ON-LINE FAULT DETECTION AND SUPERVISION IN THE CHEMICAL PROCESS INDUSTRIES 1998

A Proceedings volume from the 3rd IFAC Workshop,
IFP, Solaize (Lyon), France, 4 - 5 June 1998

Edited by

Prasad S. DHURJATI
Department of Chemical Engineering, University of Delaware,
Newark, Delaware, USA

and

Sylvie CAUVIN
IFP (Institut Français du Pétrole), France

Published for the

INTERNATIONAL FEDERATION OF AUTOMATIC CONTROL

by

PERGAMON
An Imprint of Elsevier Science

UK Elsevier Science Ltd, The Boulevard, Langford Lane, Kidlington, Oxford, OX5 1GB, UK
USA Elsevier Science Inc., 660 White Plains Road, Tarrytown, New York 10591-5153, USA
JAPAN Elsevier Science Japan, Tsunashima Building Annex, 3-20-12 Yushima, Bunkyo-ku, Tokyo 113, Japan

First edition 1998

Library of Congress Cataloging in Publication Data

A catalogue record for this book is available from the Library of Congress

British Library Cataloguing in Publication Data

A catalogue record for this book is available from the British Library

ISBN 0-08-043233 6

Transferred to Digital Printing 2009

IFAC WORKSHOP ON ON-LINE FAULT DETECTION AND SUPERVISION IN THE CHEMICAL PROCESS INDUSTRIES 1998

Sponsored by
International Federation of Automatic Control (IFAC)
- Technical Committee on Chemical Process Control

Co-sponsored by
IFAC Technical Committees on
- Fault Detection and Supervision of Technical Processes
- Safety of Computer Control Systems
AFIA (Association Française pour l'Intelligence Artificielle)
Elf
Gensym
IFP (Institut Français du Pétrole)

Organized by
IFP (Institut Français du Pétrole)

International Programme Committee (IPC)
Dhurjati, P. (USA) (Chair)
Georgakis, C. (USA) (IFAC representative)
Arzen, K.E. (S)
Blanke, M. (DK)
Cauvin, S. (F)
Corriou, J.P. (F)
Dague, P. (F)
Dote, Y. (J)
Fjellheim, R. (N)
Frank, P.M. (D)
Gertler, J. (USA)
Isermann, R. (D)
Joulia, X. (F)
Jovanovic, A. (D)
Kalitventzeff, B. (B)
Koivo, H. (SF)
Morris, J. (UK)
Patton, R.J. (UK)
Penalva, J.M. (F)
Preau, V. (AUS)
Romagnoli, J. (AUS)
Trave-Massuyes, L. (F)
Cheng Ching Yu (Taiwan)

National Organizing Committee (NOC)
Cauvin, S. (IFP) (Chair)
Couenne, N. (IFP)
Rondot, Y. (IFP)
Zaccagnino, Ph. (Elf)

PREFACE

The field of "On-Line Fault Detection and Supervision in the Chemical Process Industries" is a relatively young. Major activity in this area has taken place only in the last fifteen years. The first conference in this series of workshops with the same title was held at the University of Delaware in April 1992 although a much smaller IFAC workshop with a similar theme entitled "Fault Detection and Safety in Chemical Plants" was held in Kyoto, Japan in 1986. The second IFAC workshop in the present series was held in Newcastle, UK in 1995 and the present workshop in Solaize (Lyon), France in 1998 is the third in the series. We were encouraged by the submission of 110 abstracts and this is a reflection of the high level of international activity in this area. The reviewers helped us cut down the abstracts to 43 oral presentations and 22 poster presentations that are included in this volume of the Proceedings.

Many of the goals of the first workshop in Delaware continue to be the same: to discuss various methodologies necessary for solving industrial problems in fault diagnosis/supervision and to encourage interactions between academia and industry. This workshop also focuses on development and evaluation of methodologies for on-line fault detection and supervision in the chemical process industries. It addresses theory, application, validation, performance and evaluation of methodologies such as parameter estimation, observers, parity equations, signal analysis methods, classification, rule-based systems with probabilistic approaches, fuzzy logic and neural networks.

There are several trends that make the topic of this workshop especially relevant in today's world. The first is the tremendous advances made in automation and information technology that can potentially bring in an ever-increasing amount of information on to computer screens in the operating room of a plant. Avoiding problems of information overload and converting plant data to "on-line useful knowledge" is a key challenge. In some respects, one can draw parallels here to biological evolution where over billions of years, human beings have evolved "mental models" to interpret the huge amount of information received through their senses. In the absence of the time advantage that evolution has had, we have to rely on methodologies such as those presented in this workshop to provide assistance to operators and engineers in interpreting plant information.

A second trend that makes this field relevant in todays world is the increasing emphasis on environment and safety. Community activism and accidents such as those in Bhopal, India have caused media spotlights to be turned on the smallest of toxic releases or loss of life due to chemical accidents. The negative publicity generated by such events as well as the need to maintain the image of an environmentally conscious company make industry more sensitive to the issues of early detection of faults.

The third and perhaps most important trend that makes this field very relevant is that of the globalization of the world economy. Increasing globalization of the chemical process industry puts pressure on economic competitiveness and higher productivity. This implies reduced down-time due to faults, quick and flexible response of production to supply and demand changes, increasing reliance on automation and reduced personnel (sometimes involving loss of very experienced people).

All of these trends along with the excellent work being done in academia and industry make this field of "on-line fault diagnosis and supervision" very stimulating and relevant. We hope that the papers selected for these proceeding will contribute to the further evolution of this exciting field.

Prasad Dhurjati, University of Delaware,
Chair of the International Program Committee

Sylvie Cauvin, Institut Français du Pétrole,
Chair of the National Program Committee

CONTENTS

PLENARY PAPERS

TREND ANALYSIS

OBSERVERS

APPLICATIONS

DIFFERENT APPROACHES

NEURAL NETWORKS

STATISTICS/RECONCILIATION

QUALITATIVE

SUPERVISION/CONTROL/ALARM

POSTER PAPERS

Copyright © IFAC On-Line Fault Detection and Supervision in the
Chemical Process Industries, Lyon, France, 1998

CONTRIBUTIONS FROM THE COMMUNITY IT PROGRAMME

Patrick Corsi

Commission of the European Communities - DGIII
200 Rue de la Loi, B-1049 BRUXELLES

+32 2 296 80 88 voice, +32 2 296 83 64 fax, patrick.corsi@dg3.cec.be

Abstract : Being industrially oriented, the Information Technologies (ESPRIT) Programme is nowadays found operating at the intersection of advanced technologies and existing and new markets. The development, trialing, facilitation of the adoption and diffusion of innovative technologies by industrial and academic Esprit consortia has become the core tenet of contractual projects in all domains of ESPRIT.

This paper introduces a number of salient projects and activities which have recently been or are currently funded by the Commission of the European Communities within ESPRIT (DGIII - Industry) in relationship to advanced, intelligent techniques for industrial and business control, at the levels of systems, methods, tools, techniques, and applications with the accompanying actions to support them. A number of issues of special resonance to the Fifth Framework Programme are discussed. *Copyright © 1998 IFAC*

Keywords : Information Technologies (ESPRIT) Programme. Chemical, pharmaceutical, petroleum industries. On-line fault detection and supervision. Process control, business processes, industrial processes, distributed control systems, embedded systems, scheduling. Model-based systems. Integration of techniques: Knowledge-based systems, Neural Networks, Fuzzy Logic, Model-based Reasoning, AI and heterogeneous architectures. Human safety and security, dependability, quality.

1. THE ESPRIT CONTEXT

The Fourth Framework Programme (FP-IV) for Research and Technology Development (RTD) will conclude its financial engagements in 1998, with almost 40% being allocated to the greater field of Information and Communication Technologies (ICT).

During the years 1994-1998, the Framework Programme (FP) IV of the European Commission has consolidated the driving power of applications and the leading role of users found in its previous FPs. This was a period when applied technology - by far no longer the unique main driving force of the international business, was subsumed by the rate of change of processes and practices in all industry sectors. As a consequence, ESPRIT was forced to regard innovation as a process, and to focus on a phased view of technology assimilation within the global products and services life cycles.

This is why the past trilogy to simply fund concurrently "methods, tools and techniques" has been turned into a more industrially oriented programme, with a variation of adapted types of activities and contracts. We will outline as examples below a few more recent mechanisms that further and complement the scope of classical R&D schemes.

By definition, the initial relative immaturity of the technologies researched creates a need to nurture them towards more maturity, to promote them in hostile environments *(e.g. pilot prototypes; trial applications)*, to promote, deploy and diffuse, disseminate their results via adapted channels *(e.g. leveraging actions)*.

The trend towards user dominance lead to individual technologies, like neural networks, genetic algorithms, etc. being mixed or integrated into solutions. Overall, the convergence of information and telecommunications technologies, and the underpinning networking capability has affected applicative projects which often embody both to some degree.

The role of new mechanisms like Trial Applications (TA) and Leveraging Actions (LA)

These recent formulas aim at reducing the risks involved for organisations in deploying the technologies early enough to gain competitive advantage based on the added market value brought about by the novel technologies (ex. BOLD in drinks and other manufacturing plants).

By now, Trial Applications are a well-known model of projects. They seek to viabilise a given technology by putting it to a set of realistic tests. Often, a field prototype is thus obtained which demonstrates the usefulness of the new technical approach in that context. Normally this comes along with an understanding of the possible applicability to other domains as well.

At the moment, Leveraging Actions are being launched to support bridging the chasm between early adopters and business markets. This often regards SMEs which can't usually venture into targeted markets without having performed a consolidation of their research results by confronting them to the emerging demand. This "consolidation" may include, for instance: more detailed insights into industrial requirements, exploitation support, links with users groups, promotion of adoption of solution or product, promotion of use, internationalisation, standardisation efforts... The *raison d'être* of Leveraging Actions is to help accelerate the take up by markets of innovative products and solutions where a new economic value has been spotted that promises business benefits. This is the place where user resistance is understood and addressed and where learning processes and defensive mechanisms from the users recipient of a technology transfer can be addressed.

It is obviously not the role of ESPRIT to substitute for any pure market action, but it remains its purpose to ever more *"generate across the board socio-economic impacts facilitating penetration of IT in so pervasive a way as to ease emergence of the true Information Society anticipated for the first decade of the next century"* (The COLOMBO Report). In every single case, each Leveraging Action should seek to find out where the modern socio-economic lever is to be found. Through the leveraging act, one exits the early adopters phase definitively and hopefully enters the early maturity phase, the first phase where a profitable market lies with expected growth.

Other interesting mechanisms

Additionally, and all along the FPIV, the RTD activity was importantly accompanied and reinforced by a variety of communication and collaboration mechanisms. One can name the Networks of Excellence (ex. ERUDIT in fuzzy and other learning techniques) linking academia and early adopters, associations of suppliers and users in a given technology sector, international cooperation beyond the member states, and dissemination of results and training activities.

Organisation of this Paper

The following material briefly positions and introduces relevant and recent ESPRIT projects in the above-mentioned fields - i.e. the work of consortia essentially comprising industrial and academic organisations - whenever they emphasise process control aspects, most often in conjunction with e.g. intelligent techniques, interfaces, real-time constraints, etc.

The projects themselves may be currently running, recently terminated, or to be started soon, depending on their code number. Manifestly, they usually develop wider productions and sets of results of their own than those mentioned here, which often end up in publications, handbooks, workshops, patents, joint ventures, licensing and marketing, not mentioning their own routes to commercial exploitation. When relevant, the main application areas are indicated. For more technical data, or for information on any specific project developments, specialised readers are encouraged to initially contact the projects as listed in the documentation or on the Web site referred at the end of this paper.

2. AN ESPRIT PROJECTS SELECTION

2.1. On Methodologies and Architecture-based projects

The DIXIT project focuses on optimisation at global plant level. It is building an environment for developing distributed applications to assist plant personnel in controlling processes and emphasises safety, environmental impact and maintenance. Its architecture tends to be general and flexible for easier plant customisation. Two plants in cement manufacturing and in electricity distribution are piloting the applications. From the background research it capitalises upon, this project has the capacity to productise results in a number of continuous process control fields.

2.2. On Software Intensive, Real-Time and Related Advanced Techniques

TIGER was a seed project seeking to improve the tools available for high-speed situation assessment of complex and dynamic systems that are controlled in real-time. It aimed at detecting changes or deteriorations and predicting the development of faults in order to maintain the full availability of the gas turbine equipment of an ethylene manufacturing plant. The interfaced system developed by TIGER includes a real-time diagnostic system using numeric and qualitative simulation, with history mechanism and trend-based system to support fault prediction. The TIGER system has succeeded in making industrially viable model-based reasoning techniques (see also SHEBA-TIGER) and has been deployed internationally with success both for gas turbine manufacturers and chemical products manufacturers.

PRIDE validates the benefits of Model-Based Diagnosis (MBD) on selected industrial applications including chemicals manufacture, energy generation, water treatment, gas pumping stations and nuclear fuel production. PRIDE will contribute to methodological guidelines to enable rapid and relatively cheap applications in other companies, at other plants, and in other sectors of the process industries.

QUASI explores techniques for combined Quantum Mechanics and Molecular Mechanics simulations on a variety of High Performance Computing platforms, and applies them to industrial catalytic chemistry applications.

Some Trial Applications work by extracting a key innovative part from a previously available complex research prototype.

As one example, MARCK effectuated a technology transfer from previous advanced research in real-time knowledge-based systems (the REAKT platform) in the kiln manufacturing domain. It could show the viability of a system aimed at increasing process flexibility, process quality and reducing energy expenses. This TA resulted in a set of measures for improving an existing industrial process, and is now deployable both for kiln vendors and ceramic manufacturers.

2.3. Plant Design and Process Management

The "Integration in Manufacturing" (IiM) Unit of Esprit currently funds about 15 process industries related projects. It views "process industry" as "any industry in which production fundamentally involves chemical processes e.g. chemicals, pharmaceuticals, food and drink, energy industry." (Cluster Workshop, 1997). The "Process Industries Project Cluster" currently groups 8 R&D projects, 3 Dissemination and Awareness projects, 3 Working Groups and 2 User Group Reference Projects (UGRP).

Such UGRP co-ordinate specification of industrial IT requirements and related investments and assess their implementation. They address key issues such as: business process reengineering, methodology of system design, open systems for IT integration platforms, standards for the exchange of engineering data, tools to support plant design and construction and advanced techniques for process control. In that respect, two such groups – PRIMA and URGENT – bear special importance that are now detailed.

2.4. Special Accompanying Actions

PRIMA is a consortium composed of leading European process companies and aims at enhancing the effectiveness of investment in information and related technologies in the European process industries.

The current intense rationalisation and re-organisation within the process industries can be matched with new generations of advanced information systems and PRIMA is capitalising on these new possibilities. It privileges a business driven approach to requirements specification and investment and the results are currently being shared with vendors and industrial users. It uses a specific process modelling approach – CIMOSA - that fully supports the alignment with business objectives and enables the process industry to become more effective in its use of IT. There remain, however, blocks to achieving truly open systems (Cluster

Workshop) and wider collaborations between suppliers and users seem to be needed.

URGENT proposes to describe today's business processes and their IT component in process industries. Two examples are drilling of multilateral wells and multidimensional seismic data. The storing, managing and exploiting these data for enhanced business decisions is critical (Cluster Workshop, 1997).

SAMPLE is a PRIMA endorsed action to develop a 'Best Practice Companies' database and to provide information on the role of ICTs in supporting best practises in the process industries. The European Process Industries (EPI) can use it to assess their strategic direction and organisation structure, and the efficiency of their business processes; also to provide a basis for defining their IT requirements and future IT needs. The work will build on and contribute detailed data to the generic process industries enterprise, model which is to be developed by the PRIMA project. SAMPLE is an essential, complementary action aiming at ratifying the EPI wide generic work of PRIMA against specific sector and sub-sector needs, with the goal of enhancing/testing/tuning PRIMA results for specific applications within certain EPI sector/company profiles.

SMEs will be the key beneficiaries of this approach. The work will go beyond the current state-of-the art in supporting performance comparisons across sectors and taking account of differences in the strategic objectives of enterprises.

PROMOTER 2 follows the former Basic Research Action Working Group PROMOTER started in 1992. PROMOTER has already been instrumental in establishing a European community of software process researchers that has reached a high level of cohesion and competitiveness with respect to Japanese and American counterparts. The main objectives of PROMOTER 2 are to exploit this emerging European research community, and its strong links with industrial partners, to co-ordinate research activities, and to support dissemination of results. PROMOTER 2 will actively participate to the RENOIR Network of Excellence, to further disseminate software process technology into real industrial and business settings and to exploit feedback from the network.

2.5. *A look at relevant Longer-Term Research*

Software process technology will be the key technology for supporting workflow management, business process evolution and software quality. While the above-mentioned PROMOTER 2 action aims at consolidating and advancing the established leading European position, two Networks of Excellence operate in requirements engineering – RENOIR - and on model-based and qualitative reasoning (MONET).

The growing importance of embedded, safety critical computer systems lead to the launch of CYRF, which investigates a number of research tracks in 'synchronous languages'. Often called 'reactive systems', these are based on the abstract point of view that programs instantaneously and deterministically react to input events coming from their environment: their role is to react to events coming from their physical environment. There is a crucial need, in this domain, for clean description formalisms, design and implementation methods and tools, and formal verification. This is to be linked to project SACRES and also to the Eureka project SYNCHRON. The synchronous technology is now arousing a significant industrial demand from major European companies.

Worth to mention, VIRES is a project using formal methods for proving protocols.

The above is only a very short list of relevant projects. Section V below provides an entry point to a basic list of complementary activities of special relevance to the subject of this paper. The reader is encouraged to e.g. browse the information available on the Cordis Website.

3. SOME POINTERS TO FUTURE DIRECTIONS

3.1. *Community Policies and the coming FP-V*

The Commission has recently published its proposal for the Fifth Framework Programme (FP-V) on Research and Technological Development (see ref.), to consist of four thematic programmes under the IST Programme: Quality of Life, Information Society, Competitive growth and Energy/environment. This proposal has been submitted to the Research Council on 12 February 1998. At the moment, this proposal is still the subject of discussion with the Council of Ministers and Parliament. It is hoped that a final decision will be reached in the early summer of this year.

While an appropriate mixture of continuity and novelty will be sought and in preparing the Fifth RTD Framework Programme, the Commission has focused on a priority: integrating innovation into all the thematic programmes. The FP-V proposal discussions are continuing through 1998.

3.2. Convergence of Information & Communications Technologies (ICT)

The horizontal role of ICT in all industrial and societal activities has become a factor of crucial importance for RTD policy. The boundaries between ICT and other industrial sectors, between suppliers and users, and between the professional and consumer markets are gradually being blurred, as ICT pervades all service and production industries.

To achieve the said objectives, and leading on from the technology driven policy of the 1980s directed at a growing ICT industry, there will be firmer emphasis on a user and market led policy geared to the development of the new information infrastructure. Activities will continue to be precompetitive in nature, and will emphasise demonstration, validation, integration of technologies, specifications and standards. They will be reinforced by an effort in longer term advanced research, especially interdisciplinary research as relevant in given industrial areas.

The anticipated convergence of the three largest Programmes under FP-IV (Esprit, ACTS and Telematics) into a single integrated Programme under FP-V is a way to directly tackle the increasing complexity of systems. The Programme approach will be first and primarily led by concrete industrial and societal themes (and related objectives); synergies will be developed across technologies and in the value creation chain. Integrated and communicating systems will be favoured. RTD priorities will also go to processes in development, marketing and sales, financial services and management, and address consumer processes. The main advantage to this approach is to provide a strong framework for addressing challenges in a focused, co-ordinated way. The IST Workprogramme will be updated annually.

Other non-IST themes should also be mentioned in the context of Process Control : « products, processes, organisation for promoting competitive and sustainable growth, processes for preserving the ecosystem » (water management, global environmental change and climate, sustainable use of energy...).

Industrial Process and Control is not, in itself, such a theme. However, it could be represented as shown above and through important transversal issues like dependability, safety and quality, etc. via apposite channels in the gradual making of the Programme. Let's analyse some of the important issues at stakes.

3.3. Some further issues for consideration for Process Control activities

While Information and Communication Technologies are already underpinning our economy and business practices, new synergies need to be enacted between global R&D agents, production and advanced marketing, for the development of our economy.

One challenge of particular interest to process control industries is to merge business strategies and information strategies. Pure industrial logic's now patently underpins business processes. From a mere enabler, IT has become a key shaper and a super driver of business strategies with an enormous potential.

The resonance of business process issues also come from the greater dependence on Information as an asset. "Open" systems-of-systems begin to take over from monolithic proprietary systems. This cannot be achieved without a deeper communication, collaboration and cooperation between suppliers and users.

Will the European Process Industries meet the challenge? At a time when the point of no return from e.g. mechanical manufacturing processes to IT based processes has gone passed, very many problems reveal themselves to be analysable in terms of processes. The long-standing, huge legacy of process control bear today a general, if not generic value to all sectors of economy. This is the advocated transition from "manufacture of products to wealth created from intangible knowledge and information" (same ref.) where the economic value is to be increasingly found in Information, Contents and related Services.

Very recently, the series of four workshops for the European Dependability Initiative (see ref.) have recommended to address dependability aspects throughout the IST Programme, notably through fostering R&D, by eliciting emerging dependability issues, promote best practices in areas such as risk management, etc.; through forums to discuss changing dependability concerns, the need for standards, the promotion of synergy between security, safety and reliability; also through wider scale international collaboration. It serves as an interface to EU regulatory initiatives (security, product liability, etc) and standardisation initiatives (CEN-Information Society Standardisation System).

Openness and interoperability of systems should be high on agendas as well as: "methods for understanding risks and faults in organic (emergent) evolutionary systems (systems-of-systems), development of techniques for risk management in business processes and products, dependability in the presence of deliberate faults and intrusions..." ("Dependability Initiative", see ref.).

The complementarity between techniques and their more extensive mixed verification, validation and use must be deepened (system heterogeneity calls for more interoperable support tools and a fuller product life-cycle support).

All of this calls for multi-expertise teams, with, *inter alia,* attention to architectural models (with process and architecture integration, distribution, fault tolerance, graceful degradation) for composing complex infrastructures and for means of designing formally major systems components in view of composing systems of systems by e.g. making use of Commercial-Of-The-Shelf (COTS) components (theory of composability, quantitative evaluation, formal expressions, design, fault characterisation, design and interaction faults). And *"the massive and ubiquitous deployment of embedded systems and their link to networks is a significant trend" (same ref.).*

Integration with legacy systems into Internet-based applications, for predicting overall systems behaviours, rapid recovery strategies and global-level mechanisms for system and essential services preservation are topics to be researched.

The above evidently only constitutes a fraction of the major technological drivers underpinning the today fast evolution in process control topics. The deregulation and privatisation of operators in the Telecom networks and other sectors and services industries is also a major driving force.

The time-to-market issue is all the more critical as "competition is currently driving the move from former vertical integration towards layered models consisting of connectivity networks with services operating on top" *(same ref.).*

The pressure to produce cost-effective systems with an appropriate timing indeed further enhances the consideration of COTS systems components.

Finally, the presence of SMEs – a mandatory seed constituent for innovation at large in the fabric of the new Information, Contents and Services Society - will hopefully be facilitated through supplier/user pairs, streamlined procedures for smaller projects which were initiated during the latter part of the Fourth Framework Programme.

4. A SUBSET OF RELEVANT PROJECTS IN FRAMEWORK IV

Most of the projects listed below were launched under the Fourth Framework Programme (1994-1998). These are identifiable whenever their number is marked in the 2xxxx series. Of these, the latest were launched or are to be launched in 1998. Regularly updated synopsis and contact points are available under the given Web reference.

RTD

PRIDE 20498 **Process Industries Diagnostic Experiments,** to contribute to improving the performance of the European Process Industries, through establishing the basis for a broader application of Model-Based Diagnosis

PRIDE 21167 ESSI 2 **A Methodology for Preventing Requirements Issues from Becoming Defects,** to define and implement an experience-driven requirements methodology for: achieving faster time-to-market through reduced rework and changes during development; ensuring that requirements better reflect customer expectations and market trends; to improve the completeness, stability, and usability of requirements specifications; detecting errors in requirements, as early as possible, by tracking requirements issues over the life-cycle

PRIDE BRPR960231 BRITE/EURAM 3 **Products by Rapid Integrated Detailed Engineering,** to develop the computer based technology needed to support multidisciplinary design optimisation (MDO) and to validate the approach for the engineering component industry

SACRES 20897 **Safety Critical Embedded Systems: From Requirements to System Architecture,** embedded systems, safety-critical systems, visual programming, verification and validation, code generation, formal methods for aerospace and industrial automation

BRIDGE 22154 **Development of a Real-Time Intelligent Diagnosis Tool for Large Technical Applications,** case-based reasoning in petrochemical, transportation sectors

DIXIT 22130 **Distributed Information Technology for Strategic Multiobjective Process Control,** distributed systems, knowledge-based systems, neural networks, fuzzy logic in petrochemical, cement industry

SERENE 22187 **Safety and Risk Evaluation using Bayesian Nets,** decision support systems, bayesian belief networks, qualitative reasoning, safety-critical systems in energy sector

INRECA II 22196 **Information and Knowledge Reengineering for Reasoning from Cases,** case-based reasoning, data mining, fuzzy matching for manufacturing and car industry

PROGNOSIS 22281 **Process Diagnostics for Plant Performance Enhancement,** to demonstrate the improvements that can be achieved in plant performance monitoring, through the application of Multivariate Statistical Process Control (MSPC) to a number of production processes, and products

MONNET 22416 **Monitoring and Control of Chemical and Biochemical Processes based on the Utilisation of Neural Networks,** to develop a software package capable of assisting design engineers in the setting up and tuning of control algorithms for complex, non-linear chemical processes

PROGNOSIS II 24492 **Process Diagnostics for Plant Performance Enhancement,** to extend the PROGNOSIS technology to at-line and on-line process performance monitoring on four different processes; MSPC being now a generic tool capable of monitoring the performance of all variables on continuous or batch plant and their plant-wide interactions

CERACON 22618 **Ceramics Concept 2000,** to develop a set of computer-supported collaborative work (CSCW) tools to support process industry factories in CAD, functional design, PLC development, simulation of the designed system, production planning, optimisation and pollution control and online monitoring

CYRF 22703 **Synchronous Reactive Formalisms,** investigation of 'synchronous languages' for embedded, safety critical computer systems

HITO 22819 **Herramienta Integrada Toatal de Optimizacion,** predictive control in food and manufacturing

REFORM 22897 **A Reusable Framework for Rolling Mills,** frameworks, constraint logic programming in manufacturing

PROFES 23239 **Product Focused Improvement of Embedded Software Processes,** software process, goal oriented measurement, product quality in telecommunications, health care and petrochemical sectors

OMEGA 23245 **Object Oriented Methods and Development Environment for Geoscience Applications,** object orientation, 3D modelling, STEP, POSC in geoscience, petrochemical, mining and construction

VIRES 23498 **Verifying Industrial Reactive Systems,** formal methods on proving protocols

ORBIT 24487 **Opportunities for Re-engineering Business Processes Through Advanced IT,** a two years programme of interactive training and dissemination events in the EU and Eastern Europe designed to increase the awareness of senior executives in the process industries as to the major manufacturing advantage to be achieved by strategic investment in advanced IT solutions

QUASI 25047 **Quantum Simulation in Industry,** quantum mechanics and molecular mechanics modelling and simulation techniques for catalytic industry

SHAMASH 25491 **Standards and Hierarchical Activities Modelling; Automatic Simulation and Heuristic,** representation, analysis and optimisation of business process re-engineering in energy sector

SHEBA-TIGER 27548 **TIGER with Model Based Reasoning,** model based control monitoring, simulation and diagnostic of turbines for power generation markets

DIAMONDS 28086 **Application of advanced DIAgnostic & condition MONitoring technology to Decision Support in power plant supervision,** condition monitoring and diagnostic systems (with model-based diagnostic tool GIOTTO) for power generating plants components: steam turbines and boilers

MediaPlant 28684 **Multimedia User Interface Technologies for Plant Operating and Monitoring,** development of HMI innovative components (multimedia email system, VR plant browser, online design support system, seamless data transfer interfaces) in process control industry

AMEBICA 29172 **Auto-Adaptive Multimedia Environment Based on Intelligent Collaborative Agents,** AAM Interfaces technology with multi-agent object-oriented framework for the development of AAMI systems in industrial process control

CRISYS **Critical Instrumentation and Control Systems,** to provide design and validation methods software languages for computer and control sciences, for safe distributed control systems

ACORN BRPR950117 BRITE/EURAM 3 **Advanced Control Network,** to develop and validate a highly flexible CASE toolkit to support the management and control of the design and implementation of the next generation of open, highly distributed control systems

Trial Applications

DIAMANTA 20507 **DIADEM Methodology for MMI development Trial Application,** user interface design, man machine interface for tourism and travel, petrochemical, retail

HOD 20620 **Heterogeneous Organisation of Data,** DBMS, heterogeneous databases in petrochemical, post sectors

FLACSCOM 20786 **Preparing the FLACS-Code for Commercialisation,** risk management, simulation, structural loading, gas explosion in petrochemical sector

APPLICUS 20824 **Trial Application of Case-Based Reasoning for Customer Support,** case-based reasoning, data mining, diagnosis, maintenance, reliability in manufacturing and robotics

MARCK 20829 **Multiprocess Adaptative Real-Time Control for a Kiln,** real-time knowledge-based systems, predictive control in ceramic industry, manufacturing

REAKTANSE 20938 **Trial Application of REAKT Technology for Diagnosis in Steel Production,** real-time knowledge-based systems, diagnosis, signal processing in steel industry

AD-CAPE 24023 **Automatic Differentiation for Computer Aided Process Engineering,** Computer Aided Process Engineering codes including derivatives for FORTRAN simulation software market

Accompanying measures: Networks of Excellence

RENOIR 20800 **Requirements Engineering Network of International Cooperating Research Groups,** to provide a framework for co-ordinated joint research in requirements engineering related to industrial needs; to support the diffusion of requirements engineering research; to provide requirements engineering research training; to support technology transfer in requirements engineering

MONET 22672 **Model-Based and Qualitative Reasoning Systems Network**

Accompanying measures: Working Groups

URGENT 20337 **Oil and Gas Exploration and Production User Reference Group for Emerging New Technologies,** bridging the gap (temporal and functional) between the true, changing needs and priorities of business, and the available and emerging products and

services; to identify, prioritise and define business requirements
PRIMA II 20775 **Process Industries Manufacturing Advantage II,** an enabling infrastructure of advanced information technology for the creation of total IT systems, specifically focused on meeting the needs of the process industries: chemicals, petroleum, food, pharmaceuticals, electricity, steel, pulp & paper...
REMAFEX 20874 **Remote Maintenance for Facility Exploitation,** to define and propose new solutions with a Distributed Maintenance System which manages the global maintenance activity for manufacturing facilities
PROMOTER-2 21185 **Process Modelling Techniques Research Working Group,** to consolidate and advance the support of workflow management, business process evolution and software quality
BATCIME 21215 **Batch Control Integrated and Integrated Manufacturing Execution,** European Batch Forum aimed at setting up models and guidelines for the integration of batch control and on-line manufacturing execution functions for scheduling, quality and cost control, engineering support and production information and guidance
SAMPLE 22230 **Supportive Action to Maximise the Performances of Leading Companies of the European Industry**, to develop methods, tools and benchmark data or metrics which the European Process Industries (EPI) can use
W3USERNET 22445 **A Project to Improve the Effectiveness of European Industry by Accelerating the Commercial Use of the World-Wide Web,** World-Wide Web and Internet with W3 Consortium for health care, publishing, petrochemical and public administration
CO-MIND 22447 **Co-operative Evaluation and Dissemination of User Group Requirements by a European Technology and Education Network,** aims at supporting existing and forthcoming User Group Reference Projects through long-term requirements and their translation into sector specific scenarios, and their relevance for SMEs; to determine a mechanism that will transfer technology from the User Group Reference Projects to SMEs and Academia to influence the future
CIDIC 22667 **Computer Integrated Design of Industrial Control,** to spread the benefits of to spread the benefits of Model-Based Predictive Control

Accompanying measure: *Leveraging Action*

LEVABOLD 26496 **Exploitation support for SMEs concerning BOLD, a real-time scheduling system outcome of the BOLD trial application,** logistics, scheduling in petrochemical, textile, food, industrial automation

Other FPIII projects cited in this paper

TIGER 6862 **Real-Time Situation Assessment of Dynamic, Hard-to-Measure Systems,** integration support, real-time systems, time-dependent applications, model based reasoning in industrial machinery
ERUDIT 8193 **European Network in Uncertainty Techniques: Developments for Use in Information Technology,** an open Network of Excellence for Uncertainty Modelling and Fuzzy Technology in the European Union

5. ACKNOWLEDGEMENTS

The author wishes to thank all his colleagues from the ESPRIT Programme who contributed with their suggestions, proof reading and project inputs.

6. REFERENCES

Patrick CORSI, From Community RTD actions to European business products: Better ways through, with more modalities, IFIP'95 Conference, *First IFIP WG 8.6 - Diffusion and Adoption of Information Technology,* 14-17 October 1995, Oslo.

Patrick CORSI, Emerging Software Technologies in ESPRIT: The Way Through - yesterday, today and tomorrow; The Seventeenth International Conference on knowledge Based Systems and Applied Artificial Intelligence, Cambridge, 16-17 December 1997; and Fifth Congress of the Italian Association for Artificial Intelligence, Rome, 16-19 September 1997

"The COLOMBO Report": Preparing for the Fifth Framework Programme. Monitoring, Assessment and Evaluation of Esprit 1996-1997, http://www.cordis.lu/esprit/src/monitor.htm ; paper copies can be ordered by fax +32 2 296 83 88 or email esprit@dg3.cec.be , 1997.

"Cluster Workshop": Integration in Manufacturing: Process Industries Cluster Workshop 1997, DGIII-Esprit, Brussels, Belgium, 8-10 December 1997

An Agenda for a Dependability Initiative: Report of the Workshop on Dependability of Critical Systems and Services in the Information Society, 4-6 December 1998, Belgirate (Lake Maggiore) Italy, http://ntsta.jrc.it/dsa/workshop1297/Main.htm

Common Position on the Fifth Framework Programme, Proposition of the European Commission to the Research Council, Feb. 1998.

Further information

The IT Information Desk (European Commission, DGIII Industry) provides general and specific information on the IT Programme: Fax: +32 2 296 8388; Email: esprit@dg3.cec.be (enquiries & document orders).

Further information about Esprit and its current activities is available at Web site: http://www.cordis.lu/esprit/home.html. The Software Technologies home page that includes the EST subdomain is found at: http://www.cordis.lu/esprit/src/sthome.htm

FUTURE OF SUPERVISORY SYSTEMS IN PROCESS INDUSTRIES: LESSONS FOR DISCRETE MANUFACTURING

Ian Nimmo

Honeywell Industrial Automation and Control

Peter T. Bullemer, Ted Cochran, Steve Harp and Chris Miller

Systems and Software Technologies Department

Honeywell Technology Center

Abstract : This paper provides a brief review of the history of process control and automated systems and an overview of what we can expect in the future, together with some lessons that may be applicable to the development and operation of highly automated discrete manufacturing systems. *Copyright © 1998 IFAC*

Keywords :

Acting, ASM, DCS, Diagnostic, Evaluating, Model Based Techniques, Orienting, Process Analysis , State Estimator, Supervisory Systems,

1. SUPERVISORY SYSTEMS IN THE PAST

During the early 1970's—the beginning of the computer control era—memory and disk space were very limited, and the interface to the computer was primitive. The initial control capability provided what we describe today as Supervisory control. The computer had limited output capability, but could influence instrumentation systems by changing the setpoint to PID electronic controllers.

Early applications

Starting up a continuous polymer process is challenging: The characteristics of the product change as operating conditions change, and they also change as a function of plant throughput . These changes require different process control setpoints. In those days the operator used to manually tweak the controller setpoints but would often overcompensate and introduce new disturbances to the process. The early computers helped by automatically changing the setpoints to the controllers. Within a couple of years technology development enabled the computer to carry out full PID control using a mathematical model of a PID controller. This allowed not only automatic changes to the setpoint but changes to the tuning constants of the control loops as well. This was especially useful for non-linear processes.

However, the user interface was still very crude: Plant supervisors used to enter data by identifying the address of a variable and entering data by first entering the address and then the data, using a set of 16 piano-style keys. The process was very dangerous, because the opportunity for error was so high, especially when we consider that the data was not entered in normal base 10 numbering but in octal (base 8) or binary formats.

Soon more powerful monolithic control computers were introduced but were limited by small amounts of memory and little or no disk storage capability. The Supervisory Control Computer System held the main program and instructions for the "target" control systems. The Supervisory system downloaded the software to the control system and the supervisor initiated a booting or startup sequence. During plant operations, the target system ran its PID control algorithm and fed information about the process back to the supervisory computer system. Other information, such as batch records, historical plant data, and any abnormal condition reports was recorded manually.

The control system user interface improved slightly, allowing operators to change the control computer setpoints and alarm values directly at the control computer system. However, this required regular updates back to the Supervisory system to allow easier synchronization should the computer shut down and require reloading of the currently running program and data. This meant that the Supervisory Control Computers required several versions of the controller software, including an original or "virgin" image, a copy of the original image which had been modified during operations and which held the latest setpoints and other data parameters, and any development or test versions of the software.

During the late 1980's the Supervisory system became a very powerful system, especially on batch processes, and as Digital Corporation's VAX computers and commercial VMS operating system became available and cost effective, the use of off-the-shelf database products improved the historical data capture and replay capabilities.

As Supervisory systems continued to evolve, Honeywell introduced a more powerful Distributed Process Control Computer System (DCS) that could work independently of Supervisory systems. The DCS had better human interface, historical data capture and replay capabilities, and powerful alarm management systems. Supervisory Systems then became a place to store large amounts of historical data and for the optimization or mathematical modeling environment.

Today, the DCS works independently of any supervisory system, although some responsibility for advanced control remains delegated to more powerful supervisory systems. As new open system technologies are developed for control, it is becoming difficult to tell were the control system starts and ends and if a separate supervisory system as we know it is required anymore.

Adopting (and adapting to) new technology

The evolution of any technology has consequences for those who the technology is intended to help. The development path is rarely straight, and unexpected consequences must often be dealt with. Control systems are no exception: During the evolution of control systems, plant management and operations personnel have been required to continually reassess their abilities, progress, and needs, and adopt the technology as it becomes available. There are human and institutional consequences to this, and often the technology must be adapted to by the users as much as it is adapted for the application. The lessons learned during this process transcend the particular process and control technology involved, and provide the only hints we have available to us about what the issues of the future will be. In the sections that follow, we describe a number of examples of the sometimes unexpected consequences of the evolution of automation technology.

2. THE AEGIS STORY

AEGIS or the Abnormal Event Guidance Information System[1] was not designed in isolation by creative thoughts of individual scientists or engineers. The ASM program spent over a year identifying the problem, creating Functional Requirements and investigating available technologies. From the collective sources of information a basic design was implemented, tested and refined until a model of the solution was created. This model we call AEGIS and is being tested against known industrial problems.

During the site visits many technology problems have been identified together with at least an equal number of cultural issues. As we have shown from the earlier examples, for AEGIS to fulfill its role as a collaborative decision support system having the technological foundations to solve the problem is not enough, AEGIS must take into consideration the users' needs.

---

[1] Managing Abnormal Situations II: Collaborative Decision Support for Operations Personnel - Peter T. Bullemer, Ted Cochran, Steve Harp and Chris Miller Systems and Software Technologies Department Honeywell Technology Center Proceedings of the MVMT Workshop, Ann Arbor MI.

2.1. *Designing ASM Systems to Solve the Problem*[2]

A comprehensive approach to the design of the human-machine system interaction is needed so that operations personnel receive information appropriate to their needs, while at the same time appropriate members of the operations staff are able to collaborate to solve the problem as a team. Individual needs vary as a function of a large number of variables: the current situation, the task being performed, individual preferences and styles—and others yet to be determined. In order to serve these needs, we need to carefully assess the information requirements, not just for the current job functions present in existing plants, but for the job functions that will evolve as better decision aids become available and operators receive more support.

In the aviation community, the introduction of information technology and decision support tools resulted in nothing less than a redefinition of the jobs of the flight crew: Whereas good hand-eye coordination, excellent spatial cognition skills, and well-honed aviation intuitions used to define a large part of the repertoire of talented pilots, the current job benefits more from a thorough understanding of automation systems, good procedural discipline, and cockpit resource management skills.

If we are successful in applying user support technologies appropriately to enable effective abnormal situation management, the role of the process operations staff will also change significantly. The trick will be to introduce these technologies in a way that is simultaneously consistent with the existing culture and enabling of the new one.

The key to success is to refrain from identifying applications for the latest technological advances. Instead, we need to identify the problems that have to be solved, and only then search for answers in both sociocultural and technological terms.

The ASM team are addressing the cultural issues and must educate the industry to not accept current industrial norms that today, lead to acceptance of incidents as the "cost of doing business" and management condoning inadequate systems[3]

The delivered system must provide easy maintenance allowing the solutions to be enhanced and changed as the equipment under control also changes during the life-cycle of the plant. The solution must be capable of providing a suite of hand picked tools relevant to the plants needs. AEGIS must be seamlessly integrated with the plants existing systems, the user should have a single window to the process, support information and, to the evidence associated with diagnosing plant problems. The lessons discovered by the Formentor consortium must be fully addressed.

The Consortium are in the forth year and have officially one more funded year to complete our goals, it is too early to declare total victory over the problems that ASM presents, however, good progress has already been made. Success will depend on acceptance of the problem and willingness to implement the proposed solutions. As in our robot examples many people have already formed doubts, fears, and insecurities, and confirmed negative beliefs. What will be the reaction the first time AEGIS makes a mistake and an operator accepts bad advice and something bad happens. Will that be the "kiss of death" or will we learn, refine and move on. If we condemned Instrumentation for the same mistakes we would still be moving materials by spades and wheelbarrows. Some other issues include:-

- Integration with existing technologies
- Emergent effects of connecting previously isolated functions
- Field operations and maintenance are now in the loop again
- Paradigm shift-impact on role of DCS in operations
- Impact on the culture of operations
- Accommodation of various styles in member companies

3. COLLABORATIVE DECISION-MAKING IN PROCESS OPERATIONS ENVIRONMENT

Most oil refineries and petrochemical plants use distributed control systems (DCS) to simultaneously

---

2 Abnormal Situatio Management: NOT by New Technology Alone... Edward Cochran, Peter Bullemer

3 Adequately Address Abnormal Operations: I Nimmo Chemical Engineering Process. September'95

control thousands of process variables such as temperature and pressure. The major human role in this control is to supervise these highly automated systems. This supervisory activity requires: monitoring plant status; adjusting control parameters; executing pre-planned operations activities; and detecting, diagnosing, compensating and correcting for abnormal situations. Increased demands for higher efficiency and productivity in these industries are resulting in tremendous increases in the sophistication of process control systems through the development of advanced sensor and control technologies. However, these sensor and control technologies have not eliminated abnormal situations and will not in the future. Consequently, operations personnel continue to intervene to correct deviant process conditions.

The persistent paradox in the domain of supervisory control is that as automation technology increases in complexity and sophistication, operations professionals are faced with increasingly complex decisions in managing abnormal situations. A contributing factor to this phenomenon is that the sophistication of the user support technologies has not kept pace with the task demands imposed by abnormal situations. Thus, the focus of this program is to develop collaborative decision support technologies that will significantly improve abnormal situation management practices.

3.1. Decision making model for Operations Intervention Activities

To provide a decision making model with which to structure solutions to existing ASM problems, the ASM Consortium adopted a modified version of a model proposed by the Chemical Manufacturers Association show in Figure 1. This model provides a description of how a plant operations team responds to intervene in an abnormal situation. The left side of the figure represents the occurrence of an external abnormal situation. The operations team, represented by the dashed box, then will intervene to stabilize the process.

In the first stage of intervention (Orienting), a disturbance in the process is detected. In the Evaluating stage, the operations team develops hypotheses regarding the cause of the anomalous operating conditions. As denoted by the dashed line, which goes directly from the first to the third stages of intervention, responses to certain disturbances may be so well-rehearsed that it may appear that the Evaluating stage is skipped. Finally, the operations team must make compensatory and/or corrective action, including actions to determine if the hypotheses are correct.

All preventable abnormal situations can be traced to a breakdown in the intervention activities (For detailed discussion, see Cochran & Nimmo, 1997). For example, external inputs pertaining to process upsets can be missing, erroneous, or masked—which leads to incorrect orienting and subsequent behaviors. The fundamental goal in developing AEGIS is to improve ASM by enhancing the accuracy, completeness, and speed of the human activities in Orienting, Evaluating, and Acting.

Honeywell and the consortium can provide the technical solutions and help resolve integration and communications issues, but an industrial team is required to prepare the way for the paradigm shift--just introducing it will fail. During the ASM investigation phase we initially studied plants around the world to learn, fortunately, as we learnt we discovered some unique practices and systems that when implemented enhances operations and eliminates preventable incidents. We call these "Best Practices". By evaluating a companies processing plants against these best practices we can introduce significant benefits by helping companies achieve them. At the same time we have an opportunity to lay the foundation stone for the paradigm shift.

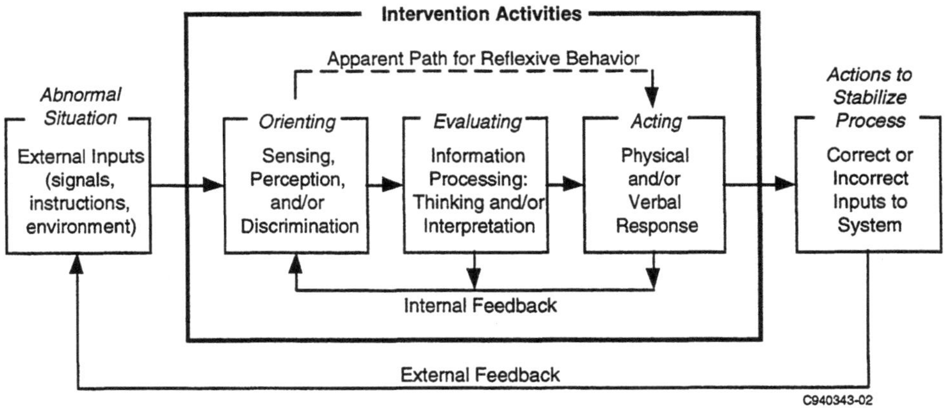

Figure 1 Decision-Making Model for Operations Intervention Activities

The purpose of AEGIS is to support the plant operations team in achieving the four high-level objectives of ASM:

- Keep the process operating normally;
- *Failing this*, restore the process to normal operation;
- *Failing this*, bring the process to a safe state;
- *Failing this*, minimize severity of any accident.

To support operations in achieving these objectives, AEGIS needs to work with the operations staff to recognize abnormal situations, correctly diagnose problems, plan the best course of action, and carry out the plan correctly and efficiently.

AEGIS's functions will be distributed over a set of six modular Roles. Collectively, these modules will provide automated assistance for the human activities of Orienting, Evaluating, and Acting, along with a mechanism for communication with plant personnel and a mechanism for modifying and improving AEGIS behavior. The AEGIS concept emphasizes human-system *collaborative* interaction; thus AEGIS will continuously interact with the plant operations team by configuring information as appropriate for the task of individual human users to facilitate the coordination of their operations. AEGIS will allocate tasks between people and computers based on the strengths of each. For example, AEGIS will improve Orienting by summarizing large amounts of data, Evaluating by resolving complex problems under pressure and formulating schedules, and Acting by managing or replacing human actions in time-critical situations. Our approach is to recognize that ASM is an integrated activity. The development of AEGIS will restructure the ASM tasks to use the strengths of the computer (automated data analysis, rapid computation, etc.) to augment the strengths of plant personnel (flexibility, creativity, common sense). The result will be a significant improvement in ASM performance in the process industries.

The AEGIS collaborative decision support vision will be presented. The vision depicts AEGIS as six major functional roles referred to as: *State Estimator*, *Goal Setter*, *Planner*, *Executor*, *Communicator*, and *Monitor*. Briefly, the State Estimator role provides a concise dynamic estimate of what is actually happening in the plant, including trend information that may be used to predict future states. The Goal Setter examines the result and proposes a prioritized agenda of high-level goals to pursue. The Planner creates a plan to achieve those goals given the current plant state. The Executor carries out the plan, ensuring that each step is executed at the right time. The Communicator manages the interface between the other Roles and the plant personnel. The Monitor examines the operation of the other Roles and regulates their activities accordingly.

The following section will describe AEGIS's State Estimator operation.

3.1.1. The State Estimator

What it Does. The State Estimator provides a concise and dynamic estimate of what is actually happening in the plant. It reduces the torrent of sensor data available on the DCS every second to a narrow stream of explanatory states we refer to as *situations*. Situations attempt to capture the necessary and sufficient information to assess the health of the process and the equipment in the plant. The scope of the estimator encompasses the complete process area supervised by a single console operator along with major interactions with upstream and downstream process areas. This would include the mechanical condition of the process equipment, the maintenance activity within the area, the process control applications and systems, the activities of the operations staff in the area, and the refining/chemical process itself. The estimator receives sensor inputs and historical information directly. Other relevant information needed as input (e.g., maintenance logs, operator observations, lab reports) can be accessed either from the communicator module (if these are available in an electronic form in different computers) or will be manually input by the operations personnel.

The goal of the State Estimator is to explain as many plant events as possible with the smallest set of *situations*. A situation may represent a root cause of an abnormal situation, such as the failure of a piece of equipment or sensor, a process disturbance forcing control changes, a trend in process variables, or an incorrect manual intervention.

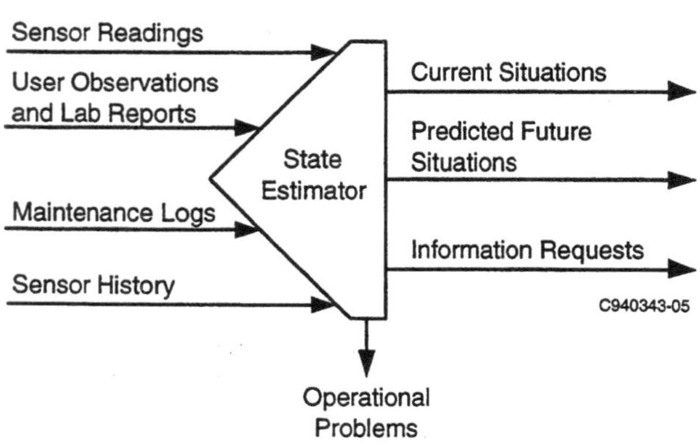

- Situations are really a generalization of the notion of diagnosis, covering a wider range of explanatory levels. Some are quite specific. For example, the situation "Valve-32 is stuck in the 24% open position" is recognizable as a specific diagnosis. Other situations represent greater degrees of uncertainty, such as "Product quality from Tower-2 has been declining for the last hour." This is better described as a high-level symptom than as a diagnosis, but it may be as much as can be deduced at an early stage of an abnormal situation. All situations derived by the state estimator should be qualified by estimates of their certainty.

- The State Estimator's vocabulary of situations is defined by the needs of the other AEGIS modules. In essence, the situations of interest to AEGIS and to plant operators are those that may make a difference in plant operations. For example, instead of continuously presenting sensor measurements from the operation of a distillation tower, the State Estimator might condense sensor information to conclude that the liquid level in the tower was "too high to support *X*" when (and only when) *X* was a planned operation—and would pass along pertinent data to other AEGIS modules for planning and communication purposes. Generally, pertinent situations such as "pressure relief valve about to release" would always be monitored. The State Estimator's vocabulary of situations and its interactions with other AEGIS modules are discussed in more depth in the section on the POCL and architecture.

- The State Estimator has several outputs. It is primarily responsible for deriving current situations—the instantaneous position of the process and equipment. However, with enough knowledge, predictions of future situations are also possible. Predictions are likewise qualified with certainty and time frame. The state estimator may be able to identify specific information it needs to make a better assessment; these are passed on as information requests. Finally, the State Estimator may be able to detect when it is failing. Such operational problems are also a form of output and are of interest to the Monitor Role.

- ***How it Helps.*** The State Estimator impact on the ASM problem will be made through the speed and accuracy of identification and diagnosis of existing or impending abnormal situations. The preventive detection and diagnosis of sources of abnormal situations will increase the time available for performing compensatory and/or corrective operational activities. Quick identification of abnormal situations and predictions of future abnormal situations allow operators to take compensatory actions before the problems escalate. A later explanation may contain the root cause description needed for a corrective action.

The AEGIS environment will provide plant situation assessments from the State Estimator to the operations team by way of the Communicator module. Suggested remedial actions will also be forwarded by the Goal Setter, Planner, and Executor.

How it Works. The AEGIS State Estimator will incorporate a synthesis of some of the most successful approaches to sensor-based diagnosis and sensor validation, as demanded by the domain. In particular, we will incorporate elements of heuristic methods as well as those based on models. Heuristic methods use the "compiled" knowledge of domain experts. Model-based methods use a model of the device or process being diagnosed along with a more generic reasoning method to derive a diagnosis.

Heuristic methods are fast and do not require a plant model, but are comparatively brittle—they cannot handle situations that were not explicitly anticipated. Model-based techniques are less brittle, but pose other problems. Most industrial chemical processes are unique. Consequently, it is difficult to build high-fidelity first-principles models of these processes and very difficult to anticipate what abnormal situations will be common. Very few processes are amenable to theoretical modeling. The work of the Advanced Process Analysis and Control System (APACS) Consortium from Canada is an elegant example of how qualitative and quantitative models can be integrated to perform process diagnosis. In APACS, a (quantitative) simulation tracks the process; deviations between sensor readings and simulated behavior trigger reasoning based on a simpler qualitative model. Hypothesized diagnoses from the qualitative reasoner are verified by running small-scale quantitative simulations to see whether the hypothesized fault(s) would generate the sensor readings observed.

A complicating factor is that the sensors and actuators used to control the plant also fail, and not infrequently. One approach to sensor failure is to use redundant sensors and select a reading based on a majority vote or some combination. This technique has been widely used in the aerospace industry. Industrial process control operates under different economic constraints, and the application of triply redundant sensors is usually unrealistic in today's plants or in designs for the near future. In the absence of physical sensor redundancy, analytical redundancy (i.e., observer-based schemes) can be used to monitor the entire system—actuators, sensors and equipment—and capitalize on redundant information available in collections of different sensors.

- Inspecting state explanations.
- Entering information not available through the sensor network.
- Directing an analysis of some specific data.
- Defining new state explanations for particular data patterns or relations.

4. CONCLUSION

The ASM Consortium based in USA and managed by Honeywell IAC have created an architecture that allows different State Estimators to plug and play in an AEGIS system. They have provided a generic State Estimator wrapper to allow different developers of State Estimators to try problem solving applications against a dynamic high-fidelity simulation of a refinery's FCCU and VRU unit.

From the initial results of the research we conclude that no single technology can address the range and complexity of problems, hence, multiple State Estimators working in harmony and being used.

We have several State Estimators monitoring, evaluating and classifying faults. The prototype AEGIS System has a plug and play feature that provides a wrapper for State Estimator Developers to map onto AEGIS using a Skeltal State Estimator that provides the connectivity to AEGIS.

This is allowing us to invite potential suppliers of diagnostic systems to try out their solutions in a real-world environment and using the simulators trainer functions to implement typical plant faults, and observe the State Estimators performance in several situations.

The time is now for diagnostic systems, the days of reacting to problems is over. We need to prevent and predict and have strategies to mitigate when we do fail to identify problems.

This conference has the unique opportunity to demonstrate to the world that the technology has finally arrived and the days of the process alarm system are numbered. Gone will be alarm floods, operators being overwhelmed by data, and panic and confusion during plant disturbances.

AEGIS promises the world more than diagnosis, more than classification, it promises orientation with plant operational and safety goals and will provide the user interface to help the operator evaluate situations and confirm corrective actions based on dynamic process and equipment conditions.

Many have tried in the past to resolve these issues and failed, we believe that collaboration between academia, industry and key suppliers is the only way forward. Sharing our valuable resource allows us to tackle high risk programs and ensures that everyone sees success from that joint investment.

In 1999 the ASM Consortium will have completed the NIST four year research program and will have spent over $20MM. Our intention is to continue without government grants and to become self supporting and continue the valuable research until we have meet our final goal which is to reduce preventable incidents by 90%. To implement an incident free environment and to save the US economy $20B/year.

Ian Nimmo

Senior Engineering Fellow – ASM Program Director

Honeywell Hi-Spec Solutions
email ian.nimmo@iac.honeywell.com

PROCESS TREND ANALYSIS USING WAVELET-BASED DE-NOISING

A. Bakhtazad [1], A. Palazoglu [2], and J. A. Romagnoli [1†],

1. ICI Laboratory of Process Systems Engineering, Department of Chemical Engineering, The University of Sydney, NSW 2006, Australia.

Tel: (61) (2) 9351 2455 Fax: (61) (2) 9351 2854

2. Department of Chemical Engineering and Materials Science, University of California, Davis, CA 95616-5294, USA.

Abstract: This paper presents a strategy to represent and classify process data for detection of abnormal operating conditions, focusing on a novel approach in de-noising of process data using wavelet coefficients. A case study features two Continuous Stirred Tank Reactors (CSTRs) in series that uses an intermediate mixer for the second feed. *Copyright © 1998 IFAC*

Keywords: Trends, Hidden Markov models, Wavelet transform, Fuzzification, Estimation theory.

1 INTRODUCTION

Today, safe and optimal operation of chemical processes require several preliminary tasks and expert operators. Analysis of measured process data, and extraction of qualitative and quantitative features of process trends are essential for the successful automation of these tasks. The trend analysis problem (Davis et al., 1995) is often decomposed into feature extraction and label assignment steps (Fig. 1). The preprocessor or data analyser is usually a numeric process to enhance the relevant signal characteristics of the process variable. The feature extractor reduces the data by taking the properties that best characterize the pattern of interest. The classifier, then, assigns the features from the extractor to one of the predefined classes or states. Here, the feature space will be partitioned into three decision regions: normal region, abnormal region and intermediate (transition) region.

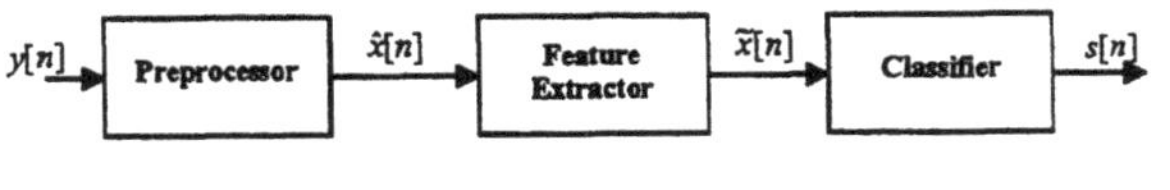

Figure 1. Diagram of data representation and classification.

In Fig. 1, $y[n]$ is the sequence of observed process data, $\hat{x}[n]$ is the de-noised sequence, $\tilde{x}[n]$ is the extracted features of the desired signal and $s[n]$ is the estimation of the behavioral trend of this process. In practice, we are given the measurement sequence of the data $y[n]=x[n]+z[n]$; that is corrupted with additional noise and/or interference $z[n]$, where $x[n]$ is the ith desired process signal. One of the most important preprocessing operations is de-noising. For many signal types, wavelets provide a compact and approximately decorrelated signal representation, which are appropriate to develop simple, but powerful algorithms for signal processing and analysis. Donoho and Johnstone (1992) proposed a method for reconstructing unknown process data where they employed thresholding in the wavelet domain and showed this to be asymptotically near optimal for a wide class of process data corrupted by additive white Gaussian noise (AWGN). Since wavelet techniques offer a different view of data than those presented by traditional techniques, wavelet analysis can often compress or de-noise process data without significant degradation, even when one wants to preserve both high and low frequency components.

In this paper, first we focus on a novel wavelet-based de-noising method based on optimal coefficients de-noising by wavelet transform. The de-noising method can be coupled with a strategy for systematically developing qualitative process information from data and capturing important process features necessary for the detection of abnormal trends. We discuss the manner in which process data are analysed or represented, by focusing on multi-resolutional wavelet analysis as a preprocessor

† To whom all correspondence should be addressed. E-mail address: jose@chem.eng.usyd.edu.au

and fuzzified triangular episodes (Wong et al., 1998) as the feature extraction method.

2 BACKGROUND

2.1 *The wavelet transform*

The discrete wavelet transform (DWT) represents a one-dimensional signal $x(t)$ in terms of shifted versions of a low-pass scaling function $\phi(t)$ and shifted and dilated versions of a prototype band-pass wavelet function $\psi(t)$. For special choices of the wavelet and scaling functions, the atoms $\psi_{J,K} \equiv 2^{-J/2}\psi(2^{-J}t-K)$, $\phi_{J,K} \equiv 2^{-J/2}\phi(2^{-J}t-K)$, $J,K \in \mathbb{Z}$ form an orthonormal basis, thus

$$x(t)=\sum_{K} u_{J_0,K}\phi_{J_0,K}(t)+\sum_{J=-\infty}^{\infty}\sum_{K} w_{J,K}\psi_{J,K}(t) \quad (1)$$

with

$$u_{J,K} \equiv \int x(t)\,\phi^{*}_{J,K}(t)\,dt\ ; \qquad w_{J,K} \equiv \int x(t)\,\psi^{*}_{J,K}(t)\,dt$$

The Wavelet coefficient $w_{J,K}$ measures the signal content around time $2^{J}K$ and frequency $2^{-J}f_0$. The scaling coefficient $u_{J,K}$ measures the local mean around time $2^{J}K$. The DWT in Eq. (1) employs scaling coefficients only at scale J_0; scaling coefficients at scales $J<J_0$ represent higher resolution approximations to the signal. Any filter bank or lifting DWT implementation produces all of the scaling coefficients $u_{J,K}$, $J<J_0$ as a natural by product. We will adopt an abstract index system for the DWT co-efficients: $u_{J,K}\rightarrow u_i$, $w_{J,K}\rightarrow w_i$ with $J(i)$ the scale of the coefficient i. We will also use $\mathbf{w}$ to denote the vector of all wavelet coefficients.

2.2 *Wavelet de-noising methods*

The wavelet expansions tend to concentrate the data energy into a relatively small number of large coefficients. This energy concentration property of the wavelet transform makes the wavelet domain appropriate for signal estimation. Let $x[n]$ be an unknown deterministic process data, and we want to reconstruct it from the set of noise-corrupted process data $y[n]=x[n]+z[n]$, where $z[n]$ is additive white Gaussian noise. Let $\mathbf{x}$, $\mathbf{y}$ and $\mathbf{z}$ denote $N\times 1$ column vectors containing the samples $x[n]$, $y[n]$ and $z[n]$, respectively and let $\mathbf{W}$ denotes $N\times N$ orthonormal wavelet transform matrix. In the wavelet domain we have $\mathbf{Y}=\mathbf{X}+\mathbf{Z}$, with $\mathbf{X}=\mathbf{Wx}$, $\mathbf{Y}=\mathbf{Wy}$ and $\mathbf{Z}=\mathbf{Wz}$. At first, we estimate the true process data wavelet coefficients $\mathbf{X}$ given the noisy observations $\mathbf{y}$. A reasonable way is to shrink the small entries of $\mathbf{Y}$ while retaining the large entries of $\mathbf{Y}$ (where the desired process data are). Since the wavelet transform approximates the Karhunen-Loeve (KL) expansion, the processing of coefficients individually stems from the fact that wavelet transform tends to decorrelate the signal. In wavelet shrinkage, the shrink/retain operation can be viewed as a diagonal filtering operation in the wavelet domain. Representing the filter by a matrix $\mathbf{H}=\mathrm{diag}\{h[0],h[1],\ldots,h[N-1]\}$, we have the process data estimate $\hat{\mathbf{x}}=\mathbf{W}^{-1}\mathbf{HWy}$. While simple, the resulting wavelet shrinkage or simply *WaveShrink* estimates are quite powerful. A block diagram for the wavelet-domain filtering is given in Fig. 2.

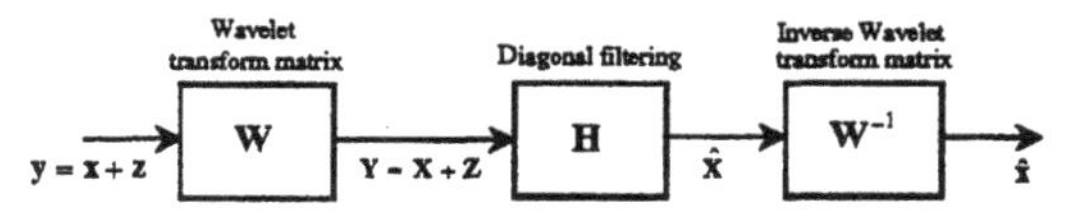

Figure 2. Wavelet-domain filtering using a diagonal weighting matrix **H**.

Nonlinear wavelet shrinkage rules can be divided into three main methods:

1. The hard threshold filter $\mathbf{H}_h$ discards coefficients below a threshold τ determined by the noise standard deviation σ and N

$$h_h[n]=\begin{cases}1 & ;\ |Y[n]|>\tau \\ 0 & ;\ \text{otherwise}\end{cases} \quad (2)$$

2. The soft threshold filter $\mathbf{H}_s$ shrinks the wavelet coefficients above the threshold

$$h_s[n]=\begin{cases}\dfrac{\mathrm{sgn}\{Y[n]\}(|Y[n]|-\tau)}{Y[n]} & ;\ |Y[n]|>\tau \\ 0 & ;\ \text{otherwise}\end{cases} \quad (3)$$

3. The optimal Wiener filter is a linear operator on the input process data that can extract the desired process data from the contaminated mixture. We can consider an approximation of the optimum Wiener filter via the wavelet transform as a nonlinear shrinkage (i.e. shaping) of wavelet coefficients. Therefore, *WienerShrink* refers to minimising the mean square error estimation of the coefficients of the desired process data among the contaminated ones by an approximation of the Wiener filter as a diagonal filter $\mathbf{H}_w$, with

$$h_w[n]=\frac{X^2[n]}{X^2[n]+\sigma^2} \quad (4)$$

It is assumed that the statistical properties of the noise and the desired process data are known. Ghael et. al. (1997) have suggested a method called empirical Wiener thresholding, to estimate σ and $X[n]$, directly from the noisy process data wavelet coefficients and then use them in Eq. (4).

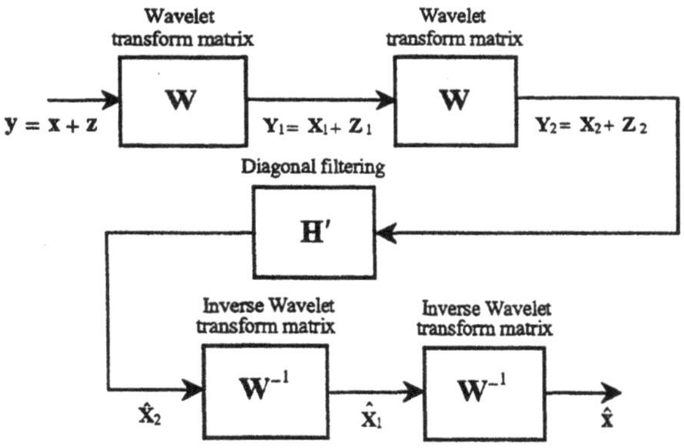

Figure 3. Wavelet-domain filtering and coefficient de-noising using a diagonal weighting matrix $\mathbf{H}'$.

In this method, we consider the block diagram of Fig. 2 as the basic structure of noise filtering. However, instead of using one stage wavelet shrinkage, (the shrink/retain operation by diagonal filtering operation in the wavelet domain), we use another similar block diagram for diagonal filter **H**. As it is clear in Fig. 2, the main purpose of using **H** is to estimate the wavelet coefficient of the desired signal. Hence, we use a similar block diagram for this new estimation problem in the wavelet domain as shown in Fig. 3, that is $\mathbf{H} = \mathbf{W}^{-1}\mathbf{H}'\mathbf{W}$, therefore we have the desired process data estimate $\hat{\mathbf{x}} = \mathbf{W}^{-1}\mathbf{W}^{-1}\mathbf{H}'\mathbf{W}\mathbf{W}\mathbf{y} = \mathbf{W}^{-2}\mathbf{H}'\mathbf{W}^{2}\mathbf{y}$.

2.3 Fuzzified Shape Analysis for Feature Extraction

A powerful method for analyzing a process trend is to characterize the time series data by the structure of the relevant process signals. In this manner, the quantitative data are reduced by converting the time series into a sequence of symbols, allowing the extraction of generic data features via generalising the shape of the signal, thereby reducing the amount of information needed for interpretation. Cheung and Stephanopoulos (1990) treated the problem of trend representation graphically by utilising a declarative language based on the notion that at the extreme or inflection points, the first or second derivatives, respectively, are zero. Thus, an *episode is* described as any part of a signal or process trend with a constant sign of the first and second derivative. This leads to the set of triangles and lines that are defined using the seven letters of the alphabet. For each episode, a triangle is created. One side of the triangle is constructed by drawing a line between the two end points of the episode. The remaining two sides are drawn by connecting the tangents of these endpoints, up to the point where the slopes intersect. This is a semi-qualitative representation because the positions of the endpoints (time, magnitude) as well as the slopes of the tangents to the curve at the endpoints are also retained. This method is discussed by Wong et al. (1998) in detail.

Fuzzy logic concept was first described by Zadeh (1965) to explain classes of objects that have no precise boundaries. The basic premise is to characterize the mapping of a set of inputs to a set of outputs by using a set of if-then rules. One of the attractive features of fuzzy logic is its ability to convert numeric data into linguistic variables. The quantitative values of magnitude and duration of a triangular episode are fuzzified into completely symbolic variables (small, medium, and large) by using two membership functions (Wong et al., 1998). With this technique, each triangle or line is transformed into nine different qualitative triangles or lines. The benefit of this technique is that the semi-qualitative representation originally using seven triangles and lines is converted into a 63 symbolic character alphabet. The new alphabet is more versatile because it allows the comparison of the sequences based on the size of the characters as well as their shape.

2.4 Gaussian mixture models

For statistical applications ranging from compression to estimation and detection, the key to successful wavelet-based algorithms is an accurate joint probability model for the wavelet coefficients of our signals of interest. A complete model for the joint probability density function $f_{\mathbf{W}}(\mathbf{w})$, with **w** the vector of wavelet coefficients, is one possibility. However, such a characterisation is intractable in practice. At the other extreme, modeling of the wavelet coefficients as statistically independent, with $f_{\mathbf{W}}(\mathbf{w}) = \prod_i f_{W_i}(w_i)$, is simple but ignores the inter-coefficient probabilistic dependencies. To obtain a trade-off, we must model the key wavelet coefficient dependencies, and only the key dependencies.

By design, wavelet-domain Hidden Markov Models (HMMs) focus on the key wavelet coefficient dependencies, learning them via maximum-likelihood-based training. The DWTs of many real-world signals tend to be sparse, with just a few non-zero coefficients containing most of the signal energy (Daubechies, 1992). Hence, the marginal density[1] $f_{W_i}(w_i)$ of each wavelet coefficient is typically described by a peaky (at $w_i = 0$) and heavy-tailed non-Gaussian density.

Such densities are well approximated by Gaussian mixture models. To each wavelet coefficient W_i, we associate a discrete hidden state S_i that takes on values $m = 1, 2, \ldots, M$ with pmf $p_{S_i}(m)$.

[1] We will use $p_S(s)$ to denote the probability mass function (pmf) of the discrete random variable S and $f_W(w)$ to denote the probability density function (pdf) of the continuous random variable W.

Conditioned on $S_i = m$, W_i is Gaussian with mean $\mu_{i,m}$ and variance $\sigma^2_{i,m}$. Hence, its marginal pdf is

$$f_{W_i}(w_i) = \sum_{m=1}^{M} p_{S_i}(m)\ f_{W_i|S_i}(w_i \mid S_i = m) \qquad (5)$$

To generate a realisation of W_i using the mixture model, we first draw a state value s_i according to $p_{S_i}(s_i)$ and then draw an observation w_i according to $f_{W_i|S_i}(w_i \mid S_i = s_i)$.

A simple way to approximate the joint density $f_{\mathbf{W}}(\mathbf{w})$ would treat the wavelet coefficients as independent Gaussian mixtures. The result (the Independent Mixture (IM) model) has proven useful for signal estimation applications. The primary motivation for this model lies in the fact that the DWT acts as an approximate K-L transform for a wide class of signals, and therefore the wavelet coefficients are approximately decorrelated. However, the wavelet coefficients of real-world signals are not statistically independent in general. Wavelet-domain Hidden Markov models (HMMs) are multidimensional mixture models in which the hidden states have a Markov dependency. The idea is to capture the dependencies in the wavelet coefficients through their hidden states. The HMT model is specified via the mixture parameters, $\mu_{i,m}, \sigma^2_{i,m}$ and transition probabilities $p_{S_i|S_{\rho(i)}}(m \mid n)$.

Before we process signals using a wavelet-domain HMM, we first must train the model to capture the Wavelet-domain properties of the signals of interest. That is, we determine the wavelet-domain HMM parameters that best characterise our observed wavelet coefficients. This standard HMM training problem can be efficiently accomplished (in linear time per iteration) using the iterative Expectation Maximisation (EM) algorithm.[2]

3 THE CASE STUDY

The case study process consists of two Continuous Stirred Tank Reactors (CSTRs) in series with an intermediate mixer for the introduction of the second feed. A single, irreversible, exothermic, first order reaction ($A \rightarrow B$) takes place in each reactor. The process has 10 variables, consisting of the input flowrates, concentrations, and temperatures (for the first CSTR and the mixer) together with the flowrates, concentrations, and temperatures of the coolant water (for both CSTRs). From the six outputs of the system, consisting of the flowrates, concentration and temperature of each CSTR, we consider the temperature of the second reactor T_2 as the output process data of this case study. Further, the coolant water flowrates are considered as the manipulated variables. The rest of the inputs can be considered constant or as disturbances to the system. To create a more practical study, measurement noise was added to each collected output. This resulted in:

$$y[n] = T_2[n] = g(m_1[n], m_2[n], T_2[n-1]) + z[n] \qquad (6)$$

where $x[n] = g(m_1[n], m_2[n], T_2[n-1])$ are the desired process data and $z[n]$ are additive measurement noises. Fig. 4 indicates the inputs and output T_2 of these two CSTRs and its input and output. In this case study, we have AWGN with zero mean and variance $\sigma = 1$.

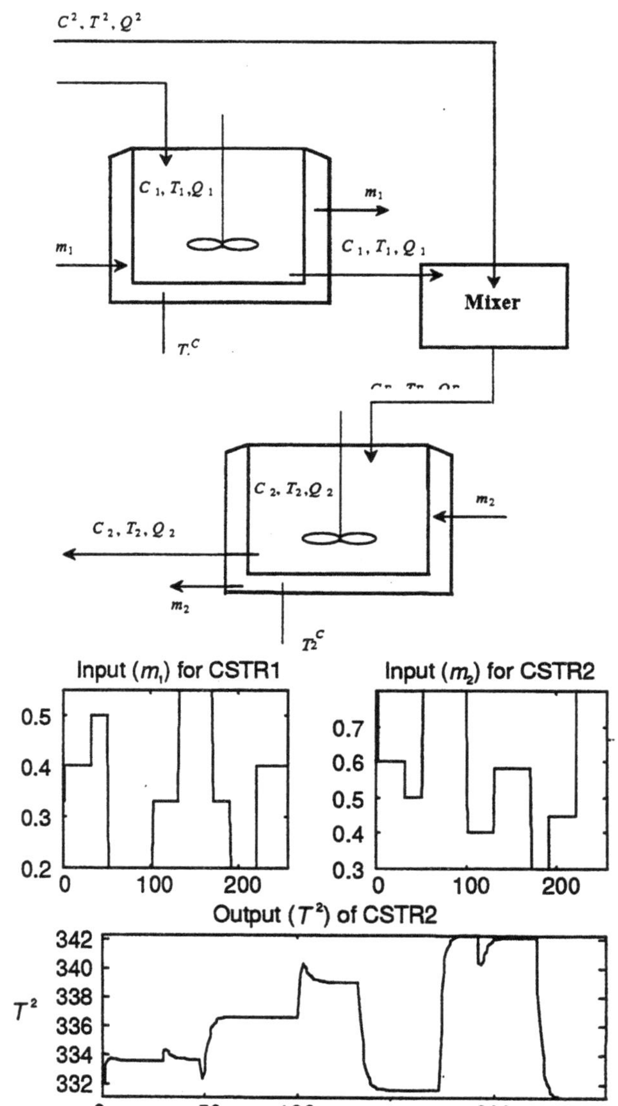

Figure 4. The double CSTR flowsheet, and the inputs and the output of these two CSTR's.

[2] EM algorithm intuition: If the values of the states S_i were known, then maximum likelihood parameter estimation would be simple. Therefore, we iterate between estimating the probabilities for the states (Expectation) and updating our model given the state probabilities (Maximisation). Under mild conditions, this iteration converges to a local maximum of the likelihood function.

Fig. 5, indicates the noisy version of the output process data and the result of our presented de-noising method to recover process data set of T^2

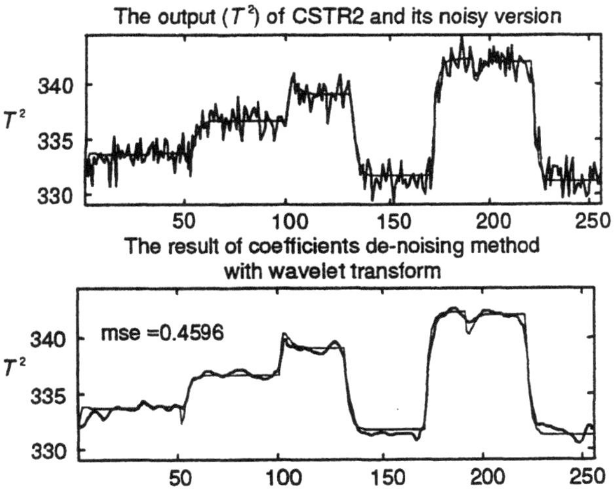

Figure 5. The noisy version of output process data and the result of our de-noising algorithms in the case study.

We showed that our de-noising method reduces the amount of mean square error estimation of the desired signal about 30% in comparison with other conventional wavelet based de-noising method. Further results will be presented at the conference.

Table 1 shows the results of fuzzification and feature extraction for a short part ($190 < n < 202$ just as an example) of the pre-processed input data as well as the classified and the estimated (most probable) working state of this system. In this table the first column denotes the input data $y[n]$, whereas the second column is the numerical values of the pre-processed data $\tilde{x}[n]$. After fuzzification and feature extraction, we have the sequence of alphabets as shown in the third column. The first prefix in the fuzzified sequence of alphabet means the result of feature extractor from the magnitude and the second one is determined by the duration of each episode. The third character in the fuzzified data column determines the type of each triangle or line. s, m, and l stand for small, medium and large, respectively. The details are presented by Wong et al. (1998). The estimated state of this system while working is in the last column. Note that this classification problem is solved by maximisation of a likelihood probability and using a trained hidden Markov model by the wavelet coefficients of the clean/original data. The classifier can detect the operational behaviours of this system, which are normal, intermediate or the transition between normal and abnormal, and abnormal states.

Table 1 The numerical values of input and pre-process data as well as, fuzzified and state estimation for a short segment in observed data.

Input data $y[n]$	Pre-processed data $\hat{x}[n]$	Fuzzified data $\tilde{x}[n]$	Estimated state $s[n]$
342.2673	341.7674	ssB	Normal ('N')
342.2660	341.5904		Normal ('N')
340.4397	341.1635	smB	Intermediate ('I')
340.2593	341.1218		Intermediate ('I')
340.5889	341.1126		Abnormal ('A')
340.9965	341.2292	smC	Intermediate ('I')
341.3423	341.4044		Intermediate ('I')
341.6000	341.5873		Normal ('N')
341.7785	341.7395	llD	Normal ('N')
341.8986	341.8551		Normal ('N')
341.9761	341.9524		Normal ('N')
342.0263	342.0175		Normal ('N')
342.0577	342.0431		Normal ('N')

The plot of this short segment of processed data is presented in fig. 6.

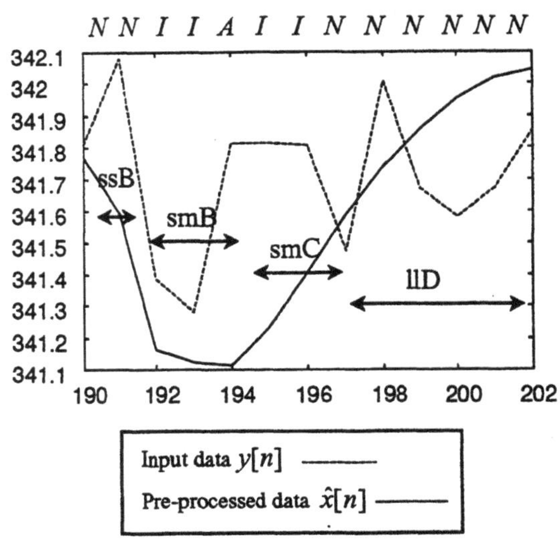

Figure 6 The result of fuzzification as well as state estimation by classifier using hidden Markov model.

REFERENCES

Cheung J.T-Y. and G. Stephanopoulos, (1990) "Representation of process trends-I. A formal representation framework," *Computers and Chemical Engineering*, v. 14, 495-510.

Daubechies, I. (1992). *Ten Lectures on Wavelets*. Society for Industrial and Applied Mathematics, Philadelphia, PA.

Davis, J. F., B.R. Bakshi, K.A. Kosanovich and M.J. Piovoso, "Process monitoring, data analysis and data interpretation," *Proc. of First International Conference on Intelligent Systems in Process Engineering*, July 9-14, (1995).

Donoho, David L. and Johnstone, Iain M. (July 1992). "Ideal Spatial Adaptation via Wavelet Shrinkage." Technical Report #400, Stanford University, CA.

Donoho, David L. (Nov. 1992). "De-Noising via Soft Thresholding." Technical Report #409, Stanford University, CA.

Ghael, S., Sayeed, A. M., and Baraniuk, R. G. (July 1997). "Improved Wavelet De-noising via Empirical Wiener Filtering." Proc. SPIE, San Diego, CA.

Wong, J. C., McDonald, K. A., and Palazoglu, A. (1998). "Classification of Process Trends Based on Fuzzified Symbolic Representation and Hidden Markov Models." *J. of Process Control*, in press.

Zadeh, L., (1965) "Fuzzy sets," *Information and Control*, v.8, 338-353.

FAULT DETECTION USING WAVELET FILTERING AND NON-LINEAR PRINCIPAL COMPONENTS ANALYSIS

R. Shao, F. Jia, E. B. Martin and A. J. Morris

Centre for Process Analysis, Chemometrics and Control
Merz Court, University of Newcastle,
Newcastle upon Tyne, NE1 7RU. UK

Abstract: A non-linear principal components analysis algorithm is proposed for process performance monitoring. Before assessing the capabilities of the monitoring scheme on an industrial dryer, the data is first pre-processed through application of a wavelet filter to remove noise and spikes. The wavelet coefficients are then used as the inputs into the non-linear algorithm. Performance monitoring charts with *non-parametric* control limits were then constructed to identify the occurrence of non-conforming operation prior to interrogating non-linear contribution plots to help identify the potential source of the fault. Encouraging results were achieved.

Keywords: Non-linear PCA, Wavelet Filtering, Wavelet Transformation, Fault Detection, Multivariate Statistical Process Control

1. INTRODUCTION

In modern process manufacturing environments, data frequently provides the basis for quality monitoring, evaluation, and control. An important step in understanding the behaviour of the process is through the extraction of the salient features encapsulated within the process data. By reducing the effective data dimensionality, it is possible to summarise the information captured in a large number of highly correlated variables by a smaller number of latent variables. The most commonly used technique is principal component analysis (PCA). Some industrial problems, however, exhibit significant non-linear behaviour and consequently the application of linear PCA may not be strictly appropriate.

In addition industrial data is frequently contaminated with process "noise" and this can have an impact on the sensitivity of the process model used for performance monitoring. It is therefore desirable to extract the "true" signal from the noise corrupted data. For data exhibiting large signal-to-noise ratios, heavy filtering may lead to significant phase-shifts if an appropriate filtering algorithm is not used. Many filters cannot efficiently handle signal spikes (Parr and Tham, 1992), whilst others require future values to implement the filter thus making it unsuitable for on-line application. Here the wavelet transform was used to pre-process the measurement signals since it exhibits better properties than traditional filters. Through application of wavelet filtering, high-frequency noise as well as sharp spikes in the data can be removed without smoothing out the important features in the process data. The discrete wavelet transform was then used to reduce the amount of data needed to construct the non-linear PCA process representation.

The non-linear PCA method is based upon the input-training network (Tan and Mavrovouniotis, 1995). It provides a parsimonious description of the process. Non-parametric control limits for the MSPC charts were calculated using kernel density estimation.

Application of the traditional approach to calculating the action and warning limits, i.e. Hotelling's T^2, is inappropriate since a non-linear mapping does not necessarily guarantee that the generated data will follow a normal distribution. Finally, non-linear contribution plots based upon model residuals and non-linear principal component scores are derived to help locate the process variable(s) that contribute to non-conforming production.

The non-linear process performance monitoring scheme is illustrated through application to an industrial drying process. After removing the high-frequency noise and sharp spikes, the non-linear PCA algorithm was applied for process monitoring and fault detection. In contrast to linear PCA, it is shown that fewer non-linear principal components are required. Finally, a non-linear contribution plot allowed the identification of the variables indicative of the fault condition.

2. WAVELET FILTERING

The wavelet transform is an extension of the Fourier transform. It projects the original signal onto the wavelet basis functions to achieve a mapping to a time-scale plane that is analogous to the time-frequency plane used in the short-time Fourier transform. These wavelet functions, localised in both the time and frequency domain, are obtained from a single prototype wavelet, called the *"mother wavelet"*, by dilation and translation (Daubechies, 1992). Through the wavelet transform, the original signal is represented in terms of a wavelet expansion using coefficients as a linear combination of the wavelet function. Thus, data operations can be performed on the corresponding wavelet coefficients. This enables a scale-invariant interpretation of the original signal. For a high scale, which corresponds to a coarse resolution, the wavelet coefficients represent the low-frequency components of the signal, i.e. the global features of the process. On the other hand, the wavelet coefficients at a low scale correspond to the high-frequency components in the signal which may be caused by either noise or local phenomenon. Thus the decomposition procedure is referred to as multiresolution analysis (Mallat, 1989).

The procedure for wavelet de-noising can be summarised as follows :-

- Apply a wavelet transform to the noisy signal and obtain the vector wavelet coefficients.
- Reduce or remove those elements in the wavelet coefficients that are believed to be attributable to noise.
- Apply the inverse wavelet transform on the remaining wavelet coefficients to obtain a de-noised function.

It is known that wavelet de-noising can remove as much of the noise as desired, but not at the expense of smoothing out real fine-scale features (Donoho and Johnstone, 1995). Generally, Wavelet de-noising methods are based upon either hard or soft thresholding approaches. Hard thresholding is where the wavelet coefficients below a certain threshold are reduced to zero. Soft thresholding reduces the value of the wavelet coefficients towards zero if they are above a certain threshold. Different methods of calculating the appropriate threshold values have been proposed (e.g. Alsberg, 1997).

3. NON-LINEAR PCA

Non-linear PCA is a generalisation of linear PCA. By projecting the predictor variables down onto curves or surfaces instead of lines or planes, a data set $X = [x_1, x_2, \dots, x_m]$ can be expressed in terms of k non-linear principal components, with $k << m$ as :

$$X = F(T) + E \qquad (1)$$

where $T = [t_1, t_2, \dots, t_k]$ is the matrix of non-linear principal component scores, F is the non-linear function equivalent to the loadings in linear PCA, and E, the residual matrix.

Tan and Mavrovouniotis (1995) proposed a non-linear PCA method based upon the concept of the Input-Training neural network (IT-net), where each input pattern is adjusted, along with internal network parameters, to reproduce a corresponding output pattern. The scaled process data X typically defines the output layer pattern and after the network has been trained, the non-linear principal scores T can be identified from the input layer. A non-linear PCA algorithm is proposed where linear PCA is first applied to compress the data, prior to applying a non-linear algorithm. This has the advantage that both second-order and higher-order correlations can be extracted. By initially applying linear PCA to the nominal data set, a new set of uncorrelated ordinates are obtained which take account of the major sources of variability due to the underlying linear features. A sub-set of these linear principal components are then retained ensuring that sufficient variability in the data has been captured. The next step involves rescaling the linear principal component scores to unit variance. Finally, the linear principal component scores are presented to the IT-network and the network optimised (trained). Adopting this procedure, both the linear and non-linear correlation structures can be captured in the non-linear principal component scores. The final non-linear mapping function can be defined as :

$$X = F(T)V^{1/2}P^T + E \qquad (2)$$

where $V(l \times l, l > k)$ is a diagonal matrix comprising the first l eigenvalues of the covariance matrix of the original observations; l is the number of linear principal component scores retained in the first step; P is the linear principal loading matrix comprising the l eigenvectors of the covariance matrix of the process observations, and E is the corresponding residual matrix.

Such an approach is not suitable for on-line process performance monitoring and fault diagnosis since it requires the application of a non-linear optimisation algorithm. However, once the non-linear function between the process observations and the non-linear principal component scores has been identified, Equation (2), the inverse function which relates the non-linear principal component scores to the process observations can be obtained. Thus it is possible to derive the non-linear principal component score for a new observation with minimal computation.

4. NON-LINEAR PCA FOR PROCESS DETECTION

Once a non-linear PCA representation of the process has been built, control limits, i.e. action and warning limits, can be calculated. Traditional multivariate control charts, such as univariate and bivariate principal component scores plots, Hotelling's T^2 and squared predicted error (the squared difference between the observed values and model predicted values) can be used to provide early warning of non-conforming operation. These control charts are derived based upon the assumption that the process data exhibits a multivariate normal distribution. However, in non-linear PCA, the principal component scores and residuals are calculated based upon a non-linear transformation of the data and thus they do not necessarily exhibit normality. Here non-parametric density estimation, a kernel density based approach, is proposed as a method for developing the non-linear control limits. This is similar to the M^2 statistic (Martin and Morris, 1996) and avoids the underlying assumption of normality.

When using the density estimation approach to calculate the control limits for the different MSPC monitoring charts, the density function of the principal component scores, the T^2 values or squared predicted errors is initially estimated. Then depending on the requirement of the confidence level, the corresponding contour or value can be calculated as the control limit for the bivariate score plot, T^2 plot and SPE plot. When the test process sample moves outside the control limits, the implication is that the process is out of statistical control. However, if the underlying distribution is normal, similar results to those based upon the assumption of normality are achieved.

When a process is recognised to be out-of-control, operator personnel need a tool to identify which variables or combination of variables, are indicative of the process malfunction. One approach is through the implementation of process variable contribution plots (Miller, 1993) which describe the change in the variables for the new observation relative to the average value calculated from the nominal process representation. Following this strategy, a contribution plot can also be derived and applied in the non-linear PCA case. The contribution plot to the SPE can be calculated in the same way as linear approach. However, since the mapping between the process variables and the non-linear principal scores is non-linear rather than a simple linear transformation, the derivation of the contribution of the process variables to the non-linear scores requires to be modified.

It is known that the partial derivative of a function with respect to a specific dimension can indicate the relative influence of the corresponding variable on that function. If the first-order partial derivatives of a multivariate function are known for a specific variable space co-ordinate, then these derivatives can be used to compare the relative influence of the different variables on the function at a particular location in variable space. The non-linear contribution plot that indicates the contribution of the process variables at a specific time point (x_0) to a non-linear score t_i can be examined as follows :-

$$X\big|_{X=X_0} \cdot \frac{\partial T}{\partial X}\bigg|_{T=t_i;X=X_0} \tag{3}$$

where $\partial T/\partial X$ is the first-order partial derivative function between T and X. This approach is also applicable for linear PCA since linearity is a particular case of non-linearity. In the linear case, the first-order partial derivatives of T to X become a constant value which in practice are the principal component loadings.

5. INDUSTRIAL APPLICATION OF FAULT DETECTION

The application of wavelet filtering and non-linear PCA for process monitoring and fault detection is illustrated by application to an industrial spray dryer. Fifteen process variables were measured on-line and a historical database representing the current operating performance was extracted from the process data base. A data set comprising 4000 samples was used as nominal data, and a second set of 1000 samples was used for model testing. An additional set of non-conforming data (500 samples) contained a number of disturbances. These occurred between samples 4-6, 306 - 310 and 370-390. This data was used for the validation of the process detection and fault diagnosis scheme.

5.1 Wavelet Filtering

Time series plots of the process variables clearly showed that most of the measurements were heavily corrupted by noise and some variables contained sharp spikes, Figure 1. Without appropriate data pre-processing, the construction of a robust nominal process model would be difficult. The first step was to remove outliers and reduce the noise levels through filtering. Wavelet filtering was used to extract the "true" signal from the noise corrupted data since it has the advantage of removing high-frequency noise without smoothing out the important features of the process.

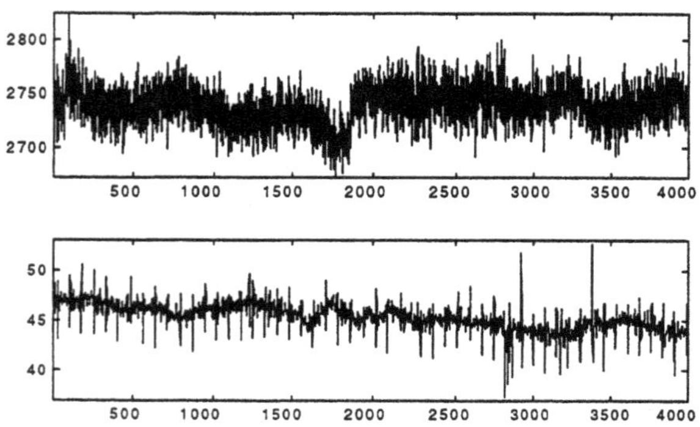

Fig. 1 Time series plot for two variables

Multiresolution analysis was first carried out for each variable to observe both the general trend and the detailed features of the process data. Figure 2 shows the results for one particular variable. Approximation coefficients at scale 4 (a4) represents the underlying trend of the signal (s) whilst the wavelet coefficients (d4~d1) show the high-frequency details. It is seen that some spikes occur more or less on a daily basis and are of either 1 to 2 minutes in duration or 4 to 5 minute duration. Discussions with plant personnel revealed that some of the peaks are the effect of automatic or manual purging of a filter and therefore these peaks should be filtered out.

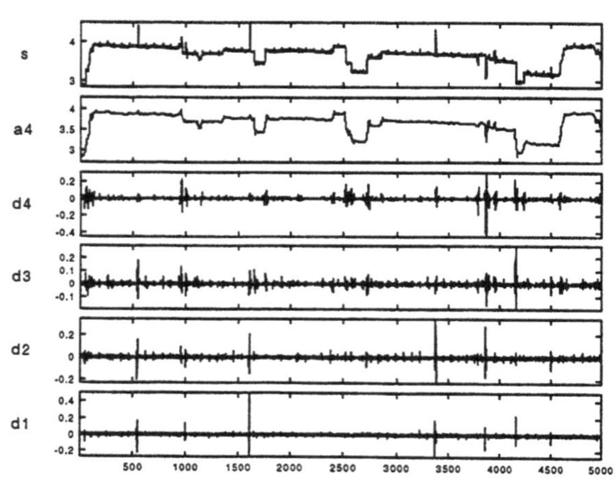

Fig. 2 Multiresolution analysis plot

Wavelet de-noising based upon hard thresholding was used to remove the high-frequency noise as well as the spikes known to be outliers or due to filter purging. Each variable was processed individually. The appropriate wavelet family was selected according to the signal properties, i.e. a long filter was adopted for smoother signals whilst a short filter was adopted for coarser signals. The level of wavelet decomposition was selected in order to achieve a balance between the genuine process changes and the level of noise removed. A number of different de-noising algorithms were investigated which have the ability to handle both white and coloured noise which may be present in the signal but which do not affect the sharp spikes. The spikes in the data, due to the purging of the filter, were handled by setting the threshold values for those variables to the maximum absolute value of the coefficients corresponding to the spikes. For other variables, histograms of wavelet coefficients were plotted to determine the noise type. The threshold values were derived from Donoho-Johnstone's fixed form threshold strategy for a specific noise type. Hard thresholding using the threshold values calculated previously was then implemented. Consequently, both noise and spikes were removed from the signal without affecting the underlying process changes, Figure 3.

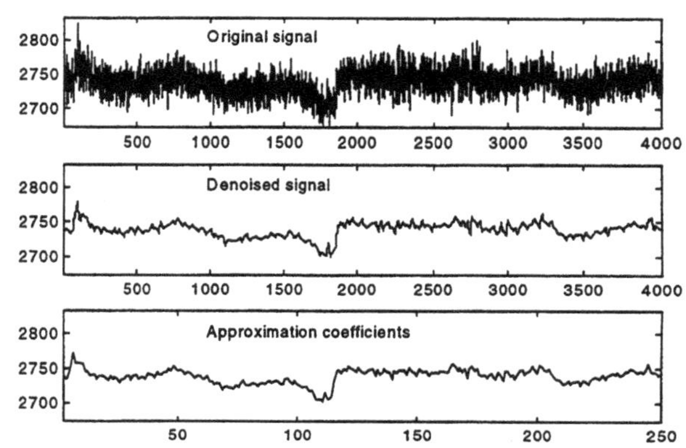

Fig. 3 Comparison between the original signal, de-noised signal and the approximation coefficients used for model derivation

The performance of wavelet de-noising was compared with both the exponential smoothing approach and the median filter. For high signal-to-noise ratios, the wavelet approach exhibited better performance. The wavelet de-noising algorithm applied to the nominal data was then used to process the validation data set and non-conforming data set. Finally, the discrete wavelet transform was applied to the de-noised nominal data. The approximation coefficients obtained (shown in Figure 3) preserve the process trend in a compact form since all high-frequency elements are omitted (Shimizu et al., 1997). These coefficients were then used to derive the non-linear PCA representation. The number of coefficients was $1/2^4$ of the original data and thus as a consequence of this reduction, a major decrease in computation time for the non-linear PCA algorithm was achieved. By removing high-frequency elements from the nominal data, the possibility of the Input-Training network to overfit is greatly reduced, consequently the generalisation ability of the network which captured the non-linear mappings is significantly improved.

5.2 Non-linear PCA

After the data matrix comprising the (250×15) wavelet coefficients was obtained following the filtering stage, the next step was to remove from the coefficients matrix any additional data points which did not correspond to nominal process operation. Finally, a data set $X(240 \times 15)$ of model coefficients was obtained and this formed the nominal data set.

Non-linear PCA was applied to the nominal data set of model coefficients. This was first transformed using linear PCA to a ten-dimensional matrix which is comparatively smaller than the original data matrix (240×15) but which still explained over 96% of the data variability. Initially a process representation was built based upon two and three non-linear principal components, but insufficient variability was captured by these two models. The next step was to build a model representation based upon four non-linear principal components. A network structure comprising four non-linear principal components, twelve nodes in the hidden layer and ten linear principal components structure was selected to identify the non-linear principal component scores. Results show the final non-linear PCA model, based upon four non-linear principal components, captured 84.5% of the variability in the nominal data set, and 78% of the data variation in the test data set comprising 1000 data points.

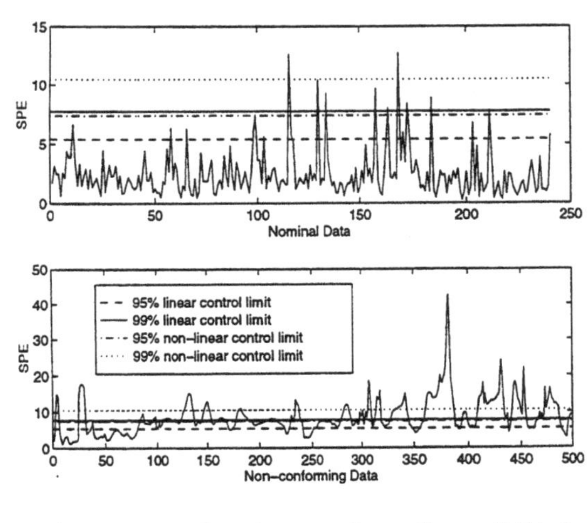

Fig. 4 Squared predicted error of non-linear PCA for nominal and non-conforming data sets.

The non-linear PCA process representation was then used to examine the non-conforming data set comprising 500 data points. Both linear and non-linear control limits were calculated for the SPE plot and the bivariate principal component scores plot in order to compare the two methods. Figure 4 illustrates the results for the squared predicted error for the nominal data set (top) and the validation data set (bottom). Also shown are the action (99%) and warning (95%) limits for the two approaches, i.e. linear and non-parametric. From the lower plot, the various disturbances are identifiable but the incident relating to samples 370-390 is more clearly apparent.

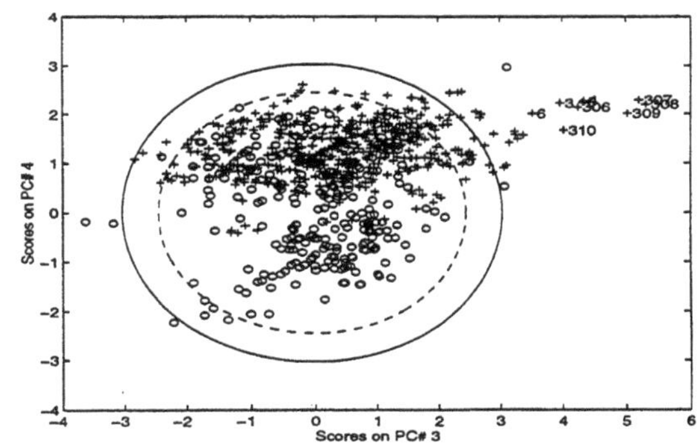

Fig. 5 Non-linear PCA with control limits calculated using Hotelling's T^2.

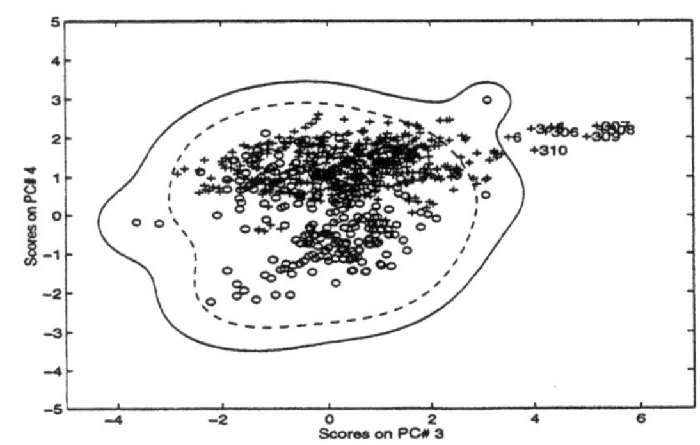

Fig. 6 Non-linear PCA with control limits calculated using non-parametric based approach.

Figure 5 describes the scores plots for non-linear principal components three versus four using the traditional approach of Hotelling's T^2 to define the control limits. Whilst Figure 6 presents the results using the non-parametric based control limits. Comparing the bivariate score plots, those samples corresponding to samples 306 - 310 ('+') lie outside the 99% non-parametric control limit. However for the monitoring plot based upon the linear limits, not only do samples 306 - 310 lie outside the action limits but also additional samples which do not correspond to periods of non-conforming behaviour. The non-parametric control limits are seen to reduce the number of false alarms and consequently the non-parametric control limits proposed in this paper exhibit more realistic and robust properties than the control limits based upon the underlying assumption of normality. The advantage of using non-linear PCA and the data driven non-parametric control limits has been clearly illustrated in this example.

Once the samples related to a process disturbance have been identified, the next step was to investigate the cause. The contribution plot to the SPE at sample 382 and the contribution to the third non-linear principal component for sample number 308 is shown in Figures 7. From Figure 7 (top plot) it can be seen that for sample 382, process variable 8 provides the strongest indication of non-conforming operation to the SPE. Whilst, from Figure 7 (bottom plot) for sample 308, process variable 4 provides the greatest contribution to the score for non-linear principal component three. The time series plot of process variable 8 and 4 for the non-conforming data is shown in Figure 8. It can be seen that there is a significant disturbance at sample 382 in variable 7 and a comparatively smaller deviation at sample 308 in variable 4 for the non-conforming data.

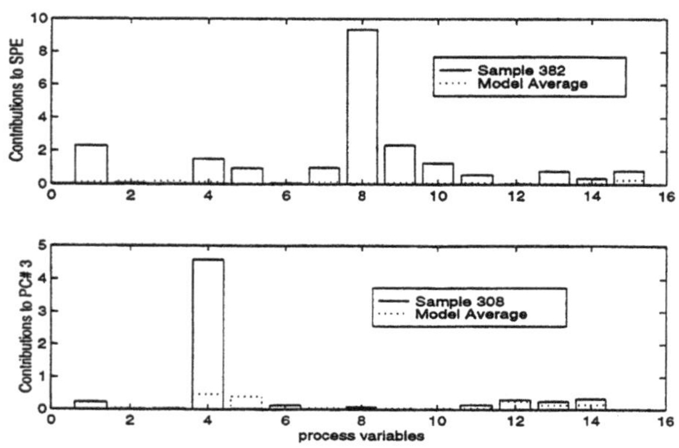

Fig. 7 Contribution Plots at the Disturbance and Minor Deviation Point

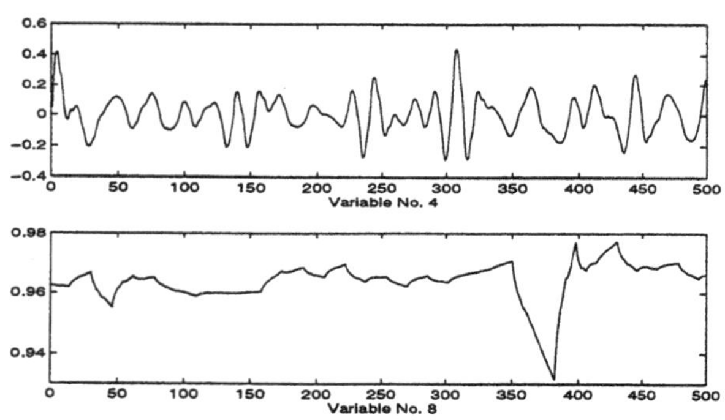

Fig. 8 Time series plot for variable 4 and variable 8 of the non-conforming data.

6. CONCLUSIONS

A non-linear PCA algorithm has been proposed for process performance monitoring and fault detection. The advantage of this method is that both linear and non-linear correlations can be extracted from the process data resulting in a simpler and more parsimonious description of the original data. Noise and spikes in the industrial data sets were eliminated through wavelet de-noising. The discrete Wavelet transform was then applied to preserve the underlying process trend in the approximation coefficients. Using these coefficients in the non-linear PCA model derivation reduces computing time. Moreover, the chance of the Input-Training network overfitting the data is significantly reduced and the generalisation ability of the network to capture the non-linear mapping is significantly improved.

Finally, data-driven non-linear control limits and non-linear contribution plots are derived to facilitate comprehensive process monitoring and fault detection where the data does not exhibit normality. The industrial application demonstrates improved performance over that of linear methods for detecting fault conditions and identifying the offending variables.

ACKNOWLEDGEMENTS

The authors acknowledge the support of the Centre for Process Analysis, Chemometrics and Control, the EU ESPRIT Project, MONNET, Number 22416 and the UK ORS Scheme for financial support of R. Shao and F. Jia.

REFERENCES

Alsberg, B. K., Woodward, A. M. and Kell, D. B. (1997). An introduction to wavelet transforms for chemometricians: A time-frequency approach. Chemometrics and Intelligent Laboratory Systems **37**, 215-239.

Daubechies, I. (1992). Ten Lectures on Wavelets, SIAM,Philadelphia.

Dong, D., MacAvoy, T. J. (1994). Nonlinear Principal Component Analysis based on Principal Curve and Neural Networks. Proc. ACC, 1284 - 1288.

Donoho, D. L. and Johnstone, I. M. (1995). Adapting to Unknown Smoothness Via Wavelet Shrinkage. J. of the American Statistical Association **90**(432), 1200-1224.

Mallat, S. G. (1989). A Theory for Multiresolution Signal Decomposition: The Wavelet Representation. IEEE Trans. on Pattern Analysis and Machine Intelligence **11**(7), 674-693.

Martin, E. B. and Morris, A. J. (1996). Non-parametric confidence bounds for process performance monitoring charts. Journal of Process Control **6**(6), 349-358.

Miller, P., Swanson, R. E. and Heckler, C. F. (1993). Contribution plots: the missing link in multivariate quality control. 37th Annual Fall Conf. ASQC, Rochester, NY.

Parr, A. and Tham, M. T. (1992). An On-line Technique for the Validation and Reconstruction of Data. IChemE Research Event.

Shimizu, H., Yasuoka, K., Uchiyama, K. And Shioya, S. (1997). On-line fault diagnosis for optimal rice alpha-amylase production process of a temperature-sensitive mutant of Saccharomyces cerevisiae by an autoassociative neural network. Journal of Fermentation and Bioengineering **83**(5), 435-442.

Tan, S. and Mavrovouniotis, M. L. (1995). Reducing Data Dimensionality through Optimising Neural Network Inputs. AIChE Journal **41**(6), 1471 - 1480.

HIERARCHICAL PLANT-WIDE MONITORING AND TRIANGULAR REPRESENTATION BASED DIAGNOSIS

Hyun-Woo Cho and Chonghun Han

Department of Chemical Engineering and Automation Research Center
Pohang University of Science and Technology
Tel: + 82 562 279 2279, Fax : + 82 562 279 2699, E-mail: Chan@postech.ac.kr

Abstract: A hierarchical plant-wide monitoring methodology is proposed based on hierarchical decomposition and principal component analysis to handle the complexity and interactions among process units. To obtain more reliable diagnostic results, pattern matching techniques are utilized along with the triangular representation of process trends in the reduced space that is defined by principal components. The proposed methodology is illustrated from the application to the hierarchical plant-wide monitoring of Tennessee Eastman benchmark process. The application has shown that the proposed methodology offers a fast and reliable monitoring and diagnosis for a large scale chemical plant. *Copyright © 1998 IFAC*

Keywords: On-line, monitoring, diagnosis, pattern recognition, hierarchical structure

1. INTRODUCTION

Maintaining high product quality in chemical manufacturing processes has become an essential task to keep the competitive edge in the increasingly severe international trade environment. The impact of abnormal situations on safety and economic aspect of process operations is enormous. Industrial statistics estimate economic impact to be about $20 billion per year in the petrochemical industries in the US alone (Mylaraswamy and Venkatasubramanian, 1997). Continuous monitoring of the measurement variables in the plant is essential to diagnose the process malfunctions and take prompt corrective actions. As the wide deployment of distributed control systems and additional process monitoring systems provides the operators with huge amount of process data on-line, the most difficult task of monitoring and diagnosis has become the efficient and reliable processing and analysis of those data to extract the information about the process (Wierda, 1994). However, this task is really challenging due to the complexities of chemical processes: hundreds or thousands of measurement variables, strong interactions among those measurements, ambiguous effects of control loops, and considerable recycle streams involved (MacGregor and Jaeckle, 1994). These situations may not only lead to the frequent occurrences of process perturbation, but also make it difficult to diagnose trouble-making causes or at most source units due to confusing candidates.

In this paper a hierarchical plant-wide monitoring methodology based on hierarchical decomposition and principal component analysis is proposed. To handle the complexity and interactions among process units, a hierarchical decomposition is employed for the monitoring. The whole plant is decomposed into several groups. Each group can be decomposed into several sub-groups. The whole plant is monitored at the top level. When a disturbance or failure is detected, the monitoring goes down to the corresponding groups where more detailed monitoring can be performed. To diagnose the causes of failure in the large complex plants, pattern matching techniques are utilized based on the triangular representation of projected values of the measurements into the smaller PC space. Because the diagnosis methodology proposed here needs only database of process measurements from past faults, it is easy to implement and maintain for the industrial problem. Case studies on Tennessee Eastman process are given to show the performance of the proposed method.

2. THEORETICAL BACKGROUND

2.1 Triangular representation of process trends

Cheung and Stephanopoulos (1990) have proposed formal descriptions of process trends in a systematic manner and represented the trends of process variables with triangular episodes. Episode can be defined for any time interval such that qualitative state is constant, and then represents an interval of uniform behavior. A triangular episode is the triangular region constructed by the intersecting lines of the initial slope, the final slope and the average slope through the boundary points as shown in Fig. 1. Triangular episodes can be considered as geometric primitives used for modeling episodes, and consequently trends. Therefore, every trend can be represented by a series of episodes, and this triangular representation of process trends permits the modeling of all the important features contained in a trend. Regardless of magnitude or time duration of fault, it can be used to recognize the fingerprints of each fault that appear differently in the processes. Rather than using correlated process data, in this work, characteristic trend of projected values of measurements in the reduced space are used to diagnose the process fault (Chen and McAvoy, 1996).

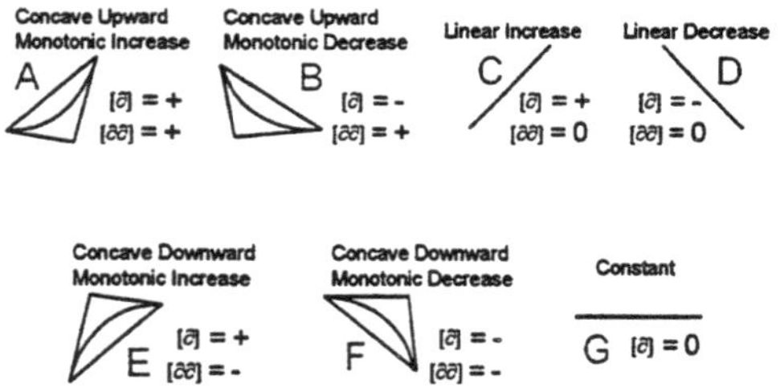

Fig. 1. Triangular components: geometrical basis of episodes (Cheung and Stephanopoulos, 1990).

2.2 K-means algorithm

K-means algorithm is based on the minimization of a performance index which is defined as the sum of the squared distances from all points in a cluster domain to the cluster center (Tou and Gonzalez, 1974). For K initial cluster centers arbitrarily chosen, at the k-th iterative step, distribute the samples x among the K cluster domains using the relation,

$$x \in S_j(k) \text{ if } \|x - z_j(k)\| < \|x - z_i(k)\| \quad (1)$$

for all $i=1,2,\ldots, K$, $i \neq j$, where $S_j(k)$ denotes the set of samples whose cluster center is $z_j(k)$. Then the new cluster centers, $z_j(k+1)$, $j=1,2,\ldots,K$, can be computed such that the performance index

$$J_j = \sum_{x \in S_j(k)} \|x - z_j(k+1)\|^2, \; j = 1,2,\ldots,K \quad (2)$$

is minimized. The $z_j(k+1)$ which minimizes this performance index is simply the sample mean of $S_j(k)$. New cluster center is given by

$$z_j(k+1) = \frac{1}{N_j} \sum_{x \in S_j(k)} x, \; j = 1,2,\ldots,K \quad (3)$$

where N_j is the number of samples in $S_j(k)$. If $z_j(k+1) = z_j(k)$ for $j=1,2,\ldots,K$, the algorithm has converged and the procedure is terminated. Otherwise, go to (2). Fault patterns that belong to the same class, in general, share common attributes, though they are not exactly same. Fault cluster shows the distribution of various patterns at specific sequence. Based on the results from K-means algorithm, this information on probable occurrences can be included in the fault prototypes. Thus K-means clustering method can be used to build reference patterns for each fault in the fault library.

2.3 Principal component analysis

PCA, one of multivariate projection methods, is a powerful tool for data compression and information extraction, which decomposes a highly correlated set of measurements into latent variables defined by the eigenvectors of the covariance of the data (Wold, 1987). This reduced set of latent variables summarizes all the relevant information by projecting the original variables down onto a low dimensional subspace. In this reduced space, the data are represented with greatly reduced collinearity explaining the variance of the original data in terms of a new set of independent principal components (Kaspar and Ray, 1992). If there are correlations and noises in the data, then several PCs are usually sufficient to describe the major variances of the data. The remaining PCs describe the variance in the noise and noise filtering effects are achieved by discarding them.

3. PROPOSED METHODOLOGY

As the complexity of process increases, it is likely

that the poor performance of on-line monitoring may be obtained. Thus new monitoring scheme different for the plat-wide purpose is required. Given the reference data under nominal operation, overall monitoring model and determine control limits are prepared. The whole plant of interest is hierarchically decomposed into several subsystems, where highly correlated variables show similar behaviors. For these variables of each subsystem, PCA is carried out to obtain a detailed individual PC model. After a tree-like hierarchical monitoring structure is constructed described as above, a fault detection function can be initiated. The monitoring and diagnosis starts from the top level. If abnormal event is monitored at this step, out-of-control state is declared, and further inspection goes down to the lower level subsystems to obtain detailed status of operation units. This procedure makes it facile to know the problem-making source region or unit in spite of considerable number of measurements. If this diagnostic advice obtained from monitoring stage are used to curtail unrelated cause candidate, diagnosis resolution may be improved. As shown in Fig. 5, hierarchical monitoring structure allows us to detect the fault earlier and diagnose the location of the cause more reliably than conventional monitoring models.

To diagnose process malfunctions of a complicated large plant, new representation scheme of fault signature is proposed. The scheme utilizes the triangular representation of projected values of original variables. This fault signature is stored in the fault library and used to match the trends of on-line measurements using a pattern matching method. To capture the patterns hidden in the sequence of episodes, the triangular representations are interpreted into a sequence of alphabet letters. Fig.2 shows the example of symbolic representations for each principal component. After measurement data are projected into model space, the triangular representation of the principal component is interpreted into a series of alphabets. Let the alphabet representation be binary valued with 0, 1 elements. Since there are seven basic representation different in nature, a binary pattern vector with seven elements can be arranged as follows.

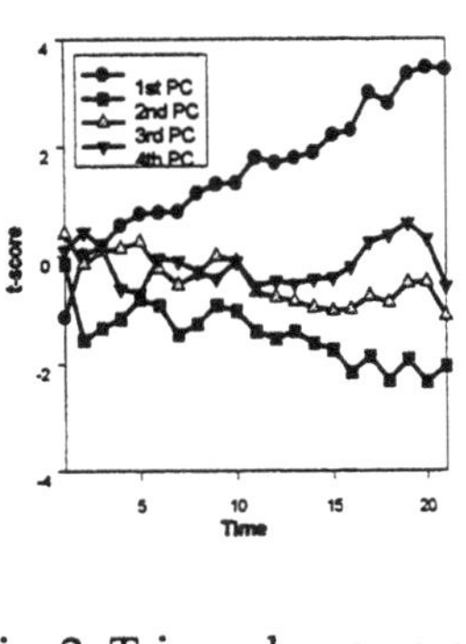

PC 1	2	3	4
E	B	B	E
E	A	A	B
E	F	B	E
E	A	A	B
A	F	F	F
B	A	B	E
A	B	B	E
A	B	A	A
A	B	A	A
E	B	F	F

Fig. 2. Triangular representation of each PC.

$$x = \left[(x_1, x_2, \dots, x_6, x_7)\right]^T \tag{4}$$

where each element x_i indicates the existence of alphabetical ith pattern. The pattern vector up to the j-th sequence of symbolic representation, $\mathbf{x}(j)$, can be constructed sequentially in the form of an augmented vector. The characteristic prototype or cluster center of N fault classes, $z_k(j)$, is described in the form of $\mathbf{x}(j)$ using K-means algorithm, and implies the average frequency of specific pattern at each sequence.

When the out-of-control state of process is identified at the monitoring stage, corresponding process measurements are projected into each PC direction retained in PC model. If on-line binary pattern vector up to sequence j is denoted as $y(j)$, its distance to each cluster center of the fault class at sequence j, $D_k(j)$, can be calculated as an Euclidean distance. When pattern vector is binary, the inverse of $D_k(j)$ gives an interpretation of the probable occurrence of the common attributes possessed by y and z: the smaller the distance, the greater the similarity. Then it can be concluded that on-line pattern vector $y(j)$, which is declared as a unusual signal vector by the on-line monitoring system, comes from k-th fault class based on

$$p_k^r(j) = \frac{\dfrac{1}{D_k^r(j)}}{\sum_{i=1}^{N} \dfrac{1}{D_i^r(j)}}, k = 1, 2, \dots, N \tag{5}$$

where $D_k^r(j) = \sum_{m=1}^{r} \sigma_m \left\| z_k^m(j) - y^m(j) \right\|$, r is the number of PCs retained, σ_m is the explained variance for m-th PC, $z_k^m(j)$ corresponds to the cluster center of m-th PC, and $y^m(j)$ is $y(j)$ of m-th PC.

$p_k^r(j)$ means diagnosis probability based on the distance to each fault class, thus fault patterns of on-line data are identified with time.

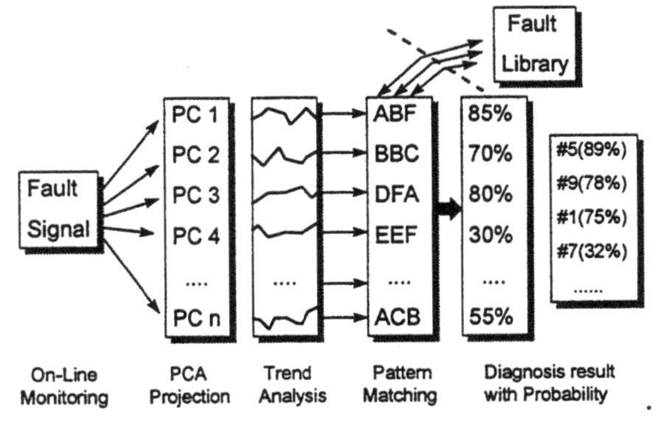

Fig.3. Diagnosis procedure using pattern matching of triangular representation.

There are many advantages that the diagnosis method based on the triangular representation of transformed variables offers. Due to the statistical characteristics and the dimension reduction effect of the proposed methodology, the proposed method is robust to measurement noises and sensor failures or missing value. The methodology is also easy to implement in real plants, because the data matrix is the only requirement for the application. Furthermore, it is independent of the fault magnitude so that it can diagnose correctly the same faults with different magnitude. Fig. 3 shows the simplified diagnosis procedure described above.

4. CASE STUDY

4.1 Description of Tennessee Eastman process

Tennessee Eastman Industrial Challenge Process was proposed as a test of alternative control and optimization strategies for continuous chemical processes (Downs and Vogel, 1993). It includes an exothermic irreversible two-phase reactor, a reactor-product condenser, a flash vapor-liquid separator, a reboiled product stripper, and a recycle compressor. There are 41 measurements and 12 manipulated variables. The process produces two products from four reactants. While there are six modes of operation at three product mass ratios, only base case of 50/50 mass ratio is considered due to the space limitation.

4.2 Hierarchical monitoring model development stage

The hierarchical reference model is constructed from the process data obtained from the process under the normal operation condition. The only data set corresponding to the specific operating region of interests should be included in statistical in-control model, from which the deviation of the process measurements are monitored. The various disturbances within some ranges were simulated to represent real situations as close as possible. From these simulated process measurement data, the simulated data that violate operation constraints and product variability specifications are excluded from the in-control model data set The reference model data should satisfy mass ratio of two products and convergence of all measurement variables. With these criteria of selecting successful operating normal conditions, statistical in-control model data matrix of 5567 by 41 was obtained after more than 24 hours of simulation for each disturbance. The model explains 78% of systematic variations with 4 PCs. Then the plant is decomposed into three subsystems so that they have minimum degree of interactions among them. The subsystems are: a reactor, a separator, and a stripper subsystems. Each subsystem includes the related variables such as the inlet, the outlet, the recycle stream, and the state variables that represent inner conditions. For each subsystem the monitoring model based on PCA is constructed. After the monitoring models have been constructed for all the subsystems, the hierarchical monitoring model is constructed from the integration of all the monitoring models.

4.3 Hierarchical monitoring stage

To test the performance of the proposed methodology in detecting the faults and the disturbances, three kind of disturbances with unusual deviation are considered: feed ratio change (case1), reactor cooling water inlet temperature change (case2), and condenser cooling water inlet temperature change (case3).

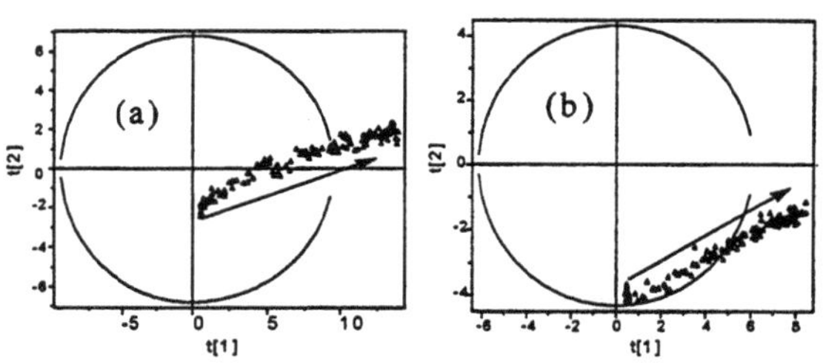

Fig. 4. Predicted t scores of case2 for (a) overall model and (b) reactor model (t[1]: 1st t score, t[2]: 2nd t score).

Fig. 4 shows the predicted t score plot of case2 for the overall model and the reactor subsystem model. As expected, the t score values of the measurements at the reduced t score space move out of normal operation region.

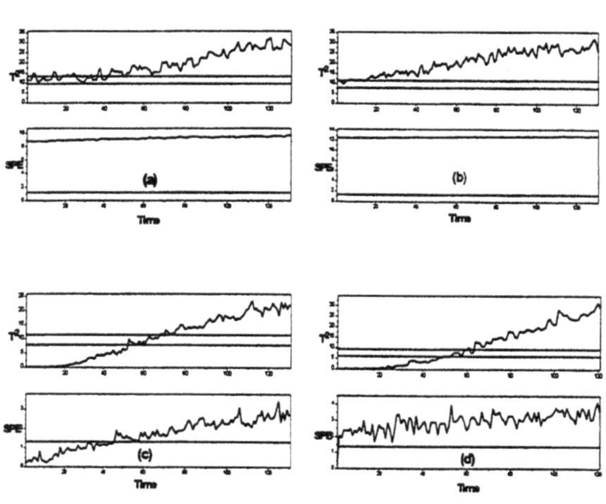

Fig. 5. Hotelling's T^2 and SPE charts for case2 based on (a) overall model; (b) reactor model; (c) separator model; (d) stripper model.

Note that the initial trajectory of the score plot in the reactor subsystem model shows more deviation from the origin than the one in the overall model. This means that subsystem model has higher resolution of detecting the out-of-control state initiated from the specific subsystem.

First, to see whether abnormality occurs at the plant level, multivariate charts for overall block are investigated. Fig. 5(a) shows Hotelling's T^2 and SPE chart based on the overall model. The upper one represents T^2 chart, and the lower SPE chart. Both show out-of-control state from the very beginning. Therefore, it is necessary to find the subsystem that caused the abnormality at the plant level. From the contribution plot at the plant level, the reactor subsystem has been identified as a possible source for the abnormality. Fig. 5(b) through 5(d) show Hotelling's T^2 and SPE chart based on each PCA model of subsystem units. Only in the control chart based on the reactor subsystem model as shown in Fig. 5(b), out-of-control state is identified. The values of T^2 and SPE in the other subsystem units tend to increase slowly over time. From the monitoring results just stated above, it can be concluded that out-of-control state stems from the reactor unit. As a result, probe should be focused on the reactor subsystem. Case1 and case3 show the result similar to case1 in the sense that out-of-control state is recognized only in the plant and reactor subsystem.

4.4 Diagnosis stage

The additional cases of IDV2, 3, 6 and 7 that are defined in Downs and Vogel (1993) are considered, including the cases mentioned in the monitoring stage above. A fault library is constructed with the simulated fault data with different magnitudes. Then reference fault patterns for each case are extracted using K-means method. Here the initial number of clusters, *K*, may be the number of faults considered. Fig. 6 shows the fault propagation trends for each PC when the fault is case1. As can be seen in the figure, strong similarity exists between standard fault signal and on-line data.

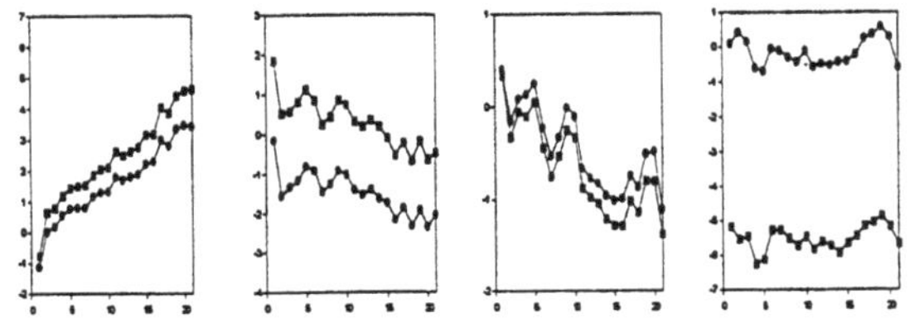

PC1 PC2 PC3 PC4

Fig. 6. Trend patterns for each PC for case1 (white dot: fault library, black dot: on-line fault data).

Table 1 Diagnosis based on the probability for case1 with time sequence

Fault Library	Time Sequence j=1	j=2	j=3	j=4	j=5
CASE1	*0.34*	*0.41*	*0.44*	*0.38*	*0.34*
CASE2	0.06	0.05	0.04	0.05	0.05
CASE3	0.11	0.14	0.15	0.18	0.17
IDV2	0.17	0.08	0.06	0.07	0.07
IDV3	0.17	0.20	0.21	0.15	0.23
IDV6	0.09	0.07	0.06	0.08	0.08
IDV7	0.06	0.05	0.04	0.07	0.06

Based on the probability value of pattern matching between the on-line data and the fault signature that is updated as the event is propagated, the fault candidate is identified from the fault library. The candidate with the largest probability is chosen to be the final cause at each sequence.

Table 1 lists the diagnosis results for case1 with time, where each value indicates the probability of the corresponding fault to be the cause. The maximum probability at each sequence is especially denoted as the underlined. Table1 shows that the diagnosis probability of the true cause is more dominant than other cause candidates.

Table 2 Diagnosis probability of three cases

Fault Library	On-line data (true cause) case1	case2	case3
CASE1	*0.34(0.27)*	0.09(0.09)	0.17(0.14)
CASE2	0.05(0.07)	*0.21(0.22)*	0.08(0.10)
CASE3	0.17(0.20)	0.12(0.12)	*0.29(0.26)*
IDV2	0.07(0.09)	0.11(0.13)	0.07(0.09)
IDV3	0.23(0.16)	0.11(0.08)	0.17(0.11)
IDV6	0.08(0.11)	0.20(0.18)	0.12(0.16)
IDV7	0.06(0.10)	0.17(0.18)	0.10(0.14)

Moreover, diagnostic results for the entire cases are listed in Table2 and Table3 with probability. The non-parenthesized values in these tables represent the probabilities at the time sequence 5, and the others at the time sequence 10.

Table 3 Diagnosis probability of other cases

Fault Library	On-line data (true cause) idv2	idv3	idv6	idv7
CASE1	0.10 (0.09)	0.16 (0.17)	0.03 (0.03)	0.10 (0.05)
CASE2	0.13 (0.16)	0.10 (0.12)	0.05 (0.07)	0.03 (0.04)
CASE3	0.08 (0.10)	0.12 (0.13)	0.06 (0.06)	0.10 (0.07)
IDV2	*0.41* *(0.33)*	0.10 (0.12)	0.03 (0.04)	0.02 (0.02)
IDV3	0.10 (0.09)	*0.27* *(0.21)*	0.03 (0.03)	0.05 (0.03)
IDV6	0.09 (0.13)	0.10 (0.12)	*0.75* *(0.73)*	0.04 (0.03)
IDV7	0.09 (0.10)	0.15 (0.15)	0.05 (0.04)	*0.66* *(0.76)*

Though the diagnosis resolution for case2 is relatively degraded especially due to the IDV6, the probability of case2 increases, and finally it is identified as a true cause. Other cases show the satisfactory diagnosis results.

5. RESULTS AND DISCUSSION

To diagnose the right cause for process abnormality from the proposed method, the problem of ambiguous time origin of a fault should be considered. It may be solved by using the recursive pattern comparison in the vicinity of uncertain time origin detected by the monitoring system. The new fault not stored in the fault library can occur in real situation due to the incompleteness of the fault library. This problem of novelty identification may be solved by updating the fault classes with the k-means algorithm. If the distance of on-line fault data to each cluster center are nearly the same, this fault data are included as a new fault class into fault classes, and the fault library is reformed.

In this work Tennessee Eastman processes are decomposed into three subsystems, and hierarchical monitoring models are constructed with two levels. However, there may be more variables and process units for the large scale chemical plants. In such cases, it takes too much time and many operators to monitor the plant. When there are much more units, more levels of hierarchy for monitoring and diagnosis will be helpful for the efficient and reliable monitoring. For the improved results, weighting each sub-block appropriately or giving sensitivity index, which takes into account the relative importance of subsystems, will be needed to handle the problem of complexity. The challenging problem of defining control limits and the frequency of the statistical model updates must be studied further.

6. CONCLUSIONS

In this paper a hierarchical monitoring and diagnosis methodology has been proposed to extract latent information from a massive amount of data and to use them to diagnose process malfunctions in plat-wide scale. To verify the performance of the proposed method for the continuous process, three kind of disturbances have been introduced to Tennessee Eastman process. Compared with the traditional monitoring method, the proposed monitoring method based on hierarchical decomposition shows much faster and more reliable results, moving the focus from the top level to more detailed level in a hierarchical manner. Diagnostic clues obtained at this step are utilized to reduce a search scope, which results in the identification of the source regions or the process units. To diagnose process malfunctions for the complicated large plant, new representation schemes of the fault signature have been proposed. They utilize the triangular representation of projected values of original variables in the reduced dimension. Due to the statistical characteristics and the dimension reduction effect of PCA, the proposed method is robust to measurement noises or sensor failures. Moreover, it is independent of the fault magnitudes so that it can diagnose correctly the same fault with different magnitudes. Several case studies have shown that the proposed diagnosis methodology has high resolution and robustness. The combined methodology offers a promising solution for monitoring and diagnosis for the large-scale industrial chemical processes.

ACKNOWLEDGMENTS

The authors deeply appreciate the financial support of the Korea Science and Engineering Foundation through the Automation Research Center at POSTECH.

REFERENCES

Chen, G. and T.J. McAvoy (1996). Process control utilizing data based multivariate statistical models. *The Can. J. of Chem. Eng.*, **74**, 1010-1024.

Cheung, J.T.-Y. and G. Stephanopoulos (1990). Representation of process trends-part I. a formal representation framework. *Comp. Chem. Eng.*, **14**, 495-510.

Downs, J.J. and E.F. Vogel (1993). A plant-wide industrial process problem. *Comp. Chem. Eng.*, **7**, 245-255.

Kaspar, M.H. and W.H. Ray (1992). Chemometric methods for process monitoring and high-performance controller design. *AIChE J.*, **38**, 1593-1607.

MacGregor, J.F. and C. Jaeckle (1994). Process monitoring and diagnosis by multiblock PLS methods. *AIChE J.*, **40**, 826-838.

Mylaraswamy, D. and V. Venkatasubramanian (1997). A hybrid framework for large scale process fault diagnosis. PSE '97-ESCAPE-7, s935-939.

Tou, J.T. and R.C. Gonzalez (1974). *Pattern recognition principles*. Chap. 3. Addison-Wesley, Massachusetts.

Wierda, S.J. (1994). Multivariate statistical process control- recent results and directions for future research. *Statistica Neerlandica*, **48**, 147-168.

Wold, S. (1987). Principal component analysis. *Chem. and Intell. Lab. Sys.*, **2**, 37-52.

A TIME-FREQUENCY ANALYSIS OF PRESSURE FLUCTUATIONS FOR THE DIAGNOSIS OF FLUIDIZED BED BOILERS

P.Simard* - H.Azov - F.Hamdi***

**umr cnrs N°6599 - Heudiasyc*
BP 20529 - 60205 Compiègne cedex
*** Cerchar*
Rue A. Dubost, BP19, 62670 Mazingarbe

Abstract: This paper presents a method developped for the precocious detection of functioning anomaly in fluidized bed boilers. This research is based on a time frequency analysis of the pressure fluctuations at the wall of a cold athmospheric fluidized bed. The analysis of these signals shows that one needs to detect and estimate transient signals mainly due to the upward movements of the bubbles. By computing precisely the parameters of the pressure transient, a possible way for the diagnosis is suggested. *Copyright © 1998 IFAC*

Keywords : Diagnosis, signal processing, time-frequency representation, fluctuations, pressure measurements, transient signals

1.INTRODUCTION

Controling complex processes such as coal boilers requires to master as much as possible the conditions of exploitation. A diagnosis tool is often appreciated to help to take a decision on the process. Supervision algorithms are only efficient if they are fed with relevant informations. The aim of this work was to find relevant informations that could help for the precocious detection of functioning anomaly in fluidized bed boilers. These informations were searched by analysing pressure fluctuations, and by relating this informations to an identified functioning anomaly.

Basically, a fluidized bed is a column of fine particles at the base of which air is blown through a grid. Numerous physical phenomena occur inside this sort of process. The pressure field can be considered as the sum of a static pressure field and a dynamic one. The first one has to be related to the difference between the air pressure at the base, and the atmospheric pressure at the top of the bed. The second one is generated by the displacements of the particles, and the upward movements of air bubbles (Aragon&Tatay 1984)

There exists a minimum speed for air supply (u_{mf}) for which the bed of particles behaves like a fluid. As this speed increases, one generally observes deeper movements generated by air bubbles. The bubbles are first generated at the grid, and then move upwards, gathering to give bigger bubbles as they raise.

We checked that the upward movements of bubbles are related to the hydrodynamic state of

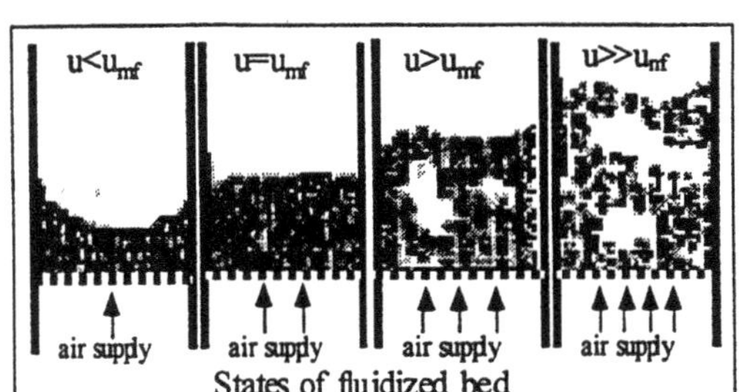

Fig. 1. The hydrodynamic state of the bed depends on the speed of the air supply. An important parameter is the minimum speed requiered to obtain the fluidization of the particles. As the speed increases, bubbles may appear.

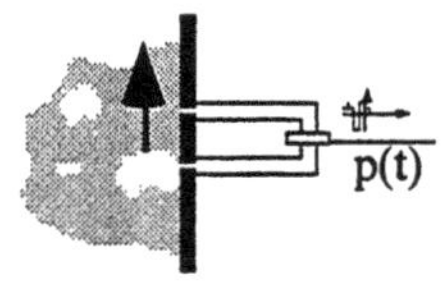

Fig. 2. The measurements are obtained by a differential pressure sensor through the wall of the bed

the bed. As a result, the proposed method can be used as a first step in a supervision algorithm.

2. TRANSIENTS DETECTION AND ESTIMATION

The measurements are carried out with differential pressure sensors through the wall of the bed. It gives very noisy signals that have to be understood as realisation of a non stationnary random process.

The mathematical model x(t) for the measurements p(t) consists of a sum of transients within a superimposed noise. These transients result from the upward movement of bubbles along the wall and close to the sensor. They give basically finite energy random signals occuring from time to time in the measurements.

2.1 Segmentation of the transients in the Wigne-Ville representation.

The first problem is a segmentation one. An algorithm based on Wigner-Ville time-frequency representation, and a model of convolution, was developped. There exist algorithms to operate a segmentation, and a few of them are able to give precisely the position. Some others are able to give a good estimate of the spectrum, and/or the amplitude function of the signal. In this paper, we propose a single method to get a segmentation, a spectrum estimate, and a signal estimate. The proposed method was compared with the SNR method proposed by Kaufman (1987) and gave definitely better results.

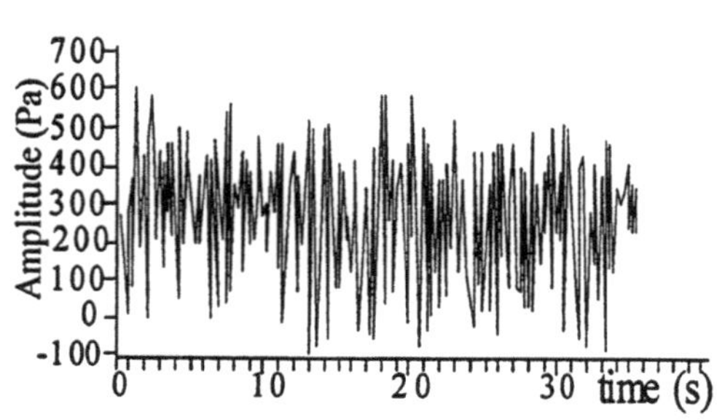

Fig. 3. Measurement signals time representation for an air speed of 1m/s.

We consider a signal $x(t)$ composed by a sum of transients $s_j(t) = a_j(t - t_j)$ in a noise:

$$x(t) = \sum_j s_j(t) + b(t) = \sum_j a_j(t - t_j) + b(t) \quad (1)$$

The transients are not necessarily all the same.
We developped a method based on the Wigner-Ville time-frequency representation $W_x(\xi,t)$ of a signal $x(t)$ (Flandrin, 1993):

$$W_x(\xi,t) = \iint x\left(t - \frac{\tau}{2}\right) x^*\left(t + \frac{\tau}{2}\right) e^{-j\xi\tau} d\tau d\xi \quad (2)$$

This transformation allows a coarse segmentation, provided that the event are distinct enough. This segmentation is computed directly in the time-frequency domain.

A sub-matrice $W_x^{(j)}(\xi,t)$ is extracted. This sub-matrice contains valuable information to compute

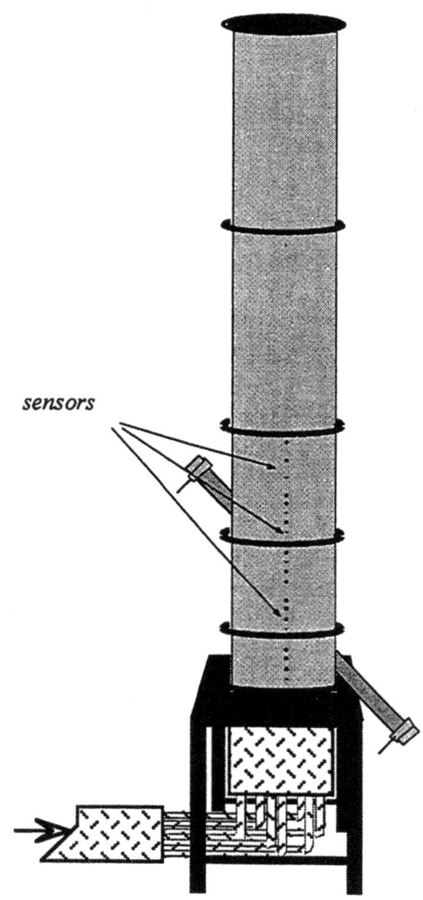

Fig. 4. The experimental device consists of a column of 3m height and a diameter of 0.45m

the spectrum of the transient signal using an averaging formulae:

$$S_j(\xi) = \int W_x^{(j)}(\xi, t) m(t) dt. \qquad (3)$$

In this expression, m(t) is a particular weighting function obtained by tracking the line of maxima in the time-frequency domain. Different other weighting functions can be used.

2.2 Estimation of the transient signal

A spectral factorization procedure is performed to get a minimum phase wavelet $\hat{s}_j(t)$ representative of the transient. This wavelet is an estimate of $a_j(t)$

The problem of reconstruction of a finite energy signal from its spectrum is known as the spectral factorization problem. Since the phase information is lost, the reconstruction can only be done under a set of hypothesis. Numerous methods have been proposed (methods of roots, Toeplitz, Kolmogorov,...) and they are generally concerned with the reconstruction of a minimum phase wavelet associated with the given spectrum (Lacoume 1985). The method choosed here was proposed by Claerbout (1985) :

i) estimate the spectrum $S_x(\xi)$
ii) Compute the inverse Fourier transform of the logarithm of $S_x(\xi)$, and take the real part.
iii) Apply a Hilbert transform, using eventually a weighting function.
iv) compute the inverse Fourier transform for the exponential of iii)
v) take the real part as the wavelet estimate $\tilde{x}(t)$

The figure compares the result with the original signal. It appears clearly that there is some phase distorsion, but this point has very little importance for the global algorithm.

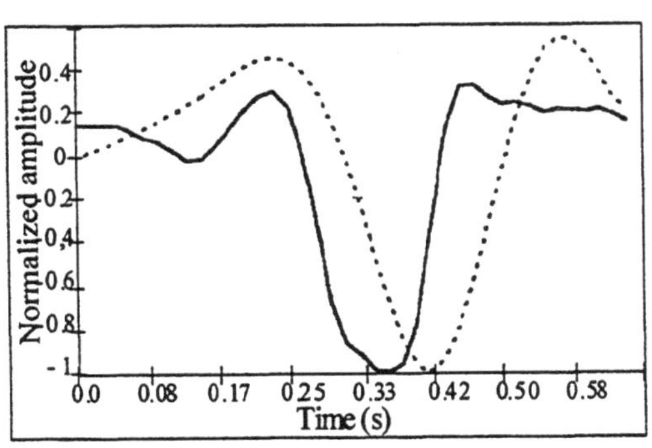

Fig. 5. The reconstruction of event wavelet give satisfactory results.

3. COUNTING BUBBLES

This following method is the first step used to count bubbles in various situations, and to measure precisely a moment for the occuring. To achieve this objective, a convolution model is used. If we assume the signal as a sum of transients, characterized by a date t_j and a shape $a_j(t)$, so that:

$$x(t) = \sum_j a_j(t - t_j) + b(t) \qquad (4)$$

The signal can also be written as:

$$x(t) = a_j(t) * \sum_j \delta(t - t_j) + b(t) \qquad (5)$$

$b(t)$ represent noise, assumed to be white.

To get the parameters of this expression, a deconvolution is performed. This operation can be achieved with a modified least square algorithm, using the information obtained by the estimation of $a_j(t)$, (Mottelet and Simard,1994). This provide a good estimation of the instants t_j.

This collection of instants is then used to compute the corresponding histogram of the time delays between two successive events.

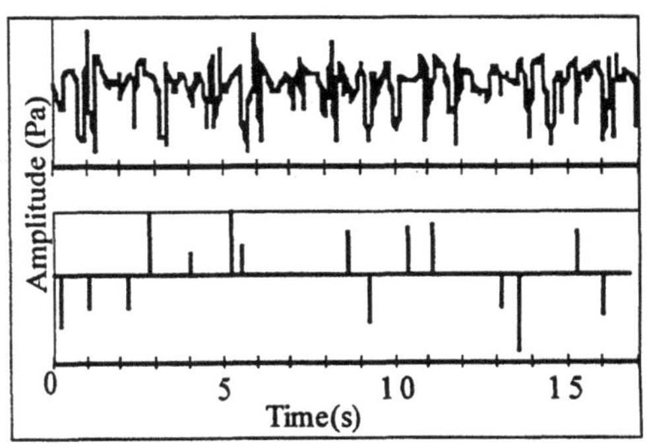

Fig. 6. The deconvolution gives a series of peaks corresponding to the instants and amplitude of the detected events.

4. EXPERIMENTAL TEST

An experimental test situation permitted to check the validity of the proposed approach. In this experiment, artificially insuflated gaz generated bubbles and the corresponding transients were recorded by the pressure sensors. This experiment was first used to check the segmentation of events, the reconstruction of the wavelets, and the estimations of the instants.

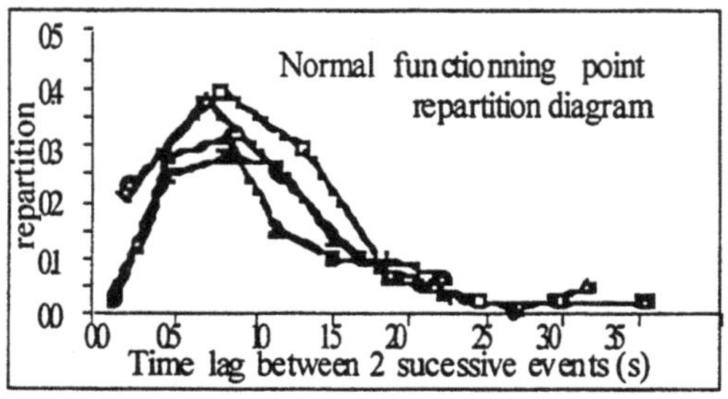

Fig. 7. The diagnosis will be held on the statistical repartition of time lags.

The limitations of the method as far as the signal to noise ratio is concerned were not studied precisely, but it is evident that the basic condition is related to the readibility of the Wigner Ville Time-Frequency plane. Anyway, it will not only depend on the signal to noise ratio, but also on the different distribution of the local spectral energy.

However, the proposed method give good results as far as the number and the instants estimations are concerned.

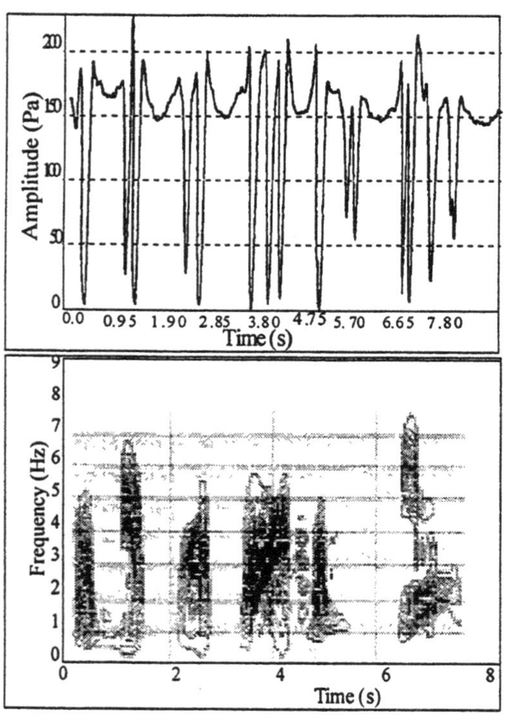

Fig. 8. Measurement signal time representation and its Wigner-Ville time frequency representation for the simulated bubbles experiment.

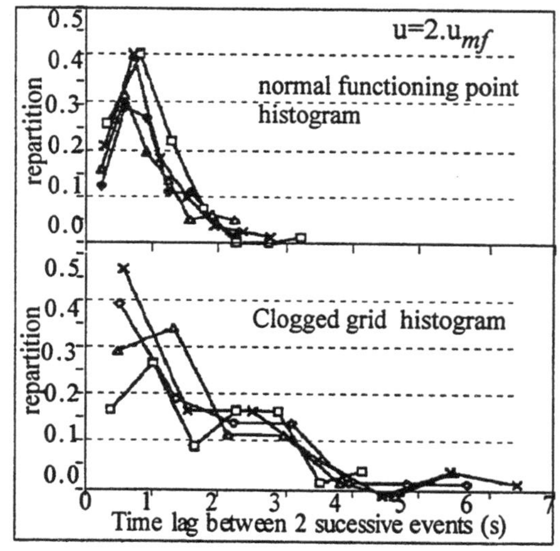

Fig. 9. Comparison of histograms for the grid clogging incident.

5. DIAGNOSIS AND PRECOCIOUS DETECTION OF FUNCTIONING ANOMALY:

The histograms appear to be different when we introduce modifications of the physical parameters of the bed.

5.1 Detection of air arrival failure.

By this way, we are able to detect a defect in the air supply through the grid, in the fluidized bed. Changes have been detected in the statistical distribution of the bubbles. This incident was due to the clogging of the grid at the base of the bed.

On figure 9, the clogging appears clearly by introducing longer time lags in the histograms.

5.2 Detection of a simulated sintering

This experiment was an attempt to simulate a bed sintering. This kind of accident is rather serious and has to be detected as soon it happens. In a 40Kg bed filled with 1mm diameter polystyrene particles, small balls with a mass of 20g were added. Their mass to volume ratio was determined precisely to keep a constant density.

6. CONCLUSION

The proposed method consists mainly of computing precisely the moments when air bubbles cross over the sensors, and to detect any change in the distribution of the time lags. The statistical test are not built yet, but it seems reasonnable to think that it can be done. The distributions looks like Paretto or χ^2 distribution.

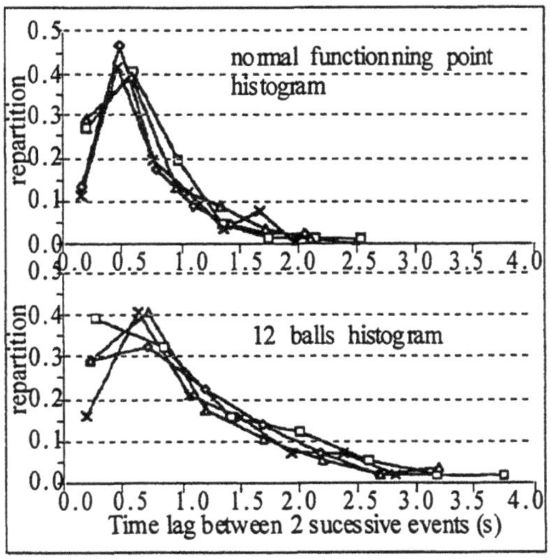

Fig. 10. Comparison of histograms for the simulated sintering experiment.

On figure 10, a change on the histogram can be noted: larger lags appear very soon. The measurements used there were obtained with 12 balls only, that is to say 0.05% of the total mass of the bed.

The experiments carried out by Cerchar were supervised by chemists, and hypotheses were made to approach as far as possible the conditions encountered in real situations. The accidental clogging of the grid was a good oppotunity to check the possibility of the method, and it was a success. However, further experiments should be carried out on industrial boilers to check these hypotheses.

These results are really encouraging. Even if the algorithm is actually rather sophisticated and has to be simplified if possible, this results show that the information does exist in the pressure fluctuation measurement.

REFERENCES

Aragon J.M. and Tatay E. (1984), Study of pressure fuctuation in fluidized beds, *Congrès Européens de Fluidisation*, Espagne.

Claerbout J.F., (1985) *Fundamentals of Geophysical Data Processing*, Ed.Blackwell sc.publ.,, pp49-63.

Flandrin P. (1993), *Temps Frequence*, ed. Hermes, Paris.

Kaufman F. (1987), Analysis of His bundle recordings, *IEEE conf. computers in cardiology*, Louvain, Belgique, pp345-348.

Lacoume J.L. et al. (1985) Factorisation spectrales, *Traitement du Signal*, vol 1, pp 35-40.

Mottelet S. and Simard P. (1994), un algorithme rapide séquentiel de déconvolution L2, *Traitement du Signal*, vol 11, n°5, pp409-416.

Semi-quantitative Model Based Chemical Process Fault Diagnosis via Episodic Fuzzy Rules

İbrahim Burak Özyurt*, Aydin K. Sunol and Lawrence O. Hall*****

**University of South Florida, Department of Chemical Engineering and Department of Computer Science & Engineering, Tampa, FL, USA. Email:ozyurt@kosh.csee.usf.edu*

***University of South Florida, Chemical Engineering Department, Tampa, FL, USA. Email:sunol@eng.usf.edu*

****University of South Florida, Computer Science & Engineering Department, Tampa, FL, USA. Email:hall@csee.usf.edu*

Abstract. A method for chemical process fault diagnosis using numerical behavior envelopes is described in this paper. A sequence of fuzzy rules is generated from numerical behavior envelopes for each fault class where any rule in a sequence is only valid within the bounds of its time interval. For each observed process variable in each fault class two sequences of episodic fuzzy rules, which provide a parsimonious, qualitative description of the trend of the variable, are automatically generated one for the lower and one for the upper numerical behavior envelope. The observed process measurements are matched against the fuzzy rules for the current time in the rule base. In case of an overlapping region defined by behavior envelopes, the introduced distance and time based fault belief scaling allows ranking of fault candidates. A novel abnormal situation will not pass the introduced system undetected due to a novel class detection mechanism. The diagnostic performance of the system is shown in two case studies. *Copyright © 1998 IFAC*

Key Words. Dynamic behavior envelopes; fault diagnosis; qualitative trend analysis; fuzzy clustering; semi-quantitative model based fault diagnosis

1. INTRODUCTION

Knowledge about how a system works facilitates the determination of why it is not working correctly, which is the idea behind model based diagnosis. In model based diagnosis, a model of the physical process is operated in parallel with the process. The outputs of the model indicate how the system should operate, while the outputs of the system indicate how it is actually operating. The discrepancy between actual output and model predictions are used as an indicator of abnormal behavior. The main objective of model based diagnostic systems (MBDS) is then determination of the fault(s) causing the discrepancy.

Semi-quantitative methods, such as fuzzy qualitative simulation (Shen and Leitch, 1993) and Q2 (Kuipers and Berleant, 1988) produce overly conservative bounds since they use a simulation time step determined by qualitative distinctions. However, the numerical behavior envelopes method (Kay and Kuipers, 1993) makes better use of available quantitative information by generating numerical envelopes on all model variables which bound them from above and below. For most of the chemical processes at least some understanding of the underlying chemical and/or physical phenomena is available, which will enable construction of a process model. However, the determination of the precise parameters for the model is usually the limiting step in the model construction process. The numerical behavior envelopes method allows use of an interval instead of a precise value for model parameters. This method, however, requires the numerical integration of two systems, to generate the upper and lower behavior envelopes for each process variable. In real time model based fault diagnosis, for a moderate to large scale process, evaluation of the numerical envelopes may be computationally intractable. The problem becomes worse, in the case that more than one fault model candidate has to be simultaneously computed in order to monitor an abnormal process behavior. In this paper, a method to convert off-line generated numerical behavior envelopes into a parsimonious sequence of fuzzy rules is described for automatic rule base generation. The generated rule base is used for real time process monitoring and fault diagnosis. Equipped with distance and time based fault belief scaling mechanisms and novel class detection capabilities, the developed MBDS minimizes the computational load problem by eliminating the evaluation of numerical behavior envelopes in real time, it allows more refined diagnosis than numerical behavior envelopes alone and provides a novel class detection mechanism to warn about the occurrence of an unknown abnormal situation. The automatic rule base generation ability facilitates knowledge base maintenance and updating. The episodic fuzzy rules not only serve as a parsimonious representation for numerical behavior envelopes, they also provide a natural language like framework to symbolically describe the behavior of the numerical behavior envelopes bounding the process sensor values via the use of linguistic fuzzy variables.

2. OVERVIEW OF SQFDiag

The system introduced in this paper, **S**emi-**Q**uantitative model based **F**ault **Diag**noser (SQFDiag), uses an episodic fuzzy rule base compiled from the behavior envelopes generated using a semi-quantitative model of the monitored system for model based fault diagnosis. A set of equations constraining the system variables is necessary for the semi-quantitative process model. However, no exact values for system parameters are needed, only an interval in which the actual value of the parameter resides is sufficient. Using interval analysis, an extremal system (Kay and Kuipers, 1993) can be constructed from the model equations and model parameter intervals. The extremal system defines a lower and upper behavior envelope (minimal and maximal systems, respectively) for each system variable, when minimal and maximal systems are integrated numerically. The resulting envelopes for each observed system variable for each fault class are then partitioned into monotonically increasing, decreasing or steady sections using the qualitative trend analysis method of Cheung and Stephanopoulos (Cheung and Stephanopoulos, 1990*a*). The geometrically distinguished points for each envelope, for each variable, for all fault classes developed in this manner are pooled and clustered by the fuzzy c-means (FCM) based partitioner to get a good estimate of the number of clusters which implicitly defines the resolution of the resulting qualitative trends (i.e. the number of episodes represented by the resulting discretization). Using the generated fuzzy discretization, the behavior envelopes are converted into sequences of fuzzy episodes which are time indexed fuzzy rules. The generation of the episodic rule base is done off-line as depicted in Figure 1a. The main objective behind converting behavior envelopes into a fuzzy episodic rule base is to facilitate real time operation, since the dynamic behavior envelope method (Kay and Kuipers, 1993) involves numerical integration of two models for upper and lower envelopes during the online monitoring, which can be intractable due to the computational load, if the system monitored is of large scale. The fuzzy

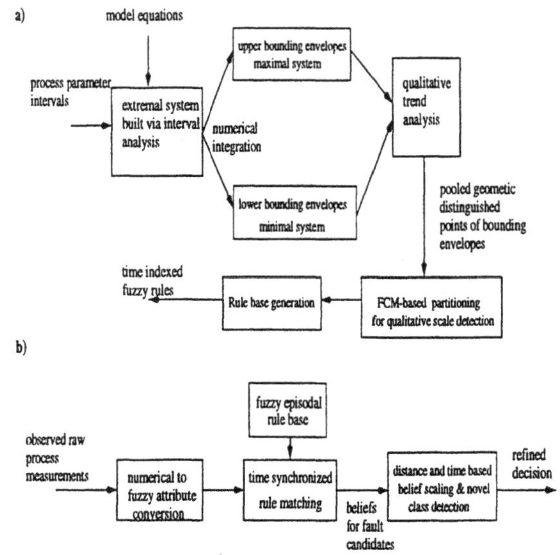

Fig. 1. SQFDiag structure: a) episodic fuzzy rule base generation (off-line mode) b) SQFDiag on-line mode.

episodic rule base is constructed automatically from the off-line simulated envelopes and it is a parsimonious representation of envelopes without real time problems. The introduced MBDS uses the fuzzy episodic rule base as a rule based expert system. The automatic knowledge base construction from a semi-quantitative deep model makes SQFDiag robust and easily maintainable. In the online mode, SQFDiag maps the observed system variables into corresponding fuzzy values and matches with the time indexed fuzzy rules in the rule base in a time synchronized manner. Since the behavior envelopes are usually conservative, significant overlapping of bounded regions for different faults may occur. In this case, to rank the possible fault candidates, distance and time based scaling mechanisms were developed. Also, a novel class detection mechanism is introduced to detect sustaining situations which are out of the bounds of all known classes. The online operation of SQFDiag is depicted in Figure 1b.

3. REPRESENTING DYNAMIC BEHAVIOR VIA EPISODIC FUZZY RULES

While dynamic behavior envelopes enable utilization of imprecise, incomplete quantitative information, for real time fault diagnosis, they still suffer from the computational load problems associated with fully quantitative methods, since they require numerical integration of the two process models for lower and upper bounds. The hypothesis-generate-test scheme used with model based diagnosis may also be a problem computationally, a set of possibly heuristic model selection rules will be necessary to narrow down fault candidates. Iteratively searching for and updating fault models (Shen and Leitch, 1995), seems intractable in real-time for industrial size processes. Rule based systems are more suitable for real-time operation than full model based systems. Compiled knowledge represented in form of rules (frames, objects etc.), with an efficient matching scheme, can be easily matched with the sensor readings in real-time. Here, a methodology for compiled knowledge representation in terms of episodic fuzzy rules using off-line generated dynamic envelopes is introduced.

The semi-quantitative process models from first principles (i.e. conservation laws) with the parameter interval values can be numerically integrated off-line to generate envelopes for each process operating mode and possible faults, building a library of envelopes. This equates to a vast amount of data. To store and use this data in a parsimonious way is of importance for real time operation. The preciseness of these envelopes is of no interest. They only conservatively bound the actual process; they provide qualitative trend behavior. Hence, representing these semi-qualitatively will be enough to use them for real time fault diagnosis. The triangular representation of trend in (Cheung and Stephanopoulos, 1990*a*; Cheung and Stephanopoulos, 1990*b*), where a trend is compactly represented by a sequence of descriptive episodes, is the qualitative representation used. The general concept of *episode* was first introduced in qualitative reasoning by Williams (Williams, 1986). An episode of a trend is defined by

1. a time interval, which is called the temporal extent of the episode,
2. a qualitative context providing meaning for the time interval.

An episode denotes uniform behavior over a time interval. Cheung and Stephanopoulos (Cheung and Stephanopoulos, 1990*a*) suggested representation by a triangular episode, which is defined as an episode, whose qualitative context is determined by the constancy of the qualitative values $[\partial\partial x]$. The qualitative values $[x]$, $[\partial x]$ and $[\partial\partial x]$ correspond to the signs of x, x' and x'', where x is 1-D function defining the trend. The main idea behind the triangular episodes is that any non-monotonic continuous function is a combination of monotonic segments. Using the first and second derivatives of a trend, geometrically distinguished time-points can be determined and used to qualitatively reconstruct the trend. For the behavior envelopes, a geometrically distinguished time-point is either a maximum ($[\partial x] = 0,\ [\partial\partial x] = -$), a minimum ($[\partial x] = 0,\ [\partial\partial x] = +$), a saddle point ($[\partial x] = 0,\ [\partial\partial x] = 0$) or the beginning or end of a monotonically increasing, decreasing or constant region. Representation of envelopes with a sequence of episodes for each observed variable for each faulty and normal operation mode still can be more than necessary to detect abnormal behavior. The notion of *scale* now becomes important. The *scale* is defined as a quantifiable parameter which characterizes the *resolution* at which the underlying physicochemical phenomena are being described by the trend (Cheung and Stephanopoulos, 1990*b*). To determine the scale of interest, which is the scale at which different operating modes of the system can be identified from the measurements, a fuzzy c-means (FCM) based partitioner explained in the following section is used. Using the pooled, representative landmarks at geometrically distinguished time points for each variable for all classes, an FCM based partitioner is used in determining the " optimal " number of fuzzy clusters. Using trapezoidal fuzzy intervals which discretize the trend, the new episodes at the scale implicitly determined using the FCM based partitioner form the base for converting the behavior envelopes into compact episodic fuzzy rules. Each fuzzy rule for an envelope, consisting of a corresponding fuzzy value for the observed variable, is time indexed, i.e. it is only valid for the duration of time defined by the time interval of the episode. The general form of an episodic fuzzy rule for the jth episode of ith fault class, valid within the time interval $[\underline{\epsilon}_{ij}, \overline{\epsilon}_{ij})$ is

$$\begin{aligned} r_{ij}:\ & t \in [\underline{\epsilon}_{ij}, \overline{\epsilon}_{ij}) \cap (\tilde{V}_1 \text{ is } \tilde{T}_1 \cap \ldots \\ & \cap \tilde{V}_k \text{ is } \tilde{T}_k \cap \ldots \cap \tilde{V}_N \text{ is } \tilde{T}_N) \end{aligned} \quad (1)$$

where $\tilde{V}_k$ is the linguistic fuzzy variable corresponding to the kth observed system variable and $\tilde{T}_k$ is the linguistic value associated with the kth variable for the jth episode of the ith fault class. The total number of observed variables is N. A system to automatically generate a fuzzy episodic rule-base from the behavior envelopes is realized in MATLAB. The systematic generation of rules from the behavior envelopes facilitates the maintenance of the rule-base. For the online real-time process monitoring and diagnosis, the generated rule-base is used to monitor the observed process variables and notify the user of the possible causes of abnormal behavior observed.

3.1. *FCM based Partitioner*

The fuzzy c-means (FCM) algorithm (Bezdek, 1981) is a well known clustering algorithm. FCM, unlike the crisp clustering algorithms, such as k-means clustering, allows soft partitioning, i.e. a pattern can partially belong to more than one cluster, according to its membership value, which is a function of the distance metric selected for clustering. The FCM based partitioner is an interactive, graphics oriented program, which provides a good estimate of the number of cluster centers and an automated mapping of raw numerical sensor data to linguistic, trapezoidal shaped fuzzy variables. In all clustering algorithms, the number of clusters (fuzzy sets), which is not known a priori, must be guessed to continue. In the FCM based partitioner, the number of clusters is estimated using two automatic stopping criteria, namely the maximum of the ratio of the slopes of the two adjacent values of FCM objective function and cluster validity index SC (Bensaid *et al.*, 1994). Accompanied by graphics, the program helps the user in selecting the appropriate number of clusters, i.e. fuzzy sets. Clustering is applied to each observed process variable to determine how many fuzzy sets should describe it. The appropriate number of fuzzy sets is not always the "optimum" number of clusters, depending on the amount of background knowledge about the process, the user can intervene, in fuzzy set selection. After the final decision, the clusters are converted to trapezoidal shaped fuzzy numbers using a least squares fit. The fuzzy values of each variable are then interactively labeled. The details of FCM based partitioner can be found in (Özyurt, 1998).

4. FAULT BELIEF CALCULATIONS

In case of overlapping fault regions, a method for updating the beliefs in occurring faults based on a distance metric is used to differentiate, as much as possible, between overlapping faults as depicted in Figure 2. The fuzzy intervals in Figure 2 constitute a FCM generated discretization of the domain of the system variable under consideration. Let's define the *α-width* of the currently active episode of a fault candidate as the α-distance between the corresponding fuzzy values of the upper and lower envelopes. Also define $\mathcal{F} = \{0, 1, \ldots, N_f\}$ as the set of process operating mode IDs, where 0 corresponds to normal operation and N_f is the number of faults. If two or more faults from the set $\mathcal{F}$ have overlapping regions and the observed abnormal behavior falls inside this region, the suggested belief scaling mechanism is used. Given overlapping fault regions i and j from $\mathcal{F}$, the scaling factor $sf_{i,k}$ for the kth variable of fault i is defined as $sf_{i,k} = \frac{a_i}{l_i}$, where a_i is the shortest distance of the observation x to the boundary of the ith fault, l_i is the α-width of fault i bounded by the envelopes. Hence, $sf_{i,k}$ is based on the ratio of the distance of the measurement from class region i. The resulting matrix of scaling factors has M rows and N columns, where M is the cardinality of the overlapping subset of faults from $\mathcal{F}$ and N is the number of the observed variables. The normalized scaling factors $\hat{sf}$

$$\hat{sf}_{ij} = \frac{sf_{ij}}{\sum_{i=1}^{M} sf_{ij}} \tag{2}$$

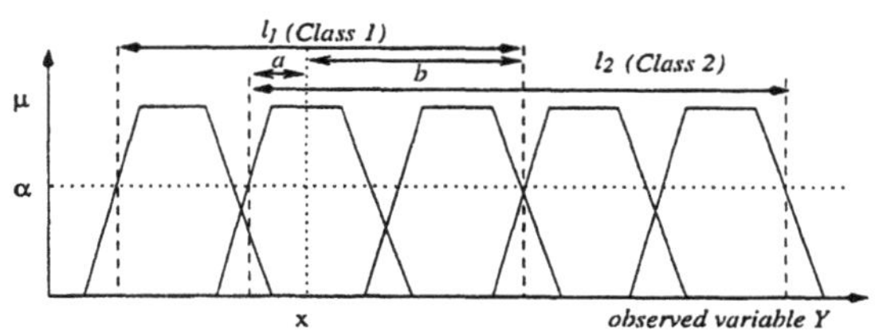

Fig. 2. Two overlapping fault regions (Class 1 and Class 2) with α-widths l_1 and l_2, respectively, for the observed variable Y. The observed value of Y at the current sampling, denoted by x, is inside the overlapping regions of the fault classes. The ratio of distance of x to the nearest boundary of each class region (segments a,b) with the α-width of classes, l_1, l_2 provides a way to scale beliefs for each class.

are used then to scale the membership values resulting from the time synchronized rule matching. The persistence of a fault during the monitoring process also increases the belief in a fault. However, the effect of time should not be monotonically increasing, it should level off, after there is no doubt, about the occurrence of the fault. For that purpose a logistic function with a user specified threshold value (this is the value of the logistic function at the time when the fault with slowest dynamics reaches quasi-steady state as known from the behavior envelopes) is used. The differences between the rate of dynamics of the faults allow us to scale fault candidates in case of a tie. The belief in the occurrence of a particular fault candidate increases more rapidly then the belief of another fault candidate with slower dynamics. The time-scaling function used for this purpose has the following form

$$f(t) = \frac{A}{1 + \exp(-c \cdot t)} + B \tag{3}$$

where the constant c is determined from the given time constant t_a and threshold value t_h for each fault using

$$c = \frac{\ln\left(\frac{1-t_h}{t_h}\right)}{t_a} \tag{4}$$

The constants A and B are used for scaling of equation 3 between desired values (between 0 and 1, in this case). The time scaling factor tf_{ij} for fault i and variable j with the constant c_{ij} determined from equation 4 using the time constant $t_{a,ij}$ at time t is

$$tf_{ij}(t) = \frac{2}{1 + \exp(-c_{ij}t)} - 1 \tag{5}$$

The final form of the belief function for fault i is

$$\text{blf}_i(t) = \frac{\sum_{j=1}^{N} \hat{sf}_{ij} \cdot tf_{ij}(t)\mu_j(t)}{\sum_{k=1}^{N_f+1} \sum_{j=1}^{N} \hat{sf}_{kj} \cdot tf_{kj}(t)\mu_j(t)} \tag{6}$$

Here $\mu_j(t)$ is the fuzzy membership value at current sampling time for the jth variable. The normalization assures that the beliefs sum up to one. The time scaling factor is bounded between 0 and 1. The monitoring system keeps track of the synchronization of the persistence durations of possible candidates. This tracking is necessary, since at every sampling time, new fault candidate(s) can enter the rule-based agenda or some old candidates can leave the agenda depending on the observed values.

The logistic function form mentioned is also used in the detection of novel faults and minimizing the occasional glitches in the monitored noisy sensors. The glitches occur if a monitored sensor measurement goes out of the α-cut bound of the currently active rule set. If this behavior is not sustaining, then to avoid early misdiagnosis, the belief in the current sustaining faults are decreased depending on the persistence time of the glitch. For this purpose the time-scaling function is used with the time constant as the quasi-steady state reaching time of the fault with slowest dynamics. If the glitch sustains, then a novel fault situation is encountered. The described diagnostic system prototype is realized in MATLAB[1].

5. GRAVITY TANK CASE STUDY

As the first case study, a simple nonlinear gravity flow tank model is considered. In this model, the outlet flow is constrained by friction. Three fault classes are considered, namely, abnormal increase in inlet flowrate, tank is leaking and outlet pipe partially plugged. Using interval arithmetic, maximal and minimal systems are generated for the three faults and normal operation, from which the behavior envelopes are generated. The data of the envelopes is then fed to the rule generator and the episodic rules are generated. For testing, noisy sensor data is simulated by adding to the actual process model simulation results Gaussian noise with zero mean and 1% and 0.6% standard deviation from the nominal value of the observed variables, tank level and outlet flowrate. For each fault, 30 cases with different additive gaussian noise are generated and tested. The process is monitored for 500 seconds.

Unlike steady-state fault diagnosis, in dynamic fault diagnosis the fault symptoms develop over time. As a performance criteria, normalized, time averaged belief in a fault is used. The performance index, PI_i for the ith fault is defined as

$$PI_i = \frac{\int_{t_0}^{t_{final}} \text{blf}_i(t)dt}{\sum_{j=1}^{N_f+1} \int_{t_0}^{t_{final}} \text{blf}_j(t)dt} \tag{7}$$

The closer the performance index of a fault is to one and/or the higher its ratio to other classes is, the better it is distinguished from the other classes. Also, since the performance index is the area under each belief curve normalized by the total of area under the belief curve of each fault, a performance index close to one indicates rapid detection. In Table 1, the performance indices averaged for 30 test cases, accompanied with standard deviations in parenthesis are listed. The diagnostic performance of the SQFDiag is rather good as indicated by the performance indices. The closeness of averaged performance indices to one for faults 2 and 3 indicates, rapid identification of these faults. Indeed the beliefs in these faults reached the value one in the first 30 seconds of 500 seconds monitoring session. The first fault is identified in the first 50 seconds of 500 seconds monitoring session, however due to the partial overlapping of fault regions after 100 seconds, belief in fault 3 increased slightly, decreasing the belief in fault 1 to a small extent.

Table 1 Performance indices for gravity tank averaged over 30 test cases.

ID	Fault 1	Fault 2	Fault 3
0	0.030 (0.002)	0.029 (0.000)	0.012 (0.003)
1	0.711 (0.011)	0.003 (0.000)	0.003 (0.001)
2	0.004 (0.001)	0.963 (0.001)	0.004 (0.001)
3	0.255 (0.011)	0.006 (0.000)	0.981 (0.004)

6. CLOSED-LOOP CSTR CASE STUDY

As the second case study, a closed-loop system, namely, a jacketed CSTR with liquid phase decomposition reaction ($A \rightarrow B$), with proportional level control and proportional integral reactor temperature control is used. Ten classes, nine faults and normal operation, are considered. Six process variable are observed. As in the previous case study, the behavior envelopes obtained from the numerical simulation of extremal systems for normal and faulty operation models, are used to generate fuzzy episodic rules, which are used in SQFDiag. Similarly, 30 test cases per fault with additive zero mean Gaussian noise and 0.3%, 1.0%, 0.017%, 0.017%, 0.82% and 0.5% standard deviations from the nominal, for variables LC, F, T, FJC, T_j and CA, respectively. The performance indexes for the best three candidates for each fault accompanied with standard deviations are summarized in Figure 3. The number on each bar indicates the corresponding candidate fault

[1] The system can be downloaded from URL http://morden.csee.usf.edu/~ozyurt/progs.html.

number. The diagnostic performance of the system is not as good as the open-loop case study, where there is no severe partial overlapping of the fault regions bounded by the envelopes. The worst performance is seen for the 4th and 6th faults. The regions for fault class 4 and 6 and the normal operating region nearly totally overlap for five of the six observed variables, for the remaining variable, the faults lie inside or very close to the overlapping region. Without the distance based scaling factor introduced in Section 4, these three classes are not distinguishable. Although the performance indexes for the other two classes are close, for these faults the system identifies the correct fault in each case. Since the available process knowledge only enables one to semi-quantitatively mimic the abnormal system behavior in the form of behavior envelopes, due to high overlapping, difficulty in identification of the actual fault is unavoidable. The introduced distance and time based scaling, however, enabled the system to diagnose the correct fault though with a small margin.

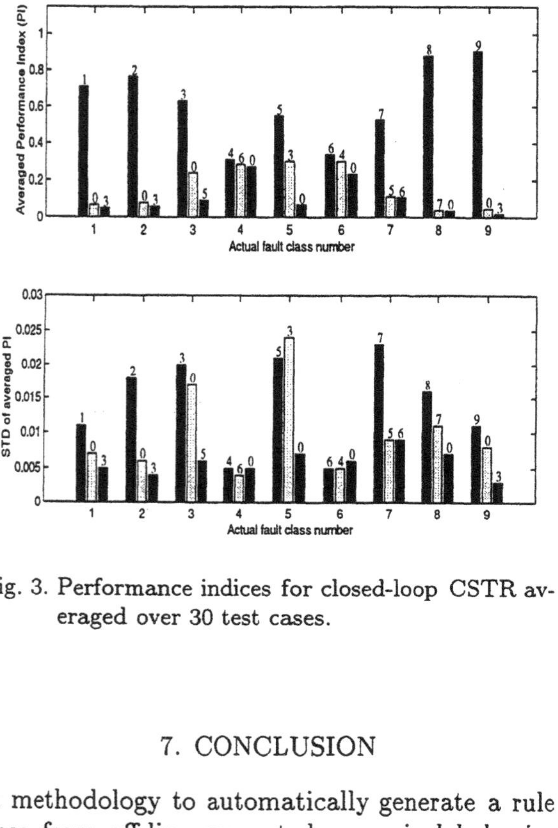

Fig. 3. Performance indices for closed-loop CSTR averaged over 30 test cases.

7. CONCLUSION

A methodology to automatically generate a rule base from off-line generated numerical behavior envelopes for normal and possible abnormal operating modes of a dynamic process is presented. The resulting fuzzy episodic rules, both provide a parsimonious representation of numerical behavior envelopes without the online numerical integration of maximal and minimal systems and also form the framework for a qualitative, natural language like description of the trend behavior of the measured and unmeasured system variables due to the use of linguistic fuzzy variables. The episodic fuzzy rule base enables time synchronized rule matching for real time process monitoring and diagnosis. The diagnostic resolution is refined by the introduced distance and time based fault belief scaling. The MBDS presented, SQFDiag, also has novel fault detection mechanism, to warn the operator about the occurrence of a new and potentially dangerous process abnormality. The automatic rule base generation enables easy maintenance and updating of the knowledge base.

8. REFERENCES

Bensaid, A. M., L. O. Hall, J. C. Bezdek and L. P. Clarke (1994). Fuzzy cluster validity in magnetic resonance images. In: *Proceedings of SPIE on Medical Imaging*. pp. 454–464.

Bezdek, J. C. (1981). *Pattern Recognition with Fuzzy Objective Function Algorithms*. Plenum. New York.

Cheung, J. T. Y. and G. Stephanopoulos (1990*a*). Representation of process trends-part i. a formal representation framework. *Computers chem. Engng* **14**, 495–510.

Cheung, J. T. Y. and G. Stephanopoulos (1990*b*). Representation of process trends-part ii. the problem of scale and qualitative scaling.. *Computers chem. Engng* **14**, 511–539.

Kay, H. and B. Kuipers (1993). Numerical behavior envelopes for qualitative models. In: *Proceedings of National Conference Artificial Intelligence*. Menlo Park, CA. pp. 606–613.

Kuipers, B. and D. Berleant (1988). Using incomplete knowledge in qualitative reasoning. In: *Proceedings of Seventh National Conference in Artificial Intelligence*. pp. 324–329.

Özyurt, I. Burak (1998). Chemical Process Fault Diagnosis using pattern recognition and semi-quantitative model based methods. PhD thesis. University of South Florida.

Shen, Q. and R. Leitch (1993). Fuzzy qualitative simulation. *IEEE Transactions on System Man and Cybernetics* **23**, 1038–1061.

Shen, Q. and R. Leitch (1995). Diagnosing continuous systems with qualitative dynamic models. *Artificial Intell. in Engng* **9**, 107–125.

Williams, B. C. (1986). Doing time:putting qualitative reasoning on firmer ground. In: *Nat. Conf. on Artificial Intell.*. Philadelphia.

EVALUATION OF STRUCTURED RESIDUALS, APPLICATION TO A GAS-LIQUID SEPARATION UNIT

Michel Kinnaert, Abdelillah Youssouf

Laboratoire d'Automatique, Université Libre de Bruxelles
CP 165, 50 Av. F. D. Roosevelt, B-1050 Brussels, BELGIUM

Abstract: A method to detect and isolate faults from structured residuals using Generalized Likelihood Ratio (GLR) tests is described. The whole residual vector is processed at once by as many GLR tests as there are faults. With this scheme, several tests can cross their threshold upon occurence of a single fault. A decision on what fault is the most likely is taken by applying a procedure based on the residual prediction error associated to each triggered test. The method is studied in simulation, and it is shown to allow fault isolation without requiring any supplementary decision tool. *Copyright © 1998 IFAC*

Keywords: Fault detection, identification and isolation, generalized likelihood ratio test, chemical process

1. INTRODUCTION

A model-based fault detection and isolation (FDI) system for monitoring of a technical process is made of two parts : a residual generator and a decision module. The first part is a filter that takes the actuator commands and the sensor outputs as inputs, and generates residuals, namely signals that are nominally equal to zero in the absence of fault, and become distinguishably different from zero upon occurrence of specific faults. The decision module monitors the set of residuals to determine whether some of them are significantly different from zero, and it deduces the most likely faulty components.

In order to perform detection and isolation of a set of faults, structured residuals can be used. In this case, only a fault specific subset of the residuals becomes non-zero in response to a single fault (Gertler, 1993). The so-called incidence matrix describes the subset of residuals which react to each fault. It is a matrix with binary entries, which can have, for instance, the form indicated below for a situation with three faults and three residuals, r_1, r_2, r_3 :

Table 1: Incidence matrix

	fault 1	fault 2	fault 3
r_1	1	0	1
r_2	1	1	0
r_3	0	1	1

Each row corresponds to a given residual and each column to a given fault. An entry equal to 1 indicates that the residual is sensitive to the corresponding fault.

A typical way of monitoring residuals is to compare them to a fixed threshold. However, such a method requires that the degrees of sensitivity of a given residual with respect to each fault be similar, and that the signal to noise ratio be large. This is not always achievable in practice. Therefore the result of a diagnosis might be very sensitive to the threshold value. This phenomenon is called diagnosis instability. Such a phenomenon can be partly alleviated by using non-Boolean inference techniques such a fuzzy set theory or the theory of evidence (Peng *et al.*, 1996). However, in those different approaches, no evaluation of the fault magnitude or the fault occurrence time is provided. This type of information is needed when temporary or successive faults must be detected. Indeed, in such situations, the residual generator must be updated by taking the fault magnitude and occurrence time into account (Willsky and Jones, 1975). To obtain this information, the residual can be evaluated by generalized likelihood ratio (GLR) tests (Basseville and Nikiforov, 1993).

Testing each residual separately by a GLR test seems to be at first a natural way of proceeding, given our background on threshold comparison. However, in this way the correlation between the different residuals is not taken into account by the statistical tests. A rule based system can be used to perform fault isolation by monitoring the relations between the ouputs of the GLR tests. Yet this system can become quite complex for realistic applications. Moreover, when scalar structured residuals are processed separately, the estimate of the fault magnitudes are often of poor quality as is illustrated later.

Therefore, the purpose of this paper is to examine the feasibility of another way to use the GLR tests: the whole residual vector is processed by as many GLR tests as there are faults, each test being tuned to detect a specific fault. The incidence matrix is thus processed columnwise instead of rowwise. The efficiency of this approach is first demonstrated on a purely theoretical example. Next, it is used to process a set of residuals obtained in (Denolin *et al.*, 1997) for FDI on a gas-liquid separation unit.

The paper is organized as follows. In section 2, a review of the GLR test is presented. The way to use it for structured residual monitoring is discussed in section 3. In section 4, simulation studies on a purely theoretical system and on the gas-liquid separation process are reported.

2. REVIEW OF THE GLR TEST

The following model of a q-dimensional residual vector is considered :

$$r(k) = G(k;\theta)\nu 1_{\{k\geq\theta\}} + r_0(k) \quad (1)$$

where r_0 is a q-dimensional zero mean Gaussian white noise sequence with known covariance matrix $V(k)$, θ is the fault occurrence time, ν is a p-dimensional vector characterizing the fault magnitude, $G(k;\theta)$, a $q \times p$ matrix, is the dynamic profile of the change for a fault starting at time θ, and $1_{\{k\geq\theta\}}$ is equal to 1 when $k \geq \theta$ and to zero otherwise. For the sake of clarity, we assume that, upon occurrence of fault i, $i = 1,\ldots p$, the ith component of ν becomes non-zero, while all other components are equal to zero. Moreover, simultaneous faults are not considered in this paper. r_0 represents the actual residual vector, in the absence of fault, and the first term in the right hand side describes the effect of a fault at time θ on the residual (at time k).

Thus $r(k)$ has the following distribution :

$\mathcal{L}(r(k)) = \mathcal{N}(0, V(k))$ in the absence of fault,

$\mathcal{L}(r(k)) = \mathcal{N}(G(k;\theta)\nu, V(k))$ after occurence of a fault.

The generalized likelihood ratio (GLR) test allows to decide between hypothesis H_0 , namely the no-fault situation ($\theta > k$), and hypothesis H_1, the faulty situation. This test amounts to computing the maximum likelihood estimates $\hat{\theta}(k)$ and $\hat{\nu}(k)$ based on $r(1),\ldots,r(k)$ and the hypothesis H_1. These values are then used for computing the log-likehood ratio, $\ell(k)$, of H_1 versus H_0, given $r(1),\ldots,r(k)$. If $\ell(k)$ is larger than a fixed threshold, hypothesis H_1 holds, otherwise hypothesis H_0 is retained. The threshold is set to assure a reasonable tradeoff between missed and false alarm rates (Willsky and Jones, 1975).

Standard developments lead to the following expressions for the test (Willsky and Jones, 1975), (Basseville and Nikiforov, 1993):

$$\hat{\nu}(k) = C^{-1}(k;\hat{\theta}(k))d(k;\hat{\theta}(k)) \quad (2)$$

where C is the deterministic matrix :

$$C(k;\theta) = \sum_{j=\theta}^{k} G^T(j;\theta)V(j)^{-1}G(j;\theta) \quad (3)$$

and d is a linear combination of the residuals :

$$d(k;\theta) = \sum_{j=\theta}^{k} G^T(j;\theta)V(j)^{-1}r(j) \quad (4)$$

The maximum likelihood estimate $\hat{\theta}(k)$ is the value of $\theta \leq k$ that maximizes :

$$\begin{aligned}\ell(k;\theta) &= \sum_{j=\theta}^{k} ln\frac{p(r(j)|H_1,\hat{\nu}(k))}{p(r(j)|H_0)} \\ &= d^T(k;\theta)C^{-1}(k;\theta)d(k;\theta) \quad (5)\end{aligned}$$

where $p(r(j)|H_0)$ ($p(r(j)|H_1,\hat{\nu}(k))$) is the probability density of $r(j)$ conditioned on H_0(on H_1 and $\hat{\nu}(k)$). The decision rule is:

$$\ell(k,\hat{\theta}(k)) \underset{H_0}{\overset{H_1}{\gtrless}} \epsilon \quad (6)$$

where ϵ denotes the decision threshold.

For implementing this test, one runs into the problem that the maximization of $\ell(k, \theta)$ must be performed over an increasing time span $\theta = 1, \ldots, k$. To avoid this problem, a finite data window is introduced. If M denotes the window length, the maximization of $\ell(k;\theta)$ is restricted to $\theta = k - M, \ldots, k$. This amounts to assuming that any fault of interest will normally trigger the test during the M time instants following its occurrence. An algorithm based on this finite window is presented in (Willsky and Jones, 1975).

3. MONITORING STRUCTURED RESIDUALS WITH THE GLR TEST

3.1 *Introduction*

Let us now turn to the evaluation of structured residuals with the GLR test. For the sake of simplicity, the discussion is based on residuals corresponding to the incidence matrix given in table 1. Yet extension to a general situation is trivial. For table 1, the residual vector can be modelled in the form (1) with

$$G(k;\theta) = \begin{bmatrix} G_{11}(k;\theta) & 0 & G_{13}(k;\theta) \\ G_{12}(k;\theta) & G_{22}(k;\theta) & 0 \\ 0 & G_{32}(k;\theta) & G_{33}(k;\theta) \end{bmatrix}$$

and $\nu = \begin{bmatrix} \nu_1 & \nu_2 & \nu_3 \end{bmatrix}^T$

We consider successively an approach where each component of the residual vector is processed by a separate GLR test (scalar residuals), and another where the whole residual vector is monitored at once.

3.2 *GLR test applied to scalar residuals*

Testing each component of the residual vector separately, as is typically done using simple thresholds, requires the use of three GLR tests. The first one is designed for a model of the form

$$r_1(k) = \begin{bmatrix} G_{11}(k;\theta) & G_{13}(k;\theta) \end{bmatrix} \begin{bmatrix} \nu_1 \\ \nu_3 \end{bmatrix} 1_{\{k \geq \theta\}} + r_{01}(k) \quad (7)$$

where $r_{01}(k)$ denotes the first entry of $r_0(k)$. Obviously, to be able to apply the GLR test on (7), the dynamic profiles of change, $G_{11}(k;\theta)$ and $G_{13}(k;\theta)$ should be different, otherwise matrix $C(k;\theta)$ as defined in (3) is singular. Should this not be true, a model of the following form could be used :

$$r_1(k) = G_{11}(k;\theta)\nu 1_{\{k \geq \theta\}} + r_{01}(k) \quad (8)$$

where $\nu = \nu_1 + \rho_{13}\nu_3/\rho_{11}$, ρ_{ij} denoting the steady state value of $G_{ij}(k;\theta)$.

However, even when $G_{11}(k;\theta)$ and $G_{13}(k;\theta)$ are distinct, it was noticed via simulations that the estimated fault magnitudes can differ significantly from their actual value, which does not allow one to judge of the severity of the fault. Moreover, r_{01} is often correlated to the other components of r_0, and this fact is not exploited when each residual is monitored separately. This motivates the search for an alternative approach.

3.3 *GLR applied to residual vector*

The GLR test will now be used to monitor directly the q-dimensional residual. Again, in the situation of table 1, this amounts to building three tests, each one being associated to one fault. For the first fault, the model to be used is :

$$r(k) = \begin{bmatrix} G_{11}(k,\theta) \\ G_{12}(k,\theta) \\ 0 \end{bmatrix} \nu_1 1_{\{k \geq \theta\}} + r_0(k) \quad (9)$$

To be able to achieve fault isolation, with that approach, it is not sufficient to simply look at which of the GLR tests was triggered. Indeed, several tests corresponding to different faults can be triggered, when one single fault has occurred. Roughly speaking, this is due to the fact that the correlation of the residual vector w.r.t. any non zero profile is non zero, and it can become sufficiently large to trigger a test based on an arbitrary profile, when the fault is large. A way to achieve isolation is to compare the quality of the estimates of the residual vector which can be derived from each test. This can be done by evaluating, for each test, a cost function based on the residual estimation error, namely :

$$J_i(\widehat{\nu}_i(\widehat{\theta}_i + M)) = \frac{1}{M+1} \sum_{j=\widehat{\theta}_i}^{\widehat{\theta}_i + M} \eta_i(j, \widehat{\theta}_i)^T \eta_i(j, \widehat{\theta}_i) \quad (10)$$

where $\eta_i(j, \widehat{\theta}_i) = r(j) - G_{.,i}(j, \widehat{\theta}_i)\widehat{\nu}_i(\widehat{\theta}_i + M)$, the index i refers to the test associated to the ith fault, and $G_{.,i}(j, \widehat{\theta}_i)$ is the ith column of $G(j, \widehat{\theta}_i)$.

$J_i(\nu)$ is precisely a cost function which is minimized by $\widehat{\nu}_i(\widehat{\theta}_i + M)$ when r is affected by fault i at time $\widehat{\theta}_i$. If several tests are triggered, the most likely fault corresponds to the cost (10) with the lowest value. Although each test is possibly triggered at the alarm time t_{ai} $(\leq \theta + M)$, $i = 1, \ldots, p$, the isolation is only completed at time $min_i\widehat{\theta}_i + 2M$ $(\geq max_i\widehat{\theta}_i + M)$. The reason is that a sufficiently large number of samples is needed to compute a reasonable estimate of the fault magnitude (and of the residual), and hence to be able to take a decision on the basis of the above criterion.

The main interest of the proposed approach is that it allows one to isolate faults without the need for supplementary decision tools. Moreover, an estimate of the fault magnitude is provided directly, while it is not clear how to get such an information when processing the residual vector componentwise, as illustrated below.

4. SIMULATION RESULTS

We first consider a purely theoretical situation where the model of the residual is perfectly known. Next application of the isolation procedure to a set of residuals generated for the supervision of a gas-liquid separation process is considered.

4.1 *Perfectly known residual model*

A residual of the form (1) is considered. r_0 is a white noise vector with covariance matrix $0.01 * I(3)$, where $I(3)$ denotes the 3×3 identity matrix, and matrix $G(k,\theta) = G(k-\theta)$ is the step response of the following transfer matrix :

$$\begin{bmatrix} \frac{0.5}{z-0.5} & 0 & \frac{0.2}{z-0.8} \\ \frac{0.3}{z-0.7} & \frac{0.6}{z-0.4} & 0 \\ 0 & \frac{0.2}{z-0.8} & \frac{0.3}{z-0.7} \end{bmatrix} \quad (11)$$

Let us first illustrate the poor estimation of the fault magnitude obtained when processing scalar residuals separately. Residual (1) is simulated with $G(k,\theta)$ given by (11), $\nu = [0.5 \;\; 0 \;\; 0]^T$ and $\theta = 500$. This corresponds to the occurence of a fault 1 of magnitude 0.5 at time instant 500. Next separate GLR tests are used to process $r_1(k), r_2(k)$ and $r_3(k)$. All GLR tests have an horizon $M = 15$ and a threshold equal to 30. The value of the horizon is chosen so that the major part of the residual transient upon occurence of a fault can be seen inside a time window of length M. The threshold was set above the value of the three test functions in the absence of faults, for typical noise realizations. An example of simulation results is now described. Tests 1 and 2, linked to r_1 and r_2, have crossed their threshold at time 504 and 506 respectively. The estimated fault occurence time are $\widehat{t}_{01} = 501$ and $\widehat{t}_{02} = 499$ and the estimated fault magnitudes are :

- For test 1, at time $\widehat{t}_{01} + M$, $[\widehat{\nu}_1 \;\; \widehat{\nu}_3] = [0.5127 \;\; -0.0243]$.
- For test 2, at time $\widehat{t}_{02} + M$, $[\widehat{\nu}_1 \;\; \widehat{\nu}_2] = [1.2812 \;\; -0.7932]$.

Although the fact that r_1 and r_2 have crossed their threshold allows one to isolate fault 1, one cannot deduce the fault magnitude, as it is not clear how to choose between the two estimates for $\widehat{\nu}_1$.

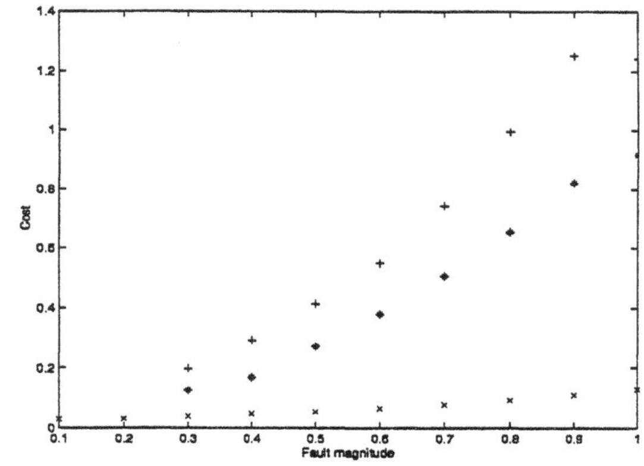

Fig.1. Cost (10) associated to test 1(x), test 2(+) and test 3(*)

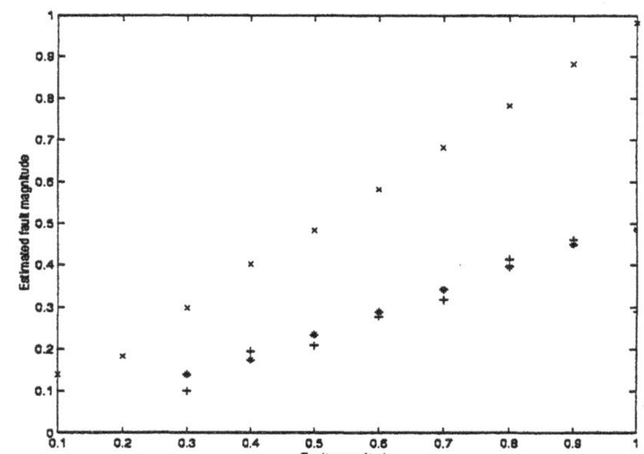

Fig.2. Estimated fault magnitude associated to test 1(x), test 2(+) and test 3(*)

Let us now illustrate the effectiveness of the approach based on processing the residual vector globally. To this end, a realisation of residual vectors of the form (1) described above with $\nu = [\alpha \; 0 \; 0]^T, \alpha = 0.1, \ldots, 1$ and $\theta = 500$ was generated. The residual vectors obtained for each fault magnitude were processed by three GLR tests associated to the three faults that can affect the residual. The tuning parameters of these tests were set as follows : horizon $M = 15$, threshold=30. Their value was set by following the guidelines mentioned previously. Figure 1 and 2 respectively show the value of the cost (10), and the estimated fault magnitude, as a function of the actual fault magnitude. Let test $i, i = 1,2,3$ be the GLR test associated to fault $i, i = 1,2,3$. From Fig. 1, it is seen that only test 1 is triggered for fault magnitudes 0.1 and 0.2. For larger fault magnitudes the value of the cost function associated to test 1 is always lower than the cost associated to test 2 and test 3 . This criterion can thus be used to deduce that fault 1 has occured. In Fig. 2, the estimated fault magnitude is seen to be close to its actual value only for test 1.

4.2 *Gas-liquid separation process*

A semi-industrial gas-liquid separation process has been modelled by a set of static and dynamic nonlinear equations of which the parameters were identified from experimental data (Vrancic *et al.*, 1995). A diagnosis system has previously been designed for the plant (Denolin *et al.*, 1997), (Daele *et al.*, 1997). The residual generation part of this system was made of static analytical redundancy relations and of fault detection filters based on nonlinear observers. A set of seven structured residuals was generated for the detection and isolation of ten faults. Each scalar residual was processed by a GLR test for which the dynamic profile of the change was assumed to be a step function. This is of course a very rough approximation. The output of each GLR test, together with extra information on the residual was converted into beliefs and timing information. The latter were processed by a causal graph in order to obtain a set of candidate faults ranked from the most to the least probable. The decision system needed with that approach was quite complex, and this is the reason why an alternative residual evaluation procedure is studied here.

In the sequel, we present preliminary results on the application of the method described in section 3.3 to evaluate the seven residuals described in (Denolin *et al.*, 1997). The response of the residual vector to step like faults of magnitude 5% occuring at time instant 200 is given in table 2 for seven possible faulty modes. f_{u_1} and f_{u_2} refer to actuator faults , namely faults in valve positions (expressed in percents of the nominal value); $f_{\ell R1}$ and $f_{\ell R2}$ stand for leaks in the two tanks (expressed as percents of the nominal output flows); c represents a clog in the injector (expressed in percents of the obstructed section), and f_{Φ_1}, f_{Φ_2} are additive bias on flow measurements Φ_1 and Φ_2(expressed in percents of the nominal flows).

The dynamic profile of the fault associated to each test must be determined for configuration of the GLR tests. This was achieved by identifying ARX models corresponding to the faulty working modes of the process. More precisely, step-like faults with magnitude 10% were simulated, and the resulting data were used for profile identification. One should notice that all the faults cannot be seen as additive changes in the physical equations describing the plant behaviour. Choosing a model of the form (1) for the residual vector is thus an approximation. Each test was adjusted so that it would not cross its threshold in the absence of fault for typical noise realizations. For a test horizon of 150 time steps, which is of the order of magnitude of the slowest dominant time constant in the dynamic profile of changes for all tests, this led to a threshold of 30.

The results presented here concern the evaluation of the residual vector depicted in column 6 of table 2. This vector was evaluated by the seven GLR tests (one test for each fault). The evolution of the test functions is plotted in Fig. 3. All the tests crossed their threshold. The associated costs (10), estimated fault occurence time, $\widehat{\theta}_i, i = 1,\ldots,7$ and estimated fault magnitude at time $\widehat{\theta}_i + M, i = 1,\ldots,7$ are indicated in table 3.

Table 3: Cost, $\widehat{\theta}_i, \widehat{\nu}_i(\widehat{\theta}_i + M)$

	Cost	$\widehat{\theta}_i$	$\widehat{\nu}_i(\widehat{\theta}_i+M)$
Test 1	0.0927	207	0.05150
Test 2	0.1656	244	-0.00007
Test 3	0.1626	213	-0.42964
Test 4	0.1616	277	-0.09749
Test 5	0.0848	210	-0.05351
Test 6	0.0105	207	0.04124
Test 7	0.1641	229	0.02277

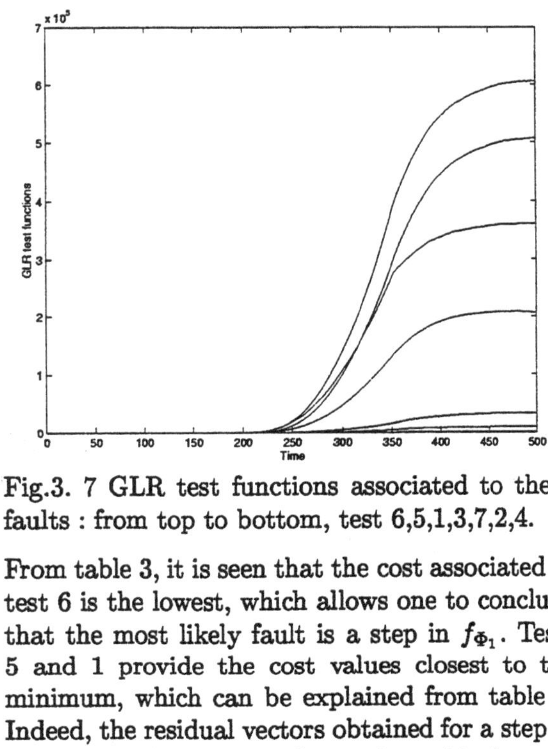

Fig.3. 7 GLR test functions associated to the 7 faults : from top to bottom, test 6,5,1,3,7,2,4.

From table 3, it is seen that the cost associated to test 6 is the lowest, which allows one to conclude that the most likely fault is a step in f_{Φ_1}. Tests 5 and 1 provide the cost values closest to the minimum, which can be explained from table 2. Indeed, the residual vectors obtained for a step in f_{u_1} or c are the most similar to the residual vector associated to f_{Φ_1}. Hence the dynamic profiles of change for tests 1 and 5 are also significantly correlated with the residual obtained upon occurence of the bias on Φ_1. This is also the reason why the estimated fault appearance time provided by tests 1, 5 and 6 are the closest to the actual fault occurence time (200). Regarding the estimate of the fault magnitude, it is seen that test 6 provides a reasonable but underestimated value (the actual fault magnitude is 0.05). It could be improved by increasing the test horizon, but this means a longer delay before a decision regarding fault isolation can be taken. More computations are also

	$f_{u_1}=5\%$	$f_{u_2}=5\%$	$f_{IR_1}=5\%$	$f_{IR_2}=5\%$	$c=5\%$	$f_{\Phi_1}=5\%$	$f_{\Phi_2}=5\%$
r_1	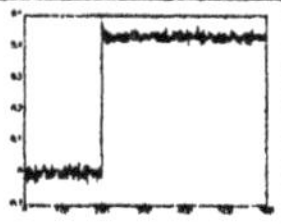	0	0	0	0	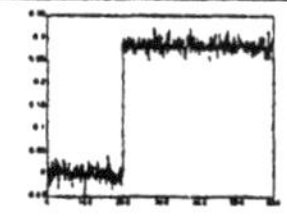	0
r_2	0	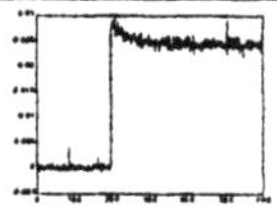	0	0	0	0	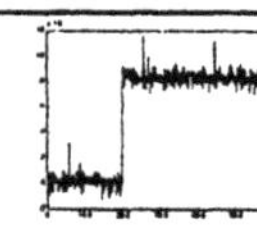
r_3	negligible transient response	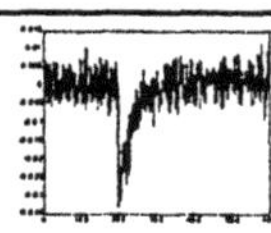	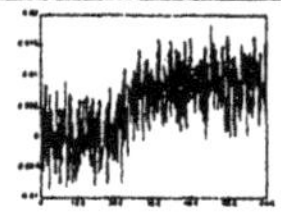	0	negligible transient response	0	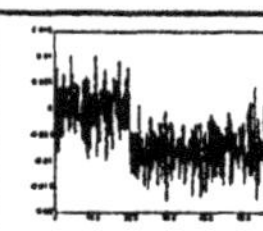
r_4	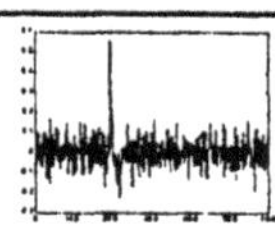	negligible transient response	negligible transient response	0	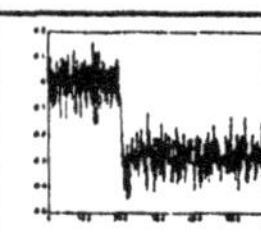	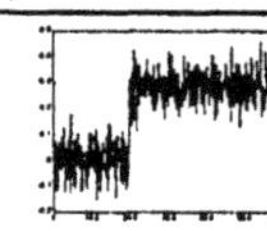	0
r_5	0	0	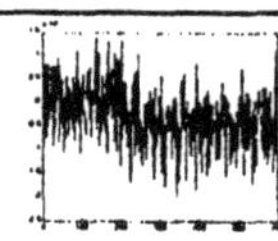	0	0	0	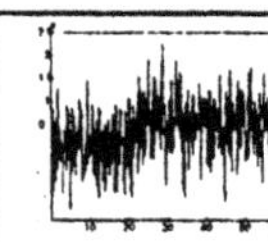
r_6	0	0	0	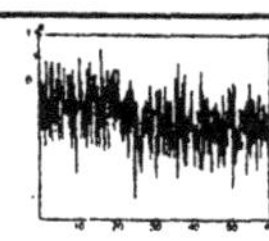	0	0	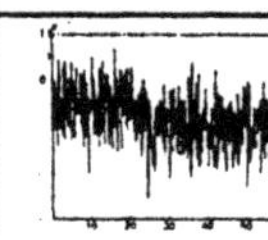
r_7	0	0	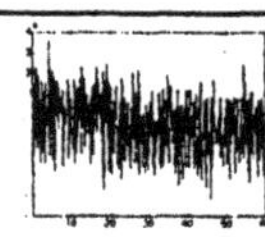	0	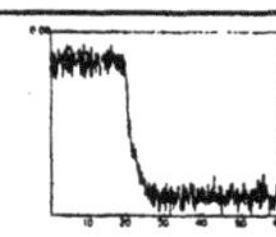	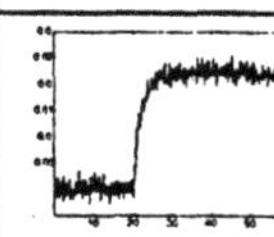	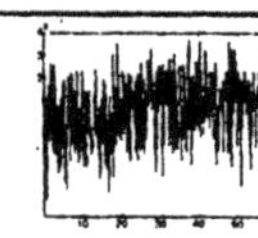

Table 2 : Rssidual patterns resulting from abrupt faults

required at each sampling time, for larger values of M. Notice that the fault estimate provided by test 1 is close to 0.05. Taking a longer horizon would however lead to a larger value for $\widehat{\nu}_1(\widehat{\theta}_1)$.

5. CONCLUSION

A method for evaluation of structured residuals by GLR tests has been presented, and tested in simulations. Further work is needed to be able to present statistical results based on a sufficiently large number of runs to account for the variability due to noise. The next step will be to adapt the approach to situations where temporary and successive faults are taking place.

6. REFERENCES

Basseville, M. and I.V. Nikiforov (1993). *Detection of Abrupt Changes: Theory and Application.* Prentice Hall. Englewood Cliffs, N.J.

Daele, R. Vanden, Y. Peng and M. Kinnaert (1997). Fault diagnosis using belief functions. *Proceedings of SAFEPROCESS'97* **1**, 546–551.

Denolin, E., D. Vrancic, M. Kinnaert, D. Juricic and J. Petrovcic (1997). Model-based diagnosis for a gas-liquid separation unit. *Proceedings of SAFEPROCESS'97* **1**, 228–233.

Gertler, J. (1993). Residual generation in model-based fault diagnosis. *Control Theory and Advanced Technology* **9(1)**, 259–285.

Peng, Y., Ph. Arte and R. Hanus (1996). Quantitative model-based fault diagnosis using belief functions. *Proceedings of CESA'96 IMACS Multiconference, Symposium on Control, Optimization and Supervision* **1**, 528–532.

Vrancic, D., D. Juricic and J. Petrovcic (1995). Semi-industrial benchmark problem: measurement and mathematical modelling of a process for separation of gas from liquid. *Internal Report, Jozef Stefan Institute.*

Willsky, A.S. and H.L. Jones (1975). A generalized likelihood ratio approach to detection and estimation of jumps in linear systems. *IEEE Trans. Automatic Control* **AC-21(1)**, 108–112.

NON-LINEAR OBSERVER BASED METHOD FOR FAULT DETECTION AND ISOLATION : Application to an industrial petrochemical process

Frédéric ARMANET, Nida OTHMAN, Nicolas COUENNE

Institut Français du Pétrole
1-4 avenue de Bois Préau
92 852 Rueil Malmaison Cédex
France

Abstract : This paper deals with the implementation of algorithmic fault diagnosis on an industrial petrochemical reactor. This process is cared out in a perfectly mixed reactor with a heat exchanger. The algorithm is based on a non linear geometric approach of the disturbance rejection. It allows to supervise three faults which may occur on the reactor with online temperature and pressure measurements. *Copyright © 1998 IFAC*

Keywords : nonlinear observer, residual filter, fault diagnosis

1. INTRODUCTION

In this paper, we present an application of residual filters synthesis to the fault diagnosis on a perfectly mixed reactor.

In the section 2, we expose the main points of the theoretical background used for the application. It is an observer based method called residual synthesis. A residual signal is a signal depending on the online measures and such that its value is sensitive to the presence of a fault. The main principle is to design partial observers which are sensitive to some faults among all the set of possible faults. If a fault occurs in the plant, the error beetwen reconstructed ouput and measure does not go to zero. Thus, a residual signal could be any function measuring this error.

The application on the industrial plant is described in the section 3. First, the physical model with the fault influence is given. Then, main methodology to design residual signals is described. Finaly, simulation results are shown.

2. THEORETICAL BACKGROUND

Non linear observer theory was extended to a large class of non linear systems. In the same time, the tools for the rejection of disturbance on non linear systems were widely studied. A typical application of these results is fault diagnosis. The extended theory presented here can be found for linear systems in (Massoumnia, et al.,1986 and 1989) and its extension to non linear systems in (Hammouri, et al., 1994).

The first part is dedicated to a description of residual generators. A residual generator is a differential system which is sensitive to one or more faults. The second part gives some results about the design of such residual generators.

2.1 residual signal properties

Non linear systems considered here are on the following form (1) :

$$\begin{cases} \dot{x} = f(x,u,e) \\ y = h(x,e) \end{cases} \qquad (1)$$

Where $x \in R^n$ is the state of the system, $u \in R^m$ the control inputs, $y \in R^p$ the measured outputs, and $e \in R^q$ is a vector which represents q independent and unknown fault signals.

The ith fault is represented as a structured disturbance of the model by a fault signal $e_i(t)$. A fault signal is non identically zero if and only if the associated fault exists.

Fault diagnosis can be made with the following method. For the ith fault, it consists in designing a differential system called residual generator (2) :

$$\begin{cases} \dot{z} = F(z,u,y) \\ r_i = H(z,u,y) \end{cases} \quad (2)$$

verifying one of the following assumptions :

- $r_i(t)$ becomes not identically zero if the signal $e_i(t)$ is not identically zero. $r_i(t)$ becomes zero if the vector $e(t)$ is identically zero. In this case, it is said that the signal r_i detects the fault e_i.

- $r_i(t)$ becomes not identically zero if and only if the signal $e_i(t)$ is not identically zero. In this case, it is said that the signal r_i isolates the fault e_i from the other fault.

$r_i(t)$ is the residual signal associated to the ith fault and the system (2) is the residual generator. The second property is stronger than the first one.

A logical analysis of each residual signal allows to conclude about fault presence among the set of all fault signals.

2.2 *residual generator design for the ith fault*

The design of a residual generator associated to the ith fault proceeds in two steps. The first one is to find a coordinate change in the state-space (3) :

$$T_i : x \rightarrow \xi = T_i(x) \quad (3)$$

and an output mapping :

$$\phi_i : y \rightarrow \bar{y} = \phi_i(y) \quad (4)$$

such that the system (2) is written in the new coordinates under the following form (5) :

$$\begin{cases} \dot{\xi}_1 = \bar{f}_1(\xi_1, \xi_2, u, e) \\ \dot{\xi}_2 = \bar{f}_2(\xi_2, u, e_i) \\ \bar{y}_1 = \bar{h}_1(\xi_1, \xi_2, e) \\ \bar{y}_2 = \bar{h}_2(\xi_2, e_i) \end{cases} \quad (5)$$

where $\xi = \begin{pmatrix} \xi_1 \\ \xi_2 \end{pmatrix}$ and $\bar{y} = \begin{pmatrix} \bar{y}_1 \\ \bar{y}_2 \end{pmatrix}$.

This coordinate change is chosen such that the sub-system (6) :

$$\begin{cases} \dot{\xi}_2 = \bar{f}_2(\xi_2, u, e_i) \\ \bar{y}_2 = \bar{h}_2(\xi_2, e_i) \end{cases} \quad (6)$$

is not perturbed by the other fault signals e_j , for $j \neq i$.

The second step in residual generator design is to find an observer for the nominal system (7) :

$$\begin{cases} \dot{\xi}_2 = \bar{f}_2(\xi_2, u, 0) \\ \bar{y}_2 = \bar{h}_2(\xi_2, 0) \end{cases} \quad (7)$$

Assuming this is possible, a residual signal associated to the fault e_i is then given in (8) :

$$r_i(t) = R_i\left(\bar{h}_2(\hat{\xi}_2) - \bar{y}_2\right) \quad (8)$$

where $\hat{\xi}_2$ is the estimated state given by the observer for the nominal subsystem (7). $R_i(v)$ is a function verifying $\forall v, R_i(v) = 0 \Leftrightarrow v = 0$.

3. APPLICATION

This part deals with the design of such fault analysis for an industrial petrochemical process. The first section describes successively the model of the process, the sensors and the different faults considered.

The second section shows the construction principles for residual generators. The first one is sensitive to only one fault. The second one is sensitive to two faults, and the third one is sensitive to all the faults.

The third and last section shows some simulation results.

3.1 *Model design*

The industrial process considered here is a dimerisation reactor. The reactions are homogeneously catalysed and take place in a perfectly mixed liquid phase. A heat exchanger evacuates the energy produced by the reaction. The model of the process is established with dynamic energy and mass balance. A scheme of the process is given in figure 1.

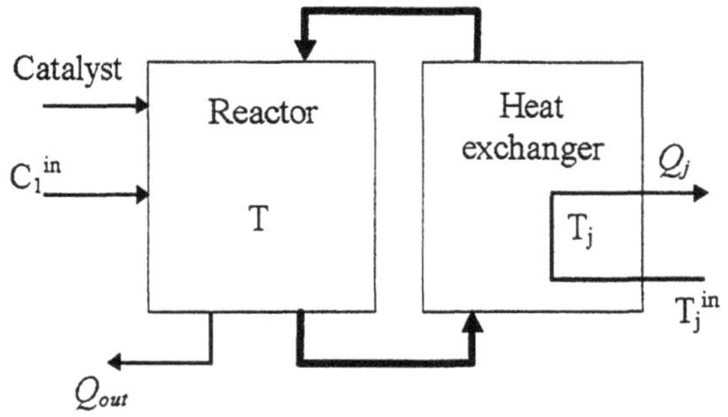

Figure 1 : scheme of the process

3.1.1 kinetic model

The reactor contains four different species which are noted respectively C_1, C_2, C_S and K_i.

The reaction scheme is :

$$C_1 \xrightarrow{r_{12}} C_2 \xrightarrow{r_{2S}} C_S$$

K_i is a catalyst element. Hence it takes place in the reaction scheme only throught its influence on the kinetic rate r_{12} and r_{2S}.

These kinetic rates follows an Arrhenius law. As there is no confusion possible, *italic* notations in the mathematical model are the representative state value associated to each of the physical variables. Mathematical expressions for kinetic rates are of the form :

$$r_{12} = \alpha_1 \exp - \frac{\lambda}{RT} C_1 K_i \qquad (9)$$

and

$$r_{2S} = \alpha_2 \exp - \frac{\lambda}{RT} C_2 K_i \qquad (10)$$

For a given domain of operating conditions, α_1, α_2 are the rate constants and λ is the activation energy. These parameters have been identified in laboratory.

There are two inlet product flows, the first one for the catalyst injection and the second one for the C_1 injection. These flow rates are respectively equal to $Q_k K_0^{in}$ and $Q_{C1} C_1^{in}$. The liquid outlet flow rate is $Q_{out}=Q_k+Q_{C1}$.

3.1.2 Heat exchange model

The temperature T in the reactor and the temperature T_j of the cooling liquid are driven by the heat exchange model. The heat exchange model is given using traditional heat balance for the mixture in the reactor and for the cooling liquid.

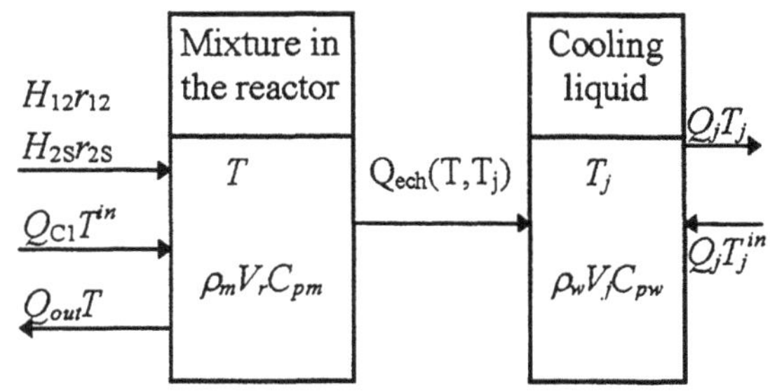

Figure 2 : Heat balance

The diagram in figure 2 shows the main flows used for the heat balance model.

The heat capacity of the cooling liquid is noted C_{pw} and $\rho_w V_j C_{pw} T_j$ represents the energy contained in the cooling liquid. The cooling liquid is injected in the heat exchanger at the known temperature T_j^{in}.

The average heat capacity of the products is noted C_{pm} and $\rho_m V_r C_{pm} T$ represents the energy contained in the reacting mixture. The enthalpies of the reactions are noted respectively H_{12} and H_{2S}. In the process, the heat quantity produced by the reaction $C_2 \rightarrow C_S$ is negligible versus the heat quantity produced by the reaction $C_1 \rightarrow C_2$.

The heat exchange between the reactor and the cooling liquid is noted Q_{ech}. It is a known function which depends essentially of T and T_j.

Finally, the inlet and outlet flow rates are taken into account in the energy balance.

3.1.3 On line sensors

Three on line measurements are available on the process.

Thermocouples measure the temperature T in the liquid phase.

Thermocouples measure temperature T_j and T_j^{in} in the heat exchanger.

Pressure sensors measure the reactor pressure. The thermodynamic properties of the mixture shows that the pressure is function only of C_1 and T in the liquid phase. Thus, the knowledge of T and the pressure at each time allows to measure the concentration C_1 in the liquid phase.

3.1.4 *Fault model*

Major faults like pump break are supervised. Nevertheless, three kinds of fault are not supervised. The first one is a temperature sensor deterioration. The second one consists in an undesirable injection of inert gas. The third and last fault is a deterioration of the catalyst.

As it was shown in paragraph 2, these three faults are represented by three unknown and independent signals noted respectively e_1, e_2 and e_3.

3.1.4.1 e_1 : *Thermocouple soot up.*

The temperature sensors in the reactor are thermocouples which are subject to important soot up. Essentially, this phenomenon is due to a polymer deposit on the thermocouple, which deteriorates the measurement precision. This uncertainty is considered as a fault. The perturbation induced by this fault on the model appears on the output equation of the temperature :

$$T_{measured} = T + e_1(t) \qquad (11)$$

3.1.4.2 e_2 : *Inert gas injection.*

The reactant C_1 is injected in the reactor as a gas. Inert gas presence in the inlet (like nitrogen) modifies the pressure in the reactor in high proportion. Then, inert gas presence induces a false measure of the concentration C_1. The output equation for C_1 is perturbed by the signal e_2 :

$$C_{1,measured} = C_1 + e_2(t) \qquad (12)$$

3.1.4.3 e_3 : *Catalyst injection*

The catalyst is a mixture of several components. The proportions for the catalyst mixture must be kept constant otherwise the selectivities are not good. A modification of these proportions is essentially translated by a modification of the kinetic rates. Hence, a perturbation of the kinetic terms with an unknown signal $e_3(t)$ reflects a fault in catalyst injection.

$$r_{12}^* = \left(1 + v_{12}(e_3(t))\right)\alpha_1 \exp -\frac{\lambda}{RT} C_1 K_i \qquad (13)$$

$$r_{2S}^* = \left(1 + v_{2S}(e_3(t))\right)\alpha_2 \exp -\frac{\lambda}{RT} C_2 K_i \qquad (14)$$

where v_{12} and v_{2S} are unknown functions of e_3. Their value is different from zero when e_3 is not equal to zero. Using these two functions is a way to represent how the fault e_3 modifies the two kinetics, although this modification is not identified.

3.1.5 *Process model*

The whole process model with fault influence is given by the following system :

$$\begin{cases}
\dot{T}_j = \dfrac{Q_{ech}(T_j, T)}{\rho_w V_j C_{pw}} + \dfrac{Q_j}{V_j}\left(T_j^{in} - T_j\right) \\
\dot{T} = -\dfrac{Q_{ech}(T_j, T)}{\rho_m V_r C_{pm}} + \dfrac{\left(Q_{C1} T^{in} - Q_{out} T\right)}{V_r} \cdots \\
\quad \cdots + \frac{H_{12}}{\rho_m V_r C_{pm}} (1 + v_{12}(e_3))\, r_{12}(C_1, T, K_i) \\
\dot{C}_1 = -(1 + v_{12}(e_3))\, r_{12}(C_1, T, K_i) \\
\quad \cdots + Q_{C1} C_1^{in} - Q_{out} C_1 \\
\dot{C}_2 = (1 + v_{12}(e_3))\, r_{12}(C_1, T, K_i) \cdots \\
\quad \cdots - (1 + v_{2S}(e_3))\, r_{2S}(C_2, T, K_i) \\
\quad \cdots - Q_{out} C_2 \\
\dot{C}_S = (1 + v_{2S}(e_3))\, r_{2S}(C_2, T, K_i) \\
\quad \cdots - Q_{out} C_S \\
\dot{K}_i = Q_k K_0^{in} - Q_{out} K_i \\
y_1 = T_j \\
y_2 = T + e_1 \\
y_3 = C_1 + e_2
\end{cases}$$

where y_1, y_2 and y_3 are the output measurements. The state variables are the temperatures T_j and T, and the concentrations of the products C_1, C_2, C_S and K_i.

3.2 *Residual generator design*

The geometric properties of this model induce that it is possible to construct three differential systems. The first one is a residual generator which isolates a disturbance on the temperature sensor from the other faults. The second one is a residual generator which detects an inert gas injection and a fault on temperature sensor. The third one is a non linear partial observer which gives an estimation of K_i. This observer can be considered as a residual generator which is sensitive to all faults.

In the following sections, the observers are high gain observers. Other observers could be used. In the same way, the residual function $R(v)$ may be any function verifying $R(v)=0 \Leftrightarrow v=0$.

3.2.1 $r_1(e_1)$: *temperature sensor fault*

For the fault on temperature sensor, it does not need a coordinate change to design the residual generator. Subsystem (15) is naturally independent from the faults e_2 and e_3.

$$\begin{cases} \dot{T}_j = \dfrac{Q_{ech}(T_j, y_2 - e_1) + Q_j \rho_w C_{pw}\left(T_j^{in} - T_j\right)}{\rho_w V_j C_{pw}} \\ y_1 = T_j \\ y_2 = T + e_1 \end{cases} \tag{15}$$

Thus, a residual generator sensitive only to e_1 is given for example by system (16) :

$$\begin{cases} \dot{\hat{T}}_j = \dfrac{Q_{ech}(\hat{T}_j, y_2) + Q_j \rho_w C_{pw}\left(T_j^{in} - \hat{T}_j\right)}{\rho_w V_j C_{pw}} \\ \quad \cdots - \theta_1\left(\hat{T}_j - y_1\right) \\ r_1(e_1) = \left(\hat{T}_j - y_1\right)^2 \end{cases} \tag{16}$$

3.2.2 $r_2(e_1, e_2)$: *Inert gas presence and temperature sensor fault*

With the following coordinate change (17) :

$$z = \rho_w V_j C_{pw} T_j + \rho_m V_r C_{pm} T + H_{12} C_1 \tag{17}$$

the influence of the catalyst fault e_3 is rejected in the subsystem (18) :

$$\begin{cases} \dot{z} = \rho_w C_{pw} Q_{C1}\left(T_j^{in} - y_1\right) \\ \quad \ldots + \rho_m C_{pm}\left(Q_{C1} T^{in} - Q_{out}(y_2 - e_1)\right) \\ \quad \ldots + H_{12}\left(Q_{C1} C_1^{in} - Q_{out}(y_3 - e_2)\right) \\ y_1 = T_j \\ y_2 = T + e_1 \\ y_3 = C_1 + e_2 \end{cases} \tag{18}$$

A residual generator which is sensitive to e_1 and e_2 but not to e_3 is given in (19) :

$$\begin{cases} \dot{\hat{z}} = \rho_m C_{pm} Q_j\left(T_j^{in} - y_1\right) \\ \quad \cdots + \rho_m C_{pm}\left(Q_{C1} T^{in} - Q_{out} y_2\right) \\ \quad \ldots + H_{12}\left(Q_{C1} C_1^{in} - Q_{out} y_3\right) \\ \quad \cdots - \theta_2(\hat{z} - \bar{y}) \\ \bar{y} = \rho_w V_j C_{pw} y_1 + \rho_m V_r C_{pm} y_2 + H_{12} y_3 \\ r_2 = (\hat{z} - \bar{y})^2 \end{cases} \tag{19}$$

It consists of an observer with output injection applied to the system (18).

3.2.3 K_i observer.

The following system is observable :

$$\begin{cases} \dot{T}_j = \dfrac{Q_{ech}(T_j, T) + Q_j \rho_w C_{pw}\left(T_j^{in} - T_j\right)}{\rho_w V_j C_{pw}} \\ \dot{T} = \dfrac{-Q_{ech}(T_j, T) + \rho_m C_{pm}\left(Q_{C1} T^{in} - Q_{out} T\right)}{\rho_m V_r C_{pm}} \\ \quad \cdots + \dfrac{H_{12}}{\rho_m V_r C_{pm}}(1 + e_3)\alpha_1 \exp - \dfrac{\lambda}{RT} C_1 K_i \\ \dot{C}_1 = -(1 + e_3)\alpha_1 \exp - \dfrac{\lambda}{RT} C_1 K_i \\ \quad \cdots + Q_{C1} C_1^{in} - Q_{out} C_1 \\ \dot{K}_i = Q_k K_0^{in} - Q_{out} K_i \\ y_1 = T_j \\ y_2 = T + e_1 \\ y_3 = C_1 + e_2 \end{cases} \tag{20}$$

Let $\hat{K}_i$ be the estimation given by a non linear observer for system (20), with e_1, e_2 and e_3 equal to zero. If there are no faults, $\hat{K}_i$ is the true estimate for catalyst concentration in the reactor. When a fault occurs, the estimate for K_i is biased. The stationary solution for the true value of K_i is known and equal to $K_i^*(t) = \dfrac{Q_k K_0^{in}}{Q_{out}}$. If the estimate $\hat{K}_i$ is very far from $K_i^*(t)$, then it means that the model is perturbed. In this case, at least one fault among e_1, e_2 or e_3 occurs.

3.3 Simulations

The three simulations show residual signal behavior in the case when one and only one fault occurs

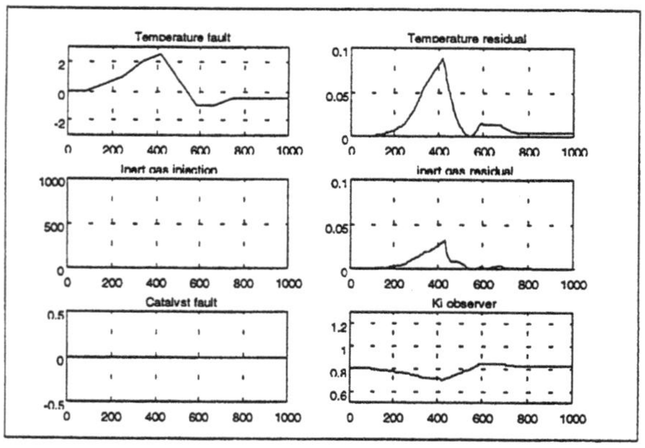

Figure 3 : Temperature fault

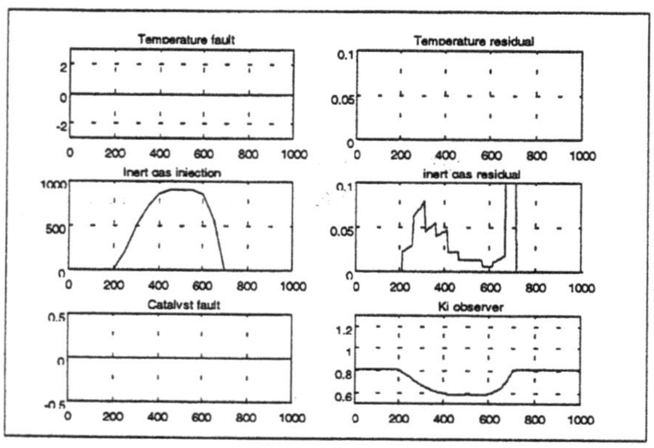

Figure 4 : Inert gas injection fault

If r_1 equal zero, at least one fault on the temperature sensors exists. If r_1 equal zero and r_2 is not equal to zero, there is no deterioration on temperature sensors and at least one fault occurs in gas injection. If r_1 and r_2 are equal to zero, there is no disturbance on temperature sensors and gas injection. So, there is a fault on catalyst injection if the K_i observation is "too far" from the expected value.

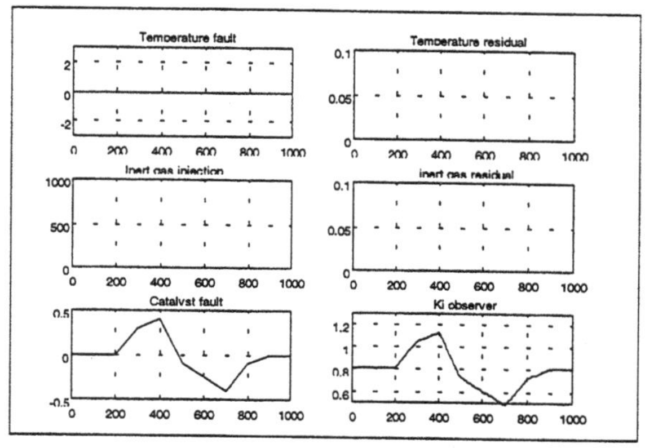

Figure 5 : catalyst fault

In the case when more than one fault simultaneously occurs, an elementary logical reasoning allows to conclude about the presence fault.

4. CONCLUSIONS

A large part of this paper is dedicated to the model of the process as the efficiency of the method is based on the quality of the mathematical model. It must be a good representation of the plant behavior. A way to do this is to keep in the model the physical means which have real influence on the measure.

For the fault model, it is sufficient to know which physical terms are disturbed by the faults. Then, observer theory and perturbation rejection are just mathematical tools to design the residuals generator. Thus, the choice for the observer is driven by considerations about feasibility, robustness or other criterion which depend on the process.

REFERENCES

H.Hammouri, M.Kinnaert, E.H. El Yaagoubi.(1994) A geometric approach to fault detection and isolation for nonlinear systems. To appear in IEEE.

M.A.Massoumnia (1986) A geometric approach to the synthesis of fault detection filters. IEEE transaction on automatic control, vol. 31, pp. 839-846, September 1986

A.S.Willsky, M.A.Massoumnia and G.C.Verghese (1989) Fault detection and isolation. IEEE transaction on automatic control, vol. 34, pp. 316-321, March 1989

P.M.Frank, (1990) Fault diagnosis in dynamic systems using analytical and knowledge based redundancy a survey and some new results. Automatica, vol. 26, pp. 459-474, March 1990

Fault Observability and Fault Reconstruction for Dynamic Systems

R.J. Patton and M. Hou

Department of Electronic Engineering
The University of Hull, Hull HU6 7RX, UK

Abstract: This paper addresses the problems of fault observability and fault reconstruction for linear time-invariant systems. Under the necessary and sufficient conditions, a simple approach to the design of schemes for fault reconstruction is developed. The design is based on a matrix pencil decomposition form which is obtainable using numerically stable orthogonal transformations. *Copyright © 1998 IFAC*

Keywords: Faults; Observability; Fault estimation

1 Introduction

The main concern in the research into fault diagnosis for dynamic systems has been focussed on fault detection and isolation. Patton *et al.* (1992) discussed the reconstruction of faults in discrete-time systems using a de-convolution approach. The faults can be uniquely reconstructed using the approach under a very restrictive condition, namely the system must have as many independent measurements as states.

The first issue arising in fault reconstruction is an evaluation of observability of faults. Roughly, fault observability means that the change of faults in a dynamic system can reflect itself in the change of measurements.

If the faults are observable, the question remaining is how to reconstruct the faults. It is worth noting that the initial condition of the system may be unknown, as it is in most applications. Hence, fault reconstruction should not assume knowledge of initial conditions of the system in most cases. The fault reconstruction problem corresponds to formulation of a reconstructor whose inputs are the measurements of the original system and whose outputs converge to the faults of the original system. The convergence speed can be specified arbitrarily. The proposed reconstructor is a generalised system as the derivatives of the measurements of the original system are normally required in the reconstruction process.

Based on the discussion on the issues of input observability and input reconstruction (Hou and Patton, 1996), the objective of this paper is to explore the equivalent issues in the context of fault observability and fault reconstruction. In the paper of Hou and Patton (1996), the relation among input reconstruction, system inverse and disturbance decoupled observers (unknown-input observers) has been investigated.

It is remarkable that, except for some very trivial cases, derivatives of the measurements are unavoidable when fault reconstructing is required. The proposed design guarantees that the lowest order of the derivatives will be involved in the fault estimator.

The organisation of this paper is as follows. Section 2 formulates the problem. Necessary and sufficient conditions for fault observability are provided in Section 3. A design procedure for fault reconstruction is given in Section 4. The design is further illustrated by an example in Section 5, followed by conclusions in Section 6.

2 Problem Formulation

Denote the time variables $f(t)$ and $y(t)$ as the fault and measurement of a process, respectively.

Definition 1 *The fault* $f(t)$ *is said to be* observable *if* $y(t) = 0$ *for* $t \geq 0$ *implies* $f(t) = 0$ *for* $t > 0$.

Definition 2 *The fault* $f(t)$ *is said to be* detectable *if* $y(t) = 0$ *for* $t \geq 0$ *implies* $f(t) \to 0$ *as* $t \to \infty$.

This paper exploits these concepts for the linear time-invariant systems of the form

$$\dot{x}(t) = A\,x(t) + B\,f(t) \quad (1)$$
$$y(t) = C\,x(t) + D\,f(t)\,. \quad (2)$$

All coefficient matrices appearing in (1-2) are known and real with compatible dimensions. The initial condition $x(0_-)$ of the system is assumed *unknown*.

For the case that the initial condition is known, fault observability is defined in the following way.

Definition 3 *The fault $f(t)$ is said to be* observable *with knowledge of $x(0_-)$ if $y(t) = 0$ for $t \geq 0$ implies $f(t) = 0$ for $t > 0$, provided $x(0_-) = 0$.*

The fault reconstruction problem is to find an estimator in the form

$$\dot{z}(t) = F_1\,z(t) + F_2\,y(t) \quad (3)$$
$$\hat{f}(t) = G_1\,z(t) + \sum_{i=0}^{p} G_{2,i}\,y^{(i)}(t) \quad (4)$$

where $\hat{f} \to f$ for arbitrary y and $x(0_-)$ of the system (1-2). $y^{(i)}$ denotes the ith derivative of y. The number p is known as the *index* of the estimator. So, the existence of the derivatives of y up to order p is assumed. The convergence speed of $\hat{f} \to u$ is required to be arbitrarily assignable. Nevertheless, a remark on the case that the convergence speed cannot arbitrarily be assigned will be given at a suitable point.

The design of the estimator consists of finding all coefficient matrices in (3-4) and determining the lowest index p.

The form of the estimator (3-4) is not unique. It is possible to design a fault estimator in which derivatives of y appear also in (3). This is so as will be shown, the existence of estimators does not depend on whether derivatives of y appear in (3) and the lowest index of possible estimator is unique. Design of an estimator in form (3-4) with the lowest index is of interest in this paper.

3 Necessary and Sufficient Conditions

Two matrices M and N of the same size with a complex variable λ may be used to construct a so-called *matrix pencil* denoted by $\lambda\,M - N$. Normally, the rank of $\lambda\,M - N$ depends on the value of λ. Nevertheless, the maximal rank or the *normal rank* (Van Dooren, 1979; Kailath, 1980) of a matrix pencil is unique and independent of λ. A complex number $\bar{\lambda}$ is said to be a finite eigenvalue of $\lambda\,M - N$, if rank $(\bar{\lambda}\,M - N) <$ normal-rank $(\lambda\,M - N)$. A matrix pencil is said to be *column unimodular*, if the pencil has full column rank for all finite λ.

Denote the set of the finite eigenvalues (single and multiple finite eigenvalues) of a matrix pencil as $\sigma(\lambda\,M - N)$. The necessary and sufficient condition of fault observability for the system (1-2) is given below.

Theorem 1 *The fault $f(t)$ in (1-2) is observable if and only if*

$$\sigma\left(\begin{bmatrix} -\lambda\,I + A & B \\ C & D \end{bmatrix}\right) = \sigma\left(\begin{bmatrix} -\lambda\,I + A \\ C \end{bmatrix}\right). \quad (5)$$

Proof: Set $y(t) = 0$ in (2). The fault $f(t)$ and the state $x(t)$ with initial condition $x(0_-)$ should satisfy the following equation in the frequency domain

$$\begin{bmatrix} -s\,I + A & B \\ C & D \end{bmatrix}\begin{bmatrix} x \\ f \end{bmatrix} = \begin{bmatrix} -x(0_-) \\ 0 \end{bmatrix}. \quad (6)$$

The objective is to show that, whatever $x(0_-)$ and therefore $x(t)$ are, Eq. (6) has a unique solution for f as $f = 0$ if and only if (5) is true.

First, the fault coefficient matrix $F = \begin{bmatrix} B \\ D \end{bmatrix}$ must be of full column rank, which is also implied in the condition (5) due to the full column normal rank of $\begin{bmatrix} -\lambda\,I + A \\ C \end{bmatrix}$. Otherwise, there exists a $f \neq 0$ satisfying $F\,f = 0$ and therefore, by letting $x(0_-) = 0$, this non-zero u satisfies (6) as well, a contradiction to Definition 1.

Thus, pre-multiplying (6) by $\begin{bmatrix} I - FF^+ \\ F^+ \end{bmatrix}$ with $F^+ = (F^T F)^{-1} F^T$ yields the equivalent equation

$$\begin{bmatrix} -s\,E_1 + A_1 & 0 \\ -s\,E_2 + A_2 & I \end{bmatrix}\begin{bmatrix} x \\ f \end{bmatrix} = \begin{bmatrix} -E_1\,x(0_-) \\ -E_2\,x(0_-) \end{bmatrix} \quad (7)$$

where

$$E_1 = (I - FF^+)\begin{bmatrix} E \\ 0 \end{bmatrix},\; A_1 = (I - FF^+)\begin{bmatrix} A \\ C \end{bmatrix} \quad (8)$$

$$E_2 = F^+\begin{bmatrix} E \\ 0 \end{bmatrix},\; A_2 = F^+\begin{bmatrix} A \\ C \end{bmatrix}. \quad (9)$$

The matrix pencil $\lambda\,E_1 - A_1$ has to be of full column normal rank, which is also implied in the condition (5) for the same reason as for the full column rank of F. Otherwise, since

$$\text{normal-rank}(\lambda\,E_1 - A_1) < \text{normal-rank}\begin{bmatrix} \lambda E_1 - A_1 \\ \lambda E_2 - A_2 \end{bmatrix} = \text{normal-rank}\begin{bmatrix} \lambda\,I - A \\ C \end{bmatrix} = \dim(x) \quad (10)$$

it is easy to show that there exists a $x(t) \neq 0$ with initial condition $x(0_-)$ satisfying $E_1 \dot{x} = A_1\,x$ but

$E_2\,\dot{x} \neq A_2\,x$. This leads to $f(t) = E_2\,\dot{x} - A_2\,x \neq 0$, a contradiction of Definition 1.

Consider the fact that $\lambda\,E_1 - A_1$ must be of full column normal rank. According to the Kronecker decomposition of a matrix pencil (Gantmacher, 1959), it can further be assumed, without loss of generality, that (7) is in the form

$$\left[\begin{array}{cc|c} -s\,I + A_f & & \\ & -sE_\varepsilon + A_\varepsilon & \\ \hline A_{21} & -s\,E_{22} + A_{22} & I \end{array}\right] \left[\begin{array}{c} x_f \\ x_\varepsilon \\ \hline f \end{array}\right] = \left[\begin{array}{c} -x_f(0_-) \\ -E_\varepsilon x_\varepsilon(0_-) \\ \hline -E_{22}\,x_\varepsilon(0_-) \end{array}\right] \quad (11)$$

where $-sE_\varepsilon + A_\varepsilon$ is column unimodular. Note the fact, see, e.g. Gantmacher (1959) and Dai (1989), that the descriptor system $E_\varepsilon\,\dot{x}_\varepsilon(t) = A_\varepsilon\,x(t)_\varepsilon$ with initial condition $E_\varepsilon\,x_\varepsilon(0_-)$ has the unique solution $x_\varepsilon(t) = 0$ for $t > 0$. Hence, since $x_f(t)$ can be arbitrary, $f(t) = 0$ for $t > 0$ if and only if $A_{21} = 0$.

It is clear that $A_{21} = 0$ if and only if $\sigma(\lambda I - A_f) = \sigma\left(\begin{bmatrix} -\lambda I + A_f \\ A_{21} \end{bmatrix}\right)$. Consider further the relations

$$\sigma\left(\begin{bmatrix} -\lambda I + A_f \\ A_{21} \end{bmatrix}\right) = \sigma\left(\begin{bmatrix} -\lambda I + A \\ C \end{bmatrix}\right) \quad (12)$$

and

$$\sigma(-\lambda I + A_f) = \sigma\left(\begin{bmatrix} -\lambda I + A & B \\ C & D \end{bmatrix}\right). \quad (13)$$

The proof is completed. □

In the above proof, it has been shown that the full column normal rank of $\lambda\,E_1 - A_1$ is a necessary condition for fault observability. The condition is equivalent to the full column normal rank of $\begin{bmatrix} -\lambda I + A & B \\ C & D \end{bmatrix}$ which is also implied in (5). Under this condition, from (12-13) it is straightforward to show that

$$\sigma\left(\begin{bmatrix} -\lambda I + A & B \\ C & D \end{bmatrix}\right) \supseteq \sigma\left(\begin{bmatrix} -\lambda I + A \\ C \end{bmatrix}\right) \quad (14)$$

which leads to the following corollary, where the notation $N_\sigma(\cdot)$ represents the total number of the elements in $\sigma(\cdot)$. It is worth noting that (14) may not hold if $\begin{bmatrix} -\lambda I + A & B \\ C & D \end{bmatrix}$ is not of full column rank.

Corollary 1 *The faul $f(t)$ in (1-2) is observable if and only if*

$$\text{normal-rank}\begin{bmatrix} -\lambda I + A & B \\ C & D \end{bmatrix} = \dim(x) + \dim(f) \quad (15)$$

and

$$N_\sigma\left(\begin{bmatrix} -\lambda I + A & B \\ C & D \end{bmatrix}\right) = N_\sigma\left(\begin{bmatrix} -\lambda I + A \\ C \end{bmatrix}\right). \quad (16)$$

This corollary is important because no eigenvalue calculation is required in checking (15) and (16). Numerically effective tests for fault observability will be detailed in Section 4.

From the above analysis and the proof of Theorem 1, there is no difficulty in deriving the following necessary and sufficient condition for fault detectability.

Theorem 2 *The fault $f(t)$ in (1-2) is detectable if and only if*

$$\sigma\left(\begin{bmatrix} -\lambda I + A & B \\ C & D \end{bmatrix}\right) - \sigma\left(\begin{bmatrix} -\lambda I + A \\ C \end{bmatrix}\right) \in \mathbf{C}^- \quad (17)$$

where $\mathbf{C}^-$ stands for the open left half complex plane.

This theorem indicates that f is detectable if and only if all the finite eigenvalues of $\begin{bmatrix} -\lambda I + A & B \\ C & D \end{bmatrix}$ which are different from those of $\begin{bmatrix} -\lambda I + A \\ C \end{bmatrix}$ must have negative real parts.

If the initial condition $x(0_-)$ is known, the condition for fault observability becomes very weak as shown below.

Theorem 3 *The fault $f(t)$ in (1-2) is observable with knowledge of initial condition $x(0_-)$ if and only if (15) holds.*

Proof: Using the same arguments as in the proof of Theorem 1, it can be shown that both F and $-\lambda\,E_1 + A_1$ have to be of full column rank and full column normal rank, respectively. These full rank properties of F and $-\lambda\,E_1 + A_1$ are equivalent to condition (15). This proves the necessity of (15). Now, assume that (15) is true. It then follows that $-\lambda\,E_1 + A_1$ has full column normal rank. According to Definition 3, set $x(0_-) = 0$ in the equation (7). Thus, $E_1\,\dot{x} = A_1\,x$ with $x(0_-) = 0$ gives the solution $x(t) = 0$ for $t > 0$ and hence, $f(t) = 0$ for $t > 0$. □

Note that that the observability assumption on the matrix pair $\{A, C\}$ is not necessary for the fault observability. It can also be proved that the controllability of the pair $\{A, B\}$ has no connection with the fault observability.

4 Fault Reconstruction

This section shows that fault observability is a necessary and sufficient condition for the existence of an estimator reconstructing faults. A systematic and simple design procedure for fault reconstruction will be provided.

Theorem 4 *There exists an estimator for fault reconstruction in the form (3-4) if and only if the system (1-2) is fault observable.*

Proof: By using arguments which are similar to those leading to the requirement for Eq. (5), the necessity can be verified. Sufficiency will be proved through completing a design procedure for fault reconstruction, given later in this section, under the condition (5) or equivalently under (15-16). □

It is convenient to use the Kronecker form for proof of fault observability conditions. But, due to the numerical unreliability of the computation (Van Dooren, 1981), this form is not suitable for developing the fault reconstruction design procedure. The staircase form (Van Dooren, 1979) which will be used in the fault reconstruction exhibits the structure of the Kronecker form and can be implemented via numerically stable orthogonal transformations.

According to Van Dooren (1979), by using orthogonal constant matrices U and V, i.e. $U^TU = I$ and $V^TV = I$, the matrix pencil $\begin{bmatrix} -\lambda I + A \\ C \end{bmatrix}$ of full column normal rank can be transformed into the form

$$U\begin{bmatrix} -\lambda I + A \\ C \end{bmatrix}V = \begin{bmatrix} \lambda E_{fAC}-A_{fAC} & \times & \times \\ & \lambda E_{\infty AC}-A_{\infty AC} & \times \\ & & \lambda E_{rAC}-A_{rAC} \end{bmatrix} \quad (18)$$

where

(a) $\lambda E_{fAC} - A_{fAC}$ is square and $|E_{fAC}| \neq 0$;

(b) $\lambda E_{\infty AC} - A_{\infty AC}$ is square, $|A_{\infty AC}| \neq 0$ and $A_{\infty AC}^{-1}E_{\infty AC}$ is a nilpotent matrix with index l. That is, $(A_{\infty AC}^{-1}E_{\infty AC})^l = 0$ but $(A_{\infty AC}^{-1}E_{\infty AC})^k \neq 0$ for $0 < k < l$, where $l \leq \dim E_{\infty AC}$.

(c) $\lambda E_{rAC} - A_{rAC}$ is column unimodular and both E_{rAC} and A_{rAC} are of full column rank;

(d) '×' denotes any pencils with appropriate dimensions, but of no particular interest.

Similarly, the matrix pencil $\begin{bmatrix} -\lambda I + A & B \\ C & D \end{bmatrix}$ which is generally not of full column normal rank can be transformed via orthogonal transformations into the form

$$\begin{bmatrix} \lambda E_{cSYS}-A_{cSYS} & \times & \times & \times \\ & \lambda E_{fSYS}-A_{fSYS} & \times & \times \\ & & \lambda E_{\infty SYS}-A_{\infty SYS} & \times \\ & & & \lambda E_{rSYS}-A_{rSYS} \end{bmatrix} \quad (19)$$

where $\lambda E_{fSYS} - A_{fSYS}$, $\lambda E_{\infty SYS} - A_{\infty SYS}$ and $\lambda E_{rSYS} - A_{rSYS}$ have the same structures as their correspondences in (18). $\lambda E_{cSYS}^T - A_{cSYS}^T$ is column unimodular and has the same structure as $\lambda E_{rSYS} - A_{rSYS}$.

By virtue of Corollary 1, the following theorem containing a numerically effective test for fault observability is immediate.

Theorem 5 *The system (1-2) is fault observable if and only if the block $\lambda E_{cSYS} - A_{cSYS}$ in (19) vanishes and $\dim(E_{fAC}) = \dim(E_{fSYS})$, where E_{fAC} and E_{fSYS} are defined in (18) and (19), respectively.*

Denote the system (1-2) as

$$\begin{bmatrix} -\rho I + A & B \\ C & D \end{bmatrix}\begin{bmatrix} x \\ f \end{bmatrix} = \begin{bmatrix} 0 \\ I \end{bmatrix} y \quad (20)$$

where ρ denotes the differential operator d/dt. According to (18), through orthogonal transformations, the following form, equivalent to Eq. (20), can be obtained as

$$\left[\begin{array}{ccc|c} -\rho E_f + A_f & -\rho E_{12} + A_{12} & -\rho E_{13} + A_{13} & \\ & -\rho E_\infty + A_\infty & -\rho E_{23} + A_{23} & \\ & & -\rho E_r + A_r & \\ \hline -\rho E_{41} + A_{41} & -\rho E_{42} + A_{42} & -\rho E_{43} + A_{43} & \bar{F} \end{array}\right] \cdot \begin{bmatrix} x_f \\ x_\infty \\ x_r \\ \hline f \end{bmatrix} = \begin{bmatrix} V_f \\ V_\infty \\ V_r \\ \hline V_4 \end{bmatrix} y \quad (21)$$

where $\bar{F}$ is non-singular, matrices E_f, A_f, E_∞, A_∞, E_r and A_r have the same structures as those in (18), the rest matrices are not important.

Eliminating E_{41} in (21) yields

$$\left[\begin{array}{ccc|c} -\rho E_f + A_f & -\rho E_{12} + A_{12} & -\rho E_{13} + A_{13} & \\ & -\rho E_\infty + A_\infty & -\rho E_{23} + A_{23} & \\ & & -\rho E_r + A_r & \\ \hline \bar{A}_{41} & -\rho \bar{E}_{42} + \bar{A}_{42} & -\rho \bar{E}_{43} + \bar{A}_{43} & \bar{F} \end{array}\right] \cdot \begin{bmatrix} x_f \\ x_\infty \\ x_r \\ \hline f \end{bmatrix} = \begin{bmatrix} V_f \\ V_\infty \\ V_r \\ \hline \bar{V}_4 \end{bmatrix} y \quad (22)$$

where

$$\bar{A}_{41} = A_{41} - E_f^{-1}A_f, \quad \bar{E}_{42} = E_{42} - E_f^{-1}E_{12} \quad (23)$$

$$\bar{A}_{42} = A_{42} - E_f^{-1}A_{12}, \quad \bar{E}_{43} = E_{43} - E_f^{-1}E_{13} \quad (24)$$

$$\bar{A}_{43} = A_{43} - E_f^{-1}A_{13}, \quad \bar{V}_4 = V_4 - E_f^{-1}V_f. \quad (25)$$

According to Theorem 4 and the proof of Theorem 1, it is easy to verify that the fault reconstruction problem is solvable if and only if $\bar{A}_{41} = 0$. Furthermore, note that, see, e.g. Dai (1989), the descriptor system $E_\infty \dot{x}_\infty(t) = A_\infty x_\infty(t) + B_\infty u(t)$ with initial condition $E_\infty x_\infty(0_-)$ has a unique solution

$$x_\infty(t) = -\sum_{i=0}^{l-1} (A_\infty^{-1}E_\infty)^i A_\infty^{-1} B_\infty u^{(i)}(t)\,, \quad t > 0\,. \quad (26)$$

It is then straightforward to provide a design procedure for fault reconstruction as follows.

Fault reconstruction design procedure:

Step 1: Test the solvability of the fault reconstruction using Theorem 5;

Step 2: Decompose the extended matrix pencil $\begin{bmatrix} -\rho I + A & B & 0 \\ C & D & I \end{bmatrix}$ into the form shown in (21) using orthogonal transformations;

Step 3: Perform simple row elementary transformation on the block $\begin{bmatrix} -\rho E_r + A_r & V_r \end{bmatrix}$ to obtain $\begin{bmatrix} -\rho I + \bar{A} & V_{r1} \\ \bar{C} & V_{r2} \end{bmatrix}$ where the pair $\{\bar{A}, \bar{C}\}$ is observable; based on this expression, design an observer for x_r defined in (21);

Step 4: Using the form (26), derive the explicit expression for $\hat{x}_\infty$ in terms of $\hat{x}_r$ (obtained from Step 3) and y (including its derivatives);

Step 5: Using the last line block in (22), obtain the explicit expression for $\hat{u}$ in terms of $\hat{x}_\infty$, $\hat{x}_r$ and y, which ends up the design procedure for fault reconstruction with the form (3-4).

Remark 1 *Step 3 can be done in many ways, for instance, through pre-multiplying* $\begin{bmatrix} -\rho E_r + A_r & V_r \end{bmatrix}$ *by the full-column-rank matrix* $\begin{bmatrix} E_r^+ \\ I - E_r E_r^+ \end{bmatrix}$ *with* $E^+ = (E_r^T E_r)^{-1} E_r^T$. *This means,* $\bar{A} = E_r^+ A_r$, $V_{r1} = E_r^+ V_r$, $\bar{C} = (I - E_r E_r^+) A_r$ *and* $V_{r2} = (I - E_r E_r^+) V_r$. *Because* $-\rho E_r + A_r$ *is column unimodular, the pair* $\{\bar{A}, \bar{C}\}$ *must be an observable pair.*

Remark 2 *If the system (1-2) is not fault observable but fault detectable, it can be proved that although* $\bar{A}_{41} \neq 0$, $\bar{A}_{41} x_f(t) \to 0$ *as* $t \to \infty$. *Hence, by setting* $\bar{A}_{41} = 0$ *formally at Step 5, the above procedure offers also a fault detector in the same form as (3-4), where* $\hat{f} \to f$ *as* $t \to \infty$. *Nevertheless, the convergence speed cannot be assigned arbitrarily because no influence can be made on the convergence of* $\bar{A}_{41} x_f(t) \to 0$.

Finally, there is a need to verify that the design procedure above yields indeed a fault estimator with the minimal index.

Consider the decomposition form in (22). From (26) and the above design procedure, the index of the fault estimator is determined through the form of the solution to x_∞. It is easy to show that the solution of x_∞ takes the form $x_\infty(t) = -\sum_{i=0}^{l-1} (A_\infty^{-1}E_\infty)^i A_\infty^{-1} B_1\, y^{(i)}(t) + B_2\, x_r(t)$ for $t > 0$, where B_1 and B_2 are some matrices which can easily been determined. Although the matrices in the 3-by-3 block of (22) are not unique, the structure of the block is unique. Assume that the 3-by-3 block along with the coefficient matrix of y is further transformed into the Kronecker form

$$\left[\begin{array}{ccc|c} -\rho I + J_f & & & \bar{V}_f \\ & -\rho J_\infty + I & & \bar{V}_\infty \\ & & -\rho \bar{E}_r + \bar{A}_r & \bar{V}_r \end{array}\right]. \quad (27)$$

It is not hard to show that $(A_\infty^{-1}E_\infty)^i A_\infty^{-1} B_1 = 0$ if and only if $J_\infty^i \bar{V}_\infty = 0$. Due to uniqueness of the Kronecker form, this means that the above design procedure does give the minimal index of the fault reconstruction process.

Remark 3 *The index p in (3-4) is the minimal number of p satisfying* $\bar{E}_{42}(A_\infty^{-1}E_\infty)^p A_\infty^{-1} B_1 = 0$. *It is clear that* $p \leq l$, *where* l *is the index of the nilpotent matrix* $A_\infty^{-1}E_\infty$.

5 Illustrative Example

This section uses an example to illustrate the design procedure of fault reconstruction. In order to avoid involvement of numerical calculations and to simplify the illustration, instead of orthogonal transformations, non-singular transformations are used in the design for these examples.

Example 1 *Consider a linear system of the form (1-2) with the following coefficient matrices*

$$A = \begin{bmatrix} 1 & 1 & 1 & 1 \\ 1 & 0 & 0 & 0 \\ 0 & 1 & 1 & 0 \\ 0 & 0 & 1 & 1 \end{bmatrix}, \quad B = \begin{bmatrix} 1 \\ 0 \\ 0 \\ 0 \end{bmatrix}$$

$$C = \begin{bmatrix} 0 & 1 & 0 & 0 \\ 0 & 0 & 0 & 1 \end{bmatrix}, \quad D = 0\,.$$

By using a non-singular matrix

$$P = \begin{bmatrix} 0 & 1 & 0 & 0 & 0 & 0 \\ 0 & 0 & 0 & 0 & 1 & 0 \\ 0 & 0 & 1 & 0 & -1 & 0 \\ 0 & 0 & 0 & 1 & 0 & 0 \\ 0 & 0 & 0 & 0 & 0 & 1 \\ 1 & 0 & 0 & 0 & 0 & 0 \end{bmatrix}, \textit{ it gives}$$

$$P\left[\begin{array}{c|c|c} -\rho E + A & B & 0 \\ C & D & I \end{array}\right] =$$

$$\left[\begin{array}{cccc|c|cc} 1 & -\rho & 0 & 0 & 0 & 0 & 0 \\ 0 & 1 & 0 & 0 & 0 & 1 & 0 \\ 0 & 0 & -\rho+1 & 0 & 0 & -1 & 0 \\ 0 & 0 & 1 & -\rho+1 & 0 & 0 & 0 \\ 0 & 0 & 0 & 1 & 0 & 0 & 1 \\ \hline -\rho+1 & 1 & 1 & 1 & 1 & 0 & 0 \end{array}\right] . \quad (28)$$

The above decomposition leads to the solution to the block $-\rho E_\infty + A_\infty$ as $x_2 = y_1$, $x_1 = \dot{y}_1$ and a reduced-order observer corresponding to the block $-\rho E_r + A_r$ as

$$\begin{aligned} \dot{z} &= (1-\alpha)z - y_1 - \alpha^2 y_2 , \qquad \alpha \in \mathbf{R},\ \alpha > 1 \\ \hat{x}_3 &= z + \alpha y_2 \\ x_4 &= y_2 . \end{aligned}$$

The last line in (28) leads to

$$\hat{f} = \ddot{y}_1 - \dot{y}_1 - y_1 - \hat{x}_3 - y_2 . \qquad (29)$$

6 Conclusion

This paper has considered the problem of fault reconstruction from the viewpoint of fault observability. A theory of fault observability has been developed. This is reflected by establishing the necessary and sufficient conditions which are explicit and testable, as well as through the approval of a systematic and numerical reliable design for fault reconstruction

For the purpose of fault reconstruction, addressing fault observability is more natural and clearer than the system invertibility. The proposed fault estimator design can be implemented by using numerically effective algorithms such as those in Van Dooren (1981).

References

Dai, L. (1989). Singular Control Systems. *Lecture Notes in Control and Information Sciences 118.* Springer, Berlin.

Gantmacher, F. R. (1959). The Theory of Matrices, Vol. II. Chelsea Publishing Co., New York.

Hou, M. and Patton, R.J. (1996). Input observability and input reconstruction. *Automatica*, to appear.

Kailath, T. (1980). Linear Systems. Prentice Hall, 1980.

Patton, R.J., Zhang, H.Y. and Chen, J. (1992). Modelling of uncertainties for robust fault diagnosis. *Proc. 31th Conf. Decision & Control*, Tucson/USA, 921-926.

Van Dooren, P. (1979). The computation of Kronecker's canonical form of a singular pencil. *Linear Algebra Appl.*, **27**, 103-140.

FAULT DETECTION FOR TIME-DELAY SYSTEMS WITHIN ROBUST RESIDUAL BASED ON FINITE MEMORY OBSERVERS.

Walter NUNINGER, Frédéric KRATZ, José RAGOT

Centre de Recherche en Automatique de Nancy
Institut National Polytechnique de Lorraine
2, Avenue de la Fôret de Haye - F 54 516 Vandoeuvre-lès-Nancy
Phone: (33) 3 83 59 59 59 - Fax: (33) 3 83 59 56 44
E-mail: { wnuninge, fkratz, jragot }@ensem.u-nancy.fr

Abstract: Fault detection is usually achieved within residuals generation (based on analytical redundancy) and evaluation. Such methods need the knowledge of a system model and therefore have to be robust to disturbances such as model uncertainties and measurement errors. In order to free oneself from the drawbacks of classic methods, as the observer performance degradation, the concept of finite memory observer has been introduced. This paper gives a short summary of the observer structure and some of its properties to further focus on the choice of the residual. In general the estimation error is taken as the residual but another residual is defined here. This residual is the difference of two state estimations at the same instant but based on two overlapping observation horizons. The relevance of this paper is to prove that this special residual is 'more' robust with respect to some model uncertainties than the estimation error if the fault detection problem is tackled. Finally, the proposed method is applied to detect sensor faults for time-delay systems. The relevance of this paper is that the observer is based on a simplified model of the time-delay system that does not take into account the delayed states. The example given is a simulated one. *Copyright © 1998 IFAC*

Key words: Time-delay systems, finite memory observer, robustness, fault detection, sliding window.

1. INTRODUCTION

This paper deals with a problem of huge interest for process control: fault detection and isolation (FDI) (Willsky, 1976) (Frank, 1990) (Patton, 1994). The diagnostic procedure has to be robust with respect to system uncertainties such as measurement noise and parameter variations. This paper only considers robust residuals generation based on observers. As all the process history is taken into account to compute the estimation (i.e. infinite memory), the model errors can be accumulated in the estimation error (the residual) that might then diverges. The finite memory observer (Medvedev and Toivonen, 1991a, 1991b) is designed to prevent such effect on the estimation error thanks to an estimation based only on a minimal number of measured data (Darouach, *et al.*, 1994). The residual proposed in this paper is defined as the difference between two estimations at the same instant but based on two overlapping horizons so that a detection window is defined (Kratz, *et al.*, 1997) (Nuninger, 1997). The relevance of this paper is to demonstrate that this special residual is better for fault detection as it is more robust with respect to model uncertainties as parameter variations for instance than the estimation error. In addition a criterion is given to chose the optimal sizes of the overlapping horizons. Therefore, no further hypothesis is required to prove the existence of the detection window. It is stressed that in this paper, time-delay systems are considered but the finite memory observers are based on simplified models (not taking into account the delayed states). As a consequence, these states stand for the model uncertainties to which ones the generated residual has to be robust.

This paper is organised as follows. First, the problem formulation and the finite memory observer structure are given. Second, the observer properties are summarised. Then, the special residual is defined and the residual sensitivity to additive sensor faults and to model uncertainties are studied. Finally, the residual generator is applied to a simulated time-delay system when only a simplified model is used to design the observer. A conclusion is also presented.

2. PROBLEM FORMULATION

2.1. System models

This paper considers the following state-space model of time-delay systems in continuous-time representation:

$$\dot{x}(t)=\sum_{s=0}^{S}A_s x(t-t_s)+\sum_{j=0}^{J}B_j u(t-t_j) \tag{1.a}$$

$$y(t)=Cx(t)+v(t) \tag{1.b}$$

where x is the state vector of dimension n, y the output measurement vector of dimension p, u the control input vector of dimension m and v the measurement noise vector (zero-mean with known covariance matrix V). Matrices A_s $(s=0,\ldots,S)$, B_j $(j=0,\ldots,J)$ and C are constant with appropriate dimensions. Integers s and j denote the number of time-delays in the state and control vectors respectively. Further, the design of the finite memory observer is based on the following simplified model of system (1) that does not take into account the influence of the delayed state and control vectors:

$$\dot{x}(t)=Ax(t)+Bu(t) \tag{2.a}$$

$$y(t)=Cx(t)+v(t) \tag{2.b}$$

where A = A_0 and B = B_0. The sensibility of the proposed residual with respect to additive sensor faults and to model uncertainties will be studied further.

2.2. Static form of the system model (2)

At the instant t, a sliding observation window is considered defined by k+1 time-delay output measurements: $y_i=y(t-\tau_i)$, $i=0,\ldots,k$. First x(t) is deduced from $x(t-\tau_i)$ by integration of equation (2.a) on the interval $[t-\tau_i, t]$. This expression is further replaced in the expression of y_i deduced from (2.b) so that the following relation is obtained:

$$y(t-\tau_i)=w_i x(t)-\bar{u}(t-\tau_i,t)+v(t-\tau_i) \tag{3.a}$$

with:

$$w_i=Ce^{-A\tau_i} \tag{3.b}$$

$$\bar{u}(t-\tau_i,t)=\int_{t-\tau_i}^{t}Ce^{A(t-\tau_i-\delta)}Bu(\delta)d\delta \tag{3.c}$$

Note that the expression (3.c) requires the knowledge of an analytical form of the input but in practice, this expression is computed numerically within the assumption that the input is constant over the considered time interval. Thanks to equation (3) which is true for each data of the sliding observation window, system (2) can be described by the following « static » form already used by Chow and Willsky (1984):

$$Z_k(t)=W_k x(t)+N_k(t) \tag{4.a}$$

where Z_k of dimension p(k+1) and the matrix W_k of dimension p(k+1)xn are defined by:

$$Z_k(t)=Y_k(t)+U_k(t) \tag{4.b}$$

$W_k=[w_i]$ with i from 1 to k

$$Y_k(t)=\begin{bmatrix}y_0\\ \vdots\\ y_k\end{bmatrix},\quad U_k(t)=\begin{bmatrix}\bar{u}(t-\tau_0,t)\\ \vdots\\ \bar{u}(t-\tau_k,t)\end{bmatrix},\quad N_k(t)=\begin{bmatrix}v(t-\tau_0)\\ \vdots\\ v(t-\tau_k)\end{bmatrix}$$

Note that N_k is the extended vector of the noise which variance is defined by:

$$R_k(t)=\mathrm{diag}(V(t-\tau_0),\ldots,V(t-\tau_k))=(Q_k^TQ_k)^{-1} \tag{5}$$

2.3. Finite memory observer

Within the use of the "static" form (4) of system (2), the state estimation is given in term of maximum likelihood by minimisation of the weighted square output estimation error other the defined observation window, i.e. the minimisation of the following criterion:

$$\phi_k=\frac{1}{2}\left\|Z_k(t)-\hat{Z}_k(t)\right\|^2_{R_k^{-1}} \tag{6.a}$$

under the constraint:

$$\hat{Z}_k(t)=\hat{Y}_k(t)+U_k(t)=W_k\hat{x}_k(t) \tag{6.b}$$

The classical solution in the least square sense is obtained:

$$\hat{x}_k(t)=\Omega_k^{-1}W_k^TR_k^{-1}Z_k(t) \tag{7}$$

where the square matrix of dimension nxn:

$$\Omega_k(t)=W_k^TR_k^{-1}(t)W_k \tag{8}$$

is assumed to be invertible. Note that the designed estimation is the product of a gain with a linear combination of the measurement z_i. It is therefore intrinsically of finite memory which size is equal to the number of data taken into account, i.e. k+1 and relies on the horizon definition $[t-\tau_k, t-\tau_0]$. The extended output estimation $\hat{Y}_k$ is deduced from (6.b), whose p first components stand for the estimation of the output at the instant t, i.e.:

$$\hat{y}_k(t)=S_{p,p(k+1)}\hat{Y}_k(t)=C\hat{x}_k(t) \tag{9.a}$$

with:

$$S_{p,p(k+1)}=\begin{bmatrix}I_p & 0_{p,pk}\end{bmatrix} \tag{9.b}$$

Only the properties of the state estimation error (10.a) are given below as the ones of the output estimation error (10.b) can be easily deduced thanks to (9).

$$e_k(t)=\hat{x}_k(t)-x(t) \tag{10.a}$$

$$\varepsilon_k(t)=\hat{y}_k(t)-y(t)=S_{p,p(k+1)}\left(\hat{Y}_k(t)-Y_k(t)\right) \tag{10.b}$$

3. PROPRIETIES OF THE FINITE MEMORY ESTIMATION

3.1. Necessary and sufficient condition

The finite memory observer exists if and only if the matrix Ω_k is invertible which is equivalent to matrix Ω_k of full column rank. As it is a symmetric matrix depending on W_k and the regular matrix R_k, a sufficient condition is that W_k is full column rank. As W_k is defined by the static form (4) of the system, it can easily be proved that this condition is equivalent to the observability condition of the model (4). Therefore, the minimal number of measurements to take into account so that the observer exists is given by the observability gramian (11) condition taking into account the k+1 data of the observation horizon (static form).

$$\Gamma_k(t) = W_k^T(t)Q_k^T(t)Q_k(t)W_k(t) \qquad (11)$$

3.2. Optimal number of measurements

The optimal number is given by the evolution of the norm of the state estimation error variance. Basing the proof on a sequential form (12) of the estimation (7) of x(t) expressed at a frozen instant t by adding one more data (z_{k+1}), it can be shown that matrix Ω_k is the solution of an algebraic Riccati equation depending on parameter k (Nuninger, 1997). Thanks to the proprieties of such equation, there exists an integer k_{opt} so that the inverse of matrix Ω_k does not change significantly for $k > k_{opt}$. This means that the new estimation of x(t), $\hat{x}_{k+1}$, based on one more data (z_{k+1}) does not differ significantly from the previous estimation $\hat{x}_k$; the adjunction of more data is therefore useless (no disturbances assumed).

$$\hat{x}_{k+1}(t) = \hat{x}_k(t) + G_{k+1}(t)\{z_{k+1}(t) - w_{k+1}\hat{x}_k(t)\} \qquad (12.a)$$
$$G_{k+1}(t) = \Omega_k^{-1}w_k^T\{r_{k+1} + w_{k+1}\Omega_k^{-1}w_{k+1}^T\} \qquad (12.b)$$

As a consequence, in practice, the optimal number of data to take into account is given by the evolution of the Euclidean matrix norm (for instance) of the estimation error variance Ω_k^{-1} or of the observability gramian Γ_k when k increases. Note that the optimal number of measurement does not rely on the τ_i that should nevertheless represent a proper virtual sampling of the data (Shannon criterion).

3.3. Estimation error proprieties

Consider additive deterministic faults d(t) on the output measurement in spite of the noise (i.e. $D_k(t) = [d(t-\tau_i)]$ for the considered horizon). Then, the output measurement of the static model is rewritten as (13) and the estimation error is (14).

$$\overline{Z}_k(t) = Z_k(t) + D_k(t) \qquad (13)$$
$$\overline{e}_k(t) = e_k(t) + \Delta e_k(t) \qquad (14.a)$$

where:

$$e_k(t) = \Omega_k^{-1}W_k^TR_k^{-1}N_k(t) \qquad (14.b)$$
$$\Delta e_k(t) = \Delta\hat{x}_k(t) = \Omega_k^{-1}W_k^TR_k^{-1}D_k(t) \qquad (14.c)$$

Note that $e_k(t)$ stands for the error when no fault occurs whereas the error differs from this value by the quantity $\Delta e_k(t)$ while faults appear. As a consequence, the estimation error is unbiased when there is no disturbances as it is proved within the estimation error mean and variance:

$$\mathrm{Esp}\{\overline{e}_k(t)\} = 0 + \mathrm{Esp}\{\Delta e_k\} = \Omega_k^{-1}W_k^TR_k^{-1}D_k \qquad (15.a)$$
$$\mathrm{Var}\{\overline{e}_k(t)\} = \mathrm{Var}\{e_k(t)\} = \Omega_k^{-1} \qquad (15.b)$$

Unfortunately, because of model uncertainties the estimation error is not a good residual to detect incipient faults. Indeed, as the real system is described by (1), the real output measurement is in fact (16.a) where ΔZ_k stands for the influence of the model uncertainties (16.b) not taken into account in the static model (3) of the system.

$$\overline{Z}_k(t) = Z_k(t) + D_k(t) + \Delta Z_k(f) \qquad (16.a)$$
$$f(x,u,t) = \sum_{s=1}^{S} A_s x(t-t_s) + \sum_{j=1}^{J} B_j u(t-t_j) \qquad (16.b)$$

As a consequence, the estimation error is rewritten as (14.a) where $e_k(t)$ is still given by (14.b) but with the variation from this value, $\Delta e_k(t)$, now defined by:

$$\Delta e_k(t) = \Omega_k^{-1}W_k^TR_k^{-1}\{D_k(t) + \Delta Z_k(f)\} \qquad (17)$$

It is obvious that both influence of faults and model uncertainties can not be distinguished as the corresponding sensitivities of the error, defined by the Jacobean matrices, are equal:

$$\frac{\partial \overline{e}_k}{\partial D_k^T} = \left[\frac{\partial \overline{e}_k(i)}{\partial D_k^T(j)}\right]_{n,(k+1)p} = \Omega_k^{-1}W_k^TR_k^{-1} = \frac{\partial \overline{e}_k}{\partial \Delta Z_k^T} \qquad (18)$$

4. NEW RESIDUAL

4.1. Definition

The finite memory estimation can be computed on-line on a fixed horizon within the recursive formulation (12). This formula (similar to the Kalman filter innovation equation) brings up the expression of the new residual. Indeed, assume a fault on the more recent data (z_{k+1}) then, from (12) it is obvious that only $\hat{x}_{k+1}$ will be wrong as it depends on the faulty z_{k+1} which is not the case of $\hat{x}_k$. Therefore, the difference between $\hat{x}_{k+1}$ and $\hat{x}_k$ is not equal to zero and the fault can be detected. On the contrary, model uncertainties influence any estimation quite the same way; making the difference can eliminate some of them so that it is less sensitive to model uncertainties. So, in order to detect incipient faults despite of model uncertainties, a robust residual (19) is defined by extension as the difference of two estimations $\hat{x}_{0.k}(t)$ and $\hat{x}_{r.k}(t)$ of the same state x(t).

$$e_{k/r}(t) = \hat{x}_{0/k}(t) - \hat{x}_{r/k}(t) \qquad (19)$$

The estimations $\hat{x}_{0,k}$ and $\hat{x}_{r,k}$ are computed on two overlapping data-sets $\{y(t-\tau_k),\ldots,y(t-\tau_0)\}$ and $\{y(t-\tau_k),\ldots,y(t-\tau_r)\}$ as shown in Figure 1.

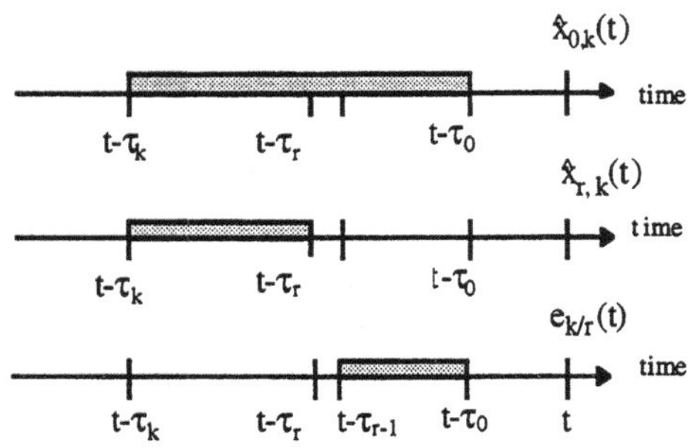

Fig. 1. Definition of the residual

It can be proved that the data of the detection window, i.e. $\{y(t-\tau_{r-1}),\ldots,y(t-\tau_0)\}$, have not the same influence than the data of the smallest horizon which is included in the largest one (these data are used for the computation of the two estimations ; this is not the case for the data of the detection window). Then, thanks to a special choice of (k,r), it is possible to give a greater sensitivity of the residual to the data of the detection window (i.e. the ones containing the information about the incipient fault) than the other data. Therefore, it is possible to design a residual which is less sensitive to some model uncertainties.

4.2. Residual formulation and statistical proprieties

Consider each estimation $\hat{x}_{0,k}(t)$ and $\hat{x}_{r,k}(t)$ deduced from (7) with the proper definitions of vectors and matrices on the given horizons denoted by the sub-indices '0,k' and 'r,k' respectively (see Figure 1). $\hat{x}_{0,k}(t)$ is given by (20) and $\hat{x}_{r,k}(t)$ satisfies the same equation with the proper notations.

$$\hat{\hat{x}}_{0,k}(t) = \hat{x}_{0,k}(t) + \Delta\hat{x}_{0,k}(t) \quad (20.a)$$

$$\hat{x}_{0,k}(t) = x(t) + \Omega_{0,k}^{-1} W_{0,k}^{T} R_{0,k}^{-1} N_{0,k}(t) \quad (20.b)$$

$$\Delta\hat{x}_{0,k}(t) = \Omega_{0,k}^{-1} W_{0,k}^{T} R_{0,k}^{-1} \{D_{0,k} + \Delta Z_{0,k}\} \quad (20.c)$$

Note that $\hat{x}_{0,k}$ (20.a) stands for the estimation while there is no faults and no model uncertainties, otherwise the estimation differs from this value by the quantity $\Delta\hat{x}_{0,k}$ (20.c). Thanks to the vectors and matrices bloc-proprieties, the estimation $\hat{x}_{0,k}$ on the bigger horizon can be expressed with respect to the vectors and matrices defining the estimation $\hat{x}_{r,k}$ based on the smaller horizon. Indeed, the following relations are satisfied (the sub-indices '0,r-1' denotes the data defined inside the detection window):

$$Y_{0,k} = \begin{bmatrix} Y_{0,r-1} \\ Y_{r,k} \end{bmatrix}, \text{ and so on for } U_{0,k}, Z_{0,k}, N_{0,k} \text{ and } D_{0,k}$$

$$R_{0,k}^{-1} = \begin{bmatrix} R_{0,r-1}^{-1} & 0_{r,k-r+1} \\ 0_{k-r+1,r} & R_{r,k}^{-1} \end{bmatrix}, \quad W_{0,k} = \begin{bmatrix} W_{0,r-1} \\ W_{r,k} \end{bmatrix}$$

$$\Omega_{0,k} = \Omega_{r,k} + \Omega_{0,r-1}$$

First assume that the model uncertainties are additive deterministic and constant parameter variations (ΔA, ΔB and ΔC) on the matrices defining the real system which is therefore given by ($A+\Delta A$, $B+\Delta B$, $C+\Delta C$). The not exact model is kept as (A,B,C). As a consequence, the output measurement Z_k can be rewritten with ΔZ_k now defined by (21) where the matrix ΔW_k and the vector ΔU_k stand for the additive and deterministic parameter variations of the matrices and vectors defining the static form of the real system. Basing the demonstration on the way W_k and U_k are designed when the indicated parameter variations occur, it is easily proved that $\Delta Z_{0,k}$ can also be rewritten with respect to $\Delta Z_{0,r-1}$ and $\Delta Z_{r,k}$ the same way $Z_{0,k}$ is expressed.

$$\Delta Z_k(t) = \Delta W_k x(t) - \Delta U_k(t) \quad (21)$$

Because of the limited number of pages required for this paper, only the final expression of the residual (22) is given (Nuninger, 1997) that leads to the statistical proprieties (23). From (22) and the further sensitivity study, it is obvious that both influence of faults and parameter variations (additive and deterministic) can be split apart thanks to the choice of the pair (k, r).

$$\bar{e}_{k/r}(t) = e_{k/r}(t) + \Delta e_{k/r}(t) \quad (22.a)$$

$$e_{k/r}(t) = \Omega_{0,k}^{-1} W_{0,r-1}^{T} R_{0,r-1}^{-1} N_{0,r-1}(t) + \left(\Omega_{0,k}^{-1} - \Omega_{r,k}^{-1}\right) W_{r,k}^{T} R_{r,k}^{-1} N_{r,k}(t) \quad (22.b)$$

$$\Delta e_{k/r}(t) = \Omega_{0,k}^{-1} W_{0,r-1}^{T} R_{0,r-1}^{-1} \left[\Delta Z_{0,r-1}(t) + D_{0,r-1}(t)\right] + \left(\Omega_{0,k}^{-1} - \Omega_{r,k}^{-1}\right) W_{r,k}^{T} R_{r,k}^{-1} \left[\Delta Z_{r,k}(t) + D_{r,k}(t)\right] \quad (22.c)$$

$$Esp\{\bar{e}_{k/r}(t)\} = Esp\{\Delta e_{k/r}(t)\} = \Delta e_{k/r}(t) \quad (23.a)$$

$$Var\{\bar{e}_{k/r}(t)\} = \Omega_{r,k}^{-1} - \Omega_{0,k}^{-1} \quad (23.b)$$

Second, the model uncertainties are due to not modelled time-delay in the real system. Therefore, $\Delta Z_{0,k}$ must be rewritten like (24) that leads to the new expression (25) of $\Delta e_{k/r}(t)$.

$$\Delta Z_{0,k}(t) = \begin{bmatrix} \Delta Z_{0,k}[0,r-1] \\ \Delta Z_{0,k}[r,k] \end{bmatrix} \quad (24)$$

$$\Delta e_{k/r}(t) = -\Omega_{r,k}^{-1} W_{r,k}^{T} R_{r,k}^{-1} \{\Delta Z_{r,k} + D_{r,k}\} + \Omega_{0,k}^{-1} W_{0,r-1}^{T} R_{0,r-1}^{-1} \{\Delta Z_{0,k}[0,r-1] + D_{0,r-1}\} + \Omega_{0,k}^{-1} W_{r,k}^{T} R_{r,k}^{-1} \{\Delta Z_{0,k}[r,k] + D_{r,k}\} \quad (25)$$

However, the proof is omitted in this paper, it can be shown that $\Delta Z_{0,k}[r,k]$ rely on $\Delta Z_{r,k}$ because of the way the static form of the real system is designed. This is quite clear for discrete time representation. Therefore, a slightly similar result as (22) can be written but in a more complex way because of coupling terms but a close conclusion could be brought out anyway. The extension for the statistical proprieties is not direct.

4.3. Residual sensitivity to parameter variations

Because of the previous remark, only the residual sensitivity to additive deterministic parameter variations is studied and compared to the residual sensitivity to additive faults. Within the previous definition of sensitivity, the following results can be easily proved from the expression (22) of the residual:

$$\frac{\partial \bar{e}_{k/r}(t)}{\partial D^T_{0,r-1}(t)} = \frac{\partial \bar{e}_{k/r}(t)}{\partial \Delta Z^T_{0,r-1}(t)} = \Omega^{-1}_{0,k} W^T_{0,r-1} R^{-1}_{0,r-1} \qquad (25.a)$$

$$\frac{\partial \bar{e}_{k/r}(t)}{\partial D^T_{r,k}(t)} = \frac{\partial \bar{e}_{k/r}(t)}{\partial \Delta Z^T_{r,k}(t)} = \left(\Omega^{-1}_{0,k} - \Omega^{-1}_{r,k}\right) W^T_{r,k} R^{-1}_{r,k} \qquad (25.b)$$

From theses results it is concluded that for a given horizon, the influence of faults and uncertainties can not be distinguished. Note that the same result applied for the state estimation error $\bar{e}_k(t)$ based on the biggest horizon. The residual sensitivity to the faults $D_{0,r-1}$ appearing on the data of the detection window (25.a) is compared to the residual sensitivity to the fault $D_{r,k}$ appearing on the data from the non-common horizon (25.b). It is obvious that both sensitivities are different (the same result applied for $\Delta Z_{0,r-1}$ and $\Delta z_{r,k}$). This remark leads to a tool for the choice of the pair (k, r). Note that in comparison with the estimation error $\bar{e}_k(t)$, the proposed residual $\bar{e}_{r/k}(t)$ has not the same sensitivity with respect to incipient faults whereas the sensitivities are equal outside the detection window:

$$\frac{\partial \bar{e}_k}{\partial D^T_{r,k}} = \Omega^{-1}_{0,k} W^T_{r,k} R^{-1}_{r,k} = \frac{\partial \bar{e}_{k/r}}{\partial D^T_{r,k}} + \Omega^{-1}_{r,k} W^T_{r,k} R^{-1}_{r,k} \qquad (26.a)$$

$$\frac{\partial \bar{e}_k(t)}{\partial D^T_{0,r-1}(t)} = \frac{\partial \bar{e}_{k/r}(t)}{\partial D^T_{0,r-1}(t)} \qquad (26.b)$$

4.4. Optimal choice of the detection window

Thanks to the previous results, it is proved that the residual $e_{r/k}(t)$ sensitivity can be improved with respect to incipient faults within a proper choice of the pair (k, r). This pair is chosen so that more influence is given to the data of the detection window with respect to the measurements outside this window. The aim is that the residual is more sensitive to incipient faults $D_{0,r-1}$ and less sensitive to some parameter variations $\Delta Z_{r,k}$ (note that the influence with respect to $\Delta Z_{0,r-1}$ can not be changed in comparison to the influence of $D_{0,r-1}$). First, the integer k is chosen so that the state estimation error is of minimal variance (the inverse of Ω_k is the solution of an algebraic Riccati equation). Second, for a fixed optimal k, integer $r < k$ is chosen so that the sensitivity of the residual with respect to the data from the detection window is greater than the sensitivity with respect to the data from the smallest horizon. In practice, such a consideration leads to the study of the following sensitivity performance index ratio $\rho D^{0,r-1}_{r,k}$ (27) when r increases. The result is a compromise between having enough data to compute the estimation and an optimal sensitivity to the incipient faults. The choice remains an heuristic one as the evolution is not simple to study in theory (especially if model uncertainties are considered instead of additive deterministic parameter variations). Anyway, the proposed residual (19) is more effective for fault detection than the classical estimation error.

$$\rho D^{0,r-1}_{r,k} = \left\| \frac{\partial \bar{e}_{k/r}}{\partial D^T_{0,r-1}} \right\|_2 \Big/ \left\| \frac{\partial \bar{e}_{k/r}}{\partial D^T_{r,k}} \right\|_2 \qquad (27)$$

5. APPLICATION ON A SIMULATED EXAMPLE

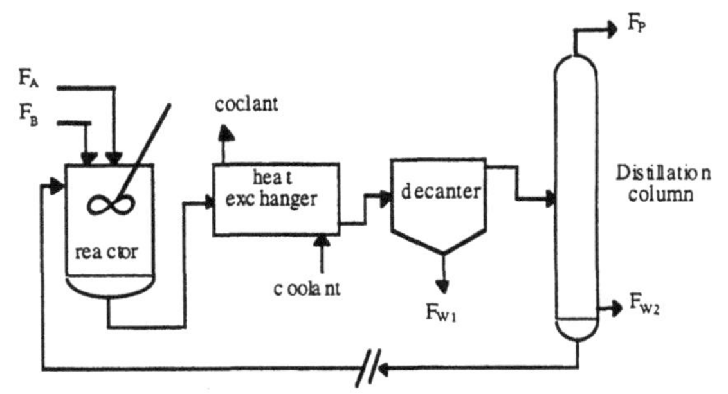

Fig. 2. Flowsheet of the chemical process (Williams et Otto, 1960).

The proposed example is a chemical process presented by (Williams and Otto, 1960). This system is a model of a refining plant (with separation and reaction involved) whose flow-sheet is shown in Figure 2 (Schoen, 1995). Sub-indices 'A' and 'B' denote the raw materials (feed rates F_A and F_B respectively), 'P' is the valuable product (F_P) whereas 'W1' stands for the by-product which is undesirable (F_{W1}) and 'W2' for the column purge lead off (F_{W2}). Note that the raw materials and by-product are recycled to the chemical reactor when reprocessed within the recycle loop that represents a significant transport lag. As a consequence, the equations of the process are non linear ones but a linearized model can be given anyway for a chosen operating point. In that case, for a recycling time of 10 minutes, the linearized model is the following (one time unit stands for 10 minutes) (Ross, 1971):

$$\dot{x}(t) = A_0 x(t) + A_1 x(t-1) + Bu(t)$$
$$y(t) = Cx(t)$$

$$A_0 = \begin{bmatrix} -4.93 & -1.01 & 0 & 0 \\ -3.20 & -5.30 & -12.8 & 0 \\ 6.40 & 0.387 & -32.5 & -1.04 \\ 0 & 0.833 & 11.0 & -3.96 \end{bmatrix}, \quad B = \begin{bmatrix} 1 & 0 \\ 0 & 1 \\ 0 & 0 \\ 0 & 0 \end{bmatrix}$$

$$A_1 = \begin{bmatrix} 1.92 & 0 & 0 & 0 \\ 0 & 1.92 & 0 & 0 \\ 0 & 0 & 1.87 & 0 \\ 0 & 0 & 0 & 0.724 \end{bmatrix}, \quad C = [0\ 0\ 1\ -1]$$

The four state components stand for the deviation in the weight composition of the materials (A, B, W1, W2 respectively). The control inputs are defined as $\partial F_A/6V_R$ and $\partial F_B/6V_R$ (V_R = 2.628m^3 is the volume of the reactor) and ∂F_A and ∂F_B are the deviations in the feed rates (in pounds per hour) from their nominal values. The proposed residual defined on the output $\varepsilon_{k/r}(t) = \hat{y}_{0,k} - \hat{y}_{r,k} = Ce_{k,r}(t)$ is designed from the evolution of the norm of the estimation error covariance matrix (k = 15) and from the evolution of the sensitivity performance index ratio $\rho D^{0,r-1}_{r,k}$ (r = 2 < k) plotted in Figure 3. In this figure the evolution of the variance norm of the residual is also plotted. In Figure 4 the residual is plotted when there is no faults (it is statistically close to zero). The output is also plotted. Two additive bias on y are assumed of magnitude 0.02, length 1.6s and 0.5s starting at instants 0.8s and 4s respectively. Two random values of magnitude 0.02 are also simulated at instants 7s and 8s. The measurement noise is of 5% magnitude.

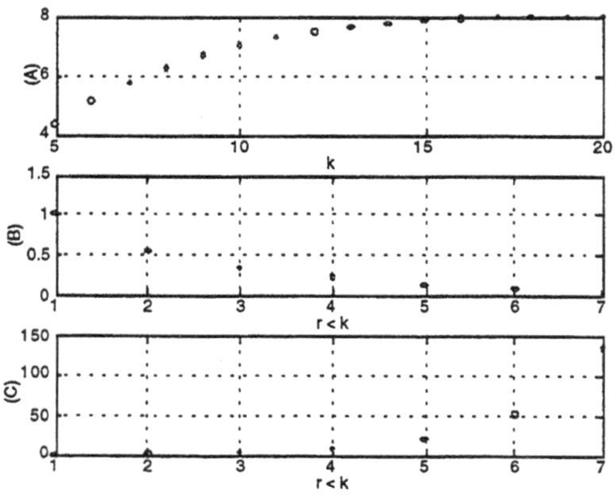

Fig. 3. Norms of the estimation error variance (A), ratio $\rho D^{0,r-1}_{r,k}$ (B) and of the residual variance (C).

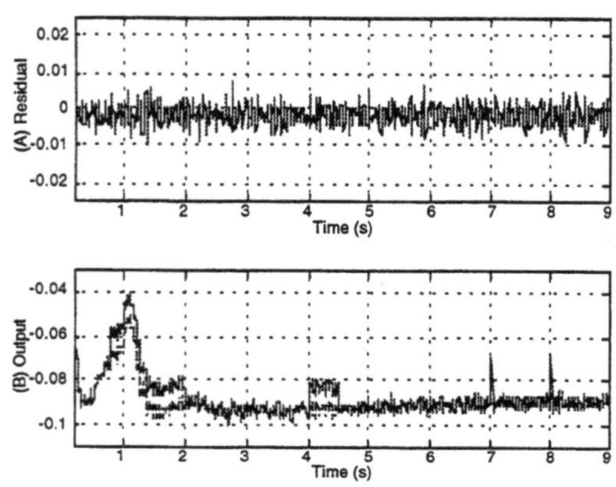

Fig. 4. (A) Residual when no faults. (B) Faulty and no-faulty output (dash) $y(t) = x_3(t) - x_2(t)$.

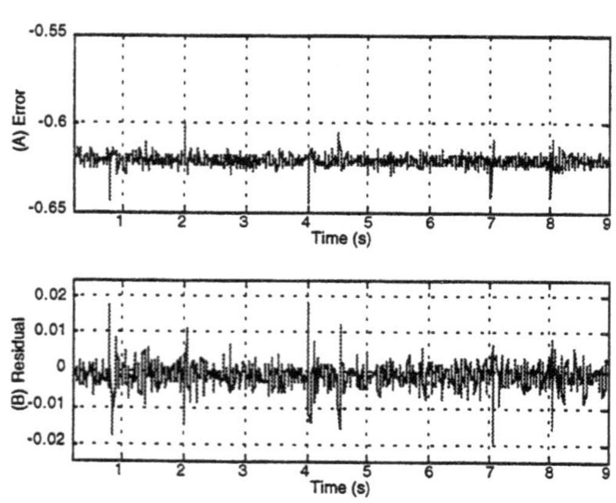

Fig. 5. Estimation error (A) and residual (B) (faults).

Finally, the residual and the estimation error are plotted in Figure 5. The error never converges to zero (even while no faults) and the evaluation procedure should be based on adaptive threshold whereas only a fixed threshold is required for the residual. It is therefore more efficient for fault detection as it is less sensitive to model uncertainties and the evaluation procedure is simpler.

6. CONCLUSION

Despite of the uncertain model (as time-delays are not taken into account to design the observer) the proposed residual is worthwhile for the detection of additive faults on the outputs. Indeed, the residual defined as the difference between two finite memory estimations, is robust with respect to some parameter uncertainties for a special choice of the parameters (k, r) that define the two overlapping horizons. Thanks to this choice, a detection window is defined and the residual is therefore more sensitive to incipient faults than to parameter uncertainties appearing in the measurements outside the detection window. This is true in comparison of the estimation error that accumulates uncertain parameters and could present a divergence phenomenon. Besides, both faults and parameter uncertainties can not be isolated. This is still true for the new residual but, it allows the isolation of incipient faults upon the detection window with respect to fault or disturbances that appeared before the detection window. Note that such a residual as also been developed and studied for discrete time representation (Kratz, *et al.*, 1997). In addition, in order to detect faults on the input, a generalised state finite memory observer is developed by Nuninger (1997). Future work will present this observer and study the residual proprieties.

REFERENCES

Chow E.Y and A.S. Willsky (1984) Analytical redundancy and the design of robust failure detection systems. *IEEE Trans. on Automatic Control*, **29 (7)**, pp. 603-614.

Darouach M., M. Zasadzinski, S.T. Xu (1994) Full-order observers for linear systems with unknown inputs. *IEEE Trans. on Automatic Control*, **39 (3)**, pp. 606-609.

Frank P.M. (1990) Fault diagnosis in dynamic systems using analytical and knowledge-based redundancy - a survey and some new results. *Automatica*, **26 (3)**, p. 459-474.

Kratz F., W. Nuninger, J. Ragot (1997) Finite memory observer based method for failure detection in dynamic systems. *SYSID'97*, **3**, pp. 1189-1194.

Medvedev A.V. and H.T. Toivonen (1991a) Three approaches towards a fault-tolerant controller structure. *Process Control Laboratory, Dept. of Chemical Engineering, Abo Akademi,* **Report 91-3.**

Medvedev A.V. and H.T. Toivonen (1991b) Investigation of a finite memory observer structure. *Process Control Laboratory, Dept. of Chemical Engineering, Abo Akademi,* **Report 91-6.**

Nuninger W. (1997) Stratégie de diagnostic robuste à l'aide de la redondance analytique. Doctorat de l'Institut National Polytechnique de Lorraine, octobre.

Patton R. J. (1994) Robust model-based fault diagnosis : the state of the art. *Proc. of IFAC, SAFEPROCESS'94*, pp. 1-24.

Ross W.D. (1971) Controller design for time lag systems via a quadratic criterion. *IEEE Trans. on Automatic Control*, **16 (6)**, pp. 664-672.

Schoen G.M. (1995) Stability and stabilization of time-delay systems. Ph. D. of Technical Sciences. Swiss Federal Institute of Technology Zurich.

Williams T.J. and R.E. Otto (1960) A generalized chemical processing model for the investigation of computer control. *Trans. of the American Institute of Electrical Engineers*, **79 (11)**, pp. 458-473.

Willsky A.S. (1976) A survey of design methods for failure detection in dynamic systems. *Automatica*, **12**, pp. 601-611.

ROBUST FAULT DETECTION FOR NONLINEAR SYSTEMS IN THE PRESENCE OF UNSTRUCTURED UNCERTAINTIES

Zhifeng Zhuang[1] **Paul M. Frank**

Fachgebiet Mess- und Regelungstechnik
Gerhard-Mercator Universitaet-GH-Duisburg
Bismarckstrasse 81BB, 47048 Duisburg, Germany
Tel: +49 203 379 2930 Fax: +49 203 379 2928
Email: (sl041zh,p.m.frank)@unidui.uni-duisburg.de

Abstract: For model-based fault detection and isolation (FDI), the residual plays an important role in indicating the occurrence of faults. The robust residual generation for nonlinear systems is in general a difficult task and the presence of unstructured uncertainties may lead to an NP-hard problem. This paper presents an alternative approach to residual generation for nonlinear systems with unstructured uncertainties by stochastic qualitative techniques. The proposed method combines stochastic modeling with the qualitative technique, forming a framework to cope with the uncertainty propagation into the system behavior as time elapses. The arising problem of spurious solutions in qualitative modeling is solved by a qualitative observer. The generated residual is normalized and less conservative compared to the adaptive threshold method. *Copyright © 1998 IFAC*

Keywords: fault detection, hidden Markov model, nonlinear systems, qualitative observer, stochastic modeling, unstructured uncertainty

1. INTRODUCTION

In model-based fault detection and isolation, the model is used to generate estimates of variables of the system. Estimates of measured variables are then compared with the actual measurements to generate error signals called residuals. The residual thus plays an important role in indicating the occurrence of faults. In the idealized cases where the system is perfectly modeled, the residual is designed to be affected only by faults. When no fault appears, it should be zero; when faults occur, it deviates from zero. However in practice, processes and their environments can not be completely known and the modeling error and unknown disturbances are unavoidable. Consequently, the residual depends not only on the faults but also on these uncertainties. An active topic in FDI research is therefore how to design the residual such that it is sensitive to faults, and on the other hand insensitive or even invariant to the effects of the uncertainties; see e.g. the survey papers (Frank, 1994; Patton, 1994).

Most early studies in robust fault detection using the observer-based approach have dealt with the structured uncertainty problem for linear systems. It has been shown that under certain restrictive conditions, the residual can be completely decoupled from the structured uncertainties and veritably reflect faults. If the conditions are not satisfied, an optimization problem has to be solved to achieve approximate disturbance decoupling. For unstructured uncertainties, a perfect disturbance decoupling is generally impossible. In this

[1] Sponsored by the German Academic Exchange Service (DAAD)

case, the generated residual should be further evaluated to decide whether the non-zero residual is indicating a fault. For this purpose, a careful selection of thresholds is necessary. In principle, a fixed threshold does not allow a satisfactory false alarm rate and an adaptive threshold seems to be superior.

An extension of the existing results for linear systems to the nonlinear systems is not direct and by no means a trivial task, because of the difficulties of estimating the state/output signals of a nonlinear system, even if the model is exactly known and no disturbances are present. Although many efforts have been made more recently, see e.g. (Vemuri and Polycarpou, 1997), the survey paper (Alcorta García and Frank, 1997) and the references contained therein, a universal solution is still an open problem .

In this paper we propose an alternative approach to residual generation for nonlinear systems with unstructured uncertainties by stochastic qualitative techniques. Our motivation for attacking this problem with the approach comes from the fact that in general the exact solutions for the problems treating unstructured uncertainties are NP hard, which can not be solved in polynomial time. A reasonable way is to solve the problems through randomized approximation, as is being done recently for robust control analysis, see e.g. (Ray and Stengel, 1993). The proposed method combines stochastic modeling with the qualitative technique, forming a framework to cope with the uncertainty propagation into the system behavior as time elapses.

The paper is organized as follows. In section 2 the behaviors of uncertain systems are reviewed in a quantized space setting which is a precondition for using randomized approximations. The emphasis is put on building the hidden Markov model that will be discussed in section 3, and on solving the problem of spurious solutions to which section 4 is dedicated. On these basis a fault detection scheme is described in section 5.

2. SYSTEM BEHAVIORS UNDER QUANTIZED SPACE

We consider the system described by the following nominal equation:

$$\dot{\mathbf{x}} = f(\mathbf{x}, \mathbf{u}, \theta)$$
$$\mathbf{y} = g(\mathbf{x}, \mathbf{u}, \theta) \qquad (1)$$

where the state $\mathbf{x} \in \Re^n$, the known input $\mathbf{u} \in \Re^r$, the output $\mathbf{y} \in \Re^m$, $\theta \in \Re^s$ is a vector of parameter uncertainties (including unknown disturbances), whose probability distribution is known. If only provided that θ belongs to the polytope described by the set

$$\Theta = \{\theta \in \Re^s, \theta^- \leq \theta \leq \theta^+\},$$

θ is then assumed to be uniformly distributed within the area.

In order to introduce the stochastic setting, we at first quantize the state space and the output space. A partition of $\Re$ leads to a set of the equivalence classes. The number of these classes is finite or countably infinite and can thus take values from the integer set $\aleph$. Each of the classes corresponds to a qualitative value q, $q = \langle x \rangle \in \aleph$ denotes the qualitative value of x. Under the certain partition σ, the equivalence classes are given by

$$[q]_\sigma = \{x \in \Re \mid \langle x \rangle_\sigma = q\}, q \in \aleph \qquad (2)$$

The quantity space $\mathcal{Q}_x$ of the qualitative values q reflects the partition σ and relates $\aleph$ to $\Re$:

$$\mathcal{Q}_x = \{[q_i] \mid \bigcup_{q_i \in \aleph} [q_i] = \Re, \quad [q_i] \cap [q_j] = \emptyset \; for \; q_i \neq q_j\} \qquad (3)$$

If $\mathbf{x}$ is a vector of dimension n and the same partition is taken for all these state variables, $\langle \mathbf{x} \rangle \in \aleph^n$ and $\mathcal{Q}_\mathbf{x} = \mathcal{Q}_x^n$. Similarly, $\mathcal{Q}_\mathbf{y} = \mathcal{Q}_y^m$. Let $[x]$ be the concise notation of $[\langle x \rangle]$ thereinafter. Suppose the range of variables at each dimension is limited and equally partitioned into N intervals, each interval with a length of Δ. The cell state $\langle \mathbf{x} \rangle \in \aleph^n$ is enumerated as $1, 2, \ldots, N^n$.

For each initial state $\mathbf{x}_0$ and fixed parameter θ, there exists a unique (quantitative) trajectory $T(\mathbf{x}_0, \theta)$ associated with the system (1). The hit state by $T(\mathbf{x}_0, \theta)$ at t is unique.

However, in a quantized space only the qualitative value $q_0 = \langle \mathbf{x}_0 \rangle$ of the initial state $\mathbf{x}_0$ is known. Furthermore, the parameter is possibly distributed in Θ. Accordingly, there are a collection of trajectories. The hit states by these trajectories at t constitute a state set, which can be represented by

$$S(q_0, \Theta, t) = \{\mathbf{x} \in \Re^n \mid T(\mathbf{x}_0, \theta)|_t = \mathbf{x}, \quad \mathbf{x}_0 \in [q_0], \theta \in \Theta\}. \qquad (4)$$

The following equality is valid for the state set on $\aleph^n$:

$$\langle S \rangle (q_0, \Theta, t) = \{\langle \mathbf{x} \rangle \in \aleph^n | S(q_0, \Theta, t) \cap [\mathbf{x}] \neq \emptyset\} \qquad (5)$$

To describe the system behavior with a randomized approximation, a probability function has to be introduced to depict the likelihood that a state is assumed. For qualitative state $q \in \aleph^n$ and $\langle \mathbf{x} \rangle(t) \in \langle S \rangle(q_0, \Theta, t)$, we define

$$p(q,t) := \Pr\{\langle \mathbf{x} \rangle(t) = q\}; \qquad (6)$$

For quantitative state $\mathbf{z} \in \Re^n$ and $\mathbf{x}(t) \in S(q_0, \Theta, t)$,

$$p(\mathbf{z},t) := \Pr\{\mathbf{x}(t) = \mathbf{z}\}. \qquad (7)$$

In this stochastic setting, we restate the hit state set by the system trajectories at t as

$$\begin{aligned} \langle S \rangle(q_0, \Theta, t) &= \{q \in \aleph^n \mid p(q,t) \neq 0\}, \\ S(q_0, \Theta, t) &= \{\mathbf{x} \in \Re^n \mid f(\mathbf{x},t) \neq 0\}, \qquad (8) \end{aligned}$$

where

$$f(\mathbf{z},t) = \frac{dp(\mathbf{z},t)}{d\mathbf{z}} \qquad (9)$$

is the probability density function of $p(\mathbf{z},t)$.

Note that $p(q,t)$ is actually the probability of $\mathbf{x}(t)$ falling into the qualitative state q and can be calculated by

$$p(q,t) = \frac{\int_{\xi \in [q] \cap S(q_0,\Theta,t)} d\xi}{\int_{\xi \in S(q_0,\Theta,t)} d\xi} \qquad (10)$$

3. HIDDEN MARKOV MODEL

The behaviors of the system with the quantized state space can approximately be represented by Markov chains (MC) (Zhuang and Frank, 1998).

Suppose the state space here is finite, with number N^n. For any $t, s \geq 0$,

$$p_{ij}(s, t+s) := \Pr\{\langle x \rangle(t+s) = j \mid \langle x \rangle(s) = i\}$$

is the transition probability of the MC. The hitting state set $S_M(\langle \cdot \rangle, t)$ and the probability $\tilde{p}(q,t)$ for assuming q at time t, of the behaviors of the MC is as follows

$$\begin{aligned} S_M(\langle \mathbf{x}_0 \rangle, 0) &= \{\langle \mathbf{x}_0 \rangle\}, \\ S_M(\langle \mathbf{x}_0 \rangle, t) &= \{q \in \aleph^n \mid \tilde{p}(q,t) > 0\}. \qquad (11) \end{aligned}$$

$\tilde{p}(q,t)$ is defined as

$$\tilde{p}(q,t) := \Pr\{\langle \mathbf{x} \rangle(t) = q\} \qquad (12)$$

and satisfies

$$\begin{aligned} &\tilde{p}(q, t+s) \\ &= \sum_{r \in \aleph} \Pr\{\langle \mathbf{x} \rangle(t+s) = q \mid \langle \mathbf{x} \rangle(s) = r\} \Pr\{\langle \mathbf{x} \rangle(s) = r\} \\ &= \sum_{r \in \aleph} p_{rq}(s, t+s) \tilde{p}(r, s), \qquad (13) \end{aligned}$$

for any $s, t \geq 0$. It is clear that the transition probability $p_{ij}(s, t+s)$ plays an important role in figuring out $\tilde{p}(q,t)$. If the input $\mathbf{u}$ of system (1) is constant, the corresponding $p_{ij}(s, t+s)$ in (13) is independent of time s, i.e., the process is stationary (homogenous), namely $p_{ij}(s, t+s) = p_{ij}(t)$. We will at first assume it is the case in our discussion.

On selecting the state transition probability of the MC that should represent the qualitative behaviors of system (1), we should let

$$\begin{aligned} S_M(\langle \mathbf{x}_0 \rangle, t) &\supseteq \langle S \rangle(\langle \mathbf{x}_0 \rangle, \Theta, t), \text{ or} \qquad (14) \\ \lfloor \tilde{p}(i,t) \rceil &\geq \lfloor p(i,t) \rceil \qquad (15) \end{aligned}$$

hold for any t and $i \in \aleph^n$, where

$$\lfloor a \rceil := \begin{cases} 1 \text{ if } a > 0 \\ 0 \text{ if } a = 0 \end{cases} \qquad (16)$$

The goal of constructing the Markov Model is to satisfy (14) and (15), at the same time to avoid an overconservative model, which generates more spurious solutions than necessary. Therefore, a reasonable choice for the transition probability is given by

$$p_{ij}(t) = \frac{\int_{\xi \in [j] \cap S(i,\Theta,t)} d\xi}{\int_{\xi \in S(i,\Theta,t)} d\xi} \qquad (17)$$

where $S(i, \Theta, t)$ is the region formed via the movement of the qualitative state i through a time interval t, and $[j] \cap S(i, \Theta, t)$ is the fraction falling into qualitative state j.

In the following we will, for the conciseness of the notations, use $X(t)$ to denote the probability function vector: $(p(\langle \mathbf{x} \rangle, t))_{N^n \times 1}$, for $\langle \mathbf{x} \rangle = 1, 2, \ldots, N^n$. According to the theory of MC (Isaacson and Madsen, 1976), $X(t)$ is governed by the equation

$$\dot{X}(t) = Q(\mathbf{u}, \theta) X(t) \qquad t \geq 0 \qquad (18)$$

where $Q = (q_{ji})$ is the intensity matrix of the Markov chain:

$$\begin{aligned} q_{ji} &= \lim_{t \to 0+} \frac{p_{ij}(t)}{t} \\ q_{jj} &= \lim_{t \to 0+} \frac{p_{jj}(t) - 1}{t}, \qquad (19) \end{aligned}$$

where $p_{ij}(t)$ is the probability of the transition from the state i at time h to state j at time $h+t$. q_{ji} is determined by $f(\mathbf{x}, \mathbf{u}, \theta)$ (Zhuang and Frank, 1998) as follows:

$$q_{k+i} = \frac{\int_{\theta^-}^{\theta^+} \int_{\mathbf{x}_{\bar{k}}^-}^{\mathbf{x}_{\bar{k}}^+} F_k^+(+) d\mathbf{x}_{\bar{k}} g(\theta) d\theta}{\Delta^{n-1}} \qquad (20)$$

$$q_{k-i} = \frac{\int_{\theta-}^{\theta+} \int_{\mathbf{x}_{\bar{k}}^-}^{\mathbf{x}_{\bar{k}}^+} F_k^-(-) d\mathbf{x}_{\bar{k}} g(\theta) d\theta}{\Delta^{n-1}}. \quad (21)$$

where $j = k^+$ represents the ith cell state transferring into the neighbouring cell state in the increasing direction at the kth dimension; $j = k^-$ corresponds to the transferring into the neighbouring cell state in the decreasing direction at the kth dimension. $\int_{\mathbf{x}_{\bar{k}}^-}^{\mathbf{x}_{\bar{k}}^+} := \int_{x_1^-}^{x_1^+} \cdots \int_{x_{k-1}^-}^{x_{k-1}^+} \int_{x_{k+1}^-}^{x_{k+1}^+} \cdots \int_{x_n^-}^{x_n^+}$, $d\mathbf{x}_{\bar{k}} := dx_1 \cdots dx_{k-1} dx_{k+1} \cdots dx_n$ and $g(\theta)$ is the probability density function of θ. We let

$$F_k^+(+) := \max[f_k(x_1, \cdots, x_{k-1}, x_k^+, x_{k+1}, \cdots, x_n, \mathbf{u}, \theta), 0]$$
$$F_k^-(-) := \max[-f_k(x_1, \cdots, x_{k-1}, x_k^-, x_{k+1}, \cdots, x_n, \mathbf{u}, \theta), 0]$$

x_k^+ and x_k^- are the boundary values adjoin to ith cell, of the k^+th and k^-th cell, respectively.

If $\mathbf{u}$ is time variable, the corresponding Markov chain is non- homogeneous and the intensity matrix $\mathbf{Q}$ is now also dependent on t.

On the other hand, let $Y(t)$ stand for the vector of the probabilities that the output signals fall into the cells in the quantized measurement space, as $X(t)$ in the case of state variables. Knowing the distribution $X(t)$ for $\mathbf{x}(t)$, we have the distribution $Y(t)$ for $\mathbf{y}(t)$ as

$$\begin{aligned} Y_i(t) &= \Pr\{\langle \mathbf{y} \rangle(t) = i\} \\ &= \sum_j \Pr\{\langle \mathbf{y} \rangle(t) = i \text{ and } \langle \mathbf{x} \rangle(t) = j\} \\ &= \sum_j \Pr\{\langle \mathbf{y} \rangle(t) = i | \ \langle \mathbf{x} \rangle(t) = j\} \Pr\{\langle \mathbf{x} \rangle(t) = j\} \\ &= \sum_j r_{ij} X_j(t), \end{aligned} \quad (22)$$

for $t \geq 0$; where

$$r_{ij} := \Pr\{\langle \mathbf{y} \rangle(t) = i | \ \langle \mathbf{x} \rangle(t) = j\} \quad (23)$$

and $Y_i(t)$ is the ith element of vector $Y(t)$. In this way, the output equation of (1) is transformed into

$$Y(t) = R(\mathbf{u}, \Theta) X(t) \qquad t \geq 0 \quad (24)$$

where $R(\mathbf{u}, \Theta)$ depends on $g(\mathbf{x}, \mathbf{u}, \theta)$ and can be calculated by the sampling method (Hsu, 1987).

The model (18) and (24) together constitute the hidden Markov model (HMM) of the continuous system (1). The HMM is actually a partially observed stochastic dynamical system model, which has been applied in the speech signal processing community (Rabiner, 1989). In our case, it is a set of linear differential equations with either the new state and the new output representing the probability distributions for the location and the likelihood for finding the original system trajectories at any time t. As pointed out in the last section, the distributions result from the partially known initial conditions and uncertain parameters.

The dimension of matrix Q is $N^n \times N^n$, but each column has only at most $2n + 1$ non-zero elements, therefore it is in general a sparse matrix. Moreover, we have for $t \geq 0$

$$\begin{aligned} l^T X(t) &= 1 \\ t^T Y(t) &= 1 \\ l^T Q(\mathbf{u}, \Theta) &= o^T \\ t^T R(\mathbf{u}, \Theta) &= l^T. \end{aligned} \quad (25)$$

where $l^T = [1 \cdots 1]$ and $t^T = [1 \cdots 1]$, the row vectors consist of N^n and N^m "1", respectively. $o^T = [0 \cdots 0]$ is the row vector with N^n "0".

Transform the model (18)(24) and (25) into the discrete-time form:

$$\begin{aligned} X(k+1) &= A(\mathbf{u}, \Theta) X(k) \qquad k \geq 0 \\ Y(k) &= C(\mathbf{u}, \Theta) X(k). \end{aligned} \quad (26)$$

For all $k \geq 0$, the following equalities hold

$$\begin{aligned} l^T X(k) &= 1 \\ t^T Y(k) &= 1 \\ l^T A(\mathbf{u}, \Theta) &= l^T \\ t^T C(\mathbf{u}, \Theta) &= l^T. \end{aligned} \quad (27)$$

The HMM retains the characteristics of state abstraction in qualitative reasoning and provides a possibility to depict the uncertain system with more detailed information than conventional qualitative modeling does.

4. SPURIOUS SOLUTIONS AND THEIR REDUCTION

In the above used Markov model, we assume that the dynamic behavioral solution set is represented by Markov chains. However, in general the equation (14) does not hold with equality sign. The residual part is called spurious solutions (SS) in the literature on qualitative reasoning. The SS have been brought in due to the quantization of the state space and the factitious assumption of Markov property. Although in some cases the effect of spurious solution can be neglected, in general the set of real solutions alone does not satisfy the equation (18). The existence of SS could deteriorate the correctness of the prediction of system behaviors described by the Markov chain.

In this section we aim at describing and reducing the effect of SS.

Assume $\mathbf{x}^*(t,\theta)$ is the solution of (1), for $t \geq 0$ and some $\theta \in \Theta$, its randomized representation is $X^*(t)$. The ith element $X_i^*(t)$ of $X^*(t)$ specifies the probability of the trajectories $\mathbf{x}^*(t,\theta)$ taking on the qualitative state i at t, given a set of initial states $[\mathbf{x}_0]$, that is

$$X_i^*(t) = p(i,t). \tag{28}$$

$X(t)$ is the solution of (18). Making use of the definition of $\lfloor \cdot \rceil$, the set of SS in qualitative space is represented as

$$\{i \in \aleph^n \mid \lfloor X_i(t) \rceil - \lfloor X_i^*(t) \rceil > 0\}. \tag{29}$$

Let $U(k)$ denote the probability distribution of the additional set caused by spurious solutions and $X^*(k)$ the probability distribution corresponding to real solutions, then

$$X(k) = X^*(k) + U(k) \qquad k \geq 0. \tag{30}$$

the state equation of HMM can be written as

$$X^*(k+1) = AX^*(k) + V(k,k+1) \tag{31}$$

for $k \geq 0$, where $V(k,k+1) = AU(k) - U(k+1)$ is taken to be disturbances or unknown inputs. Because of (27) and

$$l^T X^*(k) = 1, \tag{32}$$

we have then for all k,

$$\begin{aligned} l^T U(k) &= 0 \\ l^T V(k,k+1) &= 0. \end{aligned} \tag{33}$$

Furthermore, the assumption of knowing the initial distribution leads to

$$U(0) = [0 \cdots 0]^T. \tag{34}$$

In order to eliminate the effect of $V(k,k+1)$, we make use of the measurements to construct a filter. Because this filter is aimed at the acquirement of the correct qualitative (cell) states, we refer to it as a qualitative observer (QOB). Generally, the task of a QOB is to obtain the estimate of the state $X^*(k+1)$ in terms of the HMM and the measurement $y(k+1)$.

For the design of qualitative observers, in contrast to classical Luenberger observers or Kalman filters, there are neither precise models nor statistical characters about the disturbances available. Instead, we only know the character of SS by (33) and (34). One possible solution to the problem is to directly carry out the observation filtering, namely to check the consistence between the prediction and the measurement, as has been done in (Zhuang and Frank, 1997) at the scenario of pure qualitative models. Alternatively, we seek here an estimate $\hat{X}(k+1|k+1)$ that minimizes the following performance index:

$$J = ||\hat{X}(k+1|k+1) - \hat{X}(k+1|k)||^2 \tag{35}$$

under the constraint

$$\begin{aligned} C\hat{X}(k+1|k+1) &= Y(k+1), \\ l^T \hat{X}(k+1|k+1) &= 1, \\ \hat{X}_i(k+1|k+1) &\geq 0, \end{aligned} \tag{36}$$

for $i = 1, \cdots, N^n$. $\hat{X}(k+1|k)$ is the prediction at the $(k+1)$th step by the model,

$$\hat{X}(k+1|k) = A\hat{X}(k|k) \tag{37}$$

$Y(k+1)$ corresponds to the measurement $y(k+1)$.

Because of (27) the constraint $l^T \hat{X}(k+1|k+1) = 1$ is implied by $C\hat{X}(k+1|k+1) = Y(k+1)$ and can be removed from (36). This problem can then be formulated in general as a quadratic programming (QP) one and in turn be transformed into a linear programming problem for its solution.

5. FAULT DETECTION SCHEME

In FDI systems, it is only necessary to reconstruct the output signals and hence a qualitative output observer is required. Moreover, the design of QOB should poise the following two aspects of tasks. One is to attenuate the effect of the spurious solutions on the system behaviors, as done in the last section. In this case a measurement should be reliable, the estimate of the state is corrected according to the measurement. The other is to keep a tab on the possible abnormal behavior of measurements, which reflects fault occurrences in the system.

For this purpose, we seek an algorithm to give the estimate $\hat{Y}(k+1|k+1)$ that minimizes the following performance index:

$$\begin{aligned} J = &||\hat{Y}(k+1|k+1) - \hat{Y}(k+1|k)||^2/\sigma^2(k+1) \\ &+ ||\hat{Y}(k+1|k+1) - Y(k+1)||^2 \end{aligned} \tag{38}$$

under the constraint $\sum_i \hat{Y}_i(k+1|k+1) = 1, \hat{Y}_i(k+1|k+1) \geq 0$, for $i = 1, \cdots, N^m$. $\hat{Y}(k+1|k)$ is the prediction via the model at the $(k+1)$th step,

$$\begin{aligned} \hat{X}(k+1|k) &= A\hat{X}(k|k) \\ \hat{Y}(k+1|k) &= C\hat{X}(k+1|k) \end{aligned} \tag{39}$$

$Y(k+1)$ stems from the measurement $y(k+1)$. $\sigma^2(k+1)$ in (38) represents the degree of the influence of SS and hence reflects the credibility of the prediction $\hat{Y}(k+1|k)$. $\sigma^2(k+1)$ can be determined by the estimate of the variance of the prediction for $y(k+1)$,

$$\sigma^2(k+1) = \sum_i (y_i^c - \bar{y}(k+1))^2 \hat{Y}_i(k+1|k)$$
$$= \sum_i (y_i^c)^2 \hat{Y}_i(k+1|k) - \bar{y}^2(k+1)$$
$$\bar{y}(k+1|k) = \sum_i y_i^c \hat{Y}_i(k+1|k)$$

where $\bar{y}(k+1)$ is the mean of $\hat{y}(k+1)$, and y_i^c is the midpoint value of the ith qualitative cell and $\hat{Y}_i(k+1|k) = C_i\hat{X}(k+1|k)$, C_i is the ith row vector of C.

Again, this problem can be solved by a QP algorithm.

With the aid of the obtained estimate of qualitative outputs, a residual generator can be designed as shown in Fig.1, and

$$r(k) = 1 - \hat{Y}_i(k|k) \quad \text{if } y(k) \in [y_i^-, y_i^+] \qquad (40)$$

where $\hat{Y}_i(k|k)$ is the ith element of $\hat{Y}(k|k) = C\hat{X}(k|k)$. y_i^- and y_i^+ are the lower and upper bounds of the ith cell of outputs. This residual is automatically normalized, lying in the interval $[0,1]$. $r = 1$ corresponds to the sure occurrence of a fault. Because $\hat{Y}$ reflects the estimated outputs of system, the generation of the residual is direct and less conservative compared to adaptive threshold methods.

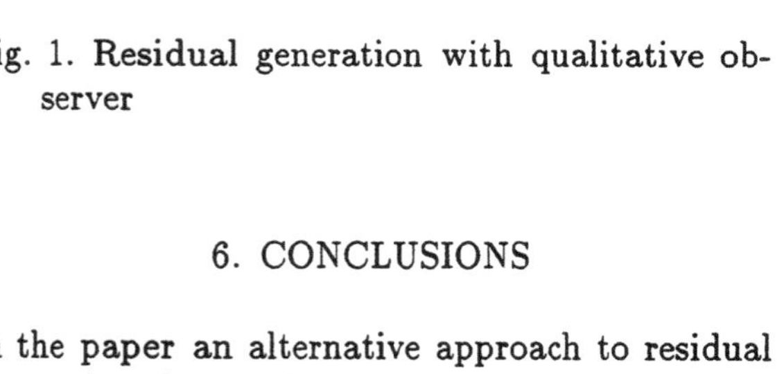

Fig. 1. Residual generation with qualitative observer

6. CONCLUSIONS

In the paper an alternative approach to residual generation for nonlinear systems with unstructured uncertainties is proposed. The behaviors of uncertain systems are rechecked in a quantized space setting. The emphasis is put on building the hidden Markov model and solving the arising problem of spurious solutions. On these bases the fault detection scheme is constructed.

As opposed to quantitative model-based methods, the system variables herein are qualitatively estimated, i.e. the estimates are intervals rather than real numbers. It can be expected that these estimates are less conservative than the existing adaptive thresholds, leading to a lower missing alarm rate. On the other hand, the modeling scheme assures the conservativeness of the estimates for variables under normal operation and thus avoids false alarm. The approach is also different from the conventional qualitative model-based techniques which do not prophesy the likelihood for taking the qualitative values for a system variable.

The study of the next steps is to evaluate and eventually improve the computation complexity.

7. REFERENCES

Alcorta García, E. and P.M. Frank (1997). Deterministic nonlinear observer-based approaches to fault diagnosis: A survey. *Control Engineering Practice* **5**(5), 663–670.

Frank, P.M. (1994). Enhancement of robustness in observer-based fault detection. *International Journal of Control* **59**(4), 955–981.

Hsu, C.S. (1987). *Cell-to-Cell Mapping – A Method of Global Analysis for Nonlinear Systems.* Springer-Verlag. New York.

Isaacson, D.L. and R. W. Madsen (1976). *Markov Chains Theory and Applications.* John Wiley &Sons. New York.

Patton, R. (1994). Robust model-based fault diagnosis: the state of the art. In: *Preprints of SAFEPROCESS'94*. Espoo, Finland.

Rabiner, L.R. (1989). A tutorial on hidden markov models and selected applications in speech recongnition. *Procedings of the IEEE* **77**(2), 257–286.

Ray, L. R. and R. F. Stengel (1993). A monte carlo approach to analysis of control system robustness. *Automatica* **29**, 229–236.

Vemuri, A.T. and M.M. Polycarpou (1997). Robust nonlinear fault diagnosis in input-output systems. *International Journal of Control* **68**(2), 343–360.

Zhuang, Z. and P.M. Frank (1997). Qualitative observer and its application to fault detection and isolation systems. *Journal of Systems and Control Engineering, Proc. of Institution of Mechanical Engineers, Pt. I* **211**(4), 253–262.

Zhuang, Z. and P.M Frank (1998). A gsmp model-framework for continuous-time dynamic systems. In: *American Control Conference 98.* Philadelphia, PA, USA.

DETECT UNEXPECTED CHANGES OF PARTICLE SIZE DISTRIBUTION IN PAPER-MAKING WHITE WATER SYSTEMS

Hong Wang

Department of Paper Science
UMIST
Manchester M60 1QD
U. K.

Abstract: This paper presents a novel approach for the change detection of the output probability density functions for dynamic stochastic systems. Using the B-splines neural network, the measured probability density functions of the system output are represented by a set of weights which are functions of the control inputs to the system. This leads to a unique expression of the dynamic characteristics of the output probability density functions for the system. An observer based fault detection is developed which detects any unexpected changes of the parameters in the dynamic part of the system. An application to the detection of unexpected changes of particle size in paper-making is included to demonstrate the use of the proposed algorithm and desired results have been obtained.

Keywords: Fault detection; observer design, papermaking; stochastic systems.

1.INTRODUCTION

It is well known that in paper-making (see Fig. 1), many different types of chemicals are used in the wet end in order to

- control paper properties in the dry end, and
- increase the efficiency of raw material usage.

Especially for the improved efficiency of raw material usage, a group of chemicals, called retention aids or polymers, are added in the thin stock stream to control the percentage of total solids retented on the paper machines after the initial formation of the paper web. As its name stands, retention is an important factor which describes the efficiency in the wet end of paper machines. It is defined as

$$y(t) = K(1 - \frac{C_w(t)}{C_h(t)}) \tag{1}$$

where

- $y(t)$ = retention value at time t;
- K= a machine dependent constant;
- $C_w(t)$ =consistency in white water;
- $C_h(t)$=consistency in head-box.

Indeed, $y(t)$ reflects the percentage of fibres and other additives (i. e., fillers and process chemicals) remaining in the final produced paper and thus indicates both the efficiency of the use of raw material and the runnability of paper machines. Due to the complicated nature of the wet end of paper machines, retention is affected by many variables. Typical examples are retention aids (polymers), stock pH values, stock flow rate, sheer forces, pulp quality (CSF), head-box slice geometry, raw material contaminants, drainage, machine speed and the structure of white water systems, etc. As a result, it is a multiple-input and one output system as shown in Fig .2 (Wang, 1997). Moreover, due to constant variations of all the input variables during paper production and the involvement of fluid dynamics during the formation phase on wire section, the system is dynamic, nonlinear and stochastic.

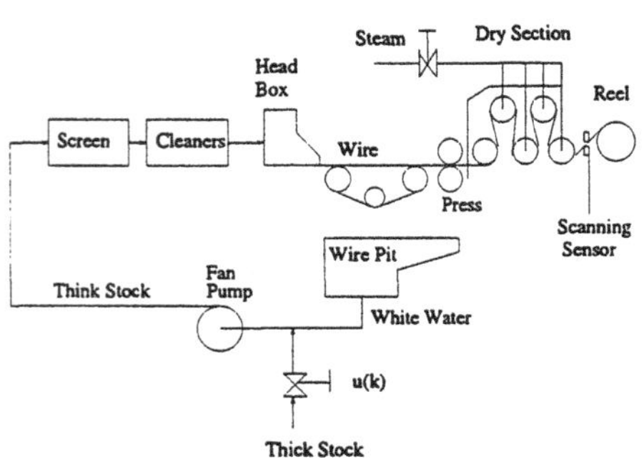

Fig. 1. A paper machine.

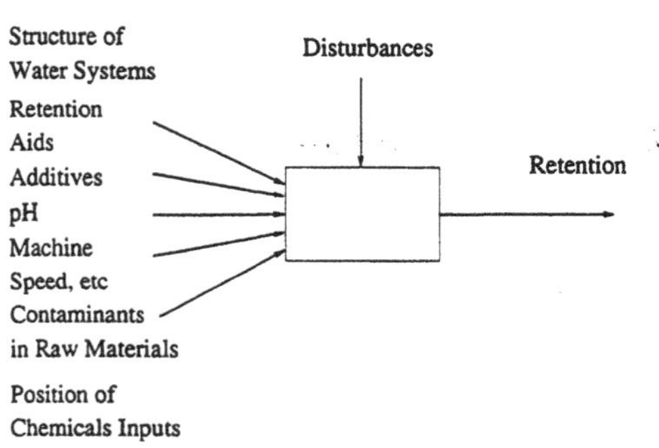

Fig. 2. Retention process.

2.THE FAULT TO BE DETECTED

In order to control retention, the consistency in the white water pit needs to be closely controlled and monitored. As such, it is desirable to detect any unexpected changes of the white water consistency at an early stage. Often, this early stage of changes can be characterised as unexpected changes (or fault) of the particle size distribution in the white water systems.

3.MODEL REPRESENTATION

Denote random variable x as the particle size in the white water system, then using the concepts of random variable, its cumulative distribution function is defined as

$$F(z) = P(x < z) \tag{2}$$

where $P(x < z)$ stands for the probability of variable x less than z. Assuming the cumulative distribution function $F(z)$ is a continuous function with respect to z, then the probability density function $f(z, u)$ is defined as

$$f(z,u) = \frac{dF(z)}{dz} + \omega(z) \tag{3}$$

where $\omega(z)$ represents the noise term on the density function, u is the measurable inputs which control the shape of the distribution function. For example, u can be regarded as retention polymers, machines speed, water drainage conditions in the wire section, headbox geometry, etc.. Assuming $z \in [a, b]$ and using the well known B-splines function theory (Brown and Harris, 1994;Girosi and Poggio, 1990), it can be seen that

$$f(z,u) = \sum_{i=1}^{N} w_i(u)B_i(z) + \omega(z) \tag{4}$$

where $w_i(u), (i = 1, 2, ..., N)$ are the weights which can be trained using the measured $f(z, u)$ and z, and $B_i(z), (i = 1, 2, ..., N)$ are pre-specified basis functions defined on the interval $z \in [a, b]$. Since all the basis functions are fixed, the unexpected changes of the particle size distribution can be regarded as to be caused by the corresponding unexpected changes in $w_i(u)$. As such, the purpose of fault detection is to use measurable $f(z, u)$ and u to detect any unexpected changes in $w_i(u)$.

Two different cases need to be considered. In the first case, we can regard the relationship between w_i and u as static. In this case, the relationship between w_i and u can be again expressed via the B-splines theory as

$$w_i(u) = \sum_{j=1}^{M} v_{ij}\phi_j(u) \tag{5}$$

where v_{ij} are the weights and ϕ_j are again the pre-specified basis functions defined on the definition domain of u. By substituting equation (5) into equation (4), it can be further obtained that

$$\begin{aligned} f(z,u) &= \sum_{i=1}^{N}\sum_{j=1}^{M} v_{ij}\phi_j(u)B_i(z) + \\ & \omega(z) = \theta^T\Phi(u,z) + \omega(z) \end{aligned} \tag{6}$$

where vector θ consists all the unknown weights v_{ij} to be trained and vector $\Phi(u, z)$ is composed by all the combined algebraic products of basis functions, B_i and ϕ_j. Since this form is exactly the same as the linear output equation widely used in parameter estimation, the standard recursive least squares estimation technique can be used to train all the weights v_{ij}. As such, the resulting residual signal from this estimation algorithm can be taken as the detection signal for fault detection purpose.

The second case is for the dynamic relationship between w_i and u (Wang, 1997, 1998). This can be further classified into linear and nonlinear dynamic relationships. For linear dynamic systems, we have

$$\dot{W} = AW + Bu \quad (7)$$

where A and B are known parameter matrices obtained from the system, and

$$W = (w_1, w_2, ..., w_N) \quad (8)$$

For nonlinear dynamic systems, this relationship can be expressed as

$$\dot{W} = h(W, u) \quad (9)$$

where $h(W, u)$ is a nonlinear vector function. By taking equation (7) or (9) as the state equation and (4) as the output equation, observer based fault detection and diagnosis can be used to perform fault detection for the system. In this study, only linear dynamic system (7) is considered, where fault detection is performed via the use of observer based techniques.

4.FAULT DETECTION

In this section, the observer-based fault detection will be discussed where the faults of the system are regarded as any unexpected changes of an additive term in the state space equation.

Since $f(z, u)$ is a probability density function which should satisfy

$$\int_a^b f(z, u)dz = 1 \quad (10)$$

for all $\forall z \in [a, b]$, by denoting

$$\begin{aligned} b_i &= \int_a^b B_i(z)dy \\ L(z) &= b_N^{-1} B_N(z) \in R^{1\times 1} \\ C(z) &= (B_1(z) - \frac{B_N(z)}{b_N} b_1, ..., \\ &\quad ..., B_{N-1}(z) - \frac{B_N(z)}{b_N} b_{N-1}) \end{aligned} \quad (11)$$

then the following minimum order dynamic relationship

$$\begin{aligned} \dot{V} &= AV + Bu + F \\ f(z, u) &= C(z)V + L(z) + \omega(z) \end{aligned} \quad (12)$$

can be used, where $V = (w_1, w_2, ..., w_n)^T$ is the state vector and $\{A, B\}$ are the known healthy parameter matrices of the system, F is an additive term representing the fault in the system, and $n = N - 1$ is the minimum order of the system. It can be shown that with this form the equality (10) is always satisfied (Wang 1997, 1998). In order to detect the fault, let us construct the following observer

$$\begin{aligned} \dot{\hat{V}} &= A\hat{V} + Bu + K\epsilon \quad (13) \\ \epsilon &= \int_a^b C^T(z)(f(z, u) - \hat{f}(z, u))dz \\ \hat{f}(z, u) &= C(x)\hat{V} + L(z) \quad (14) \end{aligned}$$

then by defining

$$e = V - \hat{V} \quad (15)$$

and using equations (12) and (13), it can be formulated that

$$\begin{aligned} \dot{e} &= AV + Bu + F - \\ &\quad A\hat{V} - Bu - K\epsilon \end{aligned} \quad (16)$$

Since it can be shown that

$$\begin{aligned} \epsilon &= \int_a^b C^T(z)C(z)(V - \hat{V})dz + \\ &\quad \int_a^b C^T(z)\omega(z)dz \end{aligned} \quad (17)$$

by denoting

$$\Sigma = \int_a^b C^T(z)C(z)dz \in R^{n\times n} \quad (18)$$

it can be further obtained that

$$\epsilon = \Sigma e + \Delta \quad (19)$$

where $\Delta = \int_a^b C^T(z)\omega(z)dz$. As a result, the error equation becomes

$$\dot{e} = (A - K\Sigma)e + F + K\Delta \quad (20)$$

Assuming that the basis functions $B_i(z), i = 1, 2, ..., N$ are selected such that the pair (A, Σ) is observable, then the gain matrix K can be selected such that all the eigenvalues of matrix $A - K\Sigma$ lie inside the lef-hand complex plane. As a result, when no fault occurs, $F = 0$ and $e(t)$ tends to be small. This means that $\epsilon(t)$ should also be small when $F = 0$. As such, the fault detection can be perform by taking $\epsilon(t)$ as the detection signal. This means that the fault can be detected through the following mechanism.

$$\|\epsilon\| > \lambda \rightarrow a \ fault \ has \ occurred. \quad (21)$$

where $\lambda > 0$ is a pre-specified threshold. In practice, the threshold should satisfy

$$\lambda > K_1 \left| \int_a^b C^T(y)\omega(z)dz \right| \tag{22}$$

to avoid any possible false alarm caused by the noise term $\omega(z)$, where

$$K_1 = \max\{\|sI - A + K\Sigma\|^{-1} K\} + 1 \tag{23}$$

Assuming that there exists a positive number δ such that

$$|\omega(z)| \leq \delta \tag{24}$$

then it can be seen that the threshold should be selected such that

$$\lambda > \delta K_1 \int_a^b \|C(z)\| \, dz \tag{25}$$

Since $C(y)$ is composed of all the basis functions, it can be seen that the threshold λ is well defined by inequality (25).

From the above formulations, it can be seen that different from the standard observer design, here the residual ϵ is related to the integration of the output probability density function $f(z,u)$. Indeed, this is very similar to the control algorithm design for the input and output form of the system (12) by Wang (1997, 1998), where the control input $u(t)$ is related to the integration of the probability density function $f(z,u)$.

5.APPLICABILITY STUDY

Several pilot plant trials have been carried out to collect the required samples, where the input of pulp and fillers are kept constant. As such, the control input u only consists of polymer dosage and the fault is reflected by unexpected changes of machine operation condition largely affected by different degrees of shear forces applied in the headbox approaching stream.

One hundred samples have been formulated and used to evaluate one hundred different probability density functions. These probability distributions are then combined together and is shown in Fig. 3 via the use of the MESH function in Matlab. From this 3-D plot it can be seen that the total number of the basis functions can be selected as $N = 10$, $a = 0$ and $b = 0.6(mm)$. This leads to the following exponential type of basis functions

$$B_i(z) = exp(-\frac{(z - z_i)^2}{\sigma_i^2}) \tag{26}$$

where z_i represents the data centre spanned by the support of the ith basis function, $B_i(z)$, and σ_i stands for the applied width of the ith basis function. Denote w_i as the weights, then the following state space model for $n = N - 1 = 9$ can be obtained.

$$\dot{w}_i = -0.083w_i + 0.78u + f_i \tag{27}$$

where u stands for the equivalent single polymer flow rate to the system and f_i represents *equivalent* abnormal conditions applied to the system. By selecting σ_i as a small positive number, the following approximation can be obtained

$$b_i = \int_0^{0.6} B_i(z)dz = \sqrt{2\pi}\sigma_i \tag{28}$$

In this case, it has been shown that the noise term $\omega(z)$ is upper bounded by $\delta = 0.16$. This leads to the threshold in equation (25) being given by $\lambda = 0.1$.

By choosing $\sigma_i = 0.03$, matrix Σ can be approximately obtained as given by

$$\Sigma = [\Sigma_1 \quad \Sigma_2] \tag{29}$$

where

$$\Sigma_1 = \begin{bmatrix} 0.042 & 0.002 & 0.0001 & 0 \\ 0.002 & 0.042 & 0.002 & 0.0001 \\ 0.0001 & 0.002 & 0.042 & 0.002 \\ 0 & 0.0001 & 0.002 & 0.042 \\ 0 & 0 & 0.0001 & 0.002 \\ 0 & 0 & 0 & 0.0001 \\ 0 & 0 & 0 & 0 \\ 0 & 0 & 0 & 0 \\ 0 & 0 & 0 & 0 \end{bmatrix} \tag{30}$$

$$\Sigma_2 = \begin{bmatrix} 0 & 0 & 0 & 0 & 0 \\ 0 & 0 & 0 & 0 & 0 \\ 0.0001 & 0 & 0 & 0 & 0 \\ 0.002 & 0.0001 & 0 & 0 & 0 \\ 0.042 & 0.002 & 0.0001 & 0 & 0 \\ 0.002 & 0.042 & 0.002 & 0.0001 & 0 \\ 0.0001 & 0.002 & 0.042 & 0.002 & 0.0001 \\ 0 & 0.0001 & 0.002 & 0.042 & 0.002 \\ 0 & 0 & 0.0001 & 0.002 & 0.042 \end{bmatrix} \tag{31}$$

Since $A = diag(-0.0083, -0.0083, ..., -0.0083)$, it can be seen that the following observer gain

$$K = 3.7 \tag{32}$$

will result in a stable matrix $A - K\Sigma$ with its eigenvalues being between $(-0.1502, -0.1784)$. From Fig. 3, it can be seen that after the sample number 30, there is a change in the machine shear condition leading to a nonzero value for $f_1 - f_3$. This is indeed reflected in Fig.4, where the response of the residual signal ϵ is shown.

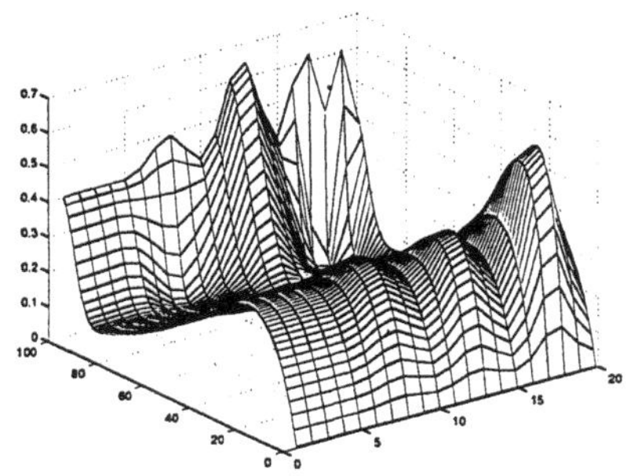

Fig. 3. The distribution mesh plot.

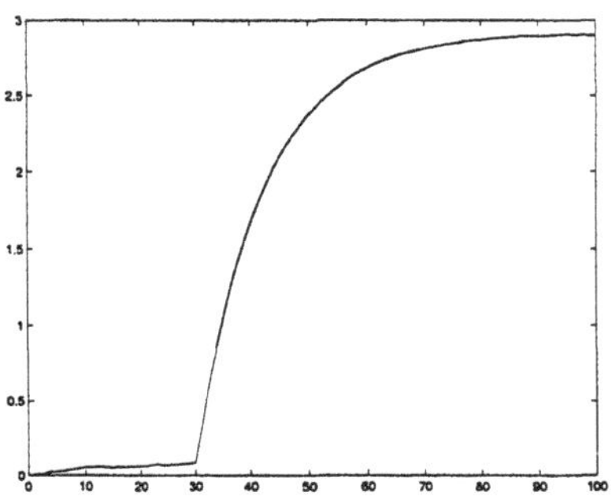

Fig. 4. The detecting signal $||\epsilon(t)||$.

6.CONCLUSIONS

In this paper, a new fault detection approach has been developed for the unexpected change detection of the probability density function of the output of stochastic systems. A new type of observer is constructed where the residual signal is calculated via the use of the weighted integration of the measured probability density function. It has been shown that under a very general condition, an observer can be constructed to perform the fault detection for the system. An applicability study to the unexpected change detection of the particle size in the white water pit of a paper machine has been made and a desired result has been obtained.

7.ACKNOWLEDGEMENT

The work is partly funded by the on-going EPRSC research grant GR/K97721. This is gratefully acknowledged.

REFERENCES

Astrom, K., J., and B. Wittenmark (1988).*Adaptive Control*, Addison Wesley.

Brown, M., and C.J. Harris (1994). *Neurofuzzy Adaptive Modelling and Control*, Prentice-Hall, Hemel-Hempstead.

Girosi, F and T. Poggio, (1990). Networks and the best approximation property, *Biol. Cybern.*, Vol. **63**, pp. 169-176.

Moore, M. W. (1997) Practical applications of a novel web monitor for paper machines, *Proc. 1997 APPITA Conference*, Melbourne.

Wang, H., M. Brown and C. J. Harris, (1996). Modelling and control of nonlinear, operating point dependent systems via associative memory networks, *Int. J. Dynamic and Control*, Vol. 5.

Wang, H., (1997). On the modelling and control of the output probability density functions for stochastic systems, submitted to *Automatica*.

Wang, H., A. P. Wang, and S. Duncan, (1997). *Advanced Process Control for Paper and Board Making*, PIRA International.

RECONCILING ACADEMIC APPROACHES AND INDUSTRIAL REALITIES: LESSONS FROM FAULT DIAGNOSIS APPLICATIONS

Prasad Dhurjati

Department of Chemical Engineering, University of Delaware,
Newark, DE 19716, USA email: dhurjati@udel.edu

Abstract: This paper reviews lessons learned from over fifteen years of industrial experience with fault diagnosis and process supervision. While a significant body of academic literature exists on methodologies for fault diagnosis, there exists a striking gap between academic approaches and industrial requirements. Reasons for this gap include availability of sensor measurements, accessibility to experts, non-availability of knowledge in the form of heuristics or models, complicated process dynamics, knowledge validation, exclusive dependence on a single methodology, inadequate software tools, difficulties in implementation, human factors, etc. The challenges in application of fault diagnosis to industrial situations will be illustrated using several examples with primary focus on the *FALCON* project.

Keywords: fault diagnosis, industrial applications, knowledge-based systems, mathematical models, process simulation

I. INTRODUCTION

In industry, considerable amount of knowledge may exist as models (steady state or dynamic), archived data, fault trees, reports after a major incident and in the form of experts who have been troubleshooting problems over long periods. Such knowledge could possibly be exploited and automated within the context of a knowledge based system leading to substantial economic benefits.

In academia, considerable progress has been made in coming up with new methodologies and developing a systematic approach towards the problem of fault diagnosis and supervision. Many research groups have developed in great depth a single methodology which is often illustrated using simulated examples.

It would be expected that many of the academic developments would have been adopted in industry and that conversely academia would have benefited from industrial feedback. Unfortunately, this has not been the case as we have discovered from our industrial interactions.

In this paper, we will start with a brief overview of some of the large scale applications. "Details" of these applications are available in other manuscripts. We will demonstrate the diversity of problems that are encountered in practice and discuss how academic approaches could be useful in the solution of these problems.

1.1 OVERVIEW OF PROJECTS

Examples of projects that were undertaken are given below:
(1) *FALCON*: diagnosing faults in a commercial scale chemical process (Rowan, 1986, Dhurjati, 1987)
(2a) *Star-resources*: optimal use of hydrogen resources in a refinery (Petti, 1992, 1993)

(2b) *Star-odor*: diagnosing odor-causing emissions in a refinery (Kordon, 1996)
(2c) *Star-upset :* analysis of archived data to anticipate plant shutdown (English, 1998)
(3) *Composites* Manufacturing: supervision of a composite process (Pillai, 1994, 1996, 1997)

In each case, the **motivation** for the project was different. (1) In the *FALCON* project, the motivation was safety and avoiding hazardous incidents. (2a) For the *Star-resource* management problem, the motivation was savings in hydrogen resources and its economic impact. (2b) For the *Star-odor* problem, the motivations were the complaints from the community and the bad publicity generated as a result . (2c) For the *Star-upset* analysis problem, the motivation was to take advantage of knowledge that existed in the archived data and use that knowledge to prevent future similar upsets that could lead to expensive plant shutdown. (3) For the *composite* process, the motivation was reproducibility in quality of parts and decrease in process time.

The amount of **information** that was available to address the problem was variable in each case and so was the nature of knowledge. For example, in *FALCON*, a limited set of measurements was initially available to the data acquisition system although we were able to increase the number of available measurements at a later stage. The initial knowledge was available in the form of a steady state simulation and access to experts. In the *Star-odor* project, a large amount of diverse information had to be exploited. This included remote data from weather monitoring systems, off-line lab measurements available once a day, information from individuals complaining about odors, about 50,000 measurements available to the data acquisition system every minute, etc. In the *composites* project, the only available measurement was the temperature of the surface and center of the part. A dynamic mathematical model was built to simulate the process but part variability from run to run made the parameter information fairly unreliable.

The **methodologies** used for each project varied depending on the goals and the amount of available on-line information and knowledge. For the *FALCON* project, the methodology was primarily mathematical (as explained later) and took advantage of residuals computed at different plant operating conditions. In the *Star-resource* problem, a coupled architecture incorporating numerical computing in a knowledge based system environment was used. One portion of the coupled system formulated an optimization problem that was subsequently solved by an external routine. The uncertainty present in some of the constraints is dealt with using a fuzzy approach to optimization. The fuzzy optimization problem is solved using standard mathematical programming techniques. In the *Star-odor* problem, knowledge from a variety of sources is integrated within a knowledge based system. For some process variables inside the refinery, a neural network model was developed that links these variables to the odor-related thresholds at the wastewater treatment plant. In the *composites* project, an off-line optimal cure profile was obtained using a dynamic mathematical simulation of the process. The optimal profile itself was obtained using a variety of techniques including genetic algorithms and heuristic guided optimization. The cure process was guided along the optimal trajectory by using process trend analysis to account for the mismatch between the model and the actual process. In the *Star-upset* analysis project, two years of archived data were analyzed to identify mathematical indicators that could warn of an impending plant shut-down.

In the remainder of this paper, we will focus primarily on the *FALCON* project and the lessons learnt from it.

II. EXAMPLES AND LESSONS LEARNED

2.1 Project Selection

In general, the first step in any fault diagnosis project is to identify an appropriate target process for the development of the knowledge based system. Important criteria for selecting the target process include availability of an on-line monitoring system, presence of real operational hazards, availability of sufficient number of sensor information and process knowledge, proprietary considerations, openness of plant management to new technologies, demonstrable economic impact, etc.

2.2 The FALCON Project

Based on many of the above criteria, the adipic acid process in Victoria, Texas was selected as the target process for the development and testing of the *FALCON* project.

The *FALCON* (FAult Analysis CONsultant) project was a joint research effort involving the University of Delaware, DuPont, and the Foxboro Company. The main objective of the *FALCON* project was to assess the feasibility of using knowledge-based systems (KBS) for on-line fault diagnosis in a commercial scale chemical process. Industrial perspectives on the project are provided by Rowan (1986, 1987, 1993) and by Shirley (1985, 1986, 1987 a, b).

A simplified schematic of the process is given in Dhurjati (1987). The process is highly exothermic and the main challenge comes from keeping this reaction within bounds. There are flow control loops on the many liquid and gas streams entering and exiting the process loop. There is also a level controller on a liquid separation unit and a cascade control system for the heat exchanger/reaction unit.

2.3 Defining objectives:

Specific objectives of the *FALCON* project were defined by DuPont personnel in terms of a minimum set of faults whose diagnosis was considered a precondition for the successful operation of the knowledge based system. The "*minimum fault set*" drove the search for diagnostic knowledge and provided the design specifications for the KBS.

A large fraction of the faults in the minimum fault set were related to the sensor, valve or controller associated with the control loops which could fail in either a "high" or a "low" mode. Other "process" faults included fouling of the heat transfer surface on the tube or shell side and failure of pumps on the process or cooling side. Faults were defined in terms of their consequences which had quantitative limits associated with it. Faults that occurred infrequently were not included in the minimum fault set or lumped with other faults.

While safety was the main criteria for selection of the faults, one of the lessons learned was that ultimately the system had to be justified in terms of *economic impact.* Sometimes the economic benefit of infrequent faults are hard to quantify. Thus for a purely economic point of view, it is easier to justify a fault diagnosis project that provides daily economic benefits in terms of productivity improvements rather than a system that identifies very infrequent faults.

2.4 Importance of measurements.

In general, there is obviously a theoretical limit to the amount of diagnostic information that can be extracted with a limited set of available measurements. One of the major limitations in the application of theories is the availability of sensor measurements. It is the single most critical factor limiting the theoretical approaches.

Most sensors are installed because of their utility for control. For fault diagnosis , one may need sensors to complete mass and energy balances or other calculations that lead to diagnostic information. At the design stage, sensors are rarely built in based on systematic fault diagnosis considerations. The number, placement and precision of the sensors on the plant largely determine the magnitude of the fault that can be detected. A sensitivity analysis based on the diagnostic methodology can be used to specify the magnitude of the faults that can be detected.

At the DESIGN stage, measurements can be built in based on fault diagnosis methodologies. This is where the theoretical developments in the field can be valuable. One can specify at the design stage of the plant, the optimal placement of the sensors and the precision needed to diagnose faults of a pre-specified magnitude. These specifications can be based on fundamental and methodological considerations. Such sensor placement eliminates the need to come up with ad-hoc methodologies based on available measurements which is unfortunately the case with most industrial applications.

In the *FALCON* project, most available sensor measurements were part of feedback control loops. A careful review of the measured process variables was performed to establish whether or not the "necessary" measurements were available to diagnose all the faults in the minimum fault set. This analysis identified 17 variables that had to be added to the list of existing measurements in order to diagnose all the faults in the minimum fault set. The triggers for fault announcement were set before the development of the diagnostic methodology. However, after evaluation of the methodology with actual plant data, it was discovered that these triggers for fault announcement cannot be set arbitrarily.

Of the 17 variables, 16 variables were either controller outputs or set points that allowed better identification of faults associated with the control loops. The other recommended measurement was the installation of a temperature sensor that made it possible to complete an energy balance calculation leading to more effective diagnosis of several faults. These recommendations were accepted after review by DuPont.

The final KBS had access to about 30 measured plant variables all of which were sampled once every 15 seconds. These variables included the set-points associated with all controllers and a number of other temperature, pressure and flow measurements.

In contrast, in the projects with the Star refinery, we were confronted with the problem of too many measurements. More than 50,000 measurements were logged in every minute in the computer monitoring system. In such a case, the problem was of identifying the information that was relevant to the diagnostic problem at hand.

2.5 Heuristics

Good heuristics are not easy to come by. In the *FALCON* project, there was an expert working in the plant for several decades who was a source of a large amount of valuable diagnostic information. In the *Star-resources* project, heuristics provided the fuzzy constraints on the optimization problem (Petti, 1992). In the *Composites* project, heuristics were generated by sensitivity analysis using the simulation (Pillai, 1997). Where available, good heuristics could however be worth a lot of theory.

The dynamic nature of a chemical process makes "knowledge engineering" very different from the process of building a "qualitative" or heuristic knowledge base derived from domain experts. The functional relationships among process variables are very complex and include a time dependency that is best expressed by a set of highly coupled mathematical expressions rather than simple heuristics. Added to the complexity is the fact that the process is operated at different production rates that may alter the functional relationships among process variables.

Heuristics are usually applicable over a limited production range and may not be able to capture the dynamic aspects of the system behavior. Also, heuristic rules are generally expressed in terms of a finite number of discrete values and information is lost in such discretized representation.

Despite this apparent complexity, performance of human experts in correctly diagnosing faults without recourse to complex mathematical relationships does have implications.

2.6 Role of mathematical models and simulation in fault diagnosis.

Mathematical models can play an important role in on-line fault diagnosis. But often, mathematical models are severely misused.

In the *FALCON* project, mathematical models were of two different types and they played two distinct roles: (1) fairly complicated first principle based models for use as plant substitutes and (2) other simple models based on measured variables for diagnosis.

First, a fairly elaborate dynamic model of the process was built using a verified steady state simulation. This model was verified and used as a plant substitute during the testing process.

In most well run plants such as the DuPont adipic acid plant, faults rarely occur and this presents a major problem in testing the expert system knowledge base. In such cases, it is not possible to exhaustively test the knowledge base using only plant data; data from the simulation of the fault is the only rigorous means available for testing and *validating* the knowledge base. An added benefit of the dynamic simulation was the accelerated rate at which the research team could gain simulated "plant operating experience" through the use of the simulation.

The dynamic simulation had the principal role of providing simulated data on infrequent faults in order to validate the knowledge base. If such infrequent fault data were available, there would have been no need for such an elaborate dynamic simulation (over 250 simultaneous differential equations and 1000 parameters).

A second but important use of the simulation was to study the reasoning process employed by human experts while attempting to diagnose faults. This use of simulation is described in greater detail in Dhurjati (1987). The second use of models was as a source of diagnostic knowledge and is described in the next section.

In the *Composites* project, a detailed dynamic simulation of the composites process was built. The dynamic simulation was used to generate off-line optimal cure profiles. The mismatch between process reality and simulation was taken into account using process trend analysis (Pillai, 1994, 1997)

2.7 Availability of model equations

Where reliable equations are available to describe the system, the best strategy is to exploit their availability (Petti, 1990, 1991). Such equations need not always be based on first principles such as conservation laws.

The *FALCON* knowledge base was a hybrid of quantitative and qualitative knowledge. The knowledge base was a quantitative core of "simple" equations (also referred to as model fragments) which are complemented by qualitative knowledge. Much of the knowledge extracted from experts by extensive interviews was already found to be contained in overall material or energy balances, controller equations or other mathematical relationships describing the system. An equation that was most useful for diagnostic purposes was an empirical equation to estimate the quantity of off-gas generated in the process.

The quantitative part of the Falcon knowledge base consisted of the following equations: material and energy balances as well as empirical relations written in terms of "measured variables", equations for controllers, valve curve correlations relating flow measurements to controller outputs and equations for calculation of heat transfer coefficients.

Material and Energy balances were usually off by around 5%. The degree of closure of these equations varied significantly depending on production levels and context. While residuals were used to detect and diagnose faults, the sensitivity of the equations to plant operating conditions was an important part of the qualitative knowledge used in reasoning with the residuals Dhurjati (1987).

2.8 Importance of Plant Operating Context

Too much emphasis in some of the academic research is on extracting information out of equations which have validity over limited plant operating range. Equations are usually valid only in a narrow plant operating context. It is important to emphasize that the main use of equations in the *FALCON* knowledge base was to examine the underlying assumptions on which the equation or model fragments were based. Most of the assumptions underlying the models fragments were also a part of the minimum fault set. In mapping engineering reality to equations, most valuable information is present in assumptions rather than in the equations. A lot of diagnostic knowledge can be extracted from assumption violation.

The plant operates in several contexts: start-up, shut down and at different production levels. Our experience was that each context had to be dealt with separately. For example, the tolerances on mass balances (both high and low) were different in each operating regime.

The violation of an assumption associated with a model fragment or equation was for us a direct and powerful way of diagnosing a fault in the minimum fault set. A large number of rules dealt with reasoning about assumption failures in the context of the plant operating state. There was a logic hierarchy for fault diagnosis based on the interdependence of assumptions.

At the highest level was diagnosis of the plant operating state. Next, there were checks for catastrophic failures and operator interactions. At the next level of fault hierarchy are the controller faults and sensor faults. This means that it is not necessary to check assumptions related to sensors in order to diagnose controller faults. Next the valve faults are checked and the last set of faults that are diagnosed are related to process failures such as frosting, fouling or restriction of off-gas flow.

2.9 Human machine interface

Good design of a human-machine interface (Mooney et al., 1988) is essential to the success of any project. The final test of the project is its implementation in the control room and its acceptance by the operators. Operators have their own criteria for accepting or rejecting the fault diagnosis system and too often it is surprisingly independent of the sophistication of the methodologies used. False indications undermine confidence in the system. One of the design criteria for *FALCON* was that "no diagnosis is better than false diagnosis!".

Integration of the fault diagnosis system with the existing software and hardware can also pose serious implementation problems. The *FALCON* system was extensively and successfully tested on simulated plant data as well as actual plant data during fault periods before being introduced into the control room for a trial period.

A lesson learnt during implementation was that changes to processes are normal and the knowledge base must have the capability of adapting to these process changes fairly easily. This requires good organization of the knowledge and ability to edit the knowledge base with ease. Some of the currently available object oriented software tools such as G-2 with a graphical interface for knowledge editing have made this more feasible.

III. CONCLUSIONS

Industrial experiences with several processes have provided important lessons for the successful application of on-line fault diagnosis and process supervision. While these lessons have been primarily illustrated using the *FALCON* project, they are fairly generic. The importance of the availability of measurements cannot be overemphasized. The right measurements can be the limiting factor for application of methodologies. Our applications show that no single methodology is usually adequate by itself, and hybrid intelligent systems can play an important role in overcoming methodological bottlenecks. Availability of knowledge and accessibility to experts are extremely important and this varies from project to project. A validated dynamic simulation that acts as a plant substitute can prove to be a valuable alternative source of obtaining process insights. Heuristics can also play an important role in limited cases but the best form of knowledge is as mathematical models that can be validated with plant data. Easy to use software tools can substantially speed up the prototyping and development phase of the project. Ease of inputting and editing knowledge and ability to integrate methodologies on a single platform are useful attributes. The validation of knowledge bases is a serious problem as a sufficient data base of actual fault situations may not exist. "Mining" archived data to provide learning from past fault cases can sometimes be very helpful. While knowledge is important in all fault diagnosis projects, human factors from operators to engineers to management can play an equally important role.

REFERENCES:

Dhurjati, P., D. E. Lamb and D. C. Chester, "Experience in the Development of an Expert System for Fault Diagnosis in a Commercial Scale Chemical Process", Foundations of Computer Aided Process Operations, 589-625, Elsevier Science Publishers, New York (1987).

English, T. J., Lars Kaiser, Frank Doyle, P. Dhurjati, "Analysis of Post-upset Plant Data for Fault Detection", Proceedings of the IFAC Workshop on "On-line Fault Detection and Supervision in the Chemical Process Industries", Solaize (Lyon), France, June 3-5, (1998)

Kordon, A. K., P. Dhurjati, and B. J. Bockrath, "On-line Expert System for Odor Complaints in a Refinery", Computers and Chemical Engineering, 20, S1449-S1454, (1996)

Mooney, D. J., D. Chester, D. Lamb, and P. Dhurjati, "Design and Operation of the *FALCON* Interface", Proceedings of ISA/88, 747-758 (1988).

Petti, T. F., J. Klein, and P. Dhurjati, "The Diagnostic Model Processor - Using Deep Knowledge for Process Fault Diagnosis", AIChE Journal, 36, 565-575 (1990).

Petti, T. F. and P. Dhurjati, "Object-Based Automated Fault Diagnosis", Chemical Engineering Communications, 102, 107-126 (1991).

Petti, T. F. and P. Dhurjati, "A Coupled Knowledge Based System Using Fuzzy Optimization for Advisory Control", AICHE Journal, 38, 1369-1378, (1992).

Petti, T. F., and P. Dhurjati, "On-Line Hydrogen Resource Management in a Refinery Using Fuzzy Optimization", in "On-Line Fault Detection and Supervision in the Chemical Process Industries", eds. P. Dhurjati and G. Stephanopoulos, IFAC Symposia Series, Number 1, pages 169-174, Pergamon Press, (1993)

Pillai, V., A. Beris, and P. Dhurjati, "Implementation of model-based optimal temperature profiles for autoclave curing of composites using a knowledge-based system", Industrial and Engineering Chemistry Research, 33, 2443-2452 (1994)

Pillai, V., A. Beris, and P. Dhurjati, "Heuristics Guided Optimization of a Batch Autoclave Curing Process", Computers and Chemical Engineering, 20, 275-294 (1996)

Pillai, V. , A. N. Beris and P. Dhurjati, "Intelligent curing of thick composites using a knowledge-based system", Journal of Composite Materials, 31, 22-51 (1997)

Rowan D. A., "Chemical Plant Fault Diagnosis Using Expert Systems Technology: A Case Study", IFAC Workshop on Fault Detection and Safety in Chemical Plants, Kyoto, Japan, 81-87, (1986)

Rowan, D. A. , "Using an Expert System for Fault Diagnosis", Control Engineering, 160-164 (1987)

Rowan D. A., "Beyond Falcon: Industrial Applications of Knowledge Based Systems", in "On-Line Fault Detection and Supervision in the Chemical Process Industries", eds. P. Dhurjati and G. Stephanopoulos, IFAC Symposia Series, Number 1, pages 249-253, Pergamon Press, (1993)

Shirley, R. S. and D. A. Fortin, " Status Report: An Expert System to Aid Process Control", Proceedings of ISA/85, 1463-1470 (1985)

Shirley, R. S., "Status Report 2: An Expert System to Aid Process Control", Proceedings of the TAPPI Engineering Conference, 425-430, (1986)

Shirley, R. S., "Status Report 3: An Expert System to Aid Process Control", Proceedings of the Annual Pulp and Paper Industry Technical Conference, 132-136, (1987)

Shirley, R. S., "Some Lessons Learned Using Expert Systems for Process Control", Proceedings of the American Control Conference, vol. 2, 1342-1346 (1987)

Industrial implementation and testing of model-based diagnosis in the PRIDE project

K.W. Mathisen*, A. Mjaavatten* and R.A.G. Kooijmans**

* *Norsk Hydro ASA, Research Centre,* P.O.Box 2560, N-3901 Porsgrunn, Norway;
** *Unilever Research Laboratory,* P.O.Box 114, 3130 AC *Vlaardingen, The Netherlands*

Abstract: The project objectives of PRIDE, PRocess Industries Diagnostic Experiments, are to design and realise a toolset for model-based diagnosis and test the toolset in full-scale industrial applications. Topology based diagnosis is one of the tools in the toolset, and the method will be implemented both in Unilever's edible oil process at Unimills in Zwijndrecht, the Netherlands and Norsk Hydro's fertiliser process in Glomfjord, Norway. Although the toolset is not yet in operation at the plants, earlier operational experience with process diagnosis, prototyping results and responses from plant operators indicate that the method will be helpful and improve productivity and efficiency. Once proven at the test sites, model-based diagnosis is particularly well suited for rapid dissemination to similar processes. *Copyright © 1998 IFAC*

Keywords: Diagnosis, Model-based, Process, Industrial, Estimation, Dynamic models

1 INTRODUCTION

Process diagnosis includes detection, isolation or localisation, identification, explanation as well as correction or remediation of abnormal process situations or faults. Process diagnosis has gained considerable academic and industrial interest during recent years, and numerous approaches have been proposed. In most cases the methods are based on models (i.e. qualitative models like sign directed graphs), observers, assumptions, process history, principal component analysis or qualitative trends (Vedam and Venkatasubramanian, 1997). In PRIDE a wide definition of model-based diagnosis is applied, the considered diagnostic methods include features typical for all the above-mentioned classes.

PRIDE, project no. 20498 in the ESPRIT-program, aims to prove the business benefits associated with Model-based Diagnostics (MBD), and the importance of the technology to the European process industries. The project started on January 1st. 1996 and has a duration of three years. The total size of the project is 46 man-years with equal financial contributions from the participating companies and the European Commission. The consortium includes six industrial end users, three technology providers and three system vendors:

- industrial end users with full-scale applications: Unilever and Norsk Hydro,
- process industry with limited testing: BG Transco, British Nuclear FL, Electricidade de Portugal - Proet and Yorkshire Water,
- technology providers: CISE SpA, KCL, Heriot Watt University and Norsk Hydro
- system vendors: Cogsys ltd , Computas and Elsag Bailey Hartmann & Braun.

PRIDE is controlled and run by the industrial end users, and Unilever is the co-ordinator.

The PRIDE toolset presently includes three methods for process diagnosis in a client-server system: Giotto based on steady-state data reconciliation, KCL-Wedge using multivariable auto-regressive modelling and Norsk Hydro topology based diagnosis. This paper describes the implementation of the latter method only.

2 TOPOLOGY BASED DIAGNOSIS

Topology based diagnosis was originally a tailored system implemented in G2/Unix and installed in the fertiliser plant in Glomfjord (Mjaavatten and Saelid 1992, Fjalestad et al. 1994). In PRIDE the method is generalised and re-implemented in JAVA as one of the tools in the PRIDE Server. Both the server and the G2 (G2, 1997) client will run on Windows NT PC's. Norsk Hydro Topology based diagnosis (NH-TBD) uses dynamic models and measurements to determine the present value of all the state variables in the plant. Measured or estimated values are compared to normal ranges. If a fault symptom is detected, possible causes are searched for using topology information. Mjaavatten and Foss, (1997) give a more thorough technical description of the diagnosis method, but the five main parts of the method is explained in the following sections.

2.1 Global flowrate calculation

The total mass flowrates in all process streams are determined by solving a linear data reconciliation problem: Given a set of mass flow measurements **y**, find the harmonised flowrates $\hat{\mathbf{y}}$ that minimise

$$J(\hat{\mathbf{y}}) = (\hat{\mathbf{y}} - \mathbf{y})^T \mathbf{R}^{-1} (\hat{\mathbf{y}} - \mathbf{y}), \quad (1)$$

subject to the mass balance condition

$$\mathbf{A}\hat{\mathbf{x}} = \mathbf{0}, \quad (2)$$

where **A** is the incidence matrix for the flow network and **x** is the full flowrate vector, consisting of both measured (**y**) and unmeasured flowrates. **R** is ideally the covariance matrix for measurement errors, but in practice a diagonal matrix of measurement variances is used.

The formulation is favourable since quadratic optimisation problems with linear constraints can be solved explicitly. All parts (streams) of the system (plant) must be observable. To achieve this, measurements of e.g. pressure drop and valve position, power, split ratios and levels of vessels are used to compute flow in streams. These measurements are denoted as indirect measurements. If parts of the system still are unobservable due to missing or faulty sensors, normal values are used. Note that by computing accumulation in vessels from level measurements main dynamic features of the system can be taken into account. For an extensive review of data reconciliation, see Crowe (1996).

2.2 Local estimators

The global flowrate calculation is supplemented by distributed dynamic estimator routines that determine the composition and temperature in all process units and streams.. The dynamic equipment models can ideally be used to predict measured variables. To resolve the resulting conflicts between measurements and prediction one may disregard (corresponding) relations, update model parameters or update measurement bias or scale-factor. The general form of the state equation is:

$$\frac{d\mathbf{x}}{dt} = \mathbf{f}(\mathbf{x}, \mathbf{u}), \quad (3)$$

$$\mathbf{y} = \mathbf{g}(\mathbf{x}, \mathbf{u}). \quad (4)$$

Here **x** is the state vector and **u** is the vector of manipulated variables. **x** will typically contain temperature and the masses of individual chemical species; **u** contains the relevant flowrates together with inlet steam composition and enthalpy from upstream local models. **y** is the vector of measured variables. Assume that the temperature and pressure in a separator are both measured. A Kalman filter, or more generally an observer, can then be used to update the state variables. The observer equations for the state prediction and state estimation update are:

$$\bar{\mathbf{x}}_{k+1} = \hat{\mathbf{x}}_k + \mathbf{f}(\hat{\mathbf{x}}_k, \mathbf{u}_k)\Delta t, \quad (5)$$

$$\hat{\mathbf{x}}_k = \bar{\mathbf{x}}_k + \mathbf{K}(\mathbf{y}_k - \mathbf{g}(\bar{\mathbf{x}}_k, \mathbf{u}_k)). \quad (6)$$

Here subscript k indicates the time step number and Δt is the length of the time step. **K** is any suitable observer feedback matrix. If a Kalman filter is used, **K** is the Kalman filter gain. The choice of **K** will reflect how we balance our confidence in the measurements to our confidence in the model. Note that errors in the inlet stream compositions and enthalpies are considered model errors in this context.

The updating of the state variables will inevitably lead to a violation of the energy balance and/or the mass balance for one or more species. The proper way to avoid this would be to include those balances in the global flowrate calculation. Such a calculation will require a complex, non-linear dynamic optimisation scheme and is beyond the scope of the present work. Alternatively, one might insist on satisfying one or more of the local balances. It is generally possible to modify one or more of the variables treated as manipulated variables in the estimator to achieve this. One might for instance attribute the error to one or more of the flowrates. The modified flowrate could be fed to a virtual flowrate measurement, which in turn could influence the global flowrate calculation.
If all balances are to be satisfied we must find a modified vector $\tilde{\mathbf{u}}_k$ of manipulated variables

that ensures that the estimated state variables satisfy the (discretised) state equation:

$$\hat{\mathbf{x}}_{k+1} = \hat{\mathbf{x}}_k + \mathbf{f}(\hat{\mathbf{x}}_k, \tilde{\mathbf{u}}_k)\Delta t \qquad (7)$$

Subtracting (5) from (7) yields

$$\hat{\mathbf{x}}_{k+1} - \bar{\mathbf{x}}_{k+1} = \mathbf{f}(\hat{\mathbf{x}}_k, \tilde{\mathbf{u}}_k)\Delta t - \mathbf{f}(\hat{\mathbf{x}}_k, \mathbf{u}_k)\Delta t \approx \mathbf{B}_k \Delta \mathbf{u}_k \qquad (8)$$

where

$$\Delta \mathbf{u}_k = \tilde{\mathbf{u}}_k - \mathbf{u}_k, \qquad (9)$$

and

$$\mathbf{B}_k = \frac{\partial \mathbf{f}(\hat{\mathbf{x}}_k, \mathbf{u}_k)}{\partial \mathbf{u}_k^T}\Delta t. \qquad (10)$$

If there are n states and m manipulated variables then (8) yields n equations to determine the m elements of $\Delta \mathbf{u}_k$. Generally, this problem will not have a unique solution and we must supply extra constraints. For instance, we might decide that the inlet flowrate and enthalpy should not be changed. In addition, we will normally insist that the total mass balance be satisfied.

Parameter estimation is achieved by augmenting the state vector with the parameter(s) to be estimated. In cases where both parameter estimation are required, averaging out high frequency disturbances and noise is usually even more important than with state estimation.

2.3 Detection monitors

Fault detection is done by monitoring key plant variables and comparing their values to the normal or allowed range. These ranges may be given by fixed constants that seldom are changed or vary with operating conditions. It is also possible to widen the ranges during planned transient operation like product type changes.

2.4 Search procedures

Fault isolation is done by a recursive search for disturbances in upstream units and streams. In process units with more than one stream the search branches to search all inlet streams. When the search reaches a unit where no inlet stream reports disturbances, this branch of the search stops. The unit with no inlet stream disturbance, but with an outlet stream disturbance is marked as a possible primary cause of the detected fault. The path from the triggered detection monitor to a possible primary cause is called a diagnostic path. Diagnostic paths may be displayed to the operator and thereby give some reasoning explanation. The procedure is illustrated in figure 1.

2.5 Internal diagnosis

Subsequent to the topology search to isolate the cause, fault identification may be performed through an internal diagnosis. This internal diagnosis depends on tailored process unit models, and typically aim at identifying the value of one or more fault parameters.

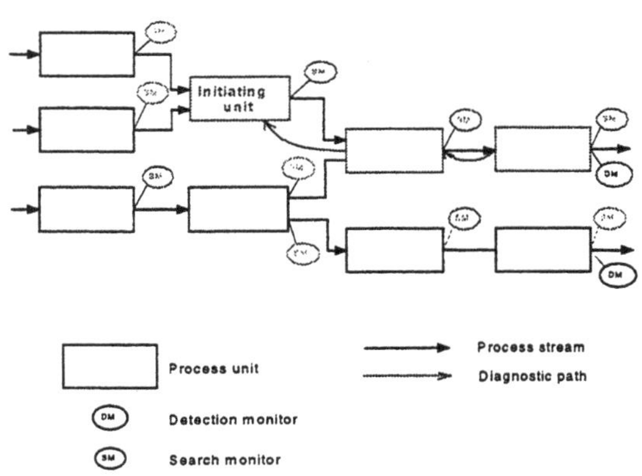

Fig. 1: Detection and search monitors illustrating the diagnostic search procedure.

3 NORSK HYDRO APPLICATION

The Norsk Hydro fertiliser plant in Glomfjord is an example of the nitrophosphate process with calcium nitrate crystallisation (Kongshaug *et al.*, 1991), see figure 2.

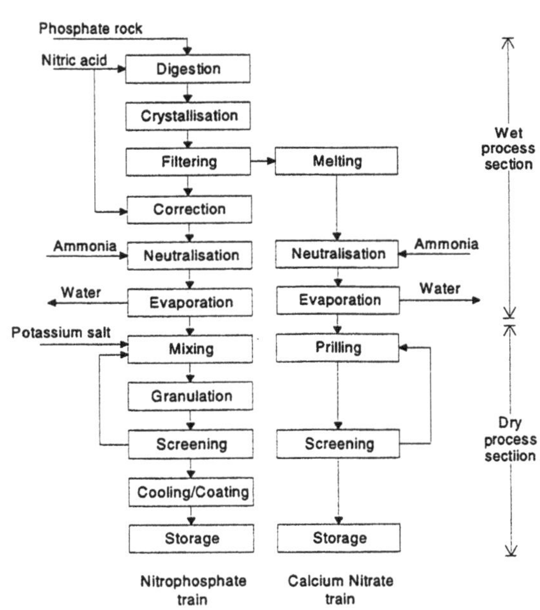

Fig. 2: The nitrophosphate process

For this application the two main objectives are:

- **Product composition**: The amount of nitrate and ammonium (N), Phosphate (P) and potassium (K) in the product are the most important product quality indicators.
- **Emissions**: Effluent streams with N and P to the sea and nitric oxides and ammonium to air are subject to strict governmental regulations. Breaking the limits may cause environmental damage and result in severe fines.

Both online process measurements and offline analysis results are used in the estimation and diagnosis. Two different process control systems exist, one for the wet process section and one for the dry process section. Tailored communication program had to be developed in order to be able to transfer data to the PC network and the main database on a Unix server machine.

Analysis results are transferred from two laboratories, the local shift laboratory and the central main laboratory. So far, estimated product concentrations are not taken into consideration when analysis results are available. One of the planned enhancements will be to remove this restriction and also put a little weight on the estimated values. However this is difficult to handle both technically and politically towards both plant management and the emission government authorities. So far, analysis are, by definition, either 100% correct or 100% wrong.

Operator usage of and interest for the existing prototype system vary a lot. System stability both for the diagnosis application and the computer network is very important for user acceptance. Since both the diagnostic server and the client program of the PRIDE toolset run on Windows NT PC's on the local area network, some earlier stability problems are eliminated.

4 UNILEVER APPLICATION

As a part of the PRIDE project, a full scale MBD application is developed for the centrifugal refining process (Hendrix, 1989) at Unimills in Zwijndrecht. The MBD application will consist of the PRIDE Toolset, together with an application manager implemented in G2 (G2, Gensym). As a part of the PRIDE Toolset the functionalities of the TBD system will be used in the MBD application. At the time of writing the MBD application for Unimills is being constructed, thus the results and conclusions presented in this part are based only on prototyping experience and development aspects related to the implementation of the MBD application. The main objectives for this application are:

- **Product composition:** The amount of soap, free fatty acids and moisture in the refined product are the most important product quality parameters. The concentration of these impurities must be kept below their maximum limits.
- **Production efficiency:** An early warning system on (possible) equipment failure will enable the operators to control the plant more tightly and without interruption. This will result in higher production efficiency.

At Unimills Zwijndrecht edible oils are refined and made suitable for human consumption. As a part of the total process the edible oils are processed in a centrifugal refining line, in order to split off the free fatty acids (FFA), phospholipids (PL), dirt and moisture from the edible oil. The centrifugal refining process consists of the following steps (see figure 3): oil supply, oil heating, phosphoric acid dosing, lye dosing, mixing, soap separation washing centrifuges and oil drying

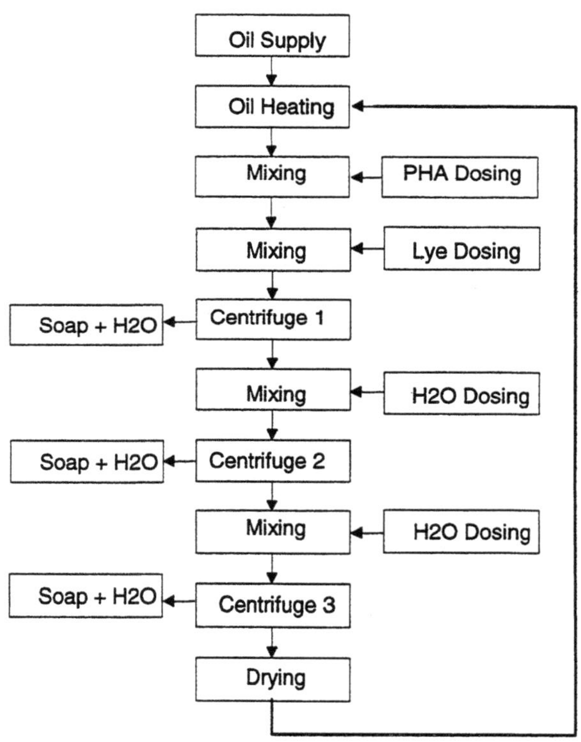

Fig. 3. The centrifugal refining process.

Coming from the oil supply the edible oil is heated to the appropriate processing temperature. In the following part phosphoric acid is dosed to agglomerate the PL and the dirt. After that lye is dosed to saponificate the FFA present in the edible oil and to neutralize the surplus of phosphoric acid. The soap, agglomerate and edible oil mixture is separated in the soap separation centrifuge. The solid phase (agglomerate) and the heavy liquid phase (soap) are split off from the light liquid phase (edible oil). After this the edible oil is washed in order to reduce the amount of soap, still present in the edible oil after the soap separation step. As the last step in the process the edible oil is dried, in order to remove the last remainders of water.

At several phases in the process measurements are taken, such as flow, temperature, pressure, valve position, power consumption and concentration. These measurements are available for the diagnostic system. One of the essential features of TBD is the ability to estimate the stream properties such as temperature, pressure, flow and concentration. Errors occurring at the front part of the process can only be propagated by means of these quantities. Although the dynamic models may be used to predict (i.e. simulate) all state variables in the process, in-line measurements of these quantities is important to get reliable estimations. At Unimills some additional measurements have been included in order to maximize the diagnostic capabilities of the MBD application.

5 DISCUSSION

The following discussion is based on Norsk Hydro experiences with the existing G2 test implementation supplemented by tool and application prototyping work in the PRIDE project performed by Norsk Hydro and Unilever.

5.1 User interface

It is important to be able to import existing process icons in order to let the MBD process mimic pictures look like the plant displays in the control rooms. Ideally, the model-based diagnosis should be completely integrated in the existing operator displays and operator procedures.

The pictures should be organised hierarchically. It is often required to be able to switch quickly between overview pictures of the whole plant and more detailed pictures of process sections and process units.

Most process operators strongly prefer to use their native language when working in the control rooms. The PRIDE Toolset native language user interface will be available through text resource file techniques.

5.2 Method limitations

TBD is a strong tool for performing detection and isolation at large plants, because of its fast back tracking of errors propagating through the plant. TBD will give the operator a clear view of the possible origin(s) of the faulty behavior of the plant. Control loops and process re-circulations limit the benefits of TBD, but the PRIDE toolset also contain KCL-WEDGE diagnostics technology. This will increase the possible benefits that may be achieved when using the full PRIDE Toolset rather than TBD.

The method is not very dependent upon the measurement density since the models are used to predict the outputs when estimation from measurements is impossible. However, with few and uncertain measurement, the diagnosis will be of less value. In the case of Unimills some extra measurements were installed to enable effective use of the TBD functionalities. However, the same remark holds for all model-based diagnostic applications in general.

The global flowrate calculation is formulated without allowed ranges for the total flowrate variables. This may yield solutions with small negative flows, also in cases where reversed flow is physically impossible. A possible solution would be to include variable range constraints and solve by e.g. successive quadratic programming.

The updating of the state variables will inevitably lead to a violation of the energy balance and/or the mass balance for one or more species. Today, updating key flowrate measurements from the local estimators reduces the problem. The proper way to avoid this would be to include those balances in the global flowrate calculation. Such a calculation will require a complex, non-linear optimisation scheme and is beyond the scope of this project. However data reconciliation tools; e.g. Massbal (Stephenson and Shewchuk, 1986; Massbal, 1996) or other diagnostic methods can be integrated in PRIDE Toolset in the future.

Most diagnostic tools are focussed on the steady state behavior of the process. The TBD tool in the PRIDE toolset however, has the capability of performing dynamic estimation. This could provide the MDB application with dynamic monitoring of key process characteristics. However, in the present version of the TBD software the allowed ranges and nominal values are restricted to steady state values. In future version of the TBD software this restriction will be eliminated and truly dynamic monitoring will be possible.

Topology based diagnosis is information demanding since the state of all streams is used. The need for phenomenological process models for all units may be limiting. Even though simple, approximate models may often be used, a future possibility of generating and including data-driven models would be advantageous.

5.3 Plant implementation

It is very important to involve the real end-users, at the plants at an early stage so that the operators have a positive attitude towards the system when it is installed. Therefore meetings have been organised in Norway and The Netherlands, bringing the operators of Norsk Hydro and Unilever together for discussion. One of the subtasks in PRIDE concentrates on the "Application Development Methodology" for how to introduce model-based diagnosis in process plants. Our experience supports the belief that such a practical handbook is needed.

REFERENCES

Crowe, C.M. (1996): "Data reconciliation – progress and challenges", *J. Proc. Cont.*, **6**, pp 89-98.

Fjalestad, K., J.A. Gravklev, A. Mjaavatten and S. Saelid (1994); *Comput. Chem. Engng.* **18** Suppl. pp S369-S373

G2 Reference Manual (1997), Gensym Corporation, 125 Cambridge Park Drive Cambridge, MA 02140 USA

Hendrix, B. (1989); Neutralization I. Theory and Practice of Conventional Caustic (NaOH) Refining, World Conference Proceedings American Oil Chemists' Society, Champaign (Illinois), pp 94-100

Kongshaug, G. *et al.* (1991). "Phosphate Fertilisers", in *Ullman's Encyclopaedia of Industrial Chemistry* Fifth edition, **A19**, pp 421-463, VCH Verlagsgesellschaft mbH, Weinheim, Germany

Massbal 3 User's Guide (1996), Honeywell Hi-Spec Solutions, 3343 Dundas Street, London, Ontario, Canada N6B 1V5

Mjaavatten, A. and B.A Foss (1997); *Comput. chem. Engng.*, **21**, pp. 1203-1218.

Mjaavatten, A. and S. Saelid (1992); *Preprints of IFAC Symposium Online Fault Detection and Supervision in the Chemical Process Industries.* Ed. P.S. Dhurjati, pp. 222-227

Stephanson, G.R and C.F. Shewchuk (1986). *AIChE J.* 32, p. 247.

Vedam, H. and V. Venkatasubramanian (1997); *Comput chem. Engng.* **21** Suppl., pp. S655-S660.

URL for ESPRIT programme: http://www.cordis.lu/esprit/home.html

PRIDE project: A general description can be found under ESPRIT (project no. 20498)

URL for the full partners in the consortium:
Unilever: http://www.unilever.com

Norsk Hydro ASA: http://www.hydro.com

Elsag Bailey Hartmann & Braun: http://www.elsag.it

CISE SpA: http://www.cise.it

Cogsys Ltd: http://www.cogsys.co.uk

Authors addresses:
K.W. Mathisen, Norsk Hydro ASA Research Centre Porsgrunn, P.O.Box 2560, N-3901 Porsgrunn, Norway. Email correspondence is preferred: Knut.Wiig.Mathisen@hre.hydro.com
Please direct all correspondence to this author.

A. Mjaavatten, see address above.

R.A.G. Kooijmans, Unilever Research Laboratory Vlaardingen, Oliver van Noortlaan 120, P.O.Box 114, 3130 AC Vlaardingen, The Netherlands. Email: Rob.Kooijmans@Unilever.com

OBSERVING THE SUGAR-BEET QUALITY USING PROCESS AND SIGNAL ANALYSING METHODS

Andrea Arenz* Michael Reimann* Eckehard Schnieder*
U. Harten**

* *Institute for Control and Automation Engineering,*
Technical University of Braunschweig,
Langer Kamp 8, D-38106 Braunschweig, Germany
** *Nordkristall GmbH,*
Verbindungschaussee, D-18273 Güstrow, Germany

Abstract: In the present paper a process supervision of the beet quality for a sugar producing plant is presented. Based on a description of the sugar process itself the occuring problem of observing the beet quality is explained. Accordingly different process and signal analysing methods are taken into consideration. The chosen observer solution uses process data and variables in order to calculate characteristic values for the beets. Next the implementation of the derived algorithms in the process control system of the sugar plant is shown. The paper concludes with a summary and a short outlook.

Keywords: Process control, Knowledge acquisition, Fault detection, Process parameter estimation, Fault identification, Filtering techniques, Observers, Signal analysis

1. INTRODUCTION

The project presented was carried out in cooperation with the sugar plant 'Nordkristall GmbH' in Güstrow (aerial view see figure 1). The plant is a part of the sugar concern 'Nordzucker' which is one of the sugar market leaders in Germany. It is located in the eastern part of Germany and was built in 1962. Due to extensive modernizations in the years '91 to '93 it is now one of the most modern sugar plants in Europe (*Geschäftsbericht*, 1996).

Fig. 1. Aerial view of the sugar plant

The process itself is mainly influenced by the natural raw product 'sugar beet'. Hence the problems occuring are in general directly caused by damaged beets.

An appropriate reaction during the process is only possible when these quality disturbances are known. Resulting from a better understanding of the process and better operating actions, the economic balance is improved.

This is especially important as the competition on the sugar market will extremely intensify due to the ending guarantees of the european quota system.

(a) Sampling of beets (b) Beet piles

Fig. 2. Delivery of beets

2. THE SUGAR PROCESS

In order to estimate the effects of the beet quality a good knowledge and understanding of the process is necessary.

The digging of beets starts at the end of september and is at the latest completed in januar. Carriers fetch the beets from the fields and deliver them at the plant where they are weighed. Samples are taken (see figure 2a) in order to reveal the actual sugar quantity which is one decisive factor for the payment of the farmers. Afterwards the beets are stored in huge piles (see figure 2b). In general three day's processing of beets is stocked up so that the plant activities are decoupled from the delivery. Resulting from the fully automated digging a high percentage of dirt is delivered together with the beets so that they need to be washed and cleansed before entering the plant.

In the 'Vorderbetrieb' (first process section) clarified juice, a liquid water-sugar solution, is processed. This is done in three steps: chip processing, extraction and juice cleansing. In order to extract sugar the surface of the beets needs to be increased. Therefore thin chips are produced in large cutting machines (see figure 3a). Now these chips seep through the extraction tower (see figure 3b) where they are steeped at constant temperature and pH-value in a water lye. During

(a) Beet chips (b) Extraction tower

Fig. 3. The 'Vorderbetrieb' (first process section)

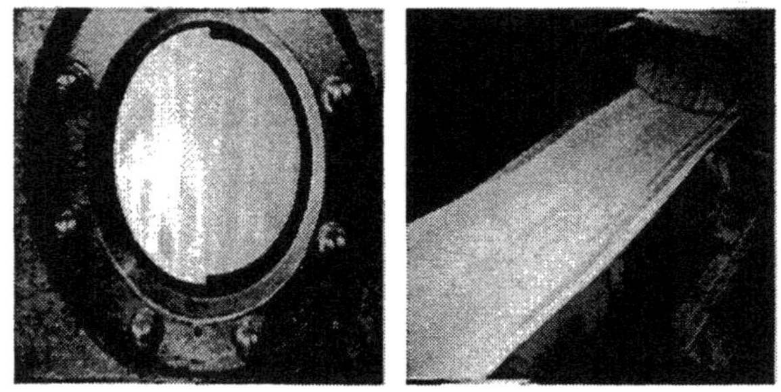

(a) Raw juice (b) Sugar

Fig. 4. The 'Zuckerhaus' (second process section)

this process up to 99 % of the sugar is extracted. As intermediate product the raw juice is obtained, which is a mixture of water, sugar, extracted dirt and solid particles. Now the non-sugar substances together with the solid particles are separated with lime-milk from the raw juice. Afterwards they are filtered out.

In the 'Zuckerhaus' (second process section) the clarified juice is concentrated in several steps. Therefore the water is vaporized (see figure 4a) until the sugar crystallizes. When the first germs arise the concentrated juice is filled in centrifuges. After a short time span the final product sugar is obtained (see figure 4b). Remaining interim products like chips and molasses are commercialized e.g. as animal food or fertilizer.

3. OCCURING PROBLEMS

The beets as raw material have many influences on the process which can not be neglected. In order to determine the quality of the beet it is very important to know when and why beets alterate (they begin to rot or are decomposed). In a next step critical process states have to be determined. This means at which points may alterated beets lead to complications within the process and may result in a stoppage of the plant.

The fresh-digged natural product sugar beet is subject to permanent changes. Biological and chemical processes change the combination of the ingredients, the cell tissue and consequently the physical properties. On the one hand these changes are due to the life of the beet itself, on the other hand external conditions have great influences. Especially the air humidity, the air exchange, the temperature and the infestation with micro organisms are very important factors. In order to preserve its ability to live the sugar beet breathes and converts saccharose; this directly decreases the sugar content. A high air exchange maintains the health of a beet but reduces the saccharose, whereas an exchange which is too low results in anaerobe decomposition. The optimal humidity is about 90 to 96 %. Below this value the beet dries out, finally dies and begins to rot. A high humidity value leads to a great increase of micro organisms. The optimal temperature to store beets is between -6 °C and +12 °C. If the temperature rises above 15 °C mould and bacterias begin to multiply. Below -10 °C the beets freeze. The critical point is the thawing when decomposition directly starts. Organisms and bacterias are millionfoldly smuggled in the process together with the dirt. They are unwanted companions which make the extraction much more difficult. If they discover good living conditions they convert sugar in non-sugar ingredients.

All these decomposition processes do not only reduce the sugar profit but furthermore result in long molecular chain structures which paste up the filter. Filter, which in general have a useful life of several days, may have to be changed after a few hours. Changing them is both costly and time intensive (about 5 to 8 workers need 8 hours for one exchange of filters). If the same happens with the substitute filter the production has to be stopped.

4. SOLUTION: PROCESS SUPERVISION

The solution of this problem of pasted filters seems very simple. If the altered beets are detected at an early stage more lime-milk or enzymes are fed to the process. These supplements have the ability to destroy the longer molecules without changing the sugar content. But for economic reasons it is of course desirable to add only as much of lime or enzymes as absolutely necessary. So the best solution is to observe the beet quality and react only when problems are likely to arise.

4.1 *Data analysis*

As direct measurements of the beet quality during the process are not possible a survey was carried out in order to obtain the actual data and information flows (Arenz *et al.*, n.d.). Based on a profound analysis four main data sources can be distinguished:

(1) workers, who observe and control the process
The whole plant is mainly run by the operators in the control room. According to the information output screen from the control system the workers observe the process and control the disturbances. The process itself is automatically (pilot) controlled by the control system.

(2) laboratory indicators
The analysis in the laboratory are very exact and give evidence of the juice contents. The methods are based on the research of several years so that correlations between results from samples and process events are prooved. Due to the long time span needed to complete a probe the minimum sample periods are about one hour. This is too long to prevent the pasting of filters.

(3) measuring points
About 1000 measuring points exist in the sugar plant. They are permanently sampled and stored in a data base. Direct correlations with the process are not given. A model of the process is not available so far and will be very complex due to the multi-input / output system with long delays. Furthermore it may be necessary to model the chemical reactions, too.

(4) weather data
The beet conditions depend strongly on the effects of the weather as described in the preceeding section. Unfortunately so far information like air pressure, humidity or temperature is not available for the time period before the beets reach the sugar plant.

Of course the different information sources have all their adavantages and disadvantages. On the one hand the workers have expert knowledge but on the other hand their information is unstructured, indirect and subjectiv. Furthermore they only see the outputs given by the control screens and e.g. can not smell if beets are altered etc. The missing communication may cause problems, too. More samples in the laboratory are both costly and the analysis will still remain too slow. This means altered beets are detected but it is too late to react adequately. Reliable weather and storing data is not available so far. Actions would be both labour intensive and costly. The machine data is easily available but characteristical values have to be generated. Direct correlations between signals and beet quality have to be found. Furthermore the measures are noisy and may be faulty. When comparing all these facts it nevertheless seems most reasonable to use the available measuring points as profound basis. Human or laboratory knowledge can be supplementary added.

4.2 *Data processing*

The raw data from the measuring points has to be processed so that in the end relevant characteristical values can be obtained. In a first step it is tried to find some correlations in the data. Afterwards it is necesary to extend the observer so that online detection of the beet quality is possible.

Starting with data from the last campaign it had to be determined when filters got pasted. This can be directly read from the time history of the filter pressure (see figure 6). Now corresponding events in other measuring points had to be found. Therefore different methods were used:

The workers were questioned what (big) events they remembered during the campaign (like stone break-in, frozen beets, pasted filters, etc.). Furthermore it was recorded which effects alterated beets have (like do they smell, are they hard to cut) and what kind of other disturbances occur with which effects.

Next the shift books were taken and searched for events. Unfortunately the entries were incomplete. An electronic shift book (Arenz *et al.*, n.d.) was proposed for the future.

Based on these informations correlation analyses were tried out. But no significant results could be found mainly because of two reasons. First of all the measurements were very noisy and disturbed not only from a bad beet quality but also from many other effects. Second and more important the data is highly coupled and effects could not really be separated.

As a lot of human knowledge is available it is rather natural to try out a fuzzy observer. Input values were different measurement points. Using the verbally described process information (expert knowledge) rules can be generated. Unfortunately this method could not be realized due to the following problems: The complexity of the system and the large number of input points lead to an enormous number of rules which are difficult to manipulate. Furthermore not all influences and correlations are known and can be included in the rule base. External influences like noises or smellings, which support the human decision process, can not be registered. Finally the data is noisy so much that even the filtered inputs have too many peaks. Besides these reasons it would be very difficult to include fuzzy set algorithms in the given environment of the control system.

The finally chosen approach uses process models in combination with process parameters. Using these models together with the measuring points as inputs, characteristical values can be determined. Those values are then processed in order to obtain a statement on the beet quality. Hence different process stages have to be defined together with the occuring material and energy flows.

4.3 *Relevant process sections*

In order to obtain well suited process models which allow the creation of reliable characteristical values certain aspects have to be kept in mind. The process itself is too complicated to be modelled. Therefore separated process sections are examined so that easier connections can be found. The observer must be fed with input variables as early as possible so that the process can be influenced in time. Hence sections too close to the filter are dropped out. Furthermore they should be carefully chosen so that as much as possible is known from the system equations. The measuring data and respectively the calculated state variables should not be too noisy and decoupled from other effects if possible. Correspondingly the following sections are closer investigated:

- beet yard
- beet cleaning
- cutting machines
- extraction tower
- raw juice processing
- filter section

On the beet yard many different conditions (like temperatur, humidity, storing influences, biological reactions) have to be evaluated. The section is therefore judged as too complicated especially because no direct measuring points exist.

In the cleansing section only indirect data (like temperature, pH-value) can be obtained from the washing water. For the given purpose this section is neglected, too.

The cutting machines seem to be well suited. From the process model a lot of facts are known. Measurement points are available. Very positive is the fact that the motor revolutions of the cutting machines are directly correlated with the beet quality (see figure 5 to 7). If altered beets are processed the knife blades blunt much faster as normal. This can be seen in the gradient of the curve. Problematic are the disturbances e.g. caused by stones or dirt which lead to a revolution increase as well. These gradients are very steep but only for a short time span. Nevertheless it may be difficult to filter them out.

In the extraction tower several measuring points are available like torque, temperature, pH-value, filling level and pressure. When processing alterated beets knots occur which lead to an increase of pressure and torque. When the speed of the tower is increased the clots break up. But as the tower is (pilot) torque controlled no exact statements can be derived for the beet quality. Furthermore the bunch of parameters result in a complex model which can not be described exactly.

From the raw juice processing section no direct measurements are available. On top of that it is at this point already too late to influence the process.

The filter section can be used in order to evaluate the results from the beet quality observer (see figure 6 and 7).

4.4 *Filtering methods*

As can be seen from figure 5 (top), the data from the measuring points is very noisy and needs to be filtered. Two different algorithms are tried out. On the one hand a Butterworth bandpass is used (see figure 5, middle). On the other hand a wavelet filter bank with (6th order) haar filters is shown (see figure 5, bottom). The results of both filtering methods are quite similar. For the use in the sugar plant the Butterworth filters are chosen as their algorithms can be easier implemented.

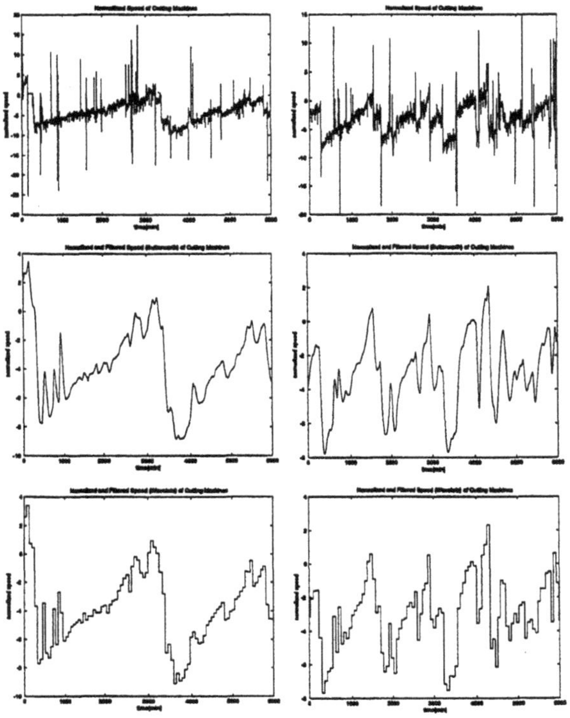

Fig. 5. Normalized speed values (left: good beet quality, right: bad beet quality, top: unfiltered, middle: Butterworth bandpass, bottom: Wavelet filter bank)

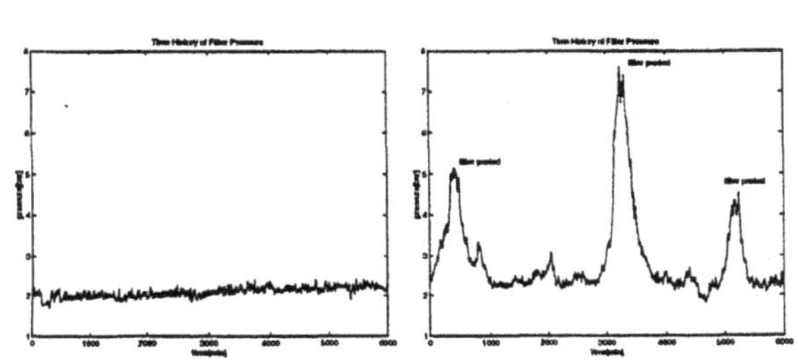

Fig. 6. Filter pressure (left: good beet quality, right: bad beet quality)

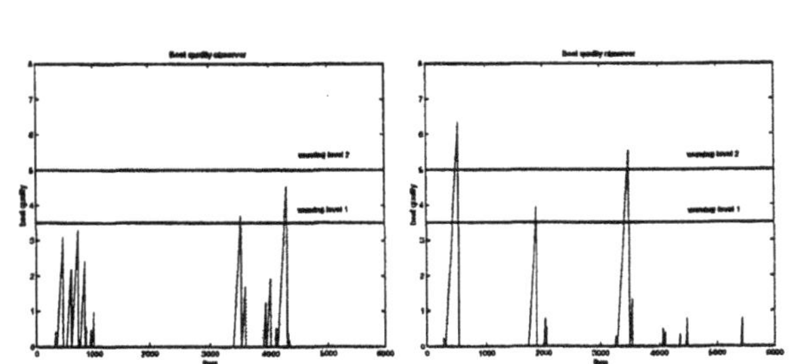

Fig. 7. Beet quality observer (left: good beet quality, right: bad beet quality)

The wavelet theory is an alternative to Fourier techniques and uses a time-frequency description. Wavelets themselves are localized waves coming from an iteration of filters (Strang and Nguyen, 1996). A filter bank is a set of filters. Its analysis results in set of averages (lowpass structure) and differences (highpass structure). Very interesting for the future research work is the ability both of denoising and discontinuity and feature detection with wavelets. E.g. runaways can be eliminated, so that the signal gets much smoother.

4.5 *The beet quality observer*

The beet quality observer works mainly with the measuring points from the cutting machines. Optionally parameters from the extraction tower can be added. In a first step the input data is filtered. Next characteristical values are defined and calculated. Now using special analyzing algorithms the quality of the beets is extracted. The whole course is described in more detail using the example of the cutting machines. The tool Matlab is used as working environment.

First a normalized speed value (see figure 5 top) is calculated from the measuring points 'motor revolutions' and 'beet quantity' so that the disturbances of the quantity control are eliminated. Next the data is filtered. Now the gradient of the curve is calculated for different time spans. Short, middle and long averaging periods are distinguished. Now the data is analyzed regarding the beet quality. Using special queries it is tried to eliminate the disturbing influences e.g. of stones, dirt or ice. The results are combined to one statement which can be displayed graphically (peaks indicate a bad quality, see figure 7). In the figures 5 to 7 these connections are shown. On the left side the beet quality was quite good whereas on the right side problems occured. The statements of the beet observer are neither very good nor bad so far. This is mainly due to the simplicity of the implemented algorithms as described in the following section. All the derived algorithms were tested with data from the last campaign. As the raw products vary widely from year to year adaptive algorithms have to be used at the beginning of each campaign in order to update the analysis functions.

5. IMPLEMENTATION

Now the algorithms had to be implemented in the sugar factory. The existing control architecture is

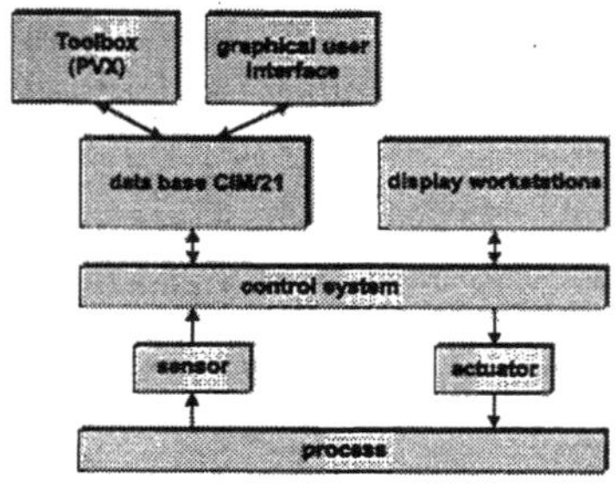

Fig. 8. Control architecture

shown in figure 8. The heart of the structure is the control system which directly controls the process. Via the graphical user interface online interactions are initiated (see figure 9a). The occuring data flows are stored in the data base tool CIM/21 which is directly linked to the control system. It provides several toolboxes for example PVX. PVX files are programmed in a very basic syntax which is similar to Assembler code. They allow the online reading and processing of measuring points. After compiling the source code the (static) PVX programs run online. When extending the system it is as well possible to include a C-code toolbox so that code, which is e.g. automatically generated by Matlab/Simulink, could be directly implemented.

When developing the evaluating algorithms the given conditions have to be kept in mind. Hence the functions have to be quite simple so that they could be easily implemented. Furthermore the functionality of the algorithms is reduced by the limited online processor capacity. In a first step the Matlab functions are (manually) converted to PVX code. Both the syntax and the results are checked. Now the source code is compiled. The program is integrated in the (online) working cycle of the CIM data base. Specified variables are stored in the data base and can be (online) displayed on the screens of the control systems (see figure 9b). Finally warnings are generated which are automatically displayed. Two different warning levels are distinguished (see figure 7). The first level which is displayed in yellow means 'pay attention, possible problems' whereas the second level in red already says 'bad beet quality, take actions.

(a) Control system　　(b) Display screen

Fig. 9. Process control

6. CONCLUSIONS

This paper deals with the derivation of an observer which gives evidence of the actual quality state of the processed sugar beets. In a first step the sugar process together with the occuring problems is described. Resulting from biological decomposition processes of alterated beets long molecular chain structures occur. They cause extreme problems by pasting up the filter system. Next different data analysing and processing methods are shortly described. The finally chosen solution uses process section models together with data from measuring points. Among the different process sections the cutting machines seem to be best suited. The gradients of the motor revolutions show a direct correlation with the beet states as alterated beets blunt the knife blades much faster than normal. The filter section is used to evaluate pasting of filters. Hence it also gives evidence whether the derived beet quality observer works well or not. As the measuring data is very noisy filter structures are needed. The results both of a Butterworth bandpass and a wavelet filter bank are shown. Next the structure of the beet quality controller is described using the example of the cutting machines. After calculating normalized values filters are applied. Then it is tried to eliminate the remaining noises and nonlinear coupling so that acceptable statements on the beet quality can be made. In order to increase the reliability of the observer different time spans are used. Resulting is a beet quality curve. It turned out that the parameters in the analyzing functions need to be adapted every year due to the changing living conditions of the beets. Finally the derived algorithms had to be implemented in the sugar plant 'Nordkristall GmbH'. Now online warnings are automatically created when a bad beet quality is detected so that the operators have more information to react adequately and in time.

REFERENCES

Arenz, A., M. Reimann, U. Harten and E. Schnieder (n.d.). Information management for process control and supervision using a hypertext system with integrated data bases. In: *Proceedings of the World Manufacturing Congress '97, New Zealand.*

Geschäftsbericht (1996). Technical report. Zucker AG Uelzen-Braunschweig.

Strang, G. and T. Nguyen (1996). *Wavelets and Filter Banks.* Wellesley-Cambridge Press.

HOW ON-LINE FAULT DETECTION AND SUPERVISION SYSTEMS CAN BENEFIT FROM THE CAPE-OPEN STANDARD INTERFACES IN PROCESS SIMULATION

Bertrand Braunschweig

Institut Français du Pétrole
Computer Science and Applied Mathematics
1 & 4 Avenue de Bois-Préau 92500 Rueil Malmaison France
bertrand.braunschweig@ifp.fr

Abstract : When fault detection and supervision systems make use of process simulation models, specific interfaces have to be developed in order to make both software modules communicate. The paper shows how this amount of work can be reduced, standardized, and made reusable by using a software component approach and by adhering to the future CAPE-OPEN[1] standard interfaces for process simulation components. The paper mentions other software standards initiatives in related application domains. *Copyright © 1998 IFAC*

Keywords : Process simulators, Standards, Software engineering

1.INTRODUCTION

Do you use process models in your supervision systems ? Do you care about the software engineering aspects of embedding a process model within a fault detection application ? Would you like your diagnosis system to be able to use several flavours of the same steady-state or dynamic model ? Would you like your process model to work seamlessly with several flavours of fault detection and supervision environments?

If you answered no to all questions, don't read this paper. But if you answered yes to one or more, take a few minutes to learn about the CAPE-OPEN initiative in process engineering software (CAPE-OPEN Consortium, 1997). You will discover how process simulation components can be re-used in different contexts without the burden of writing several hundreds of source code for interfacing them with different software. You will see examples of such possible integration with thermodynamic calculations, unit operations modules such as reactors, numerical algorithms for solving flowsheets. You will see how *middleware* such as Microsoft's COM/ActiveX (Brockschmidt, 1996) and OMG's CORBA (Object Management Group, 1997) can establish the necessary bridges between heterogeneous components originating from various sources and running on distant networked computers. You will read about two other standard development approaches, pdXi and OPC, in

[1] Brite-EuRam project BE3512 supported by the European Commission under the Industrial and Materials Technologie Programme (Brite-EuRam III) under contract BRPR-CT96-0293. Partners : BASF, BAYER, BP, DuPont, Elf, ICI, IFP, RWTH Aachen LPT, RWTH Aachen I5, Imperial College, INP Toulouse, Aspen Technology, Hyprotech, Simulation Sciences, QuantiSci.

engineering data management and online control systems.

This paper focuses on the CAPE-OPEN project, one of the many initiatives aimed at easening the development of complex applications for the 21st century computerized chemical processes, in which steady-state and dynamic simulation will play a major part. In section 2, I first address the need for defining standard interfaces in process simulation. In section 3, I give possible use of such standard interfaces in diagnosis and monitoring applications. Section 4 presents the technical aspects of using middleware code for this purpose, including an example of specification. Section 5 introduces two other initiatives in related domains. Then I conclude on how you can directly benefit from these experiences and how you can get more information.

2. INTERFACE STANDARDS FOR PROCESS SIMULATION COMPONENTS

Effective process simulation has become a critical stage of any work related to designing, building and operating chemical and petroleum processes. It is an essential tool for assessing the expected performance of a process in all conditions and at all steps of the process lifecycle. Different simulators have different strengths and to get the best results requires access to more than one simulator and integration with in-house or third-party software containing company specific information. Specifically, when dealing with fault detection and supervision systems, process models need to be integrated with a number of environments including online control systems, diagnosis tools, process databases and ERP applications. This is however a slow and costly process because of the existing diversity in interfacing simulation models in commercial systems, due to major historical differences. Think about training a neural network or a statistical model with a simulator before putting it into operation at the plant : you would either need to prepare a database of simulated conditions and transfer it to the neural network for its training ; or you could have the neural network operate the simulation model through efficient communication links, so that it would automatically produce new training and test data as needed. Think of building a model-based control system which uses a process model as the core numerical tool for predicting the state of the process under varying conditions : you could develop a specific software interface for communicating with the model, or you could rely on a standard way of accessing the principal simulation functionalities.

The CAPE-OPEN initiative aims at minimizing the development time needed for interfacing process models, or elements of process models, with each other and with third-party elements. It specifically addresses the functional exchange of process simulation at run-time, as opposed to other standards such as pdXi or STEP which are concerned by the exchange of process information through standard files and database systems.

2.1. *Open Systems Communication for Process Simulation*

CAPE-OPEN promotes the adoption of an open systems communication for process flowsheet simulation, which allows simulator users to employ elements of any simulator within any other. Specifically during the project :

- we will develop open interfaces for key simulation elements;
- we will demonstrate the effectiveness of the open system,
- we will widely disseminate the results and promote adoption of the open system by the major providers and users of process simulation.

2.2. *Scope and Technical Approach*

The types of components in simulators for which interface specifications will be produced are Thermodynamic and Physical Properties Packages, data for chemical species within the packages, individual physical properties, sequential-modular or equation-based unit operation models, numerical methods for solving dense or sparse sets of non linear algebraic and differential-algebraic systems.

The general types of simulators that will be catered for are sequential modular and equation based commercial[2] and in-house simulators; note that the interfaces are expected to be applicable to any process simulator.

The material forms that fall into the scope of the standard are uniform fluids defined as molar compositions of specific chemical species, pseudo-components, electrolyte mixtures, reacting fluids and polymer solutions and particulates.

The physical properties methods that will be provided wih standardized interfaces are the properties required for the calculation of vapour-

[2] Including Aspentech, Hyprotech and SimSci simulators

liquid-liquid-solid equilibria or subsets thereof, other thermodynamic and transport properties.

The standards will cater for both steady state and dynamic modelling of continuous and batch or mixed processes. In terms of applications, off-line simulation, optimisation, parameter estimation and data reconciliation will be covered.

The project adopts a *component software* and *object-oriented* approach which enables a complete simulation to be assembled using software components from a variety of sources. The standard interfaces will use existing *middleware*, namely *COM/ActiveX* and *CORBA*. More specifically, these interfaces will be expressed in both forms in order to be applicable to several hardware platforms and operating systems, and to be future proof. Legacy code such as FORTRAN models will be recycled by encapsulation within CAPE-OPEN compliant wrappers.

The definition of interfaces throughout the project is done following a development process based on the UML object-oriented notation, producing use cases, sequence diagrams, class diagrams and interface diagrams which will accompain the corresponding middleware implementation.

Interface specifications will be expressed both for *Microsoft's COM/ActiveX* and *OMG's CORBA* using the corresponding *Interface Definition Languages*[3].

2.3. *What the project will deliver*

When CAPE-OPEN is completed, it will be possible to assemble a simulation model from a variety of sources, such as the example on Fig. 1 in which the simulator executive and the user interface are supplied by one vendor, whereas other components (one or more unit operations, the thermodynamical and physical properties server, the solving algorithm) come from different suppliers. Extending this principle, the individual components will be able to communicate in a standard fashion with other environments such as process control and monitoring systems, or in your online fault detection application.

[3] Unfortunately, both languages are called IDL by their providers.

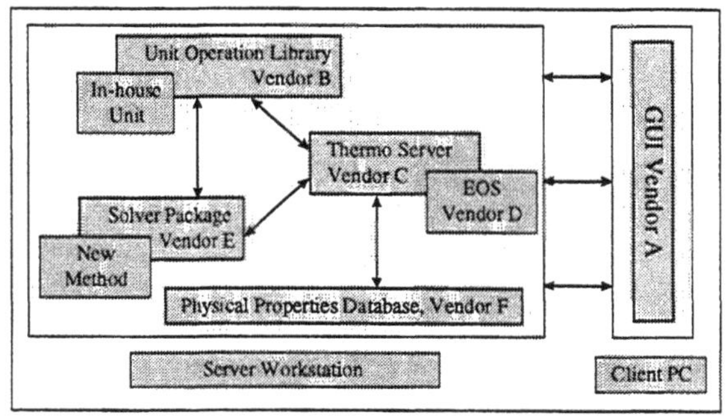

Figure 1 : A typical CAPE-OPEN simulator

3. WHERE FAULT DETECTION AND SUPERVISION SYSTEMS CAN BENEFIT

In this section I give three examples of uses of the CAPE-OPEN standard in the context of fault detection and supervision.

3.1. *Use of Thermodynamic calculations*

An intelligent process monitoring application developed with Gensym's G2® requires that a Unit Operation object calls an external thermodynamic calculation routine in order to assess its number of phases. This number of phases is passed back to G2 so that it triggers the appropriate set of rules depending on the physical state of the fluid, that is, liquid or gas/liquid. If this behaviour is attached to the top of the object hierarchy for unit operations(UOs), then specific subclasses of UOs may need to call different flash routines, possibly from different sources. As an example, a standard flash drum would use a popular thermodynamic library provided by a vendor, whereas a proprietary reactor would use a in-house thermodynamic routine. A CAPE-OPEN thermo interface introduced at the most generic object level would allow all instances of subclasses to call the thermo routine through the same set of interfaces. Moreover, taking advantage of the object-oriented definition of interfaces, the developer would only need to establish the association between the various subclasses of UOs and the corresponding thermodynamic calculations.

3.2. *Use of Unit operations*

A diagnosis application involves to calculate the theoretical equilibrium state or the dynamic behaviour of a unit operation such as a reactor, a distillation column etc. Values for output streams need to be calculated from a set of input streams and parameters describing the status and response of the unit operation. The CAPE-OPEN conceptual model introduces standard port methods through which information about the material systems traversing the unit can be accessed. Once these port methods are implemented, the diagnosis application only needs to communicate with the ports in a standardized manner.

In a program developed by IFP, a steady-state model of the reactor calculates a theoretical behaviour to be compared with the real performance of the plant. When interfaced through CAPE-OPEN ports, the reactor model can be easily replaced by another when a new reactor design is implemented; reciprocally the calling module can be implemented in several environments whereas its use of the reactor model will be unchanged. Once again here, object orientation can be used to specialize the interfaces.

3.3. *Use of numerical interface*

An online neural network classifier based on continuously acquired data needs to be retrained when the process reaches new regions of its multidimensional variable space. The training uses both real and simulated data, as the amount and coverage of the real data are not sufficient for obtaining a generic trained network able to operate in various conditions.

Instead of separately running simulations in the new region and save results onto files, a CAPE-OPEN interface would allow to take control of the simulator from the neural network, and to make it generate synthetic data when and as much as needed for the training. As a matter of fact, running the simulation would only imply to input new parameters to some process elements though the Unit Operation interface, and to place a standard call to the numerical engine in charge of solving the flowsheet. Moreover, if simulations use large amounts of CPU time, the neural network could trigger several models running on distant networked computer, in parallel: these would only need to be registered when configuring the system, the networking facilities being supplied by the middleware. The simulators themselves could be different, covering various regions of the process space, but their communications with the neural networks would remain the same.

With these three examples I have given an idea of some of the possible uses (among many others) of standard interfaces to process simulation components.

3.4. *Examples from the conference*

I now take a few examples from this conference and propose how these would benefit from using the CAPE-OPEN standard interfaces[4].

(Szederkenyi, *et al.*, 1998) use grey box models for fault detection of heat exchangers. The first principles model of the process is used by a fault model based on physical knowledge. CAPE-OPEN interfaces would allow both models to exchange functionalities in a standard way, allowing each of these models to be replaced by another one without effort.

(Cho, *et al.*, 1998) adopt a hierachical decomposition of a plant model and build local PCA estimators for each level of the decomposition. They illustrate the methodology with the Tennessee Eastman benchmark composed of five process units. CAPE-OPEN interfaces could support the runtime switching of each process units depending on the fault detection situation.

(Martinez and Wilson, 1998) reward trainable software agents by value estimates calculated using process models. With CAPE-OPEN interfaces each of these agents could have its own model, internally different, but accessed through standard calls, thus enhancing the diversity of the agents population.

(Mathisen, *et al.*, 1998) use dynamic models as local estimators for assessing the plant condition. Then in a later stage a recursive search is made though the process flowsheet in order to find disturbances. CAPE-OPEN interfaces would allow the recursive search algotihm to take control of the forward models calculations and to apply them on request.

4. WHAT IT REALLY MEANS

The CAPE-OPEN specifications will be distributed as lists of interfaces, methods and arguments together with their expressions in CORBA's IDL and in COM's IDL languages. Someone wanting to develop a CAPE-OPEN compliant component will need to integrate the same declarations in its application and use IDL compilers of both kinds to generate the corresponding instructions[5] in source language such as C, C++, Java, Smalltalk etc. Then

---

[4] All references in this part are in (Dhurjati & Cauvin 98)

[5] Mainly header files for stubs, and skeketons.

he/she will link this wrapping code with the rest of the component.

As an example, figure 2 shows the simple standard interface specification for accessing some of the ports of a Unit Operation. The method belongs to the ICAPEUNIT interface, takes as input the type and directions of ports required, and returns a pointer to a portsInterface which will then be used to communicate with the selected ports .

INTERFACE NAME	ICAPEUNIT
Method Name	GetPorts
Returns	CapeError

Return an interface to a collection containing a list of ports on the unit of a specific type and direction e.g. all input material ports.

Arguments

Name	Type	Description
[in] streamType	CapeStreamType	the type of stream required material,energy,information
[in] direction	CapeDirection	the direction of flow : input, output or any
[return] portsInterface	CapeInterface	a reference to the interface on the collection containing the specified ports

Figure 2 : a CAPE-OPEN method specification

5. OTHER STANDARDS APPROACHES

In this section I introduce two other ongoing efforts in standards for process engineering and control applications. These intiatives can be considered as complements to CAPE-OPEN in major areas : exchange of process data, and exchange of process control functionalities.

5.1. OPC : OLE for Process Control (OPC Foundation, 1997)

The OPC Foundation, a non-for-profit organization, is promoting the use of standard OLE/COM interfaces in process control. These interfaces are based on Microsoft's OLE/COM middleware. They define standard methods for accessing control systems functionalities and attributes of servers of real-time information : distributed process systems, programmable logic controllers, smart field devices and analyzers. The OPC initiative is supported by a group of more than 150 providers and users of control systems hardware and software. Prototype OPC servers have been developed and demonstrated at various occasions.

The choice of OLE/COM as unique middleware makes the OPC standard rely on Microsoft technology. Hence, Unix or VMS compatibility will only be provided through software bridges like Iona's ORBIX FOR WINDOWS or through foreign implementation of COM on these platforms. Such bridges are now being made available on specific platforms, like the ENTIREX software from Software AG (Software AG, 1997).

5.2. pdXi (Baldwin & Almeida, 1995)

The Process Design Exchange Institute (pdXi), is an initiative of the CAST (Computer and Systems Technology) Division of AIChE to promote the electronic exchange of process engineering data. pdXi have addressed the development of a generalised data model (the AP-231) and the development of a utility for accessing data that conforms to the defined model. The term process data refers to stream and equipment data and other data used to support the process engineering activity over the entire life cycle of processing facilities. Membership in pdXi is open to operating companies within the process industry, software vendors, engineering contracting firms and equipment design/manufacturing firms.

Recently an Application Programming Interface (API) has been developed by the pdXi consortium. This API allows any application to communicate with pdXi formatted files through standard calls. Fault detection and supervision applications would use the API for accessing plant data stored by pdXi-compliant applications, and reciprocally to store supervisory data that needs to be shared with other applications. Note that this is quite different from the CAPE-OPEN standard which deals with the exchange of function, where everything takes place in memory, whereas pdXi is concerned by the exchange of data through standard file mechanisms.

6.CONCLUSION

I promote the adoption of open interface standards which will encourage rather than stifle innovation in process simulation, diagnosis and control. The advent of standards in the process industry will leverage current developments as computer scientists, control systems specialists and chemical engineers will devote more time to their core activity and less to lower interest tasks such as the interfacing and re-interfacing of software.

Moreover, they will be able to use best-of-class components from commercially established firms, which will provide services that they would have to develop on their own otherwise. In summary, with standards such as CAPE-OPEN, pdXi and OPC, you will concentrate on the development of your fault detection or supervision algorithms while taking advantage of the rest of the world's developments in accompanying software.

The examples in this paper are illustrative of the kinds of benefits one could expect from standard interfaces. I am convinced that numerous additional uses of these interfaces will be proposed by others. It is your privilege to build innovative architectures by taking advantage of the core developments made by the CAPE community. It is impossible to envisage all possibilities : look at what the standard desktop bus has offered in terms hardware. Standard software interfaces will open the same perspective in allowing anyone to mix and match software components from many sources in order to develop new solutions to today's process engineering, monitoring and control problems.

The CAPE-OPEN documents are made public as they are issued and approved by the project. Goto `http:// www.quantisci.co.uk/CAPE-OPEN` for the latest information and for open interface specifications documents.

7. ACKNOWLEDGMENTS

I thank the partners of the CAPE-OPEN project and the European Commission for authorization to publish the paper. I acknowledge the partial reuse of material from some of the following references.

8. REFERENCES

Baldwin, John, and Joe de Almeida (1995), Information and data exchanger breakthrough - the new pdXi interface, AIChE 1995 Summer National Meeting, Boston .

Brockschmidt K. (1996) What OLE is Really About, Microsoft Corporation

CAPE-OPEN Consortium, Conceptual Design Document, December 1997, Adobe Acrobat PDF document downloadable from `http://www.quantisci.co.uk/CAPE-OPEN`

Dhurjati P., Cauvin S.,(1998) IFAC *Workshop on On-Line Fault Detection and Supervision in the Chemical Process Industries*, Solaize.

Cho T.H., et al., (1998) On-line process monitoring and diagnosis of chemical processes with multiple operation states. in Dhurjati P., Kramer M., Cauvin S.(1998), see reference

Mathisen K.M.,(1998) Industrial implementation and testing of model based diagnosis in the PRIDE project. in Dhurjati P., Kramer M., Cauvin S.(1998), see reference

Object Management Group (1997) *The Common Object Request Broker: Architecture and Specification.*

Opc Foundation (1997): *OLE for Process Control , Data Access Standard, Version 1.0A*

Software AG (1997) EntireX DCOM White Paper, Software AG & Microsoft

Szederkenyi G., et al., (1998) Grey box fault detection of heat exchangers, in Dhurjati P., Kramer M., Cauvin S.(1998), see reference

Wilson J.A., Martinez E.C., (1998) An agent-based approach to process supervision based on model-based reinforcement learning. in Dhurjati P., Kramer M., Cauvin S.(1998), see reference

TOWARDS AN INTEGRATED TOOL FOR CONTROL, SUPERVISION AND OPERATOR TRAINING - APPLICATION TO INDUSTRIAL WASTEWATER TREATMENT PLANTS

Konrad SZAFNICKI, Didier GRAILLOT

Ecole Nationale Supérieure des Mines de Saint-Etienne
158, cours Fauriel
F-42023 SAINT-ETIENNE Cedex 2 (France)
Tél: (+33) 4. 77.42.02.24 Fax: (+33). 4. 77.42.66.66 E-mail: konrad@emse.fr

Abstract: Chemical processes usually impose severe constraints in the fields of safety, reliability, efficiency. The know-how, necessary for their control and supervision implies a numerous and wide domains of knowledge (chemical, data treatment, process control, signal processing...) of both, quantitative and qualitative natures. An integrated implementation of the latter requires the development of dedicated tools and techniques for real-time supervision of chemical processes. The paper presents a few aspects of an expert system for real-time supervision and fault detection applied to the industrial wastewater treatment plants for surface treatment industries. The goal of the project is the design of information and decision support systems, in order to improve the operation (safety, reliability, efficiency) of waste water treatment plants as well as to provide an efficient training tool for the personnel (staff, future operators,...).

Key Words: Supervision, Diagnosis, Real-time expert systems, Water pollution, Industrial wastewaters, Computer-aided instruction.

1. INTRODUCTION

Industrial activities, and particularly metal processing activities (e.g. electroplating) use huge quantities of water during their manufacturing cycles. They thus generate significant amounts of wastewater, which usually do not satisfy environmental regulations so as to be rejected directly to the receiving water.

The regulations, however, tend to be more and more strict and limitative as far as admissible quantities of rejected pollutants (metal ions, cyanides,...) are concerned (e.g. in the new EEC-regulations). It is therefore of great interest, in order to be able to satisfy the present and future limitations, to determine and develop new, reliable means of supervision and treatment of waste-waters (Baeza *et al.*, 1998).
Moreover, Small and Medium-size Enterprises (SME) will certainly not feel very enthusiastic about the idea of installing huge numbers of additional sensors, if the idea is not properly justified. The improvement of the treatment techniques thus implies the following main directions of research efforts:

• the development of sensors which would take into account both, the quality constraints imposed by the present and future wastewater regulations and, on the other hand, the economical requirements of the concerned enterprises (Di Benedetto *et al.*, 1995),
• an "intelligent" processing of information, provided by the sensors already installed *in situ* (Szafnicki *et al.*, 1996),
• efficient on-line decision making aid,
• interactive training tools for future operators (Szafnicki *et al.*, 1997).

At present, most enterprises involved in the surface-treatments, and particularly SMEs, cannot afford a full-time agent for the supervision of the industrial waste water treatment plants. On the other hand, operators can only *note* the problems (faults, pollutions, ...) which have occured, from the Central

Monitoring Station, but still they can hardly *prevent* them, since there are no reliable industrial sensors available for the in-line simultaneous monitoring of the amounts of different metal cations (particularly: ions of Cu, Ni, Zn, Fe, Cr...) or cyanide ions in the rejected waste waters. Moreover, the supervision agent (operator) rarely has at his/her disposal any computer support, which would provide an efficient aid in the analysis of the available information and the decision making.

The problem description (administrative regulations, process complexity) is dealt with. A few specific aspects of the know-how are presented. The development of a real-time knowledge-based supervision system prototype is illustrated, which can also be used as a computer-aided operator training.

2. PROBLEM DESCRIPTION

Surface treatment industries, which mainly concern SMEs, generate toxic waste waters, containing metal ions, mainly of Ni, Cu, Zn, Al,... as well as cyanide ions. Till the end of the eighties, these industries were responsible of 30% of toxic industrial pollution of the rivers in Europe. The first laws of 1985, Table 1, imposed severe limitations, particularly on SMEs which had some difficulties inverting enough money to meet the regulations. Yet, even more restrictive limitations are being foreseen for the coming years.

Table 1: Rejection limitations (mg/l), present values and projected future evolution.

Elements	1985	1993 and future
Al	5.0	3.0-5.0
Cd	0.2	0.1-0.2
Cr^{VI}	0.1	0.1
Cr^{III}	3.0	...
Cr total	3.0	0.5-1.0
Cu	2.0	0.5-1.0
Fe	5.0	2.0-5.0
Ni	5.0	0.5-2.0
Pb	1.0	0.5
Zn	5.0	0.5-2.0
CN^-	0.1	0.1-0.2
Max. 8 *l* of waste water by m^2 of treated surface, at every rinsing stage.		

As a consequence, waste water treatment plants will have to be managed very efficiently, in particular as far as supervision and fault detection are concerned. In fact, any fault implying casual pollution may cost a lot in the near future, particularly to the SMEs. The latter are therefore interested in the development of efficient tools for real-time aid in the decision making, which would enable a « fault-free » management of the treatment plants.

The application, on which the paper will focus further on, generates waste waters which mainly contain ions of such metals as Cu, Ni, Zn,... as well as cyanide ions. A dedicated treatment plant disposes of the following main equipment (Fig. 1 & 5): head buffer bassins (alcaline and acid), decyanidation, neutralization with lime, concentrator and lamellar filter, press filter, lagoons, sand filters (duplex). This equipment is connected to a computer, which enables data recording and, if necessary, remote setpoint update.
In the following, let us focus on a few selected aspects concerning the running constraints of the different stages within the treatment plant.

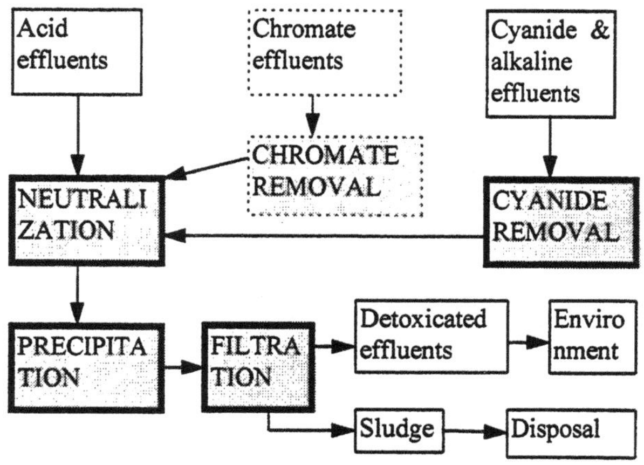

Fig. 1. A general scheme of a detoxication plant for surface treatment effluents.

• *Decyanidation:*
Cyanides oxydation in an alcaline environment theoretically consists of two successive stages: first to the form of harmless cyanates, then to nitrogen and bicarbonate. The strong oxydants that are used are: sodium hypochlorite (fr. « javel »), chlore, and monopersulfuric acid. In practice, often only the first stage is realised, due to economical reasons. This is also the case of the application, mentioned in this paper.
Decyanidation using sodium hypochlorite definitely has to occur in an alcaline environment (pH>12), otherwise the reaction speed decreases and toxic substances may be generated (e.g. CNCl), implying a fatal threat. The critical threshold is about pH 10.5.

• *Storage of NaOH:*
The solution of NaOH at a concentration of x% admits a threshold temperature T_C, under which solid phase appears due to cristallization. For example, in this application: x=50% -> T_C=10°C, Fig. 2. Therefore, the solution has to be maintained at a temperature higher than T_C, otherwise it cristallizes and may clog the tubes. For this purpose, its temperature is measured regularly and an electric resistance may be used to raise the temperature, if necessary.

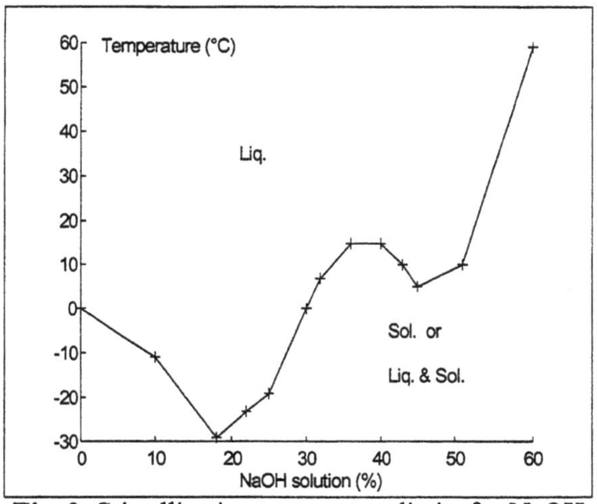

Fig. 2. Cristallization temperature limits for NaOH solutions.

• *Metal Precipitation:*
Chemical precipitation is the most common technique used for the treatment of metal-containing waters. It involves transforming a soluble compound into an insoluble form through the addition of chemicals, such that a supersaturated environment is created.

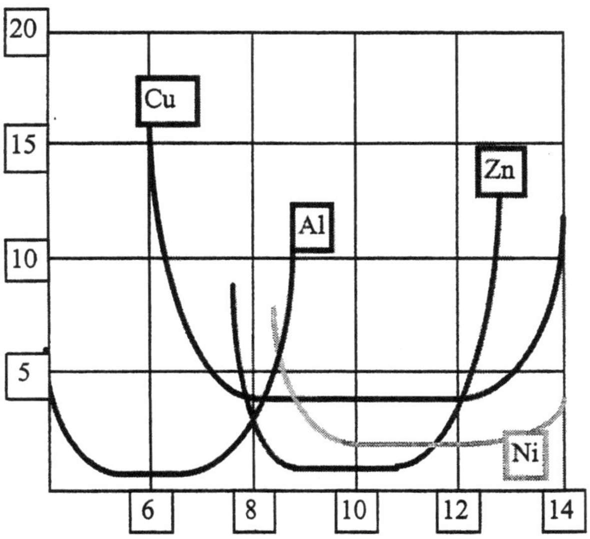

Fig. 3. Solubilities (mg/l) of a few metal hydroxides (Al^{3+}, Cu^{2+}, Zn^{2+}, Ni^{2+}) vs. pH

In hydroxide precipitation, which is used in this application, metal ions are removed by adding an alkali, such as caustic or lime, adjusting the wastewater pH to the point where the cations have a minimum solubility, followed by floculation and sedimentation - filtration operations. Usually, the solubilities of metal hydroxides in solution decrease with increasing pH to a minimum value (isoelectric point), beyond which they become more soluble because of their amphoteric nature, Fig. 3.
In case of mixtures containing several metal cations, the isoelectric points may not necessarily be close to each other. Then, a staged precipitation process can be used.

• *Sand filters*, which are run in duplex (i.e. one is used for filtration meanwhile the other is washed) should be washed regularly and optimally, i.e. before their clogging level is too high. The latter may be estimated with the help of two variables, which are measured: the loss of charge ΔP and the flow (in m^3/h). There are two parallel pumps available for the sand filters, but usually only one is run (since sufficient) to maintain the flow level. Once the maximal values of ΔP are higher than a threshold value, then filters should be permuted.

Yet, the operator's know-how remains vital for an efficient running of the plant, particularly when difficulties or faults appear. Unfortunately, experienced operators are not 100% available for this task, since they often have other responsibilities within the SME. Thus, the idea of this project has been to develop an expert tool which would be able to provide an efficient real-time decision support to the operator as well as a computer-aided instruction for future operators.

3. TOWARDS AN INTEGRATED SUPERVISION AND TRAINING TOOL.

At the first stage, knowledge has been collected and formulated from the operators who have a wide, long-term experience in dealing with the industrial waste-water treatment plants. Different scenarii of the normal run as well as of the usual faults have been put up. The latter will enable the development and implementation of the Knowledge-Base necessary for the detection of abnormal tendencies and behaviours of the plant (cross analysis of quantitative and qualitative information provided by the different measures). Subsequently, fault identification rules are to be formulated so as to provide an on-line aid in the explanation of possible causes of the detected abnormal configurations. Finally, the system is to be validated, first off-line, on real data collected on running plants, as well as by novice operators, during computer-assisted training periods, Fig. 4.

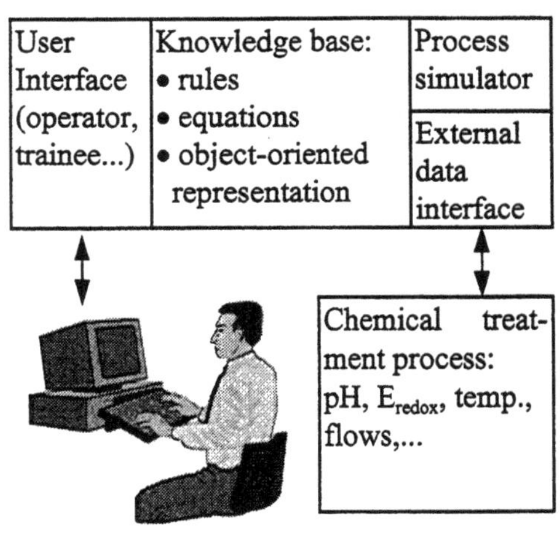

Fig. 4. General scheme of the integrated tool.

For the development of a real-time supervision and training system the G2™ real-time expert systems development tool (Gensym Corp., USA) has been chosen. The most important reasons for this choice were:

- different complementary knowledge representation possibilities: formulas, rule-bases, object-oriented representation, etc.,
- several real-time utilities; as e.g. interfaces to both, external data and the user,
- tool already validated on a number of real, industrial applications as described in e.g. (Thompson & Mertz, 1993; Szafnicki & Graillot, 1996).

The RTES-prototype has been developed and validated on real data, recorded on the treatment plant. The supervised variables are: pH of the acid and alcaline buffer bassins, temperature of NaOH solution, E_{redox} and pH of decyanidation reactor, pH of neutralization, ΔP and flow (in m^3/h) at sand filters, final pH and temperature of the rejected waste-water, Fig. 5.

The user has the possibility of simulating faults, e.g. a drop of the NaOH temperature implying a failure of the pH-control in the decyanidation reactor (for example by perturbing the « normal » run measurements). This utility enables to validate « off-line » the software's behaviour in case of occurrence of fault(s). It is also very useful for knowledge acquisition, since it enables an efficient dialogue with the experts familiar with the process as well as for training of people concerned with the maintenance or monitoring of the treatment plant.

REFERENCES

Baeza J., Gabriel D., Lafuente J. (1998). An intelligent distributed control system for a nitrogen and phosphorus removal pilot plant. *Waste Decision 98 - Intl. Workshop on Decision and Control in Waste Bio-Processing*, Narbonne, France.

Di Benedetto D., Breuil P, Poyet J.P. (1995). On-line monitors multi-metal analysers using spectrometry. *NorthWest Water Conference*, Manchester, UK.

Szafnicki K., Cournil M., Mathon A., Jullien B., Girard M.-A. (1997). Interactive Education in Control and Management for General Scope Engineering Students. *ACE'97 - The 4th IFAC Symp. On Advances in Control Education.* Istanbul, Turkey, July 14-16, 1997.

Szafnicki K., Bourgois J., Graillot D. (1996). Supervision en temps réel des stations d'épuration des effluents industriels. *DECHETS - Sciences et Techniques.* N°4, 4e trim. 1996.

Szafnicki K., Graillot D. (1996). Knowledge-based Real-time Fault Detection and Supervision of Urban Drainage Systems. *Automatica*, **Vol.32**, N° 7.

Thompson L. and Mertz G. (1993). Real-time Expert System Implementation at Monsanto-Krummrich. *Automatica*, **Vol.29**, N° 5.

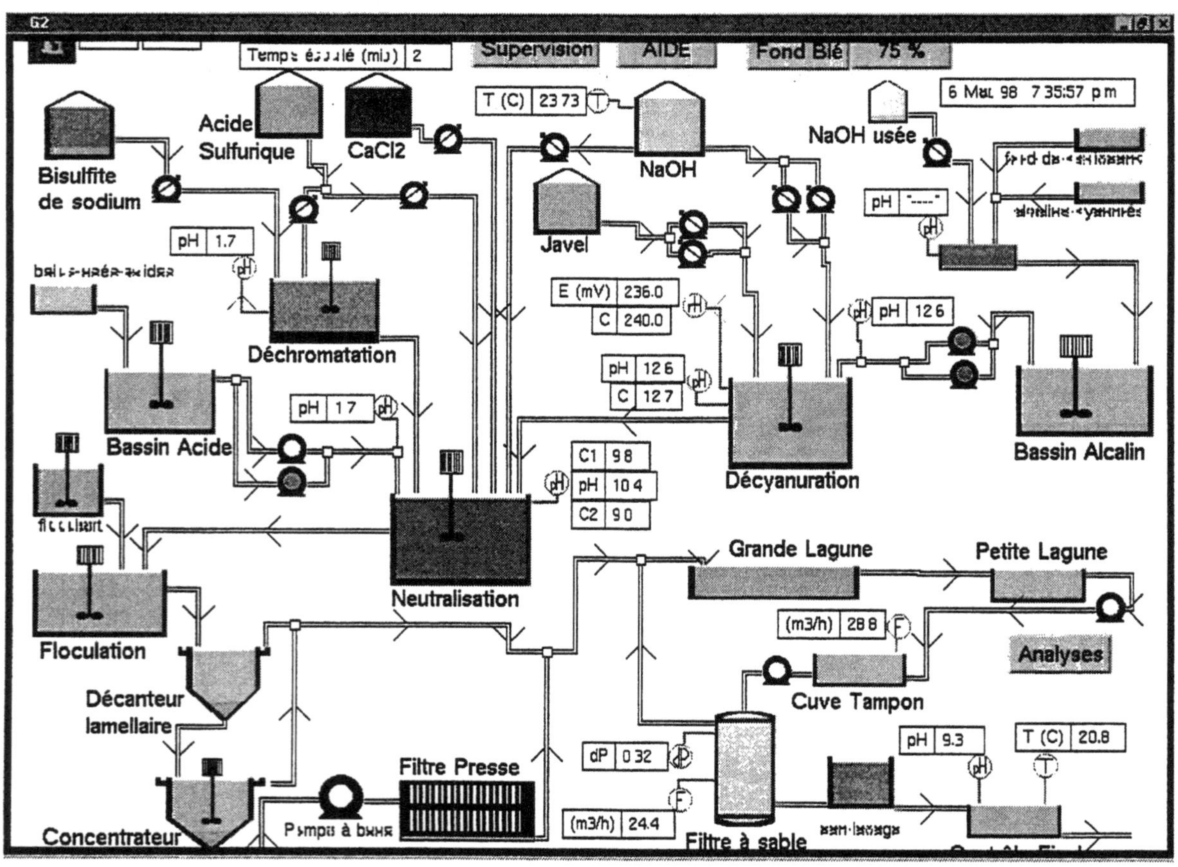

Fig. 5. Schematic overview of the treatment plant, using G2™ real-time expert system.

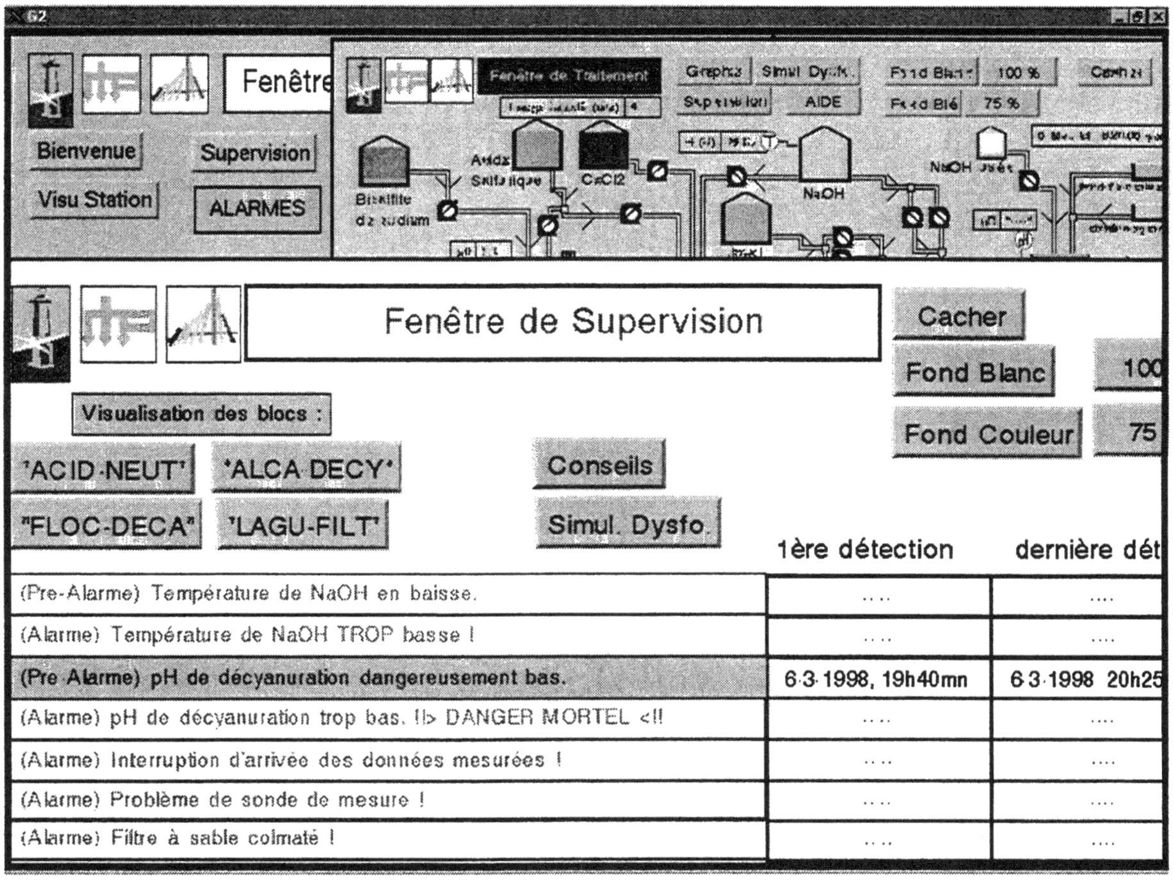

Fig. 6. Supervision and fault detection utility.

Fig. 7. Example of interactive help for process understanding. Here : some details of the detoxication hall.

PARALLEL EVOLUTIONARY APPROACH TO SYSTEM IDENTIFICATION FOR PROCESS FAULT DIAGNOSIS

Teodor Marcu and Paul M. Frank

Gerhard Mercator Universität - GH - Duisburg, FB9, FG Meß- und Regelungstechnik
Bismarckstraße 81 (BB), D-47048 Duisburg, GERMANY

Abstract: The robustness issue in model-based diagnosis of process faults is addressed by means of parameter estimation. System identification is formulated as a problem of multiobjective optimization. The solution is based on parallel evolutionary algorithms of genetic type. Process coefficients are directly identified by a generalized on-line procedure. A simplification of the stage of symptom evaluation results. Application to a laboratory process is included. A diagnosis subsystem is designed to detect incipient faults in the components of a three-tank system. *Copyright © 1998 IFAC*

Keywords: fault diagnosis; robustness; process parameter estimation; multiobjective optimization; parallel genetic algorithms; migration models; three-tank system.

1. INTRODUCTION

The most applied techniques for process fault diagnosis are based on estimation, i.e., parameter identification and observer-based methods (Isermann and Ballé, 1996). They generate analytical symptoms that characterize the state of the monitored system at a given instance of time. The diagnostic performance is dependent on the quality of the system model. Therefore, the important problems are the robustness to system uncertainty and the requirement for sensitivity to faults (Patton, 1994).

Symptom generation based on physical parameters means to use an on-line method of continuous-time parameter estimation. Due to real-time restrictions, the procedures based on generalized equation error are used (Unbehauen and Rao, 1990). Physical coefficients of the process are determined via (nonlinear) relations from the identified parameters of the mathematical model. The weak point of this approach is that, in general, no unique relationships do exist. Moreover, the identification problem may become difficult to solve in case of systems that are nonlinear in the parameters (Frank, 1992).

Stochastic search and optimization methods, based on the principles of natural biological evolution, have lately received considerable interest in engineering systems. Evolutionary algorithms of genetic type can be applied to directly identify physical parameters (Kristinsson and Dumont, 1992). Process data are used to construct an objective function of the parameters to be identified. One way to do the estimation is to find the parameters such that the area of the difference between the actual output of the process and of the model, respectively, is the smallest over a time window. The main drawback is that the predicted output of the system is obtained by means of simulation. This is not suitable for on-line applications.

The present study extends the results of the previous work of research (Marcu, 1997). It is related to the estimation of physical coefficients of a process, by means of Multiobjective Optimization based on Evolutionary Algorithms (MOEA). The paradigm of parallel Genetic Algorithms (GAs) is here accommodated. The applicability of suggested methodology is investigated with respect to the design of a model-based subsystem for diagnosing faults in the components of a three-tank system.

2. PROBLEM OF MULTIOBJECTIVE OPTIMIZATION

The methods based on generalized equation error transform a process model described by a system of differential equations into a system of linear algebraic equations. This is achieved by means of a Linear Dynamic Operator (LDO). An estimation model results. The action of LDO is usually implemented by means of a linear filter. Thus, the filtered signals of the process and their derivatives are expressed in terms of the state variables of that filter (Isermann, *et al.*, 1992).

One suggests to formulate the objective function to be minimized by the GA, in terms of the estimation model (Marcu, 1997). One illustrates this approach by considering a process model consisting of nonlinear state equations given in sensor coordinates:

$$\mathrm{d}\mathbf{x}(t)/\mathrm{d}t = \mathbf{f}(\mathbf{x}(t),\mathbf{u}(t),\mathbf{p}), \quad \mathbf{y}(t) = \mathbf{C}\cdot\mathbf{x}(t), \quad (1)$$

where $\mathbf{u}(t)$ is the input vector, $\mathbf{x}(t)$ is the state vector, $\mathbf{y}(t)$ is the output vector, $\mathbf{f}$ is a nonlinear vectorial function, $\mathbf{C}$ is a real matrix, $\mathbf{p}$ is the vector of process coefficients, and t denotes the time variable. For the sake of simplicity, one considers $\mathbf{y}(t)$=$\mathbf{x}(t)$.

Each state equation:

$$\mathrm{d}y_i(t)/\mathrm{d}t = f_i(\mathbf{y}(t),\mathbf{u}(t),\mathbf{p})$$

is transformed into an algebraic equation in terms of outputs of LDOs, namely $\mathbf{y}_{\mathrm{F},i}(t)$, $\mathbf{y}_{\mathrm{F}}(t)$ and $\mathbf{u}_{\mathrm{F}}(t)$. They correspond to $y_i(t)$, $\mathbf{y}(t)$ and $\mathbf{u}(t)$, respectively:

$$\varphi_i(\mathbf{y}_{\mathrm{F},i}(t)) = \gamma_i(\mathbf{y}_{\mathrm{F}}(t),\mathbf{u}_{\mathrm{F}}(t),\mathbf{p}). \quad (2)$$

The outputs of LDOs are sampled at given instances of time over an observation period, $t_n \in [0,T]$, n=0,1,...,k,...,N. Thus, one obtains an objective function for each equation of the model:

$$O_i(t_k,\hat{\mathbf{p}}) := \sum_{j=0}^{w-1}[\varphi_i(\mathbf{y}_{\mathrm{F},i}(t_{k-j})) - \gamma_i(\mathbf{y}_{\mathrm{F}}(t_{k-j}),\mathbf{u}_{\mathrm{F}}(t_{k-j}),\hat{\mathbf{p}})]^2, \quad (3)$$

where w denotes the window length, $k \geq w$, and $\hat{\mathbf{p}}$ represents the estimated vector of process parameters from the search space Ω. In this way, the identification of physical coefficients characterizing the model (1) represents a problem of multiobjective optimization:

$$\hat{\mathbf{p}}^*(t_k) := \min_{\hat{\mathbf{p}} \in \Omega}\{\mathbf{O}(t_k,\hat{\mathbf{p}})\}, \quad \mathbf{O}(t_k,\hat{\mathbf{p}}) := [O_i(t_k,\hat{\mathbf{p}})]_i.$$

3. PARALLEL EVOLUTIONARY ALGORITHMS

Evolutionary Algorithms (EAs) are based on computational models of fundamental processes, such as selection, recombination and mutation (Michalewicz, 1996). An algorithm begins with a collection (population) of parameter estimates, called individuals. Each one of them is evaluated for its fitness in solving a given optimization task. At each iteration (algorithm time-step), the most fit individuals are allowed to mate and bear offspring.

A MOEA procedure seeks to optimize the components of a vector-valued cost function. The solution is not a single point, but a family known as the Pareto-optimal set (Fonseca, 1995). Each point in this surface is optimal in the sense that no improvement can be achieved in one component of the objectives' vector, without degradation in at least one of the remaining components. The process of optimization is seen as the result of the interaction between an artificial selector, the Decision Maker (DM), and an evolutionary search process of genetic type. Thus, the search process generates a new set of candidate solutions, according to the utility assigned by the DM to the current set of candidates. This approach permits to accommodate goal and priority information available from problem formulation.

GAs may be parallelised in a number of ways (Chipperfield and Fleming, 1995). The migration or island models divide the population into a number of subpopulations. Each one of them is treated as a separate breeding unit under the control of a conventional GA or, more general, of MOEA. From time to time, migration of individuals occurs, such that the individuals from one (or more) subpopulation(s) are introduced into another one. The process of migration limits how much genetic diversity can occur in the global population. This pattern is determined by the number of individuals migrated, the intervals between migration, and the migration paths between subpopulations.

3.1 Parallel MOEA procedure

Following the work carried out by Fonseca (1995), the next procedure is adopted and sketched out in the sequel. It is based on a "real-coded GA" that is characterized by a direct floating-point representation of the parameters to be optimized (Michalewicz, 1996). The following stages are included:

Initialization:
- Create initial real-valued subpopulations;
- Calculate objective functions for subpopulations;
- Calculate goals for subpopulations;
- Calculate Pareto-ranking values for subpopulations;
- Select initial best individual(s);

Parallel procedure loop:
loop over a number of *epochs*
  Evolutionary loop:
  loop over a number of *MAXGEN* generations
    GA Stage:
    - Fitness assignment to whole population;
    - Select individuals from subpopulations (Stochastic Universal Sampling);
    - Recombination and mutation;

- Calculate objective functions for offspring;
- Calculate Pareto-ranking values for offspring;
- Insert best offspring in each subpopulation replacing worst parents (Elitist Strategy);

DM Stage:

- Adapt goals;
- Calculate Pareto-ranking values for new subpopulations;
- Select best individual(s) in each subpopulation;

Exit:

- Determine best individual(s) over all performed generations and subpopulations.

IF remainder(*epoch*:*MIGGEN*)=0,

Migration (fitness-based):

- Send emigrants;
- Receive imigrants;

DM Stage:

- Adapt goals;
- Calculate Pareto-ranking values for new subpopulations;
- Select best individual(s) in each subpopulation;

The main process consists of a number of iterations, called epochs. During an epoch, each subpopulation evolves under a sequential EA loop for a certain number of *MAXGEN* generations. At every *MIGGEN* epochs, each subpopulation exchanges a specific number of individuals. The worst individuals are replaced by the best imigrants. The adaptation of goal values is achieved such that the search is uniformly directed towards the middle region of the trade-off surface of objectives (Marcu, 1997).

3.2 System identification for fault diagnosis

Using a Parallel MOEA procedure (P_MOEA), one performs an on-line (recursive) identification of process coefficients. An individual consists of *P* parameters to be identified. The best individual found over all subpopulations represents the pattern that is further used in the active stage of symptom evaluation. The objective functions may have the same or different levels of priority, depending on the model to be identified. Therefore, the quality of individuals within populations and over all subpopulations, respectively, is determined with respect to the goals and assigned priorities.

The procedure may start with an initial population consisting of best individuals found in off-line (non-recursive) identification. They correspond to a normal behaviour of the process. In the latter case, the window length is $w=N$, and $t_k=T$, Eq.(3). Moreover, the procedure loop coincides with the evolutionary one. Migration occurs at every *MIGGEN* generations. For recursive identification, the evolutionary loop is run through a small number of *MAXGEN* generations (3-5), for each window of process data.

If many classes of process behaviour are available, including faulty ones, the P_MOEA procedure is of particular interest to apply. In this case, one suggests to initialize periodically each subpopulation with best individuals found in off-line identification of each known class, respectively.

All subpopulations will approximately evolve to the same solution. The comparison between the final solution and the initial subpopulations leads to fault detection and isolation. A simple criterion of decision based on distance is to be used. A fault is only detected if the final solution is different from the initial ones. For the latter case, the obtained results will constitute a new subpopulation to be used for further fault diagnosis.

4. APPLICATION

Component fault diagnosis is investigated with application to a three-tank system. One uses the SIMULINK/MATLAB programming environment (MathWorks, 1996), and the standard Genetic Algorithm Toolbox (GAT) (Chipperfield, *et al.*, 1994). An additional block library allows for the real-time communication between the programme and the plant. In order to implement the P_MOEA procedures, new functions for GAT have been developed.

Although the dynamic modelling of a three-tank system is relatively simple, the resulted nonlinear model is only a limited approximation. This is evidenced in (Wünnenberg, 1990). An advanced technique of diagnosis is there applied, namely the Unknown Input Fault Detection Observer (UIFDO). For instance, small leaks in tanks can not be detected. Such type of faults shows exactly the same response in residuals, as an inaccuracy in the description of the incoming mass flows.

Moreover, parameter estimations, by means of the well-known recursive algorithms of Discrete Square-root Filtering (Isermann, *et al.*, 1992), failed to converge to reliable results for the purpose of symptom evaluation. For a given class of process behaviour, the estimated coefficients vary with the inputs of the process.

4.1 Process description

The experimental setup "Three-Tank System" (amira, 1993) consists of three cylindrical tanks, T_1, T_3, T_2, with identical cross sections, A, being filled with water (Fig.1). The tanks are interconnected by circular pipes with identical cross sections, S_n. All three tanks are equipped with piezo-resistive transducers for measuring the level of the liquid.

An appropriate strategy implemented on a microcomputer is used to control the water inlet by two pumps. The volume flows of lateral tanks (the two process inputs, $q_1(t)$ and $q_2(t)$) are controlled such that the liquid level in the corresponding tanks (two out of three process outputs, $h_1(t)$ and $h_2(t)$) can be independently preassigned. The third output of the process, that is the level $h_3(t)$ in the middle tank, is always a resulted signal that is uncontrollable.

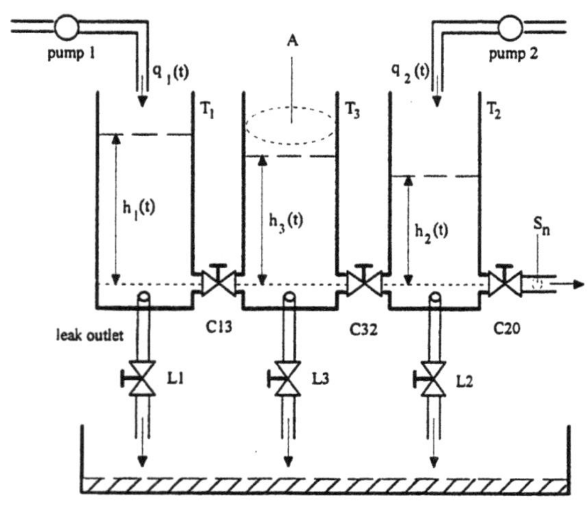

Fig. 1. The Three-Tank System DTS200.

The system can be modelled conveniently by the mass balances of the tanks. The model used by the control strategy is adopted for parameter estimation as well. The physical coefficients taken into consideration are the cross section areas of the tanks (A_T_1,A_T_3,A_T_2), the cross section areas of connecting pipes ($S_n\_T_1$,$S_n\_T_3$), and the cross section area of the outlet pipe ($S_n\_T_2$).

One considers the following notations:

$$p_1 := A\_T_1,\quad p_2 := S_n\_T_1,\quad p_3 := A\_T_3,$$
$$p_4 := S_n\_T_3,\quad p_5 := A\_T_2,\quad p_6 := S_n\_T_2;$$
$$s_1(t) := q_1(t),\quad s_2(t) := q_2(t),$$
$$s_3(t) := h_1(t),\quad s_4(t) := h_3(t),\quad s_5(t) := h_2(t),$$
$$s_6(t) := az_1 \mathrm{sign}(h_1(t) - h_3(t))\sqrt{2g|h_1(t) - h_3(t)|},$$
$$s_7(t) := az_3 \mathrm{sign}(h_3(t) - h_2(t))\sqrt{2g|h_3(t) - h_2(t)|},$$
$$s_8(t) := az_2\sqrt{2gh_2(t)},$$

where $p_i, i = \overline{1,6}$ denote the physical parameters, g represents the gravity constant, sign stands for the signum function, and $az_i, i = \overline{1,3}$ denote the outflow coefficients. The latter are determined by an experimental procedure (amira, 1993). With these notations, the model is described by the following equations:

$$\begin{aligned} ds_3(t)/dt &= (1/p_1)s_1(t) - (p_2/p_1)s_6(t), \\ ds_4(t)/dt &= (p_2/p_3)s_6(t) - (p_4/p_3)s_7(t), \\ ds_5(t)/dt &= (1/p_5)s_2(t) + (p_4/p_5)s_7(t) - (p_6/p_5)s_8(t). \end{aligned} \quad (4)$$

The connecting pipes and tanks are additionally equipped with manually adjustable valves and outlets for the purpose of simulating clogs and leaks. Seven classes of process behaviour are taken into consideration. They are the following:

<NB>: normal behaviour (valves C13,C32,C20 are open; outlets L1,L3,L2 are closed);
[L1]: leakage in tank T_1;
[L3]: leakage in tank T_3;
[L2]: leakage in tank T_2;
[C13]: clogging in pipe between tanks T_1 and T_3;
[C32]: clogging in pipe between tanks T_3 and T_2;
[C20]: clogging in outlet of tank T_2.

For the experiments, the reference values of the liquid levels are changed pulsewise with different magnitudes and durations for each controlled tank, respectively. A test period of 400s is considered. The control strategy works at a sampling rate of 0.1s. The abnormal classes correspond to incipient faults that are artificially generated at t=125s, while the reference values of the system are firstly changed at t=150s. The input-output data of the process are sampled at every 1s during the test period. The observation period for identification starts at t_0=100s.

4.2 Evolutionary system identification

One considers the approach of Poisson Moment Functional with minimal order for continuous-time models (Unbehauen and Rao, 1990). For the model given by Eq.(4), the outputs $M_0[\cdot]$ and $M_1[\cdot]$ of a cascaded filter are evaluated every 1s during the experiment. The filter has identical stages of first order. Each element has a transfer function equal to $1/(s+\lambda)$.

The nonlinear functions $s_j(t), j = \overline{1,8}$ are numerically generated, based on the measured variables of the process. The filtering is achieved by numerical integration. It is based on the discretization of state equations of the Poisson filters. Their input signals are linearly interpolated between sampling times (Isermann, *et al.*, 1992). The outputs of LDOs are used to compute recursively the objectives' values required by the P_MOEA procedure, Eqs.(2)-(4):

$$\varphi_i(t_k) := M_0[s_{i+2}(t)]_{t_k} - \lambda M_1[s_{i+2}(t)]_{t_k} - t_k e^{-\lambda t_k} s_{i+2}(t_0),$$
$$i = \overline{1,3};$$

$$\gamma_1(t_k,\mathbf{p}) := (1/p_1)M_1[s_1(t)]_{t_k} - (p_2/p_1)M_1[s_6(t)]_{t_k},$$
$$\gamma_2(t_k,\mathbf{p}) := (p_2/p_3)M_1[s_6(t)]_{t_k} - (p_4/p_3)M_1[s_7(t)]_{t_k},$$
$$\gamma_3(t_k,\mathbf{p}) := (1/p_5)M_1[s_2(t)]_{t_k} + (p_4/p_5)M_1[s_7(t)]_{t_k} - (p_6/p_5)M_1[s_8(t)]_{t_k},$$
$$\mathbf{p} := [p_i]_{i=\overline{1,6}}.$$

Thus, the following objective functions are obtained:

$$O_1(t_k,\mathbf{p}) := O_1(t_k, p_1, p_2),$$
$$O_2(t_k,\mathbf{p}) := O_2(t_k, p_2, p_3, p_4),$$
$$O_3(t_k,\mathbf{p}) := O_3(t_k, p_4, p_5, p_6).$$

On the basis of these dependencies, the following approaches to priority assignment are considered:

- all objectives have the same priority;
- objectives O_1 and O_3 have the same priority of low-level, and objective O_2 has a high-level priority;
- objective O_2 has a low-level priority, and objectives O_1 and O_3 have the same priority of high level.

For the evolutionary procedures, the following recombination operators are considered (Chipperfield, *et al.*, 1994): discrete recombination, intermediate recombination, and linear recombination. The non-uniform mutation operator of the breeder GA is further considered (Chipperfield, *et al.*, 1994; Michalewicz, 1996). The effect is akin to that of the forgetting factor in classical recursive identification.

Three types of migration topologies are under investigation. They are the ring topology, the neighbourhood migration, and the unrestricted migration (Chipperfield and Fleming, 1995). For the first approach, individuals are transferred between directionally adjacent subpopulations. A similar strategy is the second one. Migration is made only between nearest neighbours and may occur in either direction between subpopulations. Finally, in the unrestricted migration individuals may migrate from any subpopulation to another one.

4.3 Experimental results

Non-recursive identification. It has been performed, in order to obtain initial estimations of parameters corresponding to different classes of process behaviour. The initial population has been created at random. The parameter space has been defined by $A\_T_i \in [100,200]$ cm^2 and $S_n\_T_i \in [0,1]$ cm^2. The theoretical values of parameters are $A=150$ cm^2 and $S_n=0.5$ cm^2.

The best results of MOEA procedure have been obtained after 300 generations. The best results of P_MOEA procedure have been obtained after 100 generations. In the latter case, seven subpopulations have been evolved. Each subpopulation has had $P=6$ individuals. The other parameters of the evolutionary search that have led to the best results are: objective functions with the same level of priority, discrete recombination, *MIGGEN*=*P*=6, neighbourhood topology, and two emigrants/imigrants. For each faulty class, a different pattern of change has been obtained.

Recursive identification. Firstly, MOEA and P_MOEA procedures have been compared, with respect to the identification of physical parameters corresponding to the normal behaviour of the process. The window length has been chosen as w=25s. The MOEA procedure has used a single population of $2P$=12 individuals (Marcu, 1997). The P_MOEA procedure has used seven subpopulations of P=6 individuals. The initial population of both methods has been based on the best individuals found in non-recursive identification. Irrespective of the procedure used, the best results have been obtained by using objectives with different levels of priority, i.e., the second approach presented previously. Other parameters of the evolutionary search are: intermediate recombination, ring topology, *MAXGEN*=3, *MIGGEN*=6, two emigrants/imigrants. The general conclusion is that P_MOEA procedure has converged faster than MOEA.

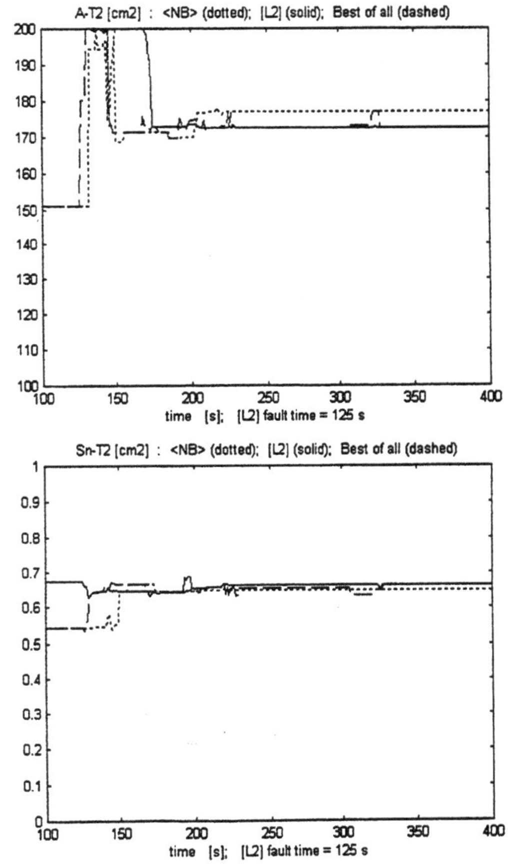

Fig. 2. Results of P_MOEA recursive identification, case of leakage in tank T_2: variables of best individual of <NB> initialized subpopulation, best individual of $[L_2]$ initialized subpopulation and best individual over all subpopulations.

Secondly, P_MOEA procedure has been used in order to perform fault detection and isolation. Each subpopulation has been initialized with identical individuals corresponding to one out of seven classes of process behaviour, respectively. For each class, the mean value of best individuals obtained in non-recursive identification has been setted. The estimated parameters have changed after the appearance of a fault. Constant values of process coefficients have been obtained after a relatively short period of time. In this way, reliable patterns are obtained for the purpose of symptom evaluation. Figures 2 and 3 show typical results corresponding to faulty behaviours of the process.

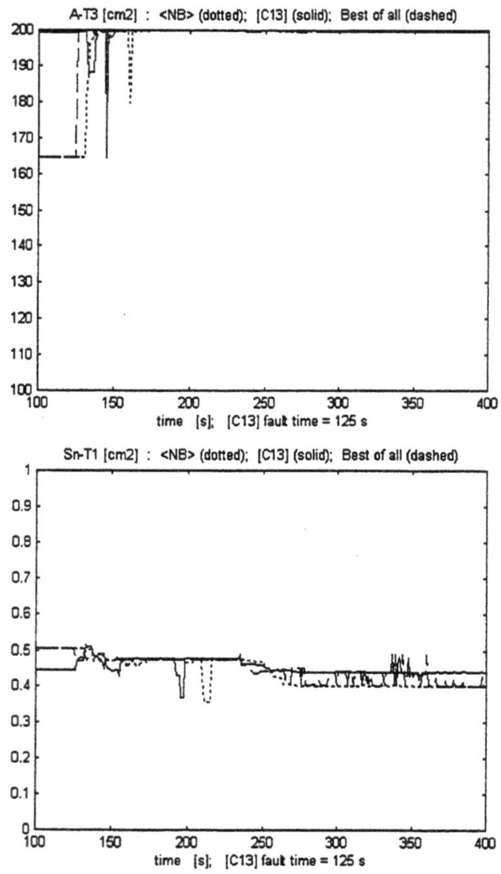

Fig. 3. Results of P_MOEA recursive identification, case of clogging between tanks T_1 and T_3: variables of best individual of <NB> initialized subpopulation, best individual of $[C_{13}]$ initialized subpopulation and best individual over all subpopulations.

5. CONCLUSIONS

The present paper investigates the applicability of evolutionary algorithms of genetic type to fault diagnosis based on parameter estimation. A generalized procedure of multiobjective optimization is suggested. This extension uses the concept of parallel genetic algorithm.

Better performances of a fault diagnosis subsystem have been obtained. Incipient faults in the components of a three-tank system are detected and isolated. They correspond to mass flows of about 15-25 ml/s. A previous approach to the robustness problem, based on a bank of UIFDOs, detected only accentuated faults (Wünnenberg, 1990). They corresponded to mass flows of about 40 ml/s.

In the suggested approach, additional parameters are introduced to control the evolutionary search. Thus, the parallelisation achieved by using the migration models is however slightly more complex to use. Further investigation has to be done to optimize the real-time implementation. It is to be realized in such a manner that the different stages of fault diagnosis are conveniently overlapped with the control of the process.

Acknowledgement. The first author gratefully acknowledges the support of this research by the Alexander von Humboldt Foundation, Germany.

REFERENCES

amira (1993). *Laboratory Experiment Three-Tank System.* amira GmbH, Duisburg, Germany.

Chipperfield, A.J., P.J. Fleming and C.M. Fonseca (1994). Genetic Algorithm Tools for Control Systems Engineering. Proc. of *1st Int. Conf. on Adaptive Computing in Engineering Design and Control*, pp.128-133, Plymouth, U.K.

Chipperfield, A.J. and P.J. Fleming (1995). Parallel Genetic Algorithms. In: *Handbook of Parallel and Distributed Copmputing* (A.Y.Zomaya (Ed.)), pp.1034-1059. McGraw Hill, New York.

Fonseca, C.M. (1995). *Multiobjective Genetic Algorithms with Application to Control Engineering Problems.* Ph.D. thesis, University of Sheffield, U.K.

Frank, P.M. (1992). Principles of Model-Based Fault Detection. Prep. of *IFAC/IFIP/IMACS Int. Symp. on Artificial Intelligence in Real-Time Control*, pp.363-370, Delft, The Netherlands.

Isermann, R., K.H. Lachmann and D. Matko (1992). *Adaptive Control Systems.* Prentice-Hall Int. Ltd., London.

Isermann, R. and P. Ballé (1996). Trends in the Application of Model Based Fault Detection and Diagnosis of Technical Processes. Prep. of *13th World Congress of IFAC*, Vol.N, pp.1-12, San Francisco, CA.

Kristinsson, K. and G.A. Dumont (1992). System Identification and Control Using Genetic Algorithms. *IEEE Trans. on Systems, Man, and Cybernetics*, **22**(5), 1033-1046.

Marcu, T. (1997). A Multiobjective Evolutionary Approach to Pattern Recognition for Robust Diagnosis of Process Faults. Prep. of *IFAC Symp. on Fault Detection, Supervision and Safety for Technical Processes*, Vol.2, pp.1193-1198, Hull, U.K.

MathWorks (1996). *SIMULINK User's Guide.* The Matworks Inc., Natick, MA.

Michalewicz, Z. (1996). *Genetic Algorithms + Data Structures = Evolution Programs.* Springer-Verlag, Berlin.

Patton, R.J. (1994). Robust Model-Based Fault Diagnosis: the State of the Art. Prep. of *IFAC Symp. on Fault Detection, Supervision and Safety for Technical Processes*, Vol.1, pp.1-24, Espoo, Finland.

Unbehauen, H. and G.P. Rao (1990). Continuous-Time Approaches to System Identification. *Automatica*, **26**(1), 23-35.

Wünnenberg, J. (1990). *Observer-Based Fault Detection in Dynamic Systems.* Fortschritt-Berichte VDI, Reihe 8, Nr.222. VDI-Verlag, Düsseldorf.

A GENERIC FAULT PROPAGATION MODELING APPROACH TO ON-LINE DIAGNOSIS AND EVENT CORRELATION

Gregory M. Stanley and Ramesh Vaidhyanathan
Gensym Corporation, 1776 Yorktown, Suite 700, Houston, TX, 77056 (USA)

Abstract: CDG (Causal Directed Graph) provides a methodology and framework for real-time fault management in large-scale systems, addressing the full life cycle of problem identification based on symptoms, diagnostic testing, and fault isolation, through recovery, as well as protecting the operator from "alarm flooding". It is based on generic fault propagation models, tied to an object-oriented domain representation and scalable algorithms. CDG combines the generality of FMEA models with on-line, asynchronous event correlation and diagnosis. The architecture of CDG is described and the modeling approach is discussed with examples. Event correlation and interactive diagnosis using CDG is illustrated through a nitric acid cooling system example. *Copyright © 1998 IFAC*

Keywords: diagnosis, fault propagation model, digraph, knowledge engineering, automation, real-time systems, failure and recovery, FMEA, test planning

1. Introduction and Motivation

Operators of large systems are constantly bombarded by events in the form of messages and alarms, often resulting from a small number of root cause problems. Operators cannot adequately analyze all these events in real-time. Once events are detected, or arrive from an external system, operators need a system to:

(1) Filter out redundant events so attention can be focused on truly new information.
(2) Correlate events - recognize and summarize or group related messages for presentation to the operator, even if the root cause is not determined.
(3) Diagnose the root causes of problems, based on incoming symptoms and test results, automatically choosing tests to be run proactively in cost order.
(4) Run automated or guided manual tests as soon as there is evidence of a problem.
(5) Execute automated or guided manual corrective actions to address symptoms or fix problems.
(6) Communicate results to users and to other computer applications.

It is evident that fault management goals for the operations environment include more than just detection and isolation. These goals are partly addressed by alarm management, event correlation, or abnormal situation management applications. However, diagnostic testing and taking corrective actions in those applications are usually ad hoc.

In addition, application developers need to provide these capabilities quickly and reliably, and build re-usable applications. For practical deployment, installation must require minimal on-site manual customization. Automatic reconfiguration is also needed, to account for changes in equipment, topology, or operating mode.

CDG (Causal Directed Graph) provides a methodology and framework for real-time fault management in large-scale systems. CDG addresses these goals with a model-based framework to standardize development using generic (class-based) fault propagation models. CDG automatically looks at secondary data or runs more elaborate tests only when problems are suspected based on symptoms. CDG manages operations in domains as diverse as process plants, communications networks, and enterprise-wide software applications.

2. Basic Modeling Elements

A fault propagation model in CDG describes the propagation of failures via potential failure paths in the system by modeling how fault events will cause other symptom events and test-result events. It is a directed graph (digraph) model, with the nodes representing fault, symptom and test events, and the arcs connecting the nodes representing the "cause and effect" relationships or dependencies among these events. An arc from an event node "A" to event node "B" means, "if event A occurs then event B will occur (after the time delay specified in the arc)" or "event A causes event B (after the time delay specified in the arc)". The events represented in the model can take the values, "true", "false" or "unknown".

A "Fault" is an underlying independent root cause problem. Faults have associated corrective or mitigation actions specified by procedures that can be started against a specific domain object. Actions can be defined to execute when a fault is suspected as a cause of the observed symptoms and when a fault is confirmed. For a fault event, a true value indicates belief that the fault has occurred and a false value indicates belief that the fault did not occur. A fault may be "suspect" if it is a possible cause of observed symptoms.

A "Symptom" is an effect of underlying faults in the monitored system. A measured symptom arrives at CDG, unsolicited asynchronously from external systems. An unmeasured symptom status is used for failed sensors, and convenience in modeling, to represent those effects that are not measurable, but are important for fault propagation. For a symptom event, a true value means that the symptom has occurred and a false value means that the symptom did not occur within the time delays specified in the arcs.

A "Test" is an observable effect of faults in the monitored system, like a measured symptom. But, unlike symptoms, the observation can be requested at any time. Test results arrive asynchronously (and possibly unsolicited), just like a symptom. Tests have an associated set of actions applied to the monitored system, specified as an external procedure name that can be started against a specific domain object. Upon request, after some indeterminate amount of time, the result of the test is returned to CDG as a truth-value. Test procedures may be fully automated, may simply be a request to an operator, or a combination of the two. The true or false value for a test indicates whether the test passes or fails, respectively. The notion of "test" is powerful and general, (Simpson and Sheppard, 1994) and can imbed arbitrarily complex analysis and actions, as long as it returns a single truth-value.

3. Limitations of Other Diagnostic Techniques

Passive Symptom Monitoring

While desirable, passive symptom monitoring is not always good enough for fault isolation. Human troubleshooters know better, performing lab tests, visual checks, or device built-in-tests. They change controller tuning, put control loops in manual mode, or run a step test. The distinction between tests and symptoms can be used to improve scalability. In large systems, it is impractical to monitor every variable regularly and often. Instead key variables are monitored often, generating symptoms. Once an initial symptom indicates a problem, additional variables can be examined as tests to complete diagnosis. Unless tests are represented in a model linking them to faults, a diagnostic system can't schedule tests, so tests are ad-hoc. In addition, tests need to be ranked based on cost factors such as disruption of the process, resource cost, or automated vs. manual testing.

Static Pattern Matching and Compiled Models

Static pattern matching (classification) includes neural net classifiers, Case Based Reasoning (CBR), and simple rule-based systems. During application development, each failure is hypothesized, and all expected symptom values are determined - a failure "signature". At run time, the pattern matcher determines the fault(s) with the signature closest to the observed symptom vector. This captures model knowledge, but is not robust enough to account for major dynamic changes in the monitored system, such as switch positions, controller modes, etc.

Unfortunately, large systems change often and developing separate pattern matchers for each possible combination of operating modes or topology wouldn't scale up. A model-based approach is required. However, if the models are compiled for use by pattern matchers, recompilation will still be needed with each change. This introduces delays, and interrupts ongoing diagnosis.

Systems requiring training from live data (e.g. neural nets or CBR) have additional problems in handling large systems, including generalization of results from failures involving symptoms in multiple objects, learning rarely-occurring faults, and deciding when old patterns are no longer relevant.

Time Window Problem

Static pattern matchers require periodic processing of a "snapshot" of data or events captured within a time window. Calculations are started fresh with each time window - results of previous analyses are not considered. Time delays between fault occurrence and symptom arrival cause problems. With only partial symptoms present, diagnostic conclusions can be wrong. Tuning the time window size to balance misdiagnosis vs. timely results is difficult. CDG processes individual events immediately as they arrive asynchronously, and event values are remembered until overwritten or timed out, without the need for a time window.

Reasoning with Missing Symptoms

While using pattern-matching techniques, the absence of a symptom event may be taken as evidence that the underlying symptom really isn't present (value of false), hence associated possible faults aren't present. That can lead to incorrect diagnosis if there are time delays between the fault and the symptom, or communications problems. Evidence based on missing symptoms should not be used until after the worst-case time delay. Additionally, when large delays are present, it is desirable to offer preliminary diagnosis before all symptoms have arrived. In that case, the only values that should be used are those for symptoms which have already arrived.

Single Failure Assumption

Many techniques assume there is a single failure. This is unrealistic in any large system where multiple faults have already occurred and previous faults continue to cause symptoms that may overlap with new problems. Static pattern matchers will fail to handle these situations, since a combination of faults leads to a pattern of symptoms far from any single fault's failure signature. Encoding all 2- or 3-failure combinations does not scale up. Instead, users must partition the system into many small diagnostic subsystems, so that only one failure is likely in a given subsystem.

Other Modeling Approaches

Process models are often based on algebraic or differential equations, or qualitative versions such as Signed Directed Graph (SDG) which represent the propagation of process variable deviations. These models do not represent test procedures, relay or PLC logic, frequency domain information, discrete operational modes (controller modes, switch positions), or messages (events) from complex software in alarm systems, shutdown systems, optimizers, or intelligent instruments. "Smart sensor error code 30", "backup power starting", or "no response" messages don't fit the paradigm. CDG shares with SDG an orientation towards asynchronous event processing and propagation of information, instead of pattern matching. Like CDG, many SDG implementations have used generic, class-level models for automating diagnosis (Finch, et al., 1990), and HAZOP analysis (Vaidhyanathan and Venkatasubramanian, 1995).

Bayesian Network (BN) is a powerful modeling technique that can represent generic fault propagation knowledge. With the additional probability information, BN might perform better in cases when there are few symptoms or tests, and where tests are expensive. But BN introduces unscalable computing complexity and is not needed in data-rich environments with inexpensive tests.

Design models for "normal" operation are often used in diagnostic systems. However, faults invalidate design model assumptions. Pattern matches on observed deviations from "normal" can be used for diagnosis, but suffer the problems noted earlier for pattern matching and compiled models. CDG can accommodate such quantitative "normal" models in the definition of the tests and symptoms.

The fault propagation models used in CDG are similar to the information flow models used for model-based design for testability and integrated diagnosis (Simpson and Sheppard, 1994). CDG extends these models for on-line event correlation, interactive diagnosis and fault mitigation. Major extensions were a strong event orientation to handle processing of asynchronous events, handling symptoms as distinct from tests, run time selection of tests dependent on previous results (instead of offline generation of one fixed fault tree for all time), specification of generic models with specific model instantiation at run time, graphical input of models and eliminating the poorly-scaling matrix calculations.

The extended real-time Failure Environment Analysis Tool (FEAT) (Malin, et al., 1992) used "failure state information flow" models which are similar to the fault propagation models used in CDG. But FEAT was limited due to passive monitoring (no tests), lack of generic modeling, the use of compiled models and a complex matrix analysis to identify single and double faults.

Scalability

Scalability (ability to scale a system up to a large number of objects) is a central issue in fault management in large-scale systems. CDG addresses this issue through linear algorithmic complexity, "management by exception" to only instantiate a specific, localized model at run time when initial symptoms indicate a possible problem, models at a high level of abstraction, and an architecture supporting distribution over multiple computers.

4. Architecture

The architecture of CDG is presented in figure 1. The input specifications required for CDG are the Generic Fault Propagation Models and the Specific Domain Model. Using these specifications, CDG correlates incoming symptom and test result events, identifies suspect faults and performs diagnosis by selecting appropriate test and mitigation actions to resolve the suspect faults. The outputs from CDG are the diagnostic conclusions and test and action requests. The various components of CDG are described in detail in the following sections:

Generic Fault Propagation Models (GFPM)

GFPMs consist of the generic cause-effect relations among the fault, symptom and test events in the domain objects. The GFPM of a domain-object class defines the propagation of failures within its instances and to other classes of domain objects via relationships. This is a generic (class-level) "library" of models, independent of any specific topology of domain objects and relationships present in any particular system. The GFPMs can be developed from first-principles models, experience-based knowledge, or FMEA results. GFPMs are specified via the developer GUI graphically.

The event nodes and the arcs in the GFPM can be made conditionally dependent on the states of the domain objects by defining appropriate conditional methods on the nodes and the arcs. These conditional methods are evaluated during event correlation and only the nodes and arcs that are valid or available for the current states of the domain objects are used for correlation. In addition, state dependent methods are defined for the event nodes to calculate the cost of running a test, the cost of fixing a fault and the cost of not fixing a fault. The costs evaluated by these methods are used to rank the candidate tests for diagnosis and the fault mitigation actions.

Specific Domain Model (SDM)

The SDM is an object-oriented model of the specific system being monitored. It might model physical equipment, as well as more abstract entities affecting system performance, such as controllers or software applications. The SDM includes objects representing the monitored domain entities, their connectivity, containment and other relationships. The SDM can be imported from external databases or files or can be specified via the developer GUI.

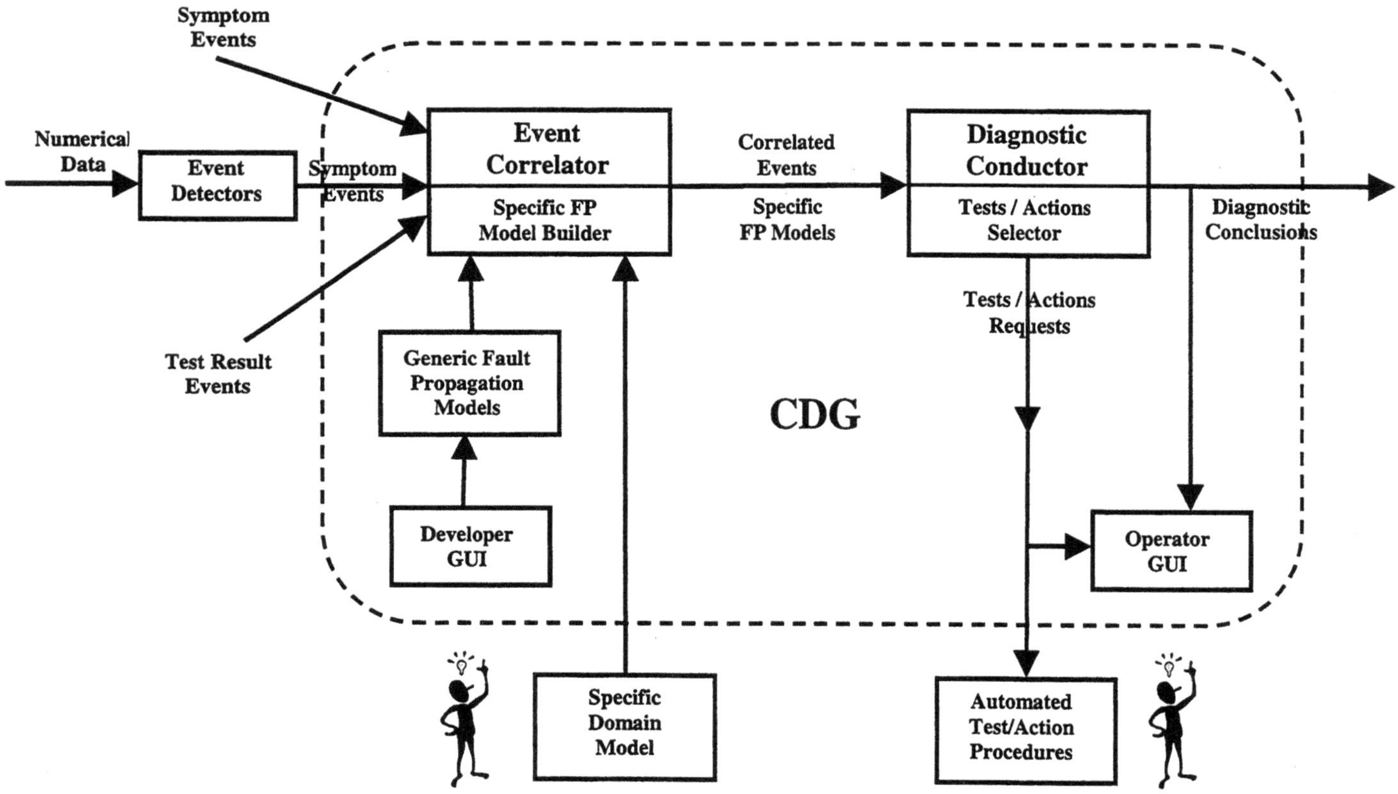

Figure 1. The Architecture of CDG

Event Correlator

The Event Correlator correlates asynchronous symptom and test result events input to CDG using Specific Fault Propagation Models (SFPMs). SFPM is a fault propagation model that describes the propagation of fault, symptom and test events within and across specific domain objects. The SFPM Builder constructs SFPMs at run time starting from the incoming events by appropriately combining the GFPMs and the SDM, just building enough event nodes to account for possible causes and effects of observed symptoms. "Event Detectors" are external applications that monitor and analyze the numerical data trend in the system, detect and generate appropriate symptom events to be input to CDG.

The event correlator recognizes that a group of events are related to each other based on their connectivity criteria in the SFPM such as the existence of a directed path or the fact that the events could be caused by common faults. Then the value of the incoming event is propagated in the SFPM to infer and predict the values of other events and to identify suspect faults. OR logic is used by default for the propagation of event values during correlation. Thus, "the value of a node is true if the value of one (or more) of its inputs is true". Conversely, "if the value of a node is false, then the value of all of its inputs ought to be false". Also using the OR logic, "if the value of an event is true, then all the upstream fault events become suspects". Hence, the event correlator identifies the suspect faults by searching upstream from the incoming symptom and test result events with a true value in the SFPM. This information is input to the Diagnostic Conductor.

During event correlation, if the value of a symptom event is predicted true by propagation and if the actual symptom with a true value does not show up within the time delays specified in the arcs, then the value of the symptom will be reset to false and re-propagated. Thus, the interim event correlation and diagnosis will be consistent with observed events at any time and the final correlation accounts for symptoms that did not occur. The default time delay is "infinite", so by default missing symptoms are not used as evidence. In addition to the OR logic, AND and NOT logic propagation relationships among the events could also be specified in the SFPM.

Diagnostic Conductor

The Diagnostic Conductor resolves the suspected faults by identifying appropriate candidate tests that when executed would provide additional information regarding those faults. These candidate tests are those which are the effects of the suspect faults. The Test Selector identifies these tests by searching downstream from the suspect faults in the SFPM. In the case of an automated test, a request is sent from CDG to execute the automated test procedure. Otherwise, the test is displayed on the operator GUI for approval before execution. The candidate tests are ranked based on cost criteria such as resource use, disruptiveness, or the information value of a test. The test results are asynchronously input back to the event correlator for further correlation to reduce the number of suspect faults.

Based on the correlation of the test results, the suspected faults are either ruled out or concluded to have occurred. Diagnostic conclusions are output from CDG to the operator and to other external systems. Whenever CDG concludes a fault as a suspect or occurred, the Actions Selector will execute appropriate mitigation actions specified for the fault. Similar to the test procedures, these mitigation actions can also be automated procedures or may require operator intervention. These mitigation actions can also be ranked based on criteria such as the failure-rate, the cost of fixing, or the cost of not fixing the faults. The test and mitigation procedures can be defined using OPAC (OPerations expert ACtions) graphical procedure development tool or G2, external to CDG.

5. GFPM of the Controller and Sensor

A part of the GFPM of the controller and sensor are shown in figures 2 and 3, respectively. Each event in the GFPMs has a category attribute that specifies the name of the event and a class attribute that specifies the class of the domain object for which the event is defined. These attribute values are displayed at the side of the event nodes. The square icon nodes represent faults and the circular icon nodes represent symptoms and tests. The smaller circular icon nodes represent symptom-views and test-views, which are the views of symptoms and tests defined for another class of domain object. A propagation-relation that specifies the relation between the domain objects of the event-view node and the event node is defined on the arc connecting these nodes.

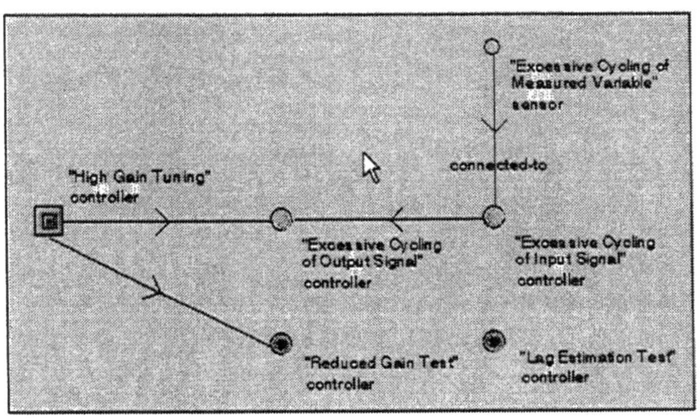

Figure 2. GFPM of the Controller

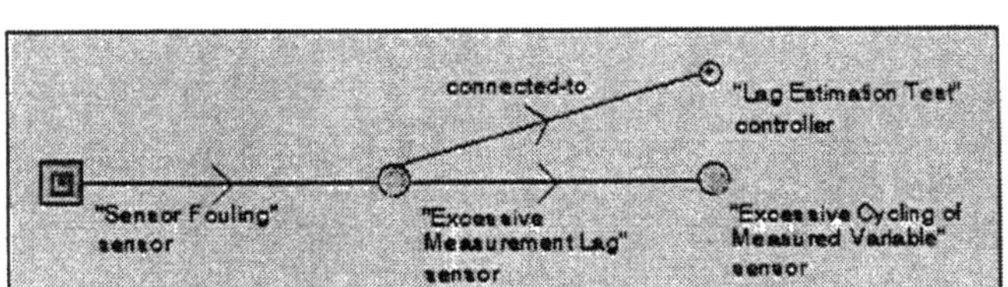

Figure 3. GFPM of the Sensor

In the GFPM of the controller, the fault "High Gain Tuning" represents the controller gain being set too high. This fault could cause the measured symptom "Excessive Cycling of Output Signal" from the controller. In addition, the measured symptom, "Excessive Cycling of Measured Variable" in a sensor connected to the controller could cause the symptom "Excessive Cycling of Input Signal" to the controller, which in turn could cause the measured symptom "Excessive Cycling of Output Signal" from the controller, due to control action.

The "Reduced Gain Test" in the controller is a candidate test for isolating the fault "High Gain Tuning" in the controller. This test involves a series of actions wherein the gain of the controller is reduced and the measured variable is monitored to determine if the cycling of the measured variable dissipates. A true or false result for this test will respectively confirm or rule out the "High Gain Tuning" in the controller as the fault that might have occurred. Mitigation actions such as "lowering the gain of the controller setting" or "switching the controller to manual mode" are defined for this fault.

In the GFPM of the sensor, the fault "Sensor Fouling" represents the accumulation of material around the thermowell of the sensor. This fault could cause the unmeasured symptom "Excessive Measurement Lag" in the sensor, which in turn could cause the measured symptom "Excessive Cycling of Measured Variable" in the sensor. The "Lag Estimation Test" in a controller connected to the sensor is a candidate test for isolating this fault. This test involves a series of actions wherein the controller is switched to manual mode from automatic mode, and then a step response test is performed by changing the manipulated variable and estimating the time-constant and the lag time by monitoring the effect on the measured variable. If the time-constant and the lag time of the sensor measurement are significantly higher than their normal values, the result of this test is true, else false. A true or false result for this test will respectively confirm or rule out the "Sensor Fouling" in the sensor as the fault that might have occurred. Mitigation actions such as "cleaning the thermowell of the sensor" or "replacing the sensor" are defined for this fault.

6. Nitric Acid Cooling System Example

In this section, event correlation and diagnosis using CDG is illustrated through a nitric acid cooling system example SDM shown in figure 4. In this system, a feedback temperature controller controls the temperature of nitric acid to a reactor by manipulating the coolant flow via a flow control valve. In addition to the GFPM of the controller and sensor described in the previous section, generic models for the flow control valve and heat exchanger are defined in CDG. The initial measured symptom, "Excessive Cycling of Measured Temperature" true, in the sensor will be detected by an event detector and this event will be input to CDG. Starting from this symptom, CDG will build the SFPM by appropriately combining the SDM and the GFPMs in the model library.

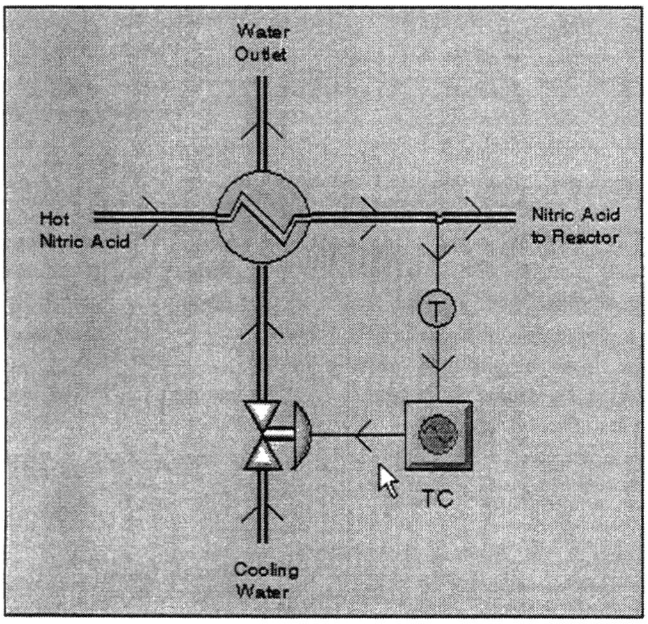

Figure 4. Nitric Acid Cooling System

By propagating the true value of the "Excessive Cycling of Measured Temperature" symptom in the SFPM, the CDG event correlator will identify the faults "High Gain Tuning" in the temperature controller and "Sensor Fouling" around the temperature sensor thermowell as the suspected root causes (assuming that the hot nitric acid inlet temperature is a measured variable and the corresponding possible cause "Excessive Cycling of Hot Nitric Acid Inlet Temperature" was already eliminated). This information will be fed to the CDG diagnostic conductor, which will then identify "Reduced Gain Test" and "Lag Estimation Test" in the controller as the respective candidate tests for resolving the suspect faults in cost order. The operator or automated procedures can then execute these test actions, and the test results are asynchronously input back to CDG. The CDG event correlator will propagate those test results in the SFPM to confirm or rule out the suspect faults.

Depending on the "Reduced Gain Test" result, the fault "High Gain Tuning" in the controller is confirmed or ruled out. Similarly, depending on the "Lag Estimation Test" result, the fault "Sensor Fouling" in the temperature sensor is confirmed or ruled out. Since CDG does not have a single fault assumption, either or both of these faults could be concluded to have occurred depending on the test results. The diagnostic conductor will inform the operator and external systems about the identification of these faults, and select appropriate mitigation actions defined for these faults for execution.

7. Conclusions

A generic fault propagation modeling approach has been developed for automating on-line event correlation and interactive diagnosis. The proposed modeling technique consists of novel concepts such as defining test and mitigation actions as part of the model and in-built state conditional dependencies. Based on this approach, a software product named CDG is currently being developed for real-time fault management in large-scale systems. CDG is part of Gensym's overall Operations Expert (OPEX) product line for managing operations environments.

CDG appropriately combines generic models with specific domain representation and builds focused specific models to investigate observed asynchronous events. Using the specific models, CDG recognizes that a group of events are correlated to each other, identifies suspected faults that could have caused the symptoms, and selects and executes candidate tests and mitigation actions to resolve the problems. CDG also provides graphical user interfaces for the development of the models, and for operator interaction. The test and mitigation actions can be defined using OPAC graphical procedure development tool or G2.

References

Finch, F. E., Oyeleye, O. O., and Kramer, M. A., "A Robust Event-Oriented Methodology for Diagnosis of Dynamic Process Systems", *Comp. and Chem. Engng.*, **14**(12), 1379 (1990).

Simpson, W.R., and Sheppard, J.W., *System Test and Diagnosis,* Kluwer Academic Publishers, Boston (1994).

Vaidhyanathan, R., and Venkatasubramanian, V., "Digraph-Based Models for Automated HAZOP Analysis", *Reliability Engineering and System Safety*, **50**, 33(1995).

Malin, J. T., Schreckenghost, D. L., and Rhoads, R. W., "Extended Real-Time FEAT", *Making Intelligent Systems Team Players: Additional Case Studies*, Section 4., NASA Technical Memorandum 104786, Johnson Space Center, Houston, TX (1992).

AN HYBRID REFERENCE MODEL FOR ON-LINE DETECTION IN DISCONTINUOUS SYSTEMS

J.C. Pascal[1], D. Andreu[2]

LAAS-CNRS 7 Av. du Colonel Roche 31077 Toulouse Cedex 4 France
[1] also at Université Paul Sabatier Toulouse France
[2] LIRMM 161 rue Ada 34192 Montpellier cedex 5 France
email: pascal@laas.fr, andreu@lirmm.fr

Abstract: The purpose of this paper is to present a model for the detection phase in discontinuous systems. This model describes the expected behaviour of the process to be controlled, in order to make a comparison with the real behaviour. The hybrid nature of the process leads to a model which supplies a discrete view describing the process configurations and a continuous view describing the continuous behaviour of the process within a given configuration. The use of this hybrid reference model is the detection of abnormal behaviour and, by the emulation principle, the verification of constraints respect before the application of a command. It can also used to analyse the consequences of a decision before making the process recovery.

Keywords: On-line detection, reference model, hybrid model, discontinuous systems

1. INTRODUCTION

The automation of discontinuous processes, like batch processes for instance, poses difficult issues because it combines continuous and discrete-event aspects (Rippin, 1993).

In the case of pure continuous system, supervisory control is in charge of monitoring the system in order to detect when it is no longer in its optimal state and to act on the local control to compensate it.

When batch processes are involved, the supervisory control has to execute the control recipe, and more precisely, to drive the execution of operation sequences described by a set of recipes. It has then to ensure the real-time management of plants resources. So, it is in charge of the dynamic configuration of the process in order to optimise its functioning according to criteria of its abstraction level. However, it has also to adapt the control when a deviation is detected. This adaptation can lead to modify process plan in interaction with the scheduling level, or to modify operating conditions in interaction with the local control levels. Consequently, it is necessary to build a monitoring function, complementary to the control one, in order to detect deviations (detection phase), to determine the cause (diagnosis phase) and to define a reaction (recovery phase). In this paper, we focus on the detection phase.

The detection of an abnormal situation of the process needs a reference model, which represents the expected behaviour of the process, allowing to make a comparison with the real behaviour. So, monitoring is based on a model of the industrial process which is run in parallel and in real-time during the operations. The same way is used in continuous systems and discrete-event systems.

In the context of continuous systems monitoring, the causal or qualitative modelling is very employed

(Leyval *et al.*, 1994; Bousson and Trave-Massuyès, 1992; Montmain 1994). The use of a process model is all the more necessary since some significant variables can not be reachable to the mesure (Féray-Beaumont and Gentil, 1991).

In the case of discrete-event systems, the monitoring based on a process model are currently used (Combacau and Courvoisier, 1990; Holloway and Krogh, 1990; Toguyeni, 1992). But in some approaches, the detection is based on the verification that the process evolution stays within the set of reachable states. This set is described on the model and it is independent of the applied control. It can not be the same for discontinuous systems monitoring. Indeed, the continuous phenomena to monitor is closely linked with the phase in which the control is.

Let us consider the example of the variation rate of a continuous variable. A great slope on the continuous variable trajectory can be considered as normal in transient regime (starting, functioning change), whereas it is a symptom of a deviation in the stationary regime (racing of an exothermic chemical reaction). Considering the evolution rate of a continuous variable independently of the active phase of the control may lead to indicate a non-existent dysfunction or to ignore an effective one. A simple example is the case of a storage tank in a loading phase whose volume is decreasing. No dysfunction can be reported if any evolution rate of the volume between the maximal input and output flows, and any values of the volume between 0 and V_{max} are considered as normal. Consequently taking only into account the reachable continuous states in order to define the detection of an abnormal behaviour is clearly insufficient. The context (the configuration) in which the control and the process are must be necessarily considered. To correctly represent this hybrid aspect, the reference model must necessarily be of hybrid nature.

2. THE HYBRID REFERENCE MODEL

We have pointed out the close link between the continuous phenomena and the discrete configuration. So then, the reference model is constituted by two linked components: one discrete, the other continuous. Such an approach of hybrid modelling, based on the cooperation of a discrete-event part and a continuous one have already been applied for control (Johnsson and Arzen, 1994) and scheduling purposes (Daubas *et al.*, 1994).

As a consequence, the model must first supply a discrete view, describing the configuration in which the system is. Indeed in a batch operated plant, the process equipment is not in continuous use, units are used sporadically with changes between active, waiting and inactive status. At different instants of time, the plant is then in different configurations, each configuration corresponding to a set of active connections between units. The supervisory control drives then the production system from dynamic configurations to dynamic configurations in order to achieve the fabrication of the product.

The model must also supply a continuous view; this view describing the continuous behaviour of the system within a given configuration. The process behaviour fluctuates depending on the unit operations being processed. So then, the hybrid model associates the set of continuous system behaviours to the set of discrete views. This model, called hybrid reference model, corresponds a qualitative view of the behaviour of the units independently from any control strategy. All the states described on the reference model must be the reachable configurations and this finite set of discrete states must not be restricted by control considerations. It can only be limited by technological constraints.

Thus, a qualitative aspect is associated to the discrete states (the configurations). According to an active configuration due to the beginning of an operation, it is possible to qualitatively determines the expected continuous phenomena evolution. For instance, in a configuration corresponding to a storage tank loading, the volume must be increasing. If it is not the case, a failure is detected.

The hybrid model of the process is then composed of a set of hybrid models: one model by unit. The discrete aspect of each model is represented by means of Petri nets. The places of the Petri net correspond to the configurations of the unit in relation to its environment (connections to others units for example). With each discrete state of the net (each place), a subset of differential algebraic equations is associated (if any operation is performed in the corresponding configuration) (Andreu *et al.*, 1996; Valette *et al.* 1995; Champagnat *et al.* 1998). This subset of equations describes the expected continuous process behaviour. The active unit configuration is characterised by the location of the token in the net (the marking) and its continuous state, described by a subset of continuous variables, is carried by the token. The set of marked places of the nets (one net by unit) corresponds to a logical description of partial states of the whole controlled system.

When the control system evolves from an operating stage to another one, the reference model simultaneously evolves. A transition firing in the control module (the control strategy is also described by means of Petri nets), which corresponds to the execution of a given operation, simultaneously results in the firing of the associated transition in the reference model. The simultaneous evolution of the reference model with the control one ensures that the automatically selected subset of equations is consistent with the operations in progress.

The use of the reference model has multiple objectives. The model must permit the detection of abnormal behaviours in comparison with the expected behaviour for the performed control. Thus, it is used to validate the execution of an operation. Furthermore it is sometimes useful to simulate the process behaviour on a short time horizon. The aim is then to verify that the command to apply will not lead to constraints violation. According to this emulation principle of the reference model, it can be envisaged to analyse the consequences of a decision before making the process recovery.

3. AN EXAMPLE OF HYBRID REFERENCE MODEL

In order to illustrate the reference model, let us consider the storage tank which is described on the next figure.

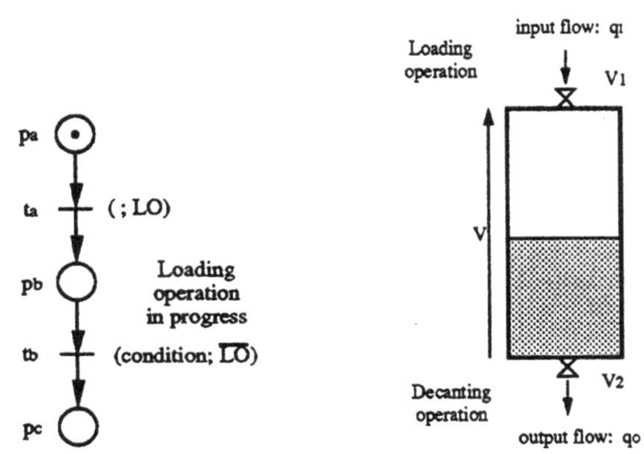

Fig 1: The storage tank example

Two activities can be performed, simultaneously or not, on the storage tank: the loading operation (expressed by LO) and the decanting operation (expressed by DO). End-of-operation conditions depend on the volume to be transferred.

The model of the storage tank (figure 2) is composed of four discrete states (the four places of the Petri net). These states correspond to the following configurations of the storage tank: inactive state (place p1), loading phase (place p2), decanting phase (place p3) and simultaneous loading and decanting phases (place p4). These four discrete states are the reachable configurations of the storage tank. If the use of the fourth configuration is not physically possible, the corresponding state must be removed. But, if this constraint stems from the control strategy, it must be specified on the control model and not on the reference one.

Equations associated to places express the variation of the storage tank volume as shown on figure 2.

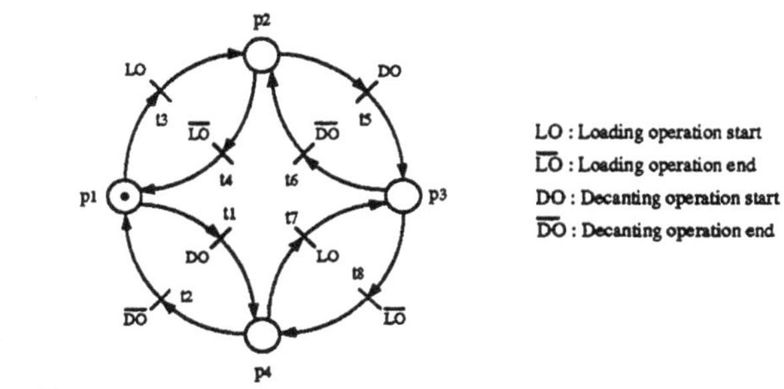

Discrete states

p1: Stable state wait (0)

p2: Loading operation in progress (1)

p3: Loading and decanting operations simultaneously in progress (2)

p4: Decanting operation in progress (3)

Associated set of equations

$$(0)\left|\begin{array}{l}\Delta V_I = 0\\ \Delta V_O = 0\\ \Delta V = 0\end{array}\right.\qquad (2)\left|\begin{array}{l}\Delta V_I = \int q_I\,dt\\ \Delta V_O = \int q_O\,dt\\ \Delta V = \Delta V_I - \Delta V_O\end{array}\right.$$

$$(1)\left|\begin{array}{l}\Delta V_I = \int q_I\,dt\\ \Delta V_O = 0\\ \Delta V = \Delta V_I\end{array}\right.\qquad (3)\left|\begin{array}{l}\Delta V_I = 0\\ \Delta V_O = \int q_O\,dt\\ \Delta V = -\Delta V_O\end{array}\right.$$

with $V(t_n) = V(t_{n-1}) + \Delta V$

Fig 2: Example of a storage tank reference model

When a transition, which corresponds to the start or the end of an operation, is fired on the control net, it synchronously results in the firing of the corresponding transition on the reference model. For instance, the start of the loading operation on the storage tank implies the firing of both transition ta, on control net, and transition t3, on the reference model one. On the latter, the place p2 becomes marked (V1 opened and V2 closed) and the subset of equations (1) is selected.

4. DETECTION OF PROCESS BEHAVIOUR DEVIATIONS

The detection is based on the comparison between the process observation and the simulation of the reference model. A process event, stemming from the process observation, is characterised by an occurrence date and a value.

By means of the reference model we obtain a temporal window which gives the possible dates for the process event occurrence. This temporal window is calculated according to parameters of the control (volume to be transferred, output flow, etc.), by means of the active set of equations describing the continuous phenomena in progress. The structure of the reference model is independent of the control strategy, but the phenomena to be monitored on the model depends on control parameters.

The reference model can not determine the expected value with an absolute precision. The active set of equations permits to determine a range of acceptable values.

The comparison mechanism consists to compare the occurrence date of the process event with the temporal window (comparison between the observed discrete state and the simulated one) and the measured value with the calculated range of values (comparison between the observed continuous state and the simulated one).

Let us consider the example of storage tank where the activity in progress is the loading operation. In control net (figure 1), the place pb is marked. Its output transition tb is enabled and the associated predicate is considered (the occurrence of a particular process event is waited, implying the loading-end-operation). The reference model determines a temporal window [d1,d2] (d1 being the earliest end-of-operation and d2 the latest one) and the expected value of the volume.

If the event occurs out of the temporal window or if the gap between the value measured on the process and the simulated value is upper than a defined threshold, the transition t4 can not be fired. The reference model and consequently the control net can not any more evolve (transition tb can not be fired). It is then possible to conclude that a failure happened. A message is then sent to a diagnosis module.

In the contrary case, it is possible to conclude that the operation has been successfully performed. The loading-end-operation produces the evolution of the control net in a new state (place pc marked). That automatically implies the change of configuration of the reference model (place p1 marked). Moreover, the reference model is updated with the value measured on the process. Thus, it keeps a faithful view of the effective state of the process.

The model used for the representation of the discrete aspect is a high level Petri net: temporal Petri nets with objects. The tokens are objects with attributes which represent a set of informations.

On the control net, the token carries information on the operation to be perform on the considered unit. For instance, on the control net shown figure 3 the pieces of information carried by the token <y> are the parameters of the operation to be executed: the volume to be transferred and the required input flow. On the reference model (figure 4), the attributes of the token <x> are pieces of information on the physical constraints of the unit (maximum volume, maximum input flow) as well as the value of the continuous state variables which characterized the unit. This token also contains attributes to select the subset of equations associated to the state in which the token is, and attributes to describe the end-of-configuration event (the considered variable and the threshold value).

Let us consider that the control net wants to start the loading operation "to transfer 2 m^3 with an input flow of 1 m^3/h". It corresponds to the firing of the transition ta. This operation is authorized if the change of configuration on the reference model is possible (existence of an enabled transition whose the firing leads to this configuration) and if the operation parameters are acceptable. On the reference model, the storage tank is in an inactive configuration, the place p1 is marked and the transition t3 is enabled. Before to fire the transition t3, the required input flow attribute of the control net token <y> is compared with the maximum input flow attribute of the reference model token <x>. If this condition is satisfied, the transition ta and t3 are simultaneously fired. The control net goes on the loading operation in progress state (place p2 marked) and it is waiting for the occurence of the end-loading process event.

On the reference model, the attribute describing the end-of-configuration event on the token <x> is updated (considered variable and volume to be transferred). The place p2 is marked (loading configuration). The subset of equations (1) and the initial and final conditions are loaded into the integrator. The expected end-of-operation date is calculated. To this date is added a temporal slack corresponding to the transient phases of the operation in order to constitute the temporal window [d1,d2]. The ouput transition t4 will stay enabled during this time interval.

The transition tb, on the control net, will be fired if the occurence date of the end-of-operation process event is within the temporal window [d1,d2] and if the measured value is consistent with the simulated one. If it is the case, the token <x> on the reference model is updated with the measured value and the place p1 becomes marked (change of configuration).

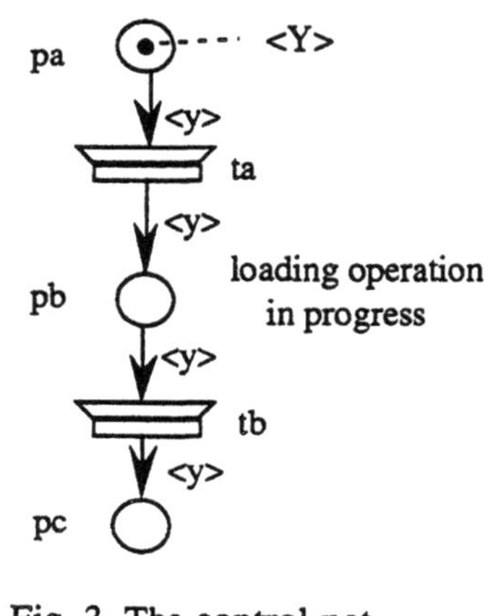

Fig. 3. The control net

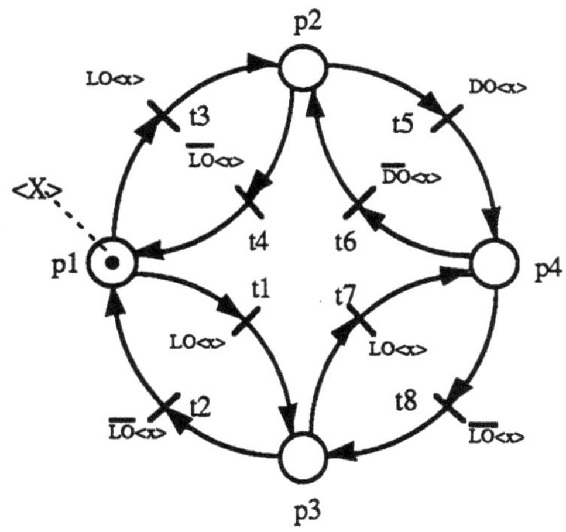

Fig. 4. The reference model.

Now, let us detail a possible mechanism to realize the simultaneous firing of transitions whose principle has been first described in Allanche *et al.*. 1984.

We have seen that the models evolution is initiated by the control net. A transition, on the control net, is broken into two transitions connected to a transition of the reference model by means of places (see the figure 5 which represents the simustaneous firing of the transition ta and t3 of the storage tank example) (Andreu 1996). These places represent channels through which information is exchanged (places containing messages). The control net sends a request and is waiting for a response from the reference model (token <r>). This mechanism guarantees that the evolution of the control net is only performed if it is authorized by the reference model. On the figure 5, the transition ta2 can only be fired if a response is sent by the reference model.

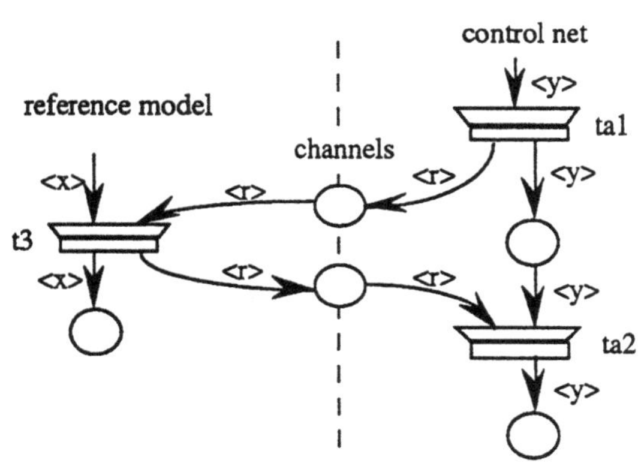

Fig. 5. Simultaneous firing mechanism

5. THE MODEL IN EMULATION

If in the previous case, the control is simultaneously applied to the process and to the reference model in order to detect abnormal behaviour of the process, the control is only applied to the model when it is used in emulation. It is a kind of temporal projection of the process state. The objective is thus to verify that the control to apply respects process constraints. For instance, we could check if the command to be applied does not violate the constraint of maximal volume of a storage tank. That allows to avoid the occurrence of a failure whose cause could have been an erroneous command.

Let us consider 2 recipes which run in parallel and use the same storage tank (figure 6 and figure 7).

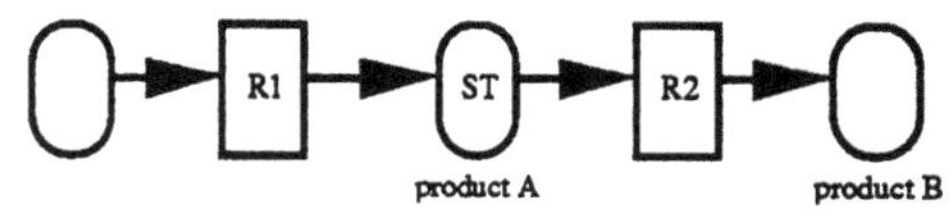

Fig. 6. The flow diagram

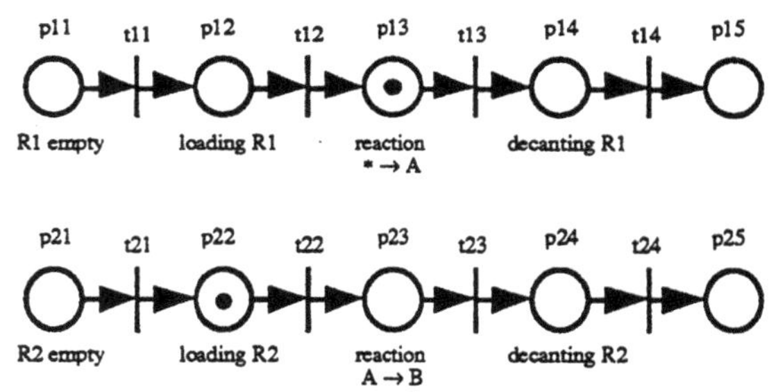

Fig. 7. The recipes

The first one products, within a reactor R1, a solution A to be stored in a storage tank ST. The second one realises a product B, within a reactor R2, by using the solution A which is contained in ST. Then, the product B is stored in an other storage tank. Let us assume that on the first recipe the reaction is going on in R1 and that on the second one the loading phase of R2 is active (figure 7). During its loading, R2 is linked to ST which contains the product. The reference model (figure 2) is in the decanting configuration (place p3 marked).

Now, let us supposed that the treatment in R1 is finished, the solution must be transferred in ST. This transfer phase depends on the continuous and discrete states of ST. This storage tank must be an available resource and the simultaneous loading-decanting configuration must be allowed. Regarding its continuous state, its available volume must permit to receive the batch. While the transition t2 is not fired, the value of the storage tank volume is not updated (the known value is that preceding the loading phase of R2). The continuous state is only known with precision when the configuration changes. So in order to make the decision, that of beginning the transfer phase from R1 to ST for instance, an emulation is necessary. The aim of this emulation is to ensure that a simultaneous loading and decanting of ST (with different flows) will not cause the violation of volume constraint.

For that kind of emulation, two types of marking are considered in the reference model. One token corresponds to the effective state of the refrence model and a second represents the emulated one.

Doing so, the reference model could be used to analyse the consequences of a decision before making the process recovery. One token corresponds to the state before the failure detection. Another one permits to test the process recovery by means of a temporal projection of the process behaviour.

6. CONCLUSION

The use of a reference model whose the nature is hybrid supplies a qualitative view of the process behaviour. Its evolution is linked to the evolution of the control discrete-event model. This ensures an automatic and consistent selection of the continuous model which represents the process behaviour in the active configuration. It permits the detection of abnormal situations. The possibility to emulate it allows the verification of the consistency of a control and the test of the process recovery.

Another use of the reference model, not presented through this paper, is the validation of a production plan. Indeed, the end-of-operation temporal window can be used by the control net to make a comparison with the temporal window supplied by the scheduling level (this latter representing the estimate duration for the operation). Such an application of the reference model before to produce, could avoid to search for the cause of a temporal constraints violation which would certainly have occurred while the production is running.

REFERENCES

Allanche P., Benzakour K., Dollé F., Gillet P., Rodrigues P. & Valette R. (1984). PSI: A Petri net based Simulator for Flexible Manufacturing Systems. *Lectures Notes in Computer Science, Advances Petri Nets*, vol. 188, p. 1-14, 1984.

Andreu D. (1996). Commande et Supervision des Procédés Discontinus: une Approche Hybride. *Thèse de Doctorat*. Université Paul Sabatier de Toulouse, novembre 1996.

Andreu D., Pascal J.C. & Valette R. (1996). Events as a key of Batch Process Control System. *CESA'96 IEEE-SMC IMACS Symposium on Discrete Events and Manufacturing Systems*, p. 297-302. Lille, July 1996.

Bousson K. & Trave-Massuyès L. (1992). A Computational Causal Model for Process Supervision. *IFAC International Symposium on Artificial Intelligence in Real-time Control*, Delft, The Netherlands, June 1992.

Champagnat R., Esteban P., Pingaud H. & Valette R. (1998). Modeling and Simulation of a Hybrid System through Pr/Tr PN-DAE Model. *ADPM'98 Conference on Automation of Mixed Process: Dynamical Hybrid Systems*. Reims, France 19-20 march 1998.

Combacau M. & Courvoisier M. (1990). A Hierarchical and Modular Structure for F.M.S. Control and Monitoring. *1st International Conference on Artificial Intelligence, Simulation and Planning in High Automaty Systems*, Tucson, Arizona, USA, march 1990.

Daubas B., Pages A. & Pingaud H. (1994). Combined Simulation, of Hybrid Processes. *IEEE International Conference on SMC*, p. 314-319, San-Antonio, USA, October 1994.

Féray-Beaumont S. & Gentil S. (1991). Modèle Qualitatif de Comportement pour un Système d'Aide à la Supervision. *APII*, vol. 5, p. 325-348.

Holloway L.E. & Krogh B.H. (1990). Fault Detection and Diagnosis in Manufacturing Systems: A Behavioral Model Approach. *2nd International Conference on Computer Integrated Manufacturing*. RPC, May 1990.

Johnsson C. & Arzen K.E. (1994). High-level Grafcet and Batch Control. *ADPM'94 Symposium on Automation of Mixed Process: Dynamical Hybrid Systems*. Brussels, p. 255-263, Belgium, november 1994.

Leyval L., Gentil S. & Féray-Beaumont S. (1994).. Model-based Causal Reasoning for Process Supervision. *Automatica*, vol. 30, no. 8, p. 1295-1306.

Montmain J. (1994). Raisonnement Approximatif et Graphes Causaux pour la Detection et la Localisation de Défaillances. *Journées d'études du GRECO Automatique "Sûreté, Surveillance, Supervision"*. Paris. November 1994.

Rippin D.W.T. (1993). Batch Process Systems Engineering: a Retrospective and Prospective Review. *Computer and Chemical Engineering*, vol. 17, p. s1-s13.

Toguyeni A-K-A. (1992). Surveillance et Diagnostic en ligne dans les Systèmes Flexibles de l'Industrie Manufacturière. *Thèse de Doctorat*, Université de Lille, novembre 1992.

Valette R., Pingaud H., Pages A., Andreu D. & . Pascal J.C. (1995). Modeling, Simulation and Control of Event-Driven Operations in Process Systems. *ETFA'95 INRIA/IEEE Symposium on Emerging Technologies and Factory Automation*, vol. 3, p. 119-128, Paris, France, october 1995.

AN ADAPTIVE FUZZY DIAGNOSIS SYSTEM FOR ON-LINE SENSOR DATA VALIDATION *

A.N. Boudaoud, M.H. Masson

Heudiasyc, U.M.R. C.N.R.S. 6599, Université de Technologie de Compiègne, B.P. 20529, 60205 Compiègne Cedex, France

e-mail : boudaoud@utc.fr

Abstract. An adaptive fuzzy diagnosis system is presented for diagnosis of incipient sensor faults in an environmental process. Simulations demonstrate the capabilities of the system to detect the occurence of slow evolutions and to localise the faulty sensor

* This work has been within the EM^{2}S esprit project p-22442. The EM^{2}S consortuim is : suez - Lyonnaise des Eaux, F, VKI Water Quality Institute, DK, Danfoss System Control, DK, Hitec, N, Computas, N and CNRS, F

Keywords. on-line, diagnostic, fuzzy, pattern recognition, detection, monitoring, drift, sensor failures, environmental, validation.

1. INTRODUCTION

For security and cost reasons, system monitoring has became increasingly an important industrial issue in which the diagnosis is a major component. The diagnosis of a system is generally described as a three steps procedure : fault detection, identification (localisation) and correction. Faults can occur at different levels of a system (sensors, controllers and process).

The diagnosis of a process consists in finding its operation state using the sensor measurements. The robustness of a diagnosis system depends on the reliability of these measurements. A sensor fault (e.g. disconnection, calibration, slow drift...) makes the measurement not valid. The use of this measurement may produce a defective diagnostic about the process operation. To garantee the reliability of the diagnostic, a monitoring sys-

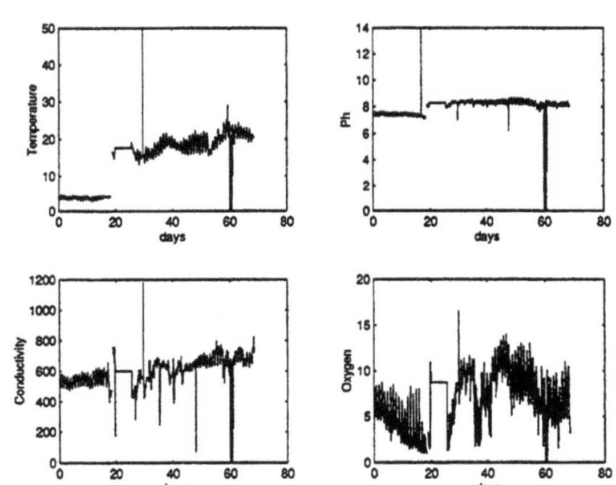

Fig. 1. Sensor measurment including invalid data : out of range measurement, outliers and disconnection of sensors are illustrated

tem must be included in order to detect and identify the sensor failures. This system is used for sensor data validation before the exploitation for the diagnostic.

This contribution concerns the sensor data validation problem in environmental systems. It is supported by the European project EM^{2}S (Environmental Monitoring and Management Systems). The main objective of this project is *the development of water quality monitoring and assessment methods for on-line quality information allowing automatic, remote and intelligent monitoring stations.* The proposed application concerns the supervision of water quality of a french river (Ouche near Dijon city) where are discharged a part of waste water and the treated water. For the monitoring, three observation stations MN1, MN2 and MN3 are installed (see figure 2). MN1 is installed before the waste water discharge point. MN2 and MN3 are respectively installed before and after the waste water treatement plant (WWTP). MN1 and MN2 provide on-line 6 parameters measurements : temperature, pH, conductivity, oxygen concentration, turbidity and level of river, MN3 provides in addition the ammonium concentration measurement. This parameter is important to assess the pollution state of the river. The relevance of the measurements depend on the reliability of the sensors. So, before any exploitation the measures given by sensors must be validated.

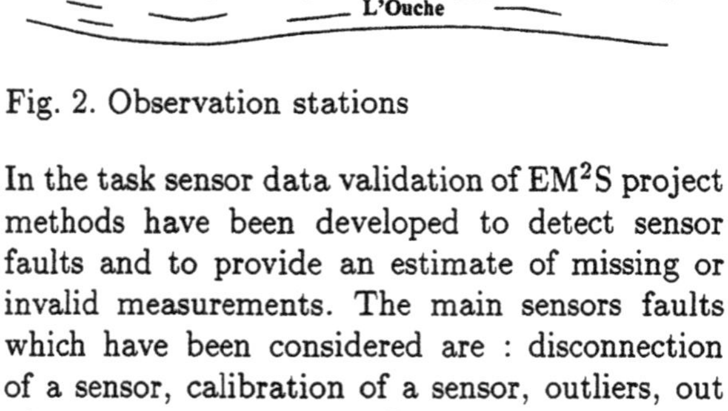

Fig. 2. Observation stations

In the task sensor data validation of EM^{2}S project methods have been developed to detect sensor faults and to provide an estimate of missing or invalid measurements. The main sensors faults which have been considered are : disconnection of a sensor, calibration of a sensor, outliers, out of range measurements, offset measurements and slow drifts of the signal (see figure 1). Unless slow drifts, the other faults are easly detected and identified. The detection of slow evolutions is a more difficult problem. An issue is presented in this paper by using an on-line fuzzy adaptive diagnosis system based on a pattern recognition approach (Boudaoud and Masson, 1997). This system is able to detect slow drifts and to localise the failed sensor.

Before describing this system, we will give some backgrounds on the pattern recognition approach for diagnosis. Results are presented to illustrate the operation and performances of the diagnosis system for the detection of incipient faults on oxygen sensors.

2. PATTERN RECOGNITION AND DIAGNOSIS

Pattern recognition provides a wide variety of methods to design diagnostic systems. The conception involves generally two main steps : analysis and exploitation.

The analysis consists firstly of extracting the relevant information from the sensor measurements (preprocessing and pattern selection). This information is used to build up a vector (*pattern vector*) in a feature space. Then a learning set is created with the past recorded pattern samples which represent the operating states encountered in the process operation (e.g. nominal, fault 1, fault 2 ...). Finally, a clustering procedure (Jain, 1988) is applied to categorize the learning set. Each category (*class*) represents an operation state for the process. To delimite these classes, one needs a discrimination rule (*a classifier*). Different approaches such as Bayes decision rule, $k-nn$ rule, neural networks, fuzzy classifiers can be used to design the classifier.

Once the classifier is designed, the exploitation of the classifier can be run for new patterns (observation). A new observation is classified into one of the known classes according to the discrimination rule.

However, this classical scheme is less effective in real situations because the learning set in often incomplete, new operating states can be encountered in the process operation (e.g. new faults). To deal with this incomplete knowledge, an adaptive approach has been introduced (Dubuisson, 1990) using the concept of distance rejection. New operation states (classes) are detected and included in the learning set. Then, the discrimination rule is adapted. This adaptation is unfortunately carried out *off-line* when the diagnostic system should react as soon as possible, and is not suitable for slow evolutive systems i.e. the system evolve slowly from a known class to a new class.

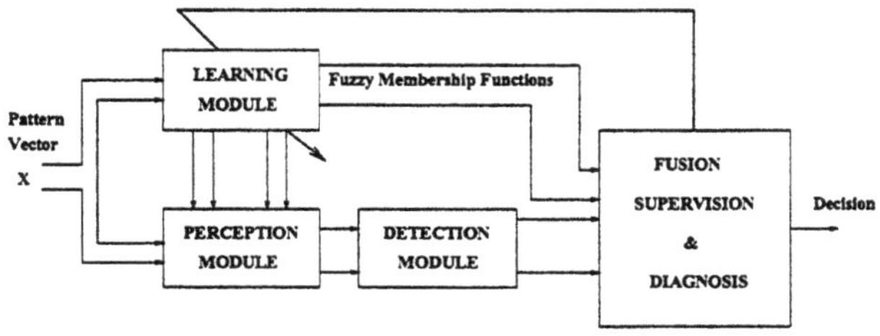

Fig. 3. On-line fuzzy diagnosis system

To deal with these constraints an *on-line* adaptive scheme has been proposed by (Boudaoud and Masson, 1997). A supervision loop is used to control the on-line adaptation of the classifier (see figure 3). It makes possible the detection of slow evolutions from a known class and stabilizations in a known or an unknown class. So, new operating states are detected and included *on-line* in the classifier.

3. DIAGNOSIS SYSTEM ARCHITECTURE

The adaptive fuzzy diagnosis system is composed of four modules :

- **learning module :** it is dedicated to the on-line elaboration of the fuzzy membership functions. The learning algorithm, fuzzy min-max clustering (Simpson, 1993), is based on a sequential agglomerative algorithm . The fuzzy membership function is constructed by agregating fuzzy prototypes.
- **perception module :** it generates a perception measure which is sensitive to the occurence of a slow evolution (gradual faults) or a stabilisation in a given class. This measure is based on a distance criterion evaluated between the current pattern vector and a known class.
- **detection module :** it used to detect the occurrence of a slow evolution or a stabilisation. This module is based on a sequential probability ratio test (SPRT) named CUSUM algorithm (Basseville and Nikiforov, 1993).
- **supervision and diagnosis module :** it is dedicated to supervise the learning module and generate a diagnosis as a symbolic inference.

These modules are well described in a previous paper (Boudaoud and Masson, 1997).

4. SIMULATIONS

The application is adressed to the diagnosis of slow drifts in sensors. Two sensors of oxygen are considered (at MN1 and MN2 obervation stations).

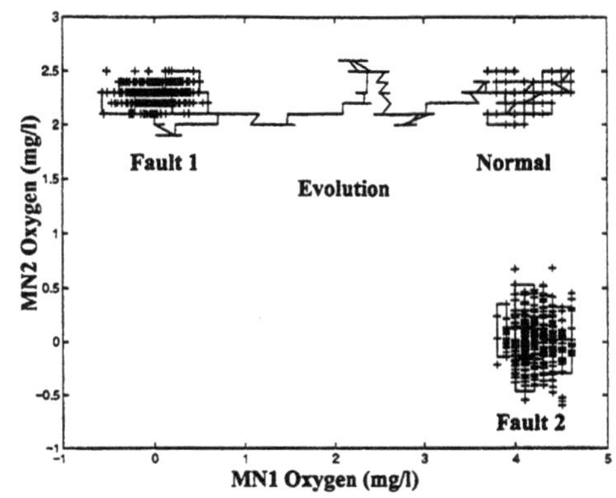

Fig. 4. Sensor data and a slow drift of sensor MN1 (speed = 0.005 mg/l/min)

For this application we consider three operation states for the sensors :

- normal state : both sensor MN1 and MN2 are in a normal operation,
- fault 1 : sensor MN1 is out of order and sensor MN2 is in a normal operation state.
- fault 2 : sensor MN1 is in a normal operation state and sensor MN2 is out of order.

We assume that a persistent drift of the sensor MN1 (resp. MN2) leads to the faulty state 1 (resp. 2).

The learning set includes three classes : normal, fault 1 and fault 2 operation states of the sensors. Data of the normal state are recorded from sensors during 20 hours (sampling period 6 minutes). The faulty states 1 and 2 are simulated during 40 hours (sampling period 6 minutes), for the fault 1(resp. 2) the measure of sensor MN1 (resp. MN2) (in a normal operation state) is recorded when the sensor MN2 (resp. MN1) measure is set to zero with a normalised gaussian noise. The pattern vector components are the measure of oxygen of each sensor.

The on-line fuzzy diagnosis system was used to learn the classes. The fuzzy membership functions of the classes (figure 5) are constructed using the

FMMC algorithm (Simpson, 1993) (leraning module). The diagnosis system is then tested to detect slow drift impemented on the sensor MN1.

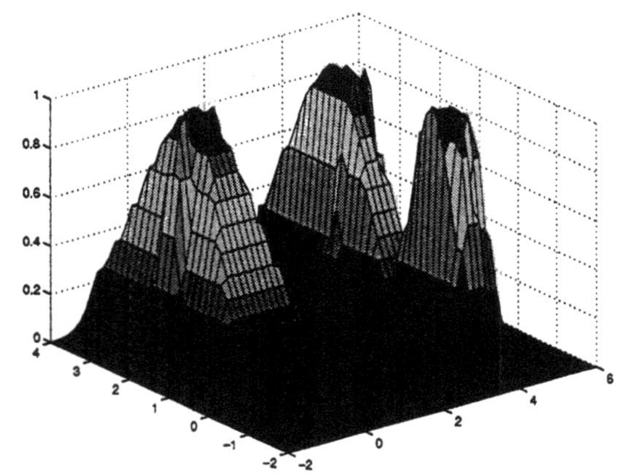

Fig. 5. Fuzzy membership functions

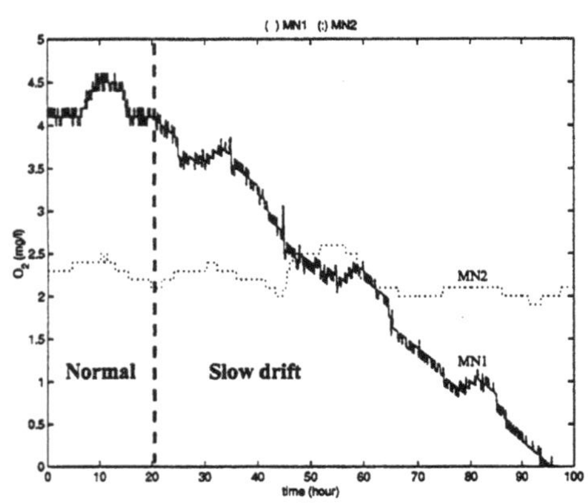

Fig. 6. Sensor measurements : drift of sensor MN1 (speed = 0.8×10^{-3} mg/l/min)

Initially, the sensor are in a normal operation state, then a linear evolution (drift) is simulated on sensor MN1 (see figure 6) for different speeds of evolution, 5 cases are considered (table 1).

Table 1. The different cases of drift

Case	speed (mg/l/min)$\times 10^{-3}$	duration(hour)
1	0.8	75
2	1.3.	48
3	1.7	42
4	3.3	21
5	8.3	8

The supervision system is used to detect and localise the faulty sensor. Tunning parameters for the detection module are fixed in order to have the minimal delay of detection with no false alarm.

The performances of detection are evaluated for different speeds of evolution (see figure 7). For a "slow" drift (speed = 0.8×10^{-3} mg/l/min) of duration of 75 hours the diagnosis system detect the drift with a delay of 21.8 hours and localise the faulty sensor after 28 hours.

Table 2. Performances : delay of detection

Diagnosis	Case 1(h)	Case 5 (h)
"normal" state ω_1		
leaving ω_1	1.6	21.8
moving toward ω_2	3.8	28.8 [1]

For a "fast" evolution (speed = 8.3×10^{-3} mg/l/min) of duration of 8 hours the evolution is detected after 1.6 hours and the faulty sensor is located with a delay of 3.8 hours (see table 2). The performances of detection are dammaged for "very slow" evolutions (see figure 7).

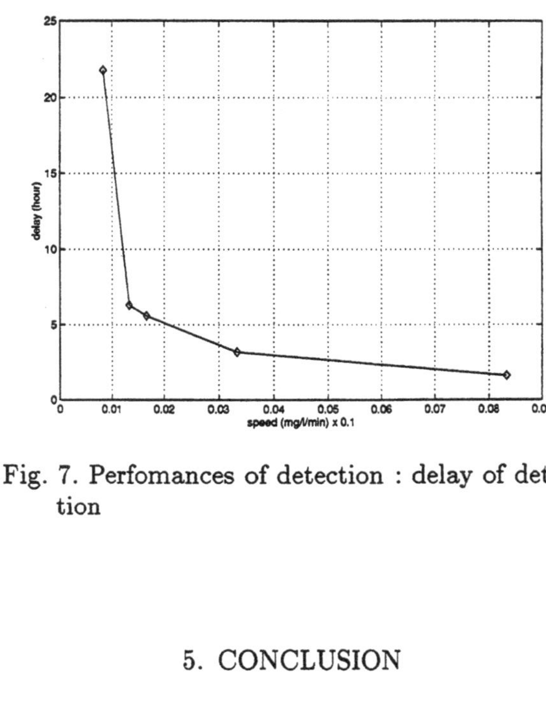

Fig. 7. Perfomances of detection : delay of detection

5. CONCLUSION

An adaptive fuzzy diagnosis system has been presented to detect slow evolutions. This system was demonstrated to detect early incipient faults and localise on-line the faulty sensor. The performances of detection are reasonably good when the rate of evolution is suficiently high. However, the delay of detection increases as the rate of drift decreases (slow evolutions). This sensitivity can be reduced by selecting for example a more relevant pattern vector or by using another perception criterion in the supervision loop. This outcome will be included in futur works.

6. REFERENCES

Basseville, M. and I.V. Nikiforov (1993). *Detection of abrupt changes : theory and application.* Prentice Hall. New Jersey.

Boudaoud, A.N. and M.H. Masson (1997). On-line adaptive fuzzy diagnosis system :a fuzzy pattern recognition approach. In: *IFAC SAFEPROCESS'97.* Kingston upon Hull, UK.

Dubuisson, B. (1990). Decision with reject options. In: *5 th European signal Processing conference.* Barcelone, Spain.

Jain, A.K. (1988). *Algorithm for clustering data.* Prentice Hall. New Jersey.

Simpson, P.K. (1993). Fuzzy min-max neural networks - part 2 : clustering. *IEEE Transactions on Fuzzy Systems* **1(1)**, 32–45.

"FAULT-TOLERANT TRACTION SYSTEM CONTROL USING FUZZY INFERENCE MODELLING"

C.J. Lopez, R.J. Patton, S. Daley

*Control & Intelligent Systems Engineering,
Department of Electronic Engineering,
The University of Hull, Cottingham Road, Hull HU6 7RX, UK
e-mail:, r.j.patton@e-eng.hull.ac.uk*

Abstract: This papers presents a new approach for the stability and design of non-linear fuzzy inference systems. The approach is based on Takagi-Sugeno (T-S) fuzzy models. The paper derives the necessary conditions for the assignability of eigenvalues to a region in the s-plane and the necessary conditions to guarantee the stability of fuzzy models. The problem is solved via the Linear Matrix Inequality (LMI) method. The paper demonstrates the application in fault-tolerant control in a railway traction system using DSP in a hardware test-ring. *Copyright © 1998 IFAC*

Keywords: Fault-tolerant control, bilinear systems, fuzzy logic, linear matrix inequality, induction motor

1. INTRODUCTION

In recent years the application of fuzzy logic to model-based fault diagnosis approaches have been gaining increasing attention in both research and application. The first methods used fuzzy set theory to express cause-effects relation in expert-systems. More recently, fuzzy logic has been combined with quantitative model-based approaches to enhance the robustness of the scheme. Typical applications included, but were not limited to, evaluation of residuals generated by observers (Frank *et al.*, 1993) or parameter estimation (Isermann; 1994), multi-valued fuzzy logic (Isermann & Ulieru, 1993; Cassar et al., 1994) for fault isolation and threshold adaptation (Schneider & Frank, 1994). In most of these existing approaches it was mainly the fuzzy-logic ability to handle vague and imprecise information which was used. Several observers had been proposed for FDI in bilinear systems using either unknown input observers, *sliding mode observers* (Edwards *et al.*, 1997), *gain-parametrised observers (*Bennett *et al.* 1996) or in Fault detection the work of Shields and co-workers (1997). In particular it is proposed to use Takagi-Sugeno fuzzy Observer, (T-S) fuzzy models for non-linear dynamic system which is described by a number of locally-linearised models. Most studies on fault diagnosis for non-linear and uncertain systems deal with two main questions. First the model is linearised at an operating point and second robust techniques are applied to generate residual signals. The strategy only works well when the linearisation does not cause a large mismatch between linear and non-linear modes and the system operates close to the specified operating point. In this paper, different operating points are self-selected with an optimisation method based on eigenvalue assignment criteria

2. TAKAGI-SUGENO FUZZY MODEL

2.1 Stability results

Several authors investigated the stability of T-S fuzzy models. By defining the open-loop system as

- Continuous Systems

$$\dot{x}(t) = \sum_{i=1}^{r} \mu_{M_i}(z(t))\big(A_i x(t)\big) \quad (1)$$

- Discrete Systems

$$x(k+1)=\sum_{i=1}^{r}\mu_{M_i}(z(k))(A_i x(k)) \qquad (2)$$

the following theorems can be derived.

Theorem 1 (Ikeda et al., 1997): *The system described by Eq. 1 is stable if a common positive matrix P exits such that the following inequality holds*

$$A_i^T P + PA_i < 0$$
$$i = 1,2,..r \qquad (3)$$

Theorem 2 : (Tanaka et al. 1990 & 1992) *The system described by Eq. 2 is stable if a common positive matrix P exists such that the following inequality holds*

$$A_i PA_i^T - P < 0$$
$$i = 1,2,..r \qquad (4)$$

Remark: The proof of theorems 1 and 2 is easy (Tanaka and Sugeno 1992). However, it is a little more involved for discrete fuzzy systems.

Inequalities 3 & 4 can be solved using linear matrix inequalities *Boyd et al. 1994*

Theorems 1, 2 only guarantee the stability, but if in addition, the system must be stable in presence of uncertainty, then the eigenvalues are required to be assigned within a specific region **S(α,r,θ)** in the s-plane or z-plane, with a robustness against the neglected frequency dynamics represented by a model uncertainty.

2.2 Pole Assignment in LMI regions

An LMI region is a useful concept to formulate pole placement objectives in LMI terms.

Definition 1: A subset D of the complex plane is called an LMI region if there exists a matrix $L=[\lambda_{ij}]\in\Re^{mxm}$ and a matrix $M = [\mu_{ij}]\in\Re^{mxm}$, such that:

$$D=\{z\in C: f_D(z)<0\} \qquad (5)$$

where

$$f_D(z) := L + Mz + M^T\bar{z} \qquad (6)$$

$\{\lambda_{ij}\}_{1\le i,j\le m}$ and $\{\mu_{ij}\}_{1\le i,j\le m}$ denote the entries of the matrices L and M defined over a region *D*

2.3 Root clustering

Theorem 4: For the system to have all its eigenvalues in *D it is sufficient to find a common positive definite matrix P, such that:*

$$M_D(A_x,P)=\left[\lambda_{ij}P+\mu_{ij}A_xP+\mu_{ji}PA_x^T\right]_{\le i,j\le m}<0$$
$$x=1....r \qquad (7)$$

The characteristic function for this LMI region is, where the poles of a matrix A are assigned to the disc, Ω_2, in the s-plane, of radius C_1 and centre $(-C_0, 0)$,

$$f_D(z)=\begin{bmatrix}-C_0 & C_1+\bar{z}\\ C_1+z & -C_0\end{bmatrix} \qquad (8)$$

Theorem 5 For a matrix A_i to have all its eigenvalues in a region D_1 *and* D_2 $(D_1\cap D_2)$, *if and only if there exists a positive definite matrix such that*

$$M_{D_1\cap D_2}(A_i,P)=Diag(M_{D_1}(A_i,P),M_{D_2}(A_i,P))$$

2.4 Solution using LMI

The problem of regional eigenvalue assignment can also be re-formulated in an LMI form.

Theorem 6: *For the system defined in Eq. 1 to have all its eigenvalues in a disc region* $f_D(z)$ *Eq. 8 it is sufficient to find a common matrix P such as the following matrix inequality holds*

$$\begin{bmatrix}-C_0P & C_1P+A_i^TP\\ C_1P+PA_i & -C_0P\end{bmatrix}<0 \;\; for\ all\ \ i \qquad (9)$$

3. T-S FUZZY OBSERVERS FOR BILINEAR SYSTEMS

3.1 Using the idea of *Parallel Distributed Compensation* (PDC), for a non-linear dynamic system represented by T-S fuzzy model, a linear time invariant observer can be associated with each rule of the fuzzy model

if ω *(t)* is M_i **then**

$$\begin{aligned}\dot{\hat{x}}(t)&=A_i\hat{x}(t)+Bu(t)+L_i(y-\hat{y}) \qquad i=1,2,...,r\\ \hat{y}&=C\hat{x}\end{aligned} \qquad (10)$$

The overall observer dynamics will then be a weighted sum of individual linear observers.

$$\begin{aligned}\dot{\hat{x}}&=\sum_{i=1}^{r}\mu_{M_i}(\omega)A_ix+Bu+L_i(y-\hat{y})\\ \hat{y}&=C\hat{x}\end{aligned} \qquad (11)$$

The schematic diagram of such an observer is shown in Fig. 1 where it can be seen that a fuzzy inference engine is used to "select" the appropriate output, from those generated by the *r* parallel observers. The transition between one model to the other depends on the operating regime defined by the angular speed, ω. The estimation error dynamics can then be shown to be given by the following differential equation

$$\dot{e}=\sum_{i=1}^{r}\mu_{M_i}(\omega(t))(A_i-L_iC)e \qquad (12)$$

Using the results presented in Section 3, it is possible to derive the conditions for a stable observer. More than that, it is also possible to assign the eigenvalues of the observer to a region in the s-plane.

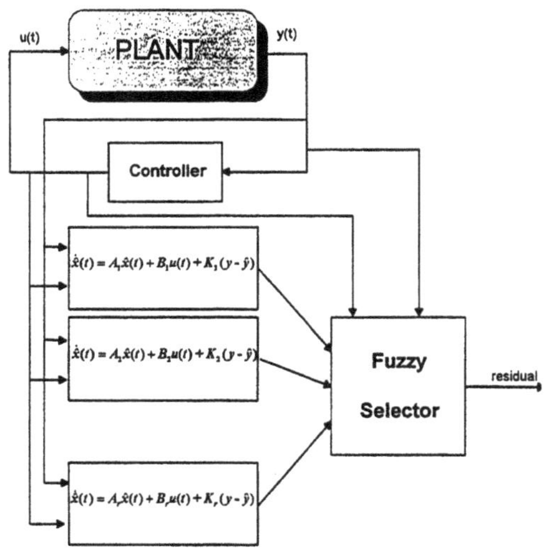

Fig. 1: Schematic diagram of the fuzzy observer

4. FAULT DIAGNOSIS OF THE INDUCTION MOTOR IN A RAIL WAY TRACTION SYSTEM: APPLICATION

The performance of the proposed observer is assessed using Matlab/Simulink simulation. The FDI forms part of a fault-tolerant control system used to regulate the torque and flux in the induction motor.

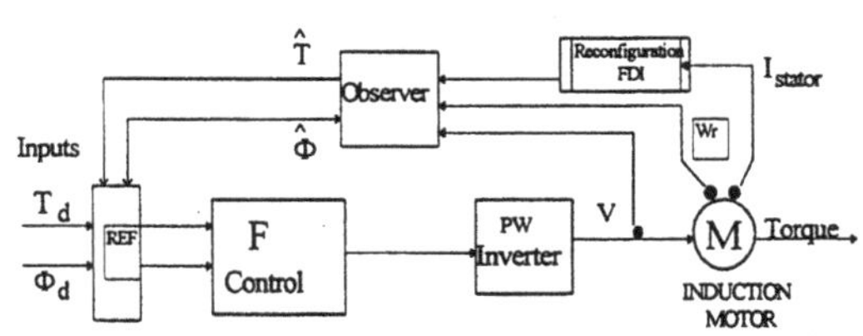

Fig. 2: Schematic diagram of the fault-tolerant control system

The overall system, shown in Fig. 2, consists of a PWM inverter which provides the necessary power to drive the machine. Since the torque and flux cannot be measured they are estimated using a bilinear observer and fed to a PI-like fuzzy controller. The following *faults* due to gain and bias changes, noise and interference pick up, signal processing effects and *disconnections* due to continuos and intermittent loss of measurement loss of can occur in the system. An FDI/Reconfiguration block, based on fuzzy observers, performs the detection and isolation of faults and takes the necessary action to maintain the controller performance. By using a power-invariant transform the complex dynamics of the 3-phase induction motor could be represented by two axes, direct and quadrature. The induction motor can be modelled by the following bilinear differential equation:

$$\dot{x} = A(\omega)x + Bu = Ax + N\omega x + Bu \qquad (13)$$
$$y = Cx$$

where

$$x = [I_{ds}\ I_{qs}\ I_{dr}\ I_{qr}]^T$$
$$y = [I_{ds}\ \ I_{dr}]^T$$
$$u = [V_{ds}\ V_{qs}\ V_{dr}\ V_{qr}]^T$$

where I and V are the current and voltage respectively. The speed ω in Eq. (13) is referred to as the bilinear input. The above mathematical model has been derived based on three main assumptions, namely: ideal air gap flux distribution, linear magnetisation characteristic and zero temperature coefficient in the windings.

The bilinear matrices used in this paper are as follows:

$$N = \begin{bmatrix} 0 & -6.5492 & 0 & -7.0314 \\ 6.5492 & 0 & 7.0314 & 0 \\ 0 & 7.0314 & 0 & 7.5492 \\ -7.0314 & 0 & -7.5492 & 0 \end{bmatrix}$$

$$A = \begin{bmatrix} -459.2987 & 0 & 233.3444 & 0 \\ 0 & -459.2987 & 0 & 233.344 \\ 427.7981 & 0 & -250.5265 & 0 \\ 0 & 427.7981 & 0 & -250.5265 \end{bmatrix}$$

$$B = \begin{bmatrix} 8.3509 & 0 \\ 0 & 8.3509 \\ -7.7781 & 0 \\ 0 & -7.7781 \end{bmatrix} \qquad C = \begin{bmatrix} 1 & 0 & 0 & 0 \\ 0 & 1 & 0 & 0 \end{bmatrix}$$

The T-S Fuzzy representation of the induction motor observer is a class of non-linear systems. The dynamics of the induction motor observer are described by the following T-S fuzzy rules.

if ω *(t)* is M_i **then**

$$\dot{\hat{x}}(t) = A_i\hat{x}(t) + Bu(t) + L_i(y - \hat{y}) \qquad i = 1,2,\ldots,r \qquad (15)$$
$$\hat{y} = C\hat{x}$$

The overall observer dynamics will then be a weighted sum of individual linear observers.

$$\dot{\hat{x}} = \sum_{i=1}^{r} \mu_{M_i}(\omega)A_i x + Bu + L_i(y - \hat{y}) \qquad (16)$$
$$\hat{y} = C\hat{x}$$

It should be emphasised that the number of models used is the one that guarantees the observer stability /pole assignment conditions derived in Section 2, which depends on the LMI region selected

Figs 3 to 5 show how the pole placement varies depending on the number of subset required .

*TS Fuzzy observer with 3 sets, (wr1=0, width=75):

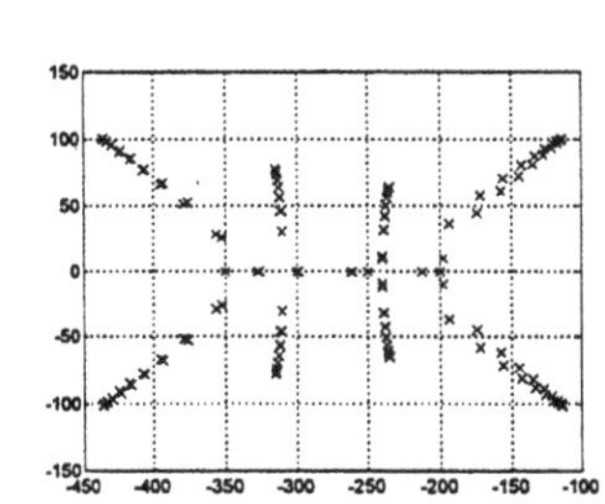

Fig. 3: Eigenvalues continuous Observer with 3sets

*TS Fuzzy observer with 5 sets, (wr1=0, width=40):

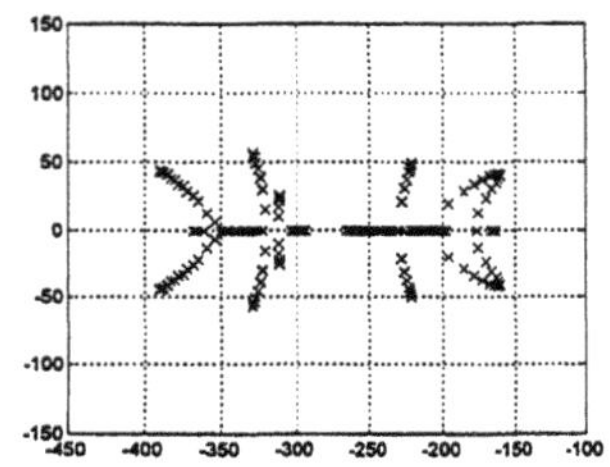

Fig. 4: Eigenvalues continuous observer with 5 sets

*TS Fuzzy observer with 10 sets,(wr1=0, width=17):

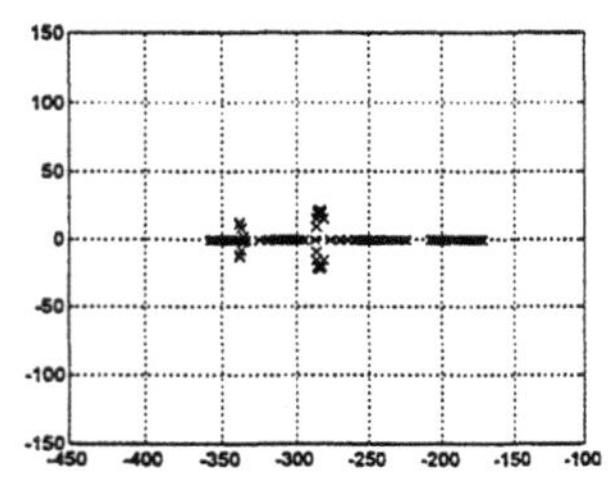

Fig. 5: Eigenvalues continuous observer (10 sets)

The required operating points w_{r1}, w_{r2},... w_{rn} with suitable widths between them are selected automatically using the high level Linear Matrix Inequalities design tool, as shown in Fig. 6. This is achieved by solving the LMI's for all the operating points.

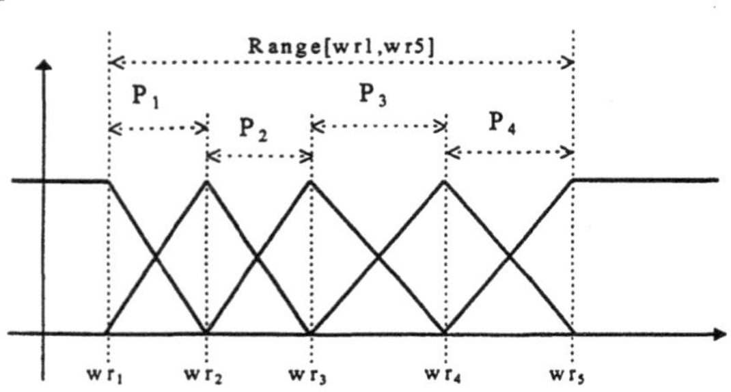

Fig. 6: Membership Fuzzy Functions

For example, assume that it is required to assign the eigenvalues of the observer to a region **S (α, r ,θ)** as shown in Fig. 7.

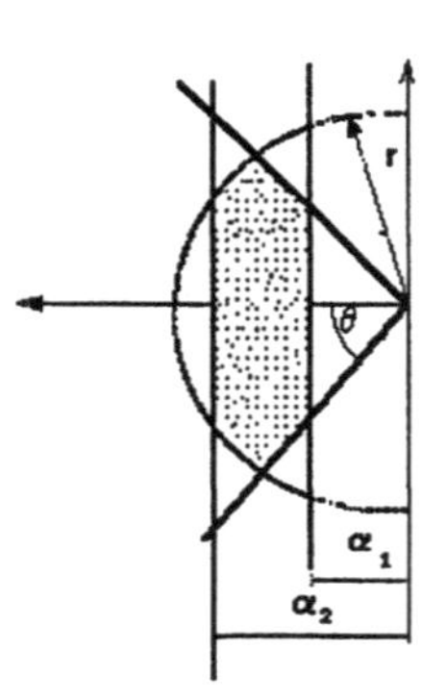

Fig. 7: Region S (α,r,θ)

The LMI region in D has the following characteristic function:

$$f_{D_1 \cap D_2 \cap D_3} = Diag(f_{D_1}, f_{D_2} f_{D_3}) \qquad (17)$$

with

*D_1 := Disc with centre at (C_0, 0) and radius C_1

$$M_{D_1} := \begin{bmatrix} -C_0 P & A_i^T P + C_1 \\ PA_i + C_1 P & -C_0 P \end{bmatrix} \qquad (18)$$

*D_2:= Conic sector centred at the origin and inner angle θ

$$M_{D_2} := \begin{bmatrix} sin\frac{\theta}{2}(A_i^T P + PA_i) & -cos\frac{\theta}{2}(A_i^T P - PA_i) \\ cos\frac{\theta}{2}(PA_i - A_i^T P) & sin\frac{\theta}{2}(A_i^T P - PA_i) \end{bmatrix} \qquad (19)$$

* D_3: = Vertical strip $h_1 \le x \le h_2$

$$M_{D_3} := \begin{bmatrix} 2\alpha_1 - (A_i P + A_i^T P) & 0 \\ 0 & (A_i P + A_i^T P) - 2\alpha_2 \end{bmatrix} \qquad (20)$$

For the observer to have all its eigenvalues in the region **S (α_1= -100, r=700 ,θ=45)**, the inequalities 18, 19, 20 must be solved. This clustering region is difficult to express with classical representation. In contrast, Theorem 5 yields the following unique LMI characterisation

$$f_D(z) := L + Mz + M^T \bar{z} \qquad (21)$$

with

$$L = \begin{bmatrix} \mathbf{200} & \mathbf{0} & \mathbf{0} & \mathbf{0} & \mathbf{0} \\ \mathbf{0} & \mathbf{-700} & \mathbf{0} & \mathbf{0} & \mathbf{0} \\ \mathbf{0} & \mathbf{0} & \mathbf{-700} & \mathbf{0} & \mathbf{0} \\ \mathbf{0} & \mathbf{0} & \mathbf{0} & \mathbf{0} & \mathbf{0} \\ \mathbf{0} & \mathbf{0} & \mathbf{0} & \mathbf{0} & \mathbf{0} \end{bmatrix}$$

$$M = \begin{bmatrix} \mathbf{1} & \mathbf{0} & \mathbf{0} & \mathbf{0} & \mathbf{0} \\ \mathbf{0} & \mathbf{0} & \mathbf{0} & \mathbf{0} & \mathbf{0} \\ \mathbf{0} & \mathbf{1} & \mathbf{0} & \mathbf{0} & \mathbf{0} \\ \mathbf{0} & \mathbf{0} & \mathbf{0} & \mathbf{0.3827} & \mathbf{-0.9239} \\ \mathbf{0} & \mathbf{0} & \mathbf{0} & \mathbf{0.9239} & \mathbf{0.3827} \end{bmatrix}$$

For the initial matrices A, B, C, D with the constraint S (α, r ,θ) the following operating points are automatically found, with 4 models and a speed range of 0-176 rad.s^{-1} is calculated using the different operating points.

$w_{r1} = 0$
$w_{r2} = 32.8125$
$w_{r3} = 86.2500$
$w_{r4} = 176$

It is simple check that all the eigenvalues lie in the region specified by Eq. 2. The eigenvalue assignment for all the rules is shown in Fig. 8.

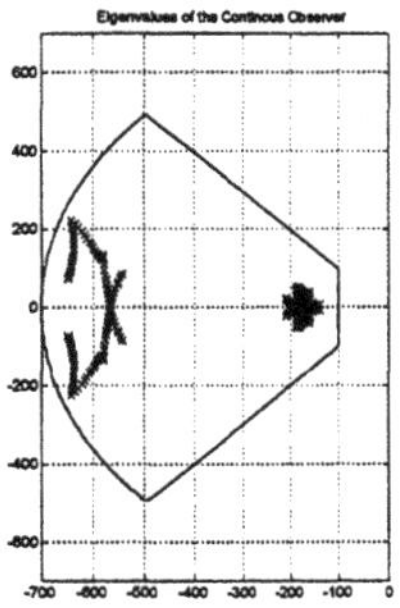

Fig. 8: Region S (α=-100 ,r =700 ,θ =45)

The FDI scheme adopted uses a bank of generalised observers (Bennett *et al.*, 1996). Each observer estimates the current in one phase of the stator from the other two sensors, and is also a residual generator for fault diagnosis. The diagnostic signal-residual is generated by the comparison the estimated and real system outputs. In so doing, the following residual signals are generated:

$$r_i = I_{phase_i} - \hat{I}_{phase_i} \qquad i = 1,2,3$$

Using LMI, it is possible to find a common matrix P, that would solve inequalities in Eqs. (18, 19 & 20) by using only 4 models. The common matrix, P is given by:

$$P3 = \begin{bmatrix} 2.5708 & 0 & -2.5026 & -0.6706 \\ 0 & 2.5708 & 0.6706 & -2.5026 \\ -2.5026 & 0.6706 & 2.7931 & 0 \\ -0.6706 & -2.5026 & 0 & 2.7931 \end{bmatrix} * 1.0e+004$$

Since only 4 models are used, the universe of discourse of ω is divided into three intervals defined by the linguistic variables *Small*, *Medium* and *High*.

Once the continuous system observer is designed, a discrete-time observer can be obtained using different discretisation methods. However, as the design is based upon pole-assignment, the eigenvalues (poles) remain the same regardless of the discretisation method chosen and the design can be made directly in discrete-time. The pole locations shown in Fig. 9 can then be used along with the first order discretisation method. This results in considerable computational saving over the use of higher order methods.

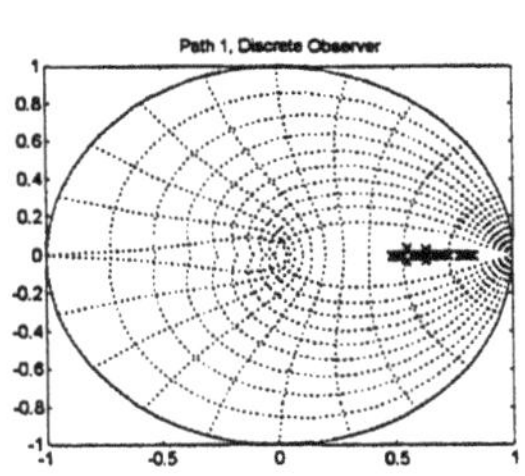

Fig. 9: Pole configuration in z-domain

Fig. 10. shows the two different ways to obtain a discrete observer. Note that for large negative valued eigenvalues, if the path 2 is selected it would be necessary to choose a more accurate method like Tustin's approximation to obtain the same results that is obtained from path 1.

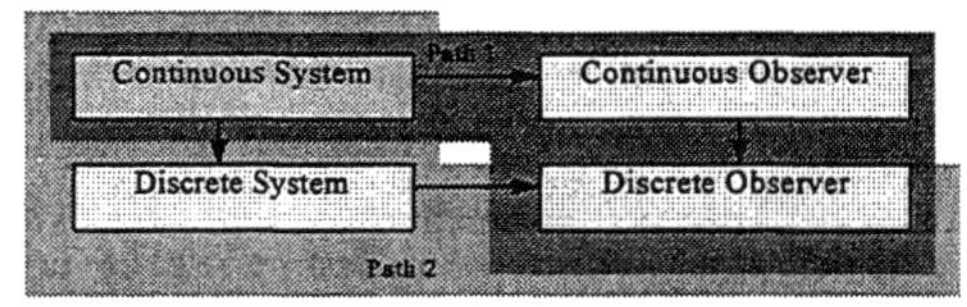

Fig. 10: Paths to obtain discrete observer

The observer is first tested against an initial perturbation, and the resulting responses are shown in Fig. 11. It can be seen that the observer presents stable dynamics, although it exhibits a relatively large overshoot. However, initial errors can be overcome, in FDI, using adaptive thresholds, especially that the transient time is very small (< 1 sec). Moreover, this error is not significant compared with that resulting from a fault in the control loop.

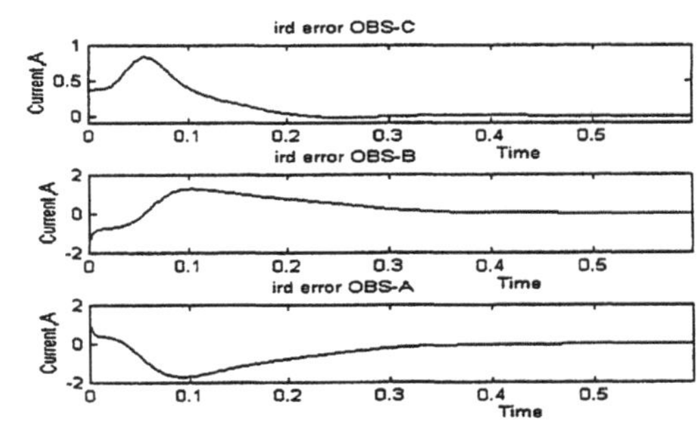

Fig. 11: Observer response to initial perturbations

4.1 DSP implementation of FDI scheme in real time

The sensor fault-tolerant control is run in real time using measurement taken from a hardware test-ring. The test-ring consists of a three phase AC induction motor driving a DC motor as load. The 3-phase voltages is supplied from a 100 KHz PWM-inverter. The induction motor selected is a 0.2 KW 3 phase 4 pole machine.

When a disconnection (fault) in phase A is injected between times t_1= 1.3 sec and t_2= 2.3 sec., the residual signal is illustrated in Fig 12. The simulation demonstrates that the fault can be easily detected using the residual generated by the fuzzy observer.

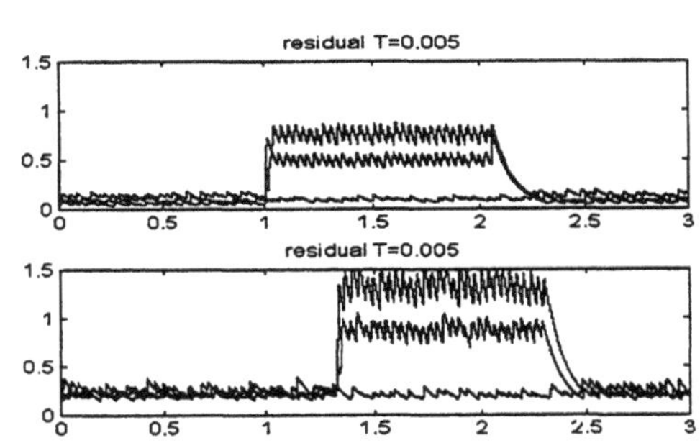

Fig. 12 : Current sensor and voltage Residual response to fault

The fuzzy observer is also tested when used in the fault-tolerant control system shown in Fig. 2. The configuration adopted for the isolation of the fault is similar to that presented by Bennett *et al.* (1996) and which is based on the generalised observer scheme. Fuzzy observers are also used to accommodate faults. To assess the performance of the overall fault-tolerant control system, intermittent faults are injected successively in each of the 3 phases. Fig. 13 shows how such faults cause the error torque to deviate from its zero value when no action is taken.

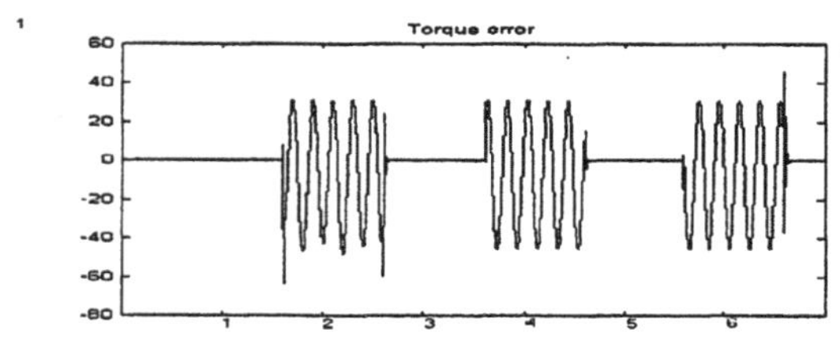

Fig. 13: Torque error (with FDI+Correction)

However, with the FDI/reconfiguration operational, it is possible to detect, isolate and obviate the effect of the three faults. This can be seen from the torque and flux responses shown in Fig. 14.

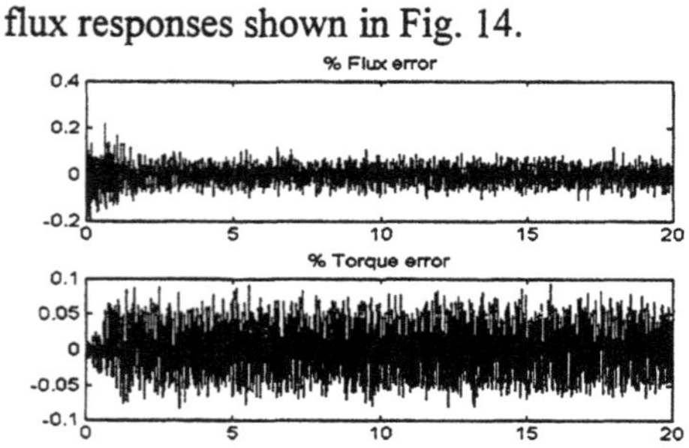

Fig. 14: Torque error (without FDI/Correction)

5. CONCLUSIONS

This paper has presented a novel approach for the stability and design of non-linear fuzzy control systems. A convex optimisation technique involving LMI has been proposed to self select the local linear models used set used in the T-S representation. The LMI-based technique guarantees the global stability of the fuzzy models. The developed scheme has been successfully applied, in real time, to a fault-tolerant control of an induction motor of a rail traction system where both the flux and torque are the regulated variables.

6. ACKNOWLEDGEMENTS

The authors are indebted to GEC-Alsthom for funding support through the provision of the induction motor test rig at the company.

REFERENCES

Bennett S M, Patton R J, Daley S & Newton D A (1996), Model Based Intermittent Fault Tolerance in an Induction Motor Drive, *CESA'96* IMACS Multiconference, Lille, July 9-12, **vol. 1,** 678-683

Bose B K (1986), *Power electronics and AC drives*, Prentice Hall

Boyd S, Ghaoui L, Feron E, & V Balakrishnan (1994), *Linear Matrix Inequalities in System and Control Theory,* SIAM. Philadelphia

Brod D M & Novontny D W (1989), Current Control of VSI-PWM Inverters. *IEEE Trans. on Ind. Applics*. **21**(4)

Cassar J P & Staroswiecki M (1994), Advanced design of the decision procedure in failure detection and isolation systems, *Proc. IFAC SAFEPROCESS'94*, Finland, **vol. 1**, 380-385

Edwards C, Spurgeon S K, Patton R J & Klotzek P (1997), Sliding mode observers for fault detection, Proc. IFAC *SAFEPROCESS'97*, Hull, 522-527

Frank P M & Kuipel B (1993), Fuzzy Supervision and application to lean production, *Int. J. Systems Sci*, **24**(10), 1935-1944

Gutman S & Jury E I (1981), A General Theory for Matrix Root-clustering in Subregions of the Complex Plane, *IEEE Trans., **AC-26**,* 853-863

Hara & Furuta K (1976), Minimal order state observer for bilinear systems, *IJC.* **24**, 705-718.

Isermann R & Ulieru M (1993), Integrated fault detection and diagnosis, *Proc. IEEE Sys., Man & Cyb.*, Le Touquet, France, 743-748.

Lorenz R D & Thomas L (1994), A. Motion Control with Induction Motors. *Proc. IEEE*, **82**

Patton R J & Frank P M & Clark R N (1989), Fault Diagnosis in Dinamic Systems: Theory and Application, Prentice Hall.

Schneider H & Frank P M (1994), Fuzzy Logic based threshold adaptation for fault detection in robots, *Proc. 3rd IEEE Conf. Control Applications*, Glasgow, UK, **vol. 2**, 1127-1132.

Takagi T & Sugeno N (1992) Stability analysis and design of fuzzy control systems, Fuzzy Sets Syst., No. **2**, 135-156.

Takagi T & Sugeno N (1985) Fuzzy identification of systems and its applications to modelling and control, IEEE Trans. SMC, **15**, 116-132

Tanaka K, Takayuki I & Wang H (1996), Design of Fuzzy Control Systems. Based on Relaxed LMI Stability Conditions. *Proc. 35th. CDC,* Kobe.

Yu D & Shields D (1995), Fault diagnosis in bilinear systems, A survey, Proc. ECC'95, Rome, 360-365

CAN CHEMICAL PROCESS INDUSTRY BENEFIT FROM ANALOG ELECTRONICS DIAGNOSIS METHODS?

Patrick Taillibert

Dassault Electronique
92214 Saint Cloud (France)
Patrick.Taillibert@dassault-elec.fr

Abstract: The present paper will **not** answer the question asked in its title! It would require a deep knowledge of both domains (chemical processes and analog circuits) that we don't have. Nonetheless, we have the feeling that the method we recently developed for diagnosing analog electronic circuits is far more general that its initial purpose. This confidence is mainly based on the fact that our method not only offers an original way to take into account differential equations -which are obviously at the basis of process modeling- but also to consider inaccuracies such as measurement errors, parameter dispersion, which are often an impediment when correctness of diagnosis is concerned. Therefore, the paper is limited to a description of our method, illustrated on very simple electronic components such as resistors, capacitors or inductors. The analogy with pipes, tanks... is obvious. Its application to process diagnosis is less obvious since many other aspects than component modeling must be taken into account, for instance: what is to be diagnosed, what ambiguity is acceptable, what are the measurements... Some of these points are also discussed.

Keywords: Fault Detection, Fault Isolation, Diagnosis, Electronics, Artificial Intelligence, Dynamic Systems. Envelopes, Intervals.

1. INTRODUCTION

The diagnosis of analog electronic circuits, and more generally of analog systems, raises many specific difficulties which justify the development of specific methods. Thus, the most significant problems are:

- quantities are defined on a continuous domain and vary over time,
- available models are differential equations which are, except in the very specific case of resistors, capacitors and inductors, piecewisely defined; moreover, parameters of these equations are subject to dispersion, which means that, at diagnostic time, one only knows the interval in which parameters are supposed to be when the component is correct, and this interval can be very large[1],
- there is a very large number of possible failures, because the domain on which parameters are defined is continuous; moreover, multiple faults are not a rare occurrence, since the failure of one component can cause the destruction of others,
- the knowledge of the nominal behavior of the correct system -e.g. for a given input- is not valuable for diagnosing since a component failure can provoke the other components (though perfectly correct) to shift well away from this nominal behavior. This is especially the case when the system contains feedback loops, which is the general case,
- measurements are scarce (at best, voltages can be measured) and inaccurate. A particular aspect of analog systems is that measurements are obtained through a sampling procedure and that, in the case

1.e.g. [50,300] for the gain of a transistor

of electronics, the sampling rate cannot always be in accordance with the measurement to do; for instance, because of the very fast switching time of components (possibly less than 100 picoseconds), it is often the case that signal variation is several order of magnitude faster than the functional time constants of the circuit. As a consequence, the Shannon theorem cannot apply and the signal cannot be safely reconstructed from the measurements. Practically, an envelope of the signal can be computed taking into account the inaccuracy of individual samples, but it is generally very difficult to compute acceptable derivative approximations.

With the model based diagnosis framework (De Kleer and Williams, 1987; Reiter, 1987), some of these difficulties have naturally been overcome, but others, for instance the inaccuracy of parameter and measurement values and the time management, have required the development of specific methods. Some purely qualitative approaches have been tested (Pan, 1984; Dague, *et al., 1987*; Ng, 1990), but the ambiguity and the inaccuracy they introduced do not allow to detect many failures.We proposed in CATS (Dague, *et al.*, 1991), a first solution to these problems (inaccuracy and time management) for systems whose observations are "repeatable" (for such systems, either the measurements of the different quantities are made simultaneously, or - as it is often the case in electronics - the same system state reoccurs periodically allowing sequential measurements) and whose fault status does not change during the diagnostic process. In CATS, physical quantities are represented by arrays of intervals, as they are issued from the sampling process. The models of correct behavior of components are described by functional relations between physical quantities. Numeric intervals represent the value of component parameters and their inaccuracy. Part of the temporal properties of components are expressed by using the first derivative of quantities. The diagnostic procedure consists in an interval propagator coupled with an Assumption-based Truth Maintenance System (ATMS) (De Kleer, 1986) which only works for approximation of derivatives to slowly varying parts of signals, with the consequence that some conflicts are not detected.

In 1992, a general evaluation of CATS, ordered by the french ministry of defense, gave very promising results. Since CATS is domain independent, it has been tested in other domains. First for the power supply system of a satellite (Devès, *et al.*, 1991) -which is not very far from electronics- and then to a spanish petrochemical plant (Alarcon, *et al.*, 1994). Both experiments were considered successful. The final industrialization of the CATS program -ordered by DGA/SPART- has started in 1997. The project -called C-DIAG/DIAMANT- will result in a library of basic modules with Application Programmatic Interfaces (API) designed to be embedded in any kind of systems and especially in Automated Test Equipments (ATE) of electronic systems. The DIAMANT part of the project consists in putting in concrete form the ability of the C-DIAG library to adapt to various environments by carrying out a diagnostic system for the Aerospatiale SESAR 400 ATE.

The story could have stopped here -at least for its research aspect- but the numerous experiments we conducted with our system let us think that improvements could be achieved and once again in the modeling techniques. In the CATS program signal representation is not sparse, component models are restricted to first order derivatives and, the last but not the least, numeric approximation of signal derivatives has to be done, which is obviously not easily achieved in a safe manner -especially when faulty circuits are considered, since any kind of signal might occur because of the fault.

That the reasons why we started in 1993 a new research project based on a representation of signals with temporal band sequences and whose results are presented in section 2. An example is given in section 3. Finally, in section 4, we discuss the limits of our approach in the perspective of its use in the diagnosis of continuous processes.

2. DIAGNOSING WITH TEMPORAL BANDS

2. 1 Modeling

The first point of our approach is that the system to be diagnosed has to be modeled as a set of equations of the form:

$$\bigwedge_i \neg ab(C_i) \supset f + g' + h'' + \ldots = 0$$

where f, g, h,... are functions of measurable quantities of the system to be diagnosed and C_i the components involved in the generation of the functions. The predicate ab(.) means abnormal. We call these equations *residuals* since they are of the same nature as the linear combination of input, output and their derivatives considered in parity space approaches used in fault detection and localization of dynamic systems (Greater and Singer, 1990). The principle of model-based diagnosis consists in finding out inconstancies in the right-hand part of the residual in order to prove, by contraposing the implication, that at least one of the C_i is faulty. In the AI community, theses sets of C_i are called *nogoods*, hitting sets of nogoods are the *candidates*, that is sets of components whose (faulty) behavior explain the observed symptoms (De Kleer and Williams, 1987; Reiter, 1987). This approach was extended to fault modes (De Kleer, 1989).

Generally, these residuals are not directly available in the usual description of physical systems. They have to be obtained by formally composing models of elementary components in eliminating the non-measurable quantities. For example, if we consider a part of a system composed of a real[1] capacitor and a real inductor in series, a residual can be derived from the models of elementary components involving current and voltage, and from Kirchhoff laws:

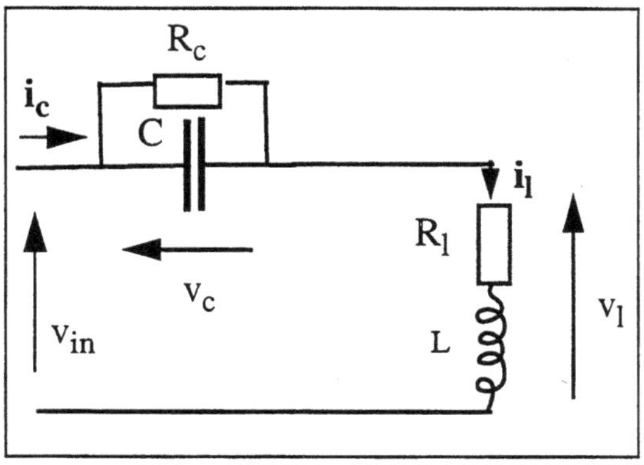

Models of elementary components

$$\neg ab(C) \supset i_c = C \cdot v_c' + v_c/R_c$$

$$\neg ab(L) \supset V_l = L \cdot i_l' + R_l \cdot i_l$$

Kirchhoff law on currents

$$i_c = i_l \quad \text{and} \quad i_c' = i_l'$$

The following residual can be derived from the previous equations:

$\neg ab(L) \wedge \neg ab(C) \Rightarrow$

$$v_l - \frac{R_l}{R_c} \cdot v_c - \left(R_l \cdot C + \frac{L}{R_c}\right) \cdot v'_c - L \cdot C \cdot v_c'' = 0$$

which corresponds to the functions:

$$\left| \begin{array}{l} f = v_l - \frac{R_l}{R_c} \cdot v_c \\ g = -\left(R_l \cdot C + \frac{L}{R_c}\right) \cdot v_c \\ h = -L \cdot C \cdot v_c \end{array} \right.$$

Our main result is that checking the consistency of such a residual is possible as soon as a piecewise polynomial envelope of the functions f, g, h... is known. We now define these envelopes, that we call "Temporal Band sequences" (TBS).

2. 2 *Temporal Band Sequences.*

We first define the concept of "temporal band" which will allow us to approximate the value of physical quantities on the temporal interval [t1, t2[.

Definition 1: "Temporal band"

A temporal band is a tuple (t_i, t_s, p_i, p_s) where t_i, t_s are real numbers, p_i, p_s are the polynomials

$$p_i(t) = \sum_{k=0}^{m} a_{ik} t^k \quad \text{and} \quad p_s(t) = \sum_{k=0}^{n} a_{sk} t^k \quad \text{such that:}$$

$$t_i < t_s \quad \wedge \forall t \in [t_i, t_s[, \quad p_i(t) \le p_s(t)$$

The interval $[t_i, t_s[$ is the *temporal extent* of the band and the highest degree of the polynomials p_i, p_s (i.e. max(m, n)) is the *degree* of the band.

We will denote by $B(t) = [p_i(t), p_s(t)]$ the value (an interval) of a band $B = (t_i, t_s, p_i, p_s)$ at the instant t.

Example: the next figures show two bands of degree 1 and 2

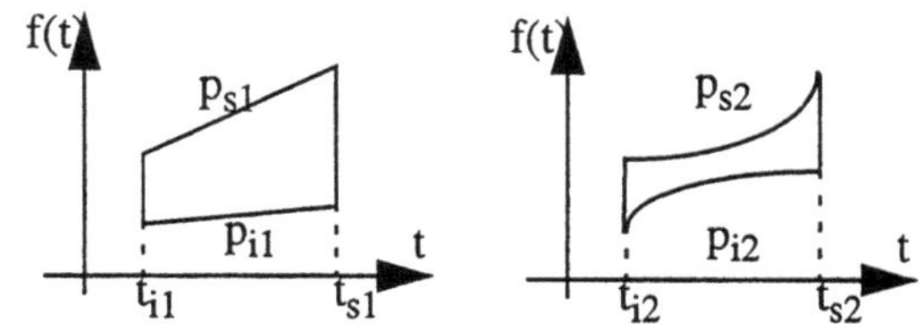

A high degree temporal band is, in general, mandatory to approximate an observation well. Nonetheless, as we will see later, our motivation to introduce polynomials of any degree in the temporal band definition is more to be able to process derivatives of higher order than to get a better approximation of an observation. That is the reason why we now define "temporal band sequences", which allows a piecewise approximation of observations:

Definition 2: "Temporal band sequence"

A temporal band sequence SB is a finite ordered set of temporal bands $\{B_1, ..., B_n\}$ such that:

$$\forall k \in [1, n-1], t_s(B_k) = t_i(B_{k+1})$$

where $t_i(B)$ is the initial instant of the band B and $t_s(B)$ the final instant.

The interval $[t_i(B_1), t_s(B_n)[$ is the *temporal extent* of the band sequence and the highest degree of the bands is the *degree* of the temporal band sequence.

We will denote by SB(t) = B(t) the value (an interval) of the temporal band sequence SB at the instant t (B is the temporal band of SB whose temporal extent contains t).

We can now define how to represent an observation:

Definition 3: "Representation of an observation"

A temporal band sequence SB whose temporal extent is $[t_1, t_2[$ is a representation of an observation v(t) over a time interval $[t_1, t_2[$ iff the observation is always encompassed within the bands, which means:

$$\forall t \in [t_1, t_2[, \exists (t_i, t_s, p_i, p_s) \in SB, \text{ such that:}$$

$$t \in [t_i, t_s[\quad \wedge p_i(t) \le v(t) \le p_s(t)$$

Example: the band sequences SB_1 and SB_2 are two possible representation of v(t)

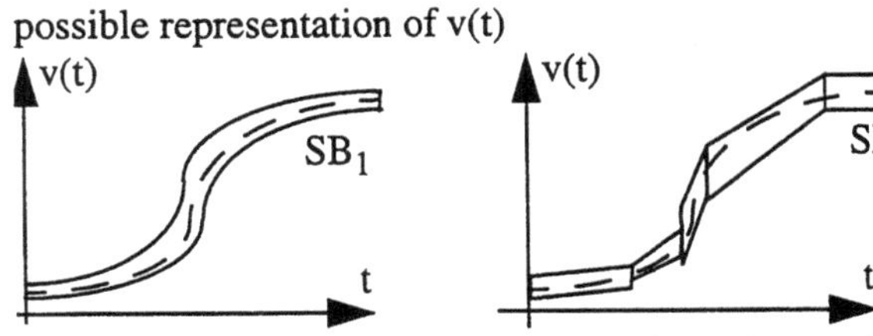

The way a TBS can be obtained depends on the knowledge available on the measurement process. For instance, if the primary measurement is a set of samples (fig a), a set of intervals can be computed, taking into account the accuracy of the device (fig. b). Then, hypothesis on the behavior of the signal between

---

1.Real means that the internal resistor is taken into account.

intervals has to be made. For instance, it might be reasonable to assume that the signal between two consecutive samples is always encompassed by the two corresponding intervals. Then polynomials are of 0th degree as it is illustrated on fig.c. Other hypotheses might give a more precise approximation (fig d).

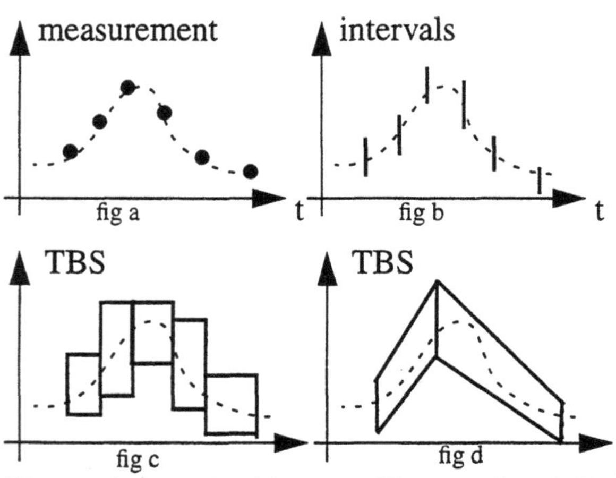

The consistency algorithm we will present later is linear in the number of bands; hence, it is preferable to compute TBS of higher degrees with less bands inside. On the contrary, since polynomial equations have to be solved, it might be interesting to stick to lower degrees in order to apply analytical methods of resolution. The experiments we did so far show that computing bands of 0th or 1^{rst} degree is a good trade off.

Various operations on TBS (sum and product between TBS and between an interval and a TBS)[1] have been defined which allow to compute the TBS representing the functions f, g, h... of a residual when a TBS is available for every observable quantity occurring in the functions and when an interval is known for every parameters. We also defined the TBS $\int_\theta SB$ as the integral of a temporal band sequence at time θ such that, if SB is an approximation of the function g(t) then $\int_\theta SB$ is an approximation of the function $\int_\theta^t g(x)dx$ as illustrated below:

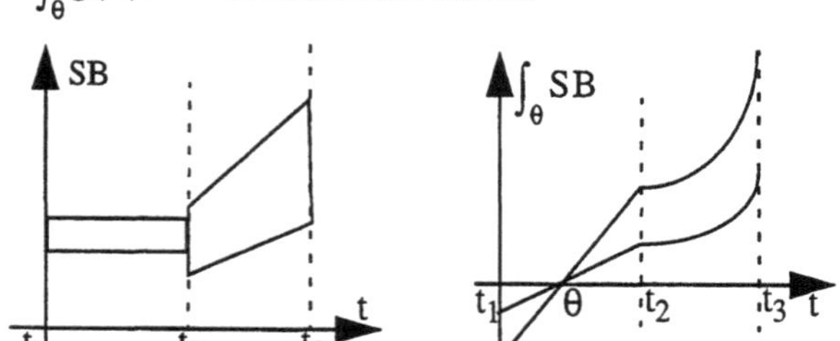

A complete description of these operations can be found in (Loiez and Taillibert, 1997; Loiez, 1997).

We now explain the way we use temporal band sequences for checking the consistency of our residuals.

1.Complete definitions and proofs can be found in (Loiez and Taillibert, 1997) and (Loiez, 1997)

2. 3 Consistency checking of residuals

We first consider <u>first-order residuals</u>.

Since or any θ the following holds:

$$\forall t,\ f(t) + g'(t) = 0 \Leftrightarrow \forall t,\ \overbrace{\int_\theta^t f(x)dx + g(t) - g(\theta)}^{k(t)} = 0 \quad ,$$

checking the consistency of a first order residual (f+g'=0) consists in:

computing SB_f and SB_g, a TBS approximation of f and g,

- choosing an integration instant θ,
- computing $K_\theta = \int_\theta SB_f + SB_g - SB_g(\theta)$ the TBS[2] which approximates k(t),
- checking whether the TBS K_θ includes 0 or not.

For <u>second order residuals</u>, the same approach can be followed, thus:

$\forall t,\ f(t) + g'(t) + h''(t) = 0 \Leftrightarrow$

$$\forall t,\ \int_\theta^t f(x)dx + g(t) - g(\theta) + h'(t) - h'(\theta) = 0$$ if

H is an interval approximating h'(θ), the right-hand side can be written:

$\forall t,\ K(t) + h'(t) = 0$

with: $K(t) = \int_\theta^t f(x)dx + g(t) - g(\theta) - H$

a TBS for K(t) can be computed when TBS for f and g are known; $\forall t,\ K(t) + h'(t) = 0$ is a first order residual which can be checked by the corresponding procedure. <u>Higher order residuals</u> are processed (recursively) in the same way.

2. 4 Example

We now illustrate this algorithm on the second order circuit presented in §2.1. The fault is supposed to be that the inductor is open. In order to simplify the numeric values, we will take R_c, R_l, C, L all equal to 1. The residual becomes:

$f = v_l - v_c$, $g = -2\ v_c$, $h = -v_c$.

Assuming a step as input signal (V_{in}), the measurements of V_c and V_l can be approximated by the two following TBS. Accuracy is purposely bad in order to illustrate the ability of our approach to find out diagnoses with imprecise measurements.

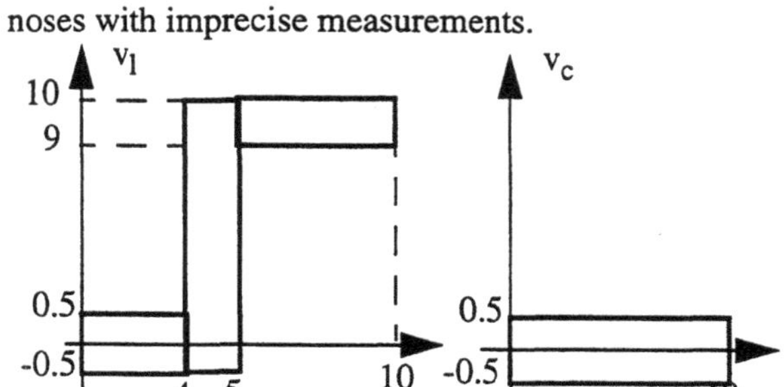

The TBS approximating f, g and h can easily be obtained by combining the TBS representing V_c and

2. $SB_g(\theta)$ is the interval value of SB_g at time θ

V_l, which gives:

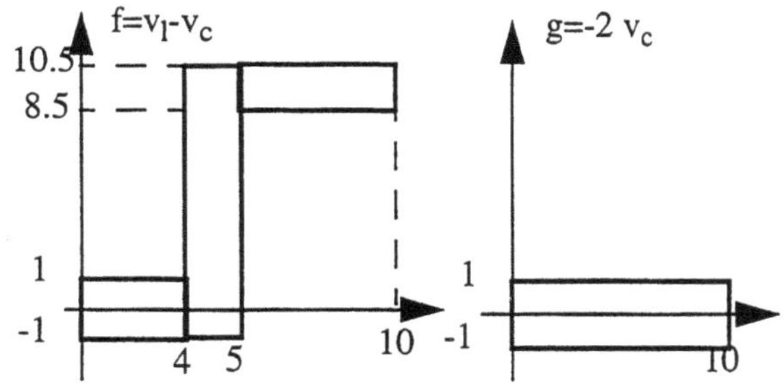

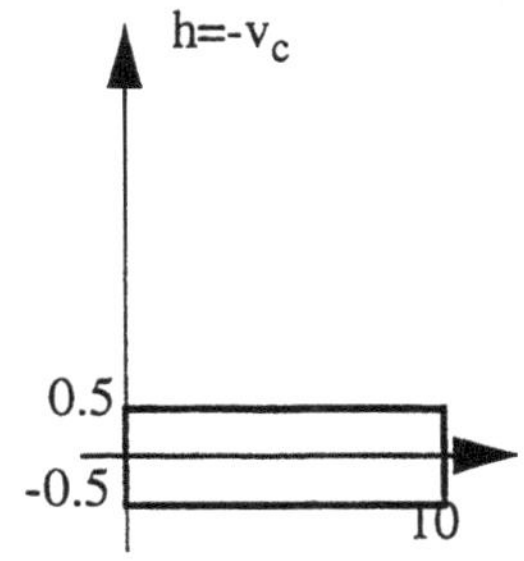

If we choose 0 as integration instant and assume that derivatives are null at 0, the consistency checking procedure consists in computing a TBS approximating: $K(t) = \int_0^t f(x)dx + g(t) - g(0) - H$ with H=0; it gives:

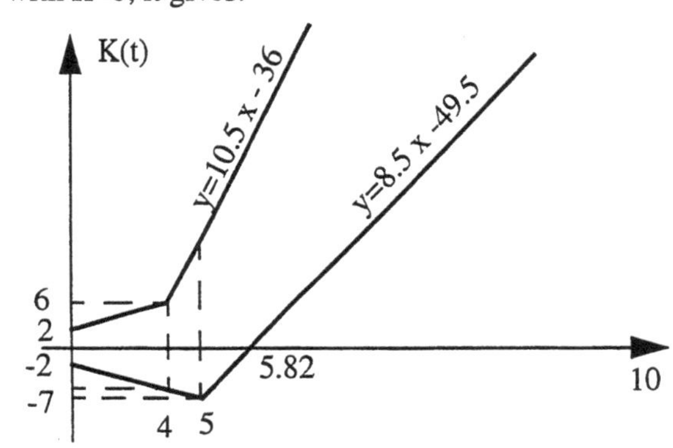

One just has to check the consistency of $\forall t, K(t) + h'(t) = 0$ which is a first order residual. Any integration instant can be chosen since no derivative approximation is anymore required. Let's choose 0 again, one has to compute the TBS

$$J_0 = \int_0 SB_K + SB_h - SB_h(0)$$

$SB_h - SB_h(0)$ is a constant TBS between -1 and 1.

As it can be seen on the TBS representing J_0, for all instants after 8.31, J_0 is not equal to 0 (the dashed part) which proves the inconsistency and that the inductor or the capacitor does not behaves correctly.

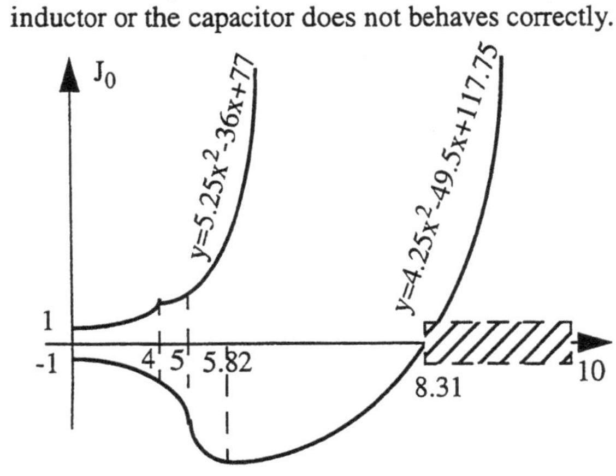

2. 5 Conclusions

This procedure has been tested on various circuits including the monostable reference problem (Dague and Taillibert, 1996) leading to results at least as good as those obtained with CATS (Dague *et al.*, 1991) without the risk of faulty diagnosis due to the computation of derivatives made by CATS in the fast-varying part of signals. Nonetheless, the procedure has two main drawbacks:

- one or several integration instants have to be chosen and this choice might lead to miss inconsistencies, because of the inaccuracy introduced by the integration process. It would be the case of the capacitor/ inductor example if the observation window was smaller (e.g. limited to 8.31).
- when second order (or beyond) residuals are considered, an approximation of the derivatives of the functions h, k... is still necessary at the integration instant (the H term above). Though this instant can be freely chosen, the only safe way to approximate derivatives when data are inaccurate is to do it in a steady part of the observation window, which does not always exist.

These drawbacks were an incentive to find out new algorithms. Recent results we obtained allow to avoid any derivative approximation (Cygan, 1997). They are based on multiple integration of the residual and also provide solutions for piecewisely defined residuals. They will be soon submitted for regular publication.

3. DISCUSSION

In spite of our poor knowledge of the problems occurring in the diagnosing of chemical processes, we will attempt a discussion about some aspects which could be an impediment to applying our method to the domain of chemical processes.

An hypothesis underlying our work is that the diagnosis is done on a temporal window and that the fault status does not change on that window. It is well adapted for fast varying dynamic systems. In the case of chemical processes, the time scales are completely different and it is the case that the fault occurs between two samples of the same measurement. The problem then consists in detecting and isolating the fault as soon as possible. A naive application of our method would be to dynamically extent the observation window and redo the consistency checking for each new sample. Although we did not investigate this point, it seems possible to devise an incremental consistency algorithm in order to avoid repeating the same calculations.

Our approach has the advantage of not requiring derivative measurements; but the residuals we consider can only refer to measurable quantities. That means that "internal" measurements of process quantities are necessary and that referring only to input/ output of the process is not realistic. The less measurements, the more ambiguous diagnosis (and the more complex residuals). In electronics, only voltages are available (at best), that is less than half the physical quantities determining the behavior of the circuit. It happened to be sufficient w.r.t. the ambiguity usually required in this domain.

Another advantage to be considered is that measurements can be carried out and processed even if the sampling rate is not strictly in accordance with the signal to be measured (the "step" in the L/C example). That might happen useful when a change of mode occurs in the process to be diagnosed, e.g. when a valve is switch on; (Mostermann and Biswas, 1997), though with other solutions, are faced to the same problem.

Finally, it is also worth to mention that our approach is strictly dedicated to diagnosis while many other methods, including those which consider objects comparable to our temporal band sequences (Berleant and Kuipers, 1997; Kay and Kuipers, 1993; Vescovi *et al.*, 1995), also provide simulation results. This is not an impediment for electronics since the huge dispersion on the value of component parameters already mentioned make simulation results of a weak interest for diagnosing. It might be different in others domains.

CONCLUSIONS

The approach of the diagnosis of analog electronic circuits we have proposed makes it possible to:

- model components with differential equations whose parameters are (strongly) inaccurate,
- take into account inaccuracies of measurements (including temporal inaccuracies),
- avoid full derivative approximation.

Recent improvements allow piecewise models of components and avoid any derivative approximation (not even at some initial instant).

These characteristics let think that it would be reasonable to experiment this approach in the context of the chemical process industry in order to define the necessary extensions (such as an incremental version of the consistency procedures) imposed by the particularities of the chemical process domain.

REFERENCES

Alarcon, I., L.B. Rodriguez-Marin, L.B. Almeida, R. Sanz, L. Fontaine, P. Gomez, X. Alaman, P. Nordin, H. Bejder and E. De Pablo(1994). Heterogenous Integration Architecture for Intelligent Control Systems. *Int. Journal of Artificial Intelligence in Engineering,* Special Issue.

Berleant, D. and B.J. Kuipers(1997). Qualitative and Quantitative Simulation, bridging the gap. *Artificial Intelligence Vol. 95 # 2, pp 215-255.*

Cygan, C. (1997). Amélioration de Procédé de Diagnostic Automatique de Systèmes Analogiques (in french). *Rapport de fin d'études de l'Ecole Centrale de Lyon.* Lyon, France.

Dague, P., P. Devès and O. Raiman (1987). Troubleshooting: when Modeling is the Trouble. *Proceedings of AAAI Conference.* Seattle, USA.

Dague, P., P. Devès, P. Luciani, O. Jehl and P. Taillibert (1991). When oscillators stop oscillating. *Proceedings of the twelfth IJCAI,* pp 1109-1115. Sydney, Australia.

Dague, P. and P. Taillibert (1996). The Monostable: A reference Problem for Analog Diagnosis. *DX'96, 7th Int. Workshop on Principles of Diagnosis.* Val Morin, Quebec, Canada..

Devès, P., C. Fischer and P. Taillibert (1991/1992). An Alternative to Expert Systems for Electrical Diagnosis. *Future Generation Computer Systems,* **Vol 7,** pp 343-352.

Kleer, J. (1986). An Assumption-based TMS. *Artificial Intelligence,* **Vol 28**, pp 127-162.

Kleer, J. and B.C. Williams (1987). Diagnosing Multiple Faults. *Artificial Intelligence,* ***Vol 32*** pp 97-130.

Kleer, J. and B.C. Williams (1989). Diagnosis with Behavioral Modes. *Proceedings of the eleventh IJCAI,* pp 104-109.Detroit, USA.

Gertler, J. and D.Singer (1990). A new structural framework for parity equation-based failure detection and isolation. *Automatica,* **Vol 26 #2**, pp 381-388.

Kay, H. and B.Kuipers (1993).Numerical Behavior Envelopes for Qualitative Models. *Proceedings of AAAI Conference, Washington, USA.*

Loiez, E. and P. Taillibert (1997). Polynomial Temporal Band Sequences for Analog Diagnosis. *Proceedings of the fifteenth IJCAI,* pp 474-479. Nagoya, Japan.

Loiez, E. (1997). Contribution au Diagnostic de Systèmes Analogiques, *Thèse de l'Université des Sciences et Technologies de Lille -in french-.* Lille, France.

Mostermann, P. and G.Biswas (1997). Formal Specifications for Hybrid Dynamical Systems. *Int. Joint Conf. on Artificial Intelligence. Nagoya, Japan.*

Ng, H.T. (1990). Model-based Multiple Fault Diagnoses of Time Varying, Continuous Physical Devices. *Proceedings of the sixth IEEE Conference on AI Applications.* Santa Barbara, USA.

Pan, Y. (1984). Qualitative reasoning with deep-level mechanism models for diagnoses of mechanism failures. *Proceedings of the first IEEE Conference on AI Applications.* Denver USA.

Reiter, R. (1987). A Theory of Diagnosis from first Principles. *Artificial Intelligence* **Vol 32**, pp 57-95.

Vescovi, M., A. Farquhar, Y. Iwasaki (1995). Numerical Interval Simulation: Combined Qualitative and Quantitative Simulation to Bound Behaviors of Non-Monotonic Systems. *Int. Joint Conf. on Artificial Intelligence.* Montreal, Canada.

KERNEL REGRESSION AND NEURAL NETWORKS FOR MODEL–FREE FAULT DIAGNOSIS

G. Fenu* T. Parisini*,[1]

* *Dept. of Electrical, Electronic and Computer Engineering, DEEI–University of Trieste, Italy*

Abstract: Correct and timely fault detection is of major importance in the field of system engineering, and constitutes a primary problem in a broad spectrum of cases, from industrial processes to high-performance systems and to mass-produced consumer equipment. A large number of methods can be found in the literature, and the recent use of non–parametric techniques and neural networks for solving fault-diagnosis problems in real industrial situations seems to be particularly promising. This paper describes a novel approach combining kernel–regression methods and neural networks to solving approximately some difficult fault-diagnosis (FD) problems. The method falls into the *model-free* class of FD techniques as no model of the plant is needed. The kernel–regressor makes it possible to detect changes in the plant dynamics, possibly due to some malfunction, whereas the neural network is used to build a decision scheme to actually diagnose the fault. A theoretical sufficient condition for fault detectability is also illustrated. *Copyright © 1998 IFAC*

Keywords: Model–free Fault Diagnosis, Kernel Smoothing, Neural Networks

1. INTRODUCTION

Monitoring complex nonlinear plants for the purpose of automated fault-detection is a task of fundamental importance in many areas of control engineering. Therefore, since the early '70s, a lot of approaches to fault–diagnosis (FD) have been developed that may be roughly classified as (i) model-based FD approaches and (ii) model-free FD approaches.

In general, model-based FD algorithms are very different from one another. Many works on such algorithms can be found in the literature (see, for instance, the classical surveys (Frank, 1990; Isermann, 1984; Willsky, 1976; Gertler, 1988), the recent special issue of *Control Engineering Practice*, vol. 5, 1997, and the references cited therein). Some of these algorithms (Frank, 1990; Willsky, 1976; Gertler, 1988) derive fault information from an optimal or suboptimal estimation of state variables. In other cases (Frank, 1990; Isermann, 1984), it is necessary to estimate the parameters of some suitable models to be identified. Other model-based methodologies rely on the construction of so-called "parity space" (Frank, 1990). Another very important class of approaches is the one based on probabilistic methods; for this class, "model–based" should be interpreted in a probabilistic sense (Basseville and Nikiforov, 1993). In some cases (Frank, 1990; Willsky, 1976), model-based algorithms exploit useful fault information derived from suitable physical redundancies. Finally, examples of methods developed on the basis of fault-detection-oriented control design can also be found (Jacobson and Nett, 1991).

It is worth noting that the applicability of model-based methods to real processes may be rather limited, in that, on one hand, these methods require realistic and accurate process models in

[1] All correspondence should be sent to Prof. T. Parisini, DEEI–University of Trieste, Via Valerio 10, 34127 Trieste, Italy, Email: thomas@dist.unige.it

order to yield reliable diagnostic signals, and, on the other hand, they are characterized by very restrictive hypotheses about process models (almost all model-based methods can be applied only to *linear* process models). The few proposed model-based methods applicable to nonlinear process models are sometimes difficult to apply because of both the presence of hard nonlinearities (hysteresis, dead zones, saturations, etc.) and the possible high dimensionality of the related real systems. Therefore, it is also evident that to use the linearized model of a system of this type for a model-based method would be of no benefit, due to the complete unreliability of the diagnostic signals that would be obtained.

On the other hand, model-free methods do not require the model of a plant to generate fault signals. This approach has been developed more recently and consists mainly in addressing a problem from the control-board operator's point of view. By comparing the measured and desired values of the controlled variables, the operator can decide on the operating mode for a system, without the need for defining an explicit dependence law. In this context, sometimes the problem can be assimilated to a pattern-recognition problem (see, for instance, (Guglielmi *et al.*, 1995) and the references cited therein).

Clearly, model-free methods have advantages and disadvantages that are opposite to the ones of model-based approaches. If a good and reliable model of a process is available, model-based techniques are clearly superior, but when such a model is not available, model-free methods are preferable.

In this paper, a novel way of using the Kernel Regression (KR) methodology in the context of model-free FD for nonlinear systems is proposed. KR is a well-known statistical tool (see, for instance, (Härdle, 1990)) that is generally used to determine smoothed interpolations of noisy scattered data. However, within the FD context, we propose a completely different way of applying the KR technique. The basic qualitative idea is the following: when a fault occurs, also some change in the smoothness characteristics of the time-behaviors of the measurable variables may occur. This change is reflected in modifications to the typical features of the kernel smoother applied over some suitable temporal batch of the measurable variables, and this could be interpreted as a *fault symptom* to be fed into the decision scheme based on a neural classifier (Lippman, 1989). The neural classifier may be trained off-line to associate the fault symptoms with some eventula critical behavior of the plant (for instance basing on some historical data).

2. THE KERNEL SMOOTHING TECHNIQUE

In this section, the kernel smoothing technique will be briefly described in the context of dynamic systems, and the statements of some basic definitions will be provided (see, for instance, (Härdle, 1990)).

Let us define a generic model of the closed-loop plant

$$\begin{cases} x_{t+1} = f(x_t, r_t, t) \\ y_t = h(x_t, t) \end{cases}, \quad t \geq 0 \qquad (1)$$

where $x_t \in \Re^n$ represents the state vector, $y_t \in \Re^p$ represents the output measurement vector, and $r_t \in R \subset \Re^q$ represents the vector of set-point signals. The time-variance of the above difference equations is introduced to take into account the possible occurrence of some malfunction.

Let us consider the batch of measurement vectors $y_{t-T}^t \triangleq \mathrm{col}\,(y_{t-T}, \ldots, y_t)$ over the moving time-window $[t-T, t]$, for $t \geq T$, and let us interpret y_{t-T}^t as the sampling of some continuous function $y(t)$ over the finite time interval $[t-T, t]$. For the sake of notational simplicity and without loss of generality, in the following we address only a scalar case, as the generalization to a non-scalar situation is only a matter of a slightly more complex notation. The kernel smoother (KS) is defined as

$$\hat{m}_\rho(\tau) = \frac{\sum_{i=t-T}^{t} y_i K_\rho(\tau - i)}{\sum_{i=t-T}^{t} K_\rho(\tau - i)}, \qquad \forall \tau : \tau \in [t-T, t] \quad (2)$$

where $K_\rho(z) \triangleq 1/\rho_t\, K(z/\rho_t)$, $K(z)$ denotes the *kernel function* and ρ_t denotes the *kernel bandwidth* at the time-instant t. The *kernel* is a continuous, bounded and symmetric real function such that $\int K(z)\mathrm{d}\,z = 1$. In qualitative terms, the function $\hat{m}_\rho(t)$ given by (2) implements an approximate smoothing of the unknown function $y(t)$ over the time-interval $[t-T, t]$. In order to better clarify how the KS is applied, we refer the reader to Fig. 1.

The choice of the kernel bandwidth ρ_t is clearly an important issue in using the KS technique. Indeed, ρ_t will play a key role in fault symptom generation. Several different criteria can be used to detemine ρ_t, and, in the present context, a useful strategy is the one of minimizing an average cost function, that is

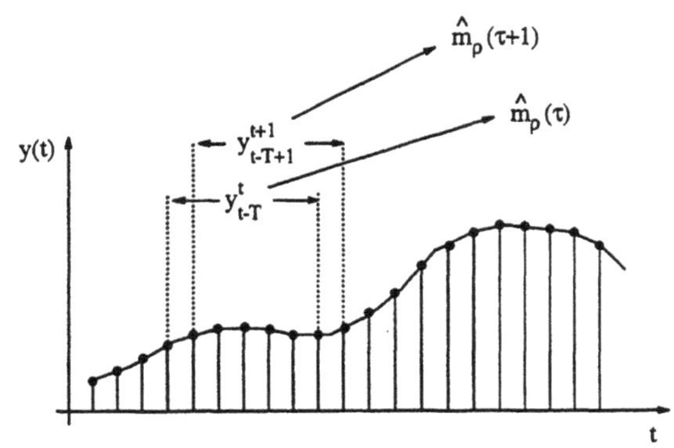

Fig. 1. Computation of $\hat{m}_\rho(\tau)$ over time.

$$\rho_t^\circ = \arg\min_{\rho_t} \int_{t-T}^{t} \mathrm{E}\,[y_\tau - \hat{m}_\rho(\tau)]^2\,\mathrm{d}\,\tau$$

Now, if we assume that the unknown function $y(t)$ can be modelled as

$$y(t) = m(t) + \varepsilon$$

where ε is a zero-mean random variable with a *constant* variance σ^2, and if we assume that the function $m(t)$ is twice-continuously differentiable, then it can be shown that

$$\rho_t^\circ = \frac{1}{N^{1/5}} \left[\frac{\sigma^2\, c_K}{d_K^2 \displaystyle\int_{t-T}^{t} [m''(z)]^2 \mathrm{d}\,z} \right]^{1/5} = \left[\frac{\sigma^2\, c_K}{N d_K^2}\right]^{1/5} \frac{1}{\left(\displaystyle\int_{t-T}^{t} [m''(z)]^2 \mathrm{d}\,z\right)^{1/5}} \quad (3)$$

where $c_K \triangleq \int K^2(z)\mathrm{d}\,z$ is the "energy" of the kernel function $K(z)$, $d_K^2 \triangleq \int z^2 K(z)\mathrm{d}\,z$ is the second moment of the kernel function $K(z)$ (interpreted as a probability density function), $m''(z)$ denotes the second derivative of the scalar function $m(\cdot)$, and N is the number of samples y_t (in our case, $N = T + 1$).

In the next section, the above basic concepts will be used for model-free fault symptom generation.

3. FAULT SYMPTOMS GENERATION

As stated in (Isermann, 1997) and also in the framework of recent meetings of the *IFAC Techn. Committee on SAFEPROCESS, a fault symptom can be defined as any change in a quantity; the change is generated on the basis of observable variables and can be related to some faulty behavior*. In this respect, the basic idea motivating the use of the KS for fault symptom generation is the following: the KS provides a smoothed interpolation of the moving batch y_{t-T}^t of measurements provided by the available sensors (see Fig. 1). The KS bandwidth ρ_t (see (2)) depends on the smoothness of the underlying function $y(t)$. If some significant change in the system dynamics occurs, this should be reflected in a change in ρ_t and thus the time-behavior of ρ_t could be interpreted as a *fault symptom*. Then, the KS block is used as a *change detector* and a diagnostic decision can be taken, that correlates the fault symptom with the current set-point vector $r_{t-T}^t \triangleq \mathrm{col}\,(r_{t-T}, \ldots, r_t)$ (see Fig. 2) and with other possible available variables. This is done by the neural classifier (block NC) that will be briefly described in Section 4.

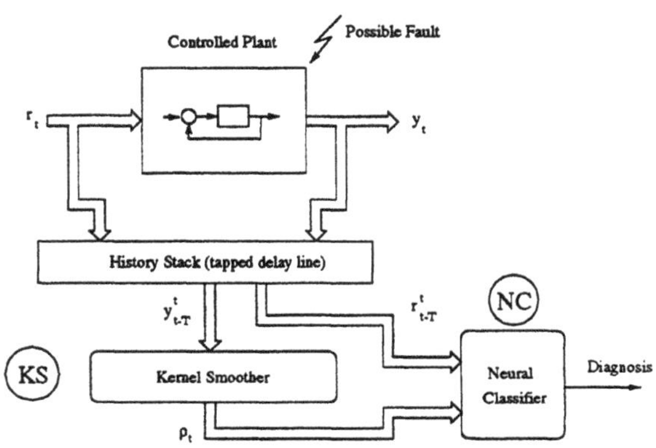

Fig. 2. Scheme of the diagnostic technique.

Now, in order to put the above idea into rather more precise terms, let us re-consider expression (3) for the optimal bandwidth ρ_t°. Clearly, ρ_t° depends on: (i) the chosen kernel function through the term c_K/d_K^2; (ii) the width N of the moving window ; (iii) the noise ε through the scalar σ^2; and (iv) the smoothness of the unknown function $m(t)$ through the term $\int_{t-T}^{t} [m''(z)]^2 \mathrm{d}\,z$. However, due to the assumed independence of time of the noise variance σ^2, and owing to the fact that the Kernel function and the time window are chosen once and hence do not depend on the time-instant t, it follows that the dependence of ρ_t° on time is only due to the term $\int_{t-T}^{t} [m''(z)]^2 \mathrm{d}\,z$.

Let $M^-_{\tau^-,\tau^+} \triangleq \min_{\tau\in[\tau^-,\tau^+]} [m''(\tau)]^2$ and $M^+_{\tau^-,\tau^+} \triangleq \max_{\tau\in[\tau^-,\tau^+]} [m''(\tau)]^2$, for some $\tau^-, \tau^+ \in \Re^+$, $\tau^- < \tau^+$. The following result can be proved (Fenu and Parisini, 1997).

Proposition. *Assume that some fault occurs at the time–instant t_f and suppose that there exist two time intervals, $[\tau_1^-, \tau_1^+]$ and $[\tau_2^-, \tau_2^+]$, with $\tau_1^+ - \tau_1^- \geq T$ and $\tau_2^+ - \tau_2^- \geq T$, and with $\tau_1^+ \leq t_f < \tau_2^-$, such that $[M^-_{\tau_1^-,\tau_1^+}, M^+_{\tau_1^-,\tau_1^+}] \bigcap [M^-_{\tau_2^-,\tau_2^+}, M^+_{\tau_2^-,\tau_2^+}] = \oslash$. Then, $\exists \Delta \in \Re, \Delta > 0$, $t^- \in [\tau_1^-, \tau_1^+]$, and $t^+ \in [\tau_2^- + T, \tau_2^+]$ such that $|\rho^o_{t^+} - \rho^o_{t^-}| \geq \Delta$.*

□

In qualitative terms, the meaning of the above result is the following (see Fig. 3): if the change in the system dynamics (due to the fault occurring at the time t_f) causes a significant variation in $m''(t)$, for $t \geq t_f$, then the KS is able to recognize it by means of a variation in the optimal bandwidth ρ^o_t within a suitable finite detection time. In this connection, the above sufficient condition can be interpreted as a *sensitivity* condition for the fault to be detected by the KS, and ρ^o_t can play the role of a *fault symptom*.

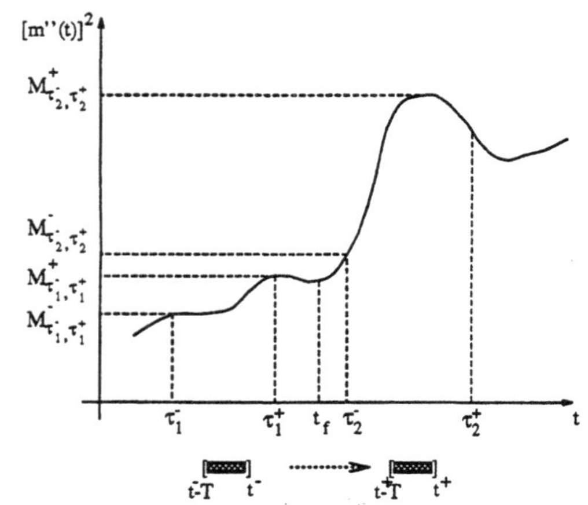

Fig. 3. Graphical interpretation of the sufficient detectability condition.

Remarks.

(1) It order that the above sufficient condition may be satisfied, a crucial role is played by the number of samples T; too large or too small values of T might result in the impossibility of finding the above time intervals $[\tau_1^-, \tau_1^+]$ and $[\tau_2^-, \tau_2^+]$. An analytical determination of T is in general impossible and experimental methods should then be used.

(2) Up to now, we have assumed the variance of the noise ε to be constant. Clearly, this may not be a reliable approximation in some case; consequently, the optimal kernel bandwidth may also change due to the non–stationarity of ε. In this case, the analysis is rather more complicated, in the sense that (3) should reflect such non–stationarity.

(3) In the above analysis, the effect of a system–dynamics change in the symptom vector has been considered. However, it is worth noting that this may not be sufficient to solve the diagnosis problem. Actually, not only faulty events might result in a significant change in ρ^o_t, and this is the reason for why it is necessary to consider also a suitable decision scheme (see Fig. 2).

3.1 *Practical issues*

Equation (3) is clearly very important from a theoretical standpoint but is difficult to apply in practice to calculate the value of the kernel bandwidth because it is true only in an asymptotic sense and involves complicated unknowns that have to be estimated from data.

The basic idea of all kernel bandwidth selection algorithms is to estimate a cost function easier to compute, like the following:

$$\frac{1}{T+1} \sum_{\tau=t-T}^{\tau=t} \mathrm{E}\,[y_\tau - \hat{m}_\rho(\tau)]^2$$

up to some constant. It would be a good result if the kernel bandwidth minimizing this estimate could also be an accurate estimate of the bandwidth minimizing the cost function itself.

One of the best known methods to find a good estimate of the "optimal bandwidth" is the *leave-one-out method* (Härdle, 1990). It is based on kernel smoothers in which one observation is left out:

$$\hat{m}_{\rho\,,\,\tau}(\tau) = \frac{\sum\limits_{i=t-T\,,\,i\neq\tau}^{t} y_i K_\rho(\tau - i)}{\sum\limits_{i=t-T\,,\,i\neq\tau}^{t} K_\rho(\tau - i)}$$

By using these modified smoothers, we can define

$$CV(\rho_t) \triangleq \frac{1}{T+1} \sum_{\tau=t-T}^{t} [y_\tau - \hat{m}_{\rho,\tau}(\tau)]^2 \quad (4)$$

The function $CV(\rho_t)$ is commonly called a *cross-validation function*, as it validates the ability to predict $\{y_\tau\}_{\tau=t-T}^{t}$ through the subsamples $\{(t, y_t)\}_{t\neq\tau}$. The estimate of the optimal bandwidth (3) is then given by

$$\hat{\rho}^o_t = \arg\min_{\rho_t} CV(\rho_t) \quad (5)$$

Remark. It is worth noting that (5) is an approximate estimate of the true bandwidth. However, early simulation examples showed that such effects are generally negligible if T is not too small.

In general, the choice of a good kernel function is not so important as the choice of the kernel bandwidth; therefore, in our FD technique, the choice criterion for the kernel function is simply the one of minimizing the computational burden associated with the specific kernel $K(z)$. According to this criterion, we use the *uniform kernel*

$$K_{unif}(z) = \begin{cases} 1/2 & \text{if } |z| \leq 1 \\ 0 & \text{otherwise} \end{cases}$$

as a kernel function.

4. THE NEURAL CLASSIFIER BLOCK

This section describes the main characteristics of the neural classifiers used for the proposed FDI methodology (for a survey of neural and nonneural classifiers, see, for instance, (Lippman, 1989)).

Let us refer again to Fig. 2. The classification task concerning block NC can be regarded as the one of developing a relation that makes it possible to decide on the membership of a set of vectors $\{\xi_j\ ;\ j = 1, \ldots, N\}$ in one of the classes belonging to a given set $\{c_i\ ;\ i = 0, 1, \ldots, M\}$. Therefore, it is necessary to determine a relation $\phi : \Re^{N_{\text{inputs}}} \rightarrow \Re^M$ such that if c_i^j is the class ξ_j belongs to, then $\phi(\xi_j) = \theta^i$, where $\theta^i \triangleq [\ 0 \cdots 0 \underset{i}{1} 0 \cdots 0\ ]^T$. In other words, the relation ϕ must partition the space $\Re^{N_{\text{inputs}}}$ into *decision regions*, with each of which a θ^i, $i = 1, \ldots, M$ has to be associated.

The determination of an approximate classifier $\tilde{\phi}$ is based on a sampling of the true relation ϕ, that is, a set of pairs $\mathcal{Tr} \triangleq \{(\xi_j, \theta_j),\ j = 1, \ldots, N\}$, where $\theta_j \in \{\theta^i, i = 1, \ldots, M\}$. $\mathcal{Tr}$ denotes the *training set*, which is used in the learning phase of the neural classifier. Approximate classifiers $\tilde{\phi}_w$ are considered, characterized by a vector w of parameters to be determined. The quality of such a classifier is strictly related to its *generalization* capabilities, that is, the capabilities of deciding correctly (i.e., according to relation ϕ) on the membership class of an input vector ξ not belonging to the training set $\mathcal{Tr}$. Clearly, the generalization capabilities depend on the type of classifier and on the number of classifier parameters that are available to satisfy the "constraints" imposed by the training set $\mathcal{Tr}$, that is, $\tilde{\phi}_w(\xi_j) = \theta_j$, $j = 1, \ldots, N$. This number is denoted by $d_w \triangleq \dim(w)$. On one hand, d_w must be sufficiently large to satisfy such constraints, at least in an approximate way. On the other hand, it is well known that better generalization performances are generally obtained by a classifier characterized by a relatively small number of parameters d_w.

In the present paper, two different neural classifiers are used: a *hyperplane classifier* based on a multilayer feedforward neural network and a *receptive fields classifier* based on a Radial Basis Functions neural network.

4.1 *The hyperplane classifier*

The multilayer neural network used in the present classification task has a single hidden layer; its structure is shown in Fig. 4.

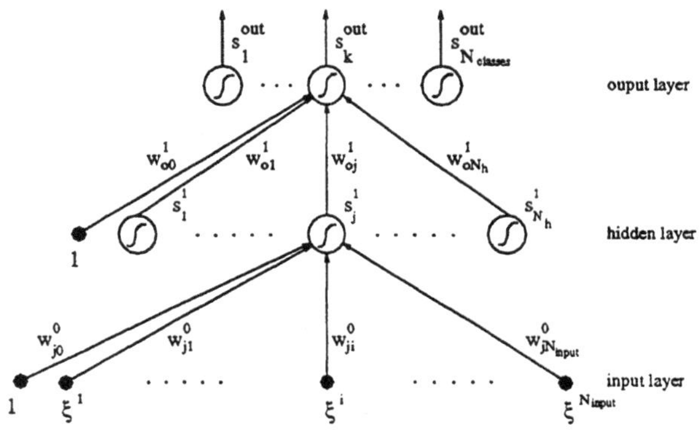

Fig. 4. One-hidden-layer feedforward neural network.

The class to which the input vector $\xi \triangleq [\xi^1, \ldots, xi^{N_{\text{inputs}}}]^T$ belongs is the class associated with the maximum output s_k^{out}. The activation functions of the neural units are sigmoidal, that is, $\text{sgm}(a) \triangleq \dfrac{1}{1+e^{-a}}$. Therefore, denoting by N_h the number of hidden units,

$$s_k^{\text{out}} = \text{sgm}\left[\sum_{j=0}^{N_h} w_{kj}^1 \cdot \text{sgm}\left(\sum_{i=0}^{N_{\text{inputs}}} w_{ji}^0 \xi^i\right)\right], \quad k = 1, \ldots, M$$

As is well known, in order to implement an arbitrary relation $\phi : \Re^{N_{\text{inputs}}} \rightarrow \Re^M$ to a certain accuracy, a single hidden layer is sufficient, provided that the number of hidden units, N_h, is large enough.

4.2 *The receptive fields classifier*

This classifier is based on a Radial Basis Functions network, whose structure is depicted in Fig. 5.

The activation functions of the neural network shown in Fig. 5 are multivariable Gaussian functions. Then, denoting by N_c the number of basis functions,

$$s_k^{\text{out}} = \sum_{j=1}^{N_c} w_{kj}\, a_j \cdot e^{-\|\xi - \rho_j\|^2_{\Sigma_j^{-1}}/2}, \quad k = 1, \ldots, N_{\text{classes}} \qquad (6)$$

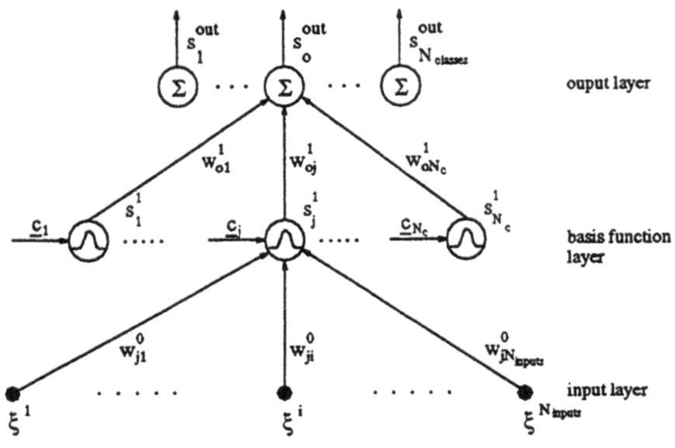

Fig. 5. Radial Basis Functions neural network.

where $a_j = \sqrt{(2\pi)^{N_{\text{inputs}}}} \cdot \sqrt{\|\Sigma_j\|}$. An RBF network is characterized by three sets of free parameters to be tuned: (i) the set $\mathcal{C} \triangleq \{\rho_j\}$ of the basis functions' centers; (ii) the set $\mathcal{V} \triangleq \{\Sigma_j\}$ of the Gaussian basis functions' covariance matrices; (iii) the weight matrix $W \triangleq [w_{kj}]$.

5. CONCLUSIONS

In the present paper, a novel approach to model-free fault diagnosis the generation of fault in nonlinear discrete-time systems has been proposed. The method does not require any model of a plant and is able to generate time-dependent symptom vectors associated with the possible faulty behavior of the plant. Moreover, a sufficient detectability condition has been stated. A neural classifier has been devised in order to take diagnostic decisions on the basis of the generated fault symptoms and the available information on the plant.

ACKNOWLEDGMENTS

This research lies in the framework of the EEC Esprit Working Group "IN-CONTROL". Moreover, the work was supported in part by the Italian Ministry for the University and Research (MURST) and by the Italian Space Agency (ASI).

6. REFERENCES

Basseville, M. and I. Nikiforov (1993). *Detection of Abrupt Changes-Theory and Applications.* Prentice-Hall, Englewood Cliffs, N.J.

Fenu, G. and T. Parisini (1997). A note on nonparametric kernel smoothing for model-free fault symptom generation. Technical Report 97/2. DEEI, University of Trieste.

Frank, P. M. (1990). Fault diagnosis in dynamic system using analytical and knowledge-based redundancy. A survey and some new results. *Automatica* **26**(3), 459–474.

Gertler, J. J. (1988). Survey of model-based failure detection and isolation in complex plants. *IEEE Control System Magazine* **8**, 3–11.

Guglielmi, G., T. Parisini and G. Rossi (1995). Fault diagnosis and neural networks: a power plant application. *Keynote Paper, IFAC Control Engineering Practice* **3**, 601–620.

Härdle, W. (1990). *Applied Nonparametric Regression.* Cambridge University Press.

Isermann, R. (1984). Process fault detection based on modeling and estimation methods–A survey. *Automatica* **20**(4), 387–404.

Isermann, R. (1997). Supervision, fault–detection and fault–diagnosis methods –An introduction. *IFAC Control Engineering Practice* **5**, 637–652.

Jacobson, C. A. and C. N. Nett (1991). An integrated approach to controls and diagnostic using the four-parameter controller. *IEEE Control System Magazine* **11**, 22–29.

Lippman, R. P. (1989). Pattern classification using neural networks. *IEEE Communications Magazine* **27**, 47–64.

Willsky, A. S. (1976). A survey of design methods for failure detection in dynamic systems. *Automatica* **12**, 601–611.

PRUNING AND EXTRACTION OF QUALITATIVE FAULT DIAGNOSIS RULES FROM A NEURAL NETWORK

J. Barry Gomm, Manori Weerasinghe and David Williams

Control Systems Research Group, School of Engineering
Liverpool John Moores University, Byrom Street, Liverpool L3 3AF, UK.
e-mail: j.b.gomm@livjm.ac.uk

Abstract: Artificial neural networks have been shown to be an effective tool for process fault diagnosis. However, a main criticism is that details of how fault diagnosis decisions are made are embedded in the complex input-output mapping performed by a network, and are in general difficult to obtain. In this paper, statistical techniques and relationships between fuzzy systems and standard radial basis function networks, are exploited to prune a trained network and to extract qualitative rules that explain the network operation for fault diagnosis. The procedures are demonstrated in an application to the diagnosis of a range of both soft and hard faults in a simulated, complex, chemical process. *Copyright © 1998 IFAC*

Keywords: rule extraction, network pruning, fault diagnosis, radial basis function networks, neuro-fuzzy networks

1. INTRODUCTION

Neural networks have successfully been applied in many applications in diverse fields. They are able to acquire knowledge from a given set of data, store it in a certain way, apply it generally to previously unseen data and predict the outcomes successfully. One of the main criticisms of neural networks, however, is the inability to explain the process of how a certain decision or outcome is reached. The potential of neural networks for classifying data for process fault diagnosis has been demonstrated (Sorsa *et al.*, 1991; Venkatasubramanian and Chan, 1989). In fault diagnosis applications, it is important to be able to explain the diagnosis process for several reasons which include the provision of additional information to facilitate decisions on the required action to be taken to isolate faults, and enhancing confidence in the diagnosis system.

Many research attempts have been projected towards extracting qualitative rules from trained neural networks to explain their behaviour. A common approach is to utilise a conventional feed-forward, multi-layer percepton network and to try to find combinations of input values to a unit that result in the unit having an activation near one (Saito & Nakano 1988, Fu 1991). Alternative network architectures have also been developed to facilitate rule extraction by relations with knowledge bases and fuzzy logic. The rule extraction algorithms developed by Towell and Shavlik (1993) require the initial network to be a Knowledge Based Neural Network (KBNN), making the method problem specific, and also made additional assumptions. A hybrid neuro-fuzzy network architecture has been used for the development of rules for diagnosis of faults in technical processes (Ayoubi, 1996). Whilst relationships between radial basis function (RBF) networks and fuzzy systems are known (Jang and Sun, 1993; Nie and Linkens, 1993) these relations have not been fully exploited to prune trained RBF networks and to extract rules from them explaining their behaviour.

In this paper, procedures for pruning and rule extraction from a trained RBF network are described. The procedures are demonstrated in an application of an RBF network to the diagnosis of faults in a simulated complex, chemical process. Results are presented to illustrate the application of the procedures and their effectiveness to reduce network complexity and extract qualitative rules explaining the network behaviour.

2. RBF NETWORK STRUCTURE

An RBF network structure is utilised and consists of n inputs, a single hidden layer of n_h non-linear processing nodes followed by a linear layer of m outputs. For fault diagnosis the network maps the input feature vector space to fault classification space. The network input-output configuration in the present study is similar to that used for other fault diagnosis applications (Himmelblau *et al.*, 1991; Sorsa *et al.*, 1991; Gomm, 1996). The input feature vector, x, comprises conditioned measurement data taken from the process at each sampling instant. The network outputs are assigned to represent each of the fault classes and the network is trained to indicate a particular fault by outputting a one on the corresponding output fault class and zeroes for the other fault class outputs. Normal condition is indicated by zeros on all network outputs.

Each node in the hidden layer computes the Euclidean distance, z_i, between the network input vector, x, and the node centre, c_i, given by:

$$z_i = \|x - c_i\|, \quad i = 1,\dots,n_h \tag{1}$$

The result from each node is passed through a Gaussian basis function, $\phi_i(z_i,\sigma_i)$, where σ_i is a positive scalar width:

$$\phi_i(z_i,\sigma_i) = \exp\left(-z_i^2/\sigma_i^2\right) \quad i = 1,\dots,n_h \tag{2}$$

The network outputs, y, are determined as a normalised weighted sum of the hidden layer outputs:

$$y_j = \frac{\sum_{i=1}^{n_h} w_{ij}\phi_i(z_i,\sigma_i)}{\sum_{i=1}^{n_h} \phi_i(z_i,\sigma_i)}, \quad j = 1,\dots,m \tag{3}$$

where w_{ij} is the weight connecting the ith hidden node to the jth network output.

The network is trained using a standard three stage training procedure (Leonard and Kramer, 1991). Initially, a batch k-means clustering algorithm is used for choosing the network centres. The hidden node width parameters are calculated using a p nearest neighbour algorithm with a unit overlap of 2. Finally, a batch least squares algorithm was used for calculating the output layer weights.

2.1 Relationships with Fuzzy Logic

Networks utilising radial basis functions have analogies with fuzzy logic systems (Jang and Sun, 1993) which is illustrated in the following. Consider a fuzzy rule base containing n_h rules with each input, x_j, fuzzified by Gaussian membership functions centred at the points c_{ij} and with activation widths σ_{ij}, $i\in[1,n_h]$, $j\in[1,n]$. Using the fuzzy product AND operator to combine the IF part of the rule base, the output of the ith rule can be written as:

$$\phi_i = \exp\left[-\frac{(x_1 - c_{i1})^2}{\sigma_{i1}^2} - \dots - \frac{(x_n - c_{in})^2}{\sigma_{in}^2}\right] \tag{4}$$

In the special case when $\sigma_i = \sigma_{i1} = \dots = \sigma_{in}$, eqn.(4) is identical to eqn.(2). Furthermore, eqn.(3) is identical to the widely used centre of gravity defuzzification technique hence, the output layer weights, w_{ij}, are equivalent to the centroids of the output fuzzy sets. Thus, each hidden node in the RBF network can be considered as a fuzzy rule activating the output fault classification:

IF x_1 is QS_{i1} AND ... AND x_n is QS_{in}
THEN Fault S_1 is w_{i1} AND ... AND Fault S_n is w_{in}

where QS_{ij} is a qualitative size, e.g. normal, positive small, negative big, associated to the jth centre element of the ith hidden node. Similar qualitative descriptors can be assigned to each of the output layer weights.

3. PROCESS DESCRIPTION AND SIMULATED FAULTS

The process under consideration is a continuous stirred tank reactor (CSTR) process (Fig. 1), which has been used in previous publications, for work carried out on fault diagnosis (Sorsa *et al.*,1991; Oyelye and Kramer,1988). The process includes a CSTR, a heat exchanger, 3 PI controllers, 2 non-linear control valves, one linear control valve, a pump and associated pipework. In the CSTR, an irreversible first-order reaction A $\Rightarrow$ B takes place. This reaction is catalytic and exothermic. The temperature of the CSTR is unstable in an open loop, and is controlled by pumping a part of the reactor outlet stream back to the reactor through the heat exchanger, where the recycle flow is cooled by an external flow. The process has three feedback control loops where discrete PI controllers are used to keep the recycle flow rate, the level of the reactor liquid and the temperature of the reactor at preset levels. These variables are controlled via the coolant liquid flow valve, the product flow valve and the reflux flow valve respectively. Fourteen process variables are measured, including the 3 controller outputs. Measurement noise was added to the process variables to simulate a more realistic process model. The model was simulated using the dynamic simulation software SPEEDUP.

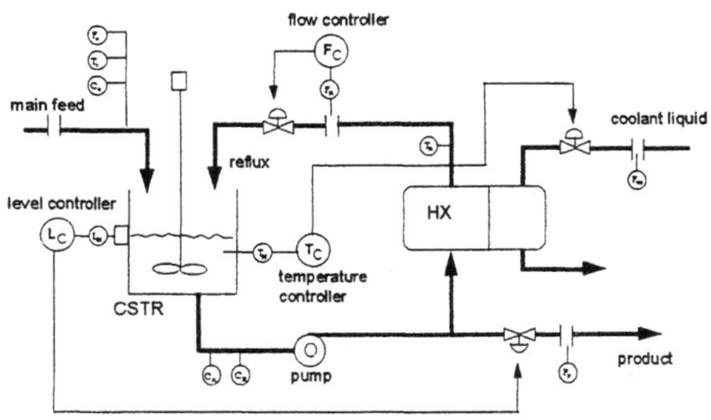

Fig. 1 Heat exchanger and CSTR process

The measured variables are as follows:

F_O	Input feed flow rate
T_O	Input feed temperature
C_O	Input feed concentration of A
C_A	Reactor outflow concentration of A
C_B	Reactor outflow concentration of B
L_M	Level of reactor
T_M	Temperature of reactor
F_R	Reflux flow rate
T_R	Reflux temperature
F_W	Coolant liquid flow rate
F_P	Product flow rate
T_C	Temperature control action signal
F_C	Flow control action signal
L_C	Level control action signal

Six faults were introduced to the process to obtain the training data for the neural networks. These were:

1. Leak in the reactor
2. Block in the input feed pipe
3. High concentration in the input feed
4. Reflux flow meter stuck high
5. Temperature control valve stuck high
6. Deactivated catalyst

Faults 4 and 5 were hard faults and the others were of 10% severity. The process was run under normal operation and under each of the fault conditions, to obtain a data set of 1400 points with 200 points on each condition.

4. NEURAL NETWORK TRAINING

The process data was conditioned prior to neural network training to facilitate the fault classification. This was carried out in a MATLAB environment. Giving each variable equal weighting is necessary in data analysis for comparison purposes, and is also a major requirement for neural network training. The data conditioning was achieved by processing the data by a standard statistical scaling method, which can be shown as:

$$x_s = \frac{x - \mu_x}{\sigma_x} \tag{5}$$

where μ_x is the mean and σ_x is the standard deviation of the process variable x under the normal operating condition. Statistical scaling gives all variables the same variance, brings them to a comparable range and amplifies the changes in the variables. The technique also provides a statistical measure of changes in individual network inputs identified by the network centres which is used later for removing unnecessary input variables.

The prepared data was then used to train the neural networks. The RBF network and training algorithms were coded in the MATLAB language for this purpose. The data set of 1400 points was partitioned into two sets of 1050 and 350 points each. The first set was used for training the networks, and the second set was used to test the trained networks. The networks trained had 14 input nodes, one for each of the measured process variables, and 6 output nodes, one for each fault. A final network with 15 centres, which gave satisfactory classification results, was selected for further investigation of the pruning and rule extraction methods. The trained network results in two matrices of main interest, the centres C (n_h x n) and the weights W (n_h x m).

5. NETWORK PRUNING

Network pruning involves creating a smaller and more efficient network from a previously trained network. This was carried out in 3 steps:

i. Removing the inputs to the network that do not contribute to the classification process.
ii. Removing the redundant centres that activate high for the same output.

iii. Removing the weights that are negligible and will not affect the activation of any output.

5.1 Removing Unnecessary Inputs

The centres matrix C, had 14 columns, one for each input, and 15 rows, one for each centre. The element values c_{ij} are the perpendicular distances from each centre to the corresponding input axis. Since the data was statistically scaled, eqn.(5), they are of zero mean ($\mu = 0$) and unit standard deviation ($\sigma = 1$) for the normal process operating condition. The centres correspond to average values of the input data in different regions of the input space. Changes in individual inputs caused by the faults which are identified by the centres can be found by examining the centre values using the probability $P(|x - \mu| < 3\sigma) \geq 0.89$ from the Chebyshev inequality. Therefore, a centre value with $| c_{ij} | < 3$ can be considered as being less than 3σ away from the point of origin, and input i does not show any significant change on centre j. Two centre input rows, corresponding to process variables T_O and L_M, showed magnitudes lower than 3 for all centres. This means that these two inputs do not vary from their normal condition values for any of the process faults. Therefore, these two input variables were removed from the network for further investigations. This reduced the neural network to a 12-15-6 (12 inputs, 15 centres, 6 outputs) architecture.

5.2. Removing Redundant Centres

The next part of the pruning procedure consisted of checking the weight matrix W to decide the centre contributions to each output. The W matrix had 6 columns, one for each output and 15 rows, one for each centre. Each element w_{jk} is the weighting between the jth centre and the kth output. A high positive weight, w_{jk}, in column k and row j, indicates that output k will be activated high by a high activation of centre j. Using this logic, it was found that while some outputs were activated by only one centre, some were activated by more than one, and is shown in Table 1.

Table 1 Identification of redundant centres

Output Node	Centres With High Weights
1	10
2	1,5,8,11
3	3,7,5
4	6
5	14
6	4,12

It was observed that for outputs with multiple high weighted centres, the centre positions were comparable. Hence, some of the centres contained redundant information and could be removed. It was decided for the outputs with redundant centres, to keep the centre with the highest weight and to remove the other centres. This resulted in the removal of centres 1, 5, 8, 3, 7 and 12.

Out of the 15 centres, only 12 centres appeared in Table 1. Centres 2, 9 and 13 did not have high positive weights for any of the faults. Centres 2 and 13 had all nearly zero weights, and showed no effect on the output node activations, and therefore it was considered that they could be safely removed. Centre 9 had a high negative weight for fault 1 and nearly zero weights for all other faults. The normal process behaviour and fault 1 nearly overlap each other in the measurement space and can be quite difficult to separate. Therefore, by considering that centre 9 represents normal behaviour, because of the high negative weight for fault 1, this node was contributing to the separation of the two classes. Thus, centre 9 was important to the fault diagnosis performance and was kept. This now results in a further reduced network of 12-7-6, which is nearly optimal.

A network with this new architecture was trained and tested at this point showing a marked improvement on the network performance. The centres retained from the above exercise were given as the initial centres for the k-means clustering algorithm. Fig. 2 shows the total mean squared error (mse) values for the original network (N_o) and the modified network (N_m).

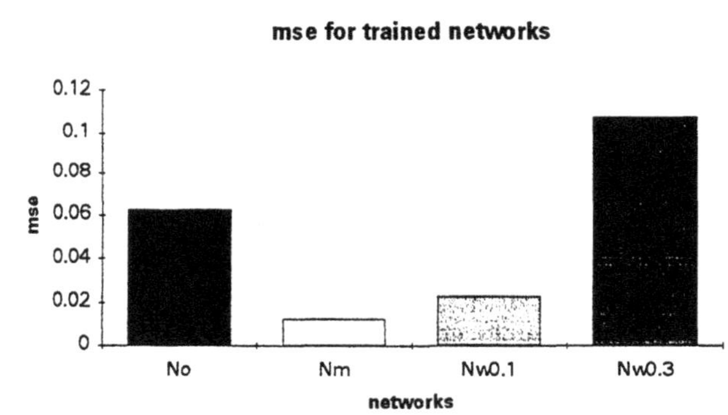

Fig. 2. The total mse values of trained networks on test data

5.3 Removing Negligible Weights

The final part of the network pruning method consists of removing the output layer weights whose magnitudes are negligible thus, not making a significant contribution to the network performance. Using the fuzzy relationship that the weights correspond to the centroids of the output fuzzy sets, weights close to zero correspond to a low network output. Hence, all weights within a defined negligible weight range can be eliminated. Presently the threshold limits of this range are chosen by a trial and error method. By removing the weights with a magnitude of less than a threshold, the network performance on a test data set was investigated. The initial full weight matrix of the modified network (N_m) consisted of 42 weights, being the number of possible links between 7 centres and 6 outputs. When a threshold of 0.1 was chosen, this resulted in a new network ($N_{w0.1}$) with only 27 weights, the other 15 weights being set to zero effectively removing these network links. For a threshold of 0.3, the new network ($N_{w0.3}$) had a weight matrix with only 10 weights.

The above networks were tested for their performance on the test data set and gave satisfactory results, all of them clearly separating each fault and normal condition. Fig. 2 compares the total mse values of the above 4 trained networks. The lowest error is given by the modified network with the full set of weights, Nm. Although the network with 10 weights ($N_{w0.3}$) has the largest mse value, it still classified the faults clearly and this network was selected to be used for the rule extraction procedure because of its reduced complexity.

6. RULE EXTRACTION

Seven rules can be extracted from the final pruned network, with each rule corresponding to one of the centres. The procedure is carried out by studying the C and W matrices for each of the centres in turn and is illustrated in an example of formulating the rule for the second network centre.

Initially, ranges for the centroids of the fuzzy sets for the network inputs and outputs are defined for different qualitative labels. In this work the following ranges were chosen:

inputs, $|c_{ij}|$: < 3 (no change, NC); [3,25] (small, S), [25,70] (medium, M) , > 70 (big, B)
outputs, $|w_{jk}|$: < 0.8 (medium, M), > 0.8 (high, H)

with the full qualitative labels being a combination of the above with the direction of the variables, positive (P) or negative (N), e.g. positive big (PB). Again, the statistical data normalisation facilitates the choice of the no change range to be within 3σ. The results of applying these qualitative lables to the pruned network parameters are given in Tables 2 and 3.

This information can now be used to extract the qualitative rules, for example for centre 2;

IF < F_O is NC AND C_O is NC AND C_A is NS AND C_B is NC AND T_M is PB AND F_R is PS AND T_R is NS AND F_W is PM AND F_P is NC AND T_C is PM AND F_C is NB AND L_C is NC >
THEN < Fault 4 is PH >
A similar procedure is applied to extract the remaining 6 rules to fully describe the network behaviour.

Table 2 Pruned neural network qualitative output layer weights

centre	Fault 1	Fault 2	Fault 2	Fault 4	Fault 5	Fault 6
1			NM			PH
2				**PH**		
3	NH	NM				NM
4	PH					
5		PH				
6					PH	
7			PH			

Table 3 Pruned neural network qualitative centres

	F_O	C_O	C_A	C_B	T_M	F_R	T_R	F_W	F_P	T_C	F_C	L_C
c1	NC	NC	PS	NC	NC	NC	NC	NC	NC	NC	NC	NC
c2	**NC**	**NC**	**NS**	**NC**	**PB**	**PS**	**NS**	**PM**	**NC**	**PM**	**NB**	**NC**
c3	NC	NC	NC	NC	NC	NC	NC	NC	NC	NC	NC	NC
c4	NC	NC	NC	NC	NC	NC	NC	NC	NS	NC	NC	NS
c5	NS	NC	NS	NC	NC	NC	NC	NC	NC	NC	NC	NC
c6	NC	NC	PS	NC	NM	NC	NS	PM	NC	NS	NC	NC
c7	NC	PS	PS	PS	NC	NC	NC	NC	NC	NC	NC	NC

7. CONCLUSIONS

The use of relations between RBF neural networks and fuzzy logic, and statistical methods, for network pruning and extracting qualitative rules were demonstrated in a fault diagnosis application to a simulated chemical process. It was shown that a pruned network can classify better than a standard fully connected network and achieve much lower output errors on test data. Increasing the pruning threshold for the output layer weights removes more links and causes the network errors to increase as expected. However, a network with a 70% reduction of parameters still exhibited a classification performance equal to a fully connected network. The extraction of qualitative rules to explain the fault diagnosis decision can facilitate enhanced confidence in a fault diagnosis system and the integration of neural networks with other methodologies.

The approach is currently being automated and its application to the diagnosis of non-catastrophic faults on a real, industrial nuclear fuel processing plant is under investigation.

ACKNOWLEDGEMENTS

The authors wish to thank the industrial and financial support received from the Springfield and Sellafield sites of British Nuclear Fuels plc, UK.

REFERENCES

Ayoubi, M. (1996) Fuzzy systems design based on a hybrid neural structure and application to the fault diagnosis of technical processes, *Control Eng. Practice*, **4**(1), 35-42.

Fu, L.M. (1991) Rule learning by searching on adapted nets. *Proceedings of the Ninth National Conference on Artificial Intelligence*, 590-595.

Gomm, J.B. (1996) On-line learning for fault classification using an adaptive neuro-fuzzy network, *Proc. IFAC World Congress*, San Francisco, USA, 30 June - 5 July, 1996, Vol. N, pp. 175-180.

Himmelblau, D.M., Barker, R.W. and Suewatanakul, W. (1991) Fault classification with the aid of artificial neural networks. *Proc. IFAC/IMACS Symp. SAFEPROCESS'91*, Baden-Baden, Germany, 10-13 Sept., 541-545.

Jang, J.S.R. and Sun, C.-T. (1993) Functional equivalence between radial basis function networks and fuzzy inference systems, *IEEE Trans. Neural Networks*, **4**(1), 156-159.

Leonard, J.A. and Kramer, M.A. (1991) Radial basis function networks for classifying process faults. *IEEE Control Systems Magazine*, **April**, 31-38.

Oyelye, O.O. and Kramer, M.A. (1988) Qualitative simulation of chemical process systems: steady state analysis, *AIChE J.*, **34**, 1441-1454.

Nie, J. and Linkens, D.A. (1993) Learning control using fuzzified self-organizing radial basis function network, *IEEE Trans. Fuzzy Systems*, **1**(4), 280-287.

Saito, K. and Nakano R. (1988) Medical Diagnostic expert system based on a pdp model. *Proceedings of IEEE International Conference on Neural Networks*, (Vol 1), 255-262.

Sorsa, T., Koivo H.K. and Koivisto H. (1991) Neural networks in process fault diagnosis. *IEEE Transactions on Systems, Man and Cybernetics*, **21**(4), 815-825.

Towell, G.G. and Shavlik, J.W. (1993) Extracting refined rules from knowledge-based neural networks. *Machine Learning*, Vol. 13, 71-101.

Venkatasubramanian, V. and Chan K., (1989) A neural network methodology for process fault diagnosis. *AIChE J.*, **35**(12), 1993-2002.

Diagnosis of Abrupt Faults Using Variable-Structure Neural Network

Da-Nian Luan, Der-Shing Chang, Ming-Ping Chang and Cheng-Ching Yu*

Department of Chemical Engineering
National Taiwan University of Science and Technology
Taipei, TAIWAN 106

Abstract: Based on these two fundamental network structures, a variable-structure neural network (VSNN) is constructed to cope with filtering and state estimation problems. It is a hybrid between the feedforward network (FFN) and the recurrent network (RecN). Similar to the Kalman filter approach, the filter gain is adjusted according the the ratio of noise and error covariance. An algorithm is proposed to approximate the error covariance and, consequently, results in a non-iterative procedure in the filter gain calculation. Moreover, when some of the states are not measurable, the VSNN naturally leads to a RecN-like architecture. A chemical reactor example is used to demonstrate the effectiveness of the VSNN state estimation scheme. Results show that the variable-structure neural network can detect and isolate incipient single as well as multiple faults in an effective manner. *Copyright © 1998 IFAC*

Keywords: fault detection, dynamic fault diagnosis, nonlinear state estimation, neural network, network architecture.

1. INTRODUCTION

The complexity of modern chemical plants and availability of inexpensive computing powers prompt us to develop automated fault diagnosis methods (Petti et al., 1990; Hsu and Yu, 1990). The last decade has seen advances in on-line monitoring and diagnosis methodologies. For diagnosis systems, some type of model is used to produce behavioral prediction. In terms of the precision of information, the model are represented in the structural, qualitative or quantitative levels. Models between the the qualitative-quantitative spectrum are also employed in several occasions.

Early detection and isolation of faults are important because operators can take remedial actions to restore normal condition or initiate the shut-down procedure. This is even more important for abrupt faults, e.g., sudden lose of cooling water, instrument malfunctions, drastic change in environmental factors etc. (Park and Himmelblau, 1983; Isermann, 1984; Willis et al., 1992). Since physical systems are described by ordinary differential equations (ODE), it is not effective to use qualitative-model based methods to detect incipient faults, despite frequently employed for on-line applications. Moreover, the imprecision in qualitative-to-quantitative model can produce rather *ambiguous* behavioral description for dynamic systems (Kramer and Palowitch, 1987). Therefore, quantitative models become an ideal candidate to detect abrupt faults.

In chemical process industries, much less work has been published for dynamic fault diagnosis than for steady-state cases (Park and Himmelblau, 1983). One obvious reason is, compared to other disciplines, chemical processes are very complex and exhibit severe nonlinearity which requires extensive engineering manpower for the first principle modeling. It is well-known that dynamic fault diagnosis problems can be formulated as filtering or state estimation problem. Various methods were published in the literature and several review papers summarize the progress. However, the modeling difficulties slow down wide-spread applications. Moreover, linear theory based techniques, e.g., Kalman filter, have to be extended to cope

*PHONE: +886-2-2737-6620
FAX: +886-2-2737-6644
E-Mail: ccyu@ch.ntust.edu.tw

with process nonlinearity. This further limits its uses.

Neural networks have emerged as useful tools in chemical engineering systems. The strength of neural networks stems from nonlinear mapping capability and roles as nonparametric models (Chen et.al., 1990; Qin et al., 1992; Willis et al., 1992; Karjala and Himmelblau, 1996; Tung et al., 1997). These features serve as an ideal model structure in fault diagnosis for chemical processes. In modeling nonlinear dynamic systems in the discrete form, the neural network architectures can be classified into two categories. One architecture is the feedforward network (FFN) where the measured outputs are used as the inputs to the neural network for the future output prediction. The other is the recurrent network structure where arbitrary feedback loops can be made in the processing elements (also called nodes). If the feedback connection is made from the *external* (or predicted) outputs to its input, the architecture, then, becomes the output recurrent network (RecN). Based on these two fundamental networks, a variable-structure neural network (VSNN) is proposed to cope with the problems of measurement noises and state estimation.

The dynamic fault diagnosis problem can be formulated as nonlinear state space equations. The VSNN architecture and algorithm can be devised to diagnose possible faults in a real-time environment. The remainder of this paper is organized as follows. The fundamental structure of the VSNN is laid in Section 2 and an algorithm is also proposed for filtering and state estimation. An chemical reactor is used to illustrate the performance of the VSNN diagnosis system in Section 3 followed by the conclusion.

2. NEURAL NETWORK FOR FAULT DIAGNOSIS

It is well-known that the FFN performs well for relatively clean data and the RecN works well at a certain noise level (Qin et al. 1992; Tung et al., 1997). One can devise a network architecture by taking advantages of both network structures: being a FFN at low noise level and becoming a RecN as the noise level increases. Therefore, an ideal neural network architecture has to adjust itself under different signal-to-noise levels.

Actually, most filters take this type of approach. For example, the Kalman filter uses the combination of the process measurement ($z(k)$) and the *a priori* estimate of the output $\hat{y}(k \mid k-1)$ in updating the output (or state) estimation. Therefore, the filter takes the following structure:

$$\hat{y}(k \mid k) = K(k)z(k) + (1 - K(k))\hat{y}(k \mid k-1) \quad (1)$$

where $\hat{y}(k \mid k)$ is the *a posterior* estimate of the output when a measurement has become available at time k. The filter gain, K, takes the value between zero and one. This is the fundamental structure any filter.

A similar structure can be applied to the neural network architecture (Figure 1). Instead of taking a distinct switch between the FFN and RecN, this new architecture uses a combination of these two for the subsequent prediction. Let us consider a set of *linear* state space equations.

$$\mathbf{x}(k+1) = \mathbf{F}\mathbf{x}(k) + \mathbf{G}\mathbf{u}(k) + \omega(k) \quad (2)$$

$$\mathbf{y}(k) = \mathbf{H}\mathbf{x}(k) \quad (3)$$

$$\mathbf{z}(k) = \mathbf{y}(k) + v(k) \quad (4)$$

with $v \in \mathcal{N}(0, R)$ and $\omega \in \mathcal{N}(0, Q)$ where Q and R are the covariance matrices of the measurement and plant noises, respectively. Let $\mathbf{H} = \mathbf{I}$ where $\mathbf{I}$ stands for the identity matrix. The Kalman filter algorithm has the following four steps: the error covariance extrapolation, the Kalman filter gain matrix calculation, state estimation update and error covariance update.

$$\mathbf{P}(k \mid k-1) = \mathbf{F}(k-1)\mathbf{P}(k-1 \mid k-1)\mathbf{F}(k-1)^T + \mathbf{Q}(k-1) \quad (5)$$

$$\mathbf{K}(k) = \mathbf{P}(k \mid k-1)\Big[\mathbf{P}(k \mid k-1) + \mathbf{R}(k \mid k-1)\Big]^{-1} \quad (6)$$

$$\hat{\mathbf{y}}(k \mid k) = \hat{\mathbf{y}}(\mathbf{k} \mid \mathbf{k}-1) + \mathbf{K}(k)\Big[\mathbf{z}(\mathbf{k}) - \hat{\mathbf{y}}(k \mid k-1)\Big] \quad (7)$$

$$\mathbf{P}(k \mid k) = \Big[\mathbf{I} - \mathbf{K}(k)\Big]\mathbf{P}(k \mid k-1) \quad (8)$$

The nonlinear version for the state estimation is the extended Kalman filter (EKF) where the algorithm performs the *linearization* procedure at each time step for the error covariance extrapolation (Park and Himmelblau, 1983; Grewal and Andrews, 1993). For large-scale nonlinear systems, the computational complexity of obtaining a solution can far exceed the already considerable burden of the standard Kalman filter algorithm. In this work, we would like to approximate the error covariance without going through extensive linearization steps. This can be done by introducing the approximated error covariance $\mathcal{P}(k)$ (Tung et al., 1997). In doing this, the covariance extrapolation and updating steps in Eqs. 5 and 8 can be eliminated. The next step is to find $\mathcal{P}(k)$. In the standard Kalman filter algorithm, the covariance matrix has the following form: $\mathbf{P}(k \mid k-1) = E[\tilde{\mathbf{y}}(k \mid k-1)\tilde{\mathbf{y}}(k \mid k-1)^T]$ which can be a full matrix where $\tilde{y}$ is the error between the true state and the estimate. However, in this work, the approximated covariance matrix is restricted to a diagonal matrix. That is, for a system with n states, we try to approximate the covariance matrix with:

$$\mathbf{P}(k \mid k-1) \approx diag\left[\tilde{y}_i(k \mid k-1)^2\right], i = 1, \cdots n \qquad (9)$$

where diag[·] stands for a diagonal matrix and $\tilde{y}_i(k \mid k-1)$ can be approximated when the corresponding $z_i(k)$ becomes available. The advantages of using Eq. 9 are two folds: (1) it can prevent the ill-conditioning in the approximated covariance matrix and (2) it simplifies significantly the feedback connections in the hybrid network. Incorporated with a moving window, the approximated covariance becomes:

$$\mathcal{P}(k) = diag\left\{ \left[\frac{1}{N_w} \sum_{j=k-N_w+1}^{k} \left(z_i(j) - \hat{y}_i(j \mid j-1)\right)\right]^2 \right\}$$
$$= diag\left\{ \mathcal{P}_{ii}(k) \right\}, \quad i = 1, \cdots n \qquad (10)$$

where $\mathcal{P}_{ii}$ is the ith diagonal element of $\mathcal{P}$. The filter gain matrix can then be calculated from:

$$\mathbf{K}(k) = \mathcal{P}(k)[\mathcal{P}(k) + \mathbf{R}(k)]^{-1} \qquad (11)$$

Notice that the measurement noises are generally uncorrelated and the filter gain matrix then becomes a diagonal matrix.

$$\mathbf{K}(k) = diag[k_{ii}] = diag\left[\frac{1}{1 + \frac{\sigma_{ii}^2}{\mathcal{P}_{ii}(k)}}\right] \qquad (12)$$

where σ_{ii}^2 is the ii^{th} entry of $\mathbf{R}$ and $\mathcal{P}_{ii}$ can be computed from Eq. 10. This implies that we have a single feedback path for each output in the hybrid network as shown in Figure 1.

Therefore, the computational procedure can be summarized as follows: (1) Update the approximated error covariance ($\mathcal{P}(k)$) from Eq. 14; (2) Compute the filter gains ($k_{ii}(k)$) using Eq. 16; (3) Compute the *a posterior* output estimation; ($y(k \mid k)$) according to Eq. 11. (4) Perform the standard feedforward neural network computing for the outputs

It is well known that the fault diagnosis problem can be formulated as a state estimation problem (Isermann, 1984). The state estimation aspect of the variable-structure neural network (VSNN) can be constructed in a straightforward manner. Following the approach of Park and Himmelblau (1983) and Karjala and Himmelblau (1996), the process faults are treated as states and assume that we have the same states and the outputs. An augmented state space equations (e.g., Eqs. 2-4) can be constructed. In doing this, the diagnosis problem becomes a state estimation problem. This problem can be handled easily using the VSNN. Consider Eq. 3 where not all the states are measurable. That implies the matrix $\mathbf{H}$ is no longer an identity matrix, i.e, we have zero's along the diagonal. Following the VSNN algorithm, it can be shown that unmeasured states j correspond to zero filter gains ($k_{jj}(k) = 0$). That implies, in VSNN, the unmeasured states follow the structure of the recurrent network.

Consider a system with four states as an example. If all states are measured except x_4, the network architecture then becomes a combination of FFN and RecN for the first three states and a truly recurrent path for the fourth state as shown in Figure 2. Notice that this architecture is embedded in the variable-structure neural network. Therefore, we can perform fault diagnosis in real-time using the VSNN.

3. APPLICATION

Since the variable-structure neural network is effective in modeling nonlinear dynamic systems and noise suppression, it is a suitable architecture for detecting abrupt faults.

3.1 Process

Without loss of generality, a continuous stirred tank reactor (CSTR) with a first order exothermic reaction (Jang et al., 1986; Karjala and Himmelblau, 1996; Tung et al., 1997) is used to illustrates filtering and gross error detection aspects several neural network structures. The state equations describe the CSTR are: $\frac{dT}{dt} = \frac{q}{V}(T_0 - T) - \frac{\Delta H_R}{\rho C_p} k_0 e^{\frac{-E_a}{T}} C - \frac{UA}{\rho C_p V}(T - T_c)$ and $\frac{dC}{dt} = \frac{q}{V}(C_0 - C) - k_0 e^{\frac{-E_a}{T}} C$. T and C stand for the reactor temperature and concentration, respectively, and T_0 and C_0 are the reactor inlet temperature and inlet concentration. Figure 3 shows that the reactor temperature (T) is maintained by adjusting the set point of the the jacket temperature (T_j). Moreover, the reactor concentration (C) is also measured. Therefore, we have two process measurements and T_j is the output of the controller. A PI controller is employed for the reactor temperature control. It is also assumed that T and C are corrupted with random noises with standard deviation of 0.05.

Five process faults (or disturbances) are assumed: (1) inlet flow rate (q), (2) inlet temperature (T_0), (3) inlet concentration (C_0), (4) pre-exponential factor in the rate constant (k_0) and (5) overall heat transfer coefficient (U). Notice that these five faults have similar *qualitative* behavior (Hsu and Yu, 1991). The signed directed graphs in Figure 4 indicate four faults show the same qualitive behavior on C and T_j. That means they are indistinguishable even at steady-state if qualitative values are employed for diagnosis.

3.2 Diagnosis

For the purpose of diagnosis, the system can be formulated as follows. First, the CSTR equations are augmented by incorporating the following equations: $\frac{dq}{dt} = \Upsilon_1$, $\frac{dT_0}{dt} = \Upsilon_2$, $\frac{dC_0}{dt} = \Upsilon_3$, $\frac{dk_0}{dt} = \Upsilon_4$ and $\frac{dU}{dt} = \Upsilon_5$. Here Υ_i denotes the change in the corresponding state variable. Therefore we have a set of nonlinear equations with 7 states where five of them are possible fault origins.

Next, the network structure is determined according to the numbers of input and outputs. Here, we have a neural network with 8 input nodes (one input (T_j), two states (T and C) and five fault origins), 7 output nodes and 9 hidden nodes. This is a network with a 8-9-7 structure (Figure 5). This VSNN has the following architecture: five outputs (faults) have completely recurrent loops and the other two outputs have partially recurrent loops as shown in Figure 5. As mentioned earlier, the architecture are determined by the availability of the output measurements.

A data set consist of 26500 data points are employed for the training. They are generated from dynamic simulation of the CSTR with random and step changes in the fault origins and measurement noises are also included. The training follows the modified back-propagation through time algorithm (Tung et al. 1997) and it takes 20000 learning steps to converged the network.

Notice that the purpose of the VSNN diagnosis system is to provide the estimation for possible fault origin. The results, Figure 6, show that the VSNN provides very good tracking of incipient fault in U. For fluctuations in T_0, good dynamic fault diagnosis is also observed as indicated in Figure 7. The results just shown are for the cases of single fault. Figure 8 shows that when faults occur simultaneous in T_0 and C_0, the VSNN gives good estimates on the faulty variables. Despite its rare probability, the case of four faults can also be correctly diagnosed using the VSNN. Notice that these results are obtained from noise corrupted measurements using the variable-structure neural network (Figures 6-9).

4. CONCLUSION

Based on these two fundamental network structures, a variable-structure neural network (VSNN) is constructed to cope with filtering and state estimation problems. It is a hybrid between the feedforward network (FFN) and the recurrent network (RecN). Similar to the Kalman filter approach, the filter gain is adjusted according the the ratio of noise and error covariances. An algorithm is proposed to approximate the error covariance and, consequently, results in a non-iterative procedure in the filter gain calculation. Moreover, when some of the states are not measurable, the VSNN naturally leads to a RecN-like architecture.

A chemical reactor example is used to demonstrate the effectiveness of the VSNN state estimation scheme. First, the dynamic fault diagnosis problem is formulated into a set of state space equations. Then, a VSNN is constructed according the availability of state measurement. Results show that the variable-structure neural network can detect and isolate incipient single as well as multiple faults in an effective manner. Moreover, this scheme is implemented in a nonparmetric manner with well-known neural network architectures.

ACKNOWLEDGEMENT

Financial support of the National Science Council of Taiwan (R.O.C.) under the grant NSC 84-2621-P011-006 is gratefully acknowledged.

REFERENCES

Chen, S.; Billings S. A.; Grant, P. M. , "Nonlinear System Identification Using Neural Networks", *Int. J. Control*, **51**, 1191-1214 (1990).

Grewal, M. S.; Andrews, A.P. , *Kalman Filtering: Theory and Practice*, Prentice-Hall: Englewood-Cliff, NJ (1993).

Hsu, Y. Y.; Yu, C. C. , "A Self-LearningFault Diagnosis System Based on Reinforcement Learning", *Ind. Eng. Chem. Res.*, **31**, 1937 (1990).

Isermann, R. , "Process Fault Detection Base on Modeling and Estimation Methods- A Survey", *Automatica*, **20**, 387 (1984).

Karjala, T. W.; Himmelblau, D. M. , "Dynamic Rectification of Data via Recurrent Neural Networks and the Extended Kalman Filter", *AIChE J.*, **42**, 2225-2239 (1996).

Kramer, M. A.; Palowitch, B. C., Jr., "A Rule-Based Approach to Fault Diagnosis sing Signed Directed Graph", *AIChE J.*, **33**, 1067 (1990).

Park, S.; Himmelblau, D. M. , "Fault Detection and Diagnosis via Parameter Estimation in Lumped Parameter Systems", *Ind. Eng. Chem. Process Des. Dev.*, **22**, 482-487 (1983).

Petti, T. F.; Klein, J.; Dhurjati, P. S., "Deep Model Processor: Using Deep Knowledge for Process Fault Detection", *AIChE J.*, **36**, 565 (1990).

Qin, S. Z.; Su, H. T.; McAvoy, T. J. , "Comparison of Four Neural Net Learning Methods for Dynamic System Identification", *IEEE Trans. Neural Networks*, **3**, 122-130 (1992).

Tung, H. H.; Luan, D. N.; Chang, D. S.; Yu, C. C. , "A Kalman Filter Type of Neural Network for Dynamic Modeling and Data Rectification", *J. Chin. Inst. Chem. Eng.*, **27**. 493-509.

Willis, M. J.; Montague, G. A.; DiMassimo, C.; Tham, M. T.; Morris, A. J. , "Artificial Neural Networks in Process Estimation and Control", *Automatica*, **28**, 1181-1187 (1992).

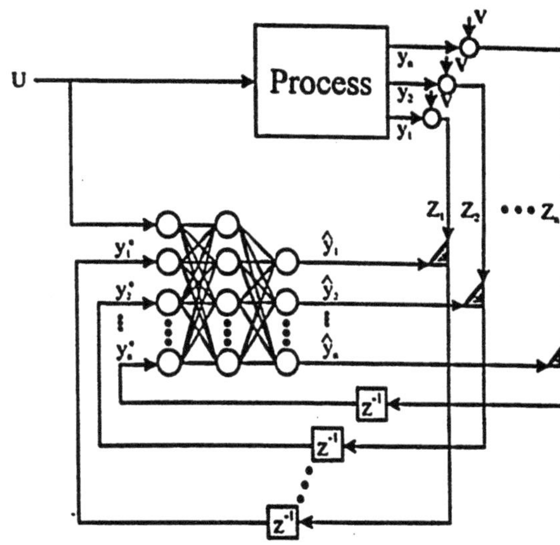

Figure 1. Architecture of variable-structure neural networks when all states measured.

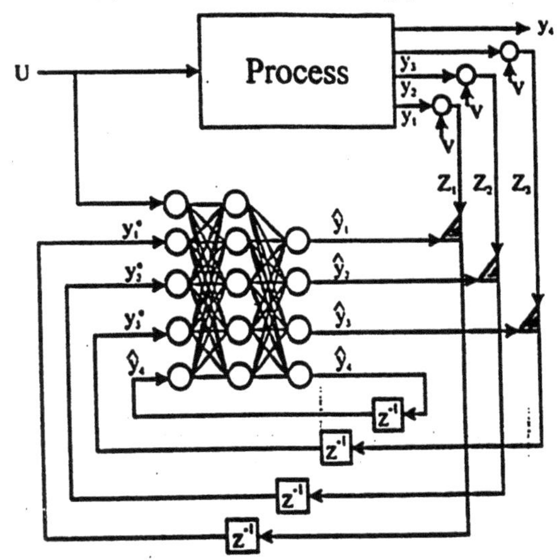

Figure 2. Architecture of variable-structure neural networks when y_4 unmeasured.

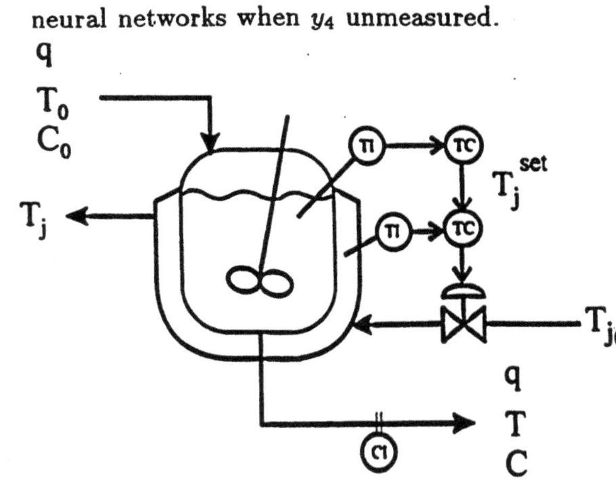

Figure 3. CSTR example.

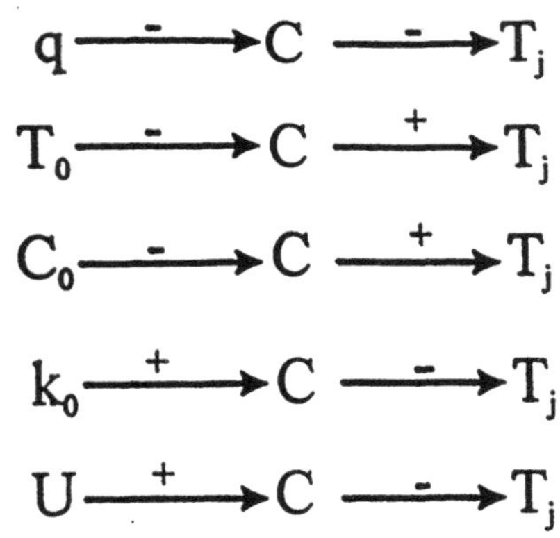

Fig. 4. Signed Directed Graph for five fault origins

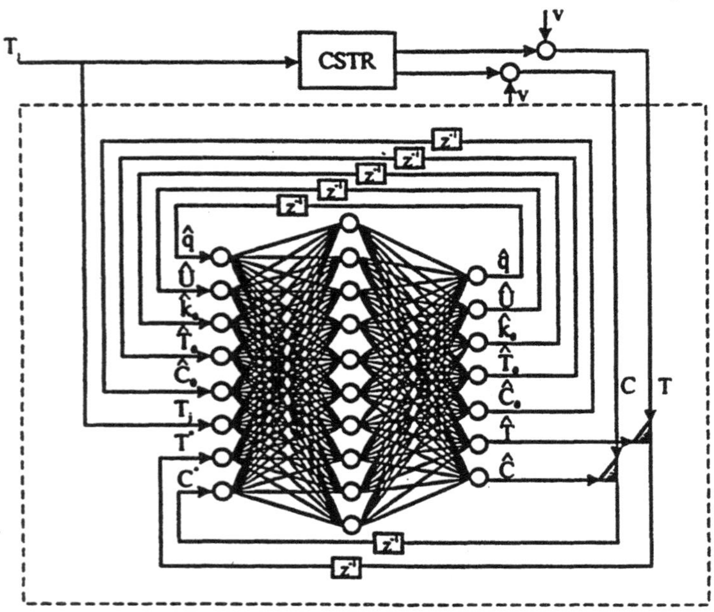

Fig.5. Variable-structure neural network fault diagnosis system for the CSTR example

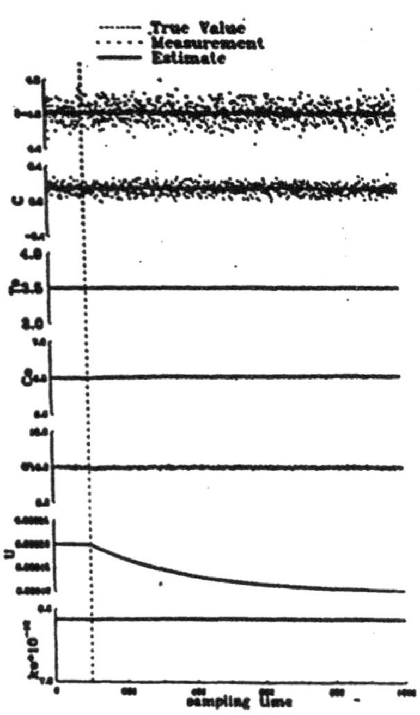

Fig.6. Results of fault diagnosis for a change in U

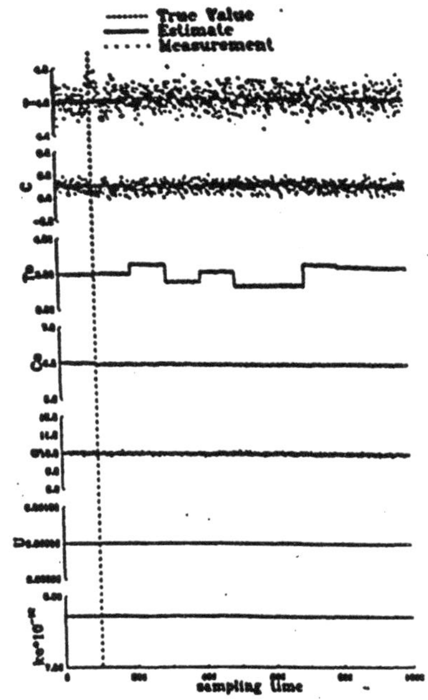

Fig.7. Results of fault diagnosis for step changes in T_0

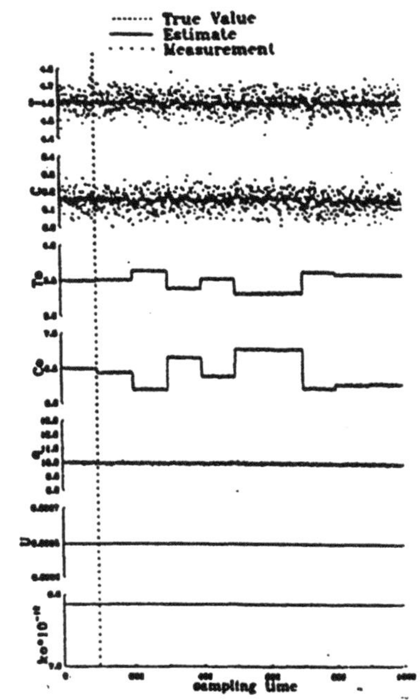

Fig.8. Results of fault diagnosis for step changes in T_0 and C_0

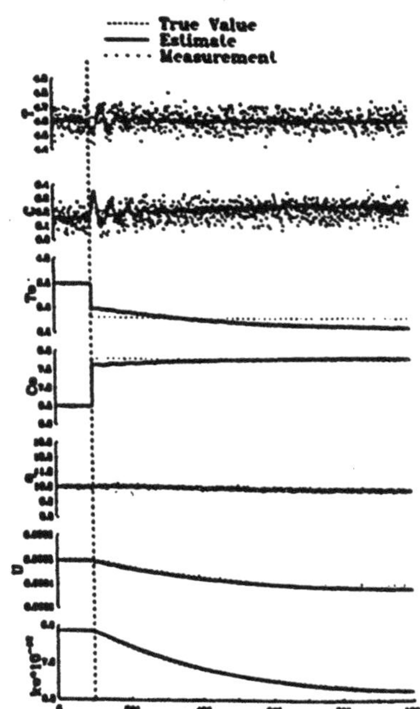

Fig.9. Results of fault diagnosis for step changes in T_0, C_0, U and k_0

FAULT DETECTION IN PAPER MAKING (A NEURAL NETWORK APPROACH)

Y. Bissessur, E. B. Martin* and A. J. Morris

Centre for Process Analysis, Chemometrics and Control
Department of Chemical and Process Engineering
**Department of Engineering Mathematics*
University of Newcastle, UK.

Abstract: In papermaking, as in other industrial activities, increasingly stringent customer requirements for product quality and consistency necessitate tighter control of the manufacturing processes. This paper describes a neural network vibration-based condition monitoring system for providing advance warnings of malfunctions in the paper machine. Spectral analysis of the raw vibration signals involves the technique of enveloping for isolating the spectral components that reflect the process defects. Feature extraction is then carried out by a suitably designed peak detection algorithm. This is followed by a neural network classifier for discriminating between the fault and the no-fault conditions. The system's performance is assessed on industrial data from a major paper manufacturer. *Copyright © 1998 IFAC*

Keywords: Fault detection, monitoring, vibration spectra, neural networks, paper industry.

1. INTRODUCTION

A typical paper machine consists of three parts, the wet end, the press section and the drier. This study concentrates on the press section since paper quality is significantly affected by the operating condition of its mechanical components. A schematic of the press section is shown in Figure 1; the arrows indicate the direction of the paper through the machine. The function of the press is to reduce the water content of the pulp by means of absorbent material called felts. A felt is a long fabric which may be grained in a variety of ways making a continuous loop and carrying the paper through the roll contact points also known as the press nips.

Whenever a felt seam becomes disjointed or uneven, a force impulse is generated as it passes through the press nips. This impulse causes the roll to bounce, almost imperceptibly, often resulting in paper flaws and leading to machine shutdowns. This can be a major problem in the paper plant due to ever increasing demands for consistent high quality product. The purpose of this work is to provide an early process malfunction warning system to improve paper quality and increase productivity through the prevention of paper deformities and breaks. Instead of the preventive approach which institutes felt replacement on a scheduled time period, the concept of predictive maintenance based on the monitoring of the vibrations at the press rolls is applied. Information on the state of the felts at any point in time can thus be obtained.

Currently, fault detection in the press section is carried out by trending the overall vibration energies (through the root-mean-square values) of the recorded spectra against time (SKF, 1994b). A developing fault is usually detected by an increase in the overall vibration levels relative to normal operation. One important drawback with this

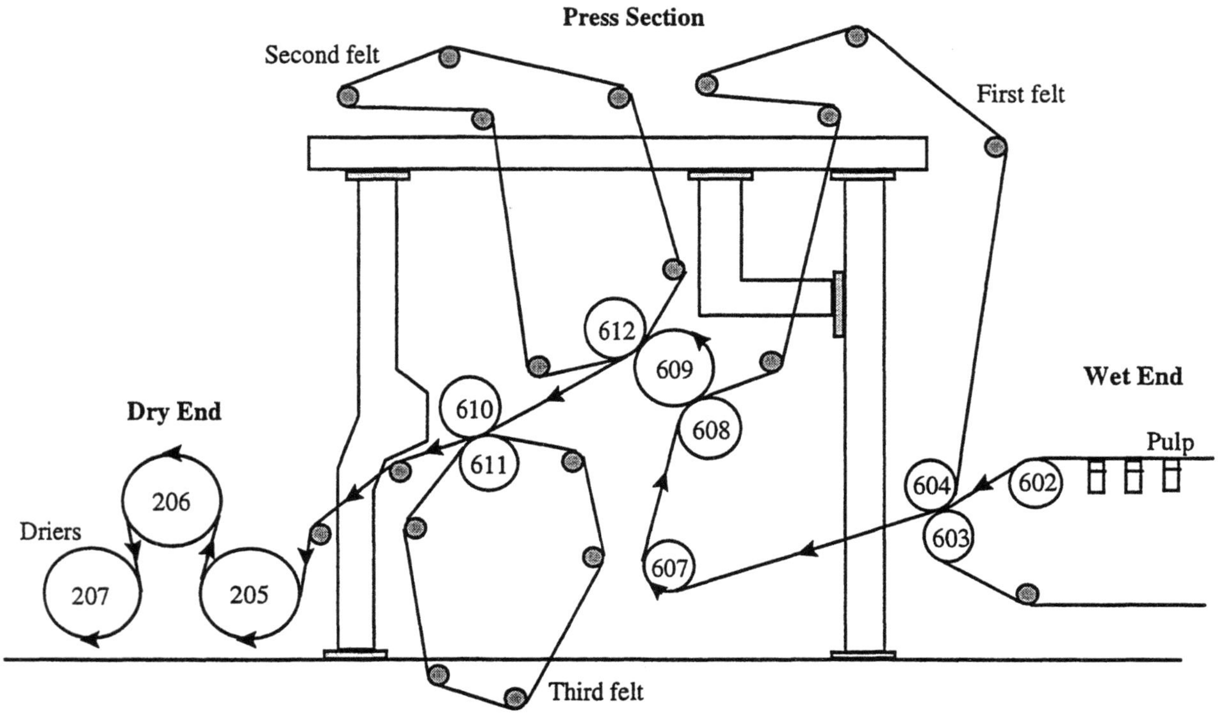

Fig. 1. Schematic of the press section.

approach is that the fault only becomes apparent at a fairly advanced stage and may have already adversely affected other components of the press section. This may lead to situations whereby the replacement of the felt does not completely remove the problem. Consequently, the paper making process is impaired over a longer period of time resulting in a degraded product. For a predictive maintenance programme to be successful, it should be capable of providing an early warning of a defect to allow corrective action that prolongs service life, maintains product consistency and schedules repairs for an optimum maintenance time interval.

This study deals with the use of robust neural network architectures for enhancing vibration-based condition monitoring programmes in the paper industry. Due to the highly non-linear nature of vibration spectra, neural networks are preferred over multivariate statistical process control approaches (Morris *et al.*, 1997; Simoglou *et al.*, 1997) which are essentially linear. Although the developed system is specifically geared towards the press felts, it can easily be extended to a wide range of other possible faults. The flowchart of Figure 2 summarises the individual stages of the diagnostic system.

The accelerometers are vibration sensors mounted at strategic points on the press rolls and yield time-domain vibration signals. These measurements are stored in data collection instruments as sampled time values together with the corresponding FFT spectra. The preprocessing stage involves demodulation and filtering techniques for isolating any defect signals present. Features are then extracted from the enveloped acceleration spectrum and used as 'cues' for discriminating between spectra from normal and damaged felts.

2. DATA AVAILABLE

Vibration data from rolls 608 and 612 (See Figure 1) was recorded over a period of 6 months. Figure 3 depicts the overall vibration levels. It is clear that problems were experienced by the first felt during two main time intervals and by the second felt during one time interval, these being indicated by arrows on the plots. This prior knowledge is used for validation of the developed system.

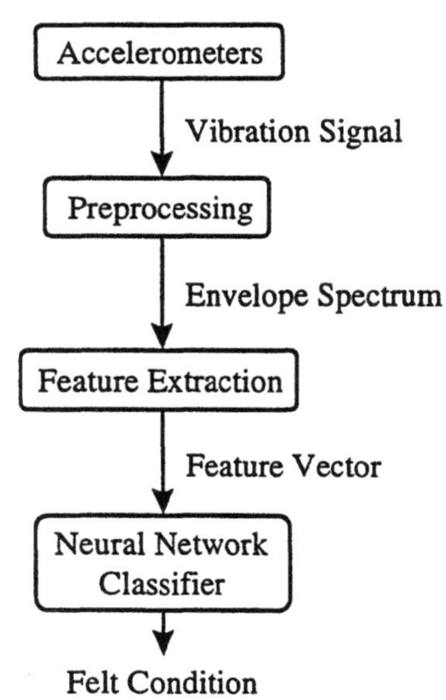

Fig. 2. Diagnostic system implementation.

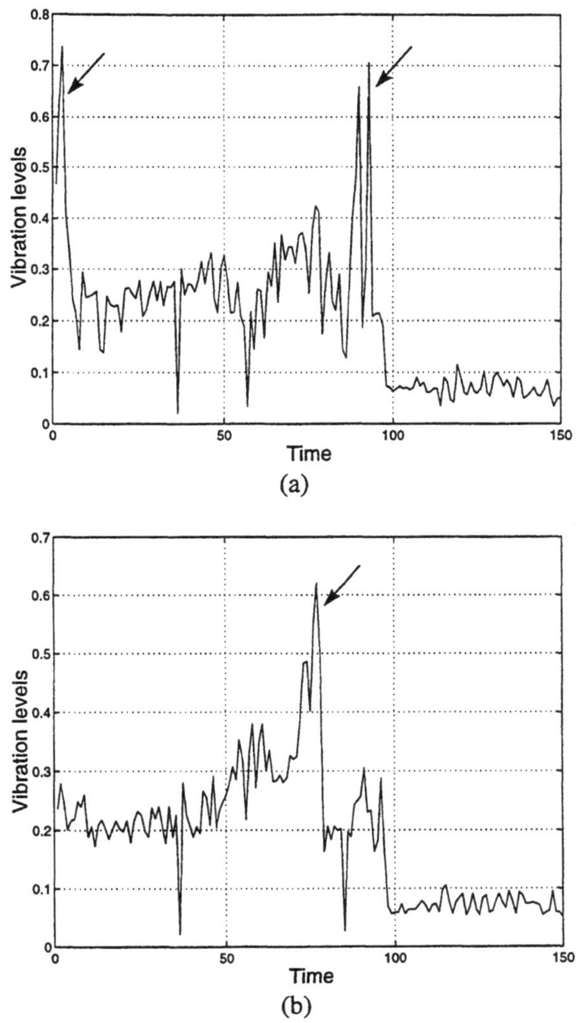

Fig. 3. Overall vibration levels from roll 608 (a) and roll 612 (b).

The raw vibration signals are filtered by 7 different bandpass filters. Each frequency band reveals information on a particular component of the roll and on its interaction with the neighbouring parts of the paper machine. Since felt anomalies manifest themselves in the 0-600 RPM range, this study concentrates on this frequency band. A total of 150 spectra, each consisting of 400 lines, are available for the two press rolls.

3. DATA PRE-PROCESSING

A typical vibration spectrum is made up of :

(i) low frequency signals from rotational components.
(ii) high frequency defect impulse signals.
(iii) machine noise.

In the time domain, the defect signals are of very short duration and, in the frequency domain, these translate into harmonic amplitudes spread over a wide frequency range and buried in noise. This makes traditional spectral vibration analysis too insensitive to detect the early defect stages.

Acceleration enveloping offers a solution to this problem. It has already received widespread acceptance in the machine condition monitoring field as an extremely powerful method for isolating defect signals to a frequency range where other noise components are not present (Smulders and Kadushin, 1994). The enveloper essentially squares the time signal in which is embedded the higher order harmonics of the defect repetitive impulse. The squaring process provides the sum and difference components of each harmonic component. The sum products are all filtered out while the difference components are vectorially summed to the 1X, 2X, 3X,... of the defect repetition rate. The enveloper, along with a low pass filter, thus transforms the high frequency components back to a base band. The result is a spectrum containing a harmonic series of amplitudes spaced at the defect frequency. One example for a felt at a normal and an advanced state of deterioration is shown in Figure 4. The amplitudes of the harmonics are usually indicative of the severity of the problem.

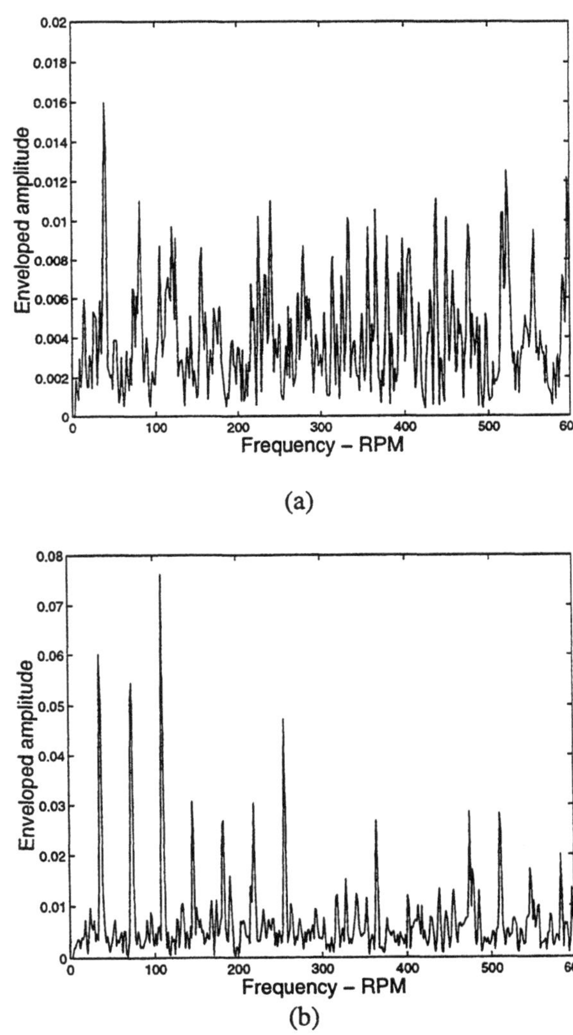

Fig. 4. Typical spectra from a normal (a) and a badly damaged felt (b).

4. FEATURE EXTRACTION

The motivation for the use of a feature extraction stage stems from the fact that large amounts of redundant information are contained within each 400 line spectrum. It is thus more beneficial to derive a reduced number of parameters for input into the neural network classifier in order to reduce the training times and network complexity. In general, an anomaly in a felt seam results in a harmonic series of peak amplitudes spaced at the repetitive rate of the felt loop (SKF, 1994a). Therefore, the harmonic amplitudes are believed to form a suitable set of discriminatory features. A change in the machine speed causes a shift in the fundamental frequencies, so the machine speed is an important parameter in selecting which harmonics need to be monitored. The fundamental is given by,

$$F = \frac{S}{L} \tag{1}$$

where S is the felt speed and L is the felt length. The features are thus made up of the amplitudes at frequencies NF. In this case, only the first 10 harmonics are considered, i.e., $N = 10$.

With only about 40% of the available spectra having the corresponding speed information, a method of determining the fundamental frequency from the envelope spectrum has therefore been devised. It is based on the fact that the first significant peak (reading the spectrum from left to right) typically represents the running speed which is equivalent to the fundamental (SKF, 1994b). This peak can be searched for and checks made for any peaks at multiples of the suspected fundamental.

The situation is complicated in many cases by the absence of a peak at the fundamental, but with peaks present at the harmonics. This is usually the case during the degeneration of a fault when some harmonics may also vanish. Therefore, software has been developed for automatically finding the fundamental and the relevant harmonics from each spectrum (Bissessur *et al.*, 1997). Briefly, the peak detection algorithm is as follows.

1. Using the minimum and the maximum values of the felt speed, calculate the upper and the lower limits for F from Equation 1, i.e., F_{max} and F_{min} respectively.
2. For $F = F_{min}$ to F_{max},
   Find the N harmonic amplitudes.
   Calculate the root mean square (RMS) value of the amplitudes.
   End
3. Determine the maximum RMS value.
4. Find F corresponding to the maximum value. This is the estimated fundamental frequency.

The performance of the peak detection algorithm has been validated using spectra for which speed information is available. The fundamental as estimated by the algorithm is consistently in close agreement with that calculated from Equation 1. This confirms its suitability for the present application.

5. CLASSIFICATION

The feedforward neural network architecture has been widely applied in industry for classification (McCormick and Nandi, 1997; Lowes and Shippen. 1996) and modelling tasks (Morris *et al.*, 1994). In this case, the network consists of 10 input units to accommodate the feature vector of 10 harmonic amplitudes, a single hidden layer and 2 output units. Outputs of [*0.9 0.1*] and [*0.1 0.9*] are arbitrarily chosen to represent *normal* and *faulty* spectra respectively during the training stage. The *Levenberg/Marquardt* algorithm (Marquardt, 1963) is used for the learning phase; its main advantage over the conventional backpropagation algorithm is a significant reduction in the training times.

A reference set of spectra from felts in normal states and in advanced stages of deterioration (10 of each encompassing the whole range of speeds of the press rolls) has been made available for training the neural network. These spectra are augmented by the addition of Gaussian noise to enhance the information content in the spectra (Conlin *et al.*, 1997) yielding a training set of 200 patterns. This technique produces a more robust classifier after training. Network robustness can additionally be significantly enhanced through stacked regression approaches (Zhang *et al.*, 1997) to overcome the problem encountered with the generalisability of single networks selected as being the 'best' trained network.

Various runs are carried out with the number of hidden units varied from 3 to 10 and with the error threshold set to 0.1. The smallest best performing network is found to be the one with 5 hidden units. All 100 normal and 100 faulty spectra are correctly classified after an average of 20 training epochs.

6. INDUSTRIAL RESULTS

The trained network is tested in turn on the envelope spectra from press rolls 612 and 608. Spectra identified as faulty are mapped onto their corresponding points in time and superimposed on the plot of overall vibration levels. They are indicated by the symbol '*' on Figure 5. It is clear that spectra with high vibration levels are correctly classified as representing faulty felts. One example of these spectra (corresponding to the point denoted by an arrow in Figure 5(a)) is shown in Figure 6(a). The presence of relatively high amplitude harmonics confirms the fact that felt deterioration has occurred

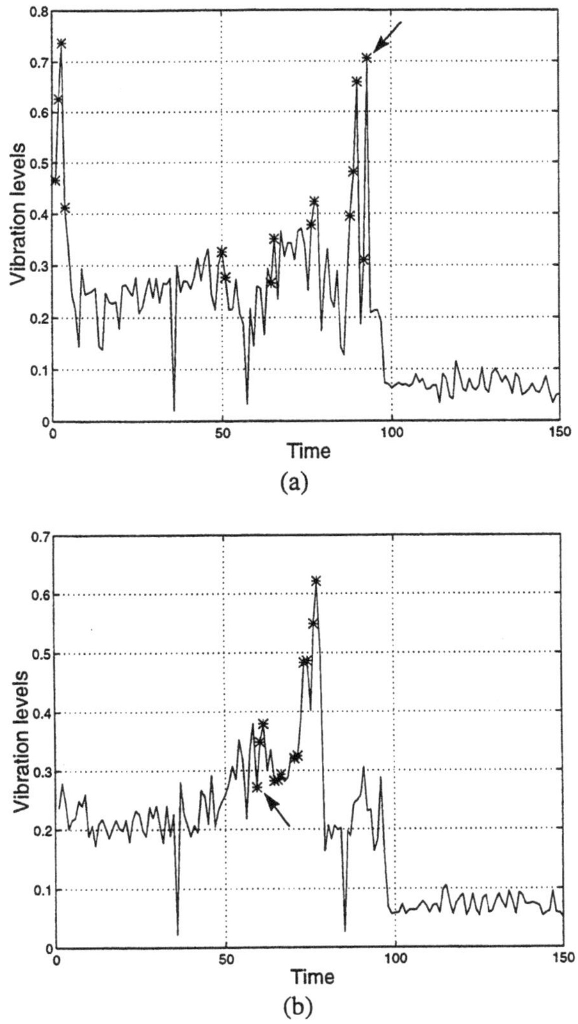

Fig. 5. Neural network results on spectra from roll 608 (a) and roll 612 (b).

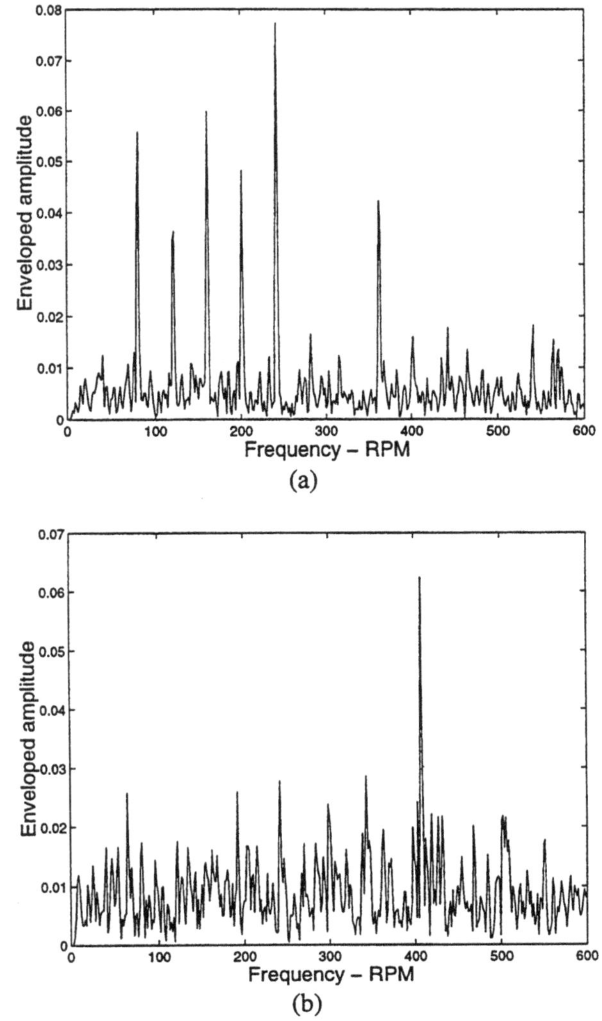

Fig. 6. Typical *faulty* spectra with high and low vibration levels, (a) and (b) respectively.

and validates the performance of the neural network classifier.

From Figure 5, some spectra with low vibration levels have also been flagged as faulty. One such spectrum (corresponding to the point denoted by an arrow in Figure 5(b)) is depicted in Figure 6(b). These results have been examined by experts from the paper mill and from a leading condition monitoring equipment manufacturer who have identified such spectra as representing roll defects (such as flat spots). This is confirmed by the high amplitude fundamental located at around 400 RPM which is the typical roll speed.

It is well known in the paper industry that such roll anomalies are one of the main causes of felt degradation. The presence of a flat spot on a roll places excessive stresses on the felt at the press nips. As shown by Figure 6(b), although this results in a high amplitude fundamental at the roll frequency, the overall vibration levels remain at relatively low values. Therefore, inspection of the vibration levels does not usually reveal a flat-spot problem at its early stage. It is only picked out when the problem has degenerated sufficiently to produce noticeable increases in the vibration levels. However, at this point, felt damage has already taken place to a large enough extent to necessitate shutdowns and repairs in the very near future.

The classification results of the neural network as represented in Figure 5 have prompted much interest from the process operators as they clearly depict roll defects (crosses at low vibration levels) prior to felt deterioration (crosses at high vibration levels). The availability of information about these roll anomalies at an early stage allows corrective action which will prolong the felt life and maintain product consistency. Repairs can then be scheduled at appropriate time intervals and thus emergency shutdowns with severe disruptions to productivity are avoided.

7. CONCLUSIONS

A system for the detection of felt deterioration in a paper machine press section has been presented. It is based on the monitoring and processing of vibration data from the rolls and consists of 3 main stages. The enveloper, which makes up the preprocessing step, isolates the defect signals from unwanted components. Features are then extracted from the envelope spectra for subsequent discriminatory tasks. This is carried out by a well-designed peak detection algorithm. The final stage of the system involves a neural network classifier for making decisions on the state of the felt.

Vibration data, made available by a major paper manufacturer, has been used for testing the developed system. The results obtained have demonstrated its capability for fault detection and for providing advance warnings of impending felt problems. Innovative contributions resulting from this study include,

(i) The development of a robust neural network classifier.
(ii) The design of a new spectral peak detection algorithm.
(iii) Application to an industrial paper manufacturing plant.

The main advantage of the approach adopted is the relative ease with which it can be extended to the on-line diagnosis of faults in a wide range of mechanical equipment in the process industries.

8. ACKNOWLEDGEMENTS

This work has been carried out within the EU BRITE/EURAM Project *IMPROQ* BE-8104/93. The authors are grateful to the industrial partners involved, in particular *KNP Fine Paper* and *SKF Condition Monitoring* in The Netherlands for providing the data and for their expert advice. The authors also acknowledge the support of colleagues from the Centre for Process Analysis, Chemometrics and Control at the University of Newcastle.

9. REFERENCES

Bissessur, Y., E.B. Martin and A.J. Morris. (1997). Felt press condition monitoring-A neural network approach. *Brite-Euram Project No. BE-8104 (IMPROQ), Technical Report No. 22,* Newcastle University.

Conlin, A., E.B. Martin and A.J. Morris. (1997). Data augmentation: An alternative approach to the analysis of spectroscopic data. Accepted for publication in *Proc. 5th Symposium on Chemometrics*, Lahti, Finland.

Lowes, S. and J.M. Shippen. (1996). Application of neural networks for diagnostic engineering of rotating machinery. *Proc. Condition Monitoring and Diagnostic Engineering Management (COMADEM'96),* pp. 113-122, Sheffield Academic Press, Sheffield, UK.

Marquardt, D. (1963). An algorithm for least squares estimation of nonlinear parameters. *SIAM Journal.of Applied Mathematics*, **11**, pp. 431-439.

McCormick, A.C. and A.K. Nandi. (1997) Classification of the rotating machine condition using artificial neural networks. *Proc. Inst. Mech. Eng.*, Part C, **211**, No.6, pp.439-450.

Morris, A.J. and E.B. Martin, (1997). Process performance monitoring and fault detection through multivariate statistical process control. *(Plenary Paper) IFAC Conference SAFEPROCESS'97*, Hull UK.

Morris, A.J., G.A. Montague and M.J. Willis. (1994). Artificial neural networks: Studies in process modelling and control. *Transactions IChemE*, **72**, Part A, pp. 3-19.

Simoglou, A., E.B. Martin, A.J. Morris, M. Wood and G.C. Jones. (1997). Multivariate statistical process control in chemicals manufacturing. *IFAC Conference SAFEPROCESS'97*, Hull, UK.

SKF Condition Monitoring. (1994a). Monitoring felt for improved paper quality. *Technical Research Bulletin 01.*

SKF Condition Monitoring. (1994b). *Vibration diagnostic guide*, Revision 1.01, San-Diego.

Smulders, A.J. and D. Kadushin. (1994). Acoustic emission technology: a viable predictive maintenance tool for the paper industry. *Tappi Journal*, **77**, No. 12, pp. 58-64.

Zhang, J., E.B. Martin, A.J. Morris and C. Kiparissides. (1997). Inferential estimation of polymer quality using stacked neural networks. (1997). *Computers and Chemical Engineering* Supp., **21**, pp. S1173-S1178.

A STUDY ON THE DEVELOPMENT OF ROBUST FAULT DIAGNOSTIC SYSTEM BASED ON NEURO-FUZZY SCHEME

Sung-Ho Kim **Kee-Sang Lee***

Department of Control & Instrumentation Eng. Faculty of Engineering, Kunsan National University, Kunsan,Korea
e-mail: shkim@ks.kunsan.ac.kr

* *Department of Electrical Engineering Faculty of Engineering Dankook University,Seoul Korea*
e-mail: keesang@ns.dankook.ac.kr

Abstract: FCM(Fuzzy Cognitive Map) is proposed for representing causal reasoning. Its structure allows systematic causal reasoning through a forward inference. By using the FCM, authors have already proposed FCM-based diagnostic algorithm. However, it is a qualitative approach in nature. Therefore, it has the same problem, i.e., lower diagnostic resolution, as the other qualitative approaches. In order to solve this problem, the quantitative knowledge is integrated into the FCM-based diagnosis by using fuzzy set theory. Generally, the integration process requires a lot of engineering efforts. For the automated integration, self-learning feature of neural network is utilized. Finally, some simulation studies are carried out to illustrate the proposed diagnostic scheme. *Copyright © 1998 IFAC*

Keywords: Fuzzy Cognitive Map, Fault Diagnosis, Neural Network

1. INTRODUCTION

In recent years, the complexity of modern industrial processes and the availability of inexpensive computer hardware prompted us to develop automated fault diagnostic systems instead of conventional diagnosis. Generally, depending on the rigorousness of the process model employed, existing fault diagnostic schemes can be classified into quantitative, qualitative and qualitative/ quantitative approaches.

Quantitative approaches which have been studied by (Clark,1989; Willsky, 1976) are based on the analytical redundancy generated by the use of estimators such as Kalman filter, state observer, and detection filters. It can accurately find out the fault origin. However, it is too time-consuming and requires as precise a model as possible to get a good diagnostic result. Qualitative approaches, on the other hand, only consider the signs of coefficients in all governing equations. The signed directed graph(SDG) is a typical example. Upon diagnosis, the consistency of the branches of a given fault origin is checked to validate this hypothesis and all possible origins are screened. In many cases, it simply gives multiple interpretations for a single event. This is an inherent limitation of the qualitative approaches. Since only qualitative knowledge is employed, the diagnostic resolution can only be improved to a certain degree.

Another one is the qualitative/quantitative approaches. (Yu and Lee,1991) integrated quantitative knowledge into a qualitative model-based diagnosis to improve diagnostic resolution.

Recently, concept of fuzzy cognitive map is proposed by (Kosko, 1986) for representing causal reasoning. Its structure allows systematic causal reasoning through a forward-evolved inference. Since forward inference behaves as temporal associative memories(TAM), it is possible to reason with FCM as we recall with TAM. By utilizing the FCM, authors have already proposed two FCM-based fault diagnostic algorithms. One can be considered as a simple transition of Shiozaki's SDG-based diagnostic approach to FCM framework. The other is an extended version of the basic diagnostic algorithm where an FCM's important property, i.e., TAM recall, is utilized. It can be effectively utilized in on-line fault diagnosis owing to its self-generated **fault FCM models** which can generate predicted pattern sequence for pattern match with observed one. However, FCM-based diagnosis is a qualitative approach in nature. Therefore, it has the same problem as the other qualitative approaches, i.e., low diagnostic resolution. As experience accumulated, some form of quantitative process knowledge is available, e.g., steady-state gain between process variables. If this information can be utilized, the diagnostic resolution of the FCM-based diagnosis can be further improved.

For this, firstly, the quantitative knowledge is integrated into the FCM-based diagnostic scheme by using fuzzy set theory and secondly **modified TAM recall procedure** is proposed. Generally, the integration of quantitative knowledge into FCM-based diagnosis requires a great deal of engineering efforts, even when all process data are available. Furthermore, quantitative knowledge is varying according as operating conditions of process and the magnitude of faults are changed. Therefore, some automated procedure for fusing the quantitative knowledge into FCM-based diagnosis is required. For this, self-learning feature of neural network is utilized. The self-learning procedure is based on reinforcement learning of the neural network.

This paper is organized as follows: Concept of FCM and FCM-based diagnostic algorithm are outlined in section 2 and 3, respectively. The detailed procedure for integration of fuzzy set theory into FCM-based diagnosis is described in section 4. In section 5, self-learning feature to the FCM-based diagnosis is considered. Finally, simulation study is given to illustrate the feasibility of the proposed diagnostic scheme.

2. FUZZY COGNITIVE MAP

FCM was proposed by Kosko to store uncertain causal knowledge. It is a fuzzy signed directed graph with feedback. FCM consists of nodes and branches. The nodes of the FCM correspond to variables and branches represent the causal influences between nodes. The influences are represented by [-1,1] on the arcs, indicating that the cause and effect variables tend to change in the same or opposite direction. Using Kosko's convention, Values -1 and 1 represent full causality, zero denotes no causal effects, and all other values correspond to different fuzzy level of causal effect. If the permissible values assigned to the directed branches are restricted only to the set {-1,0,+1}, it is called a simple FCM.

3. FCM-BASED DIAGNOSTIC ALGORITHM

Basic fault diagnostic algorithm based on the simple FCM is outlined in this section. The more detailed description can be found in (Lee,1996).

3.1 FCM-Based Fault Diagnostic Algorithm

The proposed FCM-based diagnostic algorithm is based on Shiozaki's consistent rooted trees method, which is built upon the concept of SDG(Signed Direted Graph). SDG has the same structure as FCM. Concept of Shiozaki's algorithm is as follows: Nodes which have a sign other than '0' are known as valid nodes, while branches for which the product of the signs on the initial and terminal nodes is the same as the sign of the branch are known as consistent branches. The graph which is composed of valid nodes and consistent branches is called a cause and effect(CE) graph. If there is an elementary path from a node on the SDG to all valid observed nodes, and if all the branches on these paths are consistent branches, then the tree which is composed of such a node and such consistent paths is known as a 'consistent rooted tree', and the node is its 'root'. The idea of consistent rooted-tree method is that the node which is the maximal strongly connected component of the CE graph is the candidate of the fault.

Let FCM matrix for a system, E and the observed pattern vector W be given. In what follows is the detailed FCM-based diagnostic algorithm.

Step 1. Calculating CR matrix

$$WE = Diag(W) \cdot E \cdot Diag(W) \qquad (1)$$
$$CR(i,j) = T(WE(i,j)) \qquad (2)$$

where Diag(W) represents a square matrix whose diagonal elements are those of W and all other elements off the diagonal are zero, and T is the threshold function with a threshold selected to be zero. According to Shiozaki's consistent rooted tree method, (1) can be thought of as the process of generating all possible paths. Threshold function in (2) plays the role of removing inconsistent paths from WE .

Step 2. Identifying the origin of the fault.

In order to find out the origin of the fault, we must find the maximal strongly connected node. This can be done as follows.

1. Calculate the column sum of CR matrix which represents the number of concepts causally impinging on concept C_i.

$$IN(C_i) = \sum_{k=1}^{n} CR(k, i) \quad (3)$$

2. Calculate the row sum which represents the number of concepts concept C_i causally impinges on.

$$OUT(C_i) = \sum_{k=1}^{n} CR(i, k) \quad (4)$$

3. Find the origin of fault corresponding to the maximal strongly-connected concept which has **zero column sum and non-zero row sum.**

3.2 Generation of Fault FCM Model

If an FCM and steady-state observed patterns are given, the origin of fault can be easily found out. Furthermore, if we take advantage of CR matrix, we can easily obtain fault FCM models. In general, the task of constructing fault models in other diagnostic schemes such as fault tree analysis is time-consumming and burdensome process. Detailed procedure for generating fault FCM model is as follows:

Step 1. Apply steady-state observed pattern for a known fault to the basic FCM-based diagnostic algorithm and derive its CR matrix.
Step 2. Superimpose a sign onto each non-zero elements of the CR matrix. The signs are borrowed from the original FCM matrix for the system.

Fault FCM models obtained by the above process are stored into **fault model-base** for TAM recall and pattern matching.

3.3 TAM recall Process

TAM recall is an inference process which can successively predict a future behavior. If this property is utilized in the fault diagnosis, it is possible to infer how primary fault effect is propagated through the whole system. In FCM-based fault diagnosis, hypothesis-and-test procedure is used: First, we hypothesize the non-zero element within the observed pattern vector as the origin of fault. Second, for that hypothesized fault, a corresponding fault FCM model is selected from **Fault Model-Base**. Finally, the following TAM recall process is triggered for obtaining the predicted propagation of that hypothesis.

$$PW(k+1) = PW(k) \cdot FCM_{x_i} \quad (5)$$

where FCM_{x_i} denotes a hypothesized fault FCM model with state variable x_i being an origin of fault.

4. EXTENDED FAULT FCM MODEL AND MODIFIED TAM RECALL PROCESS

Pattern matching based on TAM recall is performed between qualitative observations and qualitative predictions. In any realistic situation, quantitative observations are available. As the qualitative nature of the FCM-based diagnosis, we ignore some given information because of the limitation of the fault FCM models employed. This entails lower diagnostic resolution compared with quantitative ones. For this, a method is devised to integrate quantitative knowledge into the fault FCM models.

4.1 Extended Fault FCM Model

Consider a branch in a fault FCM model as in Fig. 1(a). The binary relationship(+1) between A and B can also be described by the ratio $\Delta B/\Delta A$ taking the value from 0^+ to infinity by using membership function as in Fig. 1(b). If some quantitative information is known, e.g., the steady-state gain between A and B, we can shape the membership function as in Fig. 1(c).

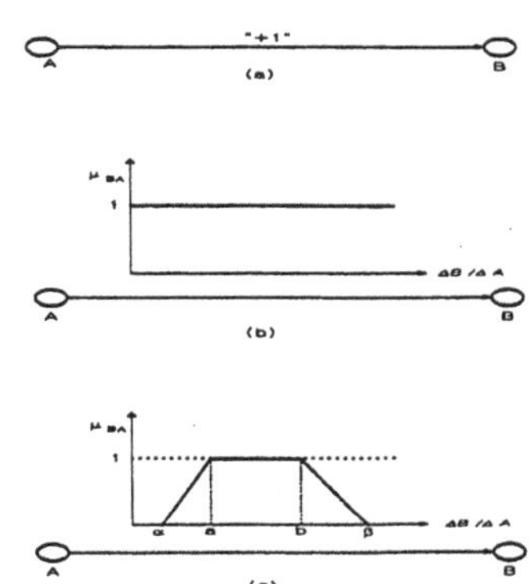

Fig. 1 Membership functions for qualitative and quantitative knowledge

The membership function of the fuzzy set theory provides a efficient way to treat the quantitative knowledge shown in Fig. 1(c). In this work, we consider the following 4-tuple fuzzy number for representing the transfer gains between process variables.

$$C = (a, b, \alpha, \beta) \quad (6)$$

If we substitute non-zero elements in an fault FCM models with the above trapezoidal fuzzy numbers, we can obtain the extended fault FCM models which can effectively represent the quantitative knowledge, e.g., transfer gains between process variables. These extended fault FCM models are stored into **fault model-base** for modified TAM recall and pattern matching.

4.2 Modified TAM Recall

Modified TAM recall based on the extended fault FCM models can be represented as in (7).

$$P(k+1) = P(k) \bigstar FCM_{x_i} \tag{7}$$

where $P(k)$ is row vector which has trapezoidal fuzzy number as its element. FCM_{x_i} is selected from the fault model-base on a hypothesis that the origin of fault is x_i. The operator $\bigstar$ represents vector and matrix multiplication. In this work, we use the following fuzzy addition ($\oplus$) and fuzzy multiplication($\otimes$) operators.

$$(a,b,\tau,\beta) \otimes (c,d,\gamma,\delta) =$$

$$\begin{cases} (ac, bd, a\gamma + c\tau - \tau\gamma, b\delta + d\beta + \beta\delta) & m > {}_00, n > {}_00 \\ (ad, bc, d\tau - a\delta + \tau\delta, -b\gamma + c\beta - \beta\gamma) & m < {}_00, n > {}_00 \\ (bc, ad, b\gamma - c\beta + \beta\gamma, -d\tau + a\delta - \tau\delta) & m > {}_00, n < {}_00 \\ (bd, ac, -b\delta - d\beta - \beta\delta, -a\gamma - c\tau + \tau\gamma) & m < {}_00, n < {}_00 \end{cases} \tag{8}$$

$$(a,b,\tau,\beta) \oplus (c,d,\gamma,\delta) = (a+c, b+d, \tau+\gamma, \beta+\delta) \tag{9}$$

The i-th element, $p_i(k+1)$ in $P(k+1)$ is calculated as follows:

$$p_i(k+1) = (p_1(k) \otimes \mu_{i1}) \oplus (p_2(k) \otimes \mu_{i2}), \cdots, (p_N(k) \otimes \mu_{iN}) \tag{10}$$

where μ_{ij} represents the ij-th trapezoidal fuzzy number in the hypothesized fault FCM model, FCM_{x_i}.

4.3 Fault Diagnosis based on Pattern matching

In this work, hypothesis-and-test scheme is adopted for fault diagnosis as before: Modified TAM recall process is applied to each extended fault FCM models to find out the origin of fault. For its verification, measure of compatibility is required to calculate the fitness between predicted and observed patterns. For this, compatibility of two fuzzy sets proposed by (Zadeh,1977) is utilized. Compatibility, τ between two fuzzy sets, A and B is defined as follows.

$$\mu_\tau(u) = \lim_{x: u = \mu_A(x)} \mu_B(x), \ \forall u \in [0,1] \tag{11}$$

By utilizing the compatibility, we can calculate the fitness of each hypothesized fault FCM models at each steps as follows.

$$\mu_{FCM_{x_i}}(k) = \min[p_{1k}(m_1(k)), \cdots, p_{Nk}(m_N(k))] \tag{12}$$

where p_{ik} represents the i-th predicted fuzzy number at the k-th step and $m_i(k)$ represents center of the i-th observed triangular fuzzy number at the k-th step. Finally, we can get the total compatibility for the hypothesized fault FCM model, e.g., FCM_{x_i} as follows.

$$\textit{Confidence}\ \deg_{FCM_{x_i}} = \min[\mu_{FCM_{x_i}}(1), \cdots, \mu_{FCM_{x_i}}(K)] \tag{13}$$

The schematic diagram for fault diagnosis based on extended fault FCM models is depicted in Fig. 2.

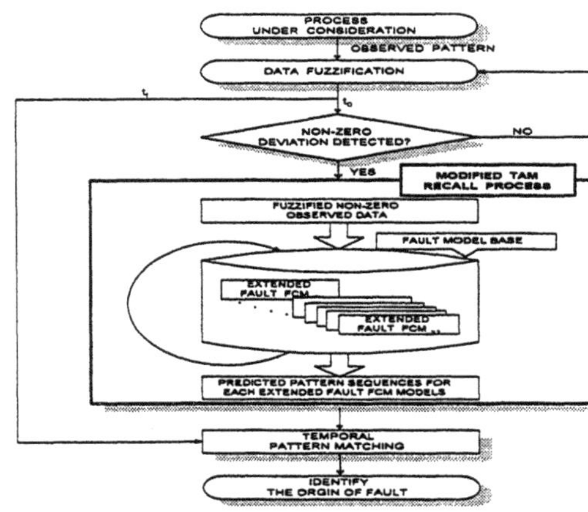

Fig.2 Block diagram for the fault diagnosis based on modified TAM recall

5. Self-Learning Feature to the FCM-based Diagnosis

The construction of the extended fault FCM model requires a great deal of engineering effort. In general, the steady-state gains between process variables need to be modified as the operating conditions and the magnitude of faults are changed. The trapezoidal fuzzy number is adequate for representing inconstant transfer gains. One important property related with the fault FCM model is that its structure remains the same under almost all possible operating conditions. Therefore, all we have to do is to tune the shape of the membership function.

5.1 Associative Reinforcement Learning

In this work, neural network scheme is employed to automatically acquire the extended fault FCM models. We consider an associative reinforcement learning scheme shown in Fig. 3.

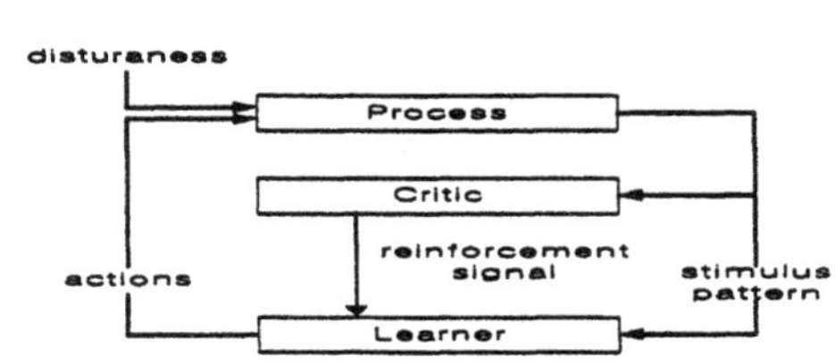

Fig. 3 Structure of reinforcement learning

In the associative reinforcement learning problem,

the learning system receives stimulus patterns as input in addition to the reinforcement signal. The optimal action on any trial depends on the stimulus patterns presented on that trial. Several associative reinforcement learning rules for neuron-like units have been studied. Fig. 4 shows a neuron-like unit called Associative Searching Element(ASE). The unit has a reinforcement input pathway, n-pathways for non-reinforcement input signals, and a single output pathway.

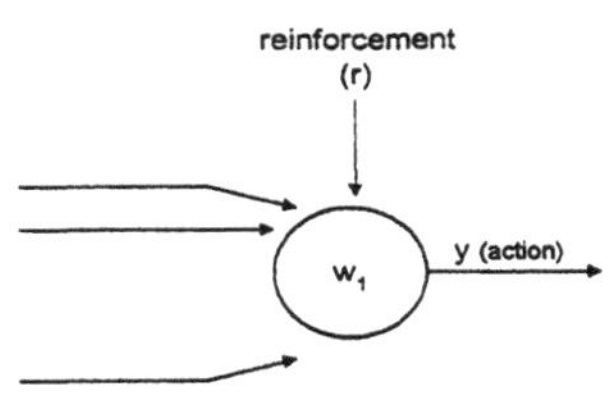

Fig. 4 Structure of ASE

The element's output $y(t)$ is determined from the input vector $X(t)=[x_1(t), x_2(t), \cdots, x_n(t)]$.

$$y(t)=f(\sum_{i=1}^{n} w_i(t)x_i(t)) \quad (14)$$

where f is the following threshold function.

$$f(x)=\begin{cases} +1 & \text{if } x\geq 0 \\ -1 & \text{if } x<0 \end{cases} \quad (15)$$

The weight w_i's are changed according to

$$w_i(t+1)=w_i(t)+\alpha r(t)e_i(t) \quad (16)$$
$$e_i(t+1)=\delta e_i(t)+(1-\delta)y(t)x_i(t) \quad (17)$$

where α = learning rate, δ = trace decay ratio, $r(t)$ = reinforcement signal at time t, $e_i(t)$= eligibility at time t of input pathway i, and $x_i(t)$ = input vector at time t.

5.2 Tuning of Membership Function Via ASE

Self-learning of extended fault FCM models via ASE is shown in Fig. 5. For a known fault origin, we can obtain the steady-state observed patterns and the steady-state transfer gains from process simulation. These measurements are fed into the corresponding extended fault FCM model selected from the fault model-base. If the compatibility between the steady-state observed pattern and the predicted steady-state pattern is not satisfactory, i.e., $\mu_{FCM_i}\neq 1$, reinforcement learning is triggered and the measurements and the diagnostic result(compatibility) are fed to the corresponding ASE. Subsequently, the shape and location of the membership function is tuned until a satisfactory diagnostic result is found. Let's consider a steady-state gain between A and B in a certain extended fault FCM model and its corresponding ASE as in Fig. 6. The shape of initial membership function is triangular and its

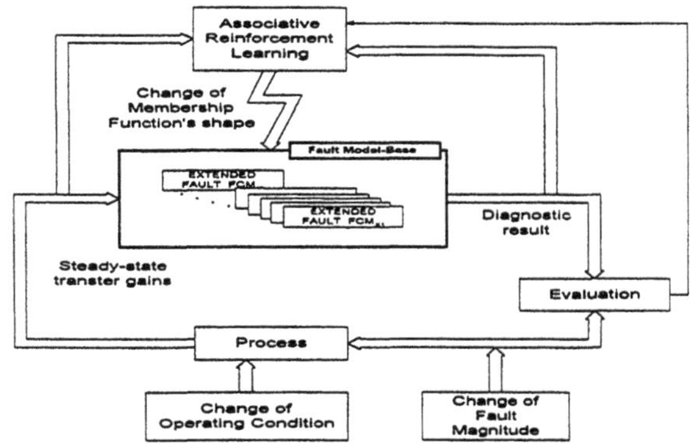

Fig. 5 Schematic diagram for self-learning of extended fault FCM models via ASE

center is located according to the measurement $\Delta B/\Delta A$. This value corresponds to the full membership, i.e., $\mu_{BA}(\Delta B/\Delta A)$=1. If another set of measurement (caused by change of operating conditions or change in magnitude of fault) is available and the result is not satisfactory, then the system responds with a reinforcement signal r=1. This indicates the location and the shape of the membership function is incorrect. Therefore, the ASE automatically tunes the shape of membership function according to new $\Delta B/\Delta A$ and compatibility of its extended fault FCM model until a right diagnosis is achieved. The membership function is relocated according to the weight change Δw. In this work, the weight w is changed as follows (Hsu and Yu,1992).

$$w(t+1)=w(t)+\alpha r(t)e(t) \quad (18)$$
$$e(t+1)=\delta e(t)+(1-\delta)(1-y(t))x(t) \quad (19)$$

The ASE adjusts the location and shape of the membership function in the following way.

$$f(x)=\begin{cases} |\delta w| & \text{if } x>b_f \\ 0 & \text{if } b_i<x<b_f \\ -|\delta w| & \text{if } b_i>x \end{cases} \quad (20)$$

where b_i and b_f are the initial value of the steady-state transfer gain satisfying y=1.

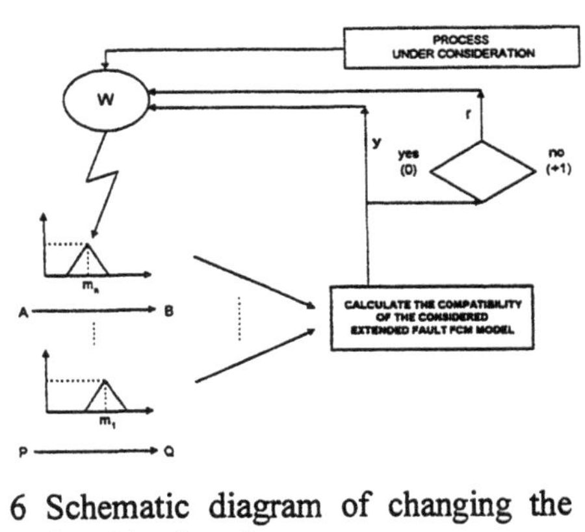

Fig. 6 Schematic diagram of changing the shape of membership function

6. SIMULATION STUDY

Tank-pipe system shown in fig.7 is used to illustrate the performance of the proposed scheme.

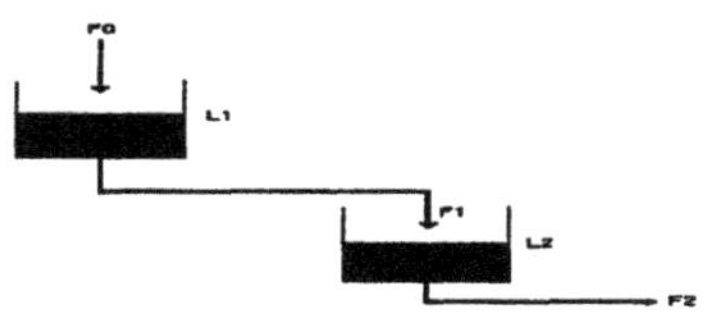

Fig. 7 Tank-pipe system

FCM for the above system is the following.

$$E=\begin{array}{c} \\ F_0 \\ L_1 \\ F_1 \\ L_2 \\ F_2 \end{array}\begin{array}{c} \begin{array}{ccccc} F_0 & L_1 & F_1 & L_2 & F_2 \end{array} \\ \begin{bmatrix} 0 & 1 & 0 & 0 & 0 \\ 0 & 0 & 1 & 0 & 0 \\ 0 & -1 & 0 & 1 & 0 \\ 0 & 0 & 0 & 0 & 1 \\ 0 & 0 & 0 & -1 & 0 \end{bmatrix} \end{array}$$

The origin of fault chosen for example is blockage in the upper tank outlet line and its steady-state observed pattern is as follows.

$$\begin{array}{cccccc} & F_0 & L_1 & F_1 & L_2 & F_2 \\ W_0=[& 0 & 1 & -1 & -1 & -1] \end{array}$$

The corresponding fault FCM model for the decrease in F_1 is obtained as follows.

$$FCM_{F1(-)}=\begin{array}{c} \\ F_0 \\ L_1 \\ F_1 \\ L_2 \\ F_2 \end{array}\begin{array}{c} \begin{array}{ccccc} F_0 & L_1 & F_1 & L_2 & F_2 \end{array} \\ \begin{bmatrix} 0 & 0 & 0 & 0 & 0 \\ 0 & 0 & 0 & 0 & 0 \\ 0 & -1 & 0 & 1 & 0 \\ 0 & 0 & 0 & 0 & 1 \\ 0 & 0 & 0 & 0 & 0 \end{bmatrix} \end{array}$$

Applying the self-learning procedure for its extended fault FCM model, i.e., $EFCM_{F_{1(-)}}$, it can be obtained as follows.

$EFCM_{F1(-)}=$

$$\begin{array}{c|ccccc|} & F_0 & L_1 & F_1 & L_2 & F_2 \\ F_0 & [0\ 0\ 0\ 0] & [0\ 0\ 0\ 0] & [0\ 0\ 0\ 0] & [0\ 0\ 0\ 0] & [0\ 0\ 0\ 0] \\ L_1 & [0\ 0\ 0\ 0] & [0\ 0\ 0\ 0] & [0\ 0\ 0\ 0] & [0\ 0\ 0\ 0] & [0\ 0\ 0\ 0] \\ F_1 & [0\ 0\ 0\ 0] & [0.1\ -1.9\ -1.3\ 0.1] & [0\ 0\ 0\ 0] & [0.1\ 1.8\ 1.9\ 0.1] & [0\ 0\ 0\ 0] \\ L_2 & [0\ 0\ 0\ 0] & [0\ 0\ 0\ 0] & [0\ 0\ 0\ 0] & [0\ 0\ 0\ 0] & [0.1\ 0.52\ 0.54\ 0.1] \\ F_2 & [0\ 0\ 0\ 0] & [0\ 0\ 0\ 0] & [0\ 0\ 0\ 0] & [0\ 0\ 0\ 0] & [0\ 0\ 0\ 0] \end{array}$$

Each membership functions in the $EFCM_{F_{1(-)}}$ are obtained via reinforcement learning by using a set of observed steady-state gains under the circumstances that both operating conditions and magnitude of faults are changed. Let's assume that primary deviation caused by the decrease in F_1 is as follows.

$$\begin{array}{cccccc} & F_0 & L_1 & F_1 & L_2 & F_2 \\ W_0=[& [0\ 0\ 0\ 0] & [0\ 0\ 0\ 0] & [0.1\ -1\ -1\ 0.1] & [0\ 0\ 0\ 0] & [0\ 0\ 0\ 0]] \end{array}$$

If we successively applies this to the modified TAM recall process, the following predicted

$$\begin{array}{lccccc} & F_0 & L_1 & F_1 & L_2 & F_2 \\ W_0 = [& [0\ 0\ 0\ 0] & [0\ 0\ 0\ 0] & [0.1\ -1\ -1\ 0.1] & [0\ 0\ 0\ 0] & [0\ 0\ 0\ 0]] \\ PW_1 = [& [0\ 0\ 0\ 0] & [0.1\ 1.3\ 1.9\ 0.1] & [0.1\ -1\ -1\ 0.1] & [0.3\ -1.9\ -1.8\ 0.27] & [0\ 0\ 0\ 0]] \\ PW_2 = [& [0\ 0\ 0\ 0] & [0.1\ 1.3\ 1.9\ 0.1] & [0.1\ -1\ -1\ 0.1] & [0.3\ -1.9\ -1.8\ 0.27] & [0.38\ -1.03\ -0.94\ 0.35]] \end{array}$$

pattern sequences are obtained.

If the magnitude of fault(decrease in F_1) is 0.6 and 0.5, respectively, steady-state observed pattern represented in triangular fuzzy number are obtained by process simulation.

$$OW_{0.6}=[[0.1\ 0\ 0.1][0.1\ 1.51\ 0.1][0.1\ -1\ 0.1][0.1\ -1.87\ 0.1][0.1\ -1\ 0.1]]$$

$$OW_{0.5}=[[0.1\ 0\ 0.1][0.1\ 1.92\ 0.1][0.1\ -1\ 0.1][0.1\ -1.90\ 0.1][0.1\ -1\ 0.1]]$$

If we compare the observed pattern with each predicted ones which are obtained from the TAM recall based on each extended fault FCM models, the following compatibilities are obtained.

	FCM_{F0}	FCM_{L1}	FCM_{F1}	FCM_{L2}	FCM_{F2}
$Compatibility_{0.6}$	0	0	1	0	0
$Compatibility_{0.5}$	0	0	0.9	0	0

From the above example, we can get a better diagnostic resolution in the face of the changes in magnitude of fault.

7. CONCLUSION

A new approach to FCM-based fault diagnosis has been proposed. By using the fuzzy set theory and neural network scheme, the proposed method can more accurately find out origin of fault in the face of the changes of operating condition and magnitude of fault. To verify its effectiveness, extensive simulation study is in progress.

REFERENCES

Clark, R.N. (1989), State estimation schemes for instrument fault detection. *In Fault diagnosis in dynamic systems* (Frank, P., ,Clark, R.N.), Prentice-Hall

Chang, C.C. and Yu, C.C.(1990), On-line Fault Diagnosis Using the Signed Directed Graph, *Ind. Eng. Chem. Res.*, **Vol.29**, No.7, pp. 1290-1299

Hsu, Y.Y. and Yu, C.C.(1992), A self-learning fault diagnosis system based on reinforcement learning, Ind. Eng. Chem. Res.**Vol.31**, pp.1937

Kosko, B.(1986), Fuzzy cognitive maps, *Int. J.Man-machine Studies*, **Vol.24**, pp.65-75

Lee, K.S. and Kim, S.H.(1995), On-line fault diagnosis by using fuzzy cognitive map, IEICE Trans., **Vol.E79-A,** No.6, pp. 921

Shiozaki, J. and Matsuyama, H.(1985), Fault dia gnosis of chemical processes by the use of signed directed graphs. Extension to the five range patterns of abnormality, *Int. Chem Eng.*, **Vol.25**, No.4, pp. 651

Willsky, A.S.(1976), A survey of design method for failure detection in dynamic systems, *Automatica*, **Vol.12**, pp. 601-611

Yu, C.C. and Lee, C.(1991), Fault diagnosis based on qualitative/quantitative process knowledge, AIChE Journal, **Vol.**37, No.4, pp. 617-628

UNIVARIATE AND MULTIVARIATE PROCESS MONITORING AND IMPROVEMENT

Mr Adrian King, The University of Manchester/WM Engineering Ltd, UK
Professor Richard Wynne, Sheffield Hallam University, UK
Dr Ian Kennedy, Managing Director, WM Engineering Ltd, Manchester, UK

Abstract: The authors demonstrate the application of traditional process monitoring and control methods such as parameter estimation, alongside Univariate Statistical Process Control techniques that have traditionally been used for quality control and monitoring. Further work is also presented on the use of Multivariate Statistical Process Control methods for process monitoring and improvement. A restriction on this work is that only the data currently used by the process monitoring and control system is considered. This is a realistic constraint, as many companies will not be able to provide a bank of additional sensors for the purposes of fault detection. This paper is of interest to researchers and practitioners alike whose ultimate aim is to apply process monitoring, fault detection and isolation in "real world" contexts rather than in the confines of the laboratory. It provides a useful step forward in the provision of guidelines for the selection of techniques for particular process types and desired outcomes.

Keywords: SPC, PCA, Process Monitoring, Fault Detection

1. INTRODUCTION

The well-documented cuts in staffing levels in all industries during recent years, as a result of recession and corporate restructuring have been a major contributor to changes in the way that plant and equipment are managed. Fewer staff on site means that any equipment defects will be less likely to be observed or corrected unless new ethos is adopted.

Interest in process fault detection techniques has naturally increased throughout the last recession and continued during recent years of growth. The opportunities for increased plant reliability as a result of the early or more confident detection of faults leading to a reduction in breakdown and an overall reduction in operating cost are clear. The authors describe a subset of related fault detection techniques and compare their application on real data.

Many fault detection and isolation methods continue to be developed and tested on simulations of plant and equipment, or on small laboratory scale equipment. The authors have concentrated on the development and application of techniques on data from operational plant and equipment. This provides a realistic test of their fault detection capabilities and includes the inherent difficulties that occur when operating in an industrial or commercial environment.

The following criteria are used in particular during the assessment and comparison process, to act as a guide for the practitioner wishing to implement a similar scheme within their own area of responsibility.

1. The simplicity of the technique and the interpretation of results.
   This is a very pertinent consideration for the industrial and commercial adoptance of the methods. Operation of them is now normally delegated to the process operators, with a minimal training commitment.

2. The ease of application of the techniques.
   Particular emphasis is placed upon their suitability for on-line automatic fault detection-obviating the need for direct operator intervention during normal operation (although interpretation of the fault condition will always be necessary and normally performed by the process operator).

3. The effectiveness of the methods.
   The success of the techniques in detecting and isolating any faults or changes that occurred during the period of monitoring was also assessed.

Univariate Statistical Process Control (SPC) has been widely used for a number of decades, chiefly in the manufacturing industries. The central aim of the approach is to bring a process into statistical stability and once this state is achieved, to reduce the variation in the process (Bird and Dale, 1994). Bird and Dale emphasise the need for ongoing commitment to ensure a successful SPC programme, and provide descriptions of two methods of implementation.

SPC is one of the techniques in the Total Quality

Management (TQM) toolbox and has, therefore, been widely used as a result of companies adopting the TQM approach to quality improvement. This widespread adoption has resulted in companies applying the techniques without consideration of the principles upon which the techniques are based. This paper provides a practical comparison of SPC with a small number of techniques that are potentially of value in process monitoring situations. This will act as a simple guide for future techniques selection.

Parameter Estimation has been widely used in the control community and has been applied to a variety of problems across industries, but in particular in the process industries.

Quality improvement is a prime consideration in process manufacturing industries as well as descrete item manufacture. Chemical process industries have traditionally used off-line quality measurements to ascertain, for example, the concentrations of the various components of the mixture and to determine whether these are within acceptable limits. This procedure continues today. However, practitioners and researchers have for some time also been utilising a group of techniques known as Multivariate Statistical Process Control (MSPC). Whereas SPC (or Univariate SPC) monitors only a single variable at a time MSPC is used to monitor the whole of the process variable space.

SPC has been widely used in manufacturing (and particularly automotive) industries whereas MSPC has seen little application outside of the chemical process industries until recently. Furthermore, the majority of research applications that are reported in the literature are for laboratory scale plant where the process conditions can be tightly controlled - compared to applications on full scale commercial and industrial plant.

Recent evidence supports the claim that practical applications are in the minority. In a review paper, with emphasis on observer, parity space and parameter estimation methods, 88% in these categories, (Isermann and Balle, 1996), report 71% of contributions being either a simulations, pilot processes or laboratory with 29% being applied to full scale industrial processes.

Univariate Techniques

A number of approaches for the use of SPC techniques have been developed. An example of the type of development is White and Schroeder, 1987, who provide a simultaneous control chart that displays information on process level and variability. The chart is successful but sacrifices ease of interpretation for the ability to combine location and variation information on one chart. Developments are also being made in the application of SPC. It has been applied to laboratory based systems by Lau and (Sebele, 1993), who use the group of techniques with on-line process analysers. (Holmberg and Pokkinen, 1996), use SPC techniques with application to a laboratory scale process and integrate the techniques with a laboratory information system.

2. SPC - THEORETICAL BASIS

Statistical Process Control techniques are based upon normal distribution theory, with the following assumptions.

The variables being observed are normally distributed with a mean of μ and a standard deviation of σ. These values are such that 95% of samples are within +/- 1.96 standard deviations from the mean value, 99% of samples are within +/- 2.58 standard deviations from the mean.

It is therefore possible to construct control limits around the mean value under the assumption under normal operating only 5% of values will be outside of 1.96 σ from the mean, with only 1% outside of 2.58 σ from the mean. These limits are then called the warning and action limits, under the assumption that any changes from normal operation will cause more and more samples to fall outside of these limits - thus breaking the guidelines set out above.

The limits can be adjusted for schemes where samples from the process population are taken at regular intervals, by applying the following limits (where n is the number in each sample). The 99% limit is $\bar{x} \pm 2.58\frac{\sigma}{\sqrt{n}}$ and the 95% limit is $\bar{x} \pm 1.96\frac{\sigma}{\sqrt{n}}$. This is the basis for SPC schemes and similar limits can be developed for the standard deviation and range of samples.

SPC is one of the simplest of process monitoring techniques to apply. The results can be easily interpreted by process operators, who are normally confident running the SPC schemes themselves. Univariate SPC should normally be used in conjunction with process capability studies, which identify the ability of the process to operate within desired tolerances, and also process variation reduction programmes (including the application of design of experiments) as part of an overall process management programme.

3. PARAMETER ESTIMATION

Parameter Estimation is also a well established procedure. (Jiang, 1996), describes the modelling procedure as identifying the parameters of the following model:

$$y(k) = a_1(k-1) + \cdots$$
$$a_n y(k-n) + b_1 u(k-1) + \cdots$$
$$b_m u(k-m) + \varepsilon(k)$$

where $a_1, a_2, \cdots, a_n, b_1, b_2, \cdots, b_m$ are the parameters of the system which are unknown, and $\varepsilon(k)$ represents modelling errors and noise.

The process data is also assumed to be distributed normally, as in the case of SPC.

Parameter estimation is most suited to a stable process where the faults manifest themselves as gradual changes in the values of the parameters

Multivariate Statistical Process Control

Recent years have seen an increase in the number and range of applications of Multivariate Statistical Process Control (MSPC). (King et al 1996), report on the use of MSPC in a Building Energy Management System application and (Bisssessur, et al, 1996), applied the techniques to a batch processing unit. Additional applications include (Veltkamp, 1993), who identified process disturbances on a liquid fed ceramic meter for the vitrification of high level nuclear waste, (Nomikos and MacGregor, 1995), and (Nomikos and MacGregor, 1994), describe the historical and on-line monitoring of a batch polymerisation reactor and (Kresta et al, 1994), provide case studies of applications to distillation columns.

There are numerous other reported applications, but the vast majority cited in the literature, relate to chemical processes and in particular small-scale laboratory processes. Although the majority are still being applied to the chemical process industries, there is now an increasing number of manufacturing applications, for example (Martin, Morris and Zhang, 1996), report on an application to parts manufacturing.

Of the range of multivariate methods, Principal Components Analysis is the most widely used. Its' popularity stems, in part, from the relative ease with which the models produced can be interrogated and understood.

The Derivation of Principal Components

The PCA technique finds combinations of variables that describe major trends, or variation, in a data set. For a data matrix, X_d, with m rows (of measurements or samples) and n columns (of variables), the covariance matrix of X_d is defined as (Wise et al, 1995):

$$cov(X_d) = \frac{X_d^T X_d}{m-1} = X$$

Where the columns of X_d have been centred to have a mean of zero and scaled to a standard deviation of one, then the result will be the correlation matrix X_d. PCA decomposes this data matrix, X_d, as the sum of the outer product of vectors, t_i and p_i plus a residual error matrix E (Wise et al, 1995):

$$X_d = t_1 p_1^T + t_2 p_2^T + \ldots + t_k p_k^T + E$$

Where t_i are the score vectors and p_i are the loadings vectors. The scores describe the relationship between the samples and the loadings the relationship between the variables.

This can be summarised as (Martin et al, 1996):

$$X_d = TP^T + E = \sum_{i=1}^{k} t_i p_i^T$$

Where k is the number of principal components retained, and is always equal to or less than the smaller dimension of X_d. In practice, k should normally be much smaller than the number of variables in the original data.

Once the principal components have been calculated, and those of interest retained, it is possible to calculate values to determine whether the process is in control or not. Although the retained principal components (or Latent Variables, LV's) with the loadings do not constitute a "model" of the system, it is possible to treat them as a model because together they describe the most important process variation and can be recombined to produce estimates of the original data matrix, X_d. Furthermore, it is possible to project a new data set onto a "model" created from normal process operation to determine whether a process disturbance has occurred during the collection of the new data set.

4. CASE STUDY APPLICATIONS - PRACTICAL ASSESSMENT

The results presented in this section were all obtained from real plant under normal operating conditions. Normal plant monitoring instrumentation was used with no attempt made to control the operating environment.

Prior to analysis the data were screened to remove outliers and then normalised to the standard normal distribution.

The case studies are based upon three different processes. A Building Energy Management system serving a single floor of a commercial office building in central London. A steel finishing process, in which the surface of the steel is sent through various pickling process to improve its appearance. A paper manufacturing plant , where the data analysed are vibration trends collected as part of an ongoing Condition Based Maintenance programme.

Univariate SPC

USPC is invaluable in processes that are stable and have predictable distributions, process locations and degrees of variation. The following plot demonstrates the application USPC to such a stable variable.

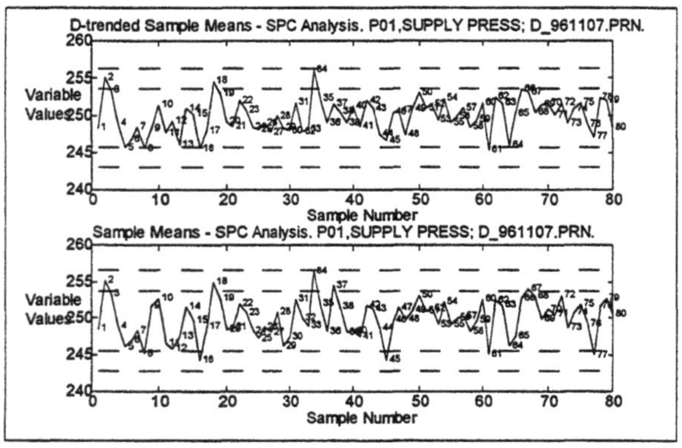

USPC is often applied inappropriately in situations where the process is not sufficiently stable, or where there is no supporting programme of variation reduction and process stabilisation. The following plot shows a typical example and the effect on the chart.

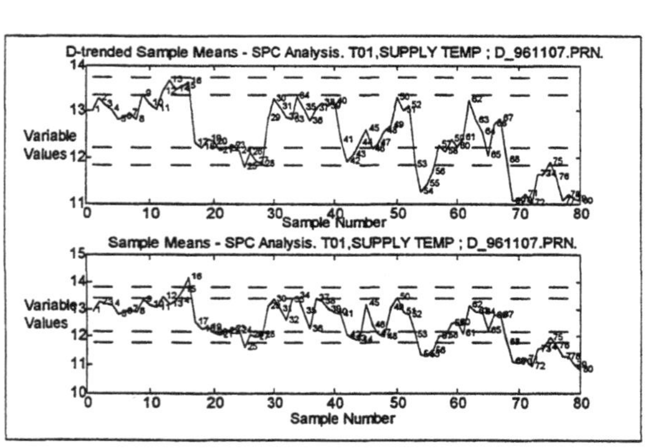

Parameter Estimation

Parameter Estimation is a useful technique in similar circumstances to USPC and can also be applied in a variety of situations where the process may not be normally distributed but follows a repeated and predictable cycle.
The next plot shows the application to an essentially cyclic process, but one in which the stability is not sufficient for successful use of the Parameter Estimation technique.

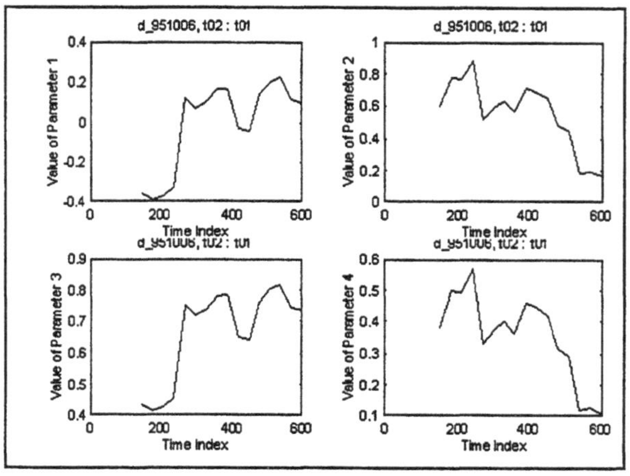

MSPC - Principal Components Analysis

The application of USPC potentially requires the simultaneous monitoring of a large number of individual plots. The application of Parameter Estimation is complicated by the large number of combinations of variables that are possible - even when a small number of process variables are under consideration. These factors can lead to the consideration of Multivariate Methods to simplify the decision making process and to ensure that important variables are not being ignored in the monitoring scheme.

For large numbers of variables, for example chemical process applications, parsimony is a prime consideration - and Multivariate Methods have significant advantages - but for a small number of variables, parsimony plays a much smaller role. The authors suggest that, regardless of the application and the number of process variables, any researchers using this group of techniques spend a large amount of time using them as investigative tools. Only when researchers are familiar with the process characteristics and those of the techniques being applied can meaningful insights be gained.

The following example shows the ability of the techniques to create apparent order from the complexity of the underlying process variable space.

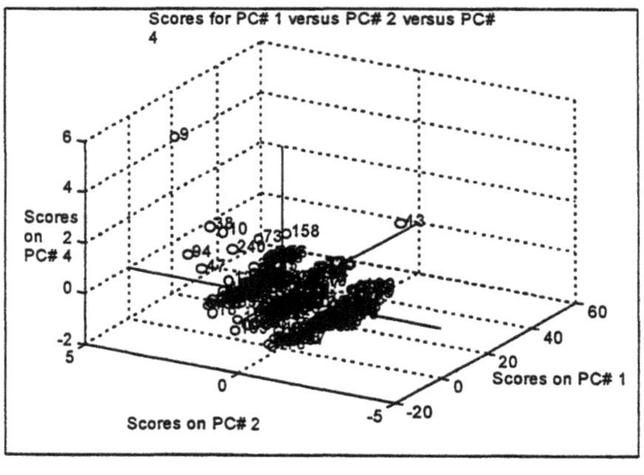

This plot will be recognisable to some researchers as indicating a possible problem with the pre-processing of the data. However, further investigation and discussion with the process engineer confirmed that PCA had been successfully used to identify three separate process operating regions from a large number of input process variables. Such regions had been suspected by the company but no quantitative evidence had been provided, until the multivariate investigations.

PCA and other multivariate techniques can be used with data that are highly correlated and can utilise the complete process variable space. They do, nevertheless, suffer from some of the drawbacks of the univariate methods. The application to non-normal processes can lead to erroneous results and conclusions and the simplification of the process investigation from possibly hundreds of variables to a relatively small number can lead to false assumptions regarding the underlying process complexity.

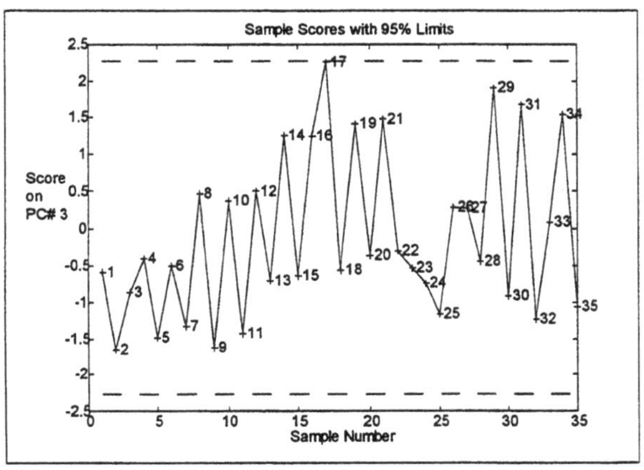

As an alternative to standard SPC methods, Exponentially Weighted Moving Averaging has been shown to be of more practical use in a relatively unstable environment such as a Building Energy Management Systems application. It is clear, however that for most effective use of any of these techniques that the process under consideration should be stabilised by removing causes of variation where possible. The output from any statistical technique will then be much more meaningful.

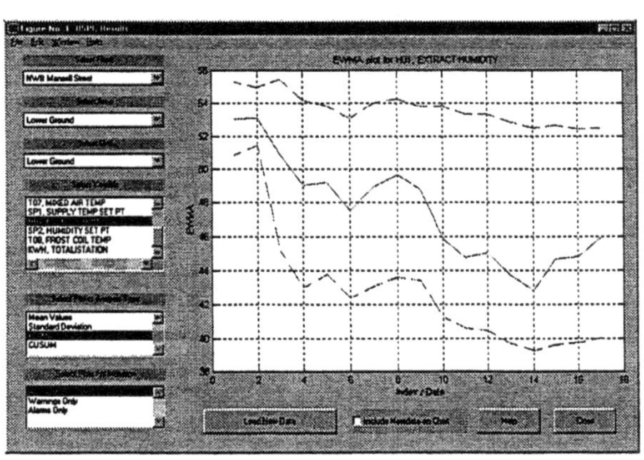

5. CONCLUSIONS

It can be seen from the above plots that although all SPC techniques can provide some insights into process operation, a degree of stability of operation is required before the techniques can be fruitfully applied.

It has been demonstrated that Univariate and Multivariate techniques can be successfully applied across a range of processes. Univariate techniques have seen most success in small-scale applications, but care must be taken not to apply the techniques in isolation. Supporting programmes and studies are necessary to obtain maximum benefit. Multivariate techniques are most successful on larger scale processes and when initially used as an investigative tool.

Experience to date has shown that knowledge of the process is essential to gain an insight into the significance of the results. This is particularly true in the case of multivariate analysis where the process parameters are not directly available.

In all cases due consideration should be given to process improvement methods - such as variation reduction techniques. There is no panacea to process monitoring and fault detection, with all techniques requiring a large amount of effort for successful application on real processes

6. ACKNOWLEDGEMENTS

This work was supported by EPSRC through the Engineering Doctorate Programme. Special thanks to National Westminster Bank Technical Services Directorate for their support throughout this programme of work.

7. REFERENCES

Bird, D. and Dale, B.G. (1994)
The Misuse and Abuse of SPC: A Case Study Examination
Int. J. of Vehicle Design, Vol. 15, Nos. ½, pp 99-107

Jiang, J. (1996)
Signal Processing Based Quantitative Schemes for Fault Detection/Diagnosis in Dynamic System
13th World Congress of IFAC International Federation of Automatic Control.
Volume N. pp139 - 144

Lau, A. and Sebele, M.G. (1993)
Statistical Quality Control procedures for the establishment and continuous validation of on-line process analyser capability
Process Control and Quality, Vol. 5, pp 87-99

White, E.M. and Schroeder, R. (1987)
A Simultaneous Control Chart
Journal of Quality Technology, Vol. 19, No. 1, pp 1-10

ISOLATION ENHANCED PRINCIPAL COMPONENT ANALYSIS*

Janos Gertler[1], Weihua Li[1], Yunbing Huang[2] and Thomas McAvoy[2]

[1]School of Information Technology, George Mason University, Fairfax, VA 22030, USA

[2]Chemical Engineering Dept, University of Maryland, College Park, MD 20742, USA

Abstract. By exposing linear dependences among the variables, Principal Component Analysis (PCA) may reduce significantly the dimensionality of plant models. While PCA is a popular tool in detecting faults in complex plants, in its original form it offers little support for fault isolation. However, by utilizing the duality between PCA and parity relations, all the powerful concepts of analytical redundancy may be transferred to PCA. Following this path, we show here how structured residuals are obtained by PCA, having the same isolation properties as analytical redundancy residuals. We demonstrate the existence conditions of such residuals and show how disturbance decoupling is implied in the method. The theoretical findings are fully supported by simulation studies performed on the Tennessee Eastman process.

Keywords: Principal component analysis; Structured residuals; Parity relations; Fault detection and isolation; Fault diagnosis.

1. INTRODUCTION

Multivariate statistical methods have recently become the tools of choice in the monitoring of complex chemical processes (MacGregor, 1989; Kresta et al, 1991; Piovoso et al, 1992a, 1992b; Wise et al, 1995). Principal Component Analysis (PCA) is used to model normal process behavior, and unusual events (faults) are then detected by referencing the observed behavior against this model. While PCA may reduce significantly the dimension of the plant model, it provides little support for fault isolation. Contribution charts (MacGregor et al, 1994) and multiblock approaches (Chen and McAvoy, 1997) have been proposed recently, but they offer only limited solutions.

As we have showed recently, a close duality exists between PCA and parity relations (Gertler and McAvoy, 1997). Parity relations belong to the analytical redundancy (AR) methods (Chow and Willsky, 1984) which rely on an explicit model of the plant. AR methods have well developed fault isolation capabilities which utilize the special (structured or directional) design of the residual set (Gertler and Singer, 1990; Gertler, 1991). By virtue of the duality, these concepts may be transferred into the PCA framework. The resulting method combines the convenience of statistical-type modeling with the powerful isolation properties of AR.

In our previuos paper, we outlined the generation of structured residuals by a set of "partial" PC models. Here we will describe a method which does this algebraically, relying on a single PC model. First we will give an introduction to parity relations. This will be followed by a brief review of the standard PCA technique, leading to the main result of the paper. We will also address the existence conditions of structured PCA residuals. It will then be shown that disturbance decoupling is included in the approach automatically, subject to the limitations well known from AR studies. Finally, we will comment on the effect and handling of control constraints.

Simulation studies, performed with the Tennessee Eastman process (Downs and Vogel, 1993), will close the paper. The results of these studies have been in complete agreement with the theory.

Note that we assume throughout this paper that (i) all the faults are related to sensors and actuators and (ii) that the system is linear. Extensions to plant faults and nonlinear systems will be treated in forthcoming papers.

2. BACKGROUND

2.1. System Description

Consider a linear static system with observed (controlled or measured) inputs $u(t)=[u_1(t) \ldots u_k(t)]'$ and observed (measured) outputs $y(t)=[y_1(t) \ldots y_m(t)]'$. These inputs and outputs are subject to actuator and sensor faults $\blacktriangle u(t)$ and $\blacktriangle y(t)$. Reserving the symbols $u(t)$ and $y(t)$ for the *observed* values of the variables, these are related to the *true* values (those acting on or arising from the plant) $u°(t)$ and $y°(t)$ as

$$u(t) = u°(t) + \blacktriangle u(t) \qquad y(t) = y°(t) + \blacktriangle y(t) \qquad (1)$$

*Research is supported by NSF under grant CTS-9714891

We assume that the variables are centered relative to their mean and that each variable and its fault are normalized with the variable's standard deviation. We make two additional assumptions here which will later be removed:

a. no disturbances act in the system;

b. no control constraints act on the variables.

Under these assumptions, the true plant variables are linked by the linear relationship

$$y^{\circ}(t) = A\, u^{\circ}(t) \qquad (2)$$

By combining u and y into a single vector

$$x(t) = [y'(t) \quad u'(t)]' \qquad (3)$$

one may write (1) and (2) as

$$x(t) = x^{\circ}(t) + \blacktriangle x(t) \qquad (4)$$

$$B\, x^{\circ}(t) = 0 \qquad where\ B = [\,I \ \ -A] \qquad (5)$$

2.2. Structured Parity Relations (Gertler&Singer, 1990)

In the analytical redundancy approach, the model matrix A is known (either from first principles or from prior identification). Parity relations rely on Eq. (5), applied to the observed variables. With (4) and (5)

$$B\, x(t) = B\, \blacktriangle x(t) = o(t) \qquad (6)$$

where $o(t)$ is the set of *primary residuals* which are computed from the observations but depend only on the faults. Further residuals are obtained by the transformation

$$r(t) = W\, o(t) \qquad (7)$$

Structured residuals. Structured residuals are so designed that each residual is sensitive to a particular subset of faults. With a suitable combination of residual structures, each fault triggers a characteristic response, also called its *fault code*. In the structure or incidence matrix, the rows are the residual structures while the columns are the fault codes. An I element means that the concerned residual does respond to the concerned fault while an O means it does not. A structure is "isolating" if each column is different and "strongly isolating" if the columns stay different from valid columns when *I*s are turned into *O*s. The following is a strongly isolating structure for four faults $\blacktriangle x_1 \ldots \blacktriangle x_4$ and four residuals $r_1 \ldots r_4$.

	$\blacktriangle x_1$	$\blacktriangle x_2$	$\blacktriangle x_3$	$\blacktriangle x_4$
r_1	*O*	*I*	*I*	*O*
r_2	*O*	*O*	*I*	*I*
r_3	*I*	*O*	*O*	*I*
r_4	*I*	*I*	*O*	*O*

When structured residuals are generated by (7) then the i-th row of W, w_i', is so chosen that the zeros assigned for the i-th row of the structure matrix be implemented. This requires that

$$w_i'\, B^i = 0 \qquad (8)$$

where B^i contains those columns of the B matrix which belong to the faults assigned for zero response in the i-th residual structure. An additional, nonzero condition is then necessary, to make the transformation defined.

Existence conditions. Clearly, if there are m primary residuals then the number of zeros in a row cannot exceed $m\text{-}1$ (more precisely, $Rank B^i \leq m\text{-}1$). Further, to assure that there are no unwanted zeros in the row structure,

$$Rank[B^i \quad b_g] = Rank B^i + 1 \qquad (9)$$

must be satisfied for all columns b_g of B outside B^i. If this latter condition is not met, the selected row-structure is not *attainable*. Finally, if any two columns of B are linearly dependent then it is not possible to construct a residual-set which distinguishes between the two faults; these faults are not *isolable*.

2.3. Standard PCA (Kresta et al, 1991)

Consider the training data-set $x(\tau)$, $\tau = 1 \ldots N$, where it is assumed that no fault is present, that is, $x(\tau) = x^{\circ}(\tau)$. Perform a singular value decomposition for

$$\Omega = [x(1), \ldots, x(N)] \qquad (10)$$

as

$$\Omega = \sum_{j=1}^{k+m} \sigma_j\, q_j\, \gamma_j' \qquad (11)$$

where q_j and γ_j are the eigenvectors of $\Omega\Omega'$ and $\Omega'\Omega$, respectively, and σ_j are the singular values (the non-negative square-roots of the eigenvalues) of $\Omega\Omega'$. In the presence of m linear relations among the variables, m of the singular values are zero, so the upper limit of the summation effectively becomes k.

It follows from (10) and (11) that any column of Ω can be expressed with the eigenvectors $q_1 \ldots q_k$ (those which belong to nonzero singular values) as

$$x(\tau) = \sum_{j=1}^{k} \sigma_j\, q_j\, \gamma_{j\tau}' = \sum_{j=1}^{k} q_j\, s_j(\tau) \qquad (12)$$

Here the eigenvectors $q_1 \ldots q_k$ are the *principal directions* and the coefficients $s_j(\tau)$ are the *scores*. The latter are the orthogonal projections of the observations on the eigenvectors which can also be computed as

$$s_j(\tau) = q_j'\, x(\tau) \qquad j = 1 \ldots k \qquad (13)$$

(12) and (13) may also be written as

$$x(\tau) = \bar{Q}\, s(\tau) \qquad s(\tau) = \bar{Q}'\, x(\tau) \qquad (14)$$

where

$$\bar{Q} = [q_1 \cdots q_k] \qquad s(\tau) = [s_1(\tau) \ldots s_k(\tau)]' \qquad (15)$$

When the above PC model is applied to an observation $x(t)$ outside the training set, this in general is not perfectly described by the expansion in the principal directions $q_1 ,,, q_k$ but an error (residual) may arise

$$x(t) - \bar{Q}\, s(t) = x(t) - \bar{Q}\,\bar{Q}'\, x(t) = \epsilon(t) \neq 0 \qquad (16)$$

3. STRUCTURED RESIDUALS BY PCA

Primary residuals. We focus our attention now on the residual $\epsilon(t)$. First we will show that

$$\epsilon(t) = Q\,Q'\,x(t) \qquad (17)$$

$$where\ Q = [q_{k+1} \cdots q_{k+m}] \qquad (18)$$

contains the eigenvectors which belong to zero singular values. To see this, consider

$$\Gamma = [\bar{Q}\quad Q] = [q_1 \cdots q_{k+m}] \qquad (19)$$

Since the eigenvectors are orthonormal, $\Gamma' \Gamma = I$, from which $\Gamma' = \Gamma^{-1}$. Therefore $\Gamma\Gamma' = \Gamma\Gamma^{-1} = I$. Thus, with (19) and (16), the result (17) follows.

We may now reinterpret the PC model as follows. The eigenvectors $q_1 \ldots q_k$ span the subspace where the training-set exists. Since the training-set is assumed to be fault-free, this is also where any future true variable-vector exists. We will refer to this as the *representation subspace.* The remaining eigenvectors $q_{k+1} \cdots q_{k+m}$ span the orthogonal complement of the representation subspace; this is where the residuals exist. This will be referred to as the *residual subspace*; it is m-dimensional. Note that any true variable is orthogonal to the residual subspace thus

$$Q'\,x^{\circ}(t) = 0 \qquad (20)$$

Notice that (17) describes the residuals in terms of the original x coordinates. It will be more advantageous to express them in terms of the $q_{k+1} \cdots q_{k+m}$ coordinates as

$$e(t) = Q'\,x(t) = Q'\,\Delta x(t) \qquad (21)$$

where we took into account (4) and (20). We will consider (21) the *primary PC residual*; it is computed from the observation $x(t)$ but depends only on the faults $\Delta x(t)$.

Relationship between Q' and B. Equation (21) shows a remarkable similarity to (6). This is not simply formal; one can show that

$$Q' = M\,B \qquad (22)$$

where M is a full-rank square matrix. To see this, observe that the m rows of both B and Q', which are $k+m$ vectors, span the orthogonal complement of the k-dimensional representation subspace. Thus (22) follows. This implies that B and Q' have the same column-rank properties. Further, since the left-hand part of B is a unit matrix, see (5), it follows that

$$M = [q_{*1} \cdots q_{*m}] \qquad (23)$$

where q_{*j} are columns of Q'.

Structured residual design. New residuals may be obtained from the primary PC residuals by the transformation

$$p(t) = V\,e(t) \qquad (24)$$

To see what the structured design involves, expand (21) as

$$e(t) = [q_{*1}\quad q_{*2} \cdots q_{*m+k}][\Delta x_1\quad \Delta x_2 \ldots \Delta x_{m+k}]' \qquad (25)$$

Thus to decouple the residual $p_i(t)$ from a fault Δx_j, the row-transformation v_i' has to be orthogonal to the column q_{*j}. That is,

$$v_i'(Q')^i = 0 \qquad (26)$$

where $(Q')^i$ contains those columns of Q' which belong to faults assigned for decoupling from the i-th residual (zero in the structure matrix). As an auxiliary condition, the length of the v_i' vector may be chosen as unity. Notice the similarity to (8).

Existence conditions follow from the column properties of the Q' matrix. For the transformation (26) to exist,

$$Rank\,(Q')^i \leq m-1 \qquad (27)$$

is required; this usually means that the number of zeroes per row must not exceed $m-1$. To avoid unintended zeroes,

$$Rank\,[(Q')^i\quad q_{*g}] = Rank\,(Q')^i + 1 \qquad (28)$$

must be satisfied for any column q_{*g} outside $(Q')^i$. Again, notice the similarity to (9). If (28) is not met, the selected row-structure is not attainable. Finally, if any two columns of Q' are linearly dependent, the respective faults are not isolable.

Design procedure. The design consists of the following steps:

1. Perform a standard PCA and obtain the Q' matrix;
2. Determine m and design a structure matrix;
3. Check Q' for rank-defects and modify the structure if necessary;
4. Implement the row-transformations by (26).

Note that the distinction between zero and nonzero singular values may not be quite clear-cut; we will comment on this in the next section.

We wish to emphasize the absolute parallelism between the PCA-based procedure and the parity relation design. In fact, one may even switch to parity relations, by computing the explicit model from the PCA model, using (22) and (23).

4. DISTURBANCE DECOUPLING

In the presence of disturbances $d(t) = [d_1(t) \ldots d_l(t)]'$, the input-output relationships (2) and (5) become

$$y^{\circ}(t) = A\,u^{\circ}(t) + F\,d(t) \qquad (29)$$

$$B\,x^{\circ}(t) = F\,d(t) \qquad (30)$$

Parity relations. In the analytical redundancy framework, F is assumed known. Decoupling the residuals from the disturbances amounts to including the latter in the incidence matrix with full zero columns. The transformation (8) is then obtained as

$$w_i' [F \quad B^i] = 0 \tag{31}$$

Clearly, the presence of F reduces the number of faults from which a residual can be decoupled by $RankF \leq l$. (31) also implies that the maximum number of disturbances from which decoupling is possible without losing fault detection is $m-1$ while that without losing isolation is $m-2$. The attainability condition (9) becomes

$$Rank[F \quad B^i \quad b_g] = Rank[F \quad B^i] + 1 \tag{32}$$

That is, the presence of the F matrix increases the possibility of non-attainability and non-isolability.

PCA. In the PCA framework, the presence of disturbances in the training data results in the reduction of the dimension of the residual space by $RankF$. Consider (30); to eliminate the disturbances apply a transformation

$$T\,B\,x^{\circ}(t) = T\,F\,d(t) = 0 \tag{33}$$

The rows of T have m elements and need to be orthogonal to $RankF$ independent vectors for $TF=0$, thus

$$RankT = m - RankF \geq m - l \tag{34}$$

which is the dimension of the residual space.

The reduced residual space can then be used to design residuals structured for the faults. (This is like first designing AR residuals decoupled only from the disturbances.) The restrictions are the same as with parity relations. Disturbances which were also present in the training data will not show up in those residuals. However, the residuals become insensitive to any fault for which (32) is violated; in general, the reduced residual space increases the possibility of non-attainability and non-isolability.

As we mentioned before, the distinction between zero and nonzero singular values may not be obvious in practice. This is because model errors (nonlinearitites) and noise also appear as disturbances and their combined number may exceed $m-1$. Therefore, the residuals need to be threshold tested. The dimension of the residual space is chosen somewhat arbitrarily, considering the trade-off between more design freedom and higher fault sensitivity.

5. CONTROL CONSTRAINTS

Recall that the equations describing plant behavior apply to the *true values* of the variables. There is another group of relationships, imposed by control actions, which act on the *observed* (measured and commanded) *values*. Such control actions include:

- feedback control keeping variables constant;
- ratio control keeping variable's linearly related.

Under feedback, it is the measured value of the controlled output which is compared to the reference, and under ratio control, a command value is linked to a measurement or another command.

The set of control constraints is described as

$$C\,x(t) = 0 \tag{35}$$

A constant variable is included as a special case since, with centering, every constant becomes zero. Eq. (35), combined with (6), leads to the residual equation

$$\begin{bmatrix} B \\ C \end{bmatrix} x(t) = \begin{bmatrix} B \\ 0 \end{bmatrix} \blacktriangle x(t) = \begin{bmatrix} o(t) \\ 0 \end{bmatrix} \tag{36}$$

Thus the constraints produce zero residuals. In AR design, where the distinction between plant and control equations is obvious, the latter are simply left out when residuals are generated.

In the PCA framework, control constraints increase the dimension of the residual subspace the same way as plant equations do. If the increased residual space is used for structured design, imperfect decoupling ensues. This is easy to see in the AR setting: assume that the control constraints were included in the design and a transformation intended to decouple from $\blacktriangle x_j$ was found by calling for

$$w_i' \begin{bmatrix} b_j \\ c_j \end{bmatrix} = 0 \tag{37}$$

Of course, the correct design condition should have been $w_i' b_j = 0$, so by applying the above w_i' to the data, decoupling would not occur. By virtue of duality, the same happens in the PCA framework, therefore the control constraints must be removed from the PC model. To achieve this, feedback loops have to be operated with (at least two) different setpoints and ratio controllers with different ratios when collecting training data.

Note that control constraints are part of a more general problem of sufficient excitation during the training of a PC model; this is the subject of ongoing research. Further issues of interest are (i) the detection of the presence of control constraints from the PC model, and (ii) the interplay between control constraints and disturbances.

6. SIMULATION RESULTS

The Tennessee Eastman Process is a simulation of a real plant that has been disguised for proprietary reasons. The process produces two products, G and H, from four reactants, A, C, D, and E. Also present are an inert, B, and a by-product, F. The process has five major units: a reactor, a product condenser, a vapour/liquid separator, a recycle compressor, and a product stripper. The gaseous reactants are fed to the reactor where they react to form liquid products. The gas phase reactions are catalyzed by a non-volatile catalyst dissolved in the liquid phase. The products leave the reactor as vapours along with unreacted feeds and the catalyst remains in the reactor. The process has 41 measurements and 12 manipulated variables. The modes of process operation are set by the G/H mass ratios. There are six modes of process operation at three different G/H mass ratios (Downs and Vogel 1993).

To simplify the demonstration, we only considered the reactor unit (Fig. 1). The reactor has 13 variables:

temperature T	feed concentrations $X_A \dots X_F$
pressure P	coolant temperature T_c
level L	coolant valve position V_c
feed flow F	

Of these, T, P, L and T_c are measured outputs, F and $X_A \dots X_F$ are measured inputs and V_c is a manipulated input. Of the 6 concentrations, at most 5 are independent.

The complete system is simulated, with 9 of the control loops operating with constant setpoints. Three variables, T, P and L, are controlled with their setpoints varied. Of these, the temperature is controlled by manipulating the coolant valve, in a cascade loop where the reference for the coolant temperature is varied inside the loop. Because of the 9 fixed setpoints, only three of the inputs to the reactor are independent, that is, the flow and one of the concentrations determine the remaining concentrations. By virtue of varying the 3 setpoints, the additional control constraints were removed. Low-level sensor and actuator noises were present throughout the simulation.

The investigations have been performed in three scenarios. In each one, the PC model is trained with a data-set of 100 samples. Then the diagnostic algorithm is designed and tested with data containing one fault at a time. In each scenario, the algorithm is designed for 6 faults, those associated with F, P, L, T, T_c and V_c.

Scenario 1. X_A is included as a measured input. For the 7 variables, PCA yields the following results:

PC #	*Eigenvalue*	*% Variance*	*% Var. total*
1	*4.09e+00*	*58.46*	*58.46*
2	*1.89e+00*	*26.95*	*85.41*
3	*1.02e+00*	*14.56*	*99.97*
4	*2.01e - 03*	*0.03*	*100.00*
5	*2.68e - 04*	*0.00*	*100.00*
6	*6.94e - 09*	*0.00*	*100.00*
7	*1.23e - 12*	*0.00*	*100.00*

The representation space has dimension 3, the residual space dimension 4. The Q' matrix is

-0.6558	*-0.3237*	*0.1479*	*-0.0507*	*-0.1539*	*0.5165*	*-0.3875*
0.5079	*-0.5492*	*-0.0126*	*0.1521*	*0.1811*	*-0.0873*	*-0.6138*
-0.2133	*-0.2937*	*-0.0066*	*-0.3192*	*-0.5405*	*-0.6864*	*-0.0546*
0.0942	*0.0831*	*0.0001*	*0.7716*	*-0.6203*	*0.0646*	*0.0027*

A possible strongly isolating structure is

	F	P	L	T	T_c	V_c
r_1	0	1	1	0	0	1
r_2	1	0	1	1	0	0
r_3	1	1	0	0	1	0
r_4	0	0	0	1	1	1

The transformation W and the fault-transfer WQ' are

-0.5625	*-0.8074*	*-0.1700*	*0.0519*
-0.4608	*0.5544*	*-0.3615*	*0.5913*
-0.0790	*-0.9685*	*0.0845*	*0.2207*
0.0106	*-0.0305*	*0.3138*	*0.9489*

0	*0.6798*	*-0.0719*	*0*	*0*	*-0.1001*	*0.7230*
0.7166	*0*	*-0.0727*	*0.6793*	*0*	*0*	*-0.1404*
-0.4373	*0.5510*	*0*	*0*	*-0.3458*	*0*	*0.6210*
0	*0*	*0*	*0.6268*	*-0.7654*	*-0.1459*	*0*

The last column of WQ', belonging to X_A, is not used. The fault responses are plotted in Fig.2. Both WQ' and the plots follow the patterns of the structure matrix.

Scenario 2. Now X_A is not taken into account as an input, thus it acts as a disturbance. The PCA results are

PC #	*Eigenvalue*	*% Variance*	*% Var. total*
1	*3.94e+00*	*65.68*	*65.68*
2	*1.04e+00*	*17.38*	*83.06*
3	*1.01e+00*	*16.91*	*99.97*
4	*1.52e - 03*	*0.30*	*100.00*
5	*2.01e - 06*	*0.00*	*100.00*
6	*1.77e - 11*	*0.00*	*100.00*

The representation space is still dimension 3 but the residual space has been reduced to dimension 3. The Q' matrix for the 6 variables is

-0.8164	*0.0297*	*0.1321*	*-0.1099*	*-0.2066*	*0.5103*
0.2551	*0.2477*	*0.0067*	*0.3675*	*0.5221*	*0.6826*
-0.0838	*-0.0687*	*0.0002*	*-0.7552*	*0.6458*	*-0.0312*

Now the submatrix formed of the first 3 columns has a rank-defect. This limits the possible structures but it is still possible to design a strongly isolating structure, as

	F	P	L	T	T_c	V_c
r_1	0	1	1	0	1	1
r_2	0	1	1	1	0	1
r_3	1	0	1	0	1	1
r_4	1	0	1	1	1	0
r_5	1	1	0	1	0	1
r_6	1	1	0	1	1	0

The fault-transfer matrix and the response plots, not shown in this paper, again follow the structure.

Scenario 3. Now an additional disturbance is applied upstream to the feed. The PCA result is

PC #	*Eigenvalue*	*% Variance*	*% Var. total*
1	*3.31e+00*	*55.13*	*55.13*
2	*1.08e+00*	*18.07*	*73.21*
3	*1.00e+00*	*16.67*	*89.88*
4	*6.07e - 01*	*10.12*	*100.00*
5	*1.29e - 04*	*0.00*	*100.00*
6	*1.43e - 07*	*0.00*	*100.00*

The residual space is further reduced to dimension 2. The Q' matrix is

0.0512	*-0.3275*	*-0.0657*	*-0.4696*	*-0.2051*	*-0.7894*
-0.0008	*0.0048*	*0.0010*	*-0.6183*	*0.7682*	*0.1660*

The first 3 columns are now pairwise linearly dependent, thus the first three faults are not isolable. A possible structure under this constraint is

	F	P	L	T	T_c	V_c
r_1	0	0	0	1	1	1
r_2	1	1	1	0	1	1
r_3	1	1	1	1	0	1
r_4	1	1	1	1	1	0

In summary, the simulation results have been found to be in complete agreement with the theory.

REFERENCES

Chen, G. and McAvoy, T. (1997). Multi-block predictive monitoring of continuous processes. *1997 IFAC ADCHEM Symposium* (Banff, Canada), pp. 73-78.

Chow, E.Y. and Willsky, A.S. (1984). Analytical redundancy and the design of robust failure detection systems. *IEEE Trans. on Automatic Control, 29*, pp. 603-614.

Downs, J.J. and Vogel, E.F. (1993). A plant-wide industrial process control problem. *Computers Chem. Eng., 17*, pp. 245-255.

Gertler, J. and Singer, D. (1990). A new structural framework for parity equation based fault detection and isolation. *Automatica, 26*, pp. 381-388.

Gertler, J. (1991). Analytical redundancy methods in FDI - Survey and synthesis. *IFAC SAFEPROCESS Symposium*, (Baden Baden, Germany), Vol. 1, pp. 9-21.

Gertler, J. and McAvoy, T.J. (1997). Principal component analysis and parity relations - a strong duality. *Preprints of IFAC SAFEPROCESS Symposium* (Hull, UK), *Vol. 2*, pp. 837-842.

Kresta, J.V., MacGregor, J.F. and Marlin, T.E. (1991). Multivariate statistical monitoring of processes. *Can. J. Chem. Eng., 69*, pp. 35-47.

MacGregor, J. (1989). Multivariate statistical methods for monitoring large datasets from chemical processes. Annual AIChE meeting, San Francisco, CA.

MacGregor, J., Jaeckle, C., Kiparissides, C. and Koutodi, M. (1994). Process monitoring and diagnosis by multiblock methods. *AIChE Journal, 40*, pp. 826-828.

Piovoso, M.J., Kosanovich, K.A. and Pearson, P.K. (1992a). Monitoring process performance in real time. *American Control Conf.* (Chicago, IL), pp. 2359-2363.

Piovoso, M.J., Kosanovich, K.A. and Yuk, J.P. (1992b). Process data chemometrica. *IEEE Tr. Instrumentation and Measurement, 41*, pp. 262-268.

Wise, B.M., Gallagher, N.B. and MacGregor, J.F. (1995). The process chemometrics approach to process monitoring and fault detection. *Proc. IFAC Workshop on On-line Fault Detection and Supervision in the Chemical Process Industries* (Newcastle, UK), pp. 1-20.

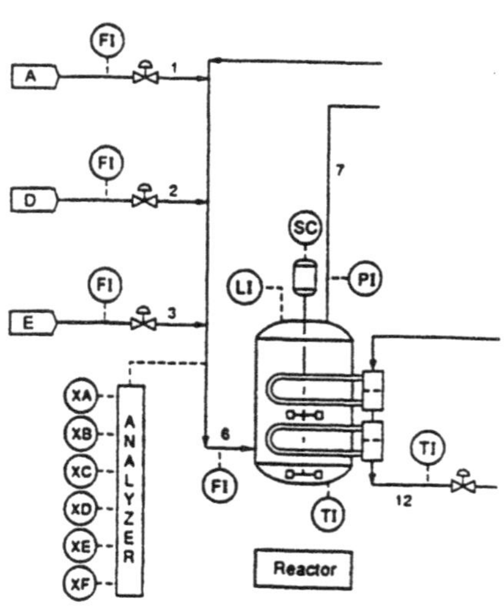

Figure 1.

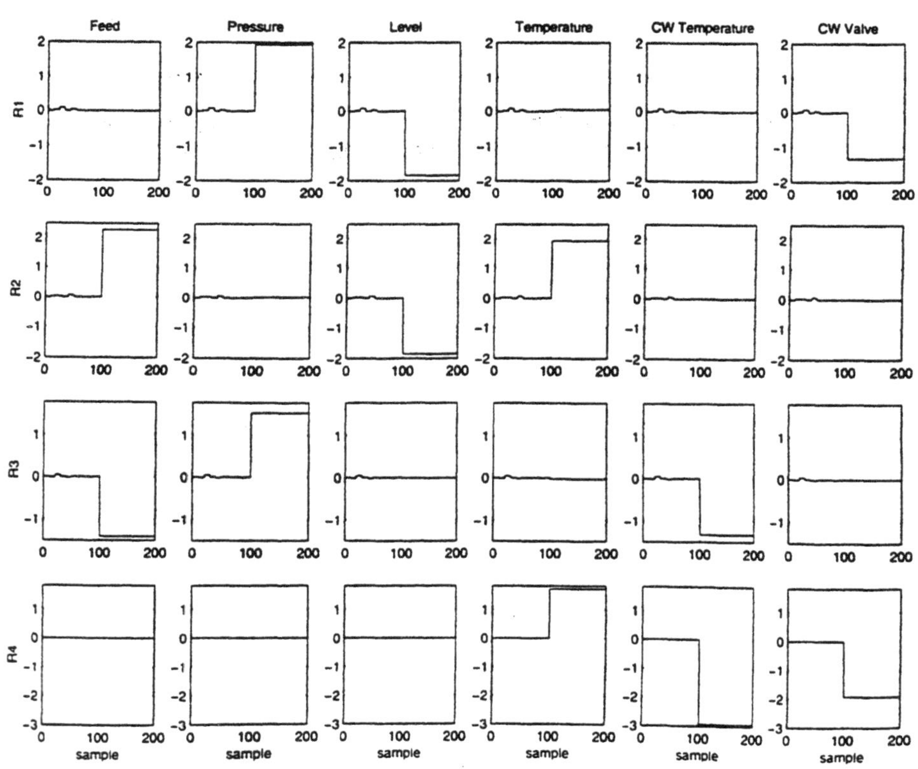

Figure 2.

REAL-TIME MONITORING FOR A PROCESS WITH MULTIPLE OPERATION MODES

Dae-Hee Hwang, Tae-Jin Ahn*, Tae-Hwa Choi, Chonghun Han**

Automation Research Center and Dept. of Chemical Eng., POSTECH
**Technology & Engineering Dept., POSCO*
Technical Research Laboratories, Iron & Steelmaking Research Team, POSCO
E-mail: chan@postech.ac.kr, Fax: (82)-562-276-2699, Tel: (82)-562-279-2279.

Abstract: Real-time monitoring of a chemical process with multiple operation modes is a challenging problem. The frequent changes of operation modes require frequent updates of monitoring models, which lead to the frequent pauses of the real-time monitoring activities. This paper proposes a monitoring methodology for a process with multiple operation modes based on hierarchical clustering and a super PCA model. The case studies have shown that the super PCA model has shown better performances than a single PCA model for all operation modes and local PCA models developed for each operation mode. *Copyright © 1998 IFAC*

Keywords: Monitoring, multilevel system, classification, multivariate quality control, fault detection.

1. INTRODUCTION

Due to increasing interests in safety and consistent product quality, real-time monitoring of the process performance has become a key issue for both productivity and safety enhancement. Multivariate statistical monitoring methods have been successful, replacing the traditional univariate chartings such as shewart chart so that one can overcome difficulties such as complexities of many measurements and strong correlations among them. Those methods assume that process variables are kept at a steady state mode (Jolliffe, 1986; Wold, *et al*., 1987).

However, such assumptions are not valid for the nostationary time varying processes or a continuous process with multiple operation modes. Multiway PCA (Nomikos and MacGregor, 1996) has been developed for the monitoring of batch processes based on the observation that all the state variables will follow statistically the same trajectories for good batches. Gallagher, *et al.* (1997) has introduced an alternative using EWMA and EWMC and Ku, *et al.* (1996) dynamic PCA for nonstationary processes.

In case of continuous processes that frequently go through mode changes due to the various changes in product specifications, feed flowrate, compositions and set points, the covariance structure changes do not allow the multivariate statistical monitoring model to be used at the monitoring at the different mode. Feasible solutions of this problem are: a) to develop a set of local monitoring models so that each of them is valid for the corresponding operation mode, b) to adaptively update the model to reflect the mode changes, and c) to develop a single monitoring model for all the different operation modes. A single monitoring model becomes too conservative and as a result leads to the lower monitoring resolution and erroneous monitoring. A local monitoring model offers the higher monitoring resolution. However, the local model may miss a portion of true covariance structure during the model development due to the incomplete operation data that are not enough to represent the process characteristics.

This paper proposes a monitoring methodology for the process that goes through several operation modes in a sequential manner. The methodology has been motivated from the observations that a) the different operation modes do not introduce a significant amount of nonlinearities in the process behavior, b) the process covariance structures for different operation modes share common characteristics that determine the process behavior. To identify the common characteristics and their minor variations from the operation data measured at different operation modes, a hierarchical clustering has been introduced as will be presented in section 2.2. After pruning and merge of clusters in the hierarchical clustering tree, the super PCA model is developed from the resulting clusters. To remove the effect of mean shift, clustering and translation are employed. As the super PCA model captures the covariance structure that is statistically average for all the clusters, it offers higher resolution and more robustness for monitoring and high adaptibility to easily accommodate new operation modes. Fig. 1 shows the overview of the proposed methodology.

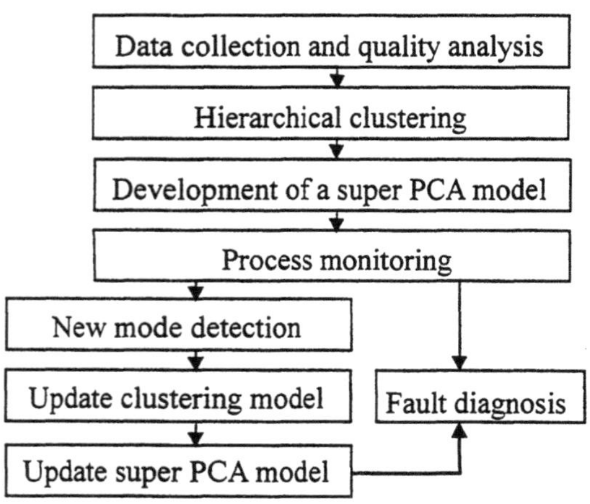

Fig. 1. Overview of the proposed methodology.

2. REAL-TIME MONITORING METHODOLOGY FOR A PROCESS WITH MULTIPLE OPERATION MODES

2.1 Data collection and analysis

The good data collection and analysis of its quality are essential to construct a good monitoring model. Variables relevant to represent the process behavior should be identified and subsequently measured to make a data matrix. When the important variables cannot be directly measured either the instrumentation or the inferential model to measure the variable is necessary. (Crowe, *et al.*, 1983). During the data processing (Söderström and Stoica, 1989; Ljung, 1987), the abnormal operation data have to be removed according to the selection algorithm (Hwang, *et al.*, 1997). The qualities of selected normal operation data such as the signal to noise ratio have to be analyzed using exploratory analysis (Santen, *et al.*, 1997).

2.2 Hierarchical clustering

Clustering is a procedure to identify the observations that belong to each operation mode. If the distribution of sample observations in n dimension is simple with a multivariate normality assumption for each cluster, nonhierarchical clustering methods such as K means and KNN clustering method can show a good clustering performance (Jonson and Wichern, 1992). However, when the observations are significantly distorted in n dimensional space with a large number of variables, the efficiency and stability of nonhierarchical clustering methods are degraded. To resolve these difficulties, a new variant of hierarchical clustering method is proposed as follows.

i) Make a data matrix that includes observation data for all the operation modes.

ii) Apply PCA to the data matrix and determine the number of clusters in the next lower level at the score plot composed of the first and the second loading vector.

iii) Apply a self-organizing map (SOM, Demuth and Beale, 1994) to the total cluster using the number of clusters identified at step ii) as an initial value and generate the smaller clusters.

iv) Calculate the prior probabilities for the generated clusters and prune out the cluster with a low value, according to equation (1).

$$p(w_i) << p(w_j) \text{ for } j=1,\ldots,n \text{ and } i \neq j \quad (1)$$

where $p(w_i)$ is a prior probability for the ith cluster.

v) Calculate the classification errors for the clusters, according to equation (2) and (3). If the error is over the threshold value of equation (4), merge the clusters into a single cluster and stop the hierarchical clustering for the corresponding branch at the current level.

$$r_{ij} = (\boldsymbol{m}_i - \boldsymbol{m}_j)^T \boldsymbol{S}^{-1}(\boldsymbol{m}_i - \boldsymbol{m}_j) \quad (2)$$

$$p(e_{ij}) = \int_{\frac{1}{2}\sqrt{r_{ij}}}^{\infty} \frac{1}{\sqrt{2\pi}} \exp(-y^2/2)dy \quad (3)$$

$$p(e_{ij}) \geq 0.1 \quad or \quad r_{ij} \leq 5 \quad (4)$$

where $p(e_{ij})$ is a probability to misclassify a w_i observation as w_j. The $\boldsymbol{m}_i$, $\boldsymbol{S}$, and r_{ij} are a sample mean of the ith cluster, a spooled sample covariance matrix and a mahalanobis distance between the ith and jth cluster, respectively.

vi) Apply the steps in a recursive manner to the clusters that have not been pruned until there are no more clusters left that are not pruned or merged.

vii) After obtaining the terminal nodes for all the branches, calculate the classification errors among the terminal nodes and make a classification error matrix.

viii) Considering the relationship between the

terminal nodes from the classification error matrix, merge the nodes whose error value is over the threshold value.

ix) Prepare the input and the target matrix to construct a back propagation neural network as a clustering model. To make the input matrix, the observation data that belong to the pruned clusters should be removed from the original data matrix. In case of the target matrix, the terminal nodes that have remained as clusters not pruned and merged are used.

x) Train a back propagation neural network as a clustering model.

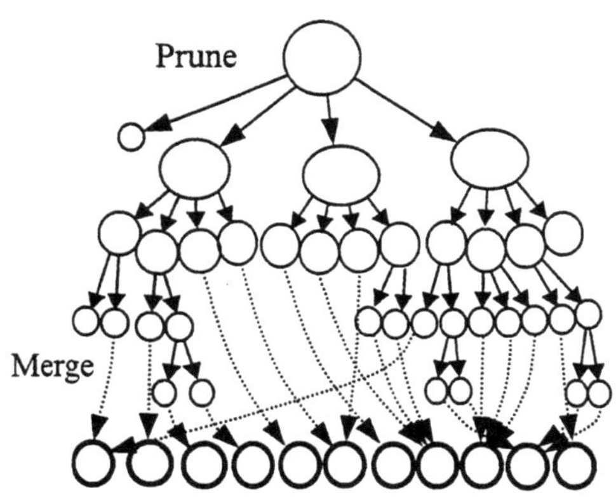

Fig. 2. Hierarchical clustering.

2.3 Super PCA modeling

A local PCA model for each clustered mode can be developed, with steady state assumption for each cluster. With new observation, we can classify it into one of clusters using a clustering model and can monitor it with its corresponding PCA model. However, as the number of clusters increase, the information stored in the database will be incredibly increased, especially for the complex systems with a large of variables. Also, local models may arise the possibility of type II error when the observations in each mode have missed some portion of true covariance structure. For example, they may have not been collected over all kind of controllable disturbances. Thus, another approach has been introduced to be more applicable for practical processes that would not stay long enough to develop a complete covariance structure at each operation mode.

The super PCA modeling has been designed to construct a more generalized PCA model, that is, more robust to type II errors than each local PCA model, due to the missed portion of a true covariance structure at each mode. This approach has a limitation that it will show better performance for the processes with similar covariance structures for the operation modes. That is, it might be not applicable for the processes operating in a wide range with highly nonlinear characteristics, so that the covariance structures are significantly different. However, in general, most of processes should be operating at the base operating conditions within a small range, due to a limitation of process capacity. Accordingly, this approach can be applicable to many practical processes, as shown in the case studies.

Super PCA modeling procedure is as follows.

i) Develop a data matrix with observation data that belong to one operation mode and consequently belong to one cluster.

ii) Standardize the cluster by centering and scaling.

iii) Translate each cluster to the origin of the entire data space by processing its standardized data matrix (Z_i) as shown in equation (5).

iv) Repeat this process i), ii), iii) for all operation modes.

v) Pile up all translated clusters to make a single super data matrix (Z_{sup}) as shown in equation (6). Piled clusters look like stacked hyper-ellipsoids for all clusters with a center in the origin.

vi) Apply SVD to the super data matrix to find PC loadings and normalized scores as shown in equation (7).

Since the super data matrix has a zero mean and a unit variance, its centering and scaling for PCA modeling are not necessary. As covariance structures are assumed similar and each hyper-ellipsoid has commonly a similar size, PC loadings (eigenvectors) for the super data matrix will be averages of all those matrices stacked in the origin, as shown in equation (9). Linear transformation in equation (5) does not change the correlation structure because it is just translation and does not introduce any topological changes.

$$Z_i = (X_i - 1m_i^T)D_i^{-1/2} \quad (5)$$

$$Z_{sup} = \left[Z_1^T Z_2^T \ldots Z_n^T\right]^T \quad (6)$$

where X_i and D_i are the ith data matrix and a standard deviation matrix. Z_i and Z_{sup} are the ith translated and standardized matrix and a super data matrix.

$$Z_{sup}^T Z_{sup} = \left[Z_1^T Z_1 + Z_2^T Z_2 + \ldots + Z_n^T Z_n\right] = V_a \Lambda_a V_a^T \quad (7)$$

$$R_1 = V_{1_a} \Lambda_{1_a} V_{1_a}^T, R_2 = V_{2_a} \Lambda_{2_a} V_{2_a}^T, \ldots, R_n = V_{n_a} \Lambda_{n_a} V_{n_a}^T \quad (8)$$

where $R_i = Z_i^T Z_i$ is a correlation matrix for the ith cluster. V_{i_a} and Λ_{i_a} is The first a eigenvectors and square rooted eigenvalues matrix for the ith cluster.

$$R_{sup} = \left[R_1 + R_2 + \ldots + R_n\right]$$
$$= \sum_{k=1}^{n} V_{k_a} \Lambda_{k_a} V_{k_a}^T \approx V_a \left(n^2 \Lambda_a\right) V_a^T$$

$$if A_{i_a} \approx A_{j_a},\ i,j=1,2,...,\quad V_a = \frac{1}{n}\sum_{k=1}^{n} V_{k_i} \quad (9)$$

2.4 Process monitoring

For process monitoring with a new operation data using a super PCA model, the procedure goes through the same process with observations of the original data matrix. The new observation will be classified into one of operation modes by the clustering model and standardized by the mean and standard deviation vector associated with its assigned cluster. Then, it will be translated to the origin as shown in equation (10).

$$z_{new} = (x_{new} - 1m_i^T)D_i^{-1/2} \quad (10)$$

The same monitoring procedure as the traditional PCA can proceed. T square and SPE values are calculated by projecting it into the PC loadings of the super data matrix and obtaining z_{new} as shown in equation (11) and (12).

$$diag(TT^T) = diag\left[(z_{new}V_a)(z_{new}V_a)^T\right] \quad (11)$$

$$SPE = \|z_{new} - z_{new}\|_2^2 \quad (12)$$

Due to the translation, this monitoring scheme will put all the monitoring responsibilities on the correlation structure alone and remove information related to shifts in the operation mode. As operation mode changes, a new mode that has never experienced can take place. In such case, the clustering model will extrapolate about a cluster, which the new observation belongs to, with old clustering information and assign it to the nearest cluster to its true cluster. As a result of translating the cluster by the amount of the mean of the nearest cluster, it is moved far away from the origin due to a wrong mean. This results in the increase of the SPE value and eventually going out of the SPE control limit. As the process data are measured and monitored continuously, the corresponding SPE values will stay out of the limit. This phenomenon indicates that a new operation mode has taken place. After determining whether the new mode has been stabilized around new steady state, the super PCA model has to be updated to incorporate new operation mode.

2.5 Model update

After operation data for the new mode have been collected for the period long enough to explain the correlation structure, the clustering model is updated by adding another terminal node to the hierarchical clustering and retraining the BP network after adding an additional output node. Target matrix will also be generated and BP will be trained using old weights as the initial values to increase the training speed. The same procedure will be repeated, whenever a new operation mode takes place.

3. CASE STUDY 1: A DISTILATION COLUMN

Methanol-water binary distillation column has been designed and simulated in HISYS to verify the methodology. Feed flow rate, reflux flowrate, and steam flowrate in the reboiler are manipulated to maintain the top and bottom product at the product specifications. The feed temperature is controlled and twenty-three variables are measured. Operation data associated with six operation modes including the base operating condition (Table 1) have been generated.

Table 1 Experimental design for six operation modes (all compositions are based on methanol)

Mode	Product spec. top	Product spec. bottom	Feed composition	Disturbance feed(%)
1	0.9	0.1	0.4	3%
2	0.94	0.06	0.4	2%
3	0.97	0.03	0.4	1%
4	0.9	0.1	0.5	3%
5	0.94	0.06	0.5	2%
6	0.97	0.03	0.5	1%

As there are no causes to contaminate the date, data quality analysis was not necessary. Due to the simple distribution structure in the data space, the proposed hierarchical clustering has been terminated in the first level, correctly resulting in six terminal modes. The super PCA modeling has been applied for five modes (mode 2-6) and the first mode left for mode change. To have the missed covariance portion, the data in mode 6 have been experimented only with 1% feed disturbance and without any random disturbances in reboiler and condenser heat duties. To check monitoring capability, the several test sets have been designed: the load disturbance in mode 4, the reflux flow sensor failure in mode 5, and one associated with the missed correlation structure in mode 6 by giving 3% feed disturbance with the disturbances in reboiler and condenser heat duty which have been never experienced in mode 6 during the experiment, but should be determined as a normal observation because it is a controllable disturbance. With the feed load disturbance, the monitoring performance of the super PCA model was compared with traditional PCA model and the local PCA model for mode 4 in Fig. 3 (calculated by Simca-P 3.01).

Fig. 3 shows that monitoring based on a traditional PCA model did not detect a load disturbance, since its control envelops cover all data distribution for five modes, even including the empty area where the fault happened. A small difference in the capability to detect the load disturbance between a super PCA model and a local PCA model indicates that both of them are comparable in the fault detection. Rather, SPE control plot of the super PCA model tells us more things about the characteristics of the fault than local PCA, considering the magnitude of abnormality, detection time and the score patterns in T square charts. For the sensor failure, the comparison was not shown, but there was no difference between them.

Fig. 4 shows that the super PCA model was sensitive to the sensor failure and gives better monitoring performance than the local PCA model, when the fault happened in the lost portion of covariance structure for mode 6. Fig. 5 shows that the super PCA model detects the mode change and the model update was verified by testing the same test set before and after model update.

4. CASE STUDY 2: A BLAST FURNACE

A *Blast Furnace* (BF) is a large reactor producing pig iron. Iron ore and cokes are charged down from the reactor top as source and fuel and Pulverized Coals (PC), secondary fuel, and hot blast are charged up from the waist level near the bottom along with oxygen. Many reactions take place in complicated ways, simultaneously with phase transitions, heat, and mass transfer inside the BF. Thus, the process has highly nonlinear characteristics and strong correlations among the variables (Omori, 1987).

The last six-month operation data from the fourth blast furnace of POSCO, an iron making company located at pohang in Korea, have been collected. The measured variables are composed of 329 temperatures, 31 pressures, and 4 gas concentrations at the top. For the selection of normal operation data, the proposed selection algorithm has been applied and data quality (exploratory data) analysis has been performed. Signal to noise ratio of most variables is higher than 10, except for pressure variables (about 3). It allows us to do a useful analysis to be possible.

Hierarchical clustering has been explored to find an insight on the data structure in 377 dimension. With the knowledge of terminal nodes in hierarchical clustering, a clustering model has been developed using BP with 377 input nodes, 120 nodes in the first hidden layer, 80 nodes in the second hidden layer and 11 output nodes. The normalized sum of squared error in the training the network was 0.0144. Fig. 2 shows the result.

For eleven clusters, standardization and translation have been done to file up them for a super data matrix. SVD of super data matrix has been performed to find the generalized PC loadings. Twenty-one PCs have been produced and the portion of total sample variance explained by them was 89%. Test sets have been designed for a sensor failure by an improper calibration in the second cluster and for a mode change. Fig. 6 shows that a super PCA model is sensitive to the fault and detects the mode change and monitoring results after model update.

5. CONCLUSIONS

The success of real-time monitoring based on multivariate statistical models depends on the following two factors: a) availability of long term operation data with a good quality that can represent the full behavior of the process, and b) how the monitoring model represent the collected data. For the process with multiple operation modes, traditional PCA models are limited for monitoring of the wide operation ranges due to the steady state assumption for process data. Without considering the changes of the covariance structure among close operation modes, the resulting monitoring model based on the PCA will lead to lower monitoring resolution. The local models developed separately for each mode may show good monitoring resolution. However, the local models may lead to unnecessarily too sensitive monitoring due to the limited amount of process data., resulting in the lower robustness. This paper has proposed a monitoring methodology based on the hierarchical clustering and the super PCA mode. The hierarchical clustering allows us to identify the common characteristics of process behavior from the hierarchical relations among clusters. From the clusters, the super PCA model is developed to capture the average covariance structure for all the different modes. The resulting super PCA model has shown much better performance in terms of resolution and robustness than a single PCA model developed for all the modes and a set of local models. The methodology offers high adaptability to easily accommodate new operation mode. However, it must be noted that the proposed methodology is valid for the modes that do not show strong nonlinearities and are not far away from other modes. To extend the current methodology to the monitoring model development of the highly nonlinear process with multiple operation modes, we think one promising idea is to develop a hierarchical tree for monitoring whose node is a super PCA model proposed in this paper.

ACKNOWLEDGMENTS

The authors deeply appreciate the financial support of the Korea Science and Engineering Foundation through the Automation Research Center at POSTECH.

REFERENCES

Söderström, T. and P. Stoica (1988). *System Identification*, Chap. 5, 12. Prentice Hall, U.K.

Ljung, L. (1987) *System Identification theory for the users*, Chap. 14. Prentice Hall, New Jersey.

Johnson, R.A. and D.W. Wichern (1992). *Applied Multivariate Statistical Analysis*, chapter 11 and 12. Prentice hall, New Jersey.

Jolliffe, I.T. (1986). *Principal Component Analysis*. Springer-Verlag, New York.

Nomikos, P. and J.F. MacGregor (1994). Monitoring batch processes using multiway principal component analysis. *AIChE J*, **40**, 1361-1375.

Wold, S., K. Esbensen and P. Geladi (1987). Principal component analysis. *Chemo. and Intell. Lab. Sys.*, **2**, 37-52.

Gallagher, B.N., B.M. Wise, S.W. Butler, S.W. White and G.G. Barna (1997). Development and benchmarking of multivariate statistical process control tools for a semiconductor etch process: improving robustness through model update. Proc. IFAC workshop on Adchem '97, Banff, Canada, 78-83.

Santen, A., G.L.M. Koot and L.C. Zullo (1997). Statistical data analysis of a chemical plant. *Com. Chem. Eng. Suppl.*, **21**, s1123-1128.

Omori, Y. (1987). *Blast Furnace Phenomena and Modeling*, Chap. 1-3. Elsevier Applied Science, London and New York.

Crowe, C.M., Y.A.G. Campos and A. Hrymak (1983). Reconciliation of process flow rates by matrix projection. *AIChE J.*, **29**, 881-888.

Demuth, H. and M. Beale (1994). *Neural Network Toolbox; User's Guide for use of MATLAB*, Chap. 8. Math Works Inc., U.S.

Hwang, D.H., T.J. Ahn and C. Han (1998). Process operation improvement based on statistical data analysis. *AIChE J.* (to be submitted).

Ku, W., R.H. Storer and C. Georgakis (1995). Disturbance detection and isolation by dynamic principal component analysis. *Chemo. and Intell. Lab. Sys.*, **30**, 179.

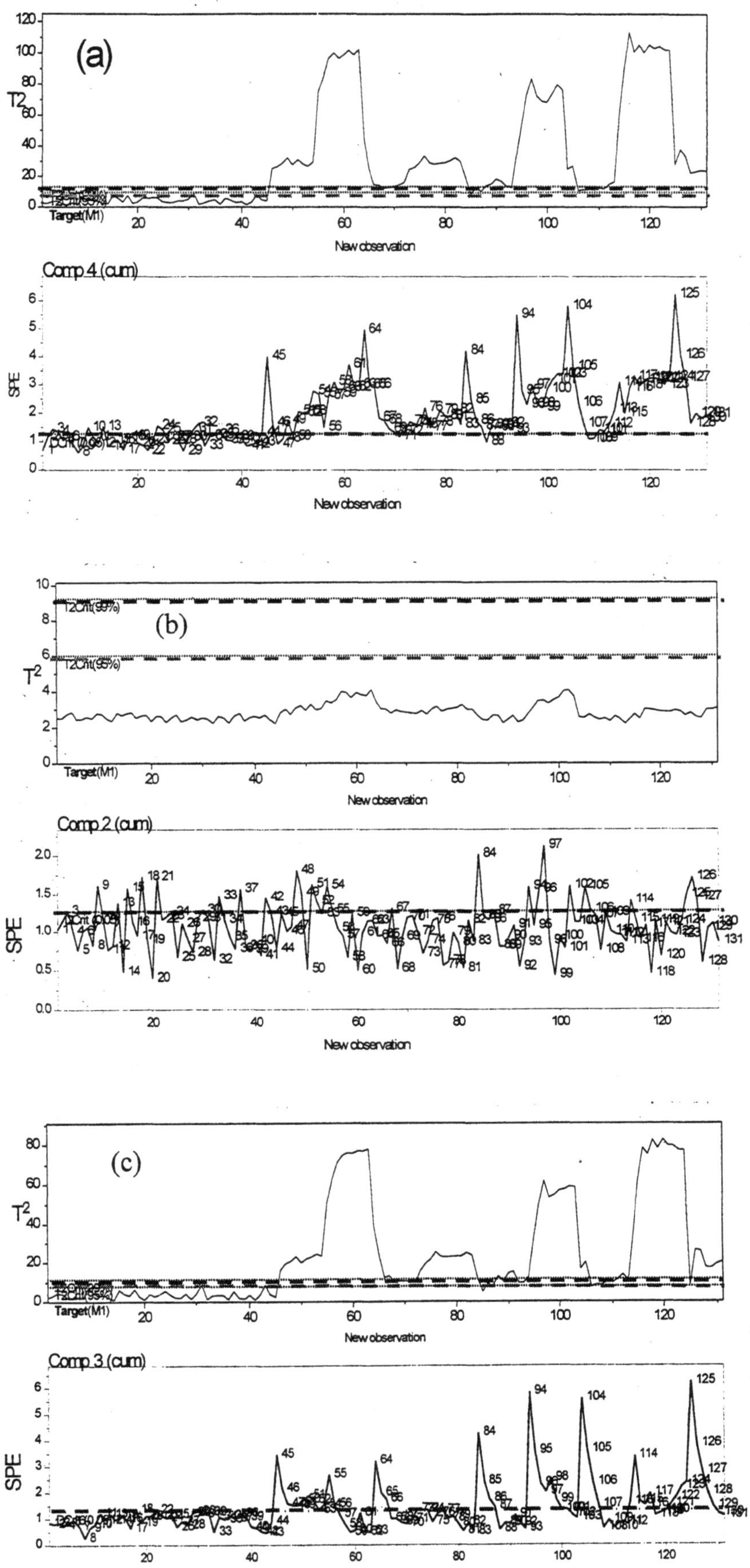

Fig. 3. The comparison of monitoring performance of super PCA model (a) with traditional PCA model (b) and a local model (c) about a feed load disturbance.

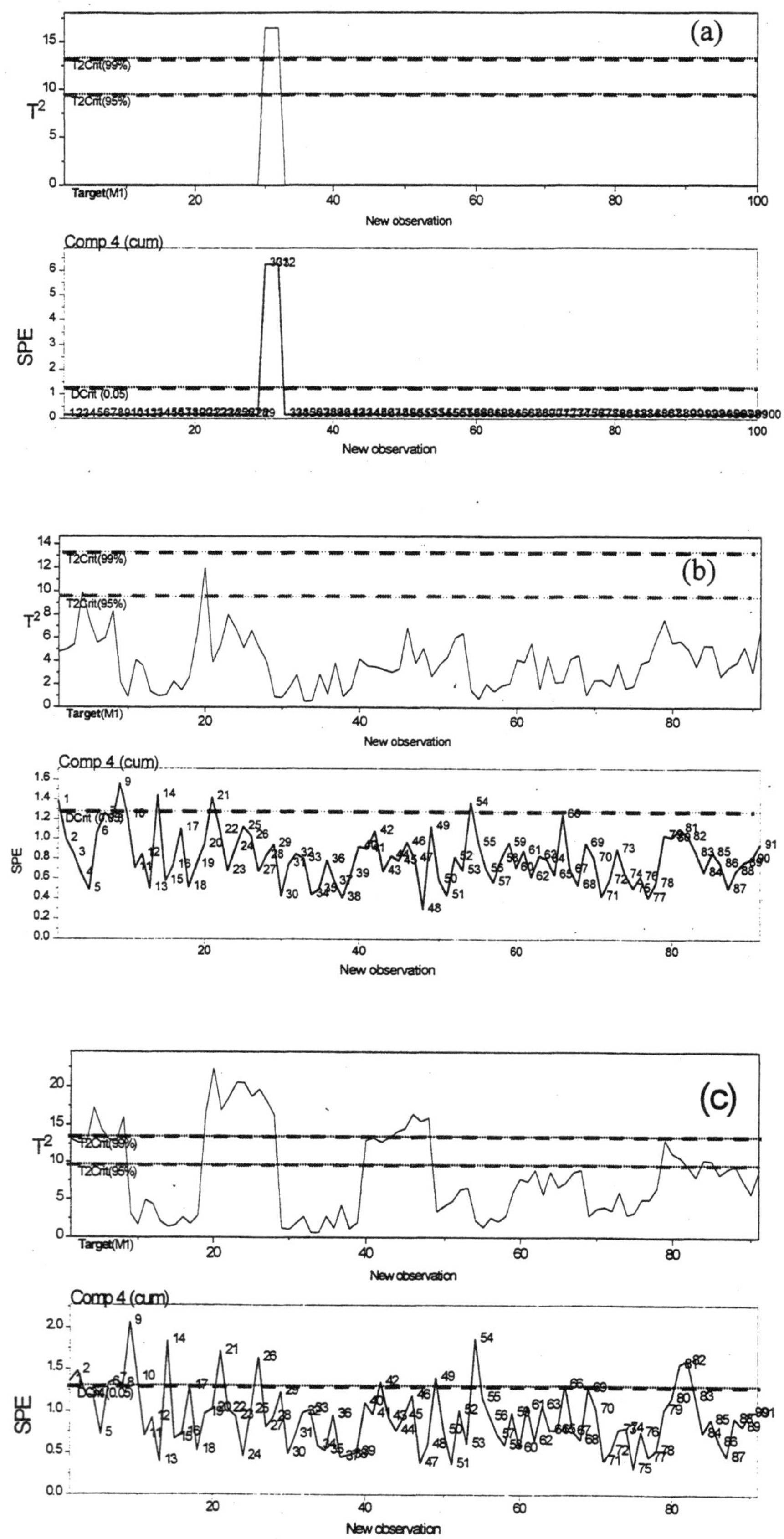

Fig. 4. Detection of a sensor failure (a). Determination of the uncollected controllable disturbance of the super PCA model in mode 6 as normal (b) and as abnormal in local PCA model (c).

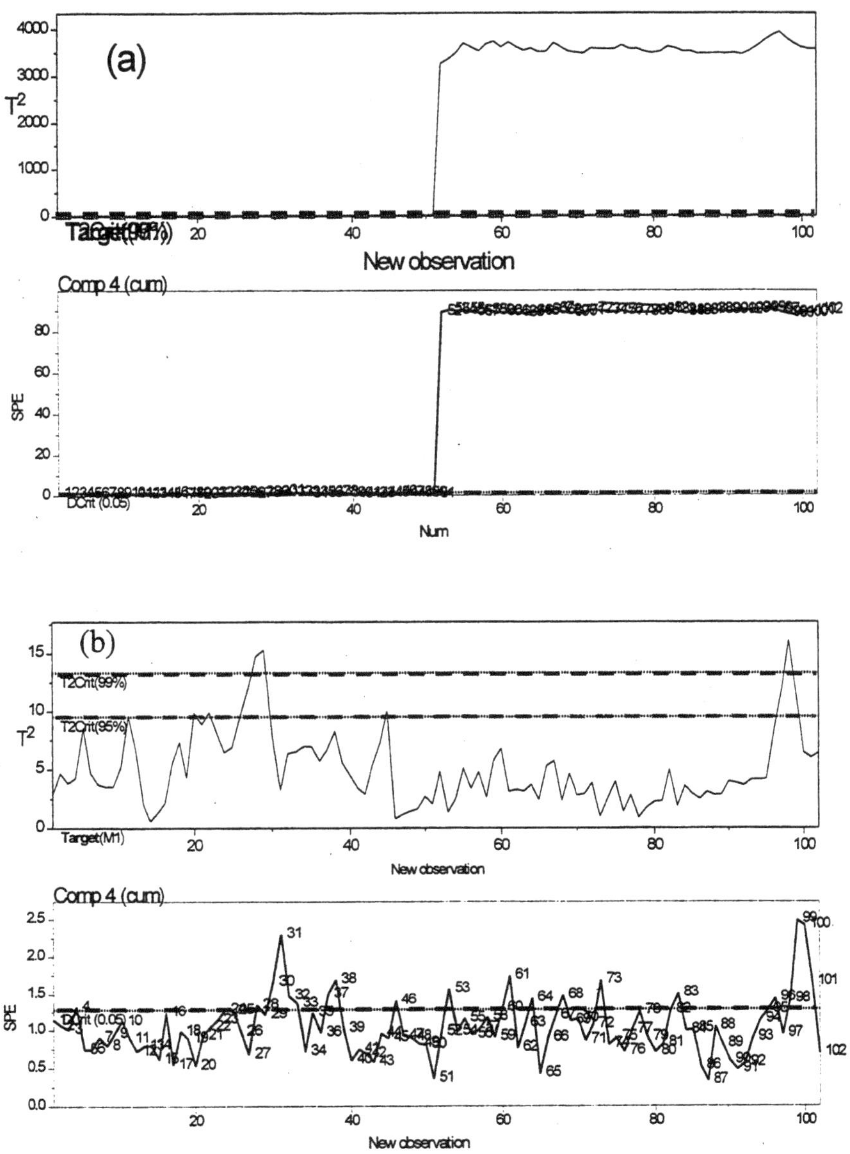

Fig. 5. Detection of a mode change (a) and monitoring results with the same test set after model update (b).

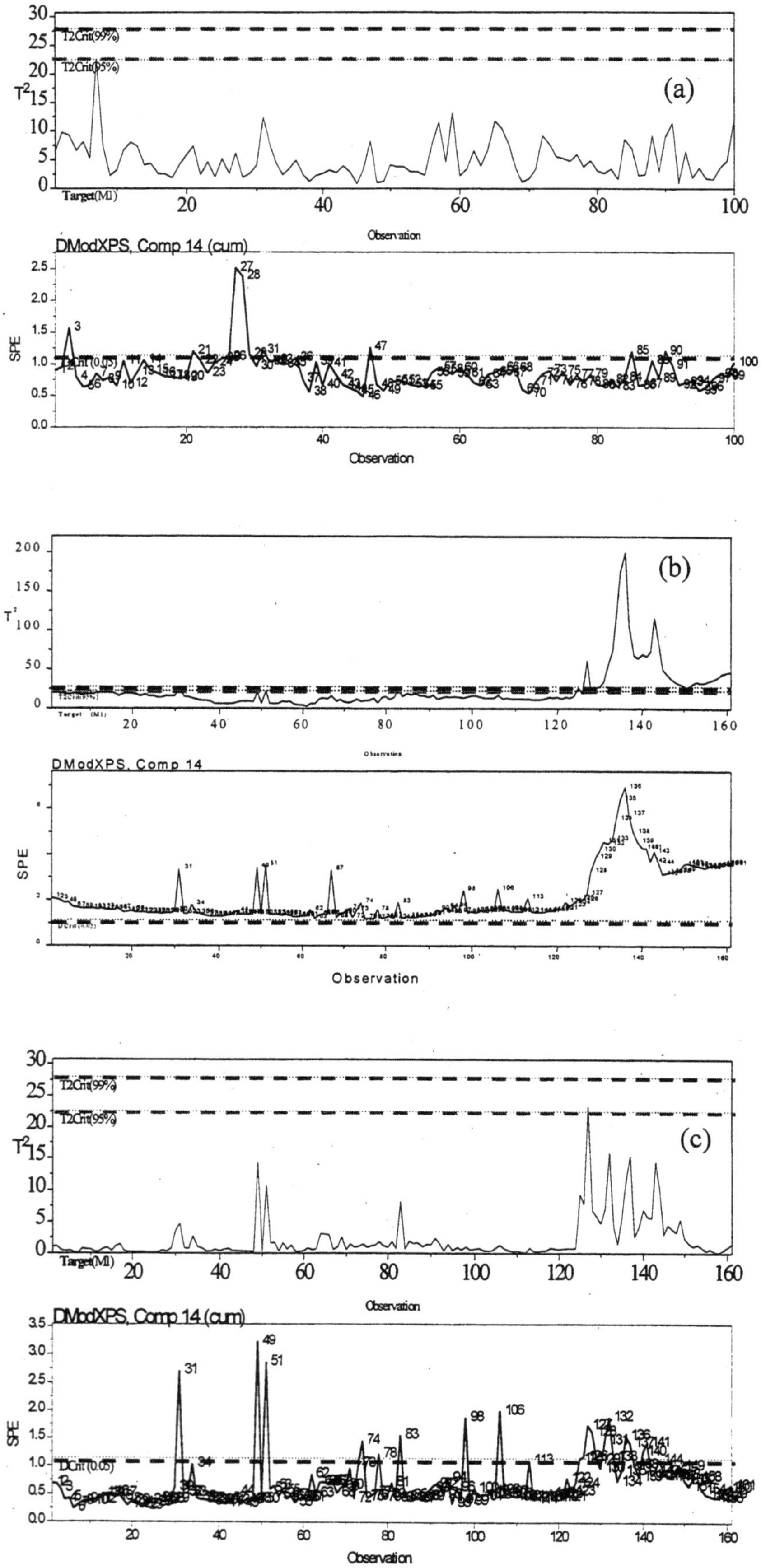

Fig. 6. Control charts for the sensor failure (a) and the mode change (b) and monitoring results after model update (c).

An Industrial Application of Principal Component Test to Fault Detection and Identification

Hongwei Tong[1] and David Bluck

Simulation Sciences Inc.
601 Valencia Ave., Suite 100, Brea, CA, U.S.A. 92823

Abstract: Process measurements do not satisfy conservation laws such as mass and energy balances because they are subject to random and sometimes gross errors. Sources of gross errors are process disturbances, leaks, departures from steady state, and malfunctioning or miscalibrated instruments. A challenging task in dealing with contaminated data is to correctly detect and identify them as such before reconciliation. Recently developed gross error detection and identification methods based on Principal Component Analysis have received attention from process industries and have been implemented in industrial software packages. This paper reviews a few cases that were solved using the Principal Component Tests with this software. *Copyright © 1998 IFAC*

Keywords: Bad data identification, Diagnostic tests, Error detection, Fault Detection, Fault identification, Normal distribution, Software tools, Statistical analysis, Validation.

1. INTRODUCTION

1.1 Background

Process measurements do not satisfy conservation laws such as material and heat balances, because they are subject to random and sometimes systematic or gross errors. Sources of gross errors are process disturbances, leaks, departures from steady state, and sensor malfunctioning or miscalibration.

Data reconciliation employs optimization and hypothesis testing techniques to detect persistent gross errors or faults and to turn inconsistent into consistent data. However, the constrained least squares procedure in data reconciliation is sensitive to gross errors. A challenging task in dealing with contaminated data is to correctly detect and identify them as such before reconciliation.

A number of methods have been used successfully in data reconciliation for gross error detection and identification. Unfortunately, they sometimes wrongly report more or fewer gross errors than are actually present. This is a problem for the following reasons. First, a gross error may propagate in data reconciliation and contaminate the reconciled data. A gross error is harmful because of its leverage in the reconciliation process. Second, a subtle though small gross error may influence quality specifications and thus must be detected and identified. Third, redundancy in a chemical or refinery process may be low compared to the size of the process. Deletion of a measurement from such a process may result in

[1] Correspondence concerning this article should be addressed to H. Tong. Email: htong@simsci.com.

additional nonredundant variables or additional unobservable variables. A sound gross error detection and identification technique should help us discard bad data while keeping good information.

1.2 Gross error detection methods

Among statistical tests frequently used in data reconciliation for detecting gross errors, the chi-square collective test compares the optimal value of the objective function in the model of data reconciliation to a threshold of chi-square values. The univariate and maximum power (MP) individual tests, on the other hand, examine each and every measurement adjustment and residual of the process constraints. Though these statistical tests have been widely used in detecting or identifying gross errors, their performance has not always been satisfactory. When these tests fail to detect gross errors, process engineers may interpret plant performance in an inadequate or misleading manner and may find it difficult to isolate gross errors.

Tong (1995) showed that the conventional statistical tests mentioned above may sometimes wrongly eliminate good while keeping contaminated data. This may complicate process simulation, optimization, and advanced process control, because these techniques rely on data to update their model parameters and it is assumed, often implicitly and without justification, that the data are of good quality.

Tong and Crowe (1995) presented the principal component (PC) transformation of the optimal measurement adjustments. The PCs are not correlated to each other, thus resulting in a sharper and less confounding statistical test for detecting and identifying gross errors.

The Principal Component Measurement Test (PCMT) was successfully applied at Simulation Sciences Inc. (SIMSCI) to industrial data reconciliation problems of various sizes. It was implemented in SIMSCI's software products ROMeo™ and DATACON™. For the cases studied in this investigation, PCMT was shown to be more acute in detecting gross errors, and to be more powerful in correctly identifying gross errors than the conventional statistical tests. An improved contribution analysis for PCMT facilitated the identification of gross errors and allowed users to isolate the errors more easily.

2. BRIEF REVIEW OF THEORY

2.1 Linear data reconciliation

For simplicity, let us consider a linear data reconciliation model. Data reconciliation of measured flow rates belongs to this category. Methods of bilinear data reconciliation can be found in Crowe (1986) and Tong (1995). Techniques of nonlinear data reconciliation are described by Swartz (1989) and Sanchez and Romagnoli (1996). A bilinear or nonlinear data reconciliation problem is often reduced to a series of linear data reconciliation problems.

Let us study a process where only flow rates are considered and all flow rates are measured. This is not a limitation since Crowe *et al.* (1983) have shown that matrix projection, and Swartz (1989) and Sanchez and Romagnoli (1996) have also shown that QR decomposition could be readily used to eliminate unmeasured variables from the process constraints. Furthermore, the technique described below can also be used to reconcile composition, temperature and pressure data.

A linear data reconciliation problem can be written as a constrained weighted least squares problem

$$\min_{\boldsymbol{a}} \; F(\boldsymbol{a}) = \boldsymbol{a}^T \Sigma_1^{-1} \boldsymbol{a} \qquad (1)$$

$$s.t. \;\; \boldsymbol{B}_1(\tilde{\boldsymbol{x}} + \boldsymbol{a}) = \boldsymbol{0} \qquad (2)$$

where $\tilde{\boldsymbol{x}}$ is an n-vector of measurements, $\boldsymbol{a}$ is the vector of adjustments to $\tilde{\boldsymbol{x}}$ so that the balances in (2) hold. $\boldsymbol{B}_1$ is an $m \times n$ matrix of rank m, consisting of the incidence matrix for the balances on species, and possibly with rows for additional process specifications. If (2) results from a linearization of a nonlinear problem, $\boldsymbol{B}_1$ consists of the Jacobian matrix of the nonlinear constraints. Σ_1 is an $n \times n$ variance-covariance matrix of $\tilde{\boldsymbol{x}}$. The optimal solution (Crowe, 1988) is

$$\boldsymbol{a} = -\Sigma_1 \boldsymbol{B}_1^T \left(\boldsymbol{B}_1 \Sigma_1 \boldsymbol{B}_1^T\right)^{-1} \boldsymbol{B}_1 \tilde{\boldsymbol{x}} \qquad (3)$$

with the corresponding $n \times n$ singular variance-covariance matrix

$$Q_1 = \text{cov}(\boldsymbol{a}) = \Sigma_1 \boldsymbol{B}_1^T \left(\boldsymbol{B}_1 \Sigma_1 \boldsymbol{B}_1^T\right)^{-1} \boldsymbol{B}_1 \Sigma_1 \qquad (4)$$

and the optimally reconciled data

$$\hat{\boldsymbol{x}} = \tilde{\boldsymbol{x}} + \boldsymbol{a} \qquad (5)$$

Without the data adjustments, the process constraints may not be closed. The residuals of the process constraints resulting from the errors in the data are expressed as

$$e = B_1 \tilde{x} \tag{6}$$

with the corresponding $m \times m$ nonsingular variance-covariance matrix

$$H_e = B_1 \Sigma_1 B_1^T \tag{7}$$

as can be seen from (2).

2.2 *Gross error detection and identification*

If $\tilde{x}$ follows a multivariate normal distribution, $\tilde{x} \sim N(x, \Sigma_1)$, where x is a vector of the unknown expectations, it can be shown that both a and e follow multivariate normal distributions, $a \sim N(0, Q_1)$ and $e \sim N(0, H_e)$, respectively, at the optimal solution. Therefore, we know *a priori* in data reconciliation that the expectations of a and e are zero if there are only random errors. This is the foundation of gross error detection.

In addition, we can show that the optimal objective function value (1) follows a chi-square distribution with m degrees of freedom

$$F^* = \min_a F(a) = a^T \Sigma_1^{-1} a = e^T H_e^{-1} e \sim \chi_m^2 \tag{8}$$

2.2.1 *Global chi-square test (Reilly and Carpani, 1963)*

We may test the data collectively by comparing the optimal objective function value (F^*) against a threshold of a chi-square value with m degrees of freedom (χ_m^2). If $F^* \geq \chi_m^2$, one or more gross errors are detected. The chi-square test had been used for detecting, but not for identifying, gross errors until Tong (1995) associated it with the Principal Component Tests.

2.2.2 *Univariate test (Mah and Tamhane, 1982)*

Since $e \sim N(0, H_e)$, we can derive a univariate constraint statistic

$$z_{e,i} = \frac{e_i}{\sqrt{(H_e)_{ii}}} \sim N(0,1),\ i = 1,2,\cdots,m \tag{9}$$

therefore, we may test each process constraint against a threshold of a uninormal variate. Similarly, we may test each measurement against a threshold of a uninormal variate, since

$$z_{a,j} = \frac{a_j}{\sqrt{(Q_1)_{jj}}} \sim N(0,1),\ j = 1,2,\cdots,n \tag{10}$$

Henceforth, we will denote (10) as UMT (Univariate Measurement Test).

2.2.3 *Principal component test (Tong and Crowe, 1995)*

We can show that a principal component constraint statistic satisfies

$$y_{e,i} = (W_e^T e)_i \sim N(0,1),\ i = 1,\cdots,m \tag{11}$$

where W_e consists of eigenvectors of H_e. Furthermore, it can be shown that $y_e \sim N(0, I)$ so that the principal components are not correlated. If there are only random errors, we know *a priori* that not only the expectations of y_e are zero but also the variance-covariance matrix is identity. As discussed earlier, we may test such a statistic in (11) against a threshold of a uninormal variate.

We should note that there is something in common in the univariate and the PC test - both compare a statistic against a uninormal variate of zero expectation and unit variance. They differ in the way in which each statistic is constructed. Since PCs are not correlated, the PC test is generally sharper and less confounding. However, a PC test requires more computations. In addition, it relies on the companion contribution analysis that maps a gross error PC into one or more constraints.

Similarly, Tong and Crowe (1995) defined the principal component measurement statistic as

$$y_{a,i} = (W_a^T a)_i \sim N(0,1),\ i = 1,\cdots,m \tag{12}$$

Henceforth, we will denote (12) as PCMT (Principal Component Measurement Test).

Crowe and Tong (1996) recently discovered the identity

$$F^* \equiv y_e^T y_e \equiv y_a^T y_a \equiv a^T \Sigma_1^{-1} a \equiv e^T H_e^{-1} e \sim \chi_m^2 \tag{13}$$

when all PCs are kept as they should be in gross error detection. This identity not only relates various matrices and vectors of theoretical importance in data reconciliation, but also serves as a guard against possible coding errors in the software development.

3. EXAMPLES

Two examples from literature highlight the difference between PCMT and UMT. Brief reviews of four more comparison case studies follow.

3.1 A process with a leak

Crowe *et al.* (1983) and Tong and Crowe (1995) used a simple process with three units in series to illustrate the linear data reconciliation algorithm and various gross error detection techniques. The diagonal variance-covariance matrix of the measured total flow rates (Σ_1) is depicted in Table 1. It is noted that measurements are independent since the matrix Σ_1 is diagonal.

Table 1 Variance-covariance matrix of the data, Σ_1, example 3.1

	Flow$_1$	*Flow*$_2$	*Flow*$_3$	*Flow*$_4$
Flow$_1$	•			
Flow$_2$		•		
Flow$_3$			•	
Flow$_4$				•

The variance-covariance matrix of the data adjustments (Q_1) is depicted in Table 2. It is interesting to note that, although Σ_1 is diagonal, Q_1 is 100% dense.

Table 2 Variance-covariance matrix of the data adjustments, Q_1, example 3.1

	Adj.$_1$	*Adj.*$_2$	*Adj.*$_3$	*Adj.*$_4$
Adj.$_1$	•	•	•	•
Adj.$_2$		•	•	•
Adj.$_3$		*symm.*	•	•
Adj.$_4$				•

PCMT fully exploits the covariance structure in Q_1, while UMT exploits only the variances in it (the diagonal). It is the covariance structure characterized by B_1 and Σ_1 and not only the variances in Q_1 that determines the outcome of data reconciliation. It is the violation to the covariance structure and not only to the variances in Q_1 that signifies the existence of gross errors. This observation is consistent with (1) through (8).

3.2 An ammonia synthesis loop

Crowe *et al.* (1983) and Tong and Crowe (1995) applied PCMT and UMT to a simple ammonia synthesis loop for gross error detection and identification. Table 3 depicts the block diagonal variance-covariance matrix of the measured species flow rates (Σ_1). Some measurements were not independent because they shared common total flow rates or were correlated by chemical reactions. Regardless of the structure of Σ_1, the resulted variance-covariance matrix of the data adjustments (Q_1) is again 100% dense (not shown). The argument regarding PCMT and UMT remains true.

Table 3 Variance-covariance matrix of the data, Σ_1, example 3.2

	$\tilde{x}_1$	$\tilde{x}_2$	$\tilde{x}_3$	$\tilde{x}_4$	$\tilde{x}_5$	$\tilde{x}_6$	$\tilde{x}_7$	$\tilde{x}_8$
$\tilde{x}_1$	•	•	•					
$\tilde{x}_2$		•	•					
$\tilde{x}_3$			•					
$\tilde{x}_4$				•	•			
$\tilde{x}_5$					•			
$\tilde{x}_6$		*symm.*				•		
$\tilde{x}_7$							•	
$\tilde{x}_8$								•

3.3 A gas plant

Data reconciliation and gross error detection in a gas plant were conducted using ROMeo™ with PCMT and UMT. As seen from Table 4, this is a large, yet very sparse problem. The number of measurements is small compared to the problem size. This is typical in many chemical and refinery processes. The nonzero pattern of the process model is shown in Fig. 1.

Fig. 2 shows the UMT on a total of 54 data adjustments. Nine of these were identified as possibly in gross error. The horizontal dashdot line represents the critical value for the UMT. Fig. 3 shows the PCMT on the same 54 adjustments, of which only two were identified as possibly in gross error. The two suspects were isolated using a variation of the contribution analysis of Tong and Crowe (1995). It is noted that PCMT was less confounding in this example. The primary suspect identified by PCMT coincides with that identified by UMT. This is a coincidence rather than a rule.

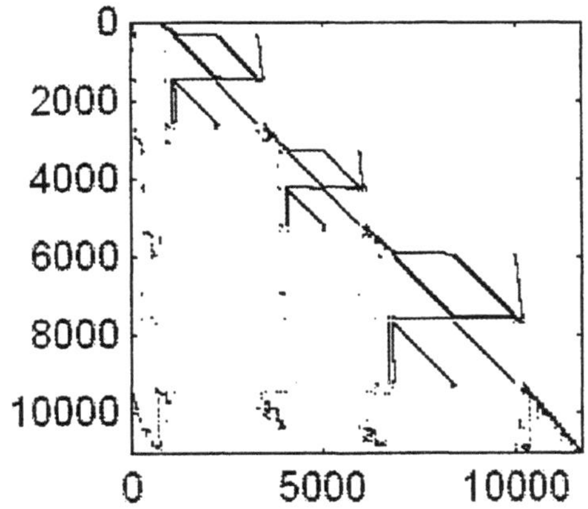

Fig. 1. Nonzeros in the gas plant model

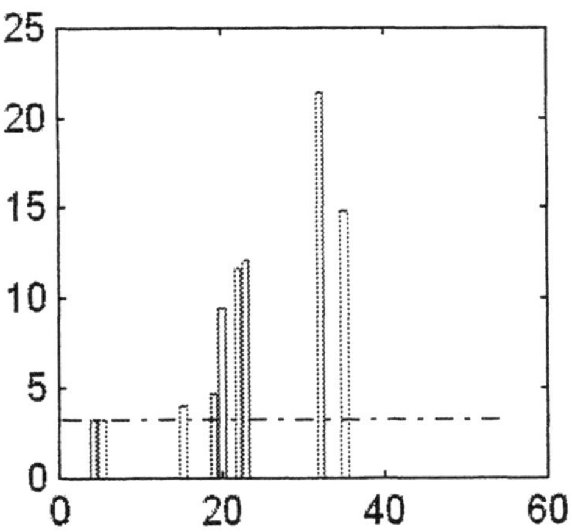

Fig. 2. UMT, example 3.3

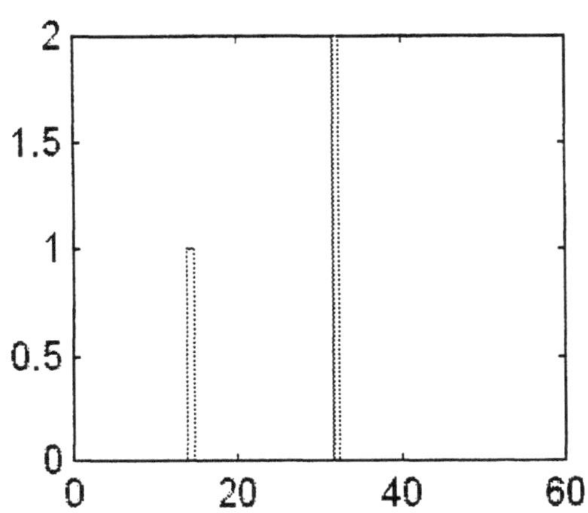

Fig. 3. PCMT, example 3.3

Table 4 Statistics of example 3.3

# of equations:	*11008*
# of nonzeros:	80639
sparsity:	0.063%
# of unmeasured variables:	10964
# of fixed variables:	703
# of measurements:	54
degrees of redundancy:	44

3.4 A refinery process

Fig. 4 shows a simplified refinery process consisting of 3 tanks, a column, and 8 streams.

An engineer who studied the problem encountered difficulty in arriving at a meaningful data reconciliation. He suspected that the accumulation in tank C was possibly in gross error. Unfortunately, neither the UMT nor the global chi-square test could support his conjecture.

The engineer's doubt about the accumulation in tank C was based on his observations. He found a 162% change from the measured to the reconciled accumulation. In fact, the measured accumulation was negative and the computed accumulation was positive. The UMT for the accumulation was smaller, yet close to its critical value. Had it been above the critical value, it would have been identified as a gross error.

The case was sent to one of the authors who used PCMT to determine that the accumulations in tanks B and C were in gross error, a finding that is in agreement with the engineer's analysis. The result is summarized in Table 5.

Table 5 PCMT vs. UMT, example 3.4

Test	Gross Err. Detected	Remark
UMT	accumulation, tank B	inadequate
PCMT	accumulation, tanks B & C	OK

3.5 Additional examples

In conducting separate data reconciliation studies for an electric power plant boiler system and for a synthesis gas process, it was found in both cases that UMTs were passed while chi-square tests were failed. Gross errors were therefore detected but could not be identified without PCMT (see 2.2.1). The problems were solved using PCMT as shown in Table 6 and Table 7.

Table 6 PCMT vs. UMT, the boiler system

Test	Gross Err. Detected	Remark
UMT (bound on)	feed water temp., mixer 2	failed χ^2
UMT (bound off)	feed flow rate, mixer 2	failed χ^2
PCMT (bound on / off)	feed water temp., mixer 2 and htr 2	OK

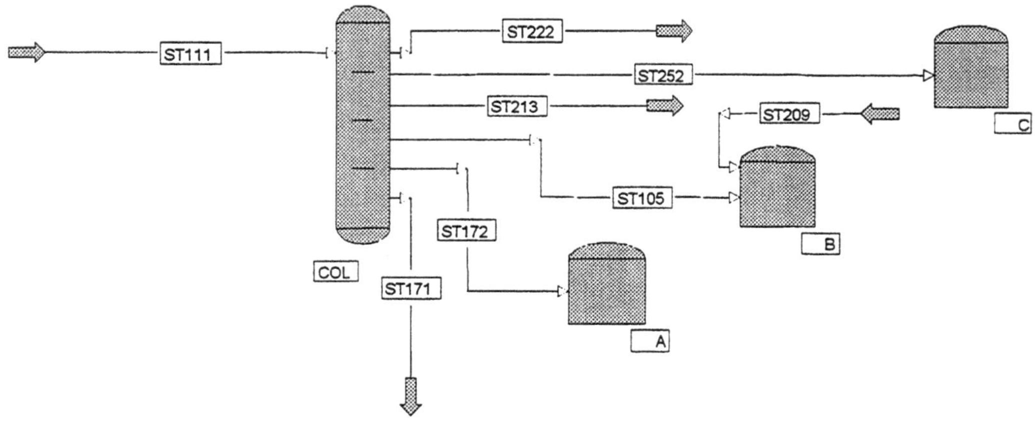

Fig. 4. Flowsheet for example 3.4

Table 7 PCMT vs. UMT, the synthesis gas process

Test	Gross Err. Detected	Remark
UMT (bound on)	CO & N2, mixer product	failed χ^2
PCMT (bound on / off)	CO & N2, mixer product; CH4, mixer feed	OK

ACKNOWLEDGMENT

H. Tong wishes to thank his colleague Bill White for his constructive comments in preparing this paper.

REFERENCES

Crowe, C. M. and H. Tong, *Private Communication*, (Oct., 1996).

Crowe, C. M., "Recursive Identification of Gross Errors in Linear Data Reconciliation," *AIChE J.*, **34**, 541, (1988).

Crowe, C. M., "Reconciliation of Process Flow Rates by Matrix Projection. II. The Nonlinear Case," *AIChE J.*, **32**, 616, (1986).

Crowe, C. M., Y. A. Garcia Campos and A. Hrymak, "Reconciliation of Process Flow Rates by Matrix Projection. I. The Linear Case," *AIChE J.*, **29**, 818, (1983).

Mah, R. S. H., and A. C. Tamhane, "Detection of Gross Errors in Process Data," *AIChE J.* **28**, 828, (1982).

Reilly, P. M., and R. E. Carpani, "Application of Statistical Theory of Adjustments to Material Balances," *13th Can. Chem. Eng. Conf.*, Montreal, Oue., (Oct., 1963).

Sanchez, M. and J. Romagnoli, "Use of Orthogonal Transformations in data classification-reconciliation," *Computers Chem. Engng*, **20**, 483, (1996).

Swartz, C. L. E., "Data Reconciliation for Generalized Flowsheet Applications," *AIChE Spring National Meeting*, Dallas, TX, (1989).

Tong, H., and C. M. Crowe, "Detection of Gross Errors in Data Reconciliation by Principal Component Analysis," *AIChE J.*, **41**, 1712, (1995).

Tong, H., "Studies in Data Reconciliation Using Principle Component Analysis," *Ph.D. Thesis*, McMaster University, Hamilton, Ont., Canada (1995).

ON-LINE MONITORING OF A BATCH PMMA REACTOR

Huei-Shyang You and Chan-Cheng Chen

Center for Industrial Safety and Health Technology
Industrial Technology Research Institute
Chutung Hsinchu, Taiwan 310, R.O.C.
E-mail: 850315@cishoa.itri.org.tw
Fax: +886-3-5820068

Abstract: A novel approach for on-line monitoring of a batch process is proposed in this article. The proposed method is mainly based on the recursive estimation of the R-statistic that considers the true distribution of the correlated noise during a batch operation. The method effectively reduces the computation loading in on-line implementation and decreases the possibility of false alarm while doing on-line monitoring of a batch operation. A case study on a pilot-plant-scale polymerization of methyl methacrylate (PMMA) simulator has demonstrated the potential benefits of this algorithm. The result is also compared with that obtained from a traditionally on-line mult-way PCA method. *Copyright © 1998 IFAC*

Keywords: Batch processes, fault detection, monitoring, on-line, process parameter estimation, statistical analysis, supervision.

1. INTRODUCTION

Conventionally, process failure is monitored and diagnosed by the operators. However, the complexity of modern chemical plants and the availability of advanced computer hardware lead us to develop automated fault detection and diagnosis. Techniques used for automated fault detection and diagnoses are very much depending on the rigorousness of the process model. Fundamental mathematical models or detailed knowledge based models such as artificial intelligence systems are usually considered. However, rather than utilizing the detailed knowledge about the process, the principal component analysis (PCA) and the projection to latent structure (PLS) methods utilize only the information contained in the historical database of past runs. As the distributed control system (DCS) is more and more popular in the process industries, such approaches seem to be more attractive for real plants. A discussion for the advantages of such methods can be found in the work of MacGregor, *et al.* (1994) and Nomikos, *et al.* (1994).

Although these techniques are further developed and applied to batch processes, for example, multi-way PCA and multi-way PLS (Wold, *et al.*, 1987), there are little works addressed the *on-line* monitoring problems (Kouri, *et al.,* 1995; Nomikos and MacGregor, 1994, 1995; Nomikos, 1996; Martin, *et al.*, 1996). In these works, normal operation data are collected batch by batch, and then multi-way PCA or multi-way PLS techniques are then applied to these data to extract the features of normal operation sets. As these features are completely constructed, a statistic test, for example, T^2 or M^2, is considered to test whether the new operation data set and the normal historical sets are of the consistent features or not. A problem in such approaches is that as the new batch evolves, the data are not complete until the end of that batch. To be an *on-line* strategy, different methods are suggested to predict or fulfill the future observations (Nomikos and MacGregor, 1995). As one already know, except for detecting of the abnormal condition between the new batch and normal batches, the main purpose of *on-line* detection is to detect the abrupt change in a new batch run, which should be described by the auto-correlation effect. However, the multi-way techniques only extract the common features between different batches, but do not catch any information about the auto-correlation effect from normal batch runs. The predicted or fulfill procedures are, indeed, to introduce an assumption about the auto-correlation effects during a batch run. Obviously, because multi-way PCA/PLS methods do not catch the auto-correlation relation from normal batch runs, the predicted or fulfill procedures are inevitable for an *on-line* detection method.

In this article, a common feature is first extracted from the historical batch runs, which considers both the cross-correlated effects between different batches and the auto-correlated effects during a batch run. A statistic criterion is then developed to on-line monitor a new batch operation. The major advantage of this approach is that, in monitoring stage, the proposed method needs not to predict or fulfill the future data, that is, the on-line monitoring is based on past observed data instead of the predicted future values, which drastically decrease the on-line computation loading. Another advantage of this approach is the real auto-correlation effects are included instead of introducing it by assumptions like the multi-way PCA/PLS approaches.

The rest of this article is organized as follows. Section 2 discusses the whole algorithm and its statistic meaning. Section 3 issues the real-time implementation algorithm. A pilot-plant-scale polymerization of methyl methacrylate (PMMA) batch reactor system is then explored to demonstrate the proposed approaches in section 4. Finally, some conclusions and discussions are made in section 5.

2. METHODOLOGIES FOR ON-LINE MONITORING

Consider a typical batch run in which $p=1,2,\cdots,P$ process variables are measured at $k=1,2,\cdots,K$ time intervals throughout a batch run. Similar data will exist on a number of similar batch runs $j=1,2,\cdots,J$. All these data are arranged into a $(J\times P\times K)$ process data matrix X. Measurements on $q=1,2,\cdots,Q$ final quality variables are taken at the end of each batch, and can be summarized in the $(J\times Q)$ quality data matrix Y.

An on-line monitoring strategy is usually developed according to the following steps. A set of *good* data is first collected and retrieved from the historical database of past batch runs. Some multivariate data analysis methods, such as the multi-way PCA/PLS methods, are then applied to the data matrices X and/or Y to extract statistic behaviors. The third step is to perform an on-line detecting of the abnormal condition according to the summarized features. In this article, a steady state identification method modified from Cao and Rhinehart (1995) is adopted by the second and the third steps. The estimated mean and standard deviation of $X_j(k)$, for $p=1,2,\cdots,P$ and for $k=1,2,\cdots,K$ are recursively updated according to the following equations:

$$\hat{X}_j(p,k)=\lambda_{1,p}X_j(p,k)+\left[1-\lambda_{1,p}\right]\hat{X}_{j-1}(p,k) \quad (1)$$

$$\hat{S}_{1,j}(p,k)=\lambda_{2,p}\left[X_j(p,k)-\hat{X}_{j-1}(p,k)\right]^2+\left[1-\lambda_{2,p}\right]\hat{S}_{1,j-1}(p,k) \quad (2)$$

$$\hat{S}_{2,j}(p,k)=\lambda_{3,p}\left[X_j(p,k)-X_{j-1}(p,k)\right]^2+\left[1-\lambda_{3,p}\right]\hat{S}_{2,j-1}(p,k) \quad (3)$$

where $\lambda_{1,p}$, $\lambda_{2,p}$, and $\lambda_{3,p}$ are tuning parameters for these low-pass estimators, respectively. Monitoring the nonstationarity behavior of a batch run with $R_j(p,k)$ statistic is then defined by

$$R_j(p,k)=\frac{\left[2-\lambda_{1,p}\right]\hat{S}_{1,j}(p,k)}{\hat{S}_{2,j}(p,k)} \quad (4)$$

If the process data $X_j(p,k)$, for $j=1,2,\cdots,J$ between different batch runs are stationary, then the process mean might be constant at every time k. Furthermore, if the noises are independent of different batch runs and the tuning parameters $\lambda_{2,p}$ and $\lambda_{3,p}$ are chosen to be equal, then $R_j(p,k)$ would be close to unity. In practice, detecting the stationary condition of a process operation would be preferable to monitor the value of $\left[2-\lambda_{1,p}\right]\hat{S}_{1,j}(p,k)$ being greater than $R_{j,crit}(p,k)\hat{S}_{2,j}(p,k)$. Where, $R_{j,crit}(p,k)$ is a threshold value of $R_j(p,k)$, and, for the simplicity, $R_{j,crit}(p,k)$ can be selected as $R_{crit}(p)$ during the whole batch and for $j=1,2,\cdots,J$.

The nonstationarity index $nssIndex_j(p,k)$ for the jth run and with respect to the pth measured variable at time k could be determined by the following rule:

IF $[2-\lambda_{1,p}]\hat{S}_{1,j}(p,k) \geq R_{crit}(p)\hat{S}_{2,j}(p,k)$

THEN $nssIndex_j(p,k)=1$;

OTHERWISE, $nssIndex_j(p,k)=0$.

Doing a weighting average on $nssIndex_j(p,k)$ for $p=1,2,\cdots,P$ and defining

$$nssFlag_j(k) = \frac{\sum_{j=1}^{k}\sum_{p=1}^{P}[Weight_j(p,i)][nssIndex_j(p,i)]}{\sum_{j=1}^{k}\sum_{p=1}^{P}Weight_j(p,i)} \tag{5}$$

and

$$Weight_j(p,k) = \frac{\left|\dfrac{X_j(p,k)-\hat{X}_{j-1}(p,k)}{\hat{S}_{1,j}(p,k)}\right|}{\sum_{p=1}^{P}\left[\dfrac{X_j(p,k)-\hat{X}_{j-1}(p,k)}{\hat{S}_{1,j}(p,k)}\right]^2} \tag{6}$$

the nonstationarity condition of a batch at time k could be determined by the empirical rule:

IF $nssFlag_j(k) \geq nssFlag_{j,crit}(k)$,

THEN *the batch j is assumed to be nonstationary at time k;*

OTHERWISE, *the batch j is assumed to be stationary up to time k.*

Practically, $nssFlag_{j,crit}(k)$, for $j=1,2,\cdots,J$ and for $k=1,2,\cdots,K$ is assigned to be a constant and can be obtained from the past runs and/or from the simulation experience.

3. REAL-TIME IMPLEMENTATION ALGORITHM

Although the theoretical derivations of the proposed method have been discussed in the previous section, on-line implementation of the monitoring scheme with recursive calculations has not been described. One way to perform the on-line monitoring of a batch operation is outlined as follows. The whole approach is summarized and illustrated in figure (1).

In the off-line implementation phase, massive amounts of process data are first collected and stored in a database. The past operating data should be clustered into normal and abnormal data sets. Note that the abnormal data sets should be removed before going to next steps, and this may be done by statistic methods or only by operator's experiences. Now according to the results of interpreting process data with statistic feature extracting methods, the parameters $\lambda_{1,p}$, $\lambda_{2,p}$, $\lambda_{3,p}$, $R_{crit}(p)$, for $p=1,2,\cdots,P$ and $nssFlag_{crit}$ (or UCL) can be determined satisfactorily. The off-line or preliminary implementation of the proposed method is thus completed in this stage.

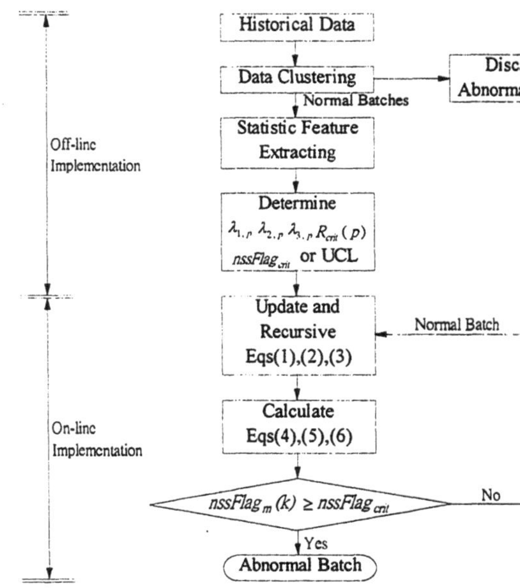

Fig 1. A schematic view of the on-line monitoring method of a batch process.

Doing an on-line implementation of the proposed method, the algorithms consist of the initialization, the recursive estimation and prediction, the decision, and the variables updating and correction steps. The resulting iterative algorithm is summarized below:

Step 1: Initialization. The sample mean $\overline{X}_0(p,k)$ and standard deviation $S_0(p,k)$, for $p=1,2,\cdots,P$ and for $k=1,2,\cdots,K$ could be estimated from a historical data set of normal batches $j=1,2,\cdots,J$:

$$\overline{X}_0(p,k) = \frac{\sum_{j=1}^{J}X_j(p,k)}{J} \tag{7}$$

$$S_0(p,k) = \frac{\sum_{j=1}^{J}[X_j(p,k)-\overline{X}_0(p,k)]^2}{J-1} \tag{8}$$

Three initial values given in equations (1), (2) and (3) are then estimated according to the following relations:

$$\hat{X}_0(p,k) = \overline{X}_0(p,k) \tag{9}$$

$$\hat{S}_{1,0}(p,k) = \frac{2}{2-\lambda_{1,p}}S_0(p,k) \tag{10}$$

$$\hat{S}_{2,0}(p,k) = 2S_0(p,k) \tag{11}$$

Practically, $\hat{X}_0(p,k)$, $\hat{S}_{1,0}(p,k)$, and $\hat{S}_{1,0}(p,k)$ should be backup and recorded in a hardware device to resume the on-line recursive algorithms while any abnormal batch is detected.

Step 2: Recursive estimation and prediction. For any batch m being operated, the values of $\hat{X}_m(p,k)$, $\hat{S}_{1,m}(p,k)$, and $\hat{S}_{2,m}(p,k)$ are updated recursively at the sampling time k, and $k = 1, 2, \cdots, K$. Equations (1), (2), and (3) are used to carried out these calculations and estimations.

Step 3: Decision. Two heuristically nonstationary identifiers $nssIndex_m(p,k)$ and $nssFlag_m(k)$ are calculated according to equations (4), (5) and (6) to highlight an abnormal condition of a batch m at time k. An empirical rule, $nssFlag_m(k) \geq nssFlag_{crit}$, is employed to identify the nonstationarity condition of a batch process. Plotting a run chart of $nssFlag_m(k)$ could make the decision method be performed very easily.

Step 4: Variables updating and correction. Update equations (1), (2), and (3). Collect new operational data at time $k+1$ to repeat the work in steps (2), (3), and (4), and do continuously.

4. SIMULATION EXAMPLE

In this section, a pilot-plant-scale PMMA batch reactor system (Chiu *et al.* 1982; Baillagou and Soong, 1985; Chang and Chen, 1992; Chang and Chung, 1995) is explored to demonstrate the proposed method. The detailed mechanisms of the PMMA reaction are described as follows.

Initiator decomposition

$$I \xrightarrow{k_d} 2R^{\bullet} \tag{12}$$

Initiation

$$R^{\bullet} + M \xrightarrow{k_i} P_1^{\bullet} \tag{13}$$

Propagation

$$P_n^{\bullet} + M \xrightarrow{k_p} P_{n+1}^{\bullet} \tag{14}$$

Chain transfer to monomer

$$P_n^{\bullet} + M \xrightarrow{k_{tm}} D_n + P_1^{\bullet} \tag{15}$$

Chain transfer to solvent

$$P_n^{\bullet} + S \xrightarrow{k_{ts}} D_n + R^{\bullet} \tag{16}$$

Termination by disproportion

$$P_n^{\bullet} + P_m^{\bullet} \xrightarrow{k_{td}} D_n + D_m \tag{17}$$

Termination by combination

$$P_n^{\bullet} + P_m^{\bullet} \xrightarrow{k_{tc}} D_{m+n} \tag{18}$$

The kinetic equations of them are also given below:

$$R_I = -k_d[I] \tag{19}$$

$$R_{R^{\bullet}} = 2fk_d[I] - k_i[R^{\bullet}][M] \tag{20}$$

$$R_M = -k_i[R^{\bullet}][M] - (k_p + k_{tm})\lambda_0[M] \tag{21}$$

$$R_{P_1} = k_i[R^{\bullet}][M] - k_p[P_1^{\bullet}] - k_t[P_1^{\bullet}]\lambda_0 + (k_{tm}[M] + k_{ts}[S])(\lambda_0 - [P_1^{\bullet}]) \tag{22}$$

$$R_{P_n} = k_p[M]([P_{n-1}^{\bullet}] - [P_n^{\bullet}]) - [P_n^{\bullet}]\sum_{n=1}^{\infty} k_t[P_m^{\bullet}] - (k_{tm}[M] + k_{ts}[S])[P_n^{\bullet}] \tag{23}$$

$$R_{\lambda_0} = k_i[R^{\bullet}][M] - k_t\lambda_0^2 \tag{24}$$

$$R_{D_n} = k_{tm}[P_n^{\bullet}][M] + k_{ts}[P_n^{\bullet}][S] + k_{td}[P_n^{\bullet}]\lambda_0 + 0.5k_{tc}\sum_{m=1}^{n-1}[P_m^{\bullet}][P_{n-m}^{\bullet}] \tag{25}$$

All parameters in above equations were reported in the literature (Chiu *et al.* 1982; Baillagou and Soong, 1985), and were checked by a pilot-plant-scale reactor system (Chang and Chen, 1992; Chang and Chung, 1995). The instrumental system of the PMMA reactor is shown in figure (2).

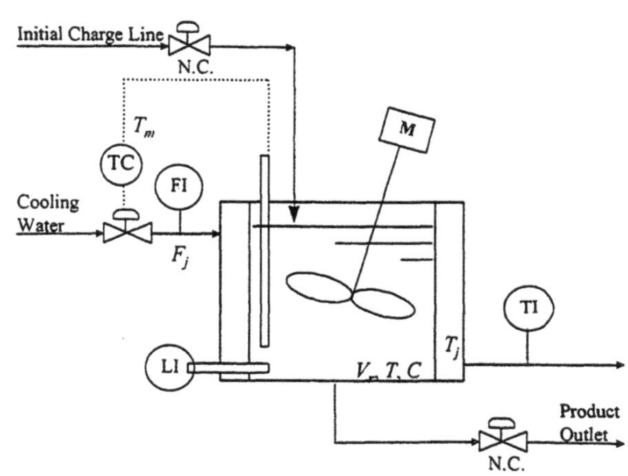

Fig 2. A PMMA batch reaction system.

In this case, there exist white noises in measurement devices, in initial charge conditions, and at the outlet temperature of cooling water. The measurable process variables $X(p,k)$, for $p = 1, 2, \cdots, 4$ and for $k = 1, 2, \cdots, 3000$ are the reactor volume $V_r(k)$, the reactor temperature $T(k)$, the flow rate of cooling water $F_j(k)$, and the outlet temperature of cooling water $T_j(k)$, respectively. All these variables are sampled per one minute.

Thirty-two normal batch run's data is collected to

construct the sample distributions of R-statistic according to equations (1) to (4). One of those R-plots is shown in figure (3). The ordered set La($\lambda_{1,p}$,$\lambda_{2,p}$,$\lambda_{3,p}$) denotes three tuning parameters adopted in the recursive estimations of sample means and standard deviations. The results have shown that $\lambda_{1,p}$ has small effects on the shape of R-plots, small $\lambda_{2,p}$ and $\lambda_{3,p}$ will make the shape of R-plots to be more narrow, and the value of R-statistic will nearly equal to unity with $\lambda_{2,p} = \lambda_{3,p}$. The threshold values of R-statistics, $R_{crit}(p)$, $p = 1, 2, \cdots, 4$ can be determined from this figure either by heuristic criterion or by statistic methods, for example, the significance level of type I error.

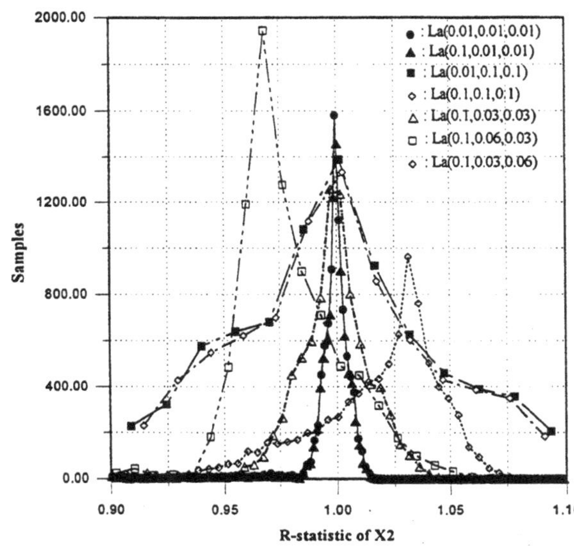

Fig 3. The sample distribution of R-statistic of $X_j(2,k)$, for $k = 1, 2, \cdots, 3000$ and for $j = 1, 2, \cdots, 32$.

Four kinds of process faults are considered to illustrate the ability of the proposed algorithm: these include three types of measurement faults and a valve chocking fault. All faults in simulation start from $k = 500$ minutes. The $nssFlag_{crit}$ or the upper control limit (UCL), in figure (4), is determined from the normal operation runs. As shown in figure (4), the proposed algorithm detects the process faults effectively.

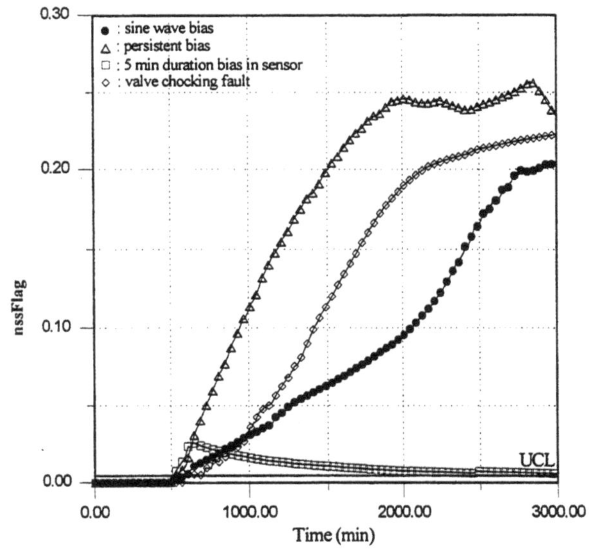

Fig 4. On-line monitoring results of the proposed method.

The sensor fault with five minutes duration bias and the valve chocking fault are also explored by the multi-way PCA approaches (Nomikos and MacGregor, 1995). The results are given in figures (5) and (6), respectively. From figure (5), this approach will have false alarms at the beginning of a batch run up to time $k = 500$ minutes. Moreover, as shown by figure (6), valve chocking fault can not be detected with the same UCL as that of sensor fault. If the UCL for valve chocking fault is relaxed to UCL(20), however, it will loss the ability to detect the sensor fault.

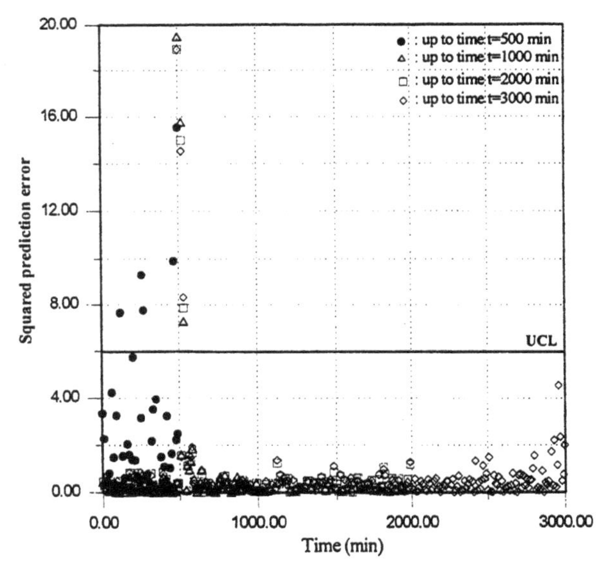

Fig 5. The monitoring results of sensor fault with five minutes duration bias using multi-way PCA method.

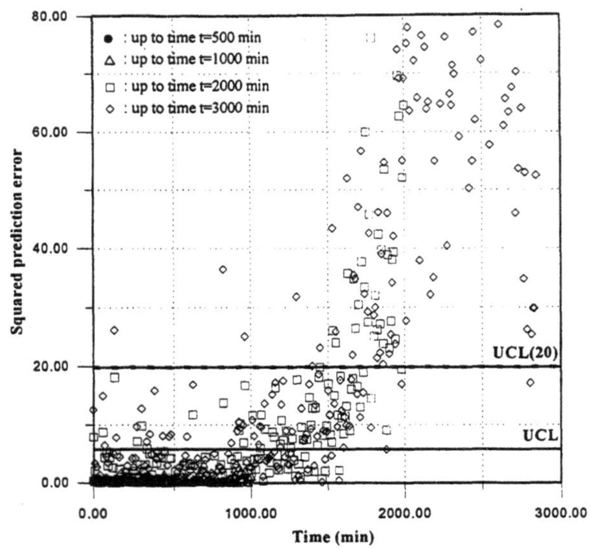

Fig 6. The monitoring results of valve chocking fault using multi-way PCA method.

5. CONCLUSIONS AND DISCUSSIONS

As reported in the literature, the multi-way PCA/PLS algorithms need to predict future process variables in the on-line application. This makes the multi-way PCA/PLS based methods have too many false alarms in the beginning of a new batch run, and this characteristic is clearly explained by figures (5) and (6). Moreover, these methods also have heavy computation loading in the monitoring stage because of the prediction procedures. In this study, a novel strategy is proposed to improve these points for an on-line monitoring of a batch process. The auto-correlation effects of normal historical batch runs are considered in the statistic feature extraction stage through the R-statistics, so it drastically decreases the computation loading in on-line monitoring stage and avoids the possibility of false alarm as shown in figures (5) and (6).

Another important characteristic of the proposed method is that the nominally normal behavior of the batch operation is updated by the forgetting factor, thus the nominal behavior has heavy weight in recent batch runs and it is robust in small changes of the nominal behaviors. However, this effect is not explained in this work and will be discussed in a future work.

ACKNOWLEDGMENT

The authors acknowledge the Ministry of Economic Affairs of the Executive Yuan for the financial support with grant number 87-EC-2-A-17-0233.

REFERENCES

Baillagou, P. E. and D. S. Soong (1985) Molecular weight distribution of products of free radical non-isothermal polymerization with gel effect-simulation of polymerization of poly(methylmethacrylate). *Chem. Eng. Sci.* **40**, 87-104.

Cao, S. and R. R. Rhinehart (1995) An efficient method for on-line identification of steady state. *J. Proc. Cont.* **5**(6), 363-374.

Chang, J. S., and S. T. Chung (1995) On-line set-point optimization and control of a batch polymerization reactor. *J. of the Chin. I. Ch. E.* **26**(4), 213-236.

Chang, J. S. and W. J. Chen (1992) Simulation of a jet-assisted circulation jacketed batch reactor for polymerization of methylmethacrylate. *Canadian J. of Chem. Eng.* **70**, 523-533.

Chiu, W. Y., G. M. Carrat and D. S. Soong (1983) A computer model for the gel effect in free-radical polymerization. *Macromolecule* **16**, 348-357.

Hoskuldsson, A. (1988) PLS regression methods. *J. of Chemometrics* **2**, 211-228.

Koutri, T., P. Nomikos and F. MacGreGor (1995) Analysis, monitoring and fault diagnosis of batch processes using multiblock and multiway PLS. *J. Proc. Cont.* **5**(4), 277-284.

Lakshminarayanan, S., S. L. Shah and K. Nandakumar (1997) Modeling and control of multivariable processes: dynamic PLS approach. *AIChE J.* **43**(9), 2307-2322.

MacGreGor, F., C. Jaeckle, C. Kiparissides and M. Koutoudi (1994) Process monitoring and diagnosis by multiblock PLS methods. *AIChE J.* **40**(5), 826-838.

Martin, E. B., A. J. Morris, M. C. Papazoglou and C. Kiparissides (1996). Batch process monitoring for consistent production. *Computers Chem. Engng* **20**, Suppl., S599-S604.

Nomikos, P. (1996) Detection and diagnosis of abnormal batch operations based on multiway principal component analysis. *ISA Transactions* **35**, 259-266.

Nomikos, P. and F. MacGreGor (1994) Monitoring batch processes using multiway principal component analysis. *AIChE J.* **40**(8), 1361-1375.

Nomikos, P. and F. MacGreGor (1995) Multivariate SPC charts for monitoring batch processes. *TECHNOMETRICS* **43**(1), 41-59.

Wold, S. P., G. K. Esbensen and J. Ohman (1987) Multiway principle components and PLS analysis. *J. of Chemometrics* **1**, 41.

SERIAL IDENTIFICATION WITH COLLECTIVE ESTIMATION OF GROSS ERRORS AND LEAKS IN PROCESS PLANT MEASUREMENT

Miguel Bagajewicz and Qiyou Jiang

University of Oklahoma, CEMS, 100 E. Boyd St, T-335, Norman, OK 73019, USA

Abstract: In this paper we review a model for collective compensation, we present the theory of gross error equivalency and a serial identification with collective compensation method (SICC) to determine gross error in steady state data reconciliation. A comparison of this strategy with other collective estimation strategies is shown. Finally, a new method based on collective identification and collective compensation using a MILP model is presented. *Copyright © 1998 IFAC*

Keywords: Data Processing, Fault Detection, Fault Isolation, Fault Location.

INTRODUCTION

The presence of gross errors invalidates the statistical basis of data reconciliation techniques, either for techniques that assume steady state or perform data reconciliation using dynamic data. For this reason, gross errors must be detected, identified and then either eliminated or corrected before a valid data reconciliation can be performed. Gross error detection has been the object of intense research with various degree of success. As methods based on steady state data reconciliation have not proven to have high power in all situations, attention has shifted lately to the use of dynamic data reconciliation based techniques.

In this paper we review a model for collective compensation recently introduced for steady state data reconciliation (Sanchez and Romagnoli, 1994; Sanchez, 1996) and independently developed by Bagajewicz and Jiang (1998) for dynamic systems. Next, we present the theory of gross error equivalency and a serial identification with collective compensation method (SICC) to determine gross error in steady state data reconciliation (Bagajewicz and Jiang, 1998). After this review a comparison of this strategy with other collective estimation strategies is shown. Finally, a new method based on collective identification and collective compensation is presented.

COLLECTIVE COMPENSATION

If leaks and measurement biases combinations are considered, both the measurement model and process constraints model are modified. The formulation for the least square problem is as follows:

$$\begin{aligned} &\underset{\varepsilon,m_p,m_b}{Min}\ \varepsilon^T \Psi^{-1} \varepsilon \\ &s.t. \\ &A(x^+ - \varepsilon - B_{rb} m_b) - B_{rp} m_p = 0 \end{aligned} \qquad (1)$$

where A is the incidence matrix of the system, x^+ is the vector of measurements, Ψ the covariance matrix and ε the vector of adjustments. In turn, B_{rp} is a (m x p) matrix with e_i column vectors indicating leak positions, m_p is the p dimensional vector of leak magnitudes, B_{rb} is a (n x s) matrix with e_j column vectors indicating the positions of proposed bias and m_b is the s dimensional vector of bias magnitudes. The solution is analytical and is omitted for space reasons.

EQUIVALENCY OF GROSS ERRORS

Two sets of gross errors are equivalent when they have the same effect in data reconciliation, that is, when simulating either one, leads to the same value of objective function. Therefore, the equivalent sets of gross errors are theoretically undistinguishable. In other words, when a set of gross errors is identified, there exists an equal possibility that the true locations of gross errors are in one of its equivalent sets. From the view of graph theory, equivalent sets exist when candidate stream/leaks form a loop in an augmented graph consisting of the original graph representing the flowsheet with the addition of environmental node. This is illustrated next through an example.

Consider the process of Figure 1 and assume that all streams are measured. Assume that the true values for each stream are 12,18,10,6,6,2. Standard deviations were taken as 2% of the true flow values. Measurements for S_1, S_3, S_6 are 12,10 and 2. For clarity, we ommit including random noise. As shown in Table 1, a bias of (-2) in S_4 and a bias of (+1) in S_5 (Case 1) can be represented by two alternative sets of two gross errors (Cases 2 and 3).

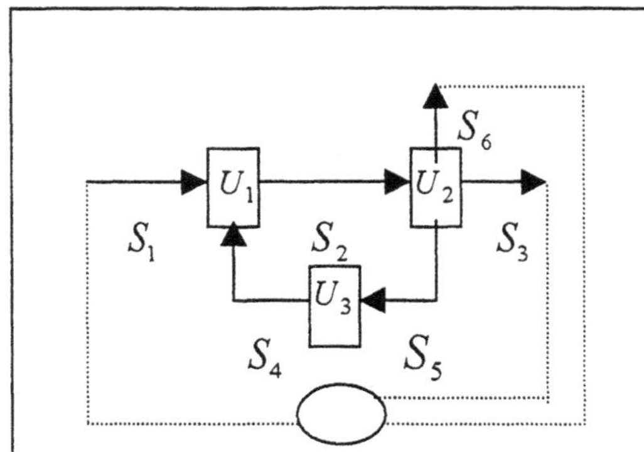

Fig. 1. Illustration of Gross Error Equivalency

Table 1 Illustration of Equivalent Sets in $\{S_2, S_4, S_5\}$ of Fig. 1

		S_2	S_4	S_5
Measurement		18	4	7
Case 1 Bias in S_4, S_5	Reconciled data	18	6	6
	Estimated biases		-2	1
Case 2 Bias in S_2, S_4	Reconciled data	19	7	7
	Estimated biases	-1	-3	
Case 3 Bias in S_2, S_5	Reconciled data	16	4	4
	Estimated biases	2		3

The above concepts are formalized using the following definitions:

Gross Error Cardinality: A set of variables has Gross Error Cardinality Γ=t when every possible set of gross errors in the system can always be equivalent to at least one set of t or less nonzero gross errors.

Basic Subset: A set of variables constitutes a Basic Subset of a system, when every set of gross errors is equivalent to a set of gross errors in the basic set.

As a corollary of this definition of basic subsets, we must note that a) for every system, more than one Basic Subset may exist, and b) all Basic Subsets are equivalent to each other, provided the right size of gross errors is used.

The above concepts implicitly suggest that a set of gross errors can be represented by different sets of gross errors, as illustrated above. In addition, the following theorem suggests a way to calculate the cardinality of a set.

Gross Error Cardinality Theorem: Let m columns $[a_1 \quad \cdots \quad a_m]$ of the system matrix A correspond to a set of flowrate variables Λ. The set has Gross Error Cardinality $\Gamma=t$ if rank $[a_1 \quad \cdots \quad a_m]=t$.

Degenerate Cases

The equivalencies above are built in the assumption that the number of gross errors identified is equal to the real number of gross errors. However, there are examples where the actual number of gross errors can be larger than the number of gross errors identified. One such example is shown in Table 2, where a set of two gross errors (Case 1) is equivalent to one gross error (Case 2). These cases are rare, as they require that the two real gross errors have equal sizes. This poses an additional uncertainty. If a certain number of gross errors are detected, there is a possibility that the actual number of gross errors is larger.

Table 2 Illustration of Degenerate Cases in $\{S_2, S_4, S_5\}$ of Fig. 1

		S_2	S_4	S_5
Measurement		18	7	7
Case 1 Bias in S_4, S_5	Reconciled data	18	6	6
	Estimated biases		1	1
Case 2 Bias in S_2	Reconciled data	19	7	7
	Estimated biases	-1		

Consider now that a certain set of gross errors was found. These variables are a basic subset of some system. The question is to determine what this system is. To perform this, the following theorem is needed.

Loop Theorem: Every loop of m streams is a set with Gross Error Cardinality $\Gamma=m-1$.

Thus the set containing all equivalent basic subsets can be found by adding one by one all variables that form a loop with the basic subset found.

We consider the environmental node for loop identification. For example, in the system of Figure 1, streams S_1, S_2, S_3 form as loop including units U_1, U_2 and the environmental node.

THE CONCEPT OF PERFECT IDENTIFICATION

As a result of the equivalency of gross errors, the issue of perfect identification of gross errors should be revised. This issue has been raised by Bagajewicz and Jiang (1998), when presenting the equivalency theory, where the concepts of gross error exact detectability and gross error correct detectability were introduced. Exact detectability is possible in principle, only when there are no equivalent sets. Thus the perfect identification performance measure (OPF), as defined by Rollins and Davis should be extended to include equivalent sets. That is, gross error identification should be considered successful if exact errors introduced are identified, or an equivalent set is identified. Degenerate cases can also be included in such counting. For example, assume a gross error is introduced in stream S_2 in the flowsheet shown in Figure 1. Assume now that a certain strategy identifies two equal and opposite sign gross errors in streams S_4 and S_5. This is a degenerate case where two gross errors are exactly equivalent to one gross error. Such a result should not be considered a failure and for this paper it will be considered a success.

Remark: These equivalencies explain why the power of gross error detection tests might have been wrongly assessed in the past, as only exact identification was considered a success

THE SICC STRATEGY

The SICC strategy (Bagajewicz and Jiang, 1998) can be summarized as follows: Determine a list of suspect gross errors using the measurement test. Run the compensation model for each suspect and add to the list of confirmed gross errors the one that gives the smallest objective function. Determine a new list of gross errors. Run the compensation method using the confirmed gross errors and adding suspect candidates one at the time. The suspect gross error that produces the smallest objective function is added to the list of confirmed gross errors. Although only biases are detected, the equivalency theory proves that leaks are equivalent to a set of biases, and therefore can be detected once the equivalent set is constructed.

COMPARISON OF SICC WITH OTHER METHODS

Several methods exist to perform gross error detection. We have conducted studies verifying the performance of our method for bias detection comparing it with other steady state collective compensation techniques like the successive application of Generalized Likelihood Ratio (Narasimhan and Mah, 1987), presented by Keller et al (1994), which we will call collective GLR (CGLR), the unbiased estimation technique (UBET) (Rollins et al, 1992) and the MSEGE method (Sanchez et al, 1998). To make these comparisons, we propose a new performance assessment scheme.

The performance measures in the literature (AVTI, OP, OPF) are based on a strategy where random errors are changed but the size of the gross error simulated is fixed. Since the solution to the compensation model is linear in the measurements one can assess the variance of the gross errors estimates analytically. Thus, we propose a different set of performance measures. This set is based on the notion that the ability of locating a gross error depends on its size and the number of gross errors. Methods may not perform the same for large or small gross errors and with different number of gross errors. Thus, our performance indexes are based on introducing several random gross errors within a certain interval [s σ, (s+Δ)σ]. We again omit including random errors in this performance assessment. Thus we define the following:

Location Performance Index (LPI):

- $LPI(i,s,\Delta)$ = *Frequency of identification of the correct location of i gross errors*

By *correct* we mean the exact location of the gross errors or their equivalent set.

Accuracy Performance Index (API):

For each run, one can define the accuracy of gross error estimation as follows:

$$\gamma = 1 - \sum_{\forall i} \frac{\left|\delta_i - \hat{\delta}_i\right|}{\delta_i} \tag{2}$$

where δ_i are the gross errors introduced and $\hat{\delta}_i$ are the values of estimated gross errors. These values of γ are then averaged for all runs to obtain the *API*. In the cases of correct identification (equivalent sets being identified or degenerate equivalent sets) the values of the originally introduced sets are calculated.

The following table shows a comparison of the method for some selected cases. In order to compare results on the same basis, the level of significance of each method was chosen such that it gives an AVTI equal to 0.1 under the null hypothesis.

Table 3 Comparison of Performance for SICC, MSEGE, UBET and CGLR

		Number of Gross Errors		
		1	2	3
LPI	SICC	0.78	0.61	0.54
	MSEGE	0.73	0.59	0.47
	UBET	0.67	0.52	0.42
	CGLR	0.67	0.59	0.48
API	SICC	0.99	0.98	0.95
	MSEGE	1.00	0.98	0.94
	UBET	1.00	0.99	0.98
	CGLR	1.00	0.99	0.98

Note: σ=2%*measurement, s=10, Δ=5

The table was constructed using all combinations of gross errors, including leaks. The variance of the gross error estimates can be obtained analytically, as they are obtained from the measurements directly. They are, however, dependent of the position of the gross error. The next table illustrates two cases.

Table 4 Comparison of Biases Estimation Results for SICC, MSEGE, UBET and CGLR

		Biased Stream	
		S_2 (1.8)	S_3 (1.0) S_5 0.84)
Standard Deviation	SICC	0.126	0.091, 0.053
	MSEGE	0.126	0.091, 0.053
	UBET	0.132	0.098, 0.127
	CGLR	0.132	0.129, 0.054

All methods provided values of the estimates fairly close to the introduced. The table shows that SICC has smaller variance. The values for SICC and MSEGE are the same, as they use the same estimators.

SPANNING TREES AND GROSS ERRORS

A spanning tree is a basic set of maximum cardinality, that is, of cardinality equal to the number of units. Thus, one can run the compensation model using the spanning tree and capturing all gross errors. Those variables in the spanning tree that show a value lower than a threshold can be disregarded. The Bonferroni test (Rollins and Davis, 1992) can be used for such purpose.

MINIMUM CARDINALITY MODEL

Assume that a spanning tree has been used to identify all gross errors in the system. This set is represented by the set of biases $\hat{\delta}$. Assuming that a minimum number of gross errors are to be determined, the following problem determines which equivalent set to $\hat{\delta}$ contains the smallest number of gross errors.

$$
\begin{aligned}
& Min \sum_{k=1}^{Ns} Y_k + \sum_{k=1}^{Nu} Z_k \\
& s.t. \\
& \quad |\delta_k| - UY_k \le 0 \\
& \quad |\delta_k| - \varepsilon_k Y_k \ge 0 \\
& \quad |l_k| - UZ_k \le 0 \\
& \quad |l_k| - \varepsilon_k Z_k \ge 0 \\
& \quad A\hat{\delta} - A\delta - l = 0
\end{aligned}
\qquad (3)
$$

The first two constraints contain a binary variable Y_k which represents the presence of a gross error of size greater than $\delta_k \ge \varepsilon_k$ in stream k (U is a large number). Setting Y_k =0, makes $\delta_k = 0$, whereas Y_k =1 allows a value larger than ε_k. The same type of constraints is used for leaks. The objective function to minimize is simply the number of gross errors. The last constraint represents the equivalency between the values of the gross errors in the spanning tree $\hat{\delta}$ and the set of new, hopefully smaller set included in the set (δ, l).

Remark: The above strategy requires solving the estimation model and the minimum location cardinality model. *It does not include the use of any test statistics. In fact, the only assumption about the probability distribution of errors is included in the objective function of the estimation model, where the maximum likelihood of a normal distribution is assumed.* If one manages to generalize the estimation model to make it distribution independent, one would have a robust estimation scheme.

To solve this problem the following is used: $|\delta| = Max(\delta, -\delta)$. In turn, the *Max operator* is replaced by a set of mixed integer linear constraints proposed by Bagajewicz and Manousiouthakis (1992). This makes the above problem an MILP problem.

ILLUSTRATION

The following table shows the results of the application of the technique for the case of figure 1. Assume that one spanning tree (S_1, S_2, S_5) is found.

Table 5 Gross Error Identification with the Minimum Cardinality Model

No.	Gross Error Introduced	Gross Error Identified with Spanning Tree	Minimum Number of Gross Errors
1	S_2 (0.5))	0.0, 0.5, 0.0	S_2 (0.5))
2	S_4 (0.5)	0.0, -0.5 , -0.5	S_4 (0.5)
3	U_2 (0.5)	0.5, 0.5, 0.0	U_2 (0.5)
4	S_1 (0.5) S_2 (1.0)	0.5, 1.0, 0.0	S_2 (0.5) S_3 (-0.5)
5	S_5 (0.5) U_2 (1.0)	1.0, 1.0, 0.5	U_2 (0.5) U_3 (0.5)
6	S_1 (0.5) S_2 (1.5) S_4 (0.5)	0.5, 1.0, -0.5	S_2 (0.5) S_3 (-0.5) S_5 (-0.5)
7	S_1 (0.5) S_2 (1.0) S_4 (0.5)	0.5, 0.5, -0.5	S_3 (-0.5) S_5 (-0.5)

In Run 1, 2 and 3, exact errors are identified. In Run 5 and 6 , one obtains equivalent sets. Run 7 is a degenerate case.

CONCLUSION

All the theory developed proves that exact identification of the location of gross errors, be it biases or leaks, is not always theoretically possible. Instead a set of measurements and possible leaks is identified, out of which many alternative sets containing the same number of gross errors identified can be selected. For this reason, successful identification is classified as exact when the real gross errors are detected, or correct when an equivalent set is identified. Thus, at the end of any successful identification, a number of equivalent alternatives exist. Our method (SICC) is capable of identifying such alternatives for both steady state and dynamic integral data reconciliation.

This paper unveils for the first time the theoretical intricacies of multiple gross error detection and presents a coherent methodology to identify leaks and biases simultaneously in steady state models.

In addition a new method is presented based on the notion that a spanning tree can be used in the compensation model to capture all gross errors and a minimum cardinality model can be used to select a basic equivalent subset containing the smallest number of gross errors.

REFERENCES

Bagajewicz M. and Q. Jiang, "An Integral Approach to Plant Linear Dynamic data Reconciliation", *AIChE J.*, 43, 2546(1997).

Bagajewicz M. and Q. Jiang, "Gross Error Modeling and Detection in Plant Linear Dynamic Reconciliation", Submitted to Computers and Chemical Engineering.(1998).

Bagajewicz M. and V. Manousiouthakis, *Mass/Heat-Exchanger Network Representation of Distillation Networks.* AIChE Journal, Vol., 38, 1769-1800, (1992).

Keller J. -Y., M. Darouach and G. Krzakala, "Fault Detection of Multiple Biases or Process Leaks in Linear Steady State Systems", 18, 1001(1994).

Narasimhan, S. and R. S. H. Mah, "Generalized Likelihood Ratio Method for Gross Error Detection", *AIChE J.*, **33**,1514 (1987).

Rollins D. and J. Davis. Unbiased Estimation of Gross Errors in Process Measurements. *AIChE J.*, **38**, 563-572 (1992)

Sánchez M., PhD Thesis, Universidad Nacional del Sur (Argentina) (1996).

Sánchez M. and J. Romagnoli. Monitoreo de Procesos Continuos: Análisis Comparativo de Técnicas de Identificación y Cálculo de Bias en los Sensores, *AADECA 94 - XIV Simposio Nacional de Control Automático* (1994).

Sánchez M., J. Romagnoli, Q. Jiang and M. Bagajewicz, "Simultaneous Estimation of Biases and Leaks in Process Plants", Submitted to Computers and Chemical Engineering (1998).

AUTOMATED INTERPRETATION OF PCA-BASED PROCESS MONITORING AND FAULT DIAGNOSIS USING SIGNED DIGRAPHS

Hiranmayee Vedam, Venkat Venkatasubramanian

Laboratory for Intelligent Process Systems
School of Chemical Engineering
Purdue University
W. Lafayette, IN 47907 - 1283, U.S.A

Abstract: Principal component analysis (PCA) has emerged as a useful tool for process monitoring and fault diagnosis. The general approach requires the user to identify the root cause by interpreting the measured variable contributions to the residual and/or principal components. This could be tedious and often impossible for a large process. It also hinders the automation of high level supervisory tasks like choice of corrective actions and their associated costs. In this paper, the interpretation of the PCA-based contributions is automated using signed directed graphs (SDGs). The implementation of the PCA-SDG based fault diagnosis algorithm is done using Gensym's expert system shell G2. Its application is illustrated on the Amoco Model IV Fluidized Catalytic Cracking Unit (FCCU). *Copyright © 1998 IFAC*

Key Words: Fault diagnosis, Multivariate Statistics, Qualitative analysis, Digraphs

1. INTRODUCTION

Abnormal situations that occur due to sensor drifts, equipment and instrument failures or changes in process parameters are manifested through changes in sensor measurements. These abnormal events, though infrequent, have significant economic and safety impact. It is estimated that the petrochemical industry in the U.S. incurs approximately $20 billion in losses due to poor abnormal situation management (ASM) (Nimmo, 1995). Since information about the root cause of the abnormality is necessary for effective ASM, early detection and diagnosis of these abnormalities has received significant attention in the recent years (Venkatasubramanian et al., 1995).

Due to the availability of significant amount of data for the normal operation of chemical processes, statistical techniques are very attractive for early detection of abnormal situations. These techniques can be broadly classified as univariate and multivariate statistical techniques. While univariate techniques are easy to implement, multivariate techniques like Principal Component Analysis (PCA) and Partial Least Squares (PLS) are more accurate as they monitor the correlation among different variables in addition to the variables themselves. Process fault detection using PCA/PLS is performed by monitoring the error in model predictions. An abnormal situation is flagged whenever this error is statistically significant. Recently, PCA/PLS based methods have been extended to perform fault diagnosis by MacGregor *et.al* (1994), Dunia *et.al* (1996) and others. These methods involve manual interpretation of the measured variables with significant contributions to the error in model prediction. However, the manual interpretation of these contributions for a large process is tedious and often impossible. In addition, manual interpretation makes the automation of supervisory tasks like the choice of corrective actions and their associated costs difficult. Hence, automated interpretation of these measured variables is desirable. In this work signed digraphs are used for this interpretation.

A signed digraph (SDG) is a qualitative model of the causal interactions between different variables in the process. Fault diagnosis using SDGs is relatively inexpensive in terms of computation and guarantees completeness, i.e., it guarantees that the actual root cause is a subset of the proposed set of root causes.

By using SDGs for the automatic interpretation of PCA-based contributions, the accuracy and ease of fault diagnosis is increased. The algorithm can also identify faults early, since the measured variable contributions reflect changes in correlation information. This gives operators more time to react to abnormal situations. The implementation of PCA-SDG based fault diagnosis algorithm is done using Gensym's expert system shell G2. Its application is illustrated on the Amoco Model IV Fluidized Catalytic Cracking Unit(FCCU) (McFarlane et al., 1993).

In Section 2, the development of PCA model and its use in performing fault diagnosis are discussed. Details of SDG based fault diagnosis follow in Section 3. The PCA-SDG based fault diagnosis algorithm is given in Section 4. The application of the algorithm to the Amoco Model IV FCCU is discussed in Section 5.

2. PCA AND FAULT DIAGNOSIS

PCA is a statistical modeling technique which finds the directions of significant variability in the data by forming combinations of variables. Consider a $m \times n$ matrix of data $\mathbf{X}$, of m samples belonging to n variables. $\mathbf{X}$ is assumed to be normally distributed with zero mean and unit variance. Define the covariance matrix of X as

$$cov(\mathbf{X}) = \frac{\mathbf{X}^T\mathbf{X}}{m-1}$$

PCA decomposes $\mathbf{X}$ as follows:

$$\mathbf{X} = \hat{\mathbf{X}} + \mathbf{E}$$

where

$$\hat{\mathbf{X}} = \sum_{i=1}^{r} \mathbf{t_i p_i} \qquad r \leq n \qquad (1)$$

and $\mathbf{E}$ is the error in model prediction. The vectors $\mathbf{t_i}$ are known as scores and contain information about the noise in the process. The vectors $\mathbf{p_i}$ are the loadings and contain information about how the variables relate to each other. They are the eigen vectors of the covariance matrix i.e.,

$$cov(\mathbf{X})\mathbf{p_i} = \lambda_i \mathbf{p_i}$$

The eigen vectors corresponding to the r largest λ_i are used in Equation 1. The scores, $\mathbf{t_i}$ are calculated as,

$$t_i = \mathbf{X}p_i$$

For a new sample $\mathbf{x}$, the error e is estimated using

$$\mathrm{e} = (\mathbf{x} - \hat{\mathbf{x}}) \qquad (2)$$

The residual R corresponding to the sample $\mathbf{x}$ is given by

$$R = \mathrm{ee}^{\mathrm{T}} \qquad (3)$$

A confidence limit on R can be calculated as discussed in Jackson and Mudholkar (1979).

Using the above equations, a PCA model of the normal operating conditions in a process can be built. To perform fault diagnosis using this model, MacGregor and coworkers (1994) suggested that whenever the scores and/or the residual, R of a new sample, $\mathbf{x}$, exceed their confidence limits, the contributions of measured variables to that deviation should be examined and interpreted manually to identify the root cause. In order to identify the fault origin, the contributions of each variable to the magnitude of the error e, i.e., the square root of the residual R are estimated using Equation 2 and interpreted manually. However, for a large process with large number of variables, this is tedious and often impossible. Further, the contribution plots often appear to be similar, thus making manual interpretation difficult. Hence, automated interpretation of these measured variables is very desirable. In this work signed digraphs are used for this interpretation. In the next section, SDGs and the algorithm to perform fault diagnosis using them are discussed.

3. SDG-BASED FAULT DIAGNOSIS

A signed directed graph is a representation of the process causal information, in which the process variables (and parameters) are represented as graph nodes and causal relations are represented by directed arcs. Nodes in the SDG assume values of (0), (+) and (-) representing the nominal steady state value, higher and lower than steady state values respectively. Directed arcs point from a cause node to its effect node. Arc signs associated with each directed arc can take values of (+) or (-) representing whether the cause and effect change in the same direction or opposite direction respectively. A SDG may also include conditional arcs which become active only if certain conditions are satisfied. For example, the arc connecting a manipulated variable to the controlled variable is active only if the controller is not in manual mode.

A SDG for the process can be developed from model equations of individual units in the process using the automated framework developed by Mylaraswamy (1994). In the absence of model equations, the procedure discussed in Vedam and Venkatasubramanian (1997) can be used.

One of the concerns in using qualitative methods like SDGs is the generation of large number of spurious solutions, leading to poor resolution. The resolution of the SDG-based diagnosis is improved using a knowledge base as discussed in Vedam and Venkatasubramanian (1997).

Iri et.al (1979) proposed an algorithm for using SDG for fault diagnosis. The inherent assumption in their approach was that a single root cause which can explain the given abnormal situation can be found. They also assume that there exists a valid causal path between the root node and the observed abnormal measurements. A root node is any node in the digraph which has at least one consistent arc connecting it to an effect node and no consistent arc connecting it to a cause node. An arc is said to be consistent if sign(cause) * sign(arc) * sign(effect) = (+). Wilcox and Himmelblau (1994) presented a new digraph-based diagnosis reasoning called possible cause-effect graph. This approach reduces the search space and hence the number of root causes generated. Based on these approaches the basic algorithm for performing single fault diagnosis using SDG can be summarized as follows :

(1) Propagate the deviation in the nodes representing process measurements (measured nodes) from effect to cause node via consistent arcs till the root nodes are identified.
(2) Use the knowledge base to screen out physically impossible root nodes.
(3) If a conflict exists in assigning a sign to any node then a voting scheme is used.
(4) For each of the root nodes a breadth-first search is performed to check if a valid causal path exists between the root node and the observed abnormal measurements.
(5) A root node is identified as the root cause for the abnormal situation if such a causal path exists.

In the SDG based diagnosis discussed above, the node deviations from steady state (signs) for the measured nodes are obtained by comparing the deviations from steady state with a preset threshold, to account for the noise. If the deviation is larger than the high threshold, a positive sign is assigned to the measured node and a negative sign is assigned if the deviation is smaller than the lower threshold. For n measured nodes $2n$ thresholds should be chosen *a priori* based on the knowledge of the process and the desired accuracy of diagnosis. This selection, which is done off-line is very time consuming and can introduce significant inaccuracies, if done incorrectly. In the next section, we will show how the number of thresholds can be reduced to one while using SDGs for the interpretation of measured variable contributions to the error associated with the PCA model.

4. SDGS FOR THE INTERPRETATION OF PCA-MODEL BASED ERROR CONTRIBUTIONS

In this Section, we will discuss the use of SDG for identifying root causes based on the contribution of measured variables to the error in PCA model prediction. The PCA model is constructed as discussed in Section 2. The process is said to have deviated from its normal behavior if the residual, R, exceeds the 95% confidence limit. In order to identify the root cause for such an abnormality, several fault diagnosis techniques can be used. These techniques could be based on historical data pertaining to known faults or on process models. Since the historical data pertaining to known faults is limited, in this work the SDG based fault diagnosis method is used. The SDG for the process can be built as discussed in (Mylaraswamy, 1996).

The measured variable deviations for the SDG-based diagnosis are obtained from their contributions to the error, $\mathbf{e}$, calculated using Equation 2 as discussed in Section 2. The error vector $\mathbf{e}$ is normalized by its magnitude to obtain the normalized contribution of each measured variable. In order to identify measured variables with significant contribution to the error, a threshold, t which specifies the meaning of "significant" is chosen. If the contribution of a measured variable is larger than the threshold, t, and positive, then the corresponding node in the SDG receives a node sign of +. If its contribution is larger than the threshold and its sign is negative, then the corresponding node sign is set to -. All measured nodes with contribution between $\pm t$ % of the magnitude of error receive a node sign of 0. The fault diagnosis algorithm described in Section 3 is then used to identify the root causes.

5. FAULT DIAGNOSIS CASE STUDY

In this Section, the application of the PCA-SDG based algorithm is illustrated on the Amoco Model IV FCCU. CATSIM (Mylaraswamy, 1996), a simulator based on FCCU model in McFarlane *et.al.* (1993), is used in this work. A number of faults can be simulated in CATSIM. These faults range from simple sensor biases to more complex process faults that affect the catalyst circulation and the delicate carbon balance in the system. There are 20 measured variables in the process.

Gaussian noise with zero mean and known standard deviations is added to the measured values.

The PCA model for the normal operating condition of the process is built using 114 samples in MATLAB PLS toolbox (Wise and Gallagher, 1996) as discussed in Section 2. The principal components and the variance explained by each principal component (PC) are shown in Table 1.

Based on variances shown in Table 1, the first three principal components are used for defining the normal region of operation.

The SDG model for the process is built from the model equations in Gensym's expert system shell G2. Figure 1 shows the SDG model for the preheater section of the FCCU. Please refer to Mylaraswamy (1996) for the SDGs of the other sections in FCCU. The threshold, t discussed in Section 4 is selected to be 20 based on the historical data available. The knowledge base used to improve the resolution of the diagnosis contains 20 rules that specify infeasible faults and faults with low probability. The PCA-SDG based fault diagnosis algorithm has been tested on 38 different fault scenarios. The algorithm has been able to correctly identify the fault in 95% of the cases tested within 1-8 samples after the introduction of the fault. The resolution of the algorithm, which specifies the number of faults identified for a given scenario, ranges from 1 to 12 faults. The results for all the faults cannot be provided here due to space constraints and only two examples that illustrate the importance of PCA-SDG based fault diagnosis algorithm are discussed below.

Example 1

The first example corresponds to 10% increase in gas oil flow rate. The FCCU operates normally for the first 14 samples at which time, the fault is introduced. The plots of the residual and the contributions of the sensors for sample number 30, normalized by the magnitude of the error is shown in Figure 2. The output of the PCA-SDG based fault diagnosis algorithm is shown in Table 2.

Example 2

In this example, a 12.5 % decrease in catalytic particle density is introduced as discussed in Example 1. The plots of the residual and the contributions of the sensors for sample number 30 normalized by the magnitude of the error is shown in Figure 3. The output of the PCA-SDG based fault diagnosis algorithm is shown in Table 3.

As shown in Figure 2 and 3, dissimilar faults can have similar contribution plots, which can lead to a wrong diagnosis if their interpretation is done manually. By using SDGs for interpreting them, the correct faults can be identified quite easily as shown in the examples above.

6. CONCLUSIONS AND FUTURE WORK

In this paper, a SDG based fault diagnosis algorithm has been used for automatic interpretation of PCA model based measured variable contributions. The application of the algorithm has been illustrated on the Amoco Model IV FCCU. Automatic interpretation of PCA-based error contributions leads to more accurate and easier diagnosis of abnormal situations. Since PCA identifies the changes in measured variables and the changes in their correlations, the PCA-SDG approach can identify the faults at an early stage. This allows the operators more time to react, thus averting their more serious consequences. The approach also reduced the number of thresholds to be chosen for SDG based diagnosis to one, making the threshold selection easy. Hence, the PCA-SDG based fault diagnosis algorithm is easier to build and can perform more accurate diagnosis than either of the methods individually.

While the PCA-SDG algorithm above performs well for single fault diagnosis, it can result in wrong or no diagnosis when multiple faults occur. Currently, we are extending the PCA-SDG based fault diagnosis algorithm to perform multiple fault diagnosis.

7. REFERENCES

Dunia, R., S. J. Qin, T. F. Edgar, and T. J. McAvoy, Identification of faulty sensors using principal component analysis. *AIChE J.*, **42**(10), 2797–2812 (1996).

Iri, M., K. Aoki, E. O'Shima, and H. Matsuyama, An algorithm for diagnosis of system failures in chemical processes. *Comput. & Chem. Engng.*, **3**, 489–493 (1979).

Jackson, J. E. and G. S. Mudholkar, Control procedures for residuals associated with principal component analysis. *Technometrics*, **21**(3), 341–349 (1979).

MacGregor, J. F., J. Christiana, K. Costas, and M. Koutoudi, Process monitoring and diagnosis by multiblock PLS methods. *AIChE J.*, **40**(5), 826–838 (1994).

McFarlane, R., R. Reineman, J. Bartee, and G. Georgakis, Dynamic simulator for a model iv fluid catalytic cracking unit. *Comput. & Chem. Engng.*, **17**(3), 275–300 (1993).

Mylaraswamy, D., *DKIT: A Blackboard-based, distributed, multi-expert environment for Abnormal Situation Management.* PhD thesis, Purdue University (1996).

Mylaraswamy, D., S. N. Kavuri, and V. Venkatasubramanian, A framework for automated development of causal models for fault diagnosis. AIChE Annual Meeting San Francisco (November, 1994).

Nimmo, I., Adequately address abnormal situation operations. *Chem. Eng. Prog.*, **91**(9), 36–45 (1995).

Vedam, H. and V. Venkatasubramanian, Signed digraph based multiple fault diagnosis. *Comput. & Chem. Engng.*, **21**(13), S655–S660 (1997).

Venkatasubramanian, V., S. N. Kavuri, and R. Rengaswamy, Process fault diagnosis - An Overview. CIPAC Tech. Report, Purdue University (1995).

Wilcox, N. A. and D. M. Himmelblau, The possible cause-effect graph model for process fault diagnosis - I. Methodology. *Comput. & Chem. Engng.*, **18**(2), 103–116 (1994).

Wise, B. M. and N. B. Gallagher, *PLS Toolbox: for use with MATLAB*. Eigenvector Technologies (1996).

Table 1. Principal Components for the Normal data of Model IV FCCU

No. of PC	% Variance Captured by this PC	% Total Variance Captured
1	89.3482	89.3482
2	8.6419	97.9901
3	1.7846	99.7747
4	0.1979	99.9726

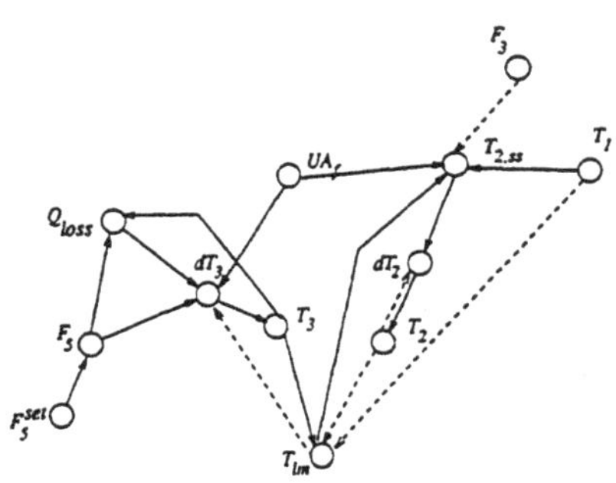

Fig. 1. SDG for Preheater Section of ModelIV FCCU

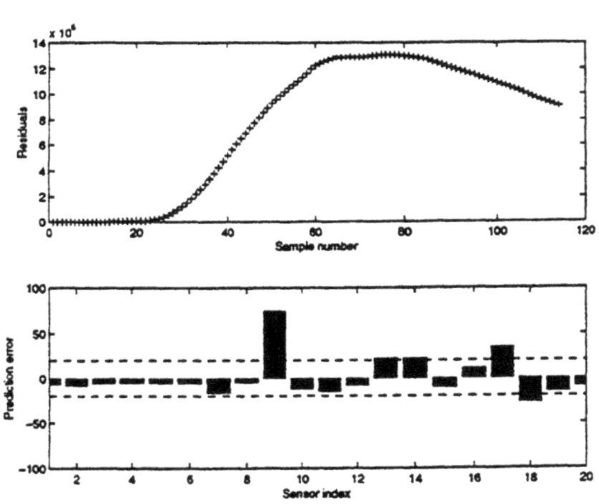

Fig. 2. Residual and Contribution Plots for 10% Increase in Gas Oil Flow Rate

Table 2. PCA-SDG results for Example 1

Fault Introduced	Faults Identified
Increase in Gas oil flow rate	Gas oil flow rate is HIGH
	Heat loss from furnace to environment is HIGH

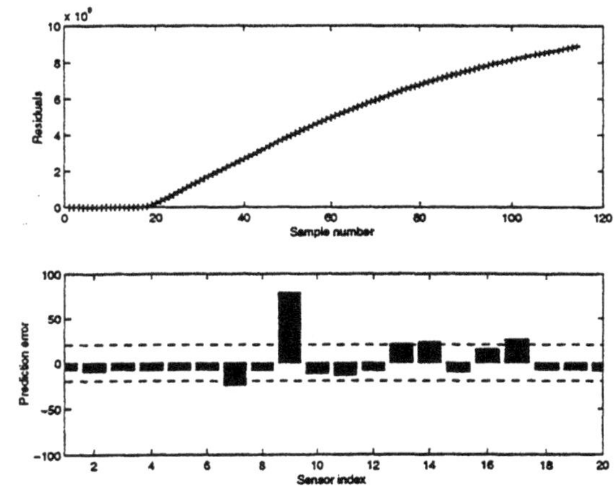

Fig. 3. Residual and Contribution Plots for 12.5 % decrease in catalytic particle density

Table 3. PCA-SDG results for Example 2

Fault Introduced	Faults Identified
Decrease in catalytic particle density	Catalytic particle density is LOW
	Atm. temp. is HIGH
	Slurry recycle Flowrate is HIGH
	Gas oil flow rate is HIGH
	Overflow factor is HIGH
	Promoter activity is LOW

Fault Diagnosis Expert System with Probability Calculations

D Leung, J Romagnoli

Laboratory of Process System Engineering
Department of Chemical Engineering
The University of Sydney
email : jose@chem.eng.usyd.edu.au

Abstract: This paper described a fault diagnosis expert system based on Possible Cause and Effect Graph methodology which is an enhanced Signed Digraph approach. This expert system incorporated Bayesian belief theorem and explanation capacity. Causal relationships in single loop and cascade controllers were re-modeled to suit the implementation. In dealing with recycle loops between process variables, a knowledge-base containing rules was employed to "break" the cyclic loop dynamically. During the diagnosis phase, the system dynamically modifies the causal network and adjusts the conditional probability using all available plant information and other processing techniques such as data reconciliation. This expert system has been implemented successfully on a pilot scale distillation column, with very promising results.

Keywords: Expert systems, fault diagnosis, inference engines, knowledge engineering, probability, rule-based system

1. INTRODUCTION

Fluctuations in the demand of chemical products of different purities and tougher environmental restrictions have resulted in more advanced and complex chemical operations. Consequently, chemical operations are now more potentially vulnerable to process upsets and hardware failure.

With the advent of artificial intelligence and computational technology, numerous researches have been directed towards designing automated fault diagnosis systems. They can be categorised as knowledge-based, model-based or pattern recognition-based (Venkatasubramanian 1994), each having their own advantages and drawbacks. Wilcox and Himmelblau (1994a, b) suggested a very effective and simple approach which is known as Possible Cause and Effect Graph ("PCEG") methodology. In this report, an implementation of an expert system on fault diagnosis of a pilot distillation process using PCEG model is described. Belief network and explanation facilities are also incorporated into this PCEG model to further enhance the diagnosis capability of the system.

2. BACKGROUND

2.1 PCEG Model

PCEG (Wilcox and Himmelblau, 1994a, b) approach is a modified Signed Digraph ("SDG") approach (Mah, 1990; Kramer and Palowitch 1987). Being a SDG based model, PCEG model shares the properties of its ancestor: ability to handle coupled systems and missing information, capability of locating all possible causes, and ease of construction from the first principle. PCEG model also has the advantage of providing more meaningful state description of the actual process, making it more user-friendly, while traditional SDG approach only uses "high", "normal" and "low" in describing the states of process variables. Another improvement in the PCEG model is the clear definition of exogenous (root) causes in the knowledge; whereas, in SDG, abnormal states can be the possible root cause of the abnormality, making the diagnosis results sometimes meaningless and incomplete (ambiguous outcomes).

3. ENHANCEMENTS OVER ORIGINAL PCEG

In our system, a few enhancements have been incorporated into the original PCEG methodology to improve the capacity of the expert system.

3.1 Bayesian Belief Network

The PCEG original implementations (Wilcox and Himmelblau, 1994b) returned a list of exogenous causes. In a large complex system, the sole provision of a list of exogenous causes is not effective in pinpointing the real causes. Since there is usually not enough manpower to investigate all possible causes simultaneously, information regarding the likelihood of all possible exogenous causes becomes very beneficial in locating the real causes in the process without delay.

In the belief network, information of operating conditions, equipment status, and theoretical and empirical equations describing the process variables is used as evidence to alter the conditional probability between nodes in the causal network.

Let $\mathbf{N_i}$ be the node and **E** be the evidence of the system. The probability of the exogenous causes is calculated using Bayesian theorem:

$$\Pr\{N_i|E\} = \frac{\Pr\{N_i\}\Pr\{E|N_i\}}{\sum_{N_i}\Pr\{N_i\}\Pr\{E|N_i\}} = \alpha \Pr\{N_i\}\Pr\{E|N_i\}$$

... Equation 3-1

Calculation of belief network is straight-forward in singlely connected network, such as the goal tree success tree ("GTST") approach (Chen and Modarres, 1994). For cyclic, multiply connected network, the problem is more complicated.

When the nodes are multiply connected, special methods such as clustering or conditioning are required (Russell and Norvig, 1995). This expert system employs depth-first search. Each branch of the causal network is calculated individually. Summation of these individual probabilities of an exogenous cause (the bottom node) is mathematically equivalent to performing conditioning.

Cyclic loops between process variables are common in causal networks for any chemical process. They pose difficulties for on-line belief network calculations. To simplify the problem, the expert knowledge comes into play. The states of process variables that are not involved in the cyclic loops can sometimes provide vital information of the possible process upset scenarios. Special rules, modelled from experience plant experts, were employed to make slight on-line adjustments (mainly arc creation and deletion) on the causal network.

When the process returns to normal operating condition, the expert system will re-establish all the causal networks for future on-line modification.

3.2 Explanation Message

A fault diagnosis system is not considered to be a complete system without explanation capability. Explanation message is generated in our system whenever an exogenous cause is found in the search. The message describes the cause-and-effect paths between the observed abnormal process variable and the diagnosed exogenous fault. The level of details of the messages can be set before the start of diagnosis.

4. PROCESS

4.1 Process Description

The pilot distillation plant is a 12-tray distillation column used to separate ethanol from water. Feed is pumped from the feed tank and is introduced into the column at Tray-6 after being preheated by feed preheater. At the top of the column, vapour is being condensed by water-cooled condenser. The condensed steam is split into distillate, which is returned into the feed tank, and reflux, which is being recycled into Tray-1 by reflux pump. Some of the liquid leaving the bottom of the column as bottom product is cooled and returned into the feed tank. The remaining liquid will be transferred into the bottom steam reboiler. The vapour is then recycled into the bottom of the column.

4.2 Instrumentation Architecture

ABB MOD 300 DCS was used for data acquisition and regulatory control. The feed flowrate was being controlled by a PID feedback controller. Two cascade controllers controlled the product specifications by manipulating Tray-5 and Tray-12 temperatures. Two other cascade controllers controlled the levels of the condenser and column bottom. The expert system gathered plant data from a VAX workstation, which acted as a data server. Figure 4.1 illustrates the system architecture.

Figure 4-1: Instrumentation architecture of the distillation process

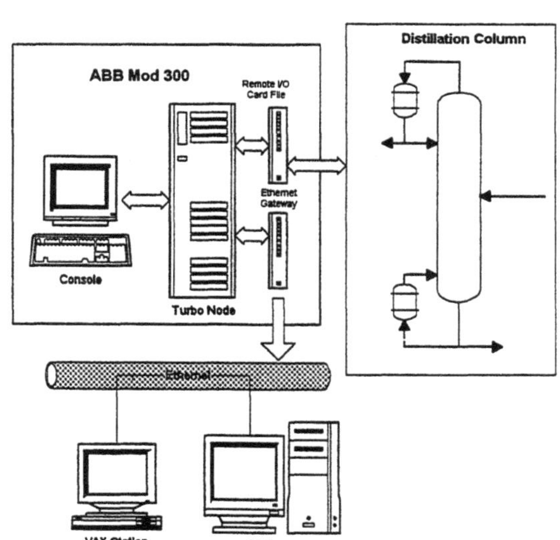

4.3 Expert System Shell

The expert system was implemented on an Intel-based PC using a real-time artificial intelligence software, *G2* by *Gensym Co. G2* employs object-oriented methodology, and has the capability to perform real-time reasoning. The programming language is similar to the domain descriptive language (DDL) used in knowledge acquisition. These features make *G2* one of the most popular expert system shells in the market.

5. MODEL OF EXPERTISE

The core of an expert system is the model of expertise. It encapsulates the problem-solving behaviour, inferencing and reasoning of human experts.

Proper modelling is important if the final KB expert system is of reasonable size. It enables re-usability of general knowledge between different KB of similar nature. In *CommonKADS* (Wielinga *et al.*, 1992), model of expertise consists of four layers: domain knowledge, inference knowledge, task knowledge and strategic knowledge. The focus of this paper is not knowledge engineering, and therefore only a simplified model of expertise is described.

5.1 Modelling of Domain Knowledge

Domain knowledge conceptualises the objects, their properties, relations, constraints and the like in the domain.

Being implemented in an object-oriented shell, the object hierarchies are critical for the purpose of programming and system management. Two "modules" are used to represent object hierarchies for plant items which appear on plant schematic and PCEGs for the purpose of fault diagnosis. In this study, only hardware failures, which include failure of utility supply such as steam or cooling water supply, are being considered as exogenous causes. Figure 5.1 shows the PCEG object hierarchy.

Not all the measured process variables are included in the domain. Only process variables that are considered to be "relatively" critical in monitoring and supervision of the distillation process are covered

Prior and conditional probabilities were also part of the domain knowledge. The prior probability of all readings had equal weighting and was set at 1.00. The conditional likelihoods were based on plant experts who operate this column.

Figure 5-1: Object hierarchy of PCEG module

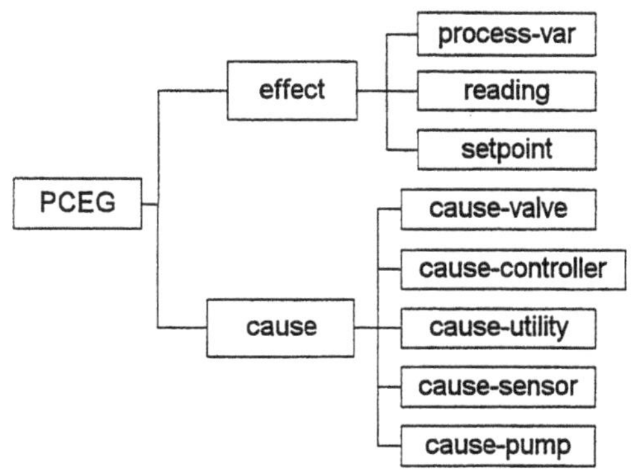

PCEG for controllers

Various controllers were used in the process, and undoubtedly, the causal relationships embedded constituted a large portion of the whole causal network.

Single Loop Feedback Controller The simplest controller used in the distillation process was single-loop feedback controller for controlling feed flowrate. When there was a deviation between the measured flowrate reading and the setpoint, the controller would attempt to overcome this upset by sending out a signal to adjust the flow control valve. Figure 5.2 shows the corresponding PCEG.

Figure 5-2: PCEG of a single-loop feedback controller

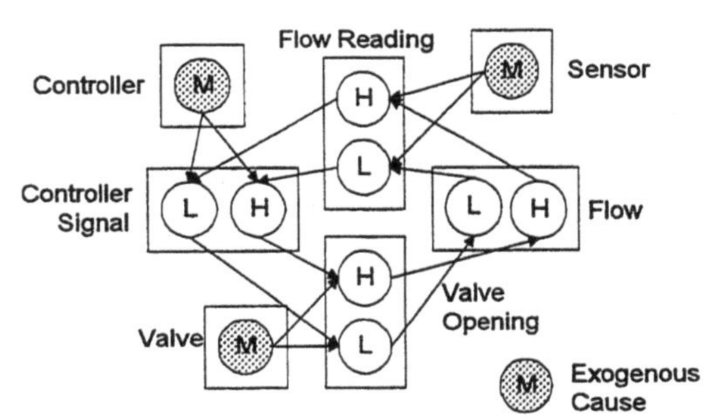

This PCEG contains very complete causal relationships for a single loop feedback controller. However, this PCEG may be too detailed when applying to model cascade controllers, resulting in decrease in computational speed and unnecessary details in the explanation. As a compromise, a simpler approach is employed (Figure 5.3).

Figure 5-3: Simplified PCEG for single loop feedback controller

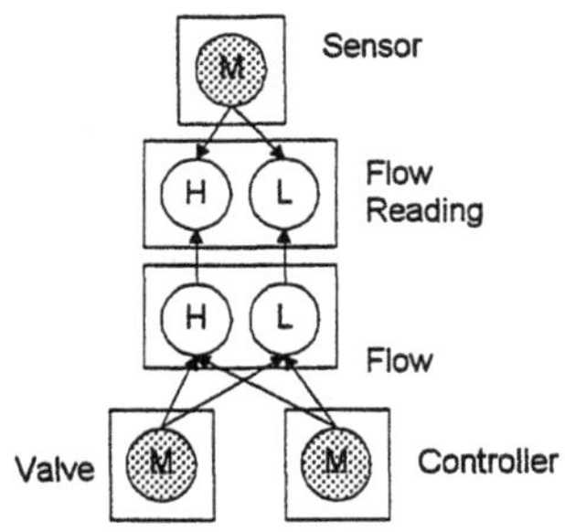

Cascade Feedback Controller Single loop feedback controller has a weakness in managing disturbances. The controller acknowledges the disturbance in manipulated variable only after a deviation in controlled variable is detected. Cascade controllers are designed to overcome this problem. A block diagram of a cascade controller is shown in Figure 5.4.

Figure 5-4: Block Diagram of a Cascade Controller

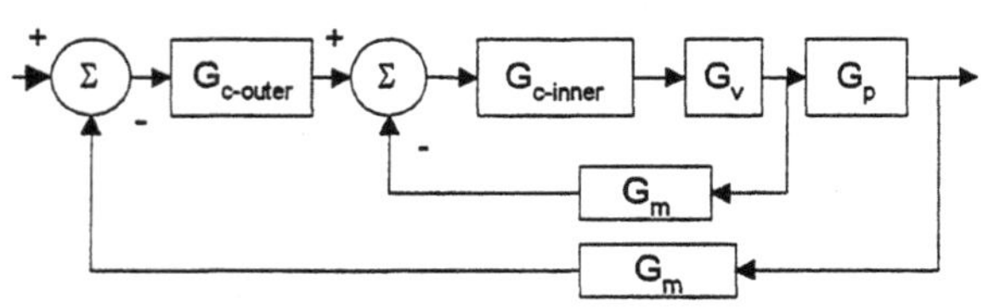

The cascade controller consists of two controller loops, an inner loop and an outer loop. The outer loop provides setpoints for the inner loop, while the inner loop removes the disturbance appeared in the manipulated variable. Figure 5.5 shows the PCEG of a cascade controller, based on the simplified single-loop feedback controller PCEG.

Figure 5-5: PCEG of a cascade controller

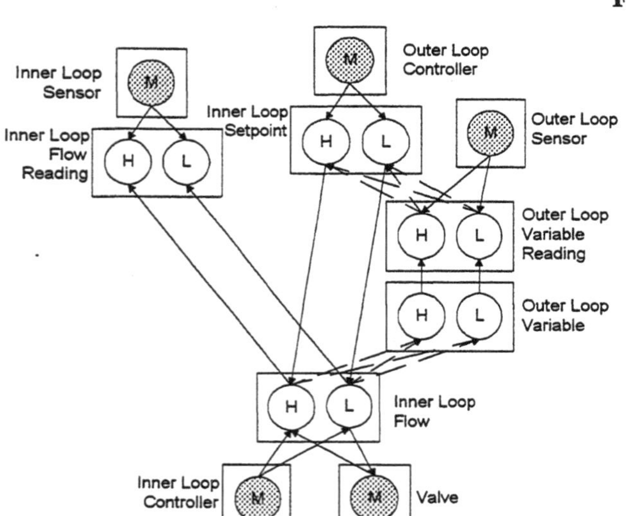

Interactions Between Process Variables As in SDG approach, interactions between process variables must be available in order to construct the PCEG. Figure 5.6 shows the interactions between all the monitored process variables.

Figure 5-6: Interactions between process variables

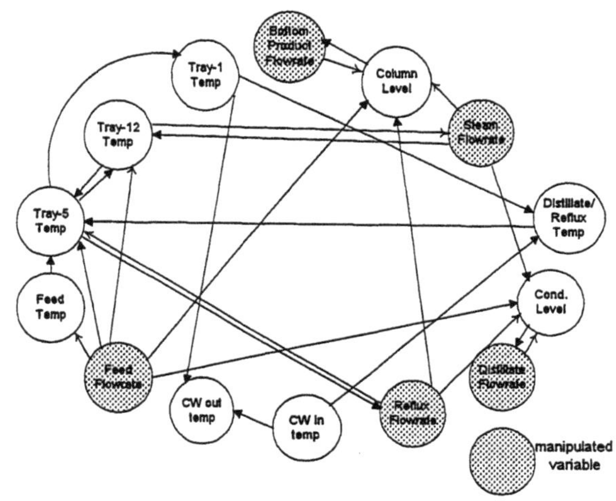

5.2 Inference Structure

The expert system acquired raw data from DCS. These values were compared against the alarm values specified in the domain knowledge. If the system detected alarms, it performed dynamic modification of the PCEG belief network depending on the alarm patterns. To avoid redundant search, which would affect the probability calculations, the alarms were then filtered to obtain the "ultimate" effect (further arc deletion). On completion, the search engine performed recursive search until the root causes were identified. The mentioned inference structure is illustrated in Figure 5.7.

Figure 5-7: Inference Structure

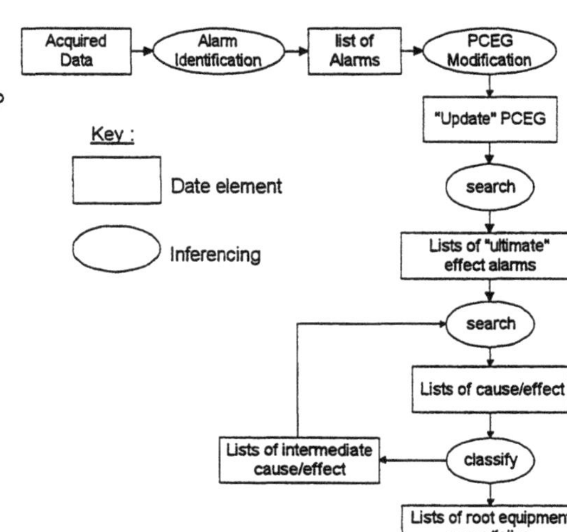

6. SAMPLE RUN

Numerous cases have been tested on this system. The following paragraphs describe a case where dynamic casual network modification is required.

The symptom for cooling water supply failure was simulated in this example. Cooling water supply failure caused high reflux temperature, which in turn caused high Tray-5 and Tray-1 temperature. This was an example of a cyclic PCEG loop of this process (between distillate temperatures, Tray-5 temperature and Tray-1 temperature). The search space of this fault symptom is illustrated in Figure 6.1.

Figure 6-1: Active PCEG search space for cooling water failure symptom

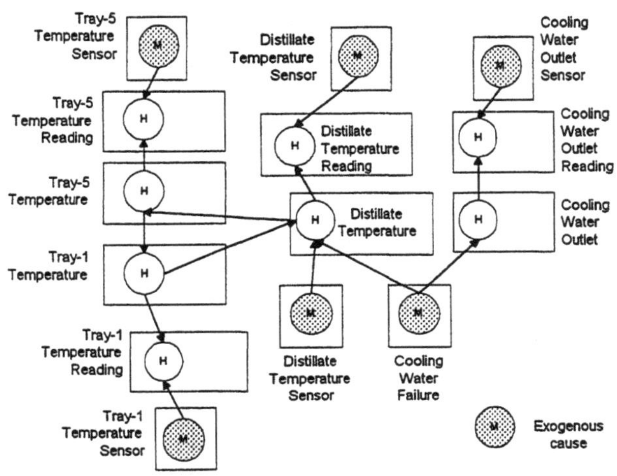

Because the Tray-12 temperature was normal, the special rules of the expert system detected that this symptom was most likely to be caused by cooling water supply failure, and not the overheating of Tray-12. It dynamically deleted the causal relationship between high Tray-1 temperature and high distillate temperature. The inference algorithms determined that high Tray-1 was the "ultimate" effect alarm. It further deleted the arc between the temperature and the temperature reading nodes of Tray-5, distillate and cooling water outlet. Figure 6.2 shows the final active PCEG network, after deletion of all the necessary causal relationship.

Results from on-line steady-state data reconciliation suggested that there is no sensor malfunction and all the sensor readings were reliable. In this setup, there is no flow sensor for cooling water. Only cooling water temperature can be used as the evidence. For instance, if the cooling water inlet temperature was normal, but the cooling water outlet temperature was high, then it is evidential that cooling water has failed. The calculation of belief network (Table 6.1) suggested that cooling water failure was in this case the root cause of the symptom.

Figure 6-2: Modified active PCEG search space for cooling water supply failure

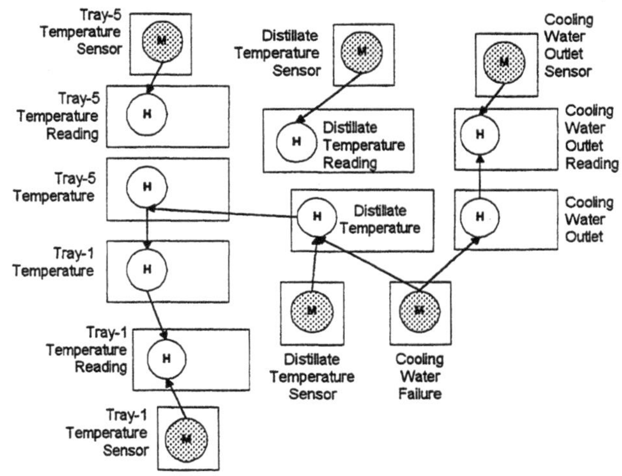

Table 6-1: Summary for evidence and conditional likelihood for sample run

Exogenous Cause	Evidence truth valve	Cond. likelihood	Norm. Probability
Tray 1 temp. sensor malfunction	False	0.05	0.025
Tray 5 temp. sensor malfunction	False	0.05	0.025
Distillate temp. sensor malfunction	False	0.05	0.025
CW outlet temp. Sensor malfunction	False	0.05	0.025
CW water supply malfunction	True	0.93	0.901

Note: Normalised probabilities are calculated based on the conditional probability of 0.98 between reading and process-var and a prior probability of 1 for each reading.

7. DISCUSSION

Unlike most of the fault diagnosis expert systems only suggest the most-likely root cause, our system provides all possible causes with probability. It did not just select the cause with the highest probability and then perform corrective action. . The operators will make the final decision and ensure the safety of the operation.

At this stage, due to the relatively small domain of the process, the causal network has not been partitioned into different plant sections. This was justified by the short time required for computation. On average, a symptom can be diagnosed within seconds on an IBM Intel 133MHz PC running on Windows NT 4.0 using 64Mb Ram. As in other SDG approaches, the search space increase exponentially with the number of "alarmed" variables Therefore, for large processes, partitioning (fault localisation) is recommended.

Figure 6-3: Human Machine Interface of the expert system.

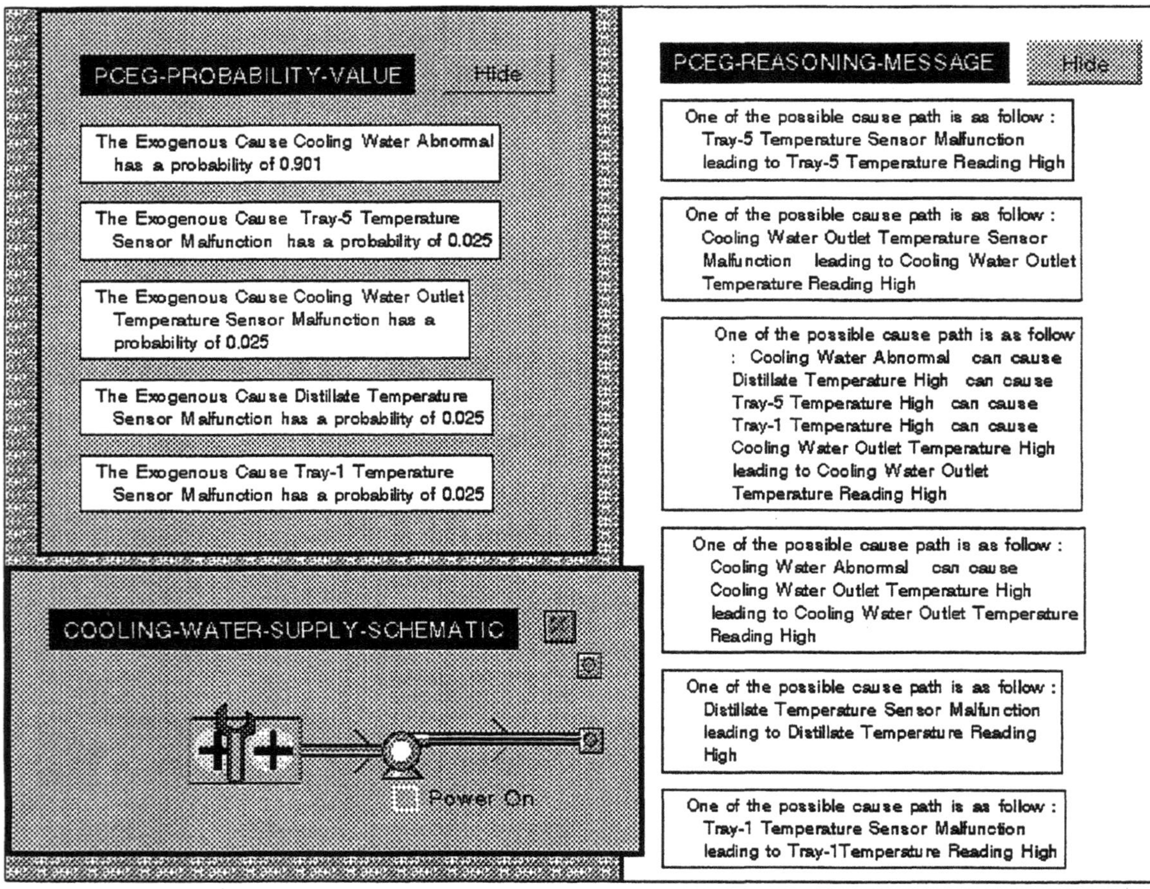

Due to the simplicity of the process, we chose not to consider time delay at this stage of development. This may be unrealistic for certain industrial process, but we have already successfully extended this dynamic approach to account for time delay. The system is at testing stage and the results so far are promising. The dynamic causal network approach has performed well within a real-time environment.

8. CONCLUSION

The complexity of today's chemical processes requires more "intelligent" system for decision support and supervision. Equipment failure can cause problems in terms of safety and operation. In this paper, a fault diagnosis expert system based on PCEG methodology is described. The breakthrough in this system is the dynamic causal network modification mechanism for handling cyclic causal loops. This enables quick probability calculation of all possible causes simultaneously by Bayesian belief theorem. Further research is undergoing to extend this dynamic approach for time delay consideration.

9. REFERENCES

Chen L W, M Modarres (1992), Hierarchical Decision Process For Fault Administration, *Comput.chem. Engng*, **16**, 425-448

Kramer, M. A., B. L. Palowitch (1987). A rule-based `approach to fault diagnosis using the signed digraph, *AICHE J.*, **33**, No.7, 1067-1078

Mah, R S H, *Chemical Process Structures and Information Flows*, Butterworths, 1990

Russell, S., P. Norvig, *Artificial Intelligence A Modern Approach*, Prentice Hall, 1995

Venkatasubramanian V. (1994). Towards Integrated Process Supervision: Current Status and Future Directions: *2nd IFAC Workshop on Computers Software Structures Integrating AIKBS Systems in Process Control*. Lund. Sweden 1994

Wielinga, B. J., A. Th. Schreiber, J. A. Breuker (1992). KADS: a modelling approach to knowledge engineering, *Knowledge Engineering*, **4**, pp 5-53

Wilcox, N. A., D. M. Himmelblau (1994a). The Possible Cause and Effect Graph (PCEG) Model For Fault Diagnosis-I. Methodology, *Comput.chem.Engng*, **18**, 103-116

Wilcox, N. A., D. M. Himmelblau (1994b). The Possible Cause and Effect Graph (PCEG) Model For Fault Diagnosis-II. Applications, *Comput.chem.Engng.*, **18**, 117-127

SCALE-SPACE FILTERING FOR QUALITATIVE INTERPRETATION OF REAL-TIME PROCESS DATA

Dongil Shin, Taesuk Chang, Yong Jin Choe[†], and En Sup Yoon

Division of Chemical Engineering, Seoul National University, Seoul 151-742, Korea
[†]LG Chemical Ltd., Taejeon 561-756, Korea

Abstract: This paper describes a qualitative interpretation method, which is used for extracting qualitative information from numeric sensor data. Firstly, whether any change has occurred in chemical process data is determined by using the CUSUM (Cumulative SUMmation) test. From the sign of the first derivatives of the process variables, sensor patterns can be classified into the seven primitives. Secondly, extraction of the trends of the data employing the modified scale-space filtering is performed. The recursive form reduces the calculation cost of the real-time scale-space filtering and solves the endpoint problem. The proposed method was tested for the artificial patterns and the simulated data of a evaporator process. *Copyright © 1998 IFAC*

Keywords: Pattern recognition, Qualitative interpretation, Change detection, Control chart, CUSUM test, Scale-space filtering

1. INTRODUCTION

The emergence of advanced control algorithms, the needs to meet the demand of energy and materials saving, and the growing community interests in the issues on the environment and safety drive chemical industries to operate processes more robustly. Even skilled operators find difficulties in identifying the conditions of processes due to great amount of process data. To deal with these problems, many researchers have been trying to devise methodologies, otherwise known as operation aid systems, that enable operators to consistently monitor the condition of processes and provide them with the required information for decision making. The fault diagnosis system is one of the core components of the operation aid systems.

There are two methods for diagnosing faults in processes: model-based (Dvorak and Kuipers, 1991; Kramer and Mah, 1993) and heuristic-based (Gujima *et al.*, 1993; Mohindra and Clark, 1993) methods. Since any fault diagnosis system depends on sensor data, whether the system manages faults qualitatively or quantitatively, its performance depends much on the way of data analysis. Especially when it comes to the fault diagnosis system using the qualitative model based on digraph, the module that converts sensor data into meaningful qualitative information plays a significant role. Many of the model-based fault diagnosis systems take advantage of qualitative models. This paper deals with the method that makes it possible to effectively extract useful qualitative information from numeric sensor data. Toward this, it is necessary to first define the characteristics of qualitative information and then develop a procedure to analyze the trends of processes in various scales by taking advantage of a traditional statistical method, the CUSUM test and the scale-space filtering.

2. MONITORING OF PROCESS VARIABLES

The point of qualitative interpretation is detecting the changes in processes. The change detection that is mentioned here means detecting the changes from steady states. While the methods such as the limit check (Isermann, 1984) and the Shewhart control chart have been applied for the change detection, this paper employs the CUSUM test which could show good performance as the average value changes with respect to time.

2.1 *CUSUM Test*

Basseville (1988) completed the CUSUM test in the viewpoint of mathematics. The CUSUM determines the occurrence of changes by employing the V-mask that contains two standard parameters. These parameters are obtained by heuristic. The occurrence of changes can be determined when outliers exist outside the V-mask on the control chart. The mathematical interpretation of the method that was converted into recursive forms is presented as follows.

$$\begin{aligned} &T = x - \mu_0 - {}^{v_m}\!/_2 + T \\ &\text{IF}(T < LM)\ LM = T \\ &GM = LM - T \end{aligned} \tag{1}$$

$$\begin{aligned} &U = x - \mu_0 - {}^{v_m}\!/_2 + U \\ &\text{IF}(U < SM)\ SM = U \\ &GP = U - SM \end{aligned} \tag{2}$$

where T and U are the cumulative sums respectively, LM is the maximum value of T, SM is the minimum value of U, GM and GP are the variables describing the differences in each cases, x is the measured value, μ_0 is the average, v_m is the size of minimum level, and λ is the threshold. The parameters v_m and λ are provided by the user.

The application of the CUSUM mentioned so far into the artificial data set is illustrated in Fig. 1a. The data set has a step change of which the size is one, from10 to 11, at time 250. Although recognizing the existence of any change and its time spot with the naked eye is next to impossible because of a Gaussian noise, the change can be detected quite readily by the use of the increase detector(g^+) and the decrease detector(g^-) of the CUSUM as shown in Fig. 2b. However, since the test depends on the analysis of the past data, the average value of the past data or an expectation are still required. With a bad average or a poor expectation that may produce a slope other than zero, the wrong determination of the trend of a variable may happen. For that reason the CUSUM test hardly identifies a linear increase, a second order increase or a series of step changes. This technique can be used only as a detector and it is necessary to devise the other method in order to classify trends. To solve this problem, the CUSUM test is applied to both a variable and its first derivative.

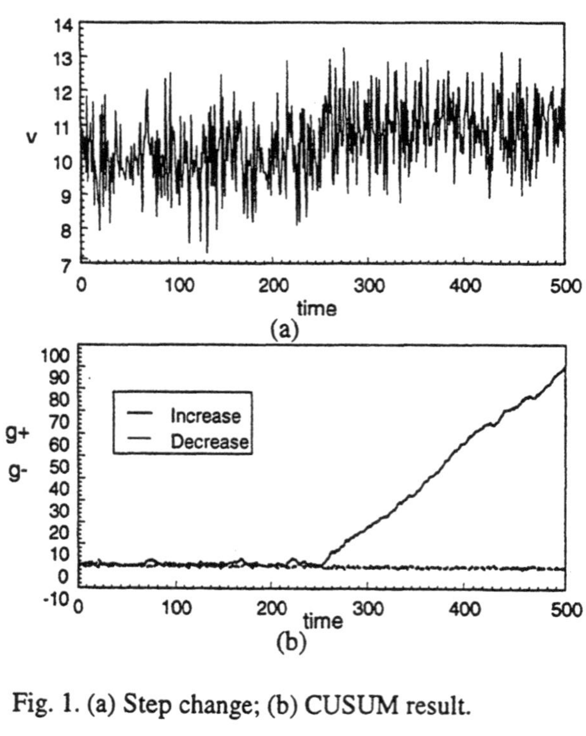

Fig. 1. (a) Step change; (b) CUSUM result.

2.2 *Definition and Analysis of Process Trends*

On the issue of the pattern recognition of process variables, Cheung and Stephanopoulos (1990a, b) classified the ranges of the same sign of the second derivatives into triangular episodes and defined the characteristics of the seven primitives. Through monitoring the signs of process variables and their first derivatives, the data can be classified into the Cheung and Stephanopoulos's seven primitives. For example, when a variable increases and its first derivative also increases, the trend is said to be a concave upward monotonic increase as shown in Table 1. There are several methods to obtain first derivative values, such as numerical differentiation methods and the direct calculation from state variable filtering. In this study, the derivative equation converted into a difference equation is selected for the convenience of computer calculation.

2.3 *Change Detection Using CUSUM Test*

An example of change detection using the CUSUM test is illustrated in Fig. 2a. After a certain period, three different trends that are distinguishable by their

Table 1 Characteristics of seven primitives

Pattern	y	y′
Constant (A)	0	0
Linear increase (B)	+	0
Linear decrease (C)	-	0
Concave upward monotonic increase (D)	+	+
Concave downward monotonic increase (E)	+	-
Concave upward monotonic decrease (F)	-	+
Concave downward monotonic decrease (G)	-	-

0 : No change + : Increase - : Decrease

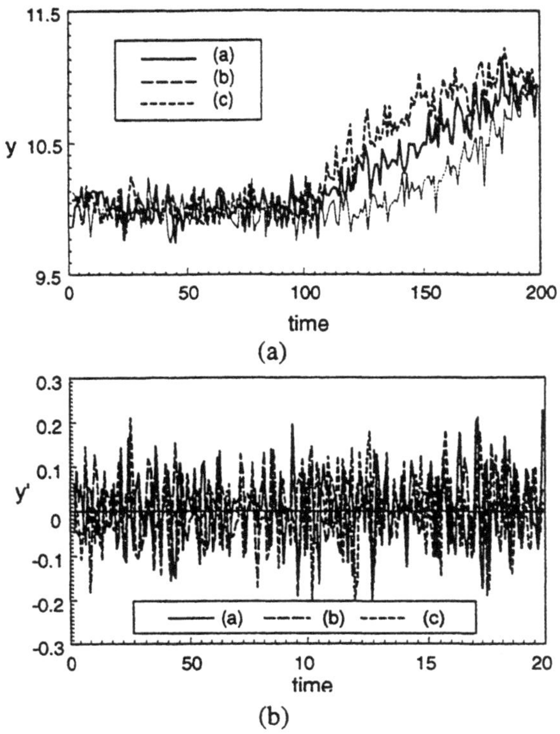

Fig. 2. (a) Increasing variables; (b) Derivatives.

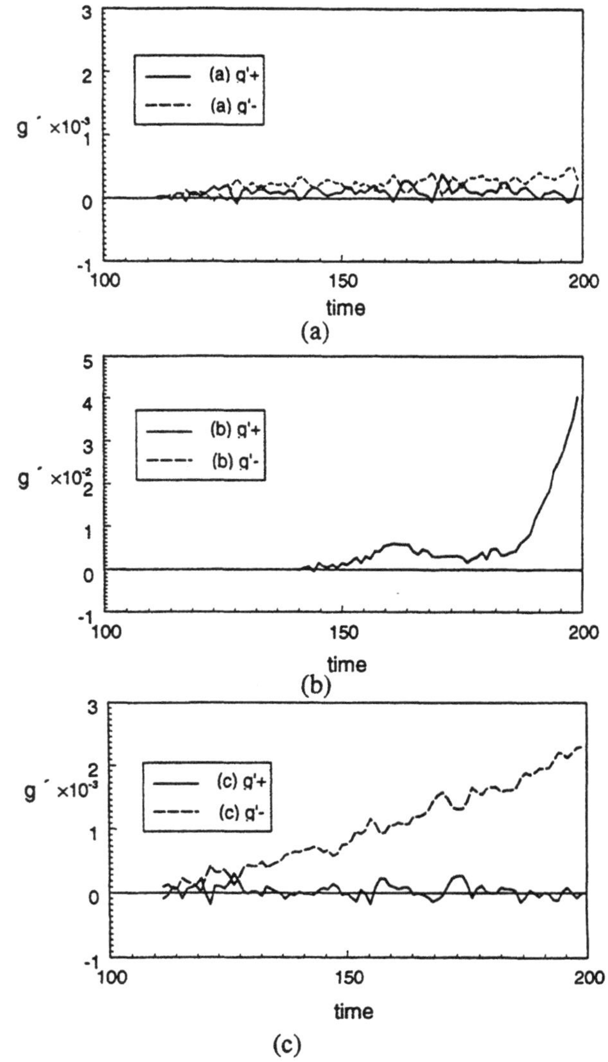

Fig. 3. (a) Linear increase; (b) Concave increase; and (c) Convex increase.

increasing proportions occur: (a) linear, (b) concave and (c) convex. The first order derivatives are also illustrated in Fig. 2b. The only conclusion that can be extracted from the original signals is that they are simply increasing but with no more obvious information. Further consideration of the first derivative values, however, gives us more detailed information that there is no change in the case (a), a decrease in (b), and an increase in (c) as illustrated in Fig. 3.

3. PROCESS TRENDS ANALYSIS WITH SCALE-SPACE FILTERING

The 'process trends' is usually understood as the behaviors of processes with respect to time that can be perceived intuitionally. The 'scale of process trends' is a qualitative parameter that accounts for the resolution of the physical and chemical phenomena described by the process trends. The larger the value of the scale of process trends, the lower the resolution. The method to convert a certain scale into other scales is highly required in order to analyze process trends in any interesting scale.

3.1 Scale-Space Filtering

The scale-space filtering is the global qualitative filtering mechanism that provides the structure that describes the process trends in various scales (Goupillaud *et al.*, 1984). A scale-space consists of the spatial coordinate and the scale axes defined by σ, the size of the Gaussian filter or the standard deviation, which is described in Eq. 3.

$$g(t,\sigma)=\frac{1}{\sigma\sqrt{2\pi}}\exp\left(\frac{-t^2}{2\sigma^2}\right) \tag{3}$$

For example, the scale-space image of a signal *f(t)* is defined as the surface in the scale-space, *F*(*t*, σ), by taking the Gaussian convolution as shown in Eq. 4 and Fig. 4.

$$F(t,\sigma)=\int_{t_0}^{t_1} f(\tau)\frac{1}{\sigma\sqrt{2\pi}}\exp\left[\frac{-(t-\tau)^2}{2\sigma^2}\right]d\tau \tag{4}$$

The scale-space filtering reflects the effect of the narrow range of adjacent data when a small value of the parameter σ is set, and vice versa. To recapitulate, with the larger value of σ, the lower resolution of filtering would be obtained. When the σ is zero, the filter would not work.

3.2 Modified Scale-space Filtering

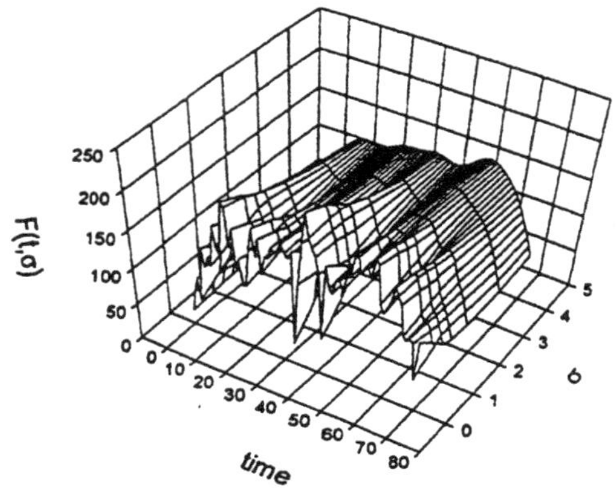

Fig. 4. Scale-space image of signal $f(t)$.

Since the scale-space filtering method extracts the time-related characteristics included in signals quite systematically, it provides the convenient structure to analyze the signals in various intuitive scales. However, the scale-space filtering has a few drawbacks at the same time. Because scale-space filtering deals with a global data set, it may distort the results of the incidents in small scales and may give rise to too much calculation load when tracking inflection points. And all data should be stored and updated with new data due to its non-recursive form. Furthermore, even though the scale-space filtering produces smoothly filtered results when used with the large value of σ, it cannot estimate the reasonable value at both ends, especially in recent data. Mokhtarian *et al.* (1986) discovered that this is because the scale-space filtering was originally developed to provide various resolutions in image processing, where the targets of filtration are mostly the closed curves that can be regarded as periodic functions. The following three modifications are proposed to make up for the listed drawbacks of the scale-space filtering method.

Calculation of Inflection Points: Inflection points are automatically tracked at the change detection step by the CUSUM test. Extra calculations other than monitoring first derivatives are not necessary.

Endpoint Problem or Edge Effect: To overcome the problems at both ends, basically the extrapolation, based on the values at both ends and the trends (or slope) acquired by the CUSUM test, is used. The data is given as follows.

$$f(t)=\begin{cases} 0 & t<t_0 \\ f_1(t) & t_0 \le t \le t_1 \\ 0 & t>t_1 \end{cases} \qquad (5)$$

Applying the scale-space filtering expressed in Eq. 4 produces Eq. 6.

$$F(t,\sigma)=\int_{t_0}^{t_1} f(\tau)\frac{1}{\sigma\sqrt{2\pi}}\exp\left[\frac{-(t-\tau)^2}{2\sigma^2}\right]d\tau \qquad (6)$$

When either t goes down below t_0 or goes up beyond t_1, $F(t,\sigma)$ approaches zero. Eq. 5 can be converted into Eq. 7 by making use of the slopes and mean values at these both ends.

$$f(t)=\begin{cases} \mu'_0(t-t_0)+\mu_0 & t<t_0 \\ f_1(t) & t_0 \le t \le t_1 \\ \mu'_1(t-t_1)+\mu_1 & t>t_1 \end{cases} \qquad (7)$$

where μ's denote the slopes at the both ends.

Since heavy noise may cause the discontinuity in the first derivative, the mean values in certain intervals are required to be applied. Applying the scale-space filtering to Eq. 7 and successive conversion into the form including error functions produces Eq. 8. In this manner the problem that the filtered values approach zero at both ends can be overcome.

$$\begin{aligned} F(t,\sigma)=&-\frac{\mu_0\sigma}{\sqrt{2\pi}}\exp\left[\frac{-(t-t_0)^2}{2\sigma^2}\right] \\ &-\frac{1}{2}\left(\mu_0(t-t_0)+\mu_0\right)\left[erf\left(\frac{t-t_0}{\sigma\sqrt{2}}\right)-1\right] \\ &+\int_{t_0}^{t_1} f_1(\tau)\frac{1}{\sigma\sqrt{2\pi}}\exp\left[\frac{-(t-\tau)^2}{2\sigma^2}\right]d\tau \qquad (8) \\ &+\frac{\mu_1\sigma}{\sqrt{2\pi}}\exp\left[\frac{-(t-\tau)^2}{2\sigma^2}\right] \\ &+\frac{1}{2}\left(\mu_1(t-t_1)+\mu_1\right)\left[erf\left(\frac{t-t_1}{\sigma\sqrt{2}}\right)+1\right] \end{aligned}$$

Calculation Load: Through the scale-space filtering, all the data have their weights and thus the calculation load increases exponentially. In this paper, the heavy calculation load is reduced by introducing the recursive form, expressed in Eq. 9.

$$\begin{aligned} \Delta F &= F_{N+1}-F_N \\ &= \frac{\Delta T}{\sigma\sqrt{2\pi}} f(t_{curr})\exp\left[\frac{-(t-t_{curr})^2}{2\sigma^2}\right] \qquad (9) \end{aligned}$$

4. CASE STUDY

The evaporator process, as illustrated in Fig. 5, is selected to illustrate the performance of the proposed methods, and the result is compared with that from the limit check which has been frequently used for detecting changes. This process has been early simulated by Newell *et al.* (1989). In the process, the feed mixed with the recycled liquid of the large flow rate is supplied to the vertical heat exchanger (evaporator), where the liquid is boiled and flows to the separator. The liquid is separated from the vapor

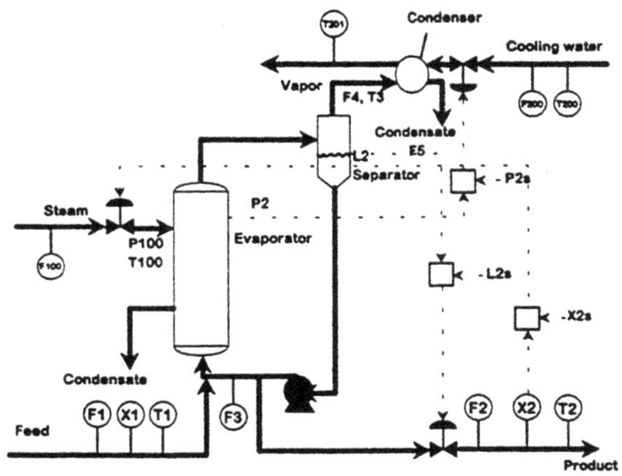

Fig. 5. Evaporator Process.

at the separator. Some of the liquid is recycled and the rest is drained as the product. The vapor part is condensed through the condenser. It is assumed that the solvent is water and the solute is a nonvolatile material. This process consists of a evaporator, a separator, and a condenser. From the mathematical viewpoint, there are basically 21 variables, three differential equations, nine algebraic equations, and three control equations. As a case study, the simulations have been carried out for more than 11 abnormal situations and three levels of process data noise(0%, 3%, and 6%). The details of each situation and the variables' changes are described in Table 2.

Table 2. Fault Simulation Parameters

Case	Fault Para.	Change From	Change To	Noise	Description
E-01-1				0 %	
E-01-2	F1	10	9	3 %	feed flow low
E-01-3				6 %	
E-02-1				0 %	
E-02-2	X1	5	8	3 %	feed conc'n high
E-02-3				6 %	
E-03-1				0 %	
E-03-2	T200	25	30	3 %	cooling water temp. high
E-03-3				6 %	
E-04-1				0 %	
E-04-2	F3	50	30	3 %	pump failure
E-04-3				6 %	
E-05-1				0 %	
E-05-2	UA2	6.84	5	3 %	heat exchanger degradation
E-05-3				6 %	
E-06-1				0 %	pressure
E-06-2	Ps	50	60	3 %	set point
E-06-3				6 %	high
E-07-1	sfault(2)	1	0	0 %	product
E-07-2	sbias(2)	0	3	3 %	flowmeter
E-07-3				6 %	stuck high
E-08-1	sfault(8)	1	0	0 %	separator level
E-08-2	sbias(8)	0	2	3 %	sensor stuck
E-08-3				6 %	high
E-09-1	vfault(1)	1	0	0 %	product stream
E-09-2	vbias(1)	0	3	3 %	valve stuck
E-09-3				6 %	high
E-10-1	sfault(18)	1	0	0 %	product conc'n
E-10-2	sbias(18)	0	30	3 %	sensor bias
E-10-3				6 %	high
E-11-1		0	0.1	0 %	evaporator
E-11-2	sigma(10)	0.05	0.1	3 %	pressure sensor
E-11-3		0.1	0.2	6 %	noise increase

In order to compare the proposed method with the limit check, the symptoms generated by the traditional 3σ method have been examined by the use of both methods. The behaviors of 21 process variables for more than 11 abnormal situation with the noise level of 3% are described in Table 3. As shown in the table, the limit check is able to determine the patterns in most cases, and also is able to identify the basic causes through pattern matching of the given symptoms. However, for the cases of E-07-2 (product flowmeter stuck high) and E-11-2 (evaporator pressure sensor noise increase), the limit check does not detect any symptoms. In view of the result, the limit check is not able to detect the faults in those two cases, and thus is not able to carry out fault diagnosis. In the case E-07-2 that the product flowmeter is stuck high due to a sensor failure, while the limit check does not detect any symptom except for the peaks in F3 and T1 as illustrated in Fig. 6 during the entire observed interval, the proposed method differentiates the symptom in F2 from the

Table 3 Symptoms at various faulty conditions

	1	2	3	4	5	6	7	8	9	10	11	12	13	14	15	16	17	18	19	20	21
E-01-2	•			•	•								•	•	•	•				•	•
E-02-2				•	•	•							•	•	•	•				•	•
E-03-2									•	•		•			•	•	•			•	•
E-04-2			•	••	••		••		••				••	•	•	••				••	•
E-05-2					••		••		•	•		•		•	•	•				•	•
E-06-2							••		•	•		•		•	•	••				••	•
E-07-2																					
E-08-2		•		•	•		•		••	•		•	•	•	•	•			•	•	•
E-09-2				•	•		•		••	•	•	•	•	•	•	•				•	•
E-10-2				•	•		•		•				•	•	•	•				•	•
E-11-2																					

1: F1 2: F2 3: F3 4: F4 5: F5 6: X1 7: X2 8: T1 9: T2 10: T3 11: L2 12: P2 13: F100 14: T100
15: P100 16: F200 17: T200 18: T201 19: F2c 20: F200c 21: P100c
▼ : decrease △ : increase ▼△ : decrease first and then increase

patterns of the other abnormal situations by using the trend of F2. Also in the case E-11-2 where there is an increase of the noise level in the sensor data from the pressure gauge of the evaporator, the limit check does not detect any symptom. The proposed method is able to determine the trend of a process variable, the data of which lie within the thresholds of 3σ, and it differentiates the specific pattern from the others.

5. CONCLUSIONS

A method has been proposed that interprets numeric sensor data into reliable qualitative information and analyzes operation data effectively and systematically. The traditional statistical method, the CUSUM test, made it possible to detect the changes in process variables effectively and to analyze the trends of a process by monitoring both process variables and their derivatives. The scale-space filtering has been modified to suit the on-line application and has been proved that it could separate the patterns in various scales. The case study performed on the evaporator process showed that the proposed method had better performance than the limit check.

ACKNOWLEDGMENTS

The authors acknowledge the financial support of the Science and Technology Policy Institute (STEPI).

REFERENCES

Dvorak, D. and B. Kuipers (1991). Process monitoring and diagnosis: a model-based approach. *IEEE Expert*. June, 67-74.

Kramer, M.A. and R.S.H. Mah (1993). Model-based monitoring. *Proceedings of 2nd Conference on Foundations of Computer Aided Process Operations*. pp. 1-24. CO, USA.

Gujima, F., B. Shibata, Y. Tsuge, J. Shiozaki, H. Matsuyama, and E. O'Shima (1993). Improvement of the accuracy of fault-diagnosis systems using signed directed graphs. *International Chemical Engineering*, **33**(4), 671-679.

Mohindra, S. and P.A. Clark (1993). A distributed fault diagnosis method based on digraph models: steady-state analysis. *Computers chem. Engng* **17**(2), 193-209.

Isermann, R. (1984). Process fault detection based on modeling and estimation methods-a survey. *Automatica*, **20**(4), 387-404.

Basseville, M. (1988). Detecting changes in signals and systems - a survey. *Automatica*, **24**(3), 309-326.

Cheung, J.T.-Y., and G. Stephanopoulos (1990a). Representation of process trends - part I. a formal representation framework. *Computers chem. Engng*, **14**(4/5), 495-510.

Cheung, J.T.-Y., and G. Stephanopoulos (1990b). Representation of process trends - part II. the problem of scale and qualitative scaling. *Computers chem. Engng*, **14**(4/5), 511-539.

Goupillaud, P., A. Grossmann and J. Morlet (1984). Cycle-octave and related transforms in seismic signal analysis. *Geoexploration*, **23**, 85-102.

Mokhtarian, F., and A. Mackworth (1986). Scale-based description and recognition of planar curves and two-dimensional shapes. *IEEE Transactions on Pattern Analysis and Machine Intelligence*, **PAMI-8**(1), 34-43.

Newell, R.B. and P.L. Lee, *Applied Process Control: A Case Study*. Prentice Hall, New York.

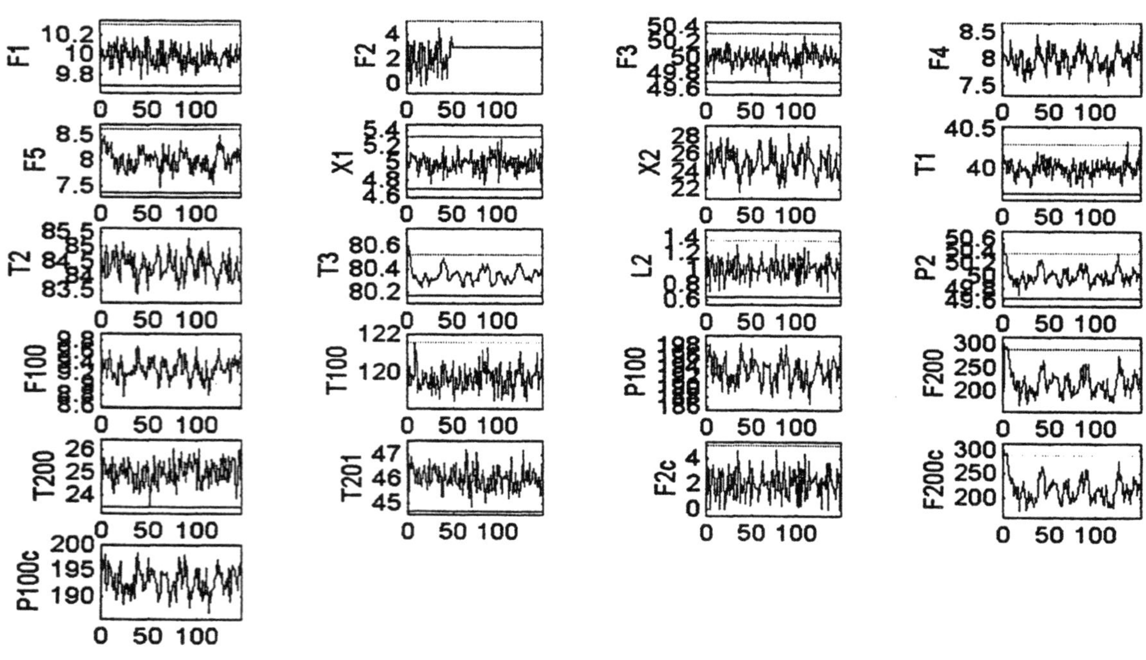

Fig. 6. Process measurement – product flowmeter stuck high.

CAUSAL MODEL BASED SUPERVISING AND TRAINING

Alexandre Evsukoff[1*], Jacky Montmain[**] and Sylviane Gentil[*]

[*]CNRS-INPG-UJF, Laboratoire d'Automatique de Grenoble, BP 46, 38402 Saint Martin d'Hères, France (evsukoff@lag.ensieg.inpg.fr, gentil@lag.ensieg.inpg.fr)
[**]CEA-CEN VALRHO, DCC/DRRV/SSP, Laboratoire d'Informatique Appliquée, BP 171, 30207 Bagnols-sur-Cèze Cedex, France (jm@concerto.valrho.cea.fr)

Abstract: In this work, causal models are proposed as a basis for knowledge representation of complex process dynamics. A prototype system has been implemented for a specific application in a nuclear fuel reprocessing plant. In this system, a causal model, built from expert knowledge, is used for operator support in process supervision and training. A fuzzy logic decision analysis provides a continuous transition from quantitative data into qualitative information and avoids threshold effects. An evaluation of the prototype system has been carried out by students of a technical school and its results are discussed in this paper. *Copyright © 1998 IFAC*

Keywords: Fault Detection and Isolation, Knowledge-Based Systems, Fuzzy Logic, Supervision System, Computer Aided Training.

1. INTRODUCTION

Safety and reliability of industrial processes are becoming the new challenge that control scientists must face. Increasing requirements on productivity and flexibility lead to operate processes in frequently varying conditions.

The process availability is guaranteed presently by skilled human operators. They have to interpret complex situations by means of a high amount of data, which changes rapidly when a malfunction occurs and propagates all along the cascaded sub-systems.

Expert operators are able to link together several apparently different problems, which are in fact related to the same primary fault, and thus make the right diagnosis. Their analysis are based on mental models, created by their past experience. Novice operators do not have such empirical knowledge and might be brought to a mistake. Their training requires quite a long time.

In complex plants, causal models can rather emulate expert operators' mental models in their diagnosis reasoning. Causal models differ from analytical models, in the sense of Russell's concepts (Rasmussen 1993), by the notion of cause and effect sequences. Hence causality is inherently related to time evolution. Causal model based reasoning has been successfully used for fault detection and isolation in complex systems (Evsukoff et al. 1997, Montmain et al. 1994, Kramer 1987).

A causal model is a qualitative description of the direct influences that some variables of a process have on other ones. Causal modelling requires to describe a complex process with a chain of elementary partial effects. Therefore the causal model provides a conceptual tool to reason on how changes propagate or how they may be explained. In this work, causal models are proposed as a basis for knowledge representation of complex process dynamics for computer aided supervising and training.

A prototype system has been implemented for a specific application in a nuclear fuel reprocessing plant The system is composed of two main modules: the detection module is based upon a numerical non - linear simulator of the process and the isolation module is based upon simple linear approximations of the numerical simulator,

[1] Supported by the Brazilian National Research Council - CNPq

structured as a causal graph, which allows the isolation of the root fault.

The fuzzy reasoning approach for both modules is necessary due to modelling imprecision and measurement noise. As a result, a compromise can be obtained between the robustness and the rapidity of the decision. Numeric - symbolic interface provided by the fuzzy reasoning allows the symbolic representation of process variables as colours in the graphical interface of the system.

The prototype system has been evaluated by students of a secondary technical school; the results of these experiments are discussed in this paper.

The paper is organised as follows: section 2 presents the application plant. Section 3 presents the basic tools for fault detection and isolation. Section 4 presents the system which has been implemented to display the fault propagation. Section 5 presents the results of an evaluation of the system with novice operators. Finally, some concluding remarks are enlightened.

2. PROCESS DESCRIPTION

This project was carried out at a nuclear fuel reprocessing plant (Fig 1). Spent fuel is reprocessed to recover the uranium and plutonium still present in the fuel rods irradiated in nuclear reactors, and to isolate the remaining fission products. The pulsed column facility considered here includes extraction columns and fission product scrubbing columns, and is designed to separate the uranium and plutonium from the fission products (FP) by selective extraction.

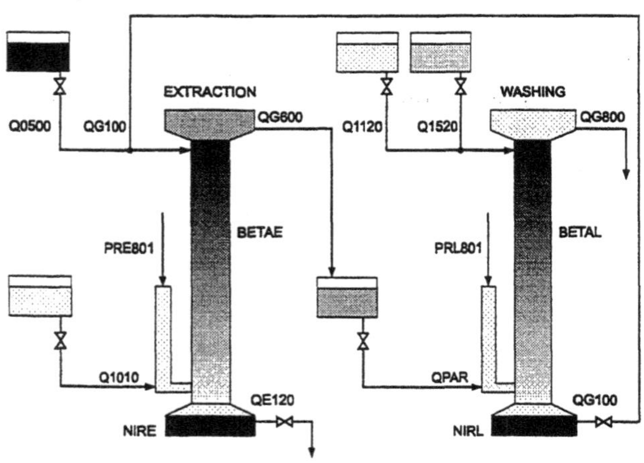

Fig 1: Schematic view of the plant

A pulsed column is a liquid-liquid extraction device. The spent fuel (comprising uranium, plutonium and FP) is dissolved in nitric acid (Q0500), and the extraction column selectively transfers the uranium and plutonium to an organic phase (inlet Q1010 and outlet QG600) consisting essentially of tributylphosphate (TBP); most of the fission products thus remain in the aqueous phase (QE120).

The extraction step requires that the aqueous and organic phases be thoroughly mixed to maximize the contact surface area between the two solvents and thereby optimize the chemical exchange phenomena. The two phases tend to separate by gravity, as nitric acid has a higher density than TBP.

In order to ensure countercurrent flow, the light phase (Q1010) is injected at the base of the column and the heavy (acid) phase at the top (Q0500 and QG100). The organic phase is the continuous phase; it initially fills the entire volume of the extraction column, and remains the predominant phase after injection of the aqueous phase: acid droplets are dispersed in the TBP. The resulting mixture is subjected to periodic pressure pulses (PRE801) to form an emulsion in order to delay the descent of the heavy phase and to mix it with the light phase. The interface – the surface physically separating the two phases – is located in the settler at the bottom of the column (NIRE), and is regulated by drawing off the aqueous phase (QE120) at a suitable rate. In the application considered here, all liquid transfers are ensured by airlifts.

The facility (Fig 1) comprises two columns coupled top-to-bottom by means of a transfer system. The extraction column described above is followed by a scrubbing column designed to enhance the decontamination of fission products from the solvent (aqueous phase: inlet Q1120 and Q1520, outlet QG100; organic phase: inlet QPAR; outlet QG800). Traces of fission products carried out from the extraction column (QG800) are rinsed from the organic phase containing the uranium and plutonium.

The columns are connected top-to-bottom (QG100) in order to recover the aqueous phase containing the fission products through a single outflow (QE120) and to increase the process efficiency by recycling the traces of uranium and plutonium in the aqueous phase from the scrubbing column.

A simulator of the process, based on a set of non-linear numerical models, is used to provide reference behaviour and also to simulate faulty behaviours (Evsukoff et al. 1997). The process simulator computes the output of fifty variables which are used by the prototype system and can simulate several kinds of faults (sensor and process faults).

3. FAULT DETECTION AND ISOLATION

Leitch and Gallanti (1992) have proposed a task classification for knowledge based system application in industrial automation. The detection and isolation modules of the prototype system can be associated to the *interpretation* and the *decision task.*

The interpretation task is a signal-to-symbol transformation that maps sensor information to process state representation. This task is performed by the detection module which transforms sensor measures into a symbolic representation of the variable, presented as colours in the graphical interface of the prototype system.

In a diagnostic system, the decision task generates fault hypotheses that satisfy system observations. Fault hypotheses are tested by the isolation module and are presented by means of a causal graph.

1.1 The detection module

For each variable, the *measurement* value $y(k)$ at time k can be written from the *reference* value $y^*(k)$ given by the numerical simulator output at the same time as:

$$y(k) = y^*(k) + \delta(k) \tag{1}$$

where $|\delta(k)| >> 0$ represents an abnormal behaviour (fault or non-measurable disturbance). The modelling errors and measurement noise are included into $\delta(k)$ and will be taken into account by fuzzification.

The *residual* $\delta(k)$ is used to compute two attributes for the signal description: the *residual mean* $x(k)$, calculated on a temporal window and the *residual mean variation* $\dot{x}(k)$, estimated on another temporal window.

Fuzzy partitions are defined over the universe of $x(k)$ and $\dot{x}(k)$ for each variable. These attributes are used to evaluate the state of the variable by a fuzzy inference system, expressed by a rule base written as follows:

$$\textit{if } x(k) \textit{ is } V \textit{ and } \dot{x}(k) \textit{ is } \dot{V} \textit{ then the state is } S \tag{2}$$

where $V \in \{NN, N, Z, P, PP\}$, $\dot{V} \in \{N, Z, P\}$ and $S \in \{OK, FA\}$ are the labels of fuzzy sets. Their meanings are the usual ones for V and $\dot{V}$. For the conclusion, *OK* accounts for normal state and *FA* for faulty state. The fuzzy sets associated to these two states are complementary.

Table 1: The detection decision table

	N	*Z*	*P*
NN	FA	FA	FA
N	FA	OK	OK
Z	OK	OK	OK
P	OK	OK	FA
PP	FA	FA	FA

The rule base is represented in table 1. For better performance, weights were associated to each rule, represented by grey scale colours in table 1. The fuzzy inference system computes at each instant the membership values of each variable to the states *OK* and *FA*. A fault is detected on a variable if the *FA* membership is significant.

The detection module provides a non-ordered list of the simultaneous faults that affect the process. The isolation module aims at converting the set of detected variables into a graph whose root is the primary fault, and whose paths illustrate the fault propagation.

1.2 The isolation module

Regarding eq. (1), the isolation problem is to decide, for each variable in the detection list, whether the detected discrepancy $\delta(k)$ is due to a local fault or a local non-measurable disturbance or due to the propagation of an upstream fault.

The basic knowledge required to relate a variable behaviour to another can be expressed in terms of causality: one variable is the cause, the other is the consequence. The causal model can be described by a digraph, whose nodes represent the variables and whose arcs represent directed relationships between variables. Note that contrary to fault-symptom trees, the causal graph represents the normal behaviour of the process (Montmain et. al. 1994).

A careful physical analysis of the process under study (Leyval et. al. 1994). has resulted in a causal graph linking input and output flows, levels and column weights (Fig 2).

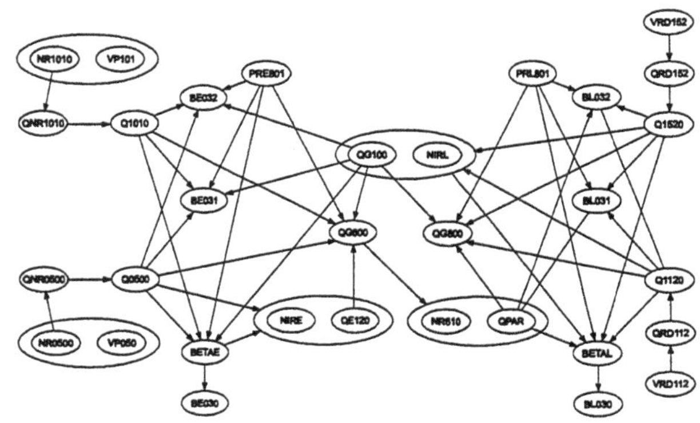

Fig 2: Causal graph

The causal links of the graph have been implemented as discrete transfer functions. For each variable, its *predicted* value $\hat{y}$ is computed from the measurements of each antecedent u_j, by the sum of the components associated to each entering arc.

The Z-transform equation of the predicted value is thus given by:

$$\hat{Y}(z) = \sum_{j=1,n} \mathrm{F}_j(z) U_j(z) \tag{3}$$

where $U_j(z)$ is the Z-transform of the measurements of each antecedent. Expressing the measured values of the antecedents according to (1):

$$\hat{Y}(z) = \sum_{j=1,n} \mathrm{F}_j(z) U_j^*(z) + \sum_{j=1,n} \mathrm{F}_j(z) \delta_j(z) . \tag{4}$$

Taking the inverse Z-transform, the predicted value of the variable can be written from (4) by means of the predicted value of its components as:

$$\hat{y}(k) = \hat{y}^*(k) + \hat{\delta}(k) \tag{5}$$

where $\hat{y}^*(k)$ is the *estimation of the reference* value and $\hat{\delta}(k)$ is the *propagated residual* that accounts for all propagated discrepancies.

The isolation procedure is carried out for each variable in the detection list. The propagated residual is the effect of upstream faults in the variable been analysed.

By the difference between (5) and (1), the following expression for the *local residual* is obtained:

$$y(k) - y^*(k) - \left(\hat{y}(k) - \hat{y}^*(k)\right) = \delta(k) - \hat{\delta}(k) \,. \qquad (6)$$

Analysing eq. (6) it can be seen that if the local residual, expressed by the right side of the equation, is approximately null, then the detected disturbance is due to the propagation of upstream faults. Otherwise the fault is local. Note that even in the presence of multiple faults, the predicted residual should isolate local faults. The rule base for the isolation module is expressed as follows :

$$\textit{if } z(k) \textit{ is } V \textit{ and } \dot{z}(k) \textit{ is } \dot{V} \textit{ then the fault is } D \qquad (7)$$

where $z(k)$ *and* $\dot{z}(k)$ are the local residual mean and its variation. $V \in \{NN, N, Z, P, PP\}$ and $\dot{V} \in \{N, Z, P\}$ are the labels of fuzzy sets in the antecedents, defined as the ones in the detection module, and thus, avoiding new parameters. The conclusion $D \in \{LO, UP\}$ is composed of two fuzzy sets where *LO* attempts for *local* faults and *UP* for *upstream* ones.

The fuzzy inference system for the isolation module uses the same rule weights as the detection module. At each instant and for each variable, the membership values to decisions *LO* and *UP* are computed in such a way that the sum of *LO* and *UP* membership values is equal to the membership to *FA*, computed in the detection module.

4. PROTOTYPE DESCRIPTION

The information about the process is presented in three different abstract layers in the prototype system. The layers present different information in the screen area. The basic frame window presents the list of the process variables in the right side. By clicking in each button the corresponding variable evolution is shown against the reference value evolution.

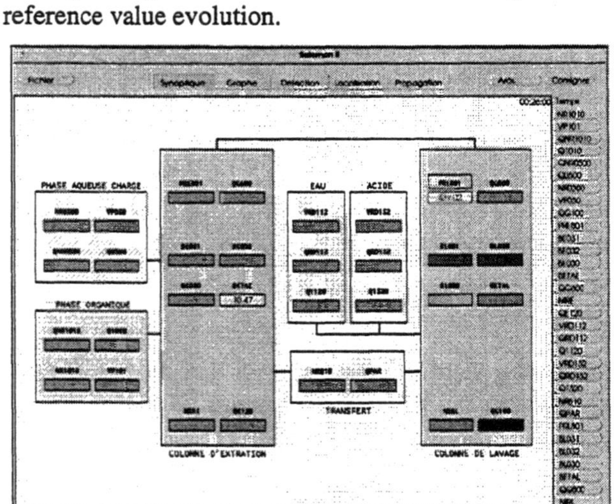

Fig 3: The first layer, structural modules of the plant.

In the first layer, the plant is represented according to a structural decomposition as in classical monitoring interfaces (Fig 3). In this view, the variables are represented by their measured value or by a colour representation obtained from the fuzzy reasoning in the detection module. The primary fault, found in the isolation module, is shown in this view by a contour around the name of the variable.

In the second layer, a first abstraction is done and the process is represented as a causal graph. The fault path is shown, with the primary fault as the root of the faulty sub-graph (Fig 4). The fault path in the causal graph can be seen as a justification module that explains the alarm chaining

The output of the fuzzy reasoning in the isolation module is represented in the graph. If the node is not red, the variable is in a normal state. If the contour of the node is in red, the alarm in the variable does not mean a local fault. If the interior of the node is red then the variable is isolated as a primary fault.

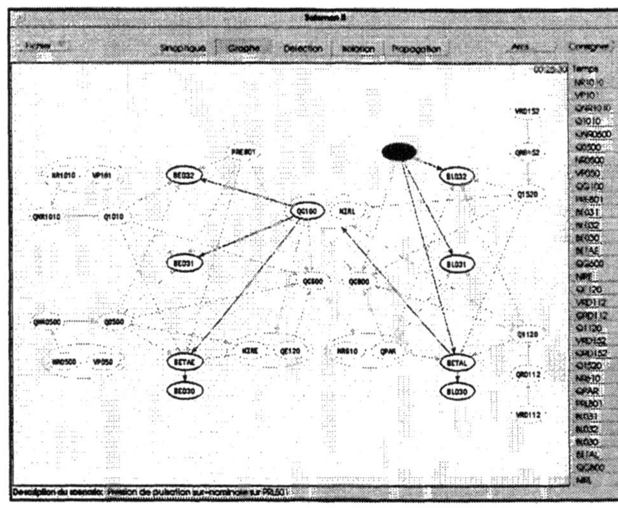

Fig 4: The second layer, causal model of the plant.

The arcs turns to red as the system is able to decide that the arc carries the propagation of some fault. If an arc between two variables in alarm is not red then two conclusions can be made: the system is not able to take a decision due to imprecision or the fault in the downstream variable is also a primary fault.

The third layer is a full abstraction layer, where only the qualitative time evolution of variables is presented, with no regard to relations between variables. The information display in the third layer was inspired by the concept of *Mass Data Display*, which has been already studied for this process (Lambert et al. 1997).

One view is dedicated to the detection module and another one is dedicated to the isolation module. In both views, the evolution of the colour code for each variable is shown in the screen area (Fig 5).

A colour code associated to the variable state is computed by the defuzzification of the output of the fuzzy reasoning in the detection and isolation modules. Each conclusion is associated to a colour. In the detection view, the conclusions of the decision are defuzzified into an index value which represents a colour variation from green (normal state) to red (faulty state). In the isolation view, the conclusions of the decision are defuzzified into an index which represents a colour variation from blue (upstream fault) to red (local fault). A colour palette composed of 64 colours is used.

In the third layer interface screen, each line is associated to a variable and each column is associated to a sampling time, where the columns close to the left side represent earlier time samples.

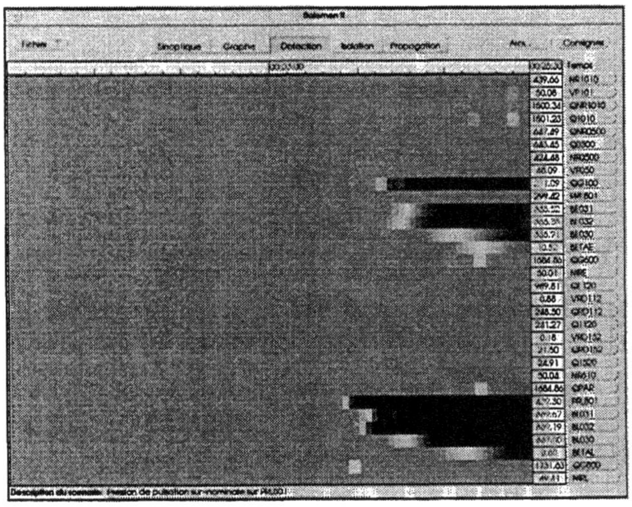

Fig 5: Detection view.

In a normal situation the whole screen is green in both detection and isolation views. In the detection view, when a fault occurs, the lines turned to red represent the variables in alarm (Fig 5). The colour degree evolution represents the abruptness of the fault.

In the isolation view, the variables in green in the detection view remain in green. The variables in red remain in red if the fault is local or turn to blue if the fault is upstream (Fig 6). By changing from detection to isolation view the effect of alarm filtering provided by the system can be evaluated.

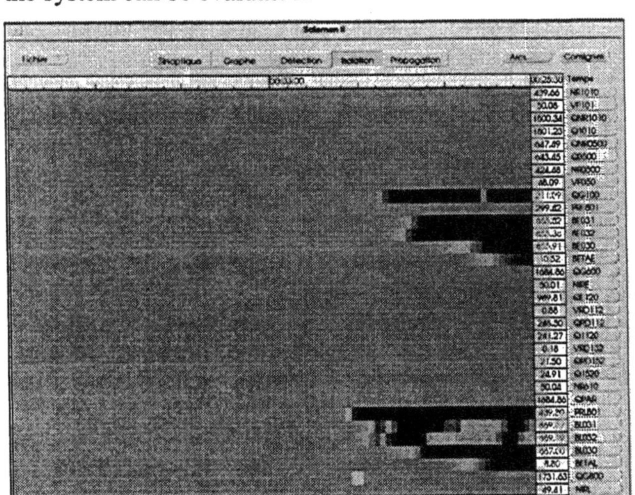

Fig 6: Isolation view.

Isolation view and causal graph representation could be concurrently used in the isolation task reasoning. With the isolation view it is very relevant to reason on the dates of appearance and disappearance of correlated alarms, on the historic of these alarms to get the right isolation reasoning. But it implies that all the relationships between variables be well assimilated by the operators.

The last view of the third layer is devoted to the predicted behaviour of the process (Fig 7). In this view, present time measures are shown in the middle of the screen. The left side screen shows past behaviour while right side screen shows the behaviour predicted by the causal model, according to the same colour representation as in the detection view.

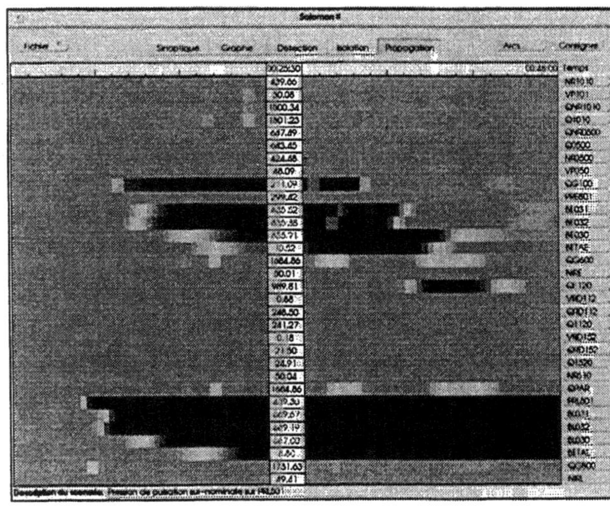

Fig 7: Propagation view.

The third layer attempts to present the time behaviour of the process state. The objective of this interface is to present a large amount of qualitative information about the fault evolution that helps the operator to analyse the alarm chaining. Moreover, isolation interface shows an image that can be seen as a graphical representation of the fault. In the propagation view, the operator is informed in the very beginning of the fault if its consequences will affect a large number of variables. In a real system this view could be used to test action alternatives.

5. EVALUATION

An evaluation of the prototype system was done with students of a technical school. Eight students trained to be process operators of technical secondary school *Lycée Jean-Baptiste Dumas* (Alès) took part into the experiments. The students were arranged into two groups with equivalent skills. Each group tried a version of the system: the first group tried the full system, while the second group had not access to the second layer with causal graph information.

A general presentation about the process and about basic notions of model based diagnostic and causal reasoning was firstly made to all the students. Experimentation were carried out with each student at a time. All students were confronted to the same four fault scenarios of growing difficulty. Each scenario took about 15 minutes for its development.

At the end of each scenario, the students were asked to explain the process evolution. At the end of each experiment, the students were asked to answer a questionnaire composed by general questions about the interface and more specific questions about the evolution of some variables in each scenario.

The aim of the experimentation was to test the way students understood the notions of model based and causal reasoning. It is clear that the too limited number of students and the time given to each of them did not allow to get precise statistics but only to get a qualitative evaluation of the relevance of the different notions for future process operators.

Classically, the main representation of the physical phenomena propagation used by students is the flow diagram. So when the fault effects can be observed on a single phase, students manage easily to achieve the right diagnosis because the fault effects propagation is perfectly equivalent to the flow diagram circulation.

When the fault is more difficult to diagnose, involving dynamical relationships that are not included in the flow diagram, students, who are not experts of liquid-liquid extractors, have had difficulties to give the right explanation. Nevertheless, those using the graph interface were better able to make a clear description of the underlying phenomena.

Consequently the graph seems to constitute a good support to understand the fault propagation through the causal dependencies as soon as the diagnosis problem is complex.

6. CONCLUSIONS

The most important difficulty reported by the students was the meaning of the names of the variables and their position in the system interface. They had to spend time to identify each variable in their diagnosis reasoning.

From the analysis of the questionnaires and from observation of the student reactions to system information, the following conclusions can be drawn:

- The students that used the version without the causal graph interface demonstrated more difficulty to understand the variables' evolution. For the others, even if the graph was, at a first seen, very complex, they all managed to explain the physical phenomena propagation using the causal relationships between the alarms.
- The representation of the process as a causal graph was well understood by the students. The teachers identified in the causal graph a powerful teaching tool complementary to the classical process representation in chemical engineering (flow diagram) and well dedicated to fault diagnosis reasoning.
- The representation of the variable states by a colour code evolving in time was easily understood by the students, once it is a very intuitive concept. As a large number of colour grades were available, abrupt changes are represented differently from slow changes. This principle was clearly perceived and approved by the students.
- The display of the variables evolution by both their measure and their reference value was well understood by the students. The concept of model based operation support was very easily assimilated.
- The causal graph view and isolation view were rather considered by the students as analysis tools while the others views were recognised as monitoring tools.

Evidently, the results shown above could not be seen as definitive once experimental conditions were far from ideal. Nevertheless, users in a first contact with the system, by a small amount of time, were able, at the end of experimentation, to understand most of complex physical behaviour of the process and to diagnose various types of faults.

AKNOWLEDGEMENTS

The authors are grateful for the kindly co-operation of the students and teachers of *Lycée Jean-Baptiste Dumas* with the evaluation described in this work.

REFERENCES

Evsukoff, A., J. Montmain and S. Gentil (1997). Dynamic model and causal knowledge-based fault detection and isolation. *IFAC SAFEPROCESS'97.*

Kramer, M. A. (1987). Malfunction diagnosis using quantitative models with non boolean reasoning in expert systems. *AiChe Journal* **33** (1).

Lambert, M., B. Riera and G. Martel (1997). Conception d'une interface de nouvelle génération pour la supervision d'un procédé de retraitement de combustibles nucleaires. *6e Congres Français de Génie des Procédés.*

Leitch, R. and M. Gallanti (1992). Task classification for knowledge-based systems in industrial automation. *IEEE Trans. on Systems Man an Cybernetics*, **22** (1), pp. 142-152.

Leyval L., S. Gentil and S. Feray-Beaumont (1994). Model based causal reasoning for process supervision. *Automatica*, **30** (8), pp. 1295-1306.

Montmain J., L. Leyval and S. Gentil (1994). Qualitative analysis for decision making in supervision of industrial continuous processes. *Mathematics and Computers in Simulation*, **36**, pp 149-163.

Rasmussen J. (1993). Diagnostic reasoning in action. *IEEE Trans. on Systems Man an Cybernetics*, **23** (4), pp. 981-991.9

SDG MODEL-BASED STRUCTURES FOR FAULT DETECTION

I.P. Gál and K.M. Hangos

Systems and Control Research Laboratory,
Computer and Automation Institute,
Hungarian Academy of Sciences, 1518 Budapest, POBox. 63, Hungary.
Phone: 36 1 1667 483, Fax: 36 1 1667 503

Abstract: A signed directed graph (SDG) model-based approach is given in this paper for fault diagnosis of process systems. The proposed method is modular, the model for fault propagation is built from the *mini SDGs* of the units driven by the process flowsheet. Thus the fault detection is decomposed into two steps: off-line generation of the SDG model and its rule set equivalent and the step of on-line fault detection. *Copyright © 1998 IFAC*

Keywords: Fault detection, Process models, Graph theoretic models, Rules, Signed Directed Graphs, Reasoning

1. INTRODUCTION

New approaches to the qualitative model-based diagnosis of industrial processes have been reported recently by Frank, Koepen-Seliger (1995), Milne (1987), Tzafestas (1989). These methodologies have also been made available to the broader diagnostic community through CAD environments with advanced man-machine interfaces by Leitch *et. al.* (1994), Venkatasubramanian *et. al.* (1994), Andow (1985).

The qualitative model-based part of the CT94-0237 Copernicus project deals with the knowledge representation and acquisition problems in the form of a qualitative model-based diagnostic toolbox and suggests a possible solution to them. The aim of the toolbox is to provide a unifying framework for the implementation and investigation of various qualitative model-based diagnostic methods. The SDG model-based diagnostic module is implemented in this toolbox reported by Gal *et al.* (1996).

The fault-detecting framework is applied in the field of chemical and technological process systems.

2. INVESTIGATIONS BASED ON SDG MODELS

2.1 The SDG model of a process

The signed directed graph (SDG) is a special type of influence graphs, which were the first type of qualitative models widely applied in chemical process modelling and analysis. SDGs are used for control and diagnosis of chemical engineering systems in particular.

The structure of an SDG consist of the following elements:

- *nodes* representing the *variables* (state, input and output) of the system to be modelled;
- *directed edges* describing *relation* between variables;
- *sign* associated to the edges describing the sign of the influence they represent.

Let us assume that we have the model of the dynamic

system in the form of non-linear ordinary differential equations being the state equations of a state-spece model:

$$\frac{d\mathbf{x}}{dt} = f_{\mathbf{x}}(\mathbf{x}, \mathbf{u}) \tag{1}$$

where

t	time;
x	vector of state (and output) variables;
f_x	vector-vector function;
u	input variables.

These set of (non-linear) ordinary differential equations can be associated to a signed directed graph (SDG). The SDG fixes the following structural properties of this ODE model.

The variable type (state, input or output) is determined by the reachability property of their associated node: whether it is intermediate (with inward and outward directed edges, i.e. state), initial (with only outward directed edges, i.e. input) or terminal (with only inward directed edges, i.e. output) node.

The number and type of variables entering the right hand side function of the ODE for the *i*th variable is given by the starting variables of the influence edges which end at the *i*th variable node.

The sign of the partial derivative of the non-linear right hand side function with respect to its independent variables is given by symbols associated to the corresponding edges.

2.2 Qualitative simulation

We can establish the effect of a given disturbance from the SDG model of a process. For example the qualitative state sgn(B) of variable B can be established in the following way:

$$\text{sgn(B)=sgn(A-B)}*\text{sgn(A)}$$

where sgn(A-B) is the sign associated to the edge A →B.

Positive difference in variable A i.e. sgn(A)=„+" and positive edge between variable A and B i.e. sgn(A-B)=„+" cause positive difference in variable B i.e. sgn(B)=„+".

If there is no edge between A and B but there is a path between them (Fig. 1.) then sgn(A-B) can be computed in the following way:

$$\text{sgn(A-B)=sgn}(x_n\text{-B})*\text{sgn}(x_{n-1}\text{-}x_n)*\ldots$$
$$\ldots*\text{sgn}(x_1\text{-}x_2)*\text{sgn(A-}x_1) \tag{2}$$

and

$$\text{sgn(B)= sgn}(x_n\text{-B})*\text{sgn}(x_{n-1}\text{-}x_n)*\ldots$$
$$\ldots*\text{sgn}(x_1\text{-}x_2)*\text{sgn(A-}x_1)*\text{sgn(A)} \tag{3}$$

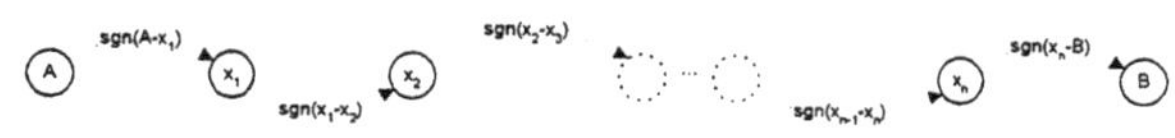

Fig. 1. Effect path trough more variables

With forward simulation on SDG models we can easily get the qualitative description of behaviour.

2.3 Fault detection

Fault detection differs in one main point from the forward simulation: the fault has to be found based on given measured data. Usually a hypothesis is made and a behaviour sample is predicted from a qualitative model. After that comparing the predicted and observed sample checks the truth of the hypothesis.

Truth tables in Table 1. and Table 2. can be used for checking hypothesises by SDG models where 'T' stands for *true* and 'F' for *false*.

Table 1. Truth table for the branches of SDG model with sign "+"

A —+→ B		B +	B 0	B -
	+	T	F	F
A	0	F	T	F
	-	F	F	T

Table 2. Truth table for the branches of SDG model with sign "-"

A —−→ B		B +	B 0	B -
	+	F	F	T
A	0	F	T	F
	-	T	F	F

A fault runs along a whole branch if the edges of the branch are consistently true based on the values of measured variables. According to that if there is positive difference in A and B, i.e. sgn(A)=sgn(B)=„+" then the fault runs further on edge A-B if the edge has positive sign, i.e. sgn(A-B)=„+". As the fault runs only on consistent edges the truth table can be used for checking the consistency of the fault hypothesis in the SDG models.

3. GRAPH CONSTRUCTION AND DEDUCTION ALGORITHM

The graph-construction algorithm builds the fault SDG of the entire system from mini SDGs on which qualitative simulation and fault detection run.

3.1 The algorithm

The underlying assumptions of the algorithm. Taken into account the restrictions of the method based on SDG models the assumptions are:

- Each fault source is given; their effects are investigated in the system.
- Only one fault appears at one time, only these faults can be detected by SDG method.
- The state of each control loop can be observed and modelled in the system. It is necessary because the states of the control loops can significantly influence the SDG model of the system, so different structures can emerge.
- Each variable goes through only one change when a fault occurs. If a variable goes through more than one change at a time then the propagation of the fault may not be identifiable with the SDG model.

Models of the process units. Taking into account the requirement of the modularity and the easy algorithmisation, the description of the SDG of process units is defined in the following way: A given type of unit is given a description once and when constructing of the flowsheet it gets an ID, i.e. it is implemented in an object-oriented way.

A *mini SDG* is a monodepth tree, which represents the effects of a given variable. Only the mini SDG-s are stored. The whole SDG model of the unit and the fault source SDG can be constructed from these mini SDG-s by the graph construction algorithm.

The main steps of the graph construction algorithm. Taking into account all possible fault sources the algorithm constructs the SDG tree by walking all the branches. Then it determines the values of the measured variables with the help of forward simulation. Then it generates the rules arranged in a ruletable, which will be used for fault detection. The detailed steps of the graph constructing algorithm are as follows:

All possible fault sources have to be enumerated for constructing the SDG tree, which contains the effects of all fault sources in case of each fault.

All possible variables should be directly measured on their available measurement point in order to compare the measured values with the effects of the assumed fault. If there is a bias in the instruments then it has to be taken into account by proper modelling of the instruments. If we assume that the measurement system is ideal we can integrate the instruments and the variables measured by them. This situation can be seen on Fig. 2.

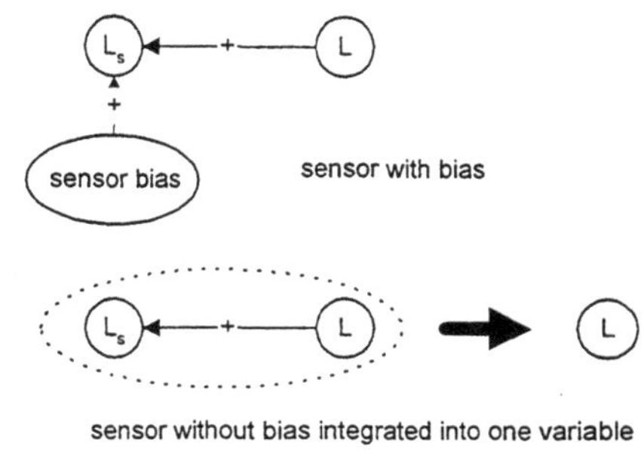

Fig. 2. Possible integration of the instruments

Construction the SDG trees for each possible fault sources. The algorithm determines the effect path for all faults so it generates a tree from the available model. Inconsistent branches and duplications are not allowed in the tree. Therefore if a variable appears again in the tree then the effect path will not be investigated from this variable because it would lead to recursion.

Critical factor: state of the control loops. How can we take into consideration the state of the control loops at fault detection? It is important because a control loop can be for example active or saturated, which affects the fault structure of the process system. The behaviour of a saturated control loop can be similar to a disconnected loop so it is essential to make a distinction between them from the viewpoint of the process. Different fault SDG trees have to be constructed taken into account the states of the control loops and their combination because the SDG of the system may change and some path may disappear from the system. Therefore different fault detecting rules have to be generated. The states of the control loops appear as a parameter in the individual fault detecting rules next to the symptoms characterised by the individual faults. We assume that the states of the control loops can be determined and can be either normal or saturated.

Connections between the units. The connections between the units are described by the edited flowsheets in the G2 environment. Two positive branches are defined vice versa between the connected units in order to construct the SDG.

If the tree-constructing algorithm reaches to input or an output of a unit it asks what kind of input or output of another unit is connected to the given unit. If this other unit is not built in the tree the algorithm will built it in the tree and will continue the constructing part of the algorithm from there.

3.2 Generation of fault detecting rules

A fault detecting rule contains:

- the name or the code of the fault;
- the states of the control loops;
- appearing symptoms, the values of the measured variables.

The constructing steps of the rules are as follows:

1. All branches are walked through and the unmeasured nodes are removed because only the measured variables can give the symptoms. After that the symptoms belonging to the fault are determined. These symptoms describe the expansion of the fault.

2. The basic rules belonging to the faults are generated. Between the symptoms which are on the same branch there is logical AND relation so if a fault expands on a branch then all variables must be true on this branch. A rule generated on a branch is a basic rule.

3. Between the individual branches there is logical OR relation, because they represent the different expansion manner of the given fault in the system.

The fault detection is performed by using these rules.

A sample of the rule table can be seen in Table 3. where CL*i* indicates the control loop status, and S*i* are the symptoms.

Table 3. Sample rule table

Fault	*CL1*	*CL2*	...	*S1*	*S2*	...
fault 1	normal	normal	...	0	+	...
fault 1	normal	normal	...	+	0	...
fault 2	normal	saturated	...	+	0	...
fault 3	normal	normal	...	0	-	...
...						

3.3 The steps of the fault detection.

After generating the rules in off-line mode the measured data are converted into qualitative values and the rules are evaluated in the following steps.

1. As the methods based on SDG-s show only the difference from the steady state (x_s) the measured qualitative data have to be converted into qualitative values compared to the steady state in order to use the fault detecting rules. It can be done in the following way:

$$\text{if } abs(x-x_s) \leq d \text{ then } \quad 0$$
$$\text{if } abs(x-x_s) > d \text{ then } \quad sgn(x-x_s)$$

where x is the measured variable, x_s is the value of the variable in the steady state, d is the half width of the zone in which the variable is not considered different from the steady state. It is for filtering out smaller environmental disturbances. An example can be seen on Fig. 3.

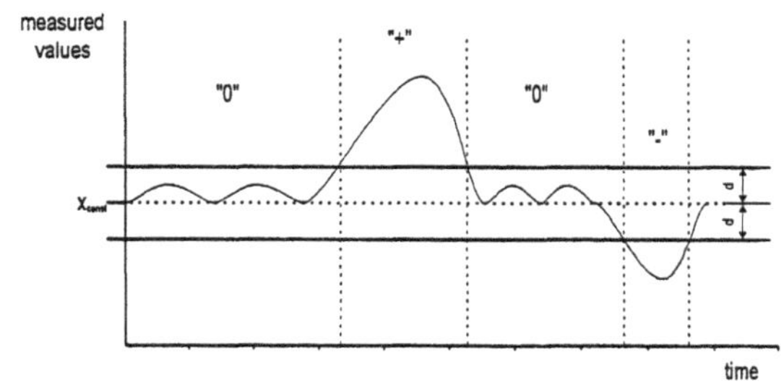

Fig. 3. Making the variables qualitative

2. The values of the measured variables are compared with the symptoms contained in the generated hypothesis and the possible appearing faults are classified.

At the evaluation of the rules according to the state of the control loops are taken into account and the hypothetical symptoms are compared with the measured variables made quantitative.

Classification of the appearing faults:

- If a subset of the symptoms of a fault is true then the fault will get into the list of the conditional faults with the appeared/all symptom factor.
- If the difference of the measured variables and the set of the fault symptoms are the same then the fault get into the list of the existing faults.
- If a symptom of an existing fault disappears then the fault gets into the list of archive faults.

4. CASE STUDY

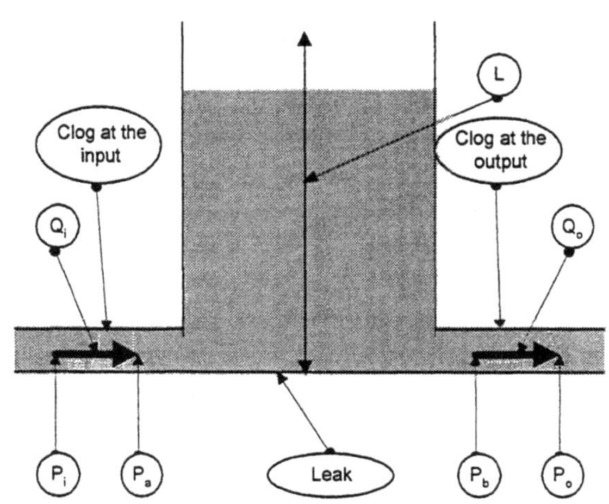

Fig. 4. The schematic of a tank

Let us consider a simple tank with an inflow-, an

outflow port and a level gauge. Let us assume that only three faults can occur:

- leaking in the tank;
- clog at the input;
- clog at the output.

The variables in the Fig. 4. are the following:

Table 3. The variables of the tank

Variable	Explanation
P_i	Pressure at the input
Q_i	Streaming speed at the input
P_a	Inlet pressure in the tank
L	Level of the liquid in the tank
P_b	Outlet pressure in the tank
Q_o	Streaming speed at the output
P_o	Pressure at the output

4.1 Construction of the fault source SDG

The SDG model of the tank with the three possible faults can be seen in Fig. 5.

The mini SDGs belonging to the SDG model of the tank can be seen in Fig. 6. It is easy to see in the figure what kind of effect path starts from each process variable. In the SDGs belonging to the faults „X" denotes the branches, which will not appear in the tree. These branches cause inconsistency or duplication. The SDGs of the tank belonging to the faults in the tank can be seen in Fig. 7., 8. and 9. The input and the output variables are denoted by the framed variables. The connections of the units is built on these framed variables.

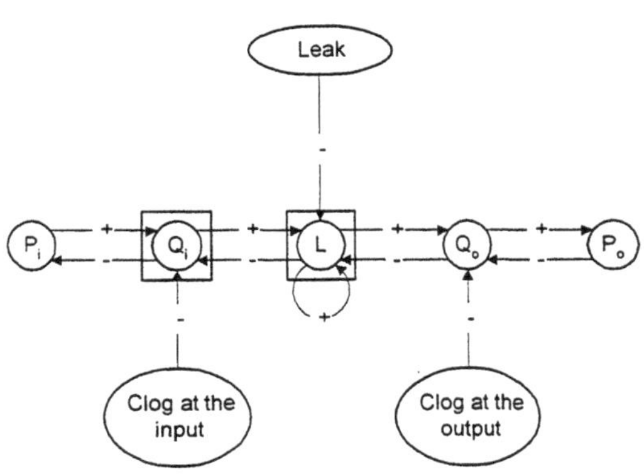

Fig. 5. The SDG model of the tank with the three possible faults

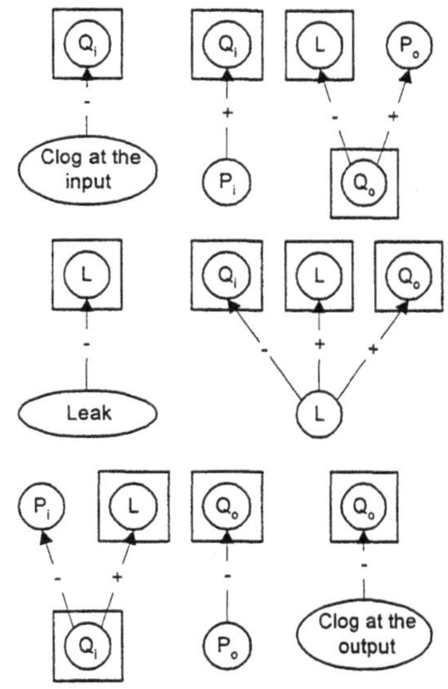

Fig. 6. The mini SDGs of the tank

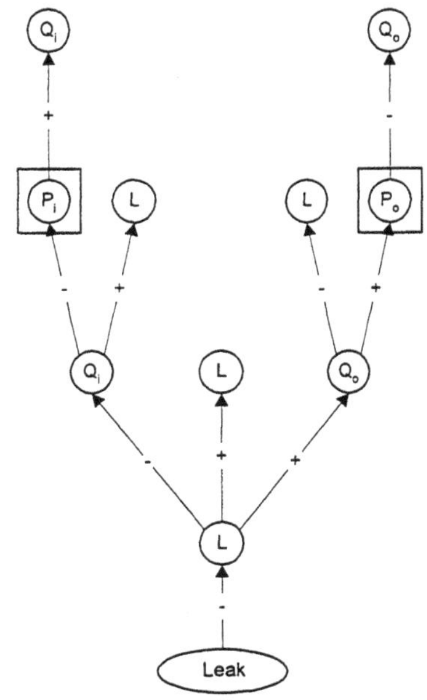

Fig. 7. The SDG tree for the leak in the tank

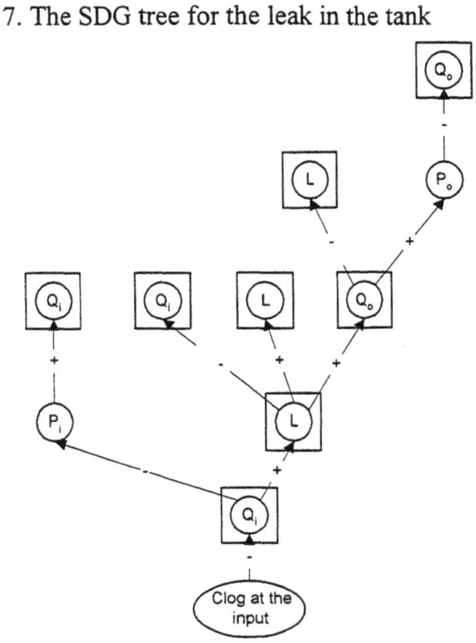

Fig. 8. The SDG tree for the clog at the input

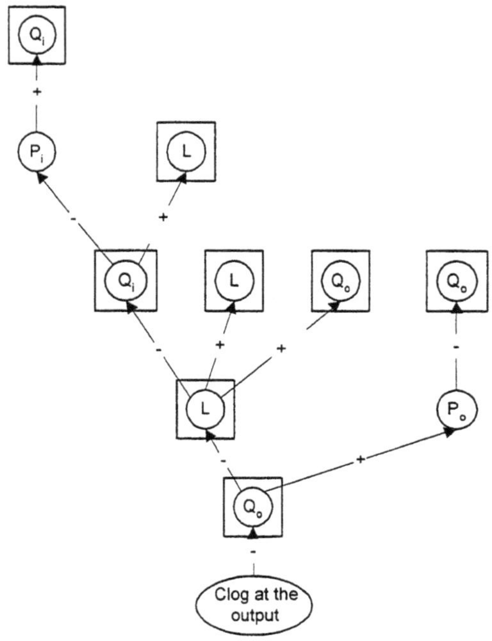

Fig. 9. The SDG tree of clog at the output in case of the tank

4.2 The ruletables

The rule set equivalent of the SDG trees of the tank corresponding to different fault sources generated by the algorithm for generation of fault detecting rules is shown in Table 4. The normalized ruletable in Table 5. is obtained therefrom by normalization, i.e. by performing the logical OR operation of the rows belonging to the same fault source (row 1 and 2 in Table 4.).

Table 4. The basic ruletable based on the SDG of the tank

Fault	Q_i	L	Q_o
leaking	0	-	-
leaking	+	-	0
clog at the output	-	+	-
clog at the input	-	-	-

Table 5. The normalised ruletable based on the SDG of the tank

Fault	Q_i	L	Q_o
leaking	-	-	-
clog at the output	+	+	-
clog at the input	-	-	-

The generated rules can be used for fault detection in any rule-based expert system and may also be manually edited to incorporate operation experience.

5. CONCLUSION

A signed directed graph (SDG) model-based approach is proposed for fault diagnosis of process systems. The proposed method is modular, the model for fault propagation is built from the *mini SDGs* of the units driven by the process flow-sheet. An algorithm for constructing the overall SDG trees of the process system for every fault source is proposed. Thereafter the SDG trees are algorithmically converted into normalised rule sets arranged in rule tables. These rules are suitable for performing the fault detection in any rule-based system.

The great advantage of this method is that the SDG description of the given units can be used in a modular way i.e. if one type of unit shows up in the system more than once the algorithm can take its model from one object.

REFERENCES

Andow, P.K. (1985). Fault Diagnosis Using intelligent knowledge based systems. *Chem. Eng. Res Des,* Vol. 63, 368-387.

Frank, P., Koepen-Selinger B. (1995). New developments using AI in fault diagnosis. *Prepr. IFAC/IMACS Workshop on Artificial Intelligence in Real-Time Control*, 1-12, Bled.

Gal, I., Znidarsic, A., Csaki, Zs., Hangos, K.M., Juricic, D., Knowledge acquisition in a qualitative diagnostic toolbox, *Symposium on Robotics and Cybernetics, CESA'96 IMACS Multiconference*, 105-109, Lille.

Leitch, R., Freitag, R. H., Stefanini, A., Tornielli, G. (1994). Using the ARTIST approach for diagnosing power transmission networks, *Intell. Syst. Eng.*, Vol. 3. No. 3, 125-37.

Milne, R. (1987). Strategies for Diagnosis. *IEEE Transaction on Systems, Man and Cybernetics*, Vol. SMC-17, No. 3, 333-339.

Tzafestas, S. (1989). System fault diagnosis using the knowledge-based methodology. In: *Fault Diagnosis in Dynamic Systems: Theory and Application.* Prentice Hall, UK

Venkatasubramanian, V., Vaidhyanathan, R. (1994). A knowledge-based framework for automating HAZOP Analysis. *AIChE Journal*, 496-505.

ACKNOWLEDGEMENT

The work has been supported by the Hungarian National Research Fund through grant T017 108 and by the EC through COPERNICUS project grant CT94-0237.

INTELLIGENT CONTROL METHODOLOGIES FOR CONTROL STRATEGY AND CONTROL STRUCTURE SELECTION

Ammar Al Khani and Heikki Koivo

Control Engineering Laboratory
Department Of Automation And Systems Technology
Helsinki University of Technology
PL. 3000, FIN - 02015 HUT, Finland
e-mail: Ammar.Al.Khani @hut.fi
Fax: 358 - 9 - 451 3333

Abstract: The purpose of this paper is to present a brief overlook and evaluation of the methods of artificial intelligence in process control that are relative to the selection of a control system and the choice of an adequate control policy of industrial processes. Consequently it discusses the selection of an adequate process control structure based on the intelligent control methodologies including neural networks, fuzzy and hybrid techniques, modeling and identification, and data analysis methods. For this goal we have elaborated a brief study and evaluation of process control methodologies for control strategy and control structure selection, and we have highlighted the needs to implement advanced adequate solutions to the existing problems. *Copyright © 1998 IFAC*

Key Words: Adaptive control, Artificial intelligence, Identification, Hierarchical Control, Intelligent Control, Fuzzy systems, Neural network.

1. INTRODUCTION

Control methods and strategies can be classified into categories in function of the temporal evolution of these techniques: The conventional, such as the PID family, the classical, such as selective controllers, the model based and predictive techniques, the new generation advanced techniques based on optimal control theory, expert systems, neural networks, fuzzy and hybrid systems.

A global overlook of the present process control systems that are implemented worldwide in various sectors of production, and ranging from chemical industries as the largest customer for these techniques, to space applications passing through medicine, shows roughly that the majority of control loops are still relying on reliable conventional controllers such as the PIDs family (Seborg, 1994; Al Khani et al, 1996).

The aim of every project is to solve the troïka relationship of Product quality, plant efficiency and investments payoff. This requires advanced modern control theory and techniques in the field to improve control quality and consequently to improve the product quality and to limit wastes of raw materials, which are leading to improve plant productivity and efficiency. The successful application of the new generation intelligent control systems has the key role to play in the technology of the next millenium.

The implementation of intelligent control theories and strategies and advanced communication networks implies, first, an evaluation study of process control methodologies and structures, which is important for understanding the problem, the demand and the adequate solution for it.
Designing a control system of a given process needs a thorough understanding of the process in order to highlight its characteristics in terms of its dynamics, its main parameters and the correlation between them, and to select an adequate model of the process. Then, the

type of the needed control be it centralized or decentralized, can be determined in function of the complexity of the process.

The choice between a direct control algorithm and a model based control algorithm remains for the designer to decide each of them having its advantages. The present study highlights some of the methods and issues related to the control scheme structure, intelligent control strategies, intelligent instrumentation, hybrid based process modeling and control, data processing, filtering and identification.

2. DISTRIBUTED INTELLIGENT CONTROL STRUCTURE

The adopted control strategy and based this discussion on, is presented in Fig. 1. It is a general intelligent control based hierarchical scheme, which is composed of three levels. This structure is much more compact than the conventional ones. It has a reduced complexity thanks to the advances in personal computers. New generation PCs permits us to build systems based on workstations and PCs, including process communication networks of Internets and Intranets. Distributed control structure has the advantage over a centralized one in de-coupling the global system into independent sub-systems, and reducing considerably the complexity and the quantity of information that the control system must deal with. Consequently, we will consider a system in a global decentralized framework. In this system each level can perform independent tasks or functions.

2.1 The Intelligent Control scheme strategy

The control policy may be based on supervised or unsupervised control and usually a supervised learning or supervised control will improve the quality of the system if the dynamics of the process is an active time-varying one. The control loops may include open loops in information regime beside the closed loop control, and this can be decided based on the observed and controlled process parameters. Each level of the intelligent control structure has its specific tasks as follows:

Level I: Here, the system comprises of sensors, data processing intelligent instrumentation and local control modules. Data processing modules perform the tasks of Filtering, elimination of noise, normalization, linearisation and discretization of data.

Level II: Includes intelligent control hierarchic module: Supervisory and observer module and the executive module. In addition, it is concerned with the process inter-communication network.

Level III: It is concerned with process modeling, process global knowledge and history, global supervision and optimisation and provides a selective bridge to extra-communication network.

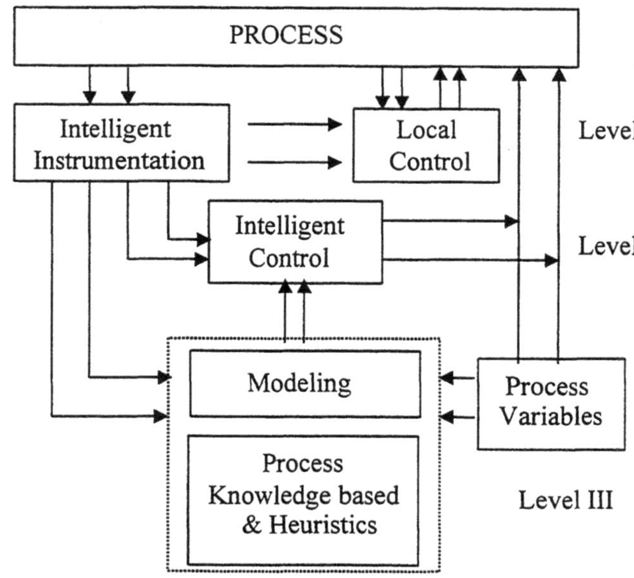

Fig. 1. A three-layer artificial intelligence based control hierarchy scheme of a complex process

3. TOOLS FOR THE DESIGN OF INTELLIGENT INSTRUMENTATION AND DATA PROCESSING

Inconsistency and noisy data and extracting knowledge from poor information are typical problems to be tackled. Their exist many techniques to help designing adequate instruments to handle such problems.

-Using neural networks as intelligent instrumentation for data processing is convenient. They can perform filtering of noisy data. Their parallel and distributed structure permits them to eliminate the noise and prevent the propagation of errors in the system. Thanks to their non-linear character, they enable to build non-linear filters based system using a system of neural networks: The approximating capability of neural networks will improve the accuracy of estimation in the case of nonlinear systems and compensate uncertainties.

-Fast Boundary Adaptation Rule: This enables performing fast adaptive scalar quantization of process data.

-Hybrid systems: Fuzzy-Neural Network: Data dependent fuzzy filter - DDF filter, where the parameters of the DDF filter are estimated using fuzzy techniques:

- Adaptive Center Weighted Average Filter
- Modified Trimmed Mean Filter

In the case of incomplete or inconsistent data, neural networks can help reconstruction of the original signal (the signals sent to and from perceptrons are continuous functions and consequently, neural network can deduce proper conclusions).

3.1 Intelligent Control Tools

By definition, an intelligent control system is a system that is able to respond to the interacting and dynamic environment of the process. It learns from the process history in order to deal with process difficulties by generating appropriate control actions. The intelligent control system is built of two modules: a supervisor and observer level and executive level for the control and identification. Techniques that are involved comprise evolutionary computing, neural networks, knowledge based and causal reasoning systems, fuzzy logic, chaos theory and hybrid systems such as neuro-fuzzy, neuro-adaptive and intelligent-conventional.

Intelligent control system that is based on neural networks is a structure which is imperatively needed when we are concerned with complex dynamic systems with their complexity, uncertainty, and present non-linearity. These systems are represented by poor analytical models and characterized by a high degree of freedom, and eventually composed of subsystems. They need multiple optimization, adding to bill a noisy data. Rectification and distillation processes are examples of such systems. As a difference from adaptive algorithms, which are used in presence of parametric uncertainties and dynamic environment, neural network based systems can perform better because they can learn from the process history the optimal values of the control parameters that correspond to different situations. Also they can be used to select adequate criteria to the subsystems, parametric identification and pattern recognition.

Let us consider a complex dynamic system. This system can be described through different models, linear and nonlinear, relative to its complex nature (Narendra et al., 1992). Consider a representation of a nonlinear system:

$$x(t+1) = (\mathbf{I} + A\Delta t)\,x(t) + f(x(t)) + (B\Delta t)\,u(t) \qquad (1)$$

After a failure, the non linear system can be expressed as:

$$x(t+1) = (\mathbf{I} + A\Delta t)\,x(t) + f_1(x(t)) + (B_1\Delta t)\,u(t) \quad (2)$$
$$x(t+1) = (\mathbf{I} + A\Delta t)\,x(t) + f_2(x(t)) + (B_2\Delta t)\,u(t) \quad (3)$$

where the non-linearities are given by functions f, f_1, f_2.

When the control objective is to maintain optimal behavior of the system in case of a failure, the supervisor level in its task of pattern recognition and detection can distinguish between the different configurations before and after a failure.

The other task of the supervisor is to decide which controller is suitable for the configuration of the system, linear or nonlinear. The recognition of the correct controller is based on the responses of the system of the possible configurations considered with the same controller. If a deviation from the desired set point is observed, it is analyzed and classified according to one of the configurations and consequently the adequate controller is selected.

To realize this structure, the authors propose a combination of two neural networks for the task of detection and pattern recognition.

3.2 The Feed Forward Scheme

This is the conventional back-propagation based neural networks (Quantrille et al., 1992). The minimum search strategy is realized by a gradient-descent sigmoid function, the objective of which is to minimize the square error between the desired y_d and calculated y_c output values of the system:

$$Er = \sum_n \varepsilon_n^2 = \sum_n (y_d - y_c)^2 \qquad (4)$$

However, the backpropagation algorithm has its limitations in real-time and on-line applications if it is not processed on parallel computing devices. The time consuming learning step may take several minutes in unsupervised mode learning for networks with many input-outputs. Many improvements have been made by combining it with other techniques in supervised mode control scheme such as recursive least squares algorithm or adaptive filters, to accelerate its convergence by providing long range prediction.

3.3 Stochastic Supervisor Learning

Consider a neural network presented by the following input-output recursive discrete time model:

$$y(k) = N[y(k-1), y(k-2), .., y(k-m+1), u(k-1), u(k-1), \ldots, u(k-m+1)] \qquad (5)$$

The stochastic supervisor can be defined by the stochastic function S, (AL Khani, 1995):

S: S: $\mathbf{Y} \rightarrow \mathbf{U}$; where Y presents the output of the Neural Network system, $\mathbf{Y} \in [0,1]$, and $\mathbf{U}$ presents the vector output of the stochastic supervisor. The stochastic supervisor can be expressed in terms of the mathematical expectation E of the conditioned penalty over the system output y (t):

$$I(t) = E[u(t) / P(t)] \qquad (6)$$

where P is the state probability distribution vector.

3.4. Model based Control

Model based control strategies comprise many techniques that have been applied on both linear and non-linear processes in different configuration and versions. They regroup families of model predictive control, internal model control, process model based control, neural networks based control and fuzzy systems. Examples of industrial and pilot application can be found in many processes in pharmaceutical, chemical and petrochemical industries. An industrial scale real comparison between seven of these techniques has been conducted by Joshi et al. (1997), in order to establish a comprehensive evaluation. The tests were carried on two processes: A Heat exchange process and a steam flowrate regime. The qualitative evaluation criteria of the controllers included: Squared error integral, controller integral of absolute value of the output changes, costs, operator and computational aspects, fault tolerance and other constraint and benefits. The subjective appreciation concluded that all the above tested techniques were difficult to handle by plant operators. However, all tested techniques have demonstrated similar performance when applied with respect to its specifications and the characteristics of the target process. A tandem master/slave between different techniques have improved the control quality by combining linear and non-linear capabilities.

3.5. Dynamic Matrix Control

This algorithm belongs to Model Predictive Control techniques (Adderrazak et al., 1996; Thibault et al., 1992). The control objective of the DMC is to predict over a future command horizon a set of M process inputs $\Delta\mathbf{I}_t$ at command intervals of $\mathbf{P}_t$ future time intervals called prediction horizon. The predicted inputs allow minimizing the sum of square deviation between the process outputs and the set points over the prediction horizon. The predicted command solution for a multivariable system by least-squares is calculated from the following equation:

$$\Delta\mathbf{I}_t = (\mathbf{A}^T\mathbf{B}^T\mathbf{B}\mathbf{A} + \mathbf{C}^T\mathbf{C})^{-1}\,\mathbf{A}^T\mathbf{B}^T\mathbf{B}\,\varepsilon_{t+1} \quad (7)$$

where $\mathbf{A}$ is the dynamic matrix of the multivariable system, $\mathbf{B}$ is a diagonal matrix of weights of controlled variables, $\mathbf{C}$ is a diagonal matrix of move suppression factors of inputs amplitudes. This is done by manipulating some control variables, and ε is a vector of deviation of the predicted outputs from the desired set points. This method has been applied successfully to many processes such as multivariable control of a polymerization reactor.

3.6. The Model Algorithmic Control

This technique is from the family of model based predictive control algorithms (Thibault et al., 1992), and its long range prediction logic is based upon the impulse response of the process. It is effective in handling process constraints. The control objective function over the prediction horizon is given by the following expression:

$$P = \sum_{d=0}^{N-1}\left(y_{t+d+1} - \hat{y}_{t+d+1}\right)^2 + \gamma\sum_{I=0}^{N-1}\left(u_{t+1} - u_{t+I+1}\right)^2 \quad (8)$$

where γ is a weight factor, $\hat{y}$ is a predicted output and y is a desired output, u is a calculated command in the horizon of prediction, by the control mechanism upon estimates made by the neural model to minimize the criteria P. The range of the prediction horizon in general is smaller than N in order to reduce calculations.

3.7. Adaptive heuristic critic

This technique is a forward-based supervised learning with reinforcement neural networks. The learning procedure in this technique is called reinforcement adaptation. In it each component receives single error estimation (Jianming et al., 1995).

3.8. Self organizing maps

Kohonen self-organizing map and self-creating map are based on a data clustering technique. They are suitable for identification and classification of process data.

3.9. Hybrid systems

These systems combine neural networks and model based control systems and they include methods such as neural model-based control, History-stake learning and a family of adaptive neural predictive control systems.

3.10. Adaptive Neural Control System

To improve the performance of learning and to accelerate the convergence of the control algorithm, we can combine the neural network based system with an adaptive controller with the objective to minimize quadratic criteria. The estimated parameters of the control law are obtained from a recursive estimator such as the least squares or the maximum of likelihood. The design of a NN system coupled with an estimation mechanism like the Kalman Filter, gives some advantages because this system combines the nonlinear dynamic with a linear filter able to deal with systems with poor information (Kang et al., 1996).

4. IDENTIFICATION

Parametric identification is crucial. Poor or rich feedback information from the process can determine our control strategy and algorithms. It is important to select those variables to which the process is most sensitive. This can be realized by different ways such as the response of the dynamics of the system to different activating inputs like impulses, step response, etc.... However, conjugating statistic methods to neural networks and fuzzy systems generates a powerful tool in process control. Sensitivity analysis is very useful to assess the estimation quality of the process parameters.

4.1 Sensitivity analysis

The frequently used techniques are Monte-Carlo, Least squares and linearization of process balance equations. For the latter method we divide the process variables into two subsets, independent variables and the others. An analysis of the variance-covariance matrices of the estimated process variables can prove the reliability of the selected variables and the correlation between them. After the selection of the model variables, we determine its type, explicit or implicit, and optimize the number of its variables. This can be done by several ways of which the criterion of Akaike:

- Akaike Criterion of Finite Prediction Error AFPE
- Akaike Information Criterion AIC

The suitable model order is that one which enables it to minimize one of Akaike's criterions above.

4.2 Neural networks and identification

The identification of neural model consists of compressing the input-output time series into a compact way to describe the plant dynamics. Many studies demonstrate that a three layer feed forward ANN is equivalent to the auto-regression moving average-ARMA dynamic models (Sjöberg, 1994; Simpson, 1990). The systems are divided into two families, linear and nonlinear. For a class of linear system the input-output behavior of the system can be described by the state relation (Asriel, 1995):

$$x\,(k+1) = \mathbf{A}x\,(k) + \mathbf{B}^{T}u\,(k) \quad (9)$$

$$y\,(k) = \mathbf{C}^{T}x\,(k) \quad (10)$$

Where $\mathbf{A}$ is nxn matrix, $\mathbf{B}$ and $\mathbf{C}$ are n-dimensional vectors. The following discrete time recursive form can represent this system:

$$y\,(k) = \Sigma_{i=1}^{n} a_i y(k-i) + \Sigma_{i=1}^{m} b_i u(k-i) \quad (11)$$

The identification task of the neural network is reduced to simple parameter estimation. However most industrial systems present some non-linearities with inaccessible states, consequently, the identified linear models will not be adequate for such systems. So, to project the adaptation capabilities of the neural network, a similar step can be taken as in linear systems:

$$x\,(k+1) = F_f[x(k)\,,\,u(k)] \quad (12)$$

$$y\hat{}(k) = F_g\,[x\,(k)] \quad (13)$$

Where F_f and F_g are non-linear functions. Let us assume that there exist a number n and a continuous function F in a time discrete recursive input-output relation of the following form:

$$y\,(k+1) = F[y(t), y(t-1),.., y(t-n+1), u(t), u(t-1),\ldots, u\,(t-n+1)] \quad (14)$$

This relation has the same input-output behavior as the original system. Consequently, a feed forward neural network with its observable inputs and outputs can express the function F, and can be trained by different techniques (Narendra et al., 1992) such as dynamic backpropagation and backpropagation Conjugate Gradient algorithm.

5. PROCESS MODELING

Intelligent control methodologies and techniques have moved modeling from analytic to post analytic phase, and considering the difficulties encountered in producing exact models of most encountered processes in the chemical industries and especially for nonlinear and in dynamic state, many mathematical methods have been proposed and applied.

5.1 Model based technique

Model based adaptive technique family is one of them. It is relying on the creation of an exact model of the process and requires its specific dynamic characteristics such as time constant, model order, delay, for each application of the controller. From industrial point of view, these requirements are time consuming, increase the costs and limit the use of the controller in different processes.

Another developed technique, the dynamic modeling technology-DMT (Gouph et al., 1997), that is based on Laguerre functions, and is able to model a given process through determining its transfer function using process observed data. This is very important for the industry because DMT-modeling can be achieved on-line and for difficult processes. This technique has been applied successfully on kraft continuous digester.

In similarity with the DMT technique, Wiener model has been used to model various non-linear processes, thanks to its structure which is composed of two parts a dynamic linear and a static non-linear one. It has been applied coupled with a neural network, to the identification and control of processes such as pH control (Duwaish et al., 1996).

Process modeling using domain knowledge technique, which is based on neural networks, presents many advantages especially when they are coupled with adaptive techniques. Neural networks behave as simple models (Quantrille et al., 1992). They have the ability of self-organizing in order to map the underlying phenomena of a given process, this property permits the neural models to replace dynamic empirical models in model based control systems. They process data by curve fitting, and the information is distributed, so it is possible to relate the qualitative effects of the neural network structure and parameters to the process parameters. The micro-feature concept and the parallel character of the neural network ensure that if the value of one variable is off, the model will not be affected greatly as in empirical models, which depend on each independent variable.

5.2 Process neural predictive model

The adaptive technique is very useful providing high control quality of processes with time-varying active dynamics (Thibault et al., 1992). In general, linear adaptive control can achieve the task. However, in practice, most of the processes have a nonlinear character, and linear identification will not produce a real reflection of the dynamics of these processes. For the class of dynamic nonlinear processes, the neural networks can perform identification with poor initial knowledge of the process.

5.3 Process Inverse Neural Model

A process neural model is a neural network that determines the estimated output of the process from the current and past values of its input and output. The plant neural controller is used in control strategies involving another neural network as a controller (Thibault et al., 1992). The task of this second neural controller is to learn the inverse dynamics of the process.
A process inverse neural model is a model representing the inverse dynamics of the process. The process inverse model can provide the estimated value of the manipulated input variable directly to obtain the direct desired output.

Some reported studies of a combination between a conventional PID controller and the inverse neural model in a practical application prove that this technique allows us to control a process characterized by correlated parameters without explicit decoupling of them. Fig. 2 presents a scheme of the specialized learning strategy. There are many proposed learning techniques to train the inverse neural model in the identification process, such as:

- Specialized learning - self learning control: Direct evaluation of the accuracy of the inverse model with respect to the output of the process
- Generalized learning: Based on off-line identification from historic database. In practice, this algorithm is used to estimate the state variables of a process, which are then used to modify the parameter of the controller to obey the control low.

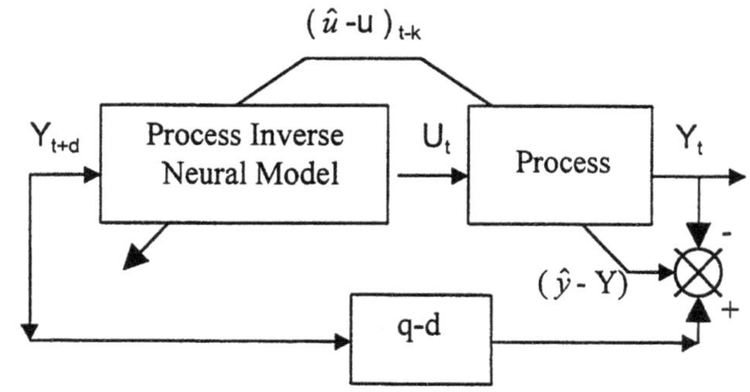

Fig. 2: Specialized Learning Strategy

5.4. Process Based Heuristic Methods

Combining the methods of artificial intelligence expert systems with neural network systems will generate a hybrid tool suitable for intelligent control and decision making and diagnostic systems. This combination will find a large field of application in industry and especially in the automation and control of complex systems (Quantrille et al., 1992). The knowledge based reasoning technique comprises the following components:

Knowledge = Facts + Rules + Heuristics

5.4.1 Rules

The rule-based problem-solving approach can be based on two strategies: causal, which is based on deep knowledge, and evidential which is based on shallow knowledge. The logic-based reasoning mechanism can be based on the IF-THEN rules containing causal relations of the input-output characteristics of the system patterns to guide the behavior and the decision taken by the neural networks or the fuzzy system.

5.4.2 Heuristics and knowledge indexing

To acquire the necessary knowledge of the process, the learning can be performed by the following methods: reasoning, past experience learning and classification.
A knowledge-based system is not able on one hand to make a decision based on pattern recognition or type of data being fed to the system. On the other hand a neural network, although it can learn from past training and process the information and classify it according to different cases, however, it does not have a reasoning mechanism to make a choice.

Consequently, a combination between the two methods will provide us with a powerful intelligent tool destined to the acquisition and organization of a considerable specific problem-solving knowledge (Mills et al., 1994). The control order then can be executed by the neural network after loading the learned coefficients from previous training, or can be executed by a second neural network in the case of a two level control structure: the first level is reasoning and decision making and the second executive module.
The difficulties can arise while translating the numeric data into symbolic information. In this case a hybrid system comprising a neural network and fuzzy system will achieve this task.

6 CONCLUSION

This study has presented a brief overview of some of the artificial intelligence methods and algorithms used in intelligent control systems of industrial processes that have been long studied and proven to be reliable and effective. Also, a briefly discussed the selection of a process representative model, and a hierarchical structure of an intelligent control of a complex process with the description of the modules it consists of and guidelines of the choice of different related algorithms.
The application of intelligent control techniques is proved to ensure better interpretation of process data, better presentation of process knowledge, improving by this the capability of the control system in reasoning and decision making, while at the same time being simple. Hybrid methods that are based on mixed conventional algorithms and artificial intelligence methodologies have better control over the process. However, intelligent control strategies and methods are for their part much more complicated than conventional ones and consequently they need tighter co-operation between academia and process industries to implement them with efficiency.

REFERENCES

Abderrazak M. Meeziou, P. B. Deshpande, C. Cozewith, N. I. Silverman and W. G. Morrison (1996). Dynamic matrix control of an ethylene-propylene diene polymerization reactor. *Ind. Eng. Chem. Res.* Vol. 35.

Al-Duwaish H., M. N. Karim, V. Chandrasekar (1996). Use of multilayer feedforward neural networks in identification and control of Wiener model. *IEE Proc.-Control Theory Appl.*, Vol. 143, No. 3, May.

AL Khani Ammar (1995). Adaptive Learning Control Scheme Based on Multilevel Stochastic Supervisor of an Artificial Neural Network. IEEE International Conference on Neural Networks, Perth, Australia.

AL Khani Ammar, Ari Nissinen, Heikki Koivo (1996). Evaluation of advanced process control methodologies and their perspective in modern industries, program ESPRIT/project SESDIP.

Asriel U. Levin, Kumpati S. Narendra (1995). Identification using feed forward networks. *Neural Computation* Vol. 7.

Dale E. Seborg (1994). A perspective on advanced strategies for process control. *Modeling, Identification And Control*, Vol.15, No.3, p. 179-189.

Gouph B. and J. T. Kay (1997). Kraft continuous digester effective alkali control. *IEEE Industry Applications Magazine*- July/August.

Jianming Tao, J. Y. Luh (1995). Reinforced learning control of Unknown dynamical systems using neural networks. *Intelligent Automation and Soft Computing*, Vol. 1, No. 2.

Joshi, N. V., Murugan P., and Rhinehart R. R. (1997). Experimental comparison of control strategies. *Control Eng. Practice*, Vol. 5, No. 7, pp. 885-896.

Kang-Ning Lou & Roland A. Perez (1996). A new system identification technique using Kalman filtering and multilayer neural networks. *Artificial Intelligence in Engineering,* **10**, p.1-8, UK.

Koivo H. N. (1995). Artificial neural networks in faults diagnosis and control. *Control Engineering Practice*.

Mills M., A. Y. Zomaya and M. O. Tadé (1994). Adaptive model-based control using neural networks. *Int. J. Control*, Vol. 60, No. 6.

Narendra K. S. and Snehasis Mukhopadhyay (1992). Intelligent Control Using Neural Networks. *IEEE Contr. Syst. Mag.*, Vol. 12, No. 2.

Quantrille Thomas E. and Y. A. Liu (1992). *Artificial Intelligence in Chemical Engineering* Academic Press.

Thibault Jules and Bernard P. A. Grandjean (1992). Neural networks in process control - A Survey. IEEE International Conference on Neural Networks.

Sjöberg, H. Hjalmarsson, L. Ljung. (1994). Neural networks in system identification. Report, Linkoping University, Sweden.

Simpson, P. K. (1990). *Artificial Neural Systems*. Pergamon Press, New York, NY.

A METHOD FOR DETECTION OF STICTION IN CONTROL VALVES

Alexander Horch, Alf J. Isaksson [1]

*S3-Process Control, Royal Institute of Technology,
S-100 44 Stockholm, Sweden*

Abstract: There has been an increasing interest in automatic control performance monitoring in recent years. The aim of such monitoring tools is to detect deterioration such as increased variability, oscillating behaviour, saturation or off-sets. Easy and efficient methods have been presented in the literature. After having detected a deteriorated control performance, one would like a monitoring tool to give hints of the possible cause of the deterioration. We will in this contribution suggest a method which allows automatic classification of one important reason for a detected control performance deterioration, namely static friction (stiction) in control valves. The suggested approach uses methodology from the fault-detection field and involves a model-based nonlinear observer and statistical hypothesis testing using a likelihood ratio test. Only little process knowledge, as time-delay, measurements of control signal and process output and some usually known valve parameters are needed. The method is evaluated on industrial data.

Keywords: Control valves, stiction, detection, performance monitoring, nonlinear filtering, hypothesis testing.

1. INTRODUCTION

Research in automatic control performance monitoring has been continuously increasing during the past decade. Several different approaches have been proposed among which the one by Harris (Harris, 1989) has received large attention in the process industry. The method relates actual output variance to minimum achievable variance. Especially in the paper industry several implementations have been reported, (Jofriet *et al.*, 1995), (Owen *et al.*, 1996). The Harris approach is very popular due to its conceptual and computational simplicity. Only the measured system output and the process time-delay have to be known. The method has been extended by several groups in order to cover a larger range of processes and process conditions. See (Tyler and Morari, 1995*b*), (Tyler and Morari, 1995*a*), (Eriksson and Isaksson, 1994) for example.

Another indication for deteriorated control performance is the presence of oscillations in the control loop. An easy and appealing method to detect oscillations was suggested in (Hägglund, 1994). Also, efforts have been made to combine the Harris performance index and Hägglund's oscillation detection algorithm, (Fu and Dumont, 1995).

Deterioration of control performance may have several reasons such as badly tuned controllers, oscillating load disturbances or nonlinearities in actuators etc. After having detected a deteriorated control performance, one would like a monitoring tool to give hints of the possible cause of the deterioration. Hägglund (Hägglund, 1994) suggests a detection scheme which, however, requires interaction with the process operator and the possibility to put the process into manual mode.

We will in this contribution suggest a method which allows automatic classification of one of the reasons for a detected control performance deterioration, namely static friction (stiction) in

[1] email: `alexander.horch, alf.isaksson}@s3.kth.se`

control valves. Stiction is known to be one of the main causes for bad control performance (Hägglund, 1994). The suggested approach makes use of a nonlinear model-based observer and statistical hypothesis testing in order to determine whether or not a control valve performs satisfactorily or not. Although the process model may not be known, it is possible to apply an observer and perform hypothesis testing by assuming a reasonable first-order plus deadtime model.

The basic idea is to utilise knowledge of the input signal to the control valve as computed by the distributed control system (DCS). This signal, together with the measured loop-output, is fed into two nonlinear filters, each representing a different valve situation. One of the filters represents the control valve with a tolerable amount of stiction whereas the other filter represents the valve with an unacceptable amount of stiction. Both predicted observer outputs are then subtracted from the real, measured, output. Given these two residuals, one can use a likelihood ratio test in order to perform hypothesis testing. This test will try to reject one of the following hypotheses:

$\mathcal{H}_0$: the control valve satisfies certain given quality specifications.

$\mathcal{H}_1$: the control valve does not satisfy certain given quality specifications.

Clearly, two questions arise at this point: Firstly, which process model shall be used in the nonlinear filters, and secondly which valve parameters shall be assumed?

Since both filters use the same process model, i.e. both observers make the same "error", it may be sufficient to assume that the process is mainly first-order with some (mostly roughly known) time-constant and a known time-delay. This assumption is often sufficient for plant dynamics in the process industry. From available product information of the control valve, one can expect to find the valve controller structure and some other valve values (e.g. mass of moving parts) which can be used in the valve model instead of general values.

The remainder of the paper is organised as follows: in Section 2 the models for the control valve and friction are presented, in Section 3, stiction in control valves is introduced and connected to performance requirements from industry. The used tools, nonlinear filtering and hypothesis testing are presented in Sections 4 and 5 respectively. Results of the method, when applied to industrial data are shown in Section 6. Finally, some conclusions (Section 7) and an overview of future work (Section 8) finish the paper.

2. MODELLING VALVE AND FRICTION

There is a large number of friction models in use for different applications. See (Olsson, 1996) for a good survey. The friction model used here to model static friction in control valves is

$$F_f = \begin{cases} F_c \cdot \mathrm{sgn}(v) & \text{if } v \neq 0 \\ F_e & \text{if } v = 0 \text{ and } |F_e| \leq F_s \\ F_s \cdot \mathrm{sgn}(F_e) & \text{if } v = 0 \text{ and } |F_e| > F_s \end{cases} \tag{1}$$

where F_c, F_e, F_f, F_s and v denote the Coulomb friction, the externally applied force, the total friction force, the static friction force and the velocity of the actuator respectively. For numerical reasons, a small deadband was added in order to avoid problems with the zero-velocity detection. The model (1) can easily be extended to include the so called Stribeck effect (Olsson, 1996). This requires, however, the determination of another unknown parameter and will not be done here. The equations of motion of the control valve are described by Newton's law,

$$m\ddot{x}(t) = F_e(t) - F_f(t) \tag{2}$$

where m and $x(t)$ are the mass and the position of the moving parts in the control valve. The force F_e is the resulting force on the piston as it is generated by the valve-inherent control mechanism. This mechanism may have either PI or PID behaviour in different implementations, depending on the valve in question. Examples are

$$F_{e1} = -K_p x(t) - K_d \dot{x}(t) + K_i \int_0^t [x_r - x] \, d\tau \tag{3}$$

$$F_{e2} = K_p(x_r - x) + K_i \int_0^t [x_r - x] \, d\tau. \tag{4}$$

The actual implementation of the control algorithm may either be known or one can use some standard PI-controller. Since our approach is to use two different nonlinear observers, it may be less important to determine all parameters exactly. However, all available knowledge should be used in order to increase the accuracy of the method.

3. STICTION

Stick-slip behaviour is a well-known problem in many control valves with high friction. To move the valve control element, the actuator force must be increased to overcome the friction in the actuator, linkages and valve body. The actuator force will continue to increase without a change in the valve position during the *stick* phase. Once the friction forces are overcome, the valve will begin to move. The energy stored in the actuator results in the valve *popping* to a new position. This is the *slip* phase. This behaviour can easily be seen when forcing the system to follow a ramp input which reverses regularly. A typical stick-slip behaviour, generated using (1) and (2) is shown in Figure 1.

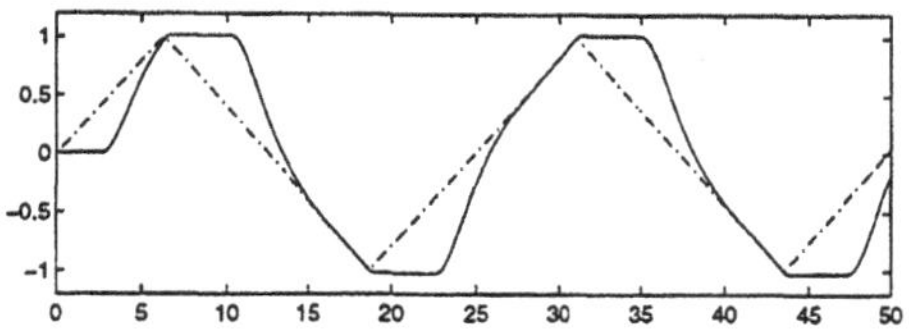

Fig. 1. Typical behaviour of a control valve with stiction. Valve position (solid) and position setpoint (dash-dotted).

It is not easy to determine generally what an "acceptable" amount of stiction is. However, a very useful standard was given by EnTech Control Engineering Inc., which can be used as a quality benchmark for control valves in the process industry (Bialkowski, 1994). EnTech's report gives specific values for different performance criteria for control valves. Using these values, one can - by simulation - determine suitable values of the friction forces F_c and F_s in (1). It is sufficient to perform a simulation of the valve model, similar to the one shown in Figure 1. The control input is changed in small bumps and the measured output is then plotted over the input. From such a plot, one can easily determine if the valve satisfies the required performance criteria. This process is described in detail in (Bialkowski, 1994). In that way, one can tune one of the valve models to represent a tolerable behaviour and another to describe deteriorated performance. This part of the proposed method represents the "tuning" part and it may be performed automatically such that the friction parameters for acceptable and unacceptable friction can be obtained without user interaction.

4. THE NONLINEAR FILTER

When dealing with the problem of predicting the states of a nonlinear model, subject to stochastic disturbances, the most popular approach is the use of an Extended Kalman Filter (EKF). One of the drawbacks of the EKF is that it requires the linearisation of the state-space equations at every time step. Even though other approaches for nonlinear filtering exist, we will here make use of the EKF.

The basic problem is to estimate the states in

$$\mathbf{x}_{k+1} = \mathbf{f}(\mathbf{x}_k, \mathbf{u}_k) + \mathbf{v}_k \tag{5}$$

$$\mathbf{z}_k = \mathbf{h}(\mathbf{x}_k, \mathbf{u}_k) + \mathbf{w}_k \tag{6}$$

where $\mathbf{x}_k$ is the n-dimensional state of the system at time k, $\mathbf{u}_k$ is the input vector, $\mathbf{v}_k$ is the q-dimensional state noise process vector due to disturbances and modelling errors, $\mathbf{z}_k$ is the m-dimensional observation vector and $\mathbf{w}_k$ is the measurement noise. It is assumed that the noise vectors $\mathbf{v}_k$ and $\mathbf{w}_k$ are zero-mean and

$$E\{\mathbf{v}_i\mathbf{v}_j^T\} = \mathbf{Q}\,\delta_{ij}, \quad E\{\mathbf{w}_i\mathbf{w}_j^T\} = \mathbf{R}\,\delta_{ij},$$
$$E\{\mathbf{v}_i\mathbf{w}_j^T\} = \mathbf{0}, \ \forall i,j.$$

The predictions of the state $\hat{\mathbf{x}}_{k+1|k}$ and the covariance matrix $\mathbf{P}_{k+1|k}$ (time update) are obtained using

$$\mathbf{A} = \frac{\partial \mathbf{f}(\hat{\mathbf{x}}_{k|k})}{\partial \mathbf{x}_k}\ ;\ \mathbf{H} = \frac{\partial \mathbf{h}(\hat{\mathbf{x}}_{k|k})}{\partial \mathbf{x}_k} \tag{7}$$

$$\hat{\mathbf{x}}_{k+1|k} = \mathbf{f}(\hat{\mathbf{x}}_{k|k}, \mathbf{u}_k) \tag{8}$$

$$\mathbf{P}_{k+1|k+1} = \mathbf{A}\mathbf{P}_{k+1|k}\mathbf{A}^T + \mathbf{Q}. \tag{9}$$

The measurement update equations are

$$\mathbf{K} = \mathbf{P}_{k|k}\mathbf{H}^T(\mathbf{H}\mathbf{P}_{k|k}\mathbf{H} + \mathbf{R})^{-1} \tag{10}$$

$$\hat{\mathbf{x}}_{k+1|k+1} = \hat{\mathbf{x}}_{k|k} + \mathbf{K}[\mathbf{z}_k - \mathbf{f}(\hat{\mathbf{x}}_{k|k})] \tag{11}$$

$$\mathbf{P}_{k+1|k+1} = (\mathbf{I} - \mathbf{K}\mathbf{H})\mathbf{P}_{k|k}. \tag{12}$$

with initial estimates for $k = 0$. This filter can be used for on-line or off-line computation.

5. HYPOTHESIS TESTING

Two of the described nonlinear filters will be used to create two different sets of residuals as described in the introduction. Using these residuals, one can apply statistical hypothesis testing in order to reject one of the two hypotheses

$\mathcal{H}_0$: the control valve satisfies certain given quality specifications.
$\mathcal{H}_1$: the control valve does not satisfy certain given quality specifications.

Mehra et al. proposed a χ^2-test for

$$V_t = \sum_{k=1}^{N} \varepsilon_k^T \left(\mathbf{P}_{k|k-1}^{\varepsilon\varepsilon}\right)^{-1} \varepsilon_k,$$

see (Mehra and Peschon, 1971). This rather simple approach, however, has some drawbacks. Therefore, a Likelihood Ratio test will be used, see for example (Willsky and Jones, 1976). For that, consider

$$S_N = 2 \ln \frac{p(\mathcal{Z}^N|\mathcal{H}_1)}{p(\mathcal{Z}^N|\mathcal{H}_0)} \tag{13}$$

where $p(\mathcal{Z}^N|\mathcal{H}_i)$ is the likelihood that the data with N samples, $\mathcal{Z}^N$, was generated with hypothesis $\mathcal{H}_i$ to be true. The factor two is used for convenience only. Since the nonlinear part of the model is located in the valve, there will be a linear measurement equation, i.e. (6) becomes simply $\mathbf{z}_k = \mathbf{H}\mathbf{x}_k + \mathbf{w}_k$. The residuals, as generated by the EKF are $\varepsilon_k^i = \mathbf{z}_k - \hat{\mathbf{z}}_k^i$, $i = 0, 1$. The likelihoods in (13) can be calculated as follows:

$$p(\mathcal{Z}^N|\mathcal{H}_i) = \prod_{k=1}^{N} p(\varepsilon_k^i|\mathcal{Z}^{k-1}) \qquad \text{with} \qquad (14)$$

$$p(\varepsilon_k^i|\mathcal{Z}^{k-1}) = \mathcal{N}(0, \Sigma_k^i)$$

$$= \frac{1}{\sqrt{2\pi}\det \Sigma_k^i} \exp\left[-\frac{1}{2}(\varepsilon_k^i)^T(\Sigma_k^i)^{-1}\varepsilon_k^i\right]$$

$$\text{where} \quad \Sigma_k^i = \mathbf{H}\mathbf{P}_{k|k-1}^i\mathbf{H}^T + \mathbf{R}$$

and $p(\varepsilon_k^i|\mathcal{Z}^{k-1})$ is the likelihood of ε_k^i, given the data $\mathcal{Z}^{k-1}$ and the index i denotes variables related to hypothesis $\mathcal{H}_i$. With (14) and using the fact that Σ_k^i is scalar (since the system is SISO), Equation (13) can be written as

$$S_N = \sum_{k=1}^{N}\left[2\left(\ln\frac{\Sigma_i^0}{\Sigma_i^1}\right) - \frac{(\varepsilon_k^1)^2}{\Sigma_i^1} + \frac{(\varepsilon_k^0)^2}{\Sigma_i^0}\right]. \quad (15)$$

Note that (15) can easily be implemented recursively for an on-line application.

In order to either accept $\mathcal{H}_0$ or $\mathcal{H}_1$, one has to choose two thresholds which can be done using

$$h \leq \ln\frac{1-\beta_1}{\beta_0}, \quad a \leq \ln\frac{1-\beta_0}{\beta_1} \qquad (16)$$

where β_i, $(i = 0, 1)$ is the probability of rejecting hypothesis $\mathcal{H}_i$ when it is true, see (Grainger *et al.*, 1995). The decision rule can then be expressed as: If $S_N \geq h$, $\mathcal{H}_1$ is accepted and if $S_N \leq -a$, $\mathcal{H}_0$ is accepted. If $-a < S_N < h$, there is not enough information to make a decision. For an on-line implementation this means that the decision is made as soon as there are sufficiently many samples to make the likelihood ratio "significant enough". Choosing $-a = -\infty$ leads to the well-known CUSUM test.

6. INDUSTRIAL EXAMPLE

The algorithm is tested using data from a pulp concentration control section in a Swedish paper mill. Pulp is diluted with water to a desired concentration. The water flow is controlled by one control valve and the pulp flow by another. The water valve had too high friction, whereas the pulp valve was performing satisfactorily. The control loop was oscillating due to the friction in the water valve. Therefore, also the pulp flow loop is oscillating in order to compensate for the fluctuations in pulp concentration. This control loop and its description using different friction models was also investigated in (Eborn and Olsson, 1995). Both control loops can be described as done in Figure 2. There, $G(s)$ denotes the dynamics of the mixing pipe and the concentration sensor, F_0 and F_1 are the EKFs to generate the residuals ε_1 and ε_2, $f(\cdot)$ is the nonlinear function (1) for the valve friction and C_v is the internal valve controller whose parameters are unknown in this case and have to be guessed. The task of the EKF is - using both signals $x_r(t)$ and $z(t)$ - to estimate the states of the system. The measured output

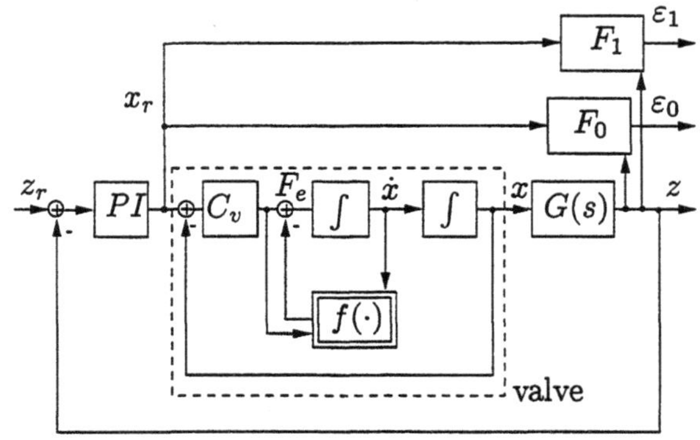

Fig. 2. Scheme of the blockdiagram which performs the nonlinear filtering and generates the residuals for hypothesis testing.

$z(t)$ and control signal $x_r(t)$ for the water valve loop are shown in Figure 3. The measurements

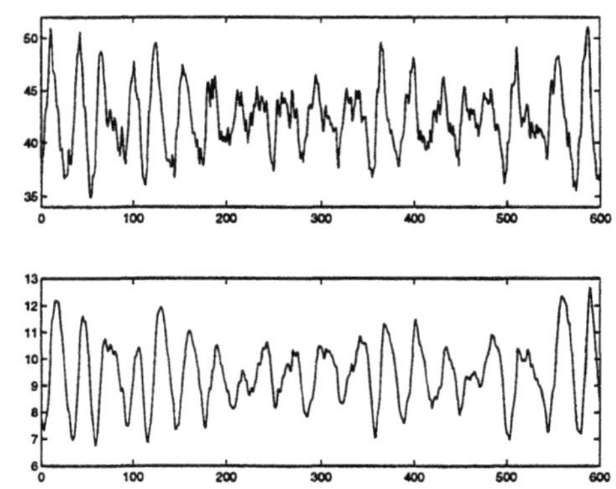

Fig. 3. Measurements from the water valve control loop. Concentration $z(t)$, (top) and valve position setpoint $x_r(t)$, (bottom). Both signals are in [%].

for the pulp valve loop are shown in Figure 4. The

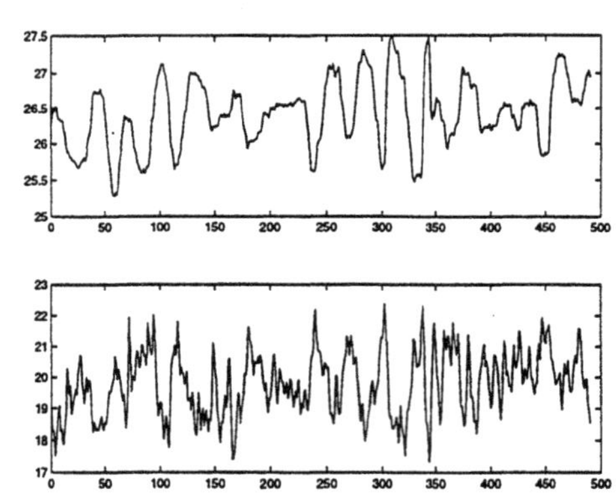

Fig. 4. Measurements from the pulp valve control loop. Concentration $z(t)$, (top) and valve position setpoint $x_r(t)$, (bottom). Both signals are in [%].

sampling time in both cases was one second. As mentioned in the introduction, the process model (i.e. mixing pipe and concentration sensor), $G(s)$, can roughly be modelled as a first order system with gain K, dead-time θ and time constant τ. Introduce then the following states:

$x_1 = z$ (measured output)
$x_2 = x$ (valve position)
$x_3 = \dot{x}$ (valve velocity)
$x_4 = \int_0^t [x_r(\tau) - x(\tau)]d\tau$

where $x_r(t)$ is the control valve setpoint as calculated by the loop PI-controller. This choice of states yields the following state-space equations whose discrete-time version is used in the EKF.

$$\begin{aligned}
\dot{x}_1 &= -\frac{1}{\tau}x_1 + \frac{K}{\tau}x_2(t-\theta) \\
\dot{x}_2 &= x_3 \\
\dot{x}_3 &= \frac{1}{m}\left[F_e(x_2, x_r) - F_f(x_2, x_3, x_r)\right] \\
\dot{x}_4 &= x_r - x_2 \\
x_r &= \text{measured input data} \\
F_e(x_2, x_r) &= K_{pv}(x_r - x_2) + K_{iv}x_4 \\
F_f(F_e, x_4) &= [\text{see Equation (1)}]
\end{aligned}$$

where m, F_e are the mass of all moving parts of the valve and the total valve force as calculated by the internal valve PI-controller (with parameters K_{pv} and K_{iv}). Note that the measured system output (concentration) is the state x_1. Hence we have an easy measurement equation $z(t) = x_1(t)$. The control structure *in* the valve is assumed to equal Equation (4). All used parameter values for the EKFs are shown in Table 1. The matrices $\mathbf{Q}$ and $\mathbf{R}$

water valve loop					
param.	K	τ	θ	m	K_{iv}
value	3	10	10	1	0.03
param.	K_{pv}	F_s^0	F_s^1	F_c^0	F_c^1
value	1	0.3	6.0	0.2	4.0
pulp valve loop					
param.	K	τ	θ	m	K_{iv}
value	0.5	10	1	1	0.03
param.	K_{pv}	F_s^0	F_s^1	F_c^0	F_c^1
value	1	0.3	3.0	0.2	2.0

Table 1. Parameters used for the consistency loop.

were chosen to be diagonal with small (0.1) default values. Their choice is a trade-off between relying on data or the model. Both measured signals, $x_r(t)$ and $z(t)$, were fed into the EKF using zero initial conditions. Hence, two sets of residuals are obtained. For the water valve loop, the model output and the data are plotted in Figure 5.

These, together with the calculated covariance matrices of the estimates are then used to calculate the likelihood ratio as described in Section 5. Choosing $\beta_0 = \beta_1 = 5\%$ in (16) yields the values as shown in Table 2. The algorithm finds that the

data	S_N	h	a	accepted $\mathcal{H}$
water valve	300.5	2.94	2.94	$\mathcal{H}_1$
pulp valve	-3.01	2.94	2.94	$\mathcal{H}_0$

Table 2. Results from the likelihood test for the industrial example.

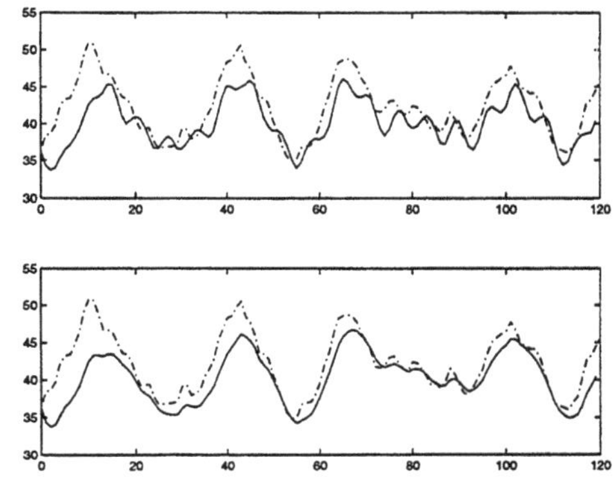

Fig. 5. Comparison of model output (solid) and data (dash-dotted) for the water valve loop. Model with acceptable friction (top) and unacceptable friction (bottom).

control valve, which is known to have too high friction, does not fulfill the given performance specifications. For the pulp valve, the null hypothesis is accepted, i.e. tolerable amount of stiction. However, the data set was almost too short to make a decision. This would be no problem with an on-line implementation. A special problem in this case is that the pulp loop is oscillating due to the oscillating water flow loop. It is hence not very easy for an operator to decide which of the valves - if a valve is the cause - has to be maintained.

Furthermore, the method is rather insensitive to errors in the assumed process transfer function and controller parameters. It is important that both EKFs use the same parameters. On the other side, it turns out that the friction parameters have to be specified as accurately as possible in order to avoid erroneous conclusions. Unfortunately, we did not have valve parameters which are usually known. Hence some assumptions had to be made.

7. CONCLUSIONS

One of the main reasons for control loop performance deterioration in process industry is stiction in control valves. A method for automatic detection of this kind of problem in a control loop was presented in this article. Desirable features of such a method are firstly the independence of tuning and secondly that the tool – once it is installed – works without further human interaction. The presented approach fulfills the second requirement, whereas it is still necessary to use available process knowledge, especially knowledge about the control valve in order to specify the limits for tolerable friction. This may be considered as a disadvantage, but the task may be automated and if process knowledge is available, it should be used by the monitoring tool. Nevertheless, there is still more research to be done in order to automate the parameter tuning.

The basic idea was to make use of two nonlinear filters to compute two different sets of residuals, each of them representing two different situations. Statistical hypothesis testing was used to determine which of the two filters seems to be more likely to having produced the actual output data.

The method was evaluated on industrial data. Even if the "correct" solution was known in this case, we did not know the real valve inherent control. Furthermore, the parameters of the friction model had to be chosen without knowing other important valve parameters. It could, however, be shown that the method can yield correct decisions even in this case.

In general, the frictions levels can be computed using published performance specifications. This may either require some simulations or one has to make use of standard values for different valve types and sizes. Such information, for example may be taken from a data base which can be created when investigating several valves. For future use, this task should be automated.

Even though an off-line algorithm was presented here, it is possible to implement the method online. Both algorithms, the EKF and the likelihood ratio can easily be implemented recursively.

8. FUTURE WORK

The presented method will be improved in future work. Since the presented friction model contains a discontinuity, a dynamic friction model may be used instead. This will simplify the nonlinear state estimation. Another important research path will be the on-line adaptation of model parameters, e.g. the process parameters K, τ or θ.

It is well known the the Extended Kalman Filter – contrary to its frequent use – has some important drawbacks since it involves the calculation of the Jacobian in every time-step. There are methods proposed in the literature which avoid differentiation, see for example (Julier *et al.*, 1995). With respect to the mentioned discontinuity in the friction model, this is an appealing choice.

Finally, the method will be tested on more industrial data sets.

ACKNOWLEDGMENT

The authors would like to thank Dr. Tore Hägglund for providing the industrial data. This work is supported by the Swedish National Board for Industrial and Technical Development (NUTEK) which is gratefully acknowledged.

9. REFERENCES

Bialkowski, W. L. (1994). Control valve dynamic specification. Technical report. EnTech Control Engineering Inc. Version 2.1.

Eborn, J. and H. Olsson (1995). Modelling and simulation of an industrial control loop with friction. In: *Proceedings of the 4th IEEE Conference on Control Applications.* Albany, New York. pp. 316–322.

Eriksson, P. G. and A. J. Isaksson (1994). "Some Aspects of Control Loop Performance Monitoring". In: *The 3rd IEEE Conference on Control Applications.* Glasgow, Scotland. pp. 1029–1034.

Fu, Y. and G. A. Dumont (1995). On-line evaluation of control loop performance. In: *4th IEEE Conference on Control Applications.* Albany, New York. pp. 655–656.

Grainger, R. W., J. Holst, A. J. Isaksson and B. M. Ninness (1995). "A parametric statistical approach to FDI for the industrial actuator benchmark". *Control Eng. Practice* **3**(12), 1757–1762.

Hägglund, T. (1994). Automatic monitoring of control loop performance. In: *Preprints of Control Systems '94.* Stockholm, Sweden. pp. 190–196.

Harris, T. J. (1989). "Assessment of Control Loop Performance". *The Canadian Journal of Chemical Engineering* **67**, 856–861.

Jofriet, P., C. Seppala, M. Harvey, B. Surgenor and T. Harris (1995). "An Expert System for Control Loop Performance Analysis". In: *81st Annual Meeting, Technical Section, CPPA.* Canadian Pulp and Paper Association. pp. B41–49.

Julier, S. J., J. K. Uhlmann and H. F. Durrant-Whyte (1995). A new approach for filtering nonlinear systems. In: *American Control Conference 95.* Seattle, Washington. pp. 1628–1632.

Mehra, R. K. and J. Peschon (1971). An innovations approach to fault detection and diagnosis in dynamic systems. *Automatica* **7**, 637–640.

Olsson, H. (1996). Control Systems with Friction. PhD thesis. Department of Automatic Control, Lund Institute of Technology.

Owen, J.G., D. Read, H. Blekkenhorst and A.A. Roche (1996). "A Mill Prototype for Automatic Monitoring of Control Loop Performance". In: *Preprints Control Systems '96.* Halifax, Canada. pp. 171–178.

Tyler, M. L. and M. Morari (1995*a*). "Performance Assessment for Unstable and Nonminimum-Phase Systems". Technical report. California Institute of Techology. IfA-Report No 95-03.

Tyler, M.L. and M. Morari (1995*b*). Performance monitoring of control systems using likelihood methods. In: *Proc. American Control Conference.* Seattle, USA. pp. 1245–1249.

Willsky, A. S. and H. L. Jones (1976). A generalized likelihood ratio approach to the detection and estimation of jumps in linear systems.. *IEEE Transactions on Automatic Control* pp. 108–112.

SUPERVISION OF COMPLEX SYSTEMS USING FUNCTIONAL, STRUCTURAL AND BOND GRAPH MODELS

B. Ould Bouamama*, E. Cherifi, M. Staroswiecki* and B. Riera****

** Laboratoire d'Automatique et d'Informatique Industrielle de Lille URA CNRS D1440*
Université des Sciences et Technologies de Lille,
Bat. EUDIL, AREMI. 59655 Villeneuve d'Ascq cedex FRANCE
belkacem.bouamama@univ-lille1.fr
*** Laboratoire d'Automatique et de Mécanique Industrielle et Humaine URA CNRS 1775*
Université de Valenciennes, BP. 311, 59304 Valenciennes cedex
elyès.cherifi@univ-valenciennes.fr

Abstract : The complexity of thermodynamic processes – large scale systems, numerous operating modes, numerous system interactions,...etc. – does not allow operators to take into account all system parameters. In order to help the operators in their tasks, the integration of decision making tools in the supervision system is necessary. This paper presents an attempt to integrate : Functional analysis, Structural analysis and Bond Graph Modeling in order to show their complementary for the design of supervision systems. The different methods are applied to a steam generator installation. It is shown how each of them contributes to help the operators in their supervision task.

Key words : Supervision, bond graphs, functional, structural, model, steam generators.

INTRODUCTION

We consider industrial thermodynamic systems, whose supervision is essential in order to insure their smooth running. The supervision consists of real time monitoring and high abstraction level coordination of the set of subsystems which constitute the process, in order to optimize the production, to diagnose faults, to perform simulations and especially to insure the safety. In normal operation, the system supervision is performed by digital control systems. However, when the operating mode changes, or in incidental or accidental situations, the presence of human operators becomes essential (Millot,1988).

Even in the case of well trained operators (which is not always true), the complexity of the process – large scale system, numerous operating modes, numerous system interactions,...etc. – does not allow them to take into account all the system parameters. In order to help the operators in their tasks, the integration of decision making tools in the supervision system is necessary.

Thermodynamic systems are characterized by the interactions of a large number of complex phenomenon which call for various types of energy. This energetic feature requires a unified modeling approach, allowing to put in evidence the physical phenomenon: energy storage, energy transformation and energy dissipation. The system operation understanding also requires a preliminary functional analysis. In addition to that, for the supervision operator, the study of the system interactions in term of coupling between the different variables of the process, presents a primary interest for the extraction of pertinent information (Cherifi,1995).

This paper presents an attempt to integrate three methods: Functional analysis, Structural analysis and Bond Graph Modeling in order to show their complementary for the design of supervision

systems. The different methods are discussed and applied to a steam generator installation.
In a first part of the paper the different methods will be presented briefly. After a description of the steam generator installation process, a second part will describe the process modeling using the different methods. Finally, their application domains and limits will be exposed and it will be shown how each of them contributes to help the operators in their supervision task.

1. BOND GRAPH, FUNCTIONAL, AND STRUCTURAL MODELING

1.1 Bond Graph modeling

At the FDI (fault detection and identification) level of the supervision system, some knowledge about the dynamic behavior of the system is required. Furthermore, to deal with the different energetic domains in the thermodynamic process, a method or language is needed. The Bond Graph language is well suited for this purpose. This approach is used because it provides a rigorous and systematic methodology for modeling the multi-energy domain embracing in our case the mechanical, hydraulic, (pump and valves), electrical (heating resistance), thermodynamic (steam production) and thermal (heat loss) elements of the steam generator process.
The key of Bond graph modeling is the representation (by a bond) of power fluxes as products of efforts and flows with elements acting between these variables and junction structures to put the system together.
Since bond graph theory has been treated rather completely elsewhere (Thoma,1975), in this paper we will deal essentially with those aspects of bond graphs peculiar to thermodynamic systems. In the thermodynamic bond graph, the coupling power (hydraulic and thermal) exchanged between two systems $\Sigma 1$ and $\Sigma 2$ is indicated by small rings around the bonds (fig. 1a). The causality is indicated by the cross-stroke position in the bond graph. In figure 1a, the cross-stroke is denoted on the right indicating that the effort acts to the right. $\Sigma 1$ impose the efforts to $\Sigma 2$ as shown by the block diagram in figure 1b.
As power variables we use as effort the temperature T and pressure P and as flow variable we use the mass flow F and the thermal flow Q.

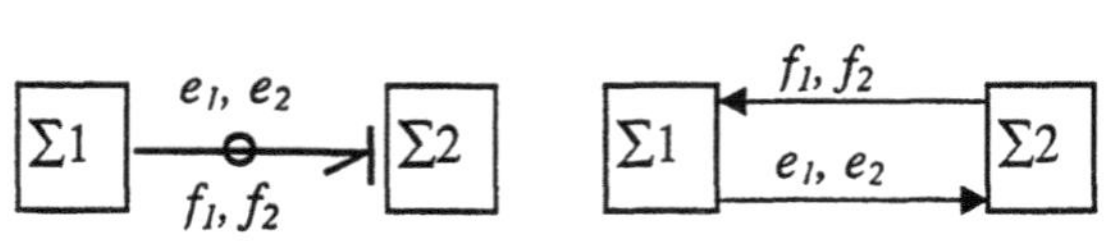

(a) (b)

Fig. 1 Thermodynamic bond graph representation (a) and causality (b)

Those variables are more natural for the operators than entropy flow for example. Really, the choice is not trivial and depends on the kind of the considered physical phenomena (Ould Bouamama and Thoma, 1997).

1.2 Functional modeling using a GTST-MPLD

Functional modeling is a technique for describing and representing complex systems. Among many recent works which contribute to modelise the functional viewpoint (Chittaro, 1993), (Féliot 1996), we chose the GTST-MPLD method.
GTST-MPLD (Goal Tree Success Tree – Master Plant Logic Diagram) presented by Modarres (1995) employs logical connectors to describe the functional hierarchy of the system. This method is composed of two part GTST and MPLD.
GTST. The GTST is divided into two parts: the Goal Tree, GT and the Success Tree, ST. The GT is concerned with the goals and functions which must be achieved by the system. The GT starts with the system's objective at the top. The objective is then decomposed into a sufficient sets of sub-goals, the achievement of which is necessary for the success of the top level "goal". All sub-goals are then described in terms of their own sub-goals which must be further decomposed until a desired level of abstraction is reached. Going further down the GT, each low-level goal or function is described in terms of a functional decomposition which explains how the goal is achieved. The first level of functions can further be decomposed into sub-functions, and the process of sub-functions definition continue again until the desired level of abstraction is reached. At the end of this process the lowest-level functions address functions performed by components. The set of all these functional decomposition form the ST. The ST is a logical structure that describes how an objective can be fulfilled by a set of components. Traditional logical connectors, such as "and", "or", and "xor", are used to describe the relationship between goals and functions.
MPLD. Because of the interrelationship between system parts, at the lowest level of GTST, the model shows multiple ways to achieve the high level functions. Since in complex systems these relationships become complicated, the MPLD is introduced to represent them in a compact and transparent fashion. The MPLD starts at the lowest level of GTST where the physical parts of the system support the lowest level of functional decomposition. The MPLD shows the interrelationships between independent parts of a system, in a hierarchical form using a dependency matrix. For each subsystem at the top of the MPLD the available success path is shown in the matrix. At the left side of the matrix the set of supporting elements needed for a success path are displayed. The application presented in this paper shows clearly how this method is used.

1.3 Structural Analysis using information theory

Structural analysis is used to represent the interaction between the different parts of the process. This is done by data analysis using information theory, it is presented here by the entropy and the transinformation concepts. These two concepts are used to calculate the degree of the transferred information between variables and sub-systems. This allows to compute indexes which represent the degree of coupling between variables. This approach is interesting because it is quite easy to apply it and to get useful results. The aim is not to get a complete dynamic model of the process which is most often impossible with regard to the complexity and the high number of variables. We intend by this analysis of the process to get information about the most important variables. Hence, it will be possible to design efficient Man Machine Interface (MMI) with useful information. In others words, we will identify the process, not in order to design a control algorithm, but to design a 'human supervision system" which will propose the right information to the human operator.

Entropy. Given a discrete variable X, having n values with the probabilities *P1, P2,.....Pn*, the entropy of the discrete variable X is defined by :

$$H(x) = -\sum_{i=1}^{N} p_i \log p_i \qquad (1)$$

with $P_i = Pr(X = X_i)$. The common logarithm used is expressed in base 2.

Let $U_1 = \{x_1, x_2, \ldots. x_p\}$ be a set of values of a discrete variable X and n(i) the occurrence number of the event $(X=xi)$. The empirical probability of the event $(X=xi)$ is defined by: $P(i)=n(i)/N$, where N is the total number of the observations. In the same way, the empirical entropy is defined by:

$$H(X) = -\sum_{i=1}^{p} p(i) \log p(i) \qquad (2)$$

For two variables X,Y, the entropy is :

$$H(X,Y) = -\sum_{i=1}^{n} \sum_{j=1}^{m} P_{ij} \log P_{ij} \qquad (3)$$

Transinformation. The transmitted information between variables X and Y or transinformation is defined by:

$$I(X:Y) = H(X) + H(Y) - H(X,Y) \qquad (4)$$

I(X,Y) measures the transmitted information between the variable *X* and *Y*.

Coupling Indexes. They measure the information link between two variables. For two variables *Xi, Xj* one possible definition is given by Conant (1972) Richetin (1975) and Toro (1982):

$$m_{ij} = \frac{I(X_i : X_j)}{\sqrt{H(X_i)H(X_j)}} \quad \text{where } 0 \le m_{ij} \le 1 \qquad (5)$$

If the variables *Xi, Xj* are statistically independent, then $m_{ij} = 0$. If they are statistically dependent then $m_{ij} = 1$. In fact this degree is just a normalization of the information index. When we compute the coupling index between all pairs of variables, we obtain what is called a coupling matrix. This matrix could be exploited in different manners for example it could be used for the decomposition of the system into a weakly coupled sub-systems, for the extraction of pertinent information (Pomorski 1996).

2. APPLICATION

2.1 Description of the pilot process

Let us consider the supervision of the pilot process schematically shown on figure 2. The test plant designed to be a scale-model of part of a power station is a complex non linear system. This installation is mainly constituted of four subsystems: a receiver with the feedwater supply system, a boiler heated by a 60 kW resistor, a steam flow system and a complex condenser coupled with a heat exchanger. The feed water flow is pressurised via the feed pump which is controlled by a relay to maintain a constant water level inside the steam generator. The heat power is determined based on the available accumulator pressure P7. The expansion of the generated steam is realised by three valves in parallel connection. V5 is a controlled valve, which allows to pass around the steam flow to the condenser. V6 is automatically controlled to maintain proper pressure P15 to the condenser. In an industrial plant, the steam flows to turbine for generating power, but at the test stand, the steam is condensed and stored in a receiver tank and returned to the steam generator.
The flow steam system and the condenser are not taken into account in this paper.

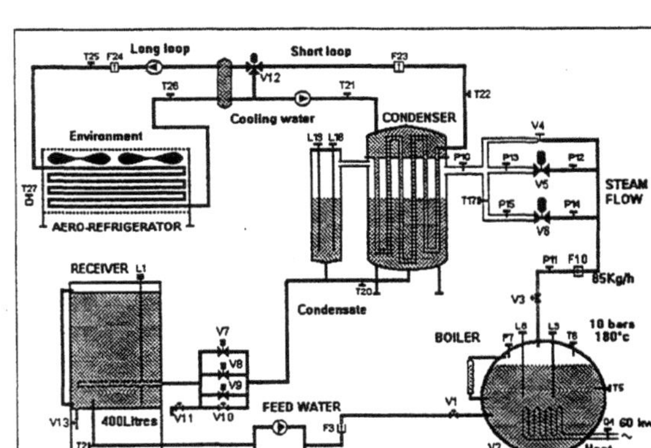

Fig. 2 : Schematic of the steam generator process

2.2 Process modeling

♦ Bond graph model of the process

The bond graph model of a part of the installation constituted by the receiver, the feed water supply system, and the boiler is given by figure 3.

In our case, the energy storage is modelled by the multiport C for coupled energy and the simple C element for single energy. Those elements represent the sate variables. Storage of coupled energy (thermal and hydraulic) occurs in the receiver (C_R), in the liquid phase (C_l) and vapour phase (C_V) in the boiler. The simple element C_{MG} represents the temperature T_{MG} stored by the metal of the boiler.

The mass exchange between the liquid and vapour phases (evaporation) is modelled by the multiport R. The pump is modelled as an effort source (Se) increasing the output pressure of the receiver $P1$ by an amount P_P. A pump characteristic is modelled by the resistive element R_{CP}. b_1 is a Boolean variable based on the water level $L8$ in the generator The transformation of electrical energy (from the voltage effort source u) into thermal energy is modelled by the RS-field multiport and the modulated transformer MTF element whose parameter b_1 (0 or 1) depends on the generator pressure signal P7. The thermal losses by conduction from the boiler to metal Q_{CG} and to atmosphere Q_{EX} are modelled respectively by resistive elements R_{GM} and R_{EX}. T_{EX} is the ambient temperature represented by the effort source Se.

The work done in pressurising the volume of vapour ($V_V \dot{P}7$) and the volume of liquid ($V_L \dot{P}5$) is modelled by the flow source Sf. As shown in the Bond graph model, this approach clearly shows up not only the energy storage, dissipation effects, and energy transformation but also a direction and causality effect of the energy flux. This is why, Bond graph methodology helps the operators to understand the basic physical principles. Furthermore, the dynamics of the process can be quantified. In fact, the seventh order state equation under non linear form. $\dot{x} = f(x,u)$ can easily be deducted from the Bond Graph. The number of state variables is equal to the number of C-elements in integral causality. They are associated to the following variables: $X=[T2\ L1\ L8\ T5\ m7\ T7\ T_{MG}]$.where L1 and $T2$ are the water level and the temperature in the receiver. *L8*, *T5*, *m7* and *T7* represent respectively the water level, the temperature, the mass of steam and temperature in the liquid and the vapour phase in the boiler. T_{MG} is the metal temperature. The causality assignment shows the operator the causality effect of energy fluxes. The causality is not a natural fact but selected by the engineer to facilitate computation. The choice is free in principle but subject to some limitation : for example Sources always impose an effort or a flow to the system. If the ambient temperature T_{EX} varies thus according to the assigned causality, the consequent effect on stored energy by the metal, the rate change of thermal energy in the boiler (T7 and T5) can be evaluated by the operator. Despite of prognostic, causality and pictorial effect presented by the bond graph model, the quantitative evaluation of the interactions between the process subsystems and variables are not taken into account. Furthermore, in terms of goals and functions the functional analysis is not described. Structural and functional analysis allow to fill those gaps.

♦ Functional description of the process

The functional approach based on GTST-MPLD method consists in modeling the goals and functions of the steam generator in a hierarchic way. First, we try to identify at the top level the objective of the process. The objective is then decomposed into a sufficient set of sub-goals, the achievement of which is necessary for the success of the top level goal. When we consider in our system just the boiler and the receiver, the objective is to produce a 85 Kg/h steam flow. This goal is decomposed into two sub goals. The first one is that the water supply is assumed in the boiler, the second one is that the energy supply is assumed in the boiler (figure 4).

These two sub-goals must be achieved together in order to achieve the top goal. In fact an "and" connector must be used, but by convention in the goal tree model, "and" connectors could be replaced by line connector. Going further down the GT, each low-level goal is described in term of functional decomposition which explain how the goal is achieved.

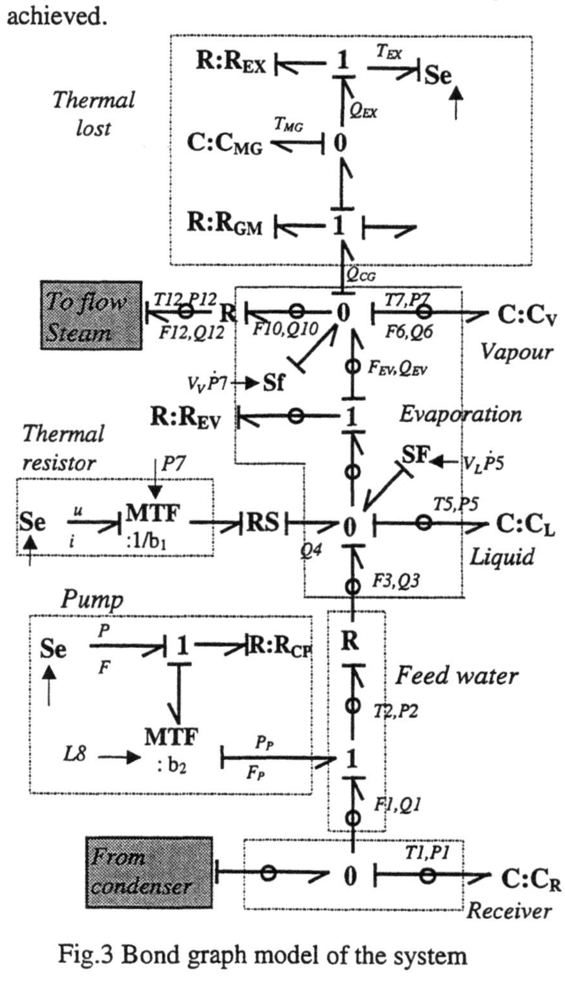

Fig.3 Bond graph model of the system

The functions are decomposed into sub-functions until the lowest level where functions address components. This figure shows clearly in hierarchical form how the goal and functions are achieved. It shows too, that different ways are possible to achieve a given goal or a function. The function "control the level of water in the boiler" can be achieved by four ways as it is shown in figure 4. Physical components can be directly represented in GTST-MPLD model. All of low level functions performed by the components of the system are shown. This characteristic makes GTST-MPLD suitable for system diagnosis, contrary to bond graph modeling. We can see that this representation is very simple to understand the functioning of the system. For the human operator, it could represents a good help in his task of supervision especially in diagnosis. Often in their supervision tasks, human operators consider the system running in terms of its functions. Concretely to cover up it, they act on its variables or components. GTST-MPLD shows the components implicated in each function, but it does not show the variables. In most cases, human operator acts on variables to change the system behavior, in order to perform and to achieve production, economical and security goals. Functions are associated implicitly to variables, that's why the integration of structural analysis becomes interesting. Structural analysis allows to get the interactions within the system. This could be used to extract the pertinent variables, then implicitly the pertinent functions. All these information are very interesting in order to design supervision system well suited to human operator.

♦ Structural description of the process

The analyzed process is started running and rich data file representing the different evolution of variables is observed in the supervision interface. The variables are very disturbed.

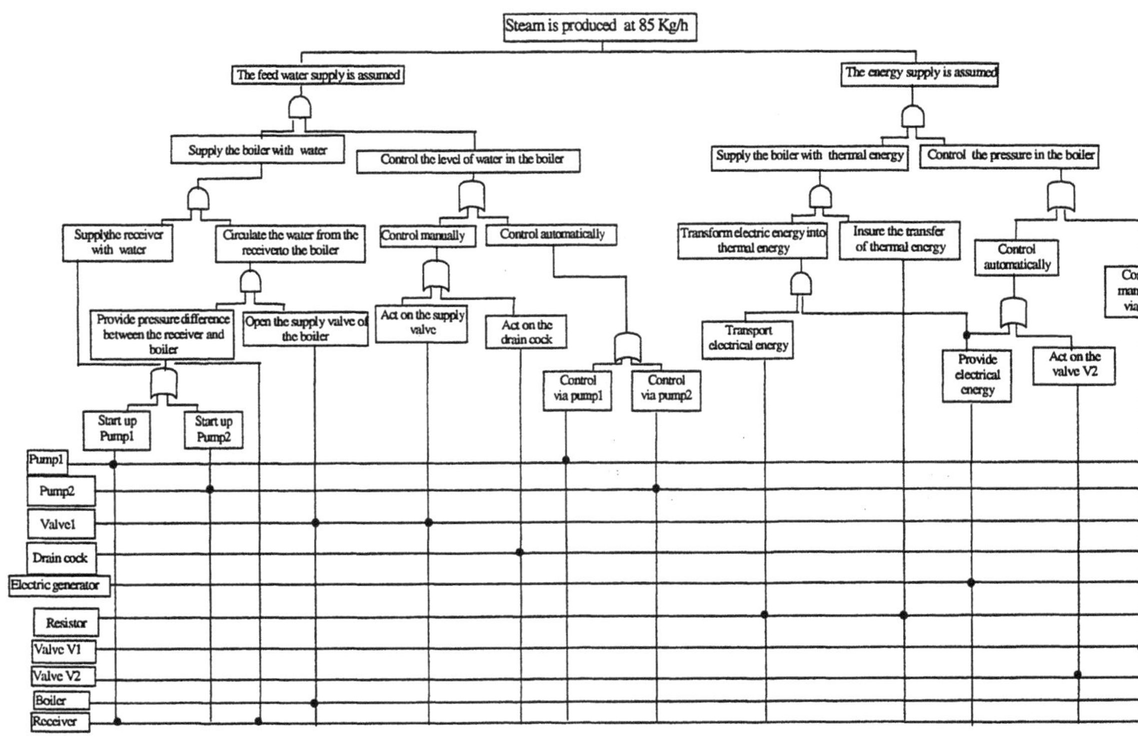

Fig 4 GTST – MPLD of the system

The filtered variables, are then divided into classes, in order to calculate a coupling matrix. The classification is necessary otherwise the calculation would take too much time. As noted previously, we present in this paper only a part of the system which is represented by 13 variables. The coupling matrix is given by the figure 6. It's shows that almost all variables are coupled each others excepted Q4 and F3 which represent respectively the electrical energy and the mass flow of the feed water. Each variable could be associated to the corresponding function(s) given by the GTST model. To set an example, let a variable P7 representing the pressure in the boiler. P7 is associated in the GTST-MPLD model to the function " control the pressure in the boiler".

In order to assure this function, the human operator must act on variable P7 which implies at the lowest level of the GTST three sub-functions. As displayed by the coupling matrix, four variables are strongly coupled each others: *P*14, *T*6, *P*7 and *P*12. In fact they represent essentially the boiler - *P*14 and *P*12 (in a parallel connection) are equal to boiler pressure *P*7 because of low hydraulic loss in the valve V3.

The associated functions to the variable P7 is then strongly linked to the associated functions of P14, P12 and T6 displayed in GTST model.

mij	P14	T5	T6	P7	L8	L9	Q4	F10	P12	T2	L1	P11	F3
P14	1	0.37	0.62	0.79	0.32	0.31	0.05	0.4	0.81	0.31	0.23	0.36	0.11
T5	0.37	1	0.41	0.39	0.41	0.35	0.09	0.35	0.38	0.38	0.2	0.41	0.07
T6	0.62	0.41	1	0.62	0.36	0.33	0.05	0.37	0.61	0.33	0.23	0.36	0.11
P7	0.79	0.39	0.62	1	0.33	0.32	0.06	0.4	0.84	0.3	0.22	0.37	0.11
L8	0.32	0.41	0.36	0.33	1	0.49	0.17	0.39	0.32	0.52	0.28	0.46	0.11
L9	0.31	0.35	0.33	0.32	0.49	1	0.09	0.33	0.31	0.48	0.24	0.4	0.11
Q4	0.05	0.09	0.05	0.06	0.17	0.09	1	0.11	0.05	0.16	0.02	0.09	0.03
F10	0.4	0.35	0.37	0.4	0.39	0.33	0.11	1	0.4	0.35	0.21	0.45	0.11
P12	0.81	0.38	0.61	0.84	0.32	0.31	0.05	0.4	1	0.31	0.22	0.37	0.1
T2	0.31	0.38	0.33	0.3	0.52	0.48	0.16	0.35	0.31	1	0.25	0.48	0.15
L1	0.23	0.2	0.23	0.22	0.28	0.24	0.02	0.21	0.22	0.25	1	0.23	0.06
P11	0.36	0.41	0.36	0.37	0.46	0.4	0.09	0.45	0.37	0.48	0.23	1	0.1
F3	0.11	0.07	0.11	0.11	0.11	0.11	0.03	0.11	0.1	0.15	0.06	0.1	1

Fig. 6 Coupling matrix of the system

From this analysis, when the operator acts on P7, he knows the functions of the system which are influenced and via the MPLD matrix, he can determine which component are implicated. This could help him to know which part of the system is implicated by acting on such variables. Furthermore, these variables (P7, P14, T6 and P12) are represented by the same sub-system: the boiler. This is an advantage of this matrix which allows to have a partition of the system. From the operator point of view, a piece of information is considered pertinent if it plays an important role in the system. In other word, if it is implicated in different system parts. When considered basic information is a variable, it is important if it is coupled with the several ones. Figure 6 shows that almost all variables are moderately coupled each others excepted Q4 and F3. Effectively these variables which represent the pressure, temperature and level of water in the boiler are very important to the operator supervising the process and they represent those for which he gives more attention. However variable Q4 representing the electrical energy function is weakly coupled with the rest of variables. This does not mean that it is not important. Indeed, from the functional point of view given by the GTST-MPLD model (fig.4), *Q4* is linked to the providing energy function situated at the bottom of the tree. This function is related to the main goal by only "and" connectors. If this function is not achieved, the main goal which is "to produce steam", could never be achieved. This case shows clearly the complementary of functional and structural approach in order to get a good understanding of the process.

3. CONCLUSION

The integration of Functional analysis, structural analysis and Bond Graph model seems of large interest in order to solve several problems in supervision of complex systems. The three descriptions have been integrated in the frame of the pilot process, and it has been shown how each of them contributes to help the operators in their supervision task. The functional analysis allows to describe the system operation in terms of goals and functions, without taking into account its physical and dynamic behavior. It provides an important help for the process operation understanding.

At the FDI (fault detection and identification) level of the supervision system, some knowledge about the dynamic behavior of the system is required. Furthermore, to deal with the different energetic domains in the thermodynamic process, a method or language is needed. The Bond Graph language is well suited for this purpose This approach clearly shows up energy storage and dissipation effects and has the advantage of an efficient transition to the computer by causality assignment and power direction. However, quantitative evaluation of the interactions between the process subsystems and variables are not taken into account by the bond graph approach. Structural analysis allows to fill this gap. In fact it studies the interactions within the system, the coupling between variables and the partition of the system into weakly coupled subsystems.

REFERENCES

Cherifi, E., E. Hais and B. Riera (1995) "Filtering Information for the Supervision of Complex Systems". *In: XVth European Annual Conference on Human Decision Making an Manual Control.* Soesterberg, The Netherlands.

Chittaro, L. G., C. Guido,. E. Tasso and E. Toppano (1993) "Functional and Teleological Knowledge in the Multimodeling Approach for Reasoning about Physical System : a Case Study in Diagnosis". In : *IEEE Trans. On Syst. Man and Cybernetics.* **Vol. 23, n°6.**

Conant, R.C (1972) "Detecting subsystems of a complex system". *IEEE. Trans. On Systems, Man and Cybernetics.* **Vol -2, n°4,** pp.550-553.

Feliot C., J.P. Cassar and M. Staroswiecki (1996) "Towards a Formal Language of Physical Systems". In: *IEEE SMC Conf., Beijing, China.*

Millot P. (1988) *"Supervision des procédés automatisés et ergonomie"* Edition Hermés, Paris.

Modarres, M. (1995) *"Functional Modeling of Complex System using a GTST-MPLD Framework".* Technical report, Department of Material and Nuclear Engineering, University of Maryland, College Park, USA.

Ould Bouamama B; J. U. Thoma and J.P. Cassar *(1997)* "Bond graph modelisation of steam condensers" In *SMC' IEEE, USA, vol.3, p. 490-2494 Olando.*

Pomorski, D and Staroswiecki M. (1996). "Analysis of Dynamical Systems Based on Information Theory". In: *WAC'96 (Word Automation Congress) - ISIAC'96,* pp. 27-30. Montpellier, France.

Richetin, M. (1975) *"Analyse strucurelle des systemes complexes en vue d'une commande hierarchisée".* Thèse de l'université Paul Sabatier de Toulouse.

Thoma, J. U., 1975, *"Introduction to Bondgraphs and their Applications",* Pergamon Press.

Toro, V. (1982) *"Contribution à l'analyse structural des systèmes complexes à l'aide de l'entropie et ses généralisation".* Thèse de doctorat, Université de Lille, France.

EARLY DETECTION OF ALARM SITUATIONS USING MODEL PREDICTIONS

Ben C. Juricek[1], Dale E. Seborg[1], Wallace E. Larimore[2]

[1]*Dept. of Chemical Engineering, University of California,Santa Barbara, CA 93106*
[2]*Adaptics Inc. 1717 Briar Ridge Rd., McLean, VA, 22101*

Abstract: A new monitoring approach is proposed for the early detection of alarm situations. The basic concept is quite simple: a dynamic model of the process is used to predict future values of key process variables as well as confidence limits for the predictions. When the model predictions exceed alarm limits with an acceptable degree of accuracy, an early warning of an alarm situation results. A specific version of this general monitoring approach is examined within this paper. It is based on simple, linear dynamic models and a Kalman filter with disturbance estimation to improve future model forecasts. Simulation results demonstrate the proposed technique works well for certain conditions. *Copyright © 1998 IFAC*

Keywords: Monitoring, Prediction, Alarm Systems

1. INTRODUCTION

It has been estimated that the U.S. petrochemical industry loses $10 billion annually due to "abnormal" situations (Nimmo, 1995), a generic term describing departures from acceptable operating conditions. Consequently, there is considerable incentive for the development of process monitoring and fault detection techniques that can detect an abnormal situation quickly and accurately. Automated aspects of process monitoring have traditionally been based on comparing current measurements with specified warning limits, e.g. Shewhart Charts. Limits are generally determined from the statistical properties of process variables during normal operation, or are based on process knowledge. Recent developments in multivariable statistical procedures (e.g., PCA) also detect faults based on statistically determined confidence limits.

Process alarm limits, unlike statistical warning limits, are defined by the process and operational or safety constraints (limits that may define a safety incident, release, or possible plant shutdown). The difference between the two limits can be seen visually in Figure 1. The Upper Control Limit (UCL) and Lower Control Limit (LCL) are the statistically defined SPC limits, and the alarm limit is much higher. Once a statistical limit has been exceeded, the concern may be what is the probability that an alarm limit might be exceeded. In this paper, a new method for the early detection of process alarm limit violations is examined. The concept is quite simple: a dynamic model of the process is used to predict future values of key process variables as well as confidence limits for the predictions. When the model predictions exceed alarm limits with an acceptable degree of accuracy, an early warning of an alarm situation results. Ideally, the warning will occur well before the actual alarm violation, so that an appropriate corrective action can be taken and undesirable situations such as unacceptable product quality or plant shutdown are avoided.

In the next section, general results from optimal prediction theory are described. The proposed method is described in Section 3, followed by three examples to demonstrate the technique, and finally a few closing remarks.

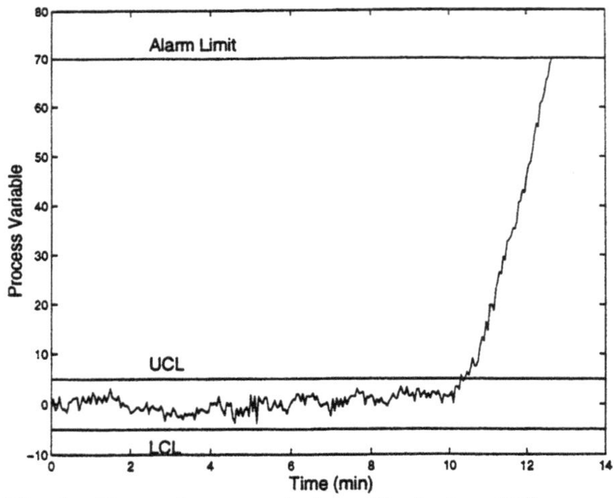

Fig. 1. Example control chart displaying differences between SPC control and alarm limits.

2. PREDICTION USING LINEAR MODELS

The following discussion is restricted to discrete, linear time invariant processes and process models; of course, the same issues will exist for nonlinear processes and models although the details will be decidedly more complicated. Any linear stochastic process model can be expressed in the standard state space form,

$$x(k+1) = Ax(k) + Bu(k) + Dd(k) + w(k) \quad (1)$$

$$y(k) = Cx(k) + v(k) \quad (2)$$

The state vector, x, is dimension n, inputs u are dimension m, outputs y are dimension p. The vector d represents unknown deterministic disturbances, and the vectors w and v represent process and measurement noise, with covariance matrices Q and R. The matrices A, B, C, and D are of the appropriate sizes. This paper examines predicting alarm violations when the term d is nonzero, and follows a simple linear trend (e.g., a step or a ramp). Assuming that the process is known except for the terms associated with d, w, and v, the process model can be expressed as,

$$\hat{x}(k+1) = A\hat{x}(k) + Bu(k) + \hat{\delta}(k) + K(y(k) - \hat{y}(k)) \quad (3)$$

$$\hat{y}(k) = C\hat{x}(k) \quad (4)$$

where K is steady-state Kalman filter gain, derived from the standard Kalman filter equations using a description of the noise covariance matrices Q and R (Gelb, 1974). The term $\hat{\delta}$ can be thought of as a "pseudo-disturbance"; it represents the combined effect of $Dd(k)$. Therefore the values of $\hat{\delta}$ have no physical meaning for describing the disturbance parameters precisely. In the normal Kalman filter formulation, this term is set to zero.

An estimate of the current state is made using past inputs, past measurements, and an estimate of the initial state. Future predictions are made using assumed values for future inputs and the current state estimate. The values for the stochastic terms are set to their expected values, zero for white noise. Let the predicted value of y at time $k+N$, using the information up to time k be denoted as $\hat{y}(k+N \mid k)$. From (3) and (4), this can be expressed as,

$$\hat{y}(k+N \mid k) = CA^N x(k) + \sum_{i=k}^{N-1} CA^{N-i-1}Bu(i) + \sum_{i=k}^{N-1} CA^{N-i-1}\delta(i) \quad (5)$$

The future values of u can be calculated using knowledge of the control system (for example, for proportional only feedback control $u(i) = -C\,y(i)$) and the future predicted values of y.

An accurate prediction in the presence of nonrandom disturbances requires quickly and accurately estimating future values of $\hat{\delta}$. Various approaches have been proposed for estimating the nonrandom bias in Kalman filtering including augmenting states that correspond to parameters of $\hat{\delta}$ (Kozub and MacGregor, 1992) and estimation via a more formal statistical framework (Willsky and Jones, 1976). The approach this paper examines is to assume a model structure for the disturbance and estimate the model parameters via least squares. For example if $\hat{\delta}$ is assumed to be a ramp disturbance,

$$\hat{\delta}(k) = \hat{\delta}(k_0) + (k - k_0)T_s\hat{\delta}_s \quad (6)$$

there are two parameters to estimate: a reference point $\hat{\delta}(k_0)$ and slope $\hat{\delta}_s$. The initial time, k_0, must be specified to be some time after the ramp has started. Substitution of the ramp model, (6), into (5) yields,

$$\hat{y}(k+N \mid k) = CA^N x(k) + \sum_{i=k}^{N-1} CA^{N-i-1}Bu(i) + \sum_{i=k}^{N-1} CA^{N-i-1}\delta(k_0) + \sum_{i=k+1}^{N-1}(i-k)CA^{N-i-1}\delta_s \quad (7)$$

The parameters of the ramp, $\hat{\delta}(k_0)$ and $\hat{\delta}_s$ are estimated from (7) using least squares regression and past values of y and u.

Confidence limits about future predictions of y can be computed using the covariance of the Kalman filter estimate, $P(k)$, the covariance of $\hat{\delta}$, $\Sigma(\hat{\delta})$, the state noise covariance and the update law for the state estimate covariance estimate (Gelb, 1974).

$$P(k+1) = AP(k)A^T + Q + \Sigma(\delta) \quad (8)$$

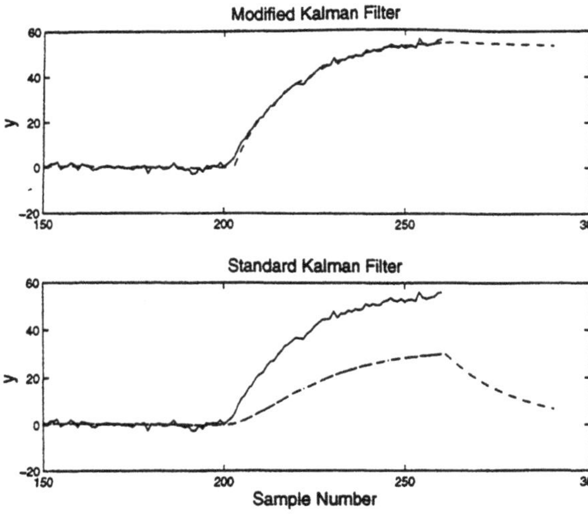

Fig. 2. Predictions with the normal and modified Kalman filters for Case I (solid line: actual values, dashed line: filter estimates, dotted line: future predictions).

The initial value for $P(k)$ is the steady-state solution of the algebraic Riccati equation. The three terms of the expression can be interpreted as the propagation of uncertainty by system dynamics, the stochastic term, and the uncertainty about $\hat{\delta}$.

3. PREDICTIONS FOR ALARM DETECTION

The mechanics of the alarm detection scheme are described in this section. Once a statistical limit has been generated (denote the time of this occurrence as t_a, as in Figure 1), the entire scheme is initiated to predict N samples ahead. The choice of N is application dependent. The estimate of the alarm limit violation time is denoted as T_A.

A ramp disturbance model form is assumed (note that a step disturbance is a subset of (6) if $\hat{\delta}_s = 0$). This choice of model structure allows for increasing disturbances, a significant advantage over the step model structure if predictions are of concern.

Once the value of $\hat{\delta}$ has been estimated as a ramp, (5) can be used for predictions. The covariance matrices from the Kalman filter and parameter estimates are used to calculate the covariance matrix of the future predictions from (8). This information can be used with the predictions to determine when a violation of a process alarm limit is predicted with acceptable accuracy. A conservative method would be to alert the operator only when the 99% confidence limits (or other value for Type I error) about future predictions exceed the alarm limit; that is, T_A is calculated from the lower confidence limit about the future predictions.

Although specifying the required accuracy about the future predictions gives a nice interpretation and method, the lower confidence limit may never exceed the alarm limit even for very low noise conditions and accurate models (this would be a common result for poorly conditioned systems). In such cases, another criterion for signaling T_A would be required, and most probably application dependent.

4. EXAMPLES

Three examples are examined in this section involving univariate and multivariate models. In the univariate case, there are two disturbances: a step that does not cause the safety limit to be exceeded and a ramp that leads to a violation. Two examples of a linearized CSTR are examined to inspect the extension to multivariable systems with different conditioning characteristics.

4.1. Univariate Example

The system in this example is given by:

$$x(k+1) = 0.95x(k) + u(k) + w(k) + d(k) \qquad (9)$$

$$y(k) = Cx(k) + v(k) \qquad (10)$$

The uncorrelated noise sequences, v and w, have variances 1 and 0.1. In Case I, $d(k)$ is a step increase at sample 201 of magnitude 3. In Case II, $d(k)$ is a ramp that begins at sample number 201 with a slope of 0.1. For this numerical example, upper and lower control limits are defined to be ± 5, and the high alarm limit is chosen to be 70.

The importance of estimating $\hat{\delta}$ for predictions and filtering is now obvious. In the absence of $\hat{\delta}$ the estimates of the standard Kalman filter will be biased, although the sign of the gain may still be correct. Future predictions will exponentially approach the bias value. Modifying the filter by including the

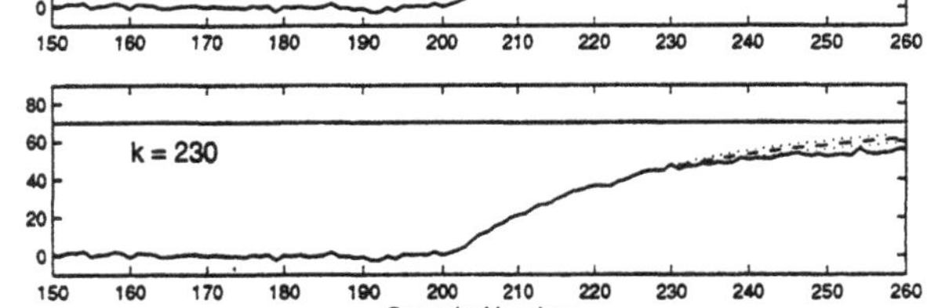

Fig. 3. Predictions for Case I (solid line: actual values, dashed line: future predictions, dotted line: 95% confidence limits)

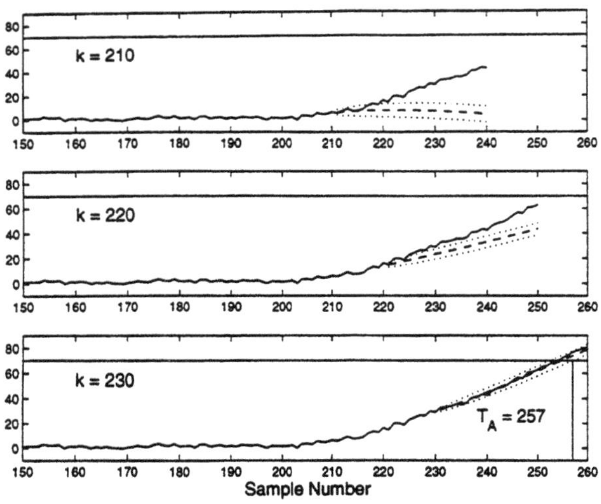

Fig. 4. Predictions for Case II (solid line: actual values, dashed line: future predictions, dotted line: 95% confidence limits)

estimate of the disturbance improves both the filtering and prediction. This is illustrated in Figure 2 for Case I.

The results for the step disturbance, shown in Figure 3, show the predictions correctly predict that no alarm violation will occur. The 30 step ahead predictions are shown in Figure 4 for Case II. For the predictions from sample number 230, the 95% confidence limits exceed the alarm limit at sample 258, four samples after the actual violation. Hence, the ramp provides a 24-sample warning of the impending violation. The plots of the parameter estimates for $\hat{\delta}$ are displayed in Figure 5 for both cases. As more data becomes available, the estimated values for $\hat{\delta}$ approach the true values. As the parameter estimates approach the true values, the accuracy of the future predictions improves.

When the estimate of $\hat{\delta}$ is poor initially, the future

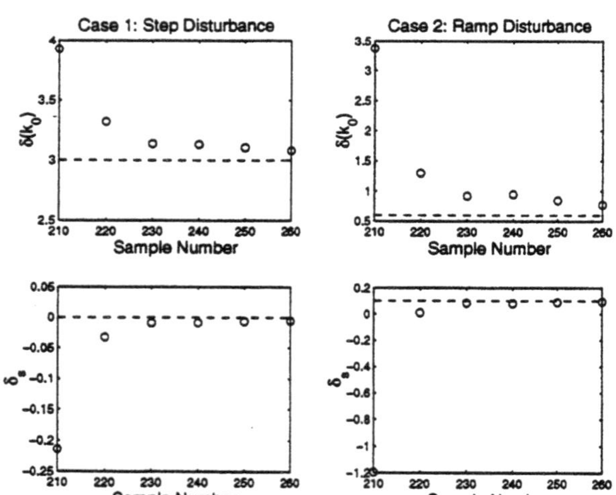

Fig. 5. Parameter estimates for univariate example (dashed line: actual values, circles: estimated values).

data points do not lie within the 95% confidence bounds, demonstrated by Figure 4 for the predictions made at k = 210 and 220.

4.2 CSTR Examples

A non-isothermal CSTR with first order reaction, $A \rightarrow B$, is described by:

$$\frac{dC_A}{dt} = \frac{q}{V}(C_{Af} - C_A) - k_0 \exp(-\frac{E}{RT})C_A \quad (12)$$

$$\frac{dT}{dt} = \frac{q}{V}(T_f - T) + \frac{-\Delta H}{\rho C_p}k_0 \exp(-\frac{E}{RT})C_A + \frac{UA}{V\rho C_p}(T_J - T) \quad (13)$$

The above equations were linearized about two operating conditions given in Table 1 and Table 2, each with a sampling period of 0.05 min. The parameters of the examples are chosen to compare systems with vastly different condition numbers.

In the following examples, C_{Af} ramps at 0.2 mol/l min, beginning at 5 min.

CSTR Example #1 The parameters in Table 1 were used to produce a well-conditioned state space model. Figures 6 and 7 display the 30 step ahead (1.5 min.) future predictions made at t = 5.5, 6.5, 7 and 7.5 min for C_A and at t = 5.5, 6.5, 7.5 and 9.25 for T. Figure 8 displays the parameter estimates at these times.

For an alarm limit of 355 K, Figure 7 shows the lower confidence bound for the predictions of T made at 9.25 min. exceeds the limit at time 10.6; i.e., T_A = 10.6. The true violation occurs at 10.25, giving a 1 min. (20 sample) warning.

As expected, the estimates of $\hat{\delta}$ approach the true

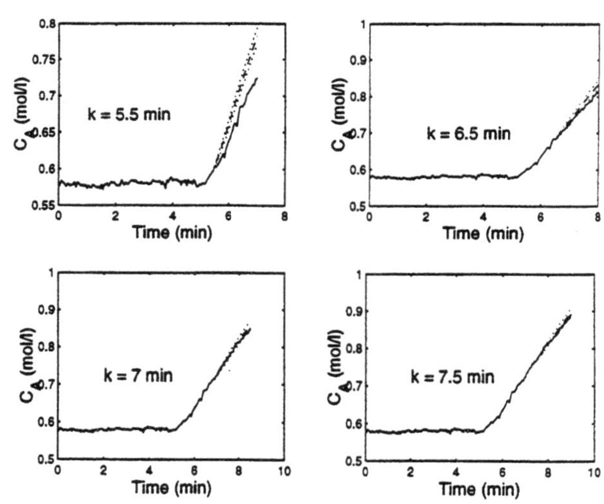

Fig. 6. CSTR example #1 future predictions for C_A (solid line: actual values, dashed line: future predictions, dotted line: 95% confidence limits).

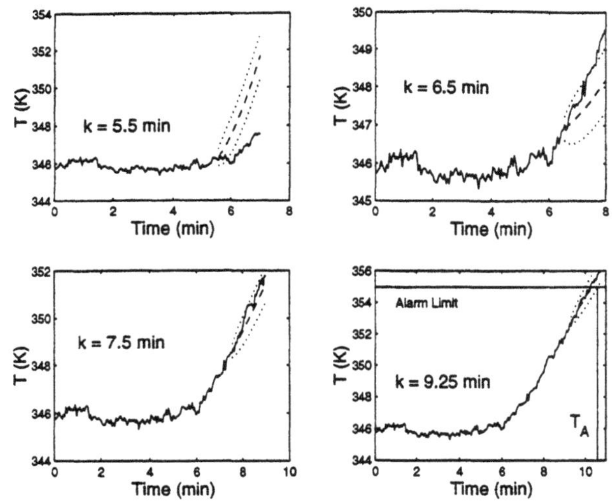

Fig. 7. CSTR example #1 future predictions for T (solid line: actual values, dashed line: future predictions, dotted line: 95% confidence limits).

values as more data is used in the least squares estimation. The true values are not precisely descriptive of the ramp disturbance properties since $\hat{\delta}$ represents the combined effect of $Dd(k)$ in (1). The accuracy of the future predictions, in terms of the error between predicted and actual values, improves as the estimates of $\hat{\delta}$ approach their true values.

CSTR Example #2 The parameters for this CSTR example are given in Table 2. Figures 9 and 10 display the 30 step ahead (1.5 min.) future predictions made at t = 5.25, 5.5, 6 and 7.5 min for C_A and T. Figure 11 displays the parameter estimates at the same prediction times.

Again, the estimates of $\hat{\delta}$ approach the true values as more data is used in the least squares estimation and thus the accuracy of the future predictions, in terms of the error between predicted and actual values, also

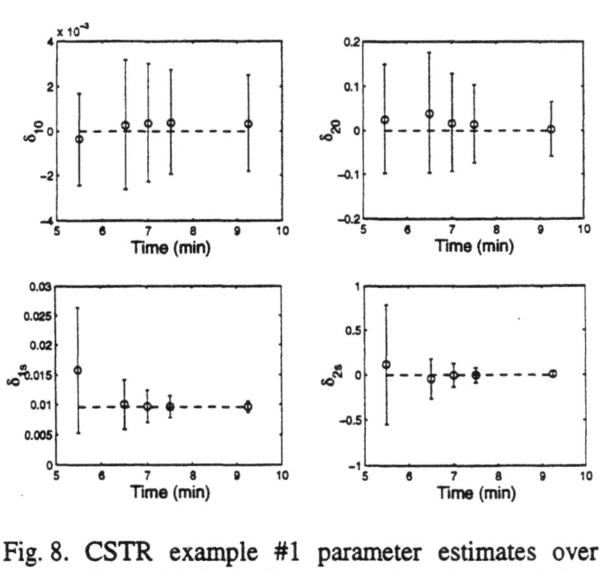

Fig. 8. CSTR example #1 parameter estimates over time (dashed line: actual values, circles: estimates, 95% error bars).

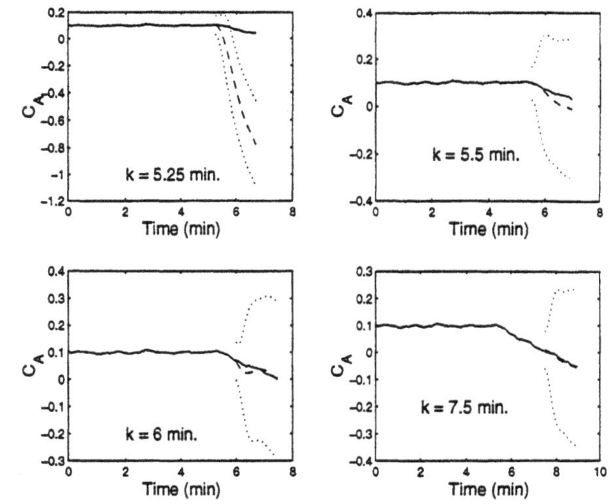

Fig. 9. CSTR example #2 future predictions for C_A (solid line: actual values, dashed line: future predictions, dotted line: 95% confidence limits).

improves.

However, using the perfect model and relatively small noise, the 95% confidence limits for C_A and T are so large they render the future predictions useless. Figure 12 shows the standard deviations of the 30-step ahead predictions for T and C_A at different prediction times. The decrease in the standard deviation using more data is due to the increased accuracy about the $\hat{\delta}$ estimates. The limiting value is almost entirely due to the system dynamics; the dominant term in (8) is APA^T. This large variance due to system dynamics arises due to the strong interaction between C_A and T (a small increase in the variability of C_A has a large effect on the variance in T) reflected by the large term (2,1) element in A. This phenomenon can also be seen by examining the confidence limits for example #1 for the much better conditioned system (the condition numbers are

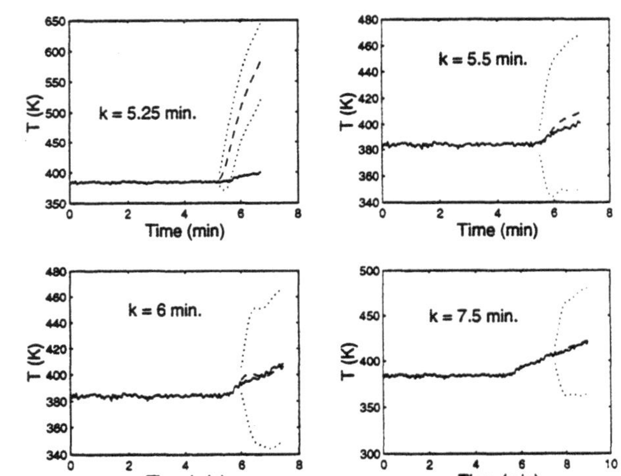

Fig. 10. CSTR example #2 future predictions for T (solid line: actual values, dashed line: future predictions, dotted line: 95% confidence limits).

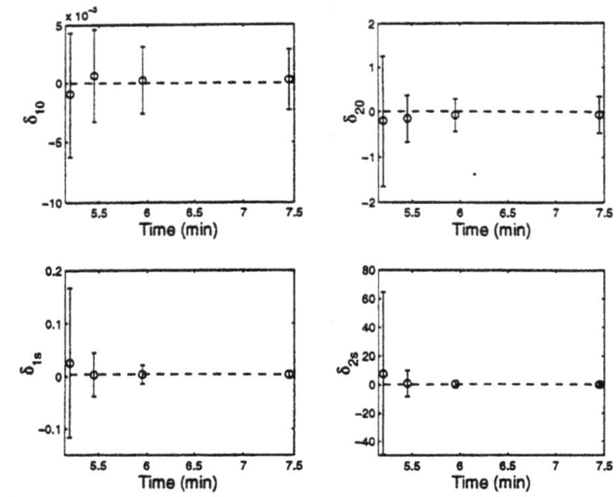

Fig. 11. CSTR example #2 parameter estimates over time (dashed line: actual values, circles: estimates, 95% error bars).

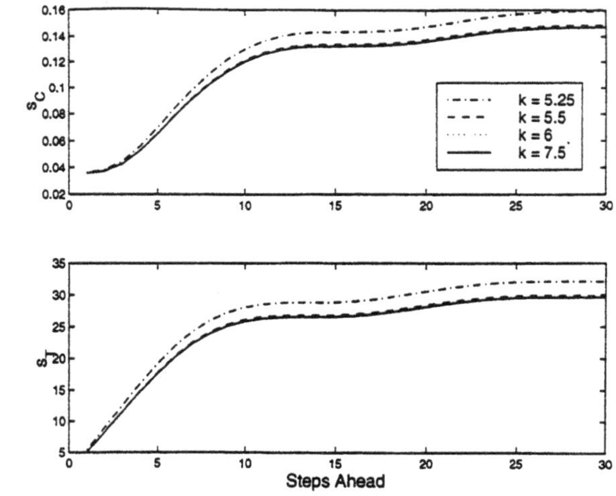

Fig. 12. Example #2 standard deviations of 30-step ahead predictions.

approximately 3 and 8000 for the two examples). Because the quantity of interest is the output covariance matrix, this phenomenon will be invariant to scaling x or y.

Note that in both CSTR examples and the numerical example the difference between the actual and predicted values is not always reflected by the theoretically determined confidence limits.

Although the covariance matrix for an optimal predictor given by (8) has a sound theoretical basis, the application may require estimating a sample covariance based on the performance of past predictions. In this way, the effect of modeling errors, the applicability of the ramp model structure for the disturbance, and the effect of the system condition can be calculated using the data rather than the theoretical values. The result may yield more easily interpretable results, at the price of theoretical precision.

5. CONCLUDING REMARKS

A novel monitoring technique has been proposed that uses predictions of process variables to determine if specified alarm limits may be exceeded. It is found that the straightforward implementation of optimal prediction theory may be useful for univariate systems or well-conditioned multivariate systems.

A sense of perspective is important in judging the usage of predictions for detecting alarm situations. It is impossible for any forecasting technique, no matter how complex, to know entirely the future behavior of disturbances (for example, a random walk would be impossible to predict with any precision). However, given a reasonably accurate process model, model predictions may be able to be used for process monitoring, analogous to using predictions for control in MPC applications.

REFERENCES

De Valliere, Ph. and D. Bonvin, Application of Estimation Techniques to Batch Reactors – III. Modeling Refinements Which Improve the Quality of State and Parameter Estimation, *Computers and Chemical Engineering*, **14**, 799-806.

Gelb, A., *Applied Optimal Estimation*, M.I.T. Press, Cambridge, MA, 1974.

Kozub, D. and J. MacGregor, State Estimation for Semi-Batch Polymerization Reactors, *Chemical Engineering Science*, **47**, 1047-1062.

Nimmo, I., Adequately Address Abnormal Operations, *Chemical Engineering Progress*, **91**, 36.

Willsky, A. and H. Jones, A Generalized Likelihood Ratio Approach to the Detection and Estimation of Jumps in Linear Systems, *IEEE Trans. Auto. Control*, **19**, 108-112.

Table 1. CSTR #1 Nominal Values and Parameters

V	100 L	E/R	8750 K
k_0	7.2×10^{10} min^{-1}	ΔH	-5×10^3 J/m$^3 \cdot$ K
ρC_p	1195 J / m$^3 \cdot$ K	UA	5×10^4 J/min $\cdot$ K
T_F	350 K	C_{Af}	1 mol. / L
T_J	309.9 K	C_A	0.42 mol. / L
T	345.8 K	q	100 L / min

Table 2. CSTR #2 Nominal Values and Parameters

V	100 L	E/R	8750 K
k_0	7.2×10^{10} min^{-1}	ΔH	-5×10^4 J/m$^3 \cdot$ K
ρC_p	239 J / m$^3 \cdot$ K	UA	5×10^4 J/min $\cdot$ K
T_F	350 K	C_{Af}	1 mol. / L
T_J	309.9 K	C_A	0.1 mol. / L
T	383.7 K	q	100 L / min

INTELLIGENT SENSOR AUDITING BY INTERFACING KNOWLEDGE-BASED SYSTEMS AND MULTIVARIATE SPM TOOLS

Eric Tatara * **Antoine Negiz** *,[2] **Aras Norvilas** *
Jeffrey DeCicco * **Ali Çinar** *,[1]

* *Department of Chemical and Environmental Engineering, Illinois Institute of Technology, 10 W. 33rd Street, Chicago, Illinois, 60616 USA*

Abstract: An intelligent sensor monitoring and auditing environment has been developed by interfacing multivariate statistical process monitoring (MSPM) techniques and knowledge-based systems (KBS) for monitoring sensor status in multivariable processes. The real-time KBS development environment G2 is integrated with an MSPM method that can monitor multivariable dynamic processes with autocorrelated data. The MSPM method is based on a canonical variate state space (CVSS) process model of a dynamic process and it is used for auditing sensor status for bias, drift and excessive noise affecting the sensors of multivariable continuous processes. Changes in the magnitudes of means and variances of residuals between measured and predicted process variables are used to detect and discriminate sensor abnormalities. The statistical model that describes the in-control variability of sensor readings is based on a CVSS model. The CV state variables obtained from the state space model are linear combinations of the past process measurements which explain the variability of the future measurements the most, and they are regarded as the principal dynamic dimensions. The method can detect and discriminate between bias change, drift, and variations in noise levels of process sensors based on the analysis of data batches. The MSPM modules are developed in Matlab, converted to C, and linked with G2. The presentation will focus on the structure and performance of the integrated system and illustration of the methodology by simulation studies. *Copyright © 1998 IFAC*

Keywords: Sensor auditing, data fusion, knowledge-based system, multivariate statistical process monitoring, canonical variate analysis, state space models.

INTRODUCTION

Misleading process information can be generated if there is a *bias* change, *drift* or high levels of *noise* in some of the sensors. Erroneous information often causes decisions and actions that are unnecessary, resulting in the deterioration of product quality, safety and profitability. Identifying failures such as a broken thermocouple is relatively easy since the signal received from the sensor has a fixed and unique value. Incipient sensor failures that cause drift, bias change or additional noise are more difficult to identify and may remain unnoticed for extended periods of time. Auditing sensor behavior can warn plant personnel about incipient sensor faults and initiate timely repair and maintenance. A new sensor auditing method based on functional redundancy generated by the CVSS space model was presented for detecting abnormal behavior in one or more sensors of multivariable processes (Negiz and Çinar 1997b, Negiz

[1] Author to whom correspondence should be addressed *cinar@charlie.iit.edu*
[2] Present address: UOP Inc., 25 East Algonquin Road, P.O. Box 5017, Des Plaines, IL 60017, USA. *anegiz@uop.com*

and Çinar 1997c). The method is integrated with a real-time KBS in G2 to enhance its decision making capabilities. The modeling and statistical monitoring algorithms developed in Matlab are compiled in C, and made callable by G2. G2 is used for coordinating the data acquisition, and inferencing about sensor status by using the refined information generated by the statistical monitoring algorithms.

Industrial continuous processes have a large number of process variables and are usually operated for extended periods at fixed operating points under closed-loop control, yielding process measurements which are autocorrelated, cross correlated, and collinear. Consequently, a method that can handle autocorrelated data from a dynamic process is necessary. The targeted application area of the proposed method is multivariable continuous processes operating at steady state and generating stationary data. If the process moves to a different state, then a new calibration model must be built from normal operation data collected when the process is operating under the new conditions without sensor failures. The method developed for *periodic* audit of process sensors is based on interpreting the magnitude of the mean and variance of the residuals between a batch of measurement data from sensors and their prediction from a process model. The sensor auditing procedure can be repeated at regular intervals to report any sensor that is operating abnormally during the period corresponding to the last data collection interval. The statistical model of the in-control variation is a *canonical variate state space* (CVSS) model (Negiz and Çinar 1997a). The model describes the dynamic behavior of the process and it is developed such that the state variables are statistically independent (orthogonal) at zero lag. The state space models and the statistical characterization of the random components are obtained directly from process data collected when the process was operating in-control. The algorithm is suitable for handling large number of variables that are autocorrelated, cross correlated, and collinear. Once an accurate statistical description of the in-control variability of a continuous process is available, the next step is the design and implementation of the sensor monitoring procedure.

DYNAMIC PROCESS MODELING WITH CVSS APPROACH

Canonical variate analysis (CVA) (Larimore 1983, Negiz and Çinar 1997a) is used to generate a state space model that describes the behavior of a process. CVA was introduced by Hotelling in 1936 and applied to dynamic systems by Akaike and Larimore (Larimore 1983). It is a subspace identification method (Overschee and de Moor 1994). Negiz and Çinar (1997) developed an SPM method based on the state variables of a CVSS model for multivariable dynamic continuous processes. The CVSS model can also be used for sensor monitoring by functional redundancy.

Consider a state space model

$$\begin{aligned} \mathbf{x}_{k+1} &= \mathbf{A}\mathbf{x}_k + \mathbf{B}\epsilon_k \\ \mathbf{y}_k &= \mathbf{C}\mathbf{x}_k + \epsilon_k \end{aligned} \qquad (1)$$

where $\mathbf{x}_k$ is the vector of n state variables at time k, $\mathbf{y}_k$ is the vector of p output measurements, ϵ_k is the innovation vector with zero mean and covariance $E(\epsilon_k \epsilon_{k+l}^T) = \mathbf{\Delta}$ if $l = 0$, and $\mathbf{0}$ otherwise. CVSS realization methods are used for determining the system matrices $\mathbf{A}$, $\mathbf{B}$, and $\mathbf{C}$, and the covariance matrix $\mathbf{\Delta}$ with the minimum number of state variables needed to capture the dynamics of the process.

CVSS realization procedure is based on a Hankel matrix that expresses the covariance between the vectors of stacked future outputs $(\mathcal{Y}_{k_J}^+)$ and past outputs $(\mathcal{Y}_{k_K}^-)$.

$$\mathcal{Y}_{k_J}^+ = \begin{bmatrix} \mathbf{y}_k \\ \mathbf{y}_{k+1} \\ \vdots \\ \mathbf{y}_{k+J-1} \end{bmatrix} \quad \text{and} \quad \mathcal{Y}_{k_K}^- = \begin{bmatrix} \mathbf{y}_{k-1} \\ \mathbf{y}_{k-2} \\ \vdots \\ \mathbf{y}_{k-K} \end{bmatrix}$$

The Hankel matrix is defined as:

$$\mathbf{H}_{KJ} = E\left(\mathcal{Y}_{k_J}^+ \mathcal{Y}_{k-1_K}^{-T}\right) \qquad (2)$$

The number of future (J) and past (K) observations are parameters that must be defined by the user. The dimensions of the state vector are selected by determining the dominant singular values of the Hankel matrix $\mathbf{H}_{JK}$ by inspection or through the use Akaike information criteria (AIC) (Negiz and Çinar 1997a).

The singular value decomposition of the scaled Hankel matrix is :

$$\bar{\mathbf{H}}_{JK} = \left(\mathbf{R}_J^+\right)^{-1/2} \mathbf{H}_{JK} \left(\mathbf{R}_K^-\right)^{-1/2} = \mathbf{U}\mathbf{\Sigma}\mathbf{V}^T \qquad (3)$$

where $\left(\mathbf{R}_J^+\right) = E\left(\mathcal{Y}_{k_J}^+ \mathcal{Y}_{k_J}^{+T}\right)$ and $\left(\mathbf{R}_K^-\right) = E\left(\mathcal{Y}_{k-1_K}^- \mathcal{Y}_{k-1_K}^{-T}\right)$ are used for scaling. The state variables can be expressed as

$$\mathbf{x}_k = \mathbf{\Sigma}^{1/2}\mathbf{V}^T \left(\mathbf{R}_k^-\right)^{-1/2} \mathcal{Y}_{k-1_K}^- . \qquad (4)$$

Once the state variables are computed, the system matrices $\mathbf{A}$, $\mathbf{B}$, and $\mathbf{C}$, and the covariance matrix $\mathbf{\Delta}$ of the innovation vector $\mathbf{e}_k = \mathbf{y}_k - E\left(\mathbf{y}_k | \mathcal{Y}_{k-1_K}^-\right)$ can be determined (Negiz and

Çinar 1997a). The state variables defined by the CVSS modeling procedure describe the variation in data such that each additional state variable describes the largest fraction of the variation remaining after the portion described by the previously defined state variables. The covariance matrix of the state variables is diagonal $E\left(\mathbf{x}_k\mathbf{x}_k^T\right) = \Sigma$ (Negiz and Çinar 1997a). These state variables are independent from each other at zero lag as shown by their crosscorrelation. Because of this property, the state variables can be used to determine the status of a multivariable process that has time dependent measurements.

Software performing canonical variate realization has been developed in Matlab (The Mathworks, Inc. 1997). The Matlab function requires a matrix of process variable data and values for the window lengths J and K. It returns the number of state variables selected and the system matrices. Canonical variate state variables describing the process are calculated from the previous K outputs, the current output, and the system matrices returned by the software.

SENSOR AUDITING WITH CV STATE VARIABLES

Information from various process sensors can be used for assessing the correctness of information generated by a sensor. This approach falls in the category of *functional redundancy*. A *multipass PLS* algorithm was developed for detecting simultaneous multiple sensor abnormalities (Negiz and Çinar 1992) as an extension of the methodology proposed by (Wise *et al.* 1989). Since the PLS regression algorithm is based on zero-lag covariance of the process measurements, the application of the *multipass PLS* algorithm is based on the assumption of serially independent observations (negligible autocorrelation). This assumption is justified for a continuous process operating at steady state and having only random noise on measurements, but is not valid if the process dynamics are significant and the data are autocorrelated. An alternative technique which handles effectively both autocorrelation and crosscorrelation in process data can be developed by using the CVSS model. Sensor audit will be based on one-step ahead residuals generated from the CVSS model. A *multipass* CVSS technique similar to the *multipass PLS* algorithm is developed for detecting multiple sensor failures. This is achieved by eliminating successively the corrupted measurements from both the calibration and test data sets and identifying a different CVSS submodel.

Assume that there are p sensors to be monitored and the *calibration* data set is of length N. The mean and the variance of the residuals for each variable is computed through the $N \times p$ residuals block matrix $\mathbf{R}$. Once the CV state space model is identified for the in-control data set, the statistics for the residuals such as mean and variance are computed. Then, a test data block of size $N_t \times p$ is formed from new process measurements. The residual statistics for the test data set are then generated by using the CV *calibration* model. The statistical test compares the statistics (mean and variance) of the residuals of the test data set with the statistics of the residuals of the *calibration* data set for detecting any significant departures.

Denote by $\mathbf{R}_{\bullet i}$ the ith $N \times 1$ residual column vector from the $N \times p$ residual block matrix $\mathbf{R}$. The statistic for testing the null hypothesis of the equality of means from two normal populations with equal and unknown variances is

$$\frac{\bar{\mathbf{R}}_{\bullet \mathbf{i}_{test}} - \bar{\mathbf{R}}_{\bullet \mathbf{i}_{model}}}{\hat{\sigma}_{p_i}\sqrt{1/N + 1/N_t}} \sim t_{N+N_t-2} \qquad (5)$$

where $\bar{\mathbf{R}}_{\bullet \mathbf{i}_{test}}$ and $\bar{\mathbf{R}}_{\bullet \mathbf{i}_{model}}$ denote the maximum likelihood estimates of the residual means for the variable i in the test sample and the *calibration* set, $\hat{\sigma}_{p_i}$ is the pooled standard deviation of the two residual populations for the i-th variable, N and N_t denote the sizes of the *calibration* and testing populations, and t_{N+N_t-2} is the t-distribution with the $N+N_t-2$ degrees of freedom (Dudewicz and Mishra 1988).

The statistic for testing the null hypothesis of the equality of variances from two normal populations with unknown means is (Dudewicz and Mishra 1988)

$$\frac{\hat{\sigma}^2_{i_{test}}}{\hat{\sigma}^2_{i_{model}}} \sim F_{N_t-1,N-1} \qquad (6)$$

where $F_{N_t-1,N-1}$ is the F distribution with respective degrees of freedom. The level of the test for each test statistic for each variable is set to 5% in our studies, but it can be changed easily depending on the process requirements. The test for the mean of the residual is two-sided, and the test for the variance is one-sided. This part of the procedure is similar to that given by (Wise *et al.* 1989).

The algorithm indicates a sensor fault when either the mean or variance of the residuals of the corresponding sensor are out of the statistical limits (based on t and F probability distributions). Since the corrupted variable affects the predictions of the remaining variables, false alarms might be generated unless the corrupted variable is taken out of both the *calibration* and the *test* data sets. The information loss due to taking a variable out of both the calibration and the test data sets is not significant since the testing procedures are based on the *iid* assumption of the residuals and not on the minimum prediction error criterion by the model. The algorithm discards the variable with the highest deviation level by looking at the residual mean and variance values. For graphical representation, a more pleasant display is obtained by plotting the ratios of residual mean and residual variance of each sensor to their respective statistical limits which are based on Eqs. 5–6. Dividing the residual mean and variance by a fixed number (statistical limit) does not change the statistical properties but makes the upper limit of the graphs unity.

The nature of the sensor fault is diagnosed by determining whether the residual mean or the residual variance or both exceed their statistical limit. If only the residual mean exceeds its limit, then the corresponding sensor is subjected to a bias change. If only the residual variance exceeds its limit, then the corresponding sensor is subjected to excessive noise. If both the residual mean and the residual variance exceed their limits, then the corresponding sensor is experiencing a drift.

Excluding variables and computing a new CV realization for the remaining variables is the key step of the *multipass* sensor auditing and fault detection algorithm. The likelihood for all of the process sensors to become simultaneously faulty is extremely small. After several successive steps, if the mean and variance of the remaining residuals still indicate significant variation, then it is more likely that a disturbance is active on the system causing the in-control variability to change significantly.

KBS STRUCTURE AND OPERATION

The G2 real-time KBS shell (Gensym 1996) is used to develop the sensor fault monitoring and diagnosis system. G2 is a hybrid system with a class-object structure for knowledge representation and rules for inferencing. The use of G2 to integrate sensor fault monitoring techniques and a KBS has several advantages. The ease of generating charts within G2, and a simple and logical language structure allow the development of a monitoring application that can be presented in a readily understandable format. G2 Standard Interface (GSI) is used to build a communication bridge between the G2 application and an external system such as process data bases or software written in C language.

The KBS for MSPM and fault diagnosis has the following components:

- G2 KBS shell,
- GSI (G2 Standard Interface), an executable program that exchanges data between G2 applications and external systems,
- Functions written in Matlab and converted to ANSI C that are called by G2 for model development.

The KBS handles data collection and storage, and when instructed by the user develops the CVSS model by using the stored data. The residuals are computed to develop the means, variances, and their limits. When sensor auditing is initiated, a new data set is collected and the residuals are computed to initiate the analysis. The KBS supervises the sensor status analysis and model development activities, and displays appropriate conclusions to the user on various screens.

MONITORING THE SENSORS OF A POLYMERIZATION REACTOR

The polymerization of vinyl acetate in a continuous stirred tank reactor as modeled by (Teymour 1989) was used to simulate and monitor a multivariable system. Four reactor input variables (feed temperature, feed initiator concentration, residence time, solvent volume fraction) were kept constant with added random noise, and the fifth variable, cooling jacket temperature, was manipulated by a PID controller to regulate the reactor temperature. The reactor simulation was run and the output data were collected for a sufficient amount of time to generate a CV state space model. Four process outputs are monitored, the reactor temperature and conversion appear to be correlated outputs, reactor initiator concentration behaved inversely to reactor temperature, and the polydispersity varies as the reactor conditions change. Various sensor faults are introduced to the system and data are collected under the faulty operating conditions. New batches of data collected under specific faulty sensor conditions are evaluated by the sensor fault monitoring software and the results of the evaluation are displayed on screen.

A case study performed to demonstrate the diagnostic procedure of the sensor fault detection routine by simulating a drift in the polydispersity measurement is presented to illustrate the method. Initially, all four of the monitored variables from the simulation are used in the construction of the CVSS calibration model. A calibration data set is collected the CVSS model is developed and the residuals between the data and model predictions are computed. At a later time, a test set of 200 samples (for example) is collected of the four process variables. The inputs to the sensor validation subroutine include the test data, the residuals of the calibration data set and system matrices computed for the calibration model. The statistical limits are computed from the residuals of the means and variances and they are shown in dashed lines in Figures 1 and 2. The subroutine also returns the ratios of the means and variances of the residuals to the statistical limits for each process variable as shown in Figures 3 and 4. In all figures, variable 1 is the initiator concentration, variable 2 is the conversion, variable 3 is the reactor temperature and variable 4 denotes the polydispersity. Since the ratio of the residual mean to the statistical limit is greater than 1 for variable number 4, the polydispersity measurement will be diagnosed as faulty. Because the residual variance for the polydispersity is also out of bounds, the sensor fault is diagnosed as a drift. The polydispersity measurement is excluded from both the calibration and the test data sets and a new CVSS model is computed by using the remaining three variables from the original calibration set. Figures 5 and 6 show the residuals for variables 1, 2, and 3, after variable 4 is excluded. The residual mean and variance statistics are all within the limits, indicating that no other faulty sensors exist in the remaining measurements. Therefore, the magnification in the residuals of the other sensors were the result of the contribution by the faulty sensor related to polydispersity measurement.

CONCLUSION

A sensor auditing method is developed for multivariable continuous processes based on functional redundancy generated by a state space process model. The method can detect and discriminate between bias change, drift, and noise. The method can be implemented to run repeatedly at frequent intervals based on data sets collected for a sufficiently long period and warn plant personnel about incipient sensor faults. The interface of the numerical procedures with a real-time knowledge-based environment such as G2 permits more in-depth diagnosis opportunities.

Acknowledgment: The authors gratefully acknowledge the ASA/NSF/NIST Research Associate appointment of A. Negiz in 1992-1993, and the research assistantships provided to A. Negiz, J. DeCicco, A. Norvilas, and E. Tatara by the National Center for Food Safety and Technology.

1. REFERENCES

Dudewicz, E. J. and S. N. Mishra (1988). *Modern Mathematical Statistics*. John Wiley. New York.

Gensym, Corporation (1996). *G2 Reference Manual.* Cambridge, MA.

Larimore, W. E. (1983). System identification, reduced-order filtering and modeling via canonical variate analysis. In: *Proc. 1983 Automatic Control Conf.*. pp. 445–451.

Negiz, A. and A. Çinar (1992). On the detection of multiple sensor abnormalities in multivariable processes. In: *Proc. 1992 American Control Conf.*. pp. 2364–2368.

Negiz, A. and A. Çinar (1997a). Statistical monitoring of multivariable continuous processes with state space models. *AIChE J.* **43**, 2002–2020.

Negiz, A. and A. Çinar (1997b). Monitoring of multivariable dynamic processes and sensor auditing. In: *Proceedings of ADCHEM'97.* pp. 55–60.

Negiz, A. and A. Çinar (1997c). State variable based statistical methods for auditing sensors of multivariable dynamic processes. In: *Proceedings of SAFEPROCESS'97.* pp. 181–186.

Overschee, P. Van and B. de Moor (1994). N4sid: Subspace algorithms for the identification of combined deterministic-stochastic systems. *Automatica* **30**, 75–93.

Teymour, Fouad (1989). The Dynamic Behavior of Free-Radical Solution Polymerization in Continuous Stirred Tank Reactors. PhD thesis. University of Wisconsin, Madison.

The Mathworks, Inc. (1997). *Matlab Reference Guide.* The Mathworks, Inc.. Natick, MA.

Wise, B. M., N. L. Ricker and D. J. Veltkamp (1989). Upset and sensor fault detection in multivariable processes. AIChE Annual Meeting, Paper 164b, San Francisco, CA.

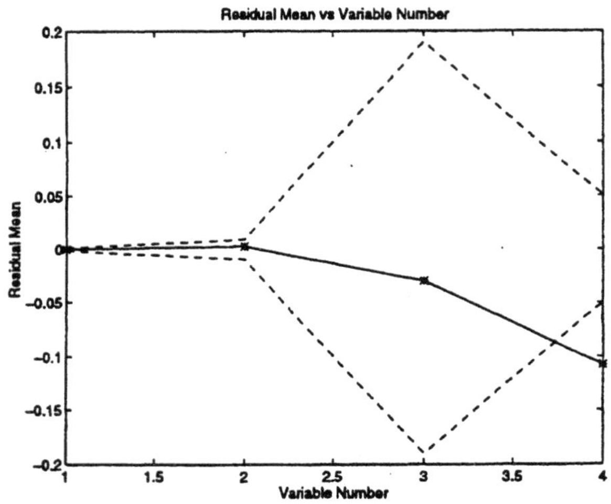

Fig. 1. The residual means for the test data set and the statistical limits (dashed lines)

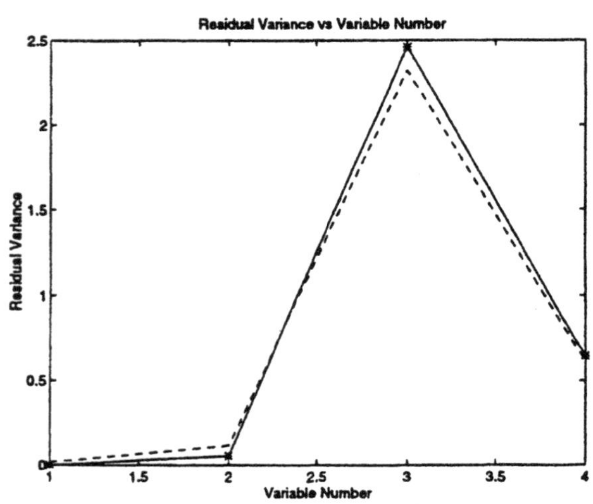

Fig. 2. The residual variancess for the test data set and the statistical limits (dashed lines)

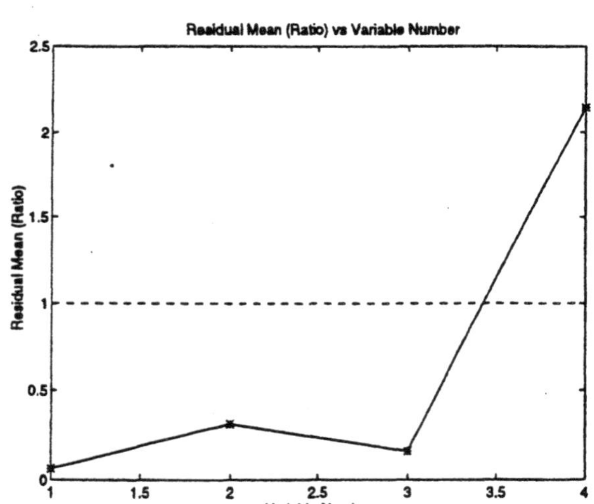

Fig. 3. The residual mean ratios for the test data set and the statistical limits (dashed lines)

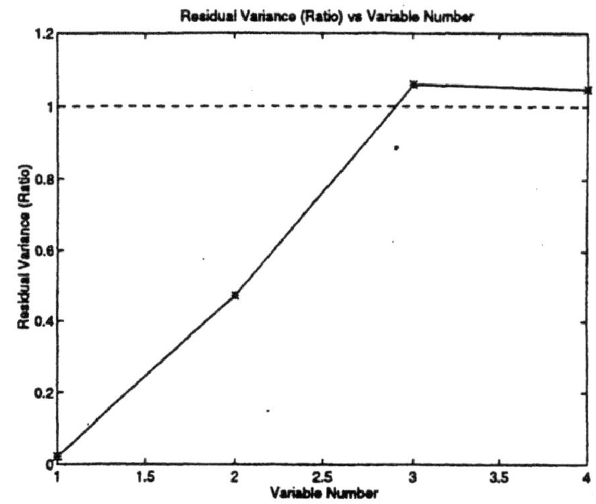

Fig. 4. The residual variance ratios for the test data set and the statistical limits (dashed lines)

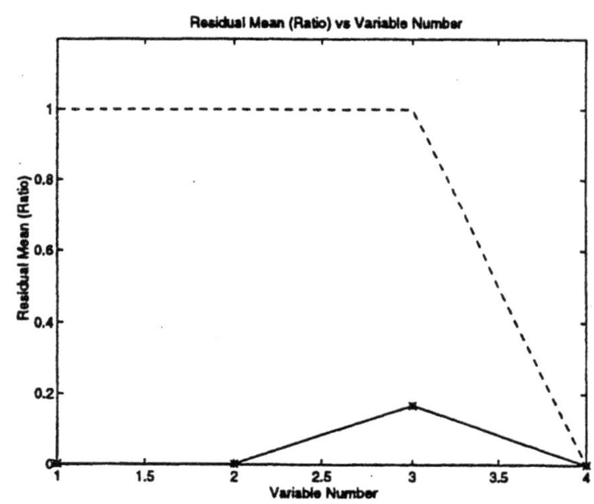

Fig. 5. The residual mean ratios for the test data set and the statistical limits (dashed lines) after sensor 4 is removed from calibration and test data sets

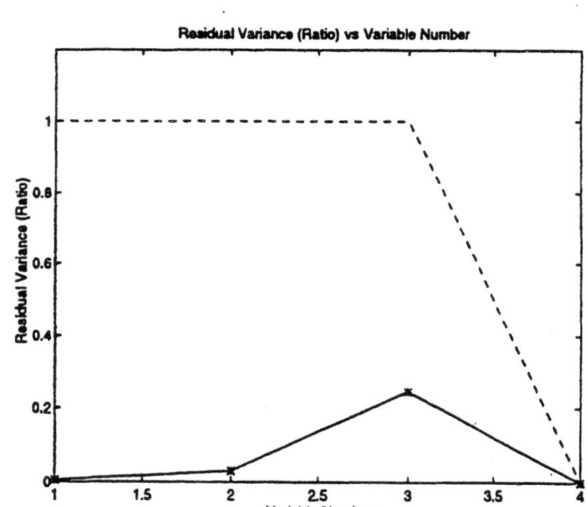

Fig. 6. The residual variance ratios for the test data set and the statistical limits (dashed lines)after sensor 4 is removed from calibration and test data sets

FAULT DETECTION AND ISOLATION ON A SYSTEM OF HEAT EXCHANGERS

J.Ph. Cassar*, C. Guernez+, M. Staroswiecki*

LAIL (CNRS 1440)
+ERASM-HEI, 13, rue de Toul, 59046 Lille Cedex, France
e-mail : caroline.guernez@hei.fupl.asso.fr
* USTL Lille 1, bât.P2, 59655 Villeneuve d'Ascq Cedex, France
e-mail : cassar@univ-lille1.fr

Abstract : This study aims at designing a Fault Detection and Isolation system for several interconnected heat exchangers. Since this kind of process is modelled by polynomial equations, this paper shows how the algebra of polynomial rings can mathematically support the FDI system design. The basic principles of FDI extract, from the model, some fault indicators, called *residuals*, that can be computed from available information. Then, to analytically get these polynomial residuals, we propose a procedure based on the algebraic approach of Elimination Theory. Groebner bases are used to perform the symbolic computation and a structural pre treatment is proposed to handle large systems of equations. Finally these ideas are applied to a system of heat exchangers. *Copyright © 1998 IFAC*

Keywords : fault detection and isolation, polynomial models, computation, large-scale systems, heat exchangers.

1. INTRODUCTION

Modern industrial systems become more complex and are subject to more frequent failures. Erroneous data make the control algorithms ineffective, and threaten the security of men and equipment. This is why monitoring systems are implemented to quickly and accurately detect and isolate failures.

Fault Detection and Isolation systems aim at the validation of the data available from the data acquisition units. They use a reference model in order to characterise the state of the system (normal or faulty) and, in the case of a failure occurrence, use specific isolation procedures to isolate the cause of the failure. The failures can occur on sensors, actuators or components of a plant.

Failure detection model based methods rest on the comparison, expressed by the value of *residuals*, between the actual behaviour of the process and a model describing its normal operating. The residuals design is derived from the analytical model of the process. The case of linear systems has been widely dealt with. Indeed linear algebra provides a well known and convenient support to the Fault Detection and Isolation (FDI) process (Iserman, *et al.*, 1996; Frank *et al.*, 1990; Gertler, 1991). One of these methods, called *Parity Space method* (Poter *et al*, 1977; Chow *et al.*, 1984), performs the direct elimination of the unknown variables of the system, thanks to projection mechanisms in linear algebra.

No investigation have been done in the non linear polynomial case. This paper proposes to apply the Theory of Elimination in the algebra of polynomial rings to eliminate the unknown variables of the system, and thus to generate residuals. Since symbolic computation tools are limited by the size of the system of equations to be processed, structural considerations are introduced in order to simplify the computational complexity.

System of heat exchangers is a good illustration of the need of such a method for FDI purpose. Indeed, the static model of an exchanger involves bilinear equations that will be easily handled by the polynomial theory. In the case of energy recovering, heat exchangers process most of the flows of the plant. Measurements performed on heat exchangers can lead to mass balances which allow the monitoring of the whole process (Cassar *et al.*, 1992). Since the mass balances are based on measurements any failure on the sensors must be previously detected and isolated. That is the purpose of this study.

The next paragraph describes the basic principles of fault diagnosis. Then the algebraic tools that allow to deal with polynomial systems (Theory of Elimination and Groebner bases) (Cox, *et al.*, 1992). are presented. The implementation of these procedures for large scale systems of equations is discussed. The results are applied on a system of heat exchangers. Finally a simulation illustrates how the isolation procedure can work.

2. FAULT DETECTION AND ISOLATION

Let us consider a static non linear system described by the following equations :

$$\underline{P}(\underline{X},\underline{U},\underline{Y}) = 0 \quad (1)$$

$$\text{where } \underline{P} : \mathbf{R}^n \times \mathbf{R}^m \times \mathbf{R}^q \rightarrow \mathbf{R}^s$$

In this model of the normal behaviour of the process, $\underline{X}$ is the state vector, $\underline{U}$ and $\underline{Y}$ are respectively the input and output vectors. Only $\underline{U}$ and $\underline{Y}$ are supposed to be known, as control and measured variables.

2.1. Residuals generation

In order to check the appropriateness of the model with the actual operation of the process, the model has to be analytically transformed in order to only contain information which are available on the process. This is why equations (1) have to be modified so that they do not contain unknown state variables $\underline{X}$ anymore. The following system, composed of the so called *Analytical Redundancy Relations(ARR)* is thus obtained :

$$\underline{\Phi}(\underline{U},\underline{Y}) = 0 \quad (2)$$

$$\text{where } \underline{\Phi} : \mathbf{R}^m \times \mathbf{R}^q \rightarrow \mathbf{R}^p$$

These ARR are satisfied in a normal operating. In the faulty case the equality to zero is assumed to be not satisfied. The equality at each time will be checked by evaluating the left hand side of equation (2), called *residuals calculation form.*

$$\underline{r} = \underline{\Phi}(\underline{U}^*,\underline{Y}^*) \quad (3)$$

where $\underline{U}$ * and $\underline{Y}$ * express the values of the measured variables.

2.2. Failure Sensitivity

Introducing in the relation (1) a model of the failures, the value of the residual vector computed by (3) can also be expressed as a function of the failures values.

$$\underline{r} = \underline{\Psi}(\underline{U}^*,\underline{Y}^*,\underline{F}^*) \quad (4)$$

For a given residual r_j, the sensitivity of a failure f_k is defined by the value of the partial derivative of r_j with respect to f_k at the operating point defined by the values of $\underline{U}$ *, $\underline{Y}$ *.

$$s_{j,k} = \left[\frac{\partial r_j}{\partial f_k}\right]_{\underline{U}^*,\underline{Y}^*} \quad (5)$$

The sensitivity identically equals zero when f_k doesn't appear in the function r_j.

2.3. Residuals structuring

In order to improve the isolation capability of the FDI system, the failures must act on the set of residuals with as different as possible effects. That leads to generate some set of residuals only sensitive to a subset of faults. The faults are considered to only affect measurements. The measurements whose failures must only affect a particular set of residuals $\underline{r}_i$ are noted $[\underline{U}_i^+,\underline{Y}_i^+]$. The corresponding residual computation form is :

$$\underline{r}_i = \underline{\Phi}_i[\underline{U}_i^+,\underline{Y}_i^+] \quad (6)$$

$[\underline{U}_i^+,\underline{Y}_i^+]$ denotes the structure of the residuals $\underline{r}_i$ (Staroswiecki *et al.*, 1993).

3. POLYNOMIAL ANALYTICAL TOOL : GROEBNER BASES

As far as polynomial systems are concerned, the ARR are obtained from the initial model (1) by a procedure of variables elimination mathematically supported by the so-called *Theory of Elimination* which is described in the following.

The models are represented by the following polynomial form :

$$\begin{cases} p_1(x_1,\dots,x_n,u_1,\dots,u_m,y_1,\dots,y_q)=0 \\ \dots \\ p_s(x_1,\dots,x_n,u_1,\dots,u_m,y_1,\dots,y_q)=0 \end{cases} \quad (7)$$

where the p_i's are polynomials which belong to the polynomial ring $\mathbf{R}(x_1,\dots,x_n,u_1,\dots,u_m,y_1,\dots,y_q)$, noted $\mathbf{R}(\underline{X},\underline{U},\underline{Y})$.

We introduce now the ideal I generated by the polynomials of the model (7) in $\mathbf{R}(\underline{X},\underline{U},\underline{Y})$: $I = \langle p_1,\dots,p_s \rangle$ contains all the polynomials that can be generated from the initial polynomials set (7) by linear combinations with polynomial coefficients in $\mathbf{R}(\underline{X}, \underline{U}, \underline{Y})$.

3.1. Residuals generation

For FDI purpose, the unknown state variables $\underline{X}$ have now to be eliminated from the system (7) to get relations of the form (2). Mathematically speaking such polynomials are elements of $I_{\underline{X}}$:

$$I_{\underline{X}} = I \cap \mathbf{R}(\underline{U}, \underline{Y}) \quad (8)$$

Let consider an order on the variables such that the n variables x_1 to x_n are dominating all the others, then $I_{\underline{X}}$ is the nth elimination ideal of I.. The following *Elimination Theorem* will allow to describe $I_{\underline{X}}$ by one of its generating bases, the Groebner Basis (Cox, *et al.*, 1992) :

Let $I \subset \mathbf{R}(\underline{X}, \underline{U}, \underline{Y})$ be an ideal and let G be a Groebner basis of I with respect to a lex order where $x_1 > \ldots > x_n > u_1 > \ldots > u_m > y_1 > \ldots > y_q$. Then, the set $G_n = G \cap \mathbf{R}(\underline{U}, \underline{Y})$ is a Groebner basis of the nth elimination ideal.

The polynomials of G_n constitute the left-hand-side members of the ARR (2) and so the calculation forms of the residuals (3).

3.2. Residuals structuring

The elimination process can also be used to obtain a desired structure defined by the sets $[\underline{U}_i^+, \underline{Y}_i^+]$. Starting from the basis of the residuals G_n, the elimination process is applied imposing an elimination order such that $[\underline{U}_i^+, \underline{Y}_i^+]$ are dominated by all the other variables : $\ldots > [\underline{U}_i^+, \underline{Y}_i^+]$

4. APPLICATION TO THE HEAT EXCHANGERS

As an example, the figure 1 exhibits the sketch of a system of heat exchangers . Three flows f_{11}, f_{21}, f_{31} at different temperatures t_{11}, t_{21}, t_{31} enter a system of six heat exchangers. The normal operation model of the system is polynomial and results from three kinds of balances relations.

$\sum_i f_i = 0$: mass balance at each node,

$\sum_i f_i . t_i = 0$: thermal balance at each node,

$\sum_{in} f_i . t_i - \sum_{out} f_j . t_j = 0$: enthalpic balance for each of the six exchangers.

4.1. Residuals generation

The model (2) consists of 34 equations which involve 54 variables. For this application, 24 measurements are performed on the system, $\underline{Y}$ = {*f12, f13, ff18, f19, f112, f11, f21, f212, f31, f33, t12, t13, t17, t18, t19, t27, t112, t11, t16, t21, t26, t212, t31, t33*}. Since no control input exist on the system : $\underline{U}$ = {}. Thus 30 variables must be eliminated to generate residuals.

For the description of algorithms generating Groebner Bases see (Buchberger, *et al.*, 19985). Since we are dealing with a large system of equations and existing computer algebra systems are not designed to handle such systems, (Guernez, *et al.*, 1997) proposed a structural pre treatment to make the implementation of FDI procedures possible.

A structural partitioning of the system into two independent subsystems is based on the minimal cut research (Berge, 1973). Without taking into account the measurements, it automatically recognises the two subsystems (E1,E2), (E5,E6) and the subsystem (E3,E4) as a link between the two previous ones. The elimination process is first applied on the subsystems instead of the global system. That aims to eliminate the variables that are unmeasured and that are not shared by several subsystems.

Applying the Groebner basis algorithm to the subsystem (E1,E2) leads to Groebner basis $\underline{G}_1$. It generates all the polynomials that do not involve the variables to be eliminated. The polynomials have been factorised to make the reading easier.

$$\underline{G}_1 \begin{cases} f_{11} - f_{16} \\ f_{11} \cdot (t_{11} - t_{16}) + f_{26} \cdot (t_{21} - t_{26}) \\ -f_{11} + f_{12} + f_{13} \\ f_{13} \cdot (t_{12} - t_{13}) - f_{11} \cdot (t_{12} - t_{16}) + f_{26} \cdot (t_{26} - t_{21}) \\ f_{21} - f_{26} \end{cases}$$

The elimination order $f_{22} > t_{22} > f_{23} > t_{23} > f_{24} > t_{24} > f_{25} > t_{25} > f_{14} > t_{14} > f_{15} > t_{15} > f_{11} > t_{11} > f_{12} > t_{12} > f_{13} > t_{13} > f_{21} > t_{21} > f_{16} > t_{16} > f_{26} > t_{26}$ has been chosen such that the unmeasured flows f_{16}, f_{26} which also belong to the subsystem (E3,E4) are not eliminated. The same process is applied to the subsystem (E5,E6) and leads to the same kind of Groebner basis $\underline{G}_2$.

$$\underline{G}_2 \begin{cases} f_{17} - f_{112} \\ f_{112} \cdot (t_{17} - t_{112}) + f_{212} \cdot (t_{27} - t_{212}) \\ -f_{112} + f_{18} + f_{19} \\ f_{19} \cdot (t_{18} - t_{19}) - f_{112} \cdot (t_{18} - t_{27}) - f_{212} \cdot (t_{27} - t_{212}) \\ f_{27} - f_{212} \end{cases}$$

The subsystem (E3,E4) can be simplified by eliminating variables which are not involved by the systems (E1,E2) or (E5,E6). Appropriate elimination orders are then such that f_{32}, t_{32} dominate all the other variables : $f_{32}>t_{32}>...$. That leads to the Groebner basis $\underline{G}_3$.

$$\underline{G}_3 \begin{cases} -f_{17}\cdot(t_{17}-t_{16})-f_{31}\cdot(t_{33}-t_{31})+f_{27}\cdot(t_{26}-t_{27}) \\ -f_{26}+f_{27} \\ f_{17}-f_{16} \\ f_{31}-f_{33} \end{cases}$$

Elimination process is then applied to the set of polynomials G1 ∪ G2 ∪ G3. This system is then reduced to 14 relations involving 27 variables among which 4 must be eliminated $(f_{16}, f_{26}, f_{17}, f_{27}.)$ A Groebner basis $\underline{G}_{\underline{X}}$ of the elimination ideal according to the unmeasured variables $I_{\underline{X}}$ is composed of 13 polynomials, which generates the polynomial residuals of the system. Annex 1 gives the expression of the corresponding residuals written in the form (3).

4.2. Structured residuals

The structure of the obtained residuals is imposed by the chosen elimination order. since f_3 is the last flow in this order, the last residual makes only appear this flow. Note that no residual can be obtained without any flow in its structure.

From the basis $\underline{G}_{\underline{X}}$, some additional structured residuals can be generated while imposing a appropriate elimination order. For example, imposing an order such that $... > t_{11} > t_{21} > t_{112} > t_{33} > t_{31} > t_{212} > f_{212} > f_{112} > f_{33}$, allows to generate the analytical redundancy relation expressing the global enthalpic balance on the whole system.

$$f_{112}\cdot(t_{11}-t_{112}) - f_{33}\cdot(t_{33}-t_{31}) - f_{212}\cdot(t_{21}-t_{212}) = 0$$

4.3. Sensitivities

If the model of the failures is defined as additive deviations, the sensitivity of a failure is equal to the partial derivative of the residual with respect to the corresponding measurement. Note that the sensitivities are polynomials in the measurements or constants. The sensitivities of failures on the flow measurements are either equal to 1 (mass balances) or defined by a polynomial that makes appear a difference of two temperatures.

For the temperatures, the sensitivities makes appear at least one flow that can be multiplied by differences of two temperatures. In the case where a difference of temperatures is a factor in the sensitivity expression, its value may be low.

To illustrate this notion consider the residual r5 which expresses the enthalpic balance on the heat exchanger E4 :

$$r5 = f_{11}\cdot(t_{17} - t_{16}) - f_{33}\cdot(t_{33} - t_{31}) - f_{212}\cdot(t_{27} - t_{26})$$

If a failure affects the temperature measurement t_{17}, the value of r5 will be : $r5 = f_{11}.\delta t_{17}$ where δt_{17} expresses the magnitude of the failure.

Following the same considerations, the sensitivity to a failure of the f_{11} measurement is $(t_{17} - t_{16})$. Given the operating conditions derived from the figure 2 in annex 3, the value of this sensitivity equals about 8 m^{-3}.

5. ILLUSTRATION

Figure 2 in annex 2 gives the evolution of some measurements. During the time interval [10,30], a bias occurs on the t_{17} measurement while a bias occurs on the measurements of t_{112} during [40,60] and t_{16} during [70,90]. The bias amplitudes are from 2 to 3 % of the nominal values.

Let observe two residuals r5, r6 to illustrate the detection and isolation processes. The structures of these residuals with respect to the failures is represented by the incidence matrix in table 1. A zero expresses that the corresponding sensitivity is identically zero. The 1's correspond to the other cases.

	t_{16}	t_{17}	t_{112}	t_{27}	...
r5	1	1	0	1	...
r6	1	0	1	0	...

Table 1. : Incidence matrix of the residuals r5, r6.

The evolutions of the residuals are given in figure 3 in annex 2 and exhibit the influences of the failures. During the time interval [10,30], the residual r5 is affected whereas r6 is not. This information is collected into a binary vector [1 0] called *observed signature*. A recognition procedure is able to match it with the signature of t_{17} . The same process is applied during the time interval [40,60] and [70,90], for t_{112} and t_{16}.

Note that two residuals are not sufficient to distinguish a failure from the other failures that may occur. Indeed, in this case, signatures of the failures are not different and the recognition procedure can only give the set of failures signatures of which correspond to the observed signature. As an example, during the time interval [10,30], the result will be $[t_{17}, t_{27}]$. That's why a set of structured residual is needed to isolate each failure in the single fault hypothesis.

6. CONCLUSION

This paper proposes to apply the FDI procedures to system described by polynomial models. The algebra of the polynomial ring provides the theoretical frame of the residual generation. This had not been done previously and opens the way to a larger range of systems to be supervised. We adapt the computer algebra systems to large systems of equations, by proposing a structural pre treatment resulting from the Graph Theory. An application illustrates our proposal. Symbolic computation was performed with the soft GB by J. C. Faugère, available by ftp (University Paris IX). The algebra allows to control the characteristics of our residuals, one of the most interesting among them is the possibility to structure the residuals in order to make them sensitive to given faults.

This procedure can be applied not only to static polynomial systems but also to dynamic ones, either discrete time or continuous time (Guernez, *et al.*, 1996). .

REFERENCES

Berge C.(1973). *Graphes et Hypergraphes*, Monographies universitaires de mathématiques, DUNOD, Paris.

Buchberger B.(1985). Groebner Bases : An Algorithmic Method. In : *Polynomial Ideal Theory, in Multidimensional Systems Theory* N.K. Bose, D. Reidel (Ed.), 184-232, Publishing Company, Dordrecht.

Cassar, J. Ph., R. Ferhati, and R. Woinet (1994). Fault Detection and Isolation System Design for a Refinery Unit. *Proceedings of Safe Process'94 IFAC Symposium on Fault Detection, Supervision and Safety for Technical Processes*, Espoo, Finland, June 13-15. pp 183-188.

Chow E.Y., Willsky A.S. (1984). Analytical Redundancy and the Design of Robust Failure Detection Systems, *IEEE Trans. on Automatic Control*, AC-**29** n°7.

Cox D., Little J., O'Shea D. (1992). *Ideals, Varieties, and Algorithms.* Undergraduate Texts in Mathematics. Springer-Verlag. London.

Frank P.M. (1990). Fault Diagnosis in Dynamic Systems Using Analytical and Knowledge-Based Redundancy - A Survey, *Automatica*, **26**, n°3, 459-474.

Gertler J.J. (1991). Analytical Redundancy Methods in Fault Detection and Isolation - A Survey and Synthesis, *Proceedings of the SAFEPROCESS'91 IFAC/IMACS Symposium*, Baden Baden, Germany, September 10-13, vol.1, pp.9-21.

Guernez C. , Cassar J.P. Staroswiecki M. (1996). Fault Detection and Isolation on Polynomial Systems : Analytical Redundancy Relations Generation and Residuals Structuration. *Proceedings of CESA'96 IMACS Multi conference, Symposium on Control, Optimization and Supervision*, July 9-12, , Lille, France, pp.690-695

Guernez C., M. Petitot, J.P. Cassar, M. Staroswiecki (1997). Fault Detection and Isolation on Non Linear Polynomial Systems. *Proceedings of 15th IMACS World Congress on Scientific Computation, Modelling and Applied Mathematics*, Berlin, Germany, August 24 - 29. vol. IV. pp. 67-73.

Iserman R.(1996). Ballé Peter, Trends in the Application of Model Based Fault Detection and Diagnosis of Technical Processes, *Proceedings of 13th IFAC World Congress*, San Francisco, USA, 30 June-5 July.

Potter J.E., M. C. Shuman (1977), Tresholdless redundancy management with array of skewed instruments. *Electronic Flight Control Systems, Agadograh*, **224**, 15-25

Staroswiecki M., Cassar J. Ph., Cocquempot V. (1993). A General Approach for Multicriteria Optimization of Structured Residuals. *Proceeding of TOOLDIAG'93 conference on Fault Diagnosis*, Toulouse, France, April 5-7, pp 800-807.

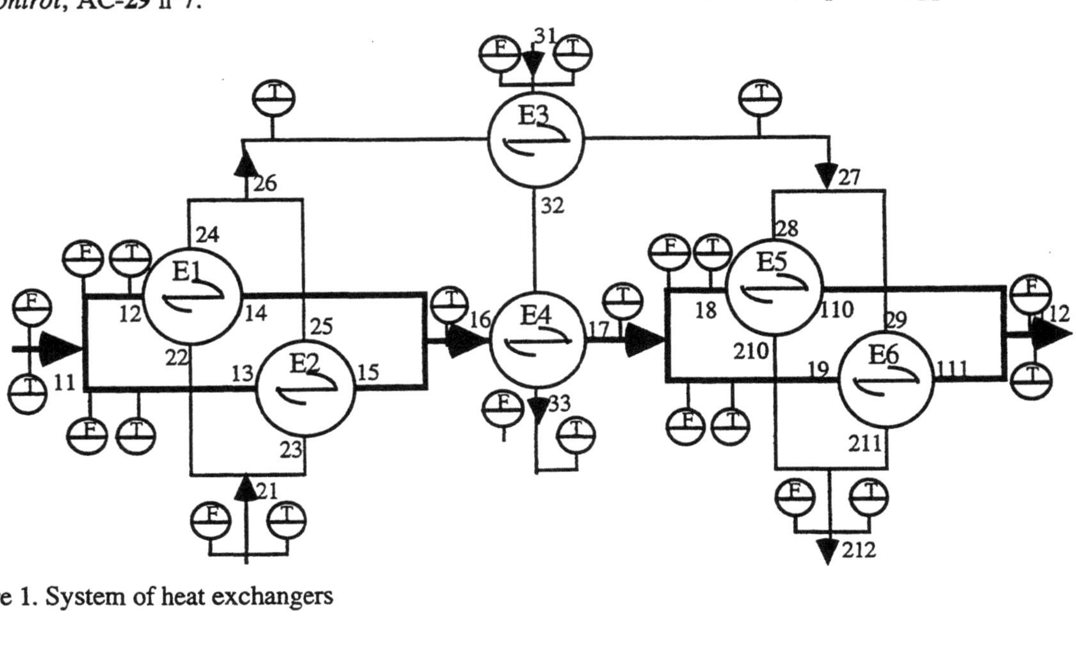

Figure 1. System of heat exchangers

ANNEX 1

$$r1 = -f_{11} + f_{12} + f_{13}$$
$$r2 = f_{13}.(t_{12} - t_{13}) - f_{212}.(t_{21} + t_{26})$$
$$r3 = -f_{112} + f_{18} + f_{19})$$
$$r4 = f_{11} - f_{112}$$
$$r5 = f_{11}.(t_{17} -.t_{16}) - f_{33}.(t_{33} - t_{31}) - f_{212}.(t_{27} - t_{26})$$
$$r6 = f_{11}.(t_{112} -.t_{16}) - f_{33}.(t_{33} - t_{31}) - f_{212}.(t_{212} - t_{26})$$
$$r7 = f_{11}.(t_{11} - t_{16}) + f_{212}.(t_{21} - t_{26})$$
$$r8 = f_{21} - f_{212}$$
$$r9 = f_{212}.((t_{17}-t_{16}).(t_{21}-t_{212})+(t_{11}-t_{112}).(t_{26}-t_{27}))+f_{33}.(t_{31}-t_{33}).(t_{17}-t_{112}).(t_{16}-t_{11})$$
$$r10 = f_{212}.(-t_{16}.(t_{27}-t_{212})+t_{112}(t_{27}-t_{26})+t_{17}.(t_{26}-t_{212}))+f_{33}.(t_{31}-t_{33}).(t_{17}-t_{112})$$
$$r11 = f_{212}.(-t_{16}.(t_{21}-t_{212})+t_{112}.(t_{21}-t_{26})+t_{11}.(t_{26}-t_{212}))+f_{33}.(t_{31}-t_{33}).(t_{11}-t_{16})$$
$$r12 = f_{31} - f_{33}$$
$$r13 = f_{33}.(t_{31}-t_{33})(t_{21}.(t_{17}.-t_{112})+t_{27}.(t_{16}.-t_{11})+t_{26}.(t_{112}.-t_{17})+t_{212}.(t_{212}.-t_{16}))$$

ANNEX 2

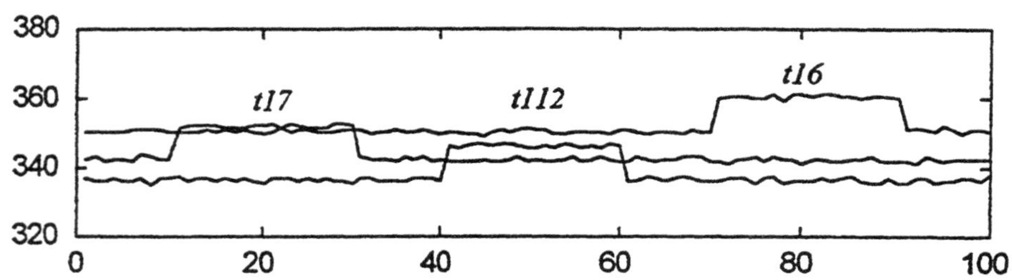

Figure 2 : Temperature measurements

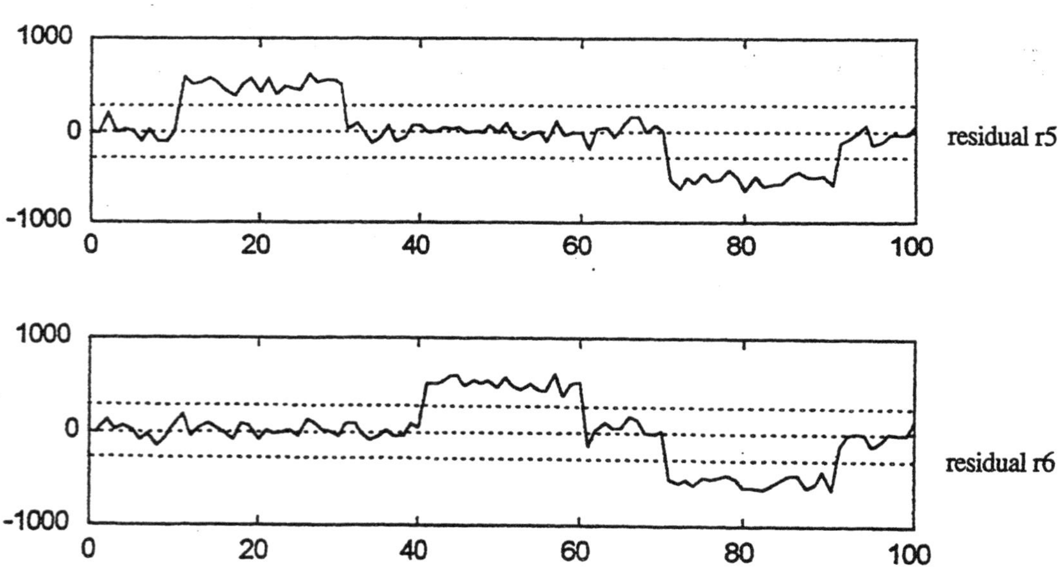

Figure 3 : Residuals r5 and r6 evolution

ROBUST DESIGN OF A FAULT DIAGNOSTIC SYSTEM FOR A SUPERSONIC WIND TUNNEL

Daw-Shang Hwang[+], Pau-Lo Hsu[*] and Shao-Kung Chang[**]

[+] *Department of Telecommunication Engineering, the China Junior College of Marine Technology, Taipei, Taiwan, Republic of China.*
[*] *Institute of Control Engineering, National Chiao-Tung University, Hsinchu, Taiwan. 300 Republic of China.*
[**] *Mechanical Industry Research Laboratories, Industrial Technology Research Institute, Hsinchu, Taiwan. 300 Republic of China.*

Abstract: By using the unknown input decoupling technique, this paper presents a robust fault detection and isolation (FDI) system design for a supersonic wind tunnel. A decoupled fault/residual map is designed in the sensor FDI system to achieve multiple fault isolation for principal sensors. Since the relations between actuator faults and the generated residuals are insignificant, a decoupled fault/residual map for actuator is not achievable. Nevertheless, based on the fault specific directions, single fault isolation can still be achieved for actuators with the proposed FDI system. *Copyright © 1998 IFAC*

Keywords: Robust fault detection and isolation; unknown input decoupling; wind tunnel.

1. INTRODUCTION

The high speed wind tunnel systems are commonly used to test the performance of high-speed flight vehicles under various conditions (Pope and Goin, 1965; Hwang and Hsu, 1992). In this paper, an intermittent blowdown-type wind tunnel is considered as shown in Fig. 1. With a chosen nozzle and the referred Standard Atmosphere Table (NOAA, 1976), the experimental conditions of the Mach number and altitude can be transformed into the conditions of the setting pressure and temperature for the wind tunnel system. They are achieved via proper operation of the air and fuel control valves.

In addition to the control system, the fault diagnostic system is also crucial for the wind tunnel system to maintain a reliable test environment. The fault diagnostic system, mainly including fault detection and isolation (FDI) tasks, is usually designed with a linear process model known as the model-based FDI system. Most of the processes, however, are nonlinear and hence modeling errors are inevitable. It may cause false alarms and deteriorate performance of the model-based FDI system. The fundamental objectives of the present diagnostic system design are as follows:

(1) To announce faults correctly (either sensor fault or actuator fault) and further verification of diagnostic results can proceed to help engineers to correct the fault in real-time to maintain a reliable test environment.

(2) The diagnostic system should be robust in the sense that it is insensitive to the noises, disturbances and uncertainty but sensitive to faults only.

Recently, robust approaches to design FDI systems have gained much attention, as reported in Watanabe and Himmelblau (1982), Viswannadham and Srichander (1987), Frank and Wünnenberg (1989), Patton, Frank and Clark (1989), Saif and Guan (1993), Frank (1994). The "unknown input decoupling" approach is an important robust technique for designing FDI systems by concerning the structured uncertainty. Most unknown uncertainties such as modeling error and the disturbances to a system can be modeled as unknown input, and their structural characteristics are summarized in the unknown input distribution matrix

(Patton and Chen, 1992). Based on this assumption, it is desirable to make the FDI system independent of all unknown inputs in order to obtain robust diagnostic results. Unknown input decoupling designs can be achieved by using the unknown input observer (UIO) (Watanabe and Himmelblau 1982, Viswannadham and Srichander 1987, Frank and Wünnenberg 1989, Saif and Guan 1993, Frank 1994, Chang and Hsu 1995) or eigenstructure assignment (Patton and Kangethe 1989, Patton and Chen 1993). Such designs generally involve great computational complexity. Similar to the UIO design, Hwang et al. (1997) proposed a more straightforward method to design the FDI system efficiently. The corresponding design algorithm is briefly introduced as in the following Section.

2. DESIGN ALGORITHM

Consider the discrete-time state-space model of a dynamic system with input vector $\mathbf{u}$ ($\mathbf{u} \in \mathbf{R}^p$) and measurement output vector $\mathbf{y}$ ($\mathbf{y} \in \mathbf{R}^m$) as

$$\mathbf{x}(k+1) = A\,\mathbf{x}(k) + B\,\mathbf{u}(k) + E\,\mathbf{d}(k) + K\,\mathbf{f}(k) \quad (1a)$$
$$\mathbf{y}(k) = C\,\mathbf{x}(k) + G\,\mathbf{f}(k) \quad (1b)$$

where $\mathbf{x} \in \mathbf{R}^n$ is the state variable, $\mathbf{f} \in \mathbf{R}^q$ represents the effect of actuator faults, component faults, and sensor (instrument) faults, $K\mathbf{f}(k)$ and $G\mathbf{f}(k)$ model the actuator, component, and sensor faults with the distribution matrix K and G whose elements depend on how they are mounted in the system. $\mathbf{d} \in \mathbf{R}^r$ characterizes the uncertainties implicitly in the system and the term $E\mathbf{d}(t)$ models the system disturbance either in structured form or unstructured form (Watanabe and Himmelblau 1982, Frank and Wünnenberg 1989). Based on the unknown input observer design method. The necessary condition for unknown input decoupling is that

$$\mathrm{rank}(CE) = \mathrm{rank}(E)$$

Then, a projection matrix T,

$$T = I - E\,(CE)^{+}\,C \quad (2)$$

which satisfies $TE = 0$, is employed in the proposed design where $(CE)^{+}$ is the generalized inverse of the matrix (CE):

$$(CE)^{+} = [(CE)^{T}(CE)]^{-1}(CE)^{T} \quad (3)$$

An equivalent system free of unknown inputs can thus be obtained as

$$\mathbf{z}(k+1) = \tilde{A}\,\mathbf{z}(k) + \tilde{B}\,\mathbf{u}(k) + [\tilde{K} - \tilde{A}\,E(CE)^{+}G]\,\mathbf{f}(k) + \tilde{A}\,E(CE)^{+}\mathbf{y}(k) \quad (4a)$$
$$\mathbf{y}(k) = C\,\mathbf{z}(k) + CE(CE)^{+}\mathbf{y}(k) + (I_m - CE(CE)^{+})\,G\,\mathbf{f}(k) \quad (4b)$$

where

$$\tilde{A} = TA, \qquad \tilde{B} = TB, \qquad \tilde{K} = TK. \quad (5)$$

By eq (4), a basic fault detection filter can be formed as

$$\mathbf{e}(z) = L_y(z)\,\mathbf{y}(z) - L_u(z)\,\mathbf{u}(z) \quad (6a)$$
$$= L_f(z)\,\mathbf{f}(z). \quad (6b)$$

where

$$L_y(z) = (I_m - CE(CE)^{+}) - C(zI_n - \tilde{A})^{-1}\tilde{A}\,E(CE)^{+} \quad (7a)$$
$$L_u(z) = C\,(zI_n - \tilde{A})^{-1}\tilde{B} \quad (7b)$$
$$L_f(z) = (I_m - CE(CE)^{+})G + C\,(zI_n - \tilde{A})^{-1} \times (\tilde{K} - \tilde{A}\,E(CE)^{+}G) \quad (7c)$$

Now, the filter output is independent of unknown inputs and is sensitive to the fault only. To increase the design freedom and to meet the required isolation performance, a dynamic weighting matrix $W(z)$ is further appended to the basic filter of eq. (6). A robust residual generator in the final form is obtained as

$$\mathbf{r}(z) = W(z)\,\mathbf{e}(z) \quad (8a)$$
$$= H_y(z)\,\mathbf{y}(z) - H_u(z)\,\mathbf{u}(z) \quad (8b)$$
$$= H_f(z)\,\mathbf{f}(z) \quad (8c)$$

where

$$H_y(z) = W(z)\,L_y(z), \quad H_u(z) = W(z)\,L_u(z)$$
$$H_f(z) = W(z)\,L_f(z) \quad (9)$$

Noted that in the proposed method, based on eqs (2) and (5), basic filter matrices of $L_y(z)$, $L_u(z)$ and $L_f(z)$ are directly obtained (eq. (7)). The required design procedure in the method is simply to choose a suitable $W(z)$ to achieve desirable diagnostic performance. A basic requirement for $W(z)$ is to make the filter matrices $H_y(z)$ and $H_u(z)$ to be realizable and stable. If the basic fault/residual map $L_f(z)$ in (eq. 7c) has full column rank (rank q), one can easily find a matrix $W_0 \in \mathbf{R}^{q\times m}$ so that $[W_0\,L_f(z)] \in \mathbf{R}^{q\times q}$ is square and of full rank, i.e. $[W_0\,L_f(z)]$ is invertable. The weighting matrix can be designed as

$$W(z) = H_f(z)\,(W_0\,L_f(z))^{-1}\,W_0 \quad (10)$$

In this manner, diagnostic performance of the FDI system can be easily assigned by using $H_f(s)$. It can be assigned in a diagonal form such that the i-th fault will reflect at the i-th residual, hence, mutual fault isolation is easily achieved. On the other hand, if $L_f(z)$ in eq. (7c) is not in a full column rank, then mutual isolation of multiple faults is not achievable. One way of isolating different faults is to generate a residual vector $\mathbf{r}(k)$ in the structured form (e.g. Frank and Wünnenberg 1989, Hwang et al., 1997). Detailed procedures are outlined in Chang et al. (1995).

3. MODEL EQUATIONS

Based on the mass and energy balance principles, nonlinear governing equations for the present supersonic wind tunnel system were derived by Hwang and Hsu (1992) and in Hwang (1997). For a sampling instance of $\Delta T = 0.4$ second, a discrete-time linearized state-space model of the wind tunnel is

$$A = \begin{bmatrix} 0.0041 & -0.0003 & 0 & 0 \\ -0.1197 & 0.0086 & 0 & 0 \\ 0.0069 & -0.0004 & 0.0070 & -0.0001 \\ -1.7718 & 0.1066 & -1.6193 & 0.0120 \end{bmatrix}$$

$$B = \begin{bmatrix} 0.3253 & 0 \\ 0.7508 & 0 \\ 0.0636 & 0.0320 \\ -9.9491 & 15.6063 \end{bmatrix} \quad (11)$$

The state $\mathbf{x}$ and input $\mathbf{u}$ are

$$\mathbf{x}(k) = \begin{bmatrix} P_1(k) - P_{10} \\ T_1(k) - T_{10} \\ P_2(k) - P_{20} \\ T_2(k) - T_{20} \end{bmatrix}$$

$$\mathbf{u}(k - k_d) = \begin{bmatrix} u_a(k - k_{da}) - u_{a0} \\ u_f(k - k_{df}) - u_{f0} \end{bmatrix} \quad (12)$$

where P_i is the pressure in the i-th plenum (bar), T_i is the temperature in the i-th plenum (K), u_a is the air control valve opening (0-100%), and u_f is the fuel control valve opening (0-100%). Based on the testing flight conditions of the Mach number 3.3 and the altitude 19.82 Km, the corresponding operating conditions are as in the following (Hwang, 1997): $P_{10} = 10.36$ bar, $T_{10} = 298$ K, $P_{20} = 3.25$ bar, $T_{20} = 623$ K, $u_{a0} = 32.5$ opening, and $u_{f0} = 32.5$ opening. The distributed characteristics of the process and the time lag involved in the combustion process are summarized as input delays T_{da} and T_{df} for pressure and temperature control, respectively, as

$$T_{da} = k_{da}\Delta T \approx 0.8 \text{ sec.}, \quad T_{df} = k_{df}\Delta T \approx 2.0 \text{ sec.}$$

Moreover, by using the unknown input decoupling technique, the basic assumption is that the unknown input distribution matrix E is available. Therefore, the matrix E should be estimated first. Patton and Chen (1993) proposed a useful method for computing the unknown input distribution matrix E for practical nonlinear plants by re-identifying plant parameters from different operating points. Based on the linearized model of eq. (10) and follow the method as described in Patton and Chen (1993), the resulting E is obtained as

$$E = [E_1 \; E_2] = \begin{bmatrix} 0.0231 & -0.2610 \\ -0.0314 & -0.9377 \\ 0.0071 & -0.2284 \\ -0.9992 & 0.0219 \end{bmatrix} \quad (13)$$

Note that the principle unknown input direction of E_1 is very similar to the directions of input matrix B:

$$\angle(B_1, E_1) = 6.1393\,°, \quad \angle(B_2, E_1) = 2.2958\,°.$$

where B_i is the i-th column vector of B matrix. It indicates that most of uncertainties are approximately reflected at process inputs. Practically, uncertainties such as the nonlinear and time-varying flow characteristics of the control valves, the decrease in the supply air pressure and the supply fuel tank pressure, and the nonlinear and time-varying combustion conditions involved in the heating system are indeed reflected at input of the wind tunnel system. To decouple the unknown inputs, it also results in a weak relationship between the actuating signal (including actuator faults) and the generated residuals. This will cause problems on designing the actuator FDI system and will be discussed in Section 5.

4. THE SENSOR FDI

To achieve a reliable test environment, tracking variables of P_1 and T_2 are also assigned as diagnostic variables (residuals) to monitor possible faults in these two measurements. It is noted that, if either of these two measurements are faulty, the established test environment is erroneous. It may lead to incorrect design results for the test object, which can cause serious problem on designing high performance flight vehicles in real applications. Since all the states of the process are measurable, redundant sensors of T_1 and P_2 are also included to increase the design freedom of the FDI· system. Hence, to design the sensor FDI system, corresponding matrices are defined as

$$C = I_4, \qquad K = 0_{4\times 2} \qquad G = \begin{bmatrix} 1 & 0 \\ 0 & 0 \\ 0 & 0 \\ 0 & 1 \end{bmatrix} \quad (14)$$

Here, redundant sensors of T_1 and P_2 are assumed to be reliable to provide useful information for the FDI system. Furthermore, because the resulting fault/residual map $L_f(z)$ is nonsingular, a weighting matrix W_0 can be easily selected as

$$W_0 = \begin{bmatrix} 1 & 0 & 0 & 0 \\ 0 & 1 & 0 & 0 \end{bmatrix}$$

The final form of the fault/residual map $H_f(z)$ is

assigned as

$$H_f(z) = \begin{bmatrix} \dfrac{0.7981z}{z-0.2019} & 0 \\ 0 & \dfrac{0.7981z}{z-0.2019} \end{bmatrix} \quad (15)$$

i.e. the i-th faults will be shown at the i-th residual in the first-order lag dynamics with a time constant of 0.25 second, it corresponds to the continuous poles of -4, mapped according to $z = e^{s\Delta T}$ with the sampling period of $\Delta T = 0.4$ second. By using eqs. (8)-(10), the resulting diagnostic filter is realized in a state-space form as follows:

$H_y(z)$:

$$A_{Hy} = \begin{bmatrix} -0.0384 & -0.0023 & 0 & 0 \\ -0.0749 & 0.0065 & 0 & 0 \\ 1.6076 & 0.0443 & 0.2019 & 0 \\ -0.0049 & 14.4017 & 0 & 0.2019 \end{bmatrix}$$

$$B_{Hy} = \begin{bmatrix} 0 & -0.0018 & 0.0137 & 0 \\ 0 & -0.0012 & -0.0038 & 0 \\ 0.1995 & 0.0494 & -0.4306 & 0 \\ 0 & 3.2905 & -13.5257 & -0.1995 \end{bmatrix}$$

$$C_{Hy} = \begin{bmatrix} 0 & 0 & 4.0000 & 0 \\ 0 & 0 & 0 & -4.0000 \end{bmatrix}$$

$$D_{Hy} = 0_{2*4} \quad (16)$$

$H_u(z)$:

$$A_{Hu} = \begin{bmatrix} -0.0384 & -0.0023 & 0 & 0 \\ -0.0749 & 0.0065 & 0 & 0 \\ 1.6076 & 0.0443 & 0.2019 & 0 \\ -0.0049 & 14.4017 & 0 & 0.2019 \end{bmatrix}$$

$$B_{Hu} = \begin{bmatrix} 0.0395 & -0.0018 \\ 0.2497 & -0.2463 \\ 0 & 0 \\ 0 & 0 \end{bmatrix}$$

$$C_{Hu} = \begin{bmatrix} 0 & 0 & 4.0000 & 0 \\ 0 & 0 & 0 & -4.0000 \end{bmatrix}$$

$$D_{Hu} = 0_{2*2} \quad (17)$$

The present FDI system was verified through the nonlinear simulation program which obeys the mass and energy conservation laws (Hwang, 1997). Fig. 2 and Fig. 3 show the diagnostic results of the proposed sensor FDI in two different cases, one with a bias fault 1.0 bar in P_1 issued at t = 24 seconds, and the other with a bias fault of 20C in T_2 also issued at t = 24 seconds. Setpoint changes in pressure and temperature have been considered in both cases. As shown in the figures, for the design using the same fault/residual map without considering the robustness, the diagnostic results are sensitive to model uncertainties and the setpoint changes such that faults in sensors cannot be detected correctly. This deficiency can be overcome with the present FDI system. Results show that fault occurrence in the sensors, include the time and the size, can be properly identified with good decoupled performance. Note that a small error in the residuals of T_2 which indicates the unstructured uncertainty effects can be neglected by selecting an appropriate threshold.

5. THE ACTUATOR FDI

To design the actuator FDI system, $K = B$ and $G = 0_{4\times 2}$ are defined. The corresponding basic fault/residual map $L_f(z)$, however, is ill-conditioned due to the similarity between the principle directions of E_1 and B, as described in Section 3. Therefore, the proposed robust actuator FDI design as described in Section 2 cannot be directly applied to this special case. However, it is noted that when the sensor FDI system in eq. (8) is dealing with the actuator fault, the residuals becomes

$$\mathbf{r}(z) = -H_u(z)\,\mathbf{f}_a(z) \quad (18)$$

The column vectors of the steady state gain matrix $-H_u(1)$ reveal the specific fault directions in steady state of the actuator fault $\mathbf{f}_a$ corresponding to the residuals $\mathbf{r}$. In other words, by checking the direction of the residuals, the same FDI scheme can also be applied for detecting actuator fault. A decision logic, includes flag function and decision function, is employed to complete the task of fault isolation:

$$\psi(t) = \left[\frac{1}{p}\sum_{i=1}^{p}\mathbf{r}_i^2(t)\right]^{\frac{1}{2}} \quad (19a)$$

$$\theta_i(t) = \cos^{-1}\left(\frac{|<\mathbf{r}(t), H_{fi}>|}{|\mathbf{r}(t)||H_{fi}|}\right) \text{ for } i = 1,2, \dots, p. \quad (19b)$$

where $\psi(t)$ is the magnitude of $\mathbf{r}(t)$ served as a flag function indicating whether any failure occurs. When $\psi(t)$ becomes large, it indicates the presence of certain failures. The $\theta_i(t)$ is the angle between the residual vector and the i-th failure direction H_{fi} where H_{fi} is the i-th column of the steady-state gain of $-H_u(1)$. It is used as the decision fuction that when $\theta_i(t)$ is close to zero, the i-th actuator is then declared to be faulty and thus fault isolation is achieved. However, as multiple faults occur, its fault specific direction is changing with fault size as

$$\mathbf{r}_{ss} = \sum_{i=1}^{p} H_{fi}\,\mathbf{f}_{ai}$$

where $\mathbf{r}_{ss}$ is the steady-state value of residual vector $\mathbf{r}$. This fault specific direction $\mathbf{r}_{ss}$ is no longer suitable for fault isolation. Therefore, in this scheme, only single fault is permitted at a time for fault isolation.

It is noted that by using the same FDI system for both sensor and actuator, it is further assumed that the sensor and actuator faults will not occur simultaneously and specific fault directions for sensors and for actuators are different. Fig. 4 and Fig. 5 are the responses of the flag function and the decision function when fault exists in the air control valve or in the fuel control valve. As shown in the figures, the flag functions provide good indication of the fault occurrence. Either fault in the air control valve or in the fuel control valve can be isolated by checking decision functions as they approach to zero.

6. CONCLUSIONS

A robust FDI system for a supersonic wind tunnel system is proposed in this paper. Discrepancies between the linearized nominal model and the actual process are summarized as the unknown input effects. They are estimated to form the unknown input distriution matrix for the system. Similarity between the principal unknown input direction and the directions of the input matrix indicates the main uncertainties occured the at process inputs.

By using the unknown input decoupling technique, the proposed FDI system for the wind tunnel system is insensitive to the estimated unknown inputs but sensitive to faults only. In the sensor FDI system design by using redundant sensors, the proposed sensor FDI successfully achieves mutual isolation of multiple faults for principal sensors. Because unknown input direction of the wind tunnel process is similar to the directions of the input matrix, it leads to a singular fault/residual map for actuator such that the proposed FDI system for actuators cannot be directly designed. However, by checking the angle between fault specific directions and generated residuals of the sensor FDI, the robust FDI task can be still achieved for the actuator. with single fault isolation. Simulation results show that in this wind tunnel applications, faults in sensors and actuators can be identified satisfactorily in spite of the existing significant model uncertainties of the process.

REFERENCES

Chang, S. K., P. L. Hsu, and K. L. Lin (1995). A parametric transfer matrix approach to fault identification filter design and threshold selection. *Int. J. of Systems Sci.*, **26**, 741-754.

Chang, S. K., and P. L. Hsu (1995). A novel design for the unknown input fault detection observer. *Control-Theory and Advanced Technology*, **10**, 1029-1051.

Frank, P. M. (1994). Enhancement of robustness in observer-based fault detection. *Int. J. of Control*, **59**, 955-981.

Frank, P. M., and J. Wünnenberg (1989). Robust fault diagnosis using unknown input observer scheme. in *Fault Diagnosis in Dynamic Systems: Theory and Application.* Prentice Hall, Englewood Cliffs, Chapter 3, pp. 47-98.

Hwang, D. S. (1997). *Robust Control and Diagnostic System Design for the Supersonic Wind Tunnel.* Ph. D. Disertation, National Chiao-Tung University, Hsinchu, Taiwan.

Hwang, D. S. and P. L. Hsu (1992). The LQG/LTR controller design for the wind tunnel. *Proc. of 11th Conference on Modeling, Identification and Control*, Innsbruck, pp. 371-374.

Hwang, D. S., S. K. Chang, and P. L. Hsu (1997). A practical design for the robust fault detection and isolation systems. *Int. J. Sys. Sci.*, **28**, 265-275.

NOAA, NASA, and USUF (1976). *U.S. Standard Atmosphere.* U.S. Goverment Printing Office, Wahington.

Patton, R. J., and J. Chen (1992). Robust fault detection of jet engine sensor systems using eigenstructure assignment. *AIAA J. of Guidance, Control and Dynamics*, **15**, 1491-1497.

Patton, R. J., and J. Chen (1993). Optimal unknown input distribution matrix selection in robust fault diagnosis. *Automatica*, **29**, 837-841.

Patton, R. J., P. M. Frank and R. N. Clark (1989). *Fault Diagnosis in Dynamic Systems: Theory and Application.* Prentice Hall, Englewood Cliffs

Patton, R. J., and S. M. Kangethe (1989). Robust fault diagnosis using eigenstructure assignment of observers. In *Fault Diagnosis in Dynamic Systems*, Prentice Hall. Chapter 4, pp. 99-154.

Pope, A. and K. L. Goin (1965). *High Speed Wind Tunnel Testing.* John Wiley & Sons, New York.

Saif, M. and Y. Guan (1993). A new approach to robust fault detection and identification. *IEEE Trans. on Aerospace and Electronic Systems*, **29**, 685-695.

Viswannadham, N. and R. Srichander (1987). Fault detection using unknown input observers. *Control-Theory and Advanced Technology*, **3**, 91-101.

Watanabe, K. and D. M. Himmelblau (1982). Instrument fault detection in systems with uncertainties. *Int. J. of Systems Sci.*, **13**, 137-158.

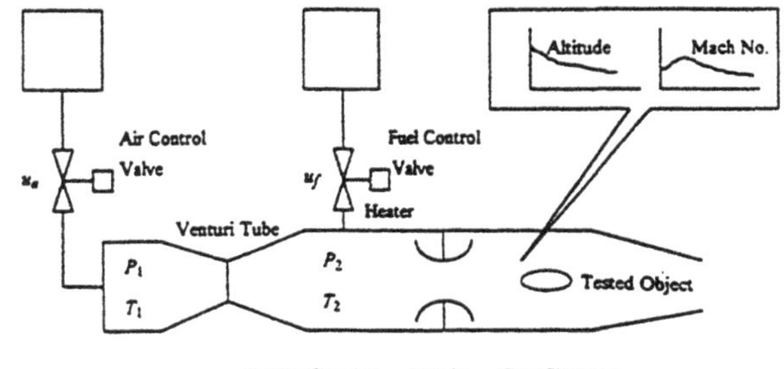

Fig. 1. Diagram of the supersonic wind tunnel.

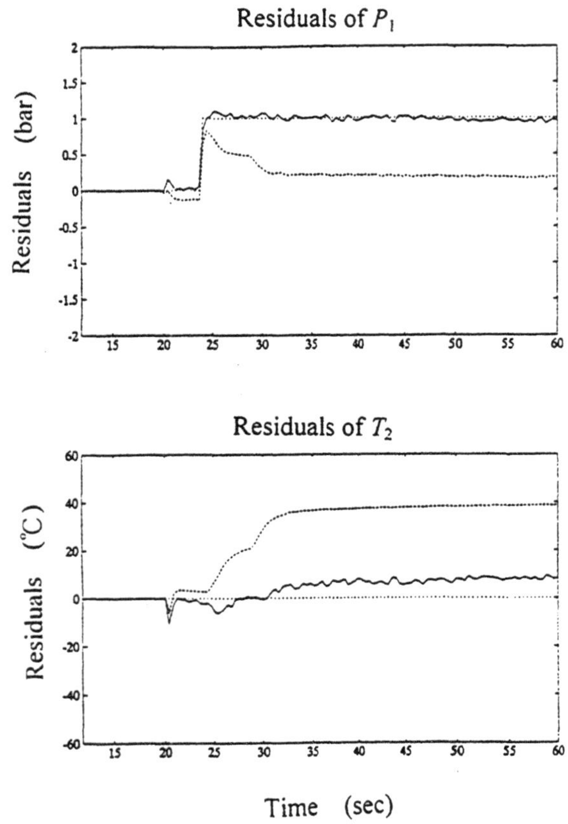

Fig. 2. Residual responses of the FDI with (solid) and without (dashed) robust design, for a 1.0 bar bias fault on sensor of P_1.

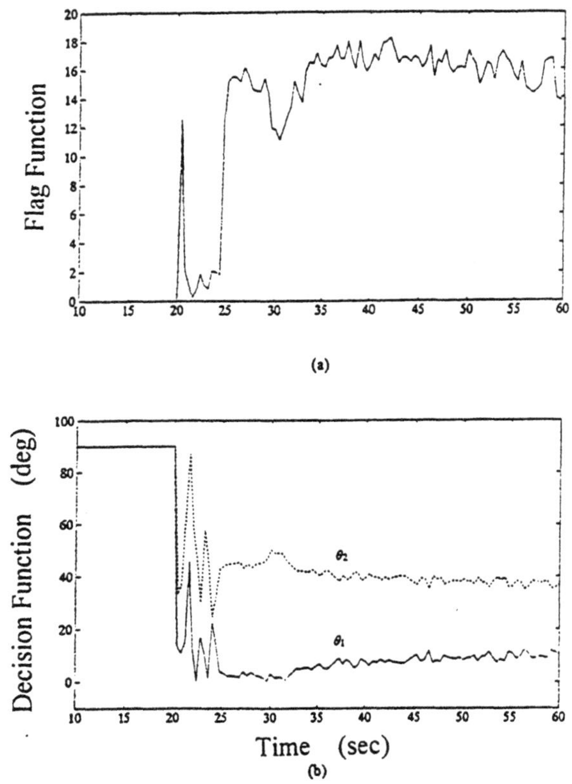

Fig. 4. Residual responses of the actuator FDI corresponding to 0.1 bias fault on the air control valve (a) flag function and (b) decision function.

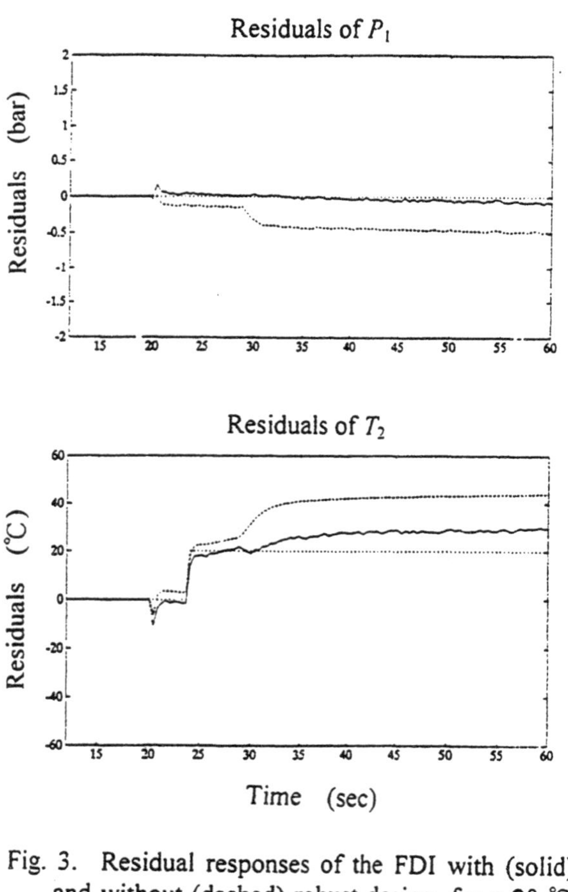

Fig. 3. Residual responses of the FDI with (solid) and without (dashed) robust design, for a 20 ℃ bias fault on sensor of T_2.

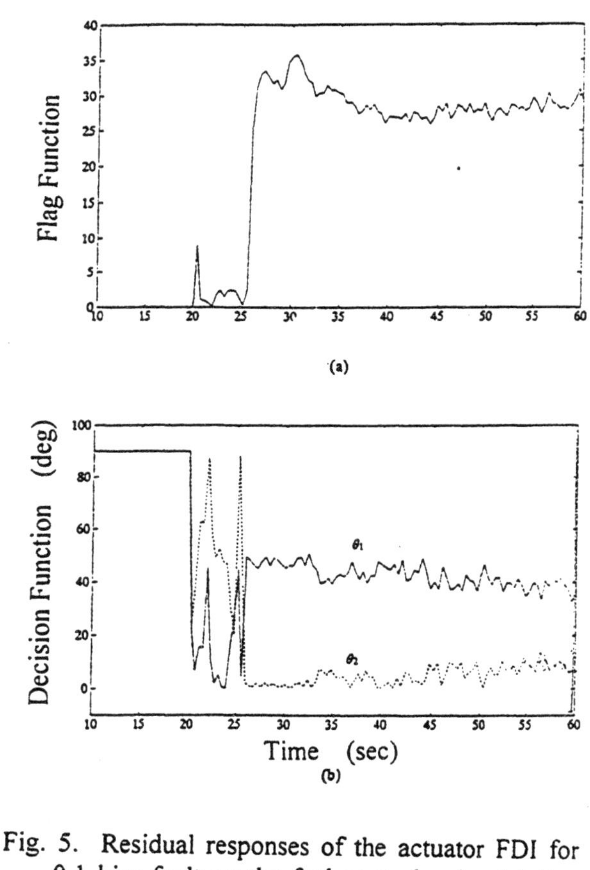

Fig. 5. Residual responses of the actuator FDI for 0.1 bias fault on the fuel control valve (a) flag function and (b) decision function.

ON-LINE SENSOR FAULT DIAGNOSIS AND ACCOMMODATION : APPLICATION TO A THREE-TANK-SYSTEM

D. Theilliol, H. Noura, D. Sauter, N. Pezzin

CRAN - ESA CNRS 7039 - Université Henri Poincaré - Nancy 1, BP 239
54506 Vandœuvre Cedex - France
Phone: (33) 3 83.91.27.01, Fax: (33) 3 83.91.20.30, e-mail: theillio@cran.u-nancy.fr

Abstract : In this paper, a fault-tolerant control method is developed. This method makes possible the compensation of additive or multiplicative sensor faults and failures. Its principle is based on the on-line detection, localization and estimation of the faulty element when a fault occurs on the system. Then, a new control law is computed in order to compensate the fault effect on the system. The performances of this method are tested in simulation on a pilot plant. *Copyright © 1998 IFAC*

Keywords: Fault accommodation, Fault-tolerant control, On-line Fault Diagnosis, Unknown Input Observers.

1. INTRODUCTION

The role of a human operator is to preserve normal operating conditions according to several plant parameters, measurements and observations. But complex automated industrial systems are vulnerable to faults in instrumentation as sensors, actuators or controllers. This leads to a reduction in the safety and the availability of these plants. The aim of the reconfigurable control systems or fault accommodation techniques is to adjust or to modify on-line the nominal control laws in order to maintain the safety of the operators and the reliability of the process.

The different studies dealing with this problem are based on hardware or analytical redundancy. The hardware redundancy technique consists in switching from the failed part of the process to another achieving the same task (Napolitano and Swaim, 1989). Analytical redundancy is an alternative to solve the reconfiguration problem avoiding the disadvantages of the hardware redundancy such as high cost and space.

During the last decades, many researchers have dealt with the fault-tolerant control problem. The survey papers of Patton (1997) and Zemlyakov (1996) give the state of the art in the field of reconfiguration and fault-tolerant control (FTC).

The classical way to achieve the fault-tolerant control task, summarized in Fig. 1, consists of the following steps:

- nominal control law
- fault detection and isolation (FDI)
- fault estimation
- fault compensation.

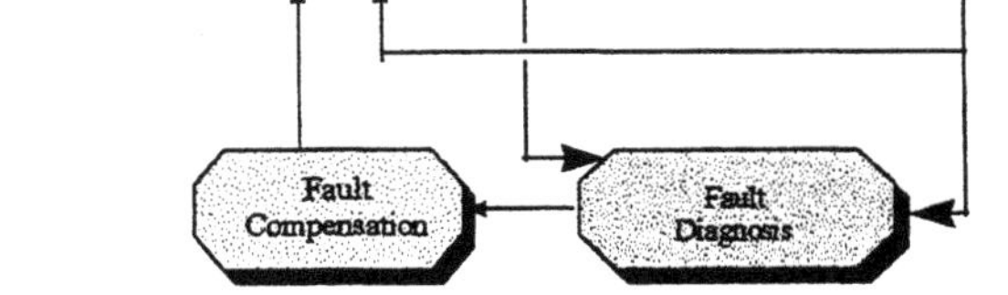

Fig. 1. General scheme for FTC

The paper is organized as follows. The section 2 presents the three-tank system used to illustrate the performances of the fault tolerant control method, developed here, in presence of sensor faults. In section 3, the principle and the objectives of the presented fault-tolerant control method are given not only in regulation but also in the tracking control. The next section is devoted to the strategy of fault diagnosis based on unknown input observers used to detect and isolate the faults. The new control law which aims to reduce or annihilate the fault effect on the system outputs is discussed. The section 4 covers the comments on the simulation results of the fault-tolerant control method applied to the three-tank-system. Finally, concluding remarks are given.

2. THREE-TANK SYSTEM

2.1 Process description

The approach presented in this paper has been applied to the simulated process, shown in Fig.2. The process is composed of three cylindrical tanks with identical cross section S. The tanks are coupled by two connecting cylindrical pipes with a cross section S_n and an outflow coefficient μ13. The nominal outflow is located at the tank 2, it also has a circular cross section S_n and an outflow coefficient μ_2. Two pumps driven by DC motors supply the tanks 1 and 2. The flow rates through these pumps are defined by the calculation of flow per rotation. All three tanks are equipped with sensors for measuring the level of the liquid (h_1, h_2, h_3).

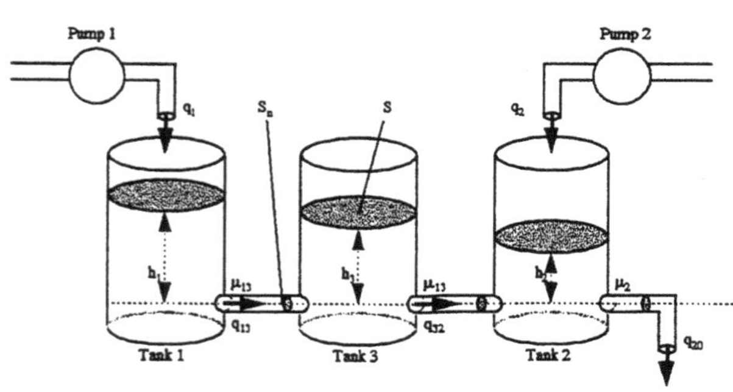

Fig. 2. Synoptic of the Three-tank-system

$S_n = 5*10^{-5} m^2$: cross section of the pipes

$g = 9.81 m/s^2$: earth acceleration

$\mu_{13} = 0.5$: outflow coefficient of tanks 1 - 3

$\mu_2 = 0.6$: outflow coefficient of tank 2

$S = 0.0154\ m^2$: cross section of the tanks

The system can be simulated conveniently using Matlab/Simulink by means of the following non-linear mass balance equations:

$$\begin{aligned} S\frac{dh_1(t)}{dt} &= q_1(t) - q_{13}(t) \\ S\frac{dh_2(t)}{dt} &= q_2(t) + q_{32}(t) - q_{20}(t) \\ S\frac{dh_3(t)}{dt} &= q_{13}(t) - q_{32}(t) \end{aligned} \tag{1}$$

where:

$$q_{13}(t) = \mu_{13} S_n \operatorname{sign}(h_1(t) - h_3(t))\sqrt{2g|h_1(t) - h_3(t)|}$$

water flow rate from tank 1 to 3

$$q_{32}(t) = \mu_{13} S_n \operatorname{sign}(h_3(t) - h_2(t))\sqrt{2g|h_3(t) - h_2(t)|}$$

water flow rate from tank 3 to 2

$$q_{20}(t) = \mu_2 S_n \sqrt{2gh_2(t)} \qquad \text{outflow rate.}$$

As all the three liquid levels are measured by level sensors, the output vector is $\underline{y} = [h_1 \quad h_2 \quad h_3]^T$. The control input vector is $\underline{u} = [q_1 \quad q_2]^T$. The purpose is to track small control the system around an operating point. Thus, it has been linearised around this operating point which is given by:

$$\underline{y}_0 = [0.4 \quad 0.2 \quad 0.3]^T \text{ (m)}$$

$$\underline{u}_0 = [0.35 \quad 0.33]^T * 10^{-4} \text{ (m}^3\text{/s).}$$

Using the Torricelli rule, for $h_1 > h_3 > h_2$, the linearised system can then be described by the discrete state space representation with a sampling period $T_s = 1s$:

$$\begin{aligned} \underline{x}(k+1) &= A\underline{x}(k) + B\underline{u}(k) \\ \underline{y}(k) &= C\underline{x}(k) \end{aligned} \tag{2}$$

with:

$$A = \begin{bmatrix} 0.988 & 0.0001 & 0.0112 \\ 0.0001 & 0.9781 & 0.0111 \\ 0.0112 & 0.0111 & 0.9776 \end{bmatrix}$$

$$B = \begin{bmatrix} 64.568 & 0.0014 \\ 0.0014 & 64.22 \\ 0.3650 & 0.3637 \end{bmatrix}$$

and C is the identity matrix.

2.2 Tracking System Design

The fault-tolerant control approach developed in this paper is suitable, not only in regulation, but also in tracking problem where outputs are required to track reference inputs.

In tracking control problem, the number of outputs which have to follow a reference input vector, $\underline{y}_r$, must be less than or equal to the number of the control inputs (D'Azzo and Houpis, 1995). Thus, in the system given by (2), the output equation can be rewritten as:

$$\underline{y}(k) = C\underline{x}(k) = \begin{bmatrix} E_1 \\ E_2 \end{bmatrix} \underline{x}(k) = \begin{bmatrix} \underline{y}_1(k) \\ \underline{y}_2(k) \end{bmatrix} \quad (3)$$

where $\underline{y}_1 \in \Re^p$ represents the vector of p outputs that are required to follow the reference input vector $\underline{y}_r$. The feedback controller is required to cause the output vector $\underline{y}_1$ to track the reference input vector in the sense that in steady-state:

$$\underline{y}_r(k) - \underline{y}_1(k) = \underline{0} \quad (4)$$

To achieve this task, a comparator and integrator vector $\underline{z}$ is added in order to satisfy the following relation:

$$\begin{aligned} \underline{z}(k+1) &= \underline{z}(k) + T_s\left(\underline{y}_r(k) - \underline{y}_1(k)\right) \\ &= \underline{z}(k) + T_s\left(\underline{y}_r(k) - E_1\underline{x}(k)\right) \end{aligned} \quad (5)$$

with T_s the sampling period. Therefore, the open-loop system is governed by the augmented state and output equations where I_p is an identity matrix of dimension p:

$$\begin{aligned} \begin{bmatrix} \underline{x}(k+1) \\ \underline{z}(k+1) \end{bmatrix} &= \begin{bmatrix} A & 0 \\ -T_s E_1 & I_p \end{bmatrix} \begin{bmatrix} \underline{x}(k) \\ \underline{z}(k) \end{bmatrix} + \begin{bmatrix} B \\ 0 \end{bmatrix} \underline{u}(k) \\ &\quad + \begin{bmatrix} 0 \\ T_s I_p \end{bmatrix} \underline{y}_r(k) + \begin{bmatrix} F \\ 0 \end{bmatrix} \underline{f}(k) \\ \underline{y}(k) &= \begin{bmatrix} C & 0 \end{bmatrix} \begin{bmatrix} \underline{x}(k) \\ \underline{z}(k) \end{bmatrix} \end{aligned} \quad (6)$$

The nominal feedback control law of this system can be computed by:

$$\underline{u}(k) = -K\underline{\tilde{x}}(k) = -\begin{bmatrix} K_1 & K_2 \end{bmatrix} \begin{bmatrix} \underline{x}(k) \\ \underline{z}(k) \end{bmatrix} \quad (7)$$

with $\underline{\tilde{x}} = \begin{bmatrix} \underline{x}^T & \underline{z}^T \end{bmatrix}^T$ and $K = \begin{bmatrix} K_1 & K_2 \end{bmatrix}$ is the feedback gain matrix obtained, for instance, by pole assignment, Linear Quadratic optimization To achieve this control law, it is assumed that the state variables are available for measurement. Moreover, we consider the state space where the outputs are the state variables (C is an identity matrix) and the outputs h_1 and h_2 are controlled.

3. FAULT-TOLERANT CONTROL

3.1 Problem formulation

The approach developed in this paper represents a general scheme for designing a system fault-tolerant control in presence of sensors faults.

Consider a classical discrete linear system under nominal conditions described by the following state space representation :

$$\begin{aligned} \underline{x}(k+1) &= A\underline{x}(k) + B\underline{u}(k) \\ \underline{y}(k) &= C\underline{x}(k) \end{aligned} \quad (8)$$

where $A \in \Re^{n \times n}$, $B \in \Re^{n \times p}$ and $C \in \Re^{q \times n}$ are respectively the state, the control and the output matrices. $\underline{x} \in \Re^n$ is the state vector, $\underline{u} \in \Re^p$ is the control input vector and $\underline{y} \in \Re^q$ is the output vector.

A sensor fault can affect the system in additive and/or multiplicative way (Rao *et al.*, 1994):

$$y_j^f(k) = \beta_j y_j(k) + y_{j0} \quad (9)$$

where y_j and y_j^f represent the jth nominal and fault sensor. y_{j0} is a bias and $0 \le \beta_j \le 1$ corresponds to a gain degradation factor of the jth sensor.

According to (9), in presence of additive and multiplicative faults, the system described by (8) can be represented by the general formulation :

$$\begin{aligned} \underline{x}(k+1) &= A\underline{x}(k) + B\underline{u}(k) \\ \underline{y}(k) &= (C + \delta C)\underline{x}(k) + \delta\underline{y}_0 \end{aligned} \quad (10)$$

where $\delta C \in \Re^{q \times n}$ represents the sensor multiplicative faults and $\delta\underline{y}_0 \in \Re^{n \times 1}$ represents the sensor additive faults.

The equation (10) can also be rewritten as follows :

$$\begin{aligned} \underline{x}(k+1) &= A\underline{x}(k) + B\underline{u}(k) \\ \underline{y}(k) &= C\underline{x}(k) + \underline{f}(k) \end{aligned} \quad (11)$$

where $\underline{f} \in \Re^n$ represents the fault vector. This fault vector and its direction are assumed to be completely unknown.

In closed-loop, when a sensor fault occurs, the control law try to cancel the static error created by the corrupted output. Consequently the real output is different from the reference input. The consequence of a bias of 20% on sensor h_1 can be observed to the Fig. 6.

3.2 Solution

A solution could consist in estimating the fault magnitude and apply the control law using ($\underline{y}$-$\underline{f}$) instead of $\underline{y}$. Another solution is considered here in order to realize an extension of our work to component and/or actuator faults : a Fault Detection and Isolation (FDI) module is used to localize the faulty sensor y_j and once this fault is localized, the control law is computed using the estimation of the faulty element. A robust FDI method based on the unknown input observer scheme (Frank, 1990) is used here. The estimation of the faulty element is realized by the other measurements according to the generalized observer scheme (Frank, 1990).

3.3 Fault Detection and Isolation

The aim of a diagnosis procedure is to solve two main decision tasks: fault detection which consists in deciding whether a fault has occurred or not and isolation which is to decide which element of the system has failed. The general procedure is composed of the three following steps:

- Residuals generation: It consists in associating, to the pair model-observation, features which allows to evaluate a difference with respect to normal operating conditions.
- Residuals evaluation: The residuals are compared to some predefined threshold according to a test and, at this stage, symptoms are produced.
- Decision making: The role of the decision maker is to decide according to the symptoms, which elements are faulty, that is isolation.

Faults in systems are usually diagnosed using analytical redundancy when comparing measured and estimated outputs of the system. While a single residual is sufficient to detect a fault, a set of structured residuals is required for fault isolation. Several approaches proposed by Frank (1990), Patton (1993) and Gertler (1991) have been suggested to generate structured residuals.

Here, the observer scheme approach is considered. The basic idea of this approach is to reconstruct the outputs of the system from the measurements or subsets of measurements by means of observers according to a dynamical model of the plant. To isolate fault sensor, the Unknown Input Observer (U.I.O.) based on observer schemes approach is used. In the U.I.O. technique, each estimator must be sensitive to faults and not sensitive to a set of unknown inputs that affect the system. For the fault detection and isolation, the objective is to build a bank of U.I.O. that each of them is driven by one output and all inputs of the system as presented in Fig. 3. The ith residual r_i is only sensitive to the ith sensor fault. The other p-1 output signals are considered as unknown inputs. Each residual is made robust to these missing signals. To satisfy the existence of U.I.O., the optimal time domain approximation is considered (Frank, 1990).

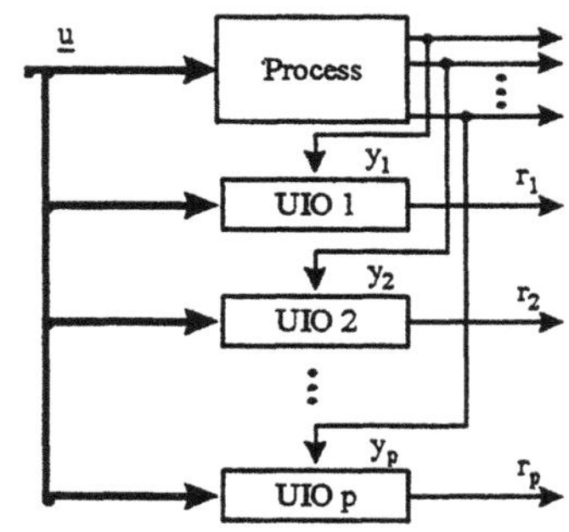

Fig. 3. Unknown Input Observer Scheme

Residual evaluation is based on the assumption that if a fault occurs, the statistical characteristic of the residuals is modified. The residual evaluation involves statistical tests such as limit checking test, generalized likelihood ratio test, Hinkley test, trend analysis test... The output of the statistical test can be built according to a test applied to each residual (Basseville, 1988).

Decision making is realized according to an elementary logic (Leonhardt and Ayoubi, 1997). A simple logic test is used to isolate the faulty actuator as each residual is only sensitive to one sensor fault.

3.4 Output Estimation

To estimate the corrupted output, a bank of observer is built following the generalized observer scheme. Each observer is driven by all inputs and all outputs except the ith measured variable. y_i is not used in the ith observer due to the fact that y_i is assumed to be corrupted by the fault and therefore does not carry information about the system (see Fig. 4.).

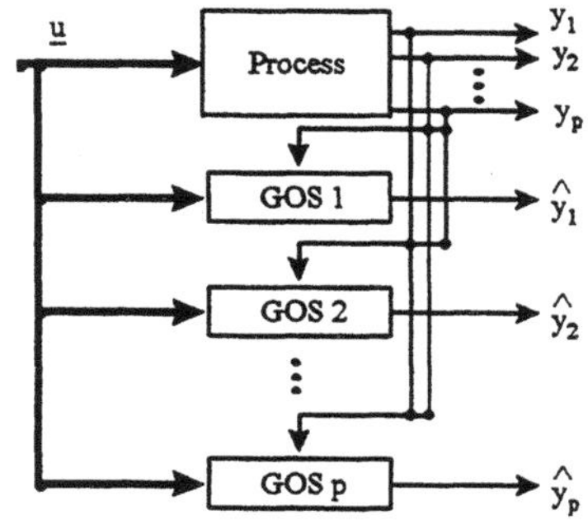

Fig. 4. Output Estimation Scheme

3.5 *General scheme*

The general concept of this approach can be summarized by Fig. 5.

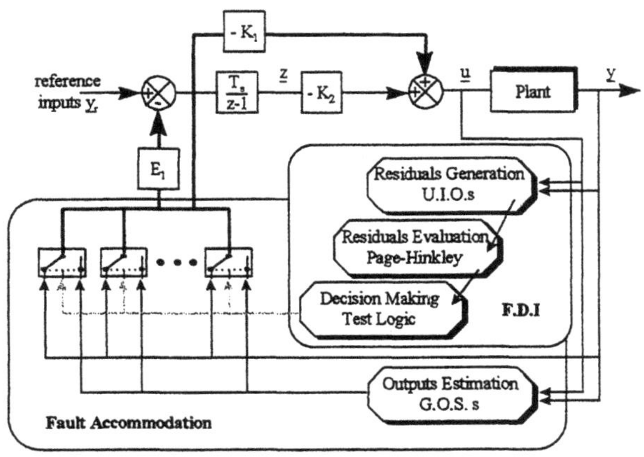

Fig. 5. Fault-tolerant control scheme

4. RESULTS AND COMMENTS

The FTC method developed in this paper has been applied to the simulation of the non linear process depicted in section 2. The simulation results shown in the following figures are step responses with respect to set-point changes, obtained first with a classical control law then with the FTC. The reference inputs correspond to a step of 5cm for h_1 and a step of 2.5cm for h_2. All figures are illustrated for a range of 1500s.

Firstly a bias of 20% on level sensor h_1 occurs at instant 1000s. The Fig.6. shows the result of residual generation on the first output. The other residuals are insensitive to faults. At instant 1000s, the residual presents some variation due to the presence of sensor fault. The residual evaluation detects this evolution at instant 1002s and consequently the control law switches to the estimated output h_1. In Fig.7., it can be noticed that with the FTC method the real level follows the set-point, it is not the case for the classical control law. The same remarks can be made when a bias of 20% on level sensor h2 occurs at instant 700s (see Fig. 8 and Fig. 9).

Our approach is also tested when the level sensor h_1 is out of order at instant 500s. With the FTC method the level h_1 follows the set-point but with a classical control law, the level increases dangerously and reaches the physical limitations of the tank (at 60cm) (see Fig. 10). The consequence on level h_2 can be seen on Fig. 11. With the FTC method, the level h_2 follows the reference value. The Fig. 12 and Fig. 13 represent the flow rates q_1 and q_2. It can be noticed that these control inputs are saturated without using the FTC method.

5. CONCLUSIONS

The method developed in this paper emphasizes the importance of the fault-tolerant control in high automated systems. This method is suitable for sensor faults or failures. The performances of this approach are tested in simulation on a three-tank system. The results are shown and commented. It has been shown that the faults are compensated. The application of this method to the real pilot plant is on going and the compensation of a complete loss of a component or actuator will be studied.

6. REFERENCES

Basseville, M. and I. Nikiforov (1993). Detection of abrupt changes: theory and application. In: Information and System Sciences Series, Prentice Hall International (UK) Ltd.

D'Azzo, J. and C.H. Houpis (1995). Linear control system analysis and design, conventional and modern. In: McGraw-Hill Series in Electrical and Computer Engineering.

Gertler, J. (1991). Analytical redundancy methods in fault detection and isolation - Survey and synthesis. Safeprocess'91 IFAC/IMACS Symposium, Baden Baden, Germany, Vol. 1, pp. 9-21.

Frank, P.M. (1990). Fault diagnosis in dynamic systems using analytical and knowledge-based redundancy - A survey and some new results, Automatica, **vol. 26**, n°3, pp 459-474.

Leonhardt, S. and M. Ayoubi (1997). Methods of fault diagnosis. Control Engineering Practice, **Vol. 5**, pp. 683-692.

Napolitano, M.R. and R.L. Swaim (1989). A new technique for aircraft flight control reconfiguration, Proc. AIAA Guidance, Navigation and Control Conf., Pt. 1.

Patton, R.J. and J. Chen (1991). A review of parity space approaches to fault diagnosis, SAFEPROCESS'91, Baden Baden, **vol. 1**, pp. 239-255.

Patton, R.J. (1997). Fault-tolerant control: the 1997 situation. Safeprocess'97 IFAC Symposium, Kingston Upon Hull, United Kingdom, **Vol. 2**, pp. 1033-1055.

Rao M., Xia Q., and Ying Y., (1994) "Modeling and advanced control for process industries - Applications to paper making processes". Ed. Springer-Verlag Advances in Industrial Control. Chap. 6, pp 157-191.

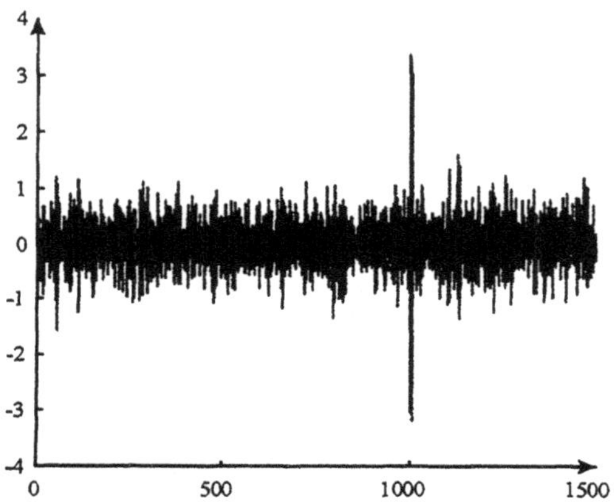

Fig. 6. Residual r_1 with a bias of 20% on h_1

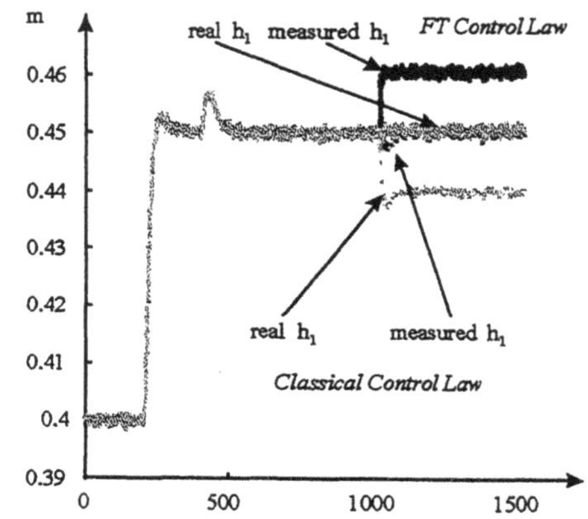

Fig. 7. Level h_1 with a bias of 20% on h_1

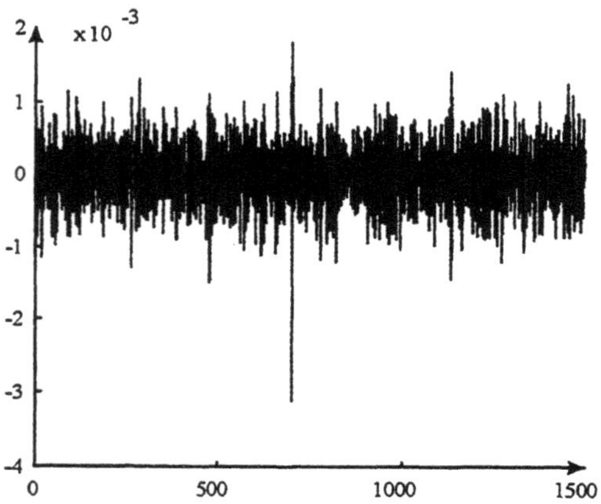

Fig. 8. Residual r_2 with a bias of 20% on h_2

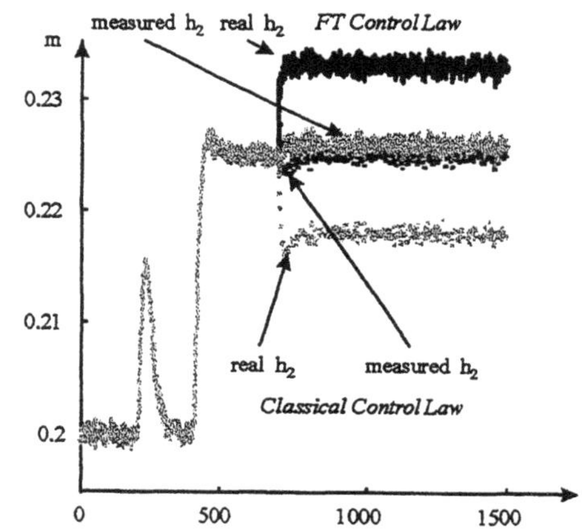

Fig.

Fig. 9. Level h_2 with a bias of 20% on h_2

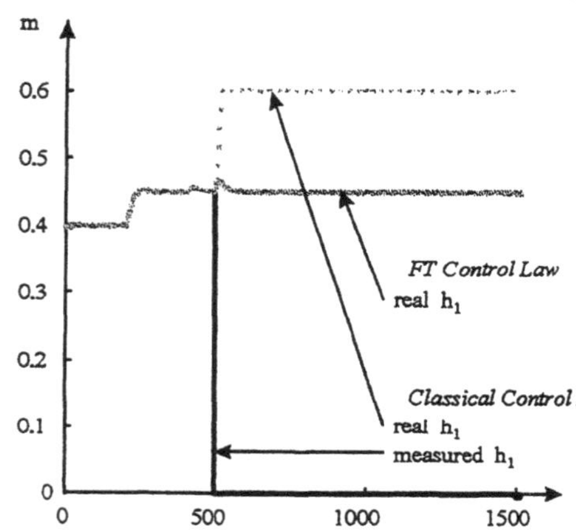

Fig. 10. Level h_1 with level sensor h_1 out of order

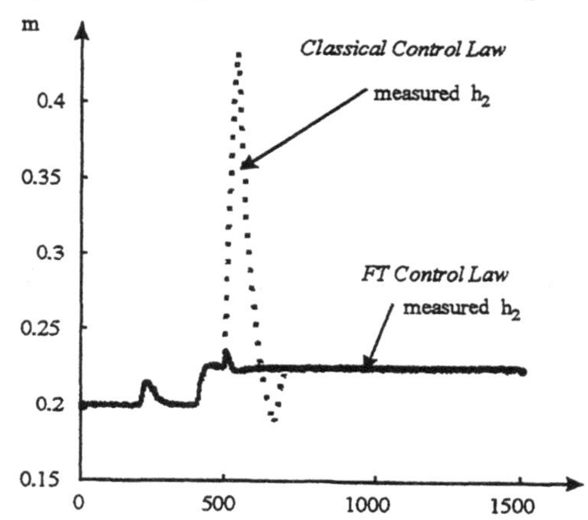

Fig. 11. Level h_2 with level sensor h_1 out of order

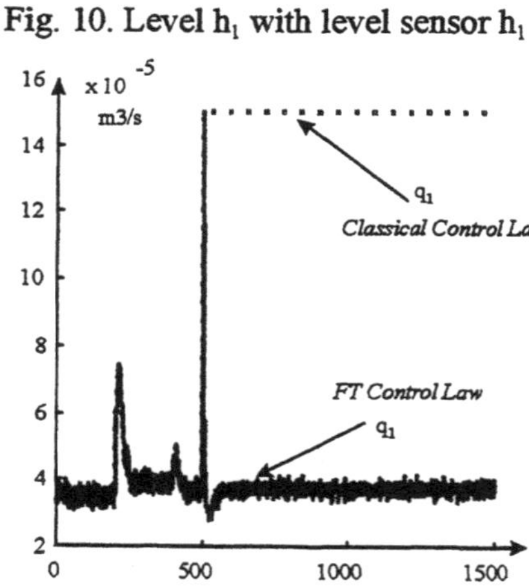

Fig. 12. Flow rate q_1 with level sensor h_1 out of order

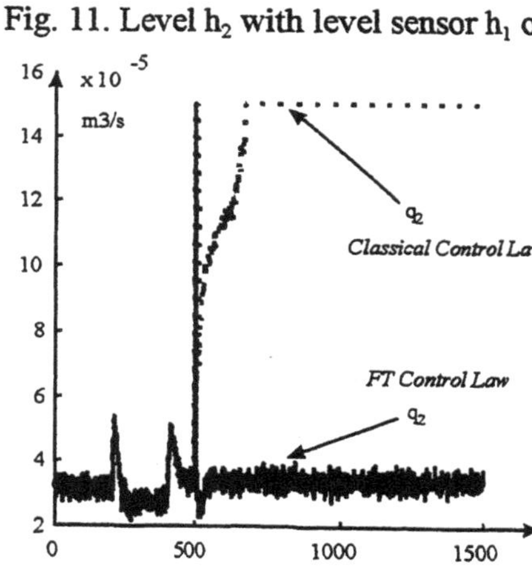

Fig. 13. Flow rate q_2 with level sensor h_1 out of order

SUBOPTIMAL CONDITIONS FOR LEAKAGE DETECTABILITY IN PIPELINES

Cristina Verde *,[1] **Fausto Ibinarriaga** **

* *Instituto de Ingenieria UNAM PB 70-472 Coyoacan DF Mexico*
verde@servidor.unam.mx
** *Facultad de Ingenieria UNAM*

Abstract: This paper deals with the multi-leak detection and location in pipelines. The proposed approach is based on the continuous parity equations which describe the fluid dynamic behaviour in n sections when only the states at the pipeline extremes are measured. Since the perfect decoupling conditions for multi-leak are not satisfied, a criterium with equality constraint is formulated. This allows us to minimize the effects of the state and of one leak in the residual and at the same time to maximize the effect of the rest of the leaks. To address the location problem of multi-leak a comparative scheme of the residuals with bias is proposed. It is shown by simulation that even small leak can be located and if two leaks appear, the scheme can identify both of them.

Keywords: Multi-leak in pipelines, residual generation, detectability conditions, dedicated observer system, parity equations.

1. INTRODUCTION

Pipelines integrity is an important and constant concern in the operation of fluids distribution networks. Although pipelines are protected against damage, the external impacts, internal pressures, corrosion, and natural events surge and as consequence leaks, dents and breaks can occur in the system. On the other hand the location of large leak or pipeline breaks can be easy detected by traditional mass balance method, but as the leak gets smaller it becomes much difficult to detect and locate its. Moreover, the mass balance method is useless to distinguish multi-leak. In general the main objective of the fault detection and isolation task, FDI, for the pipelines is to obtain, as soon as possible, the location of leaks with few instrumentation and high precision.

The particular problem of leak detectability based on model knowledge and available measurements of the pipeline has been tackled in the past using different approaches. Billman *et al* (Billman and Isermann, 1984) proposed a FDI scheme based on parameter estimation and on sensitive static observer to get the errors between model and process; however the method is incapable to locate two leaks. Recently, Liou (Liou, 1996) developed a method in which a pseudo random binary disturbance and pressure data are used to obtain the impulse function response and the change of the time transient of the fluid under faults. In this case steady flow conditions are required to detect and locate a leak or a dent. Other authors (Mactaggart and Myers, 1996) reported successful results using a practical approach based on volume balance in the small Bonito Pipeline in the Gulf of Mexico. Practical engineers has been reported that the ultrasonic flow sensor is a potential tool to locate leaks (Baumoel, 1994); the main limitation with this sensor is its high cost.

In general, the model based fault detection and isolation issue has been solved using different ap-

[1] Partially supported by DGAPA

proaches: parity space (Chow and Willsky, 1984), algebraic methods in frequency domain (Frank and Ding, 1994), in state space (Patton and Chen, 1991) and by geometric concepts (Massoumnia. *et al.*, 1989). It was recognized, during the Pre-Congress Tutorial of the 13th IFAC World Congress (Frank, 1996), the variety of applications in which the above approaches have shown success; nevertheless, they have not be applied to the location of leak and dent in pipeline.

In this paper the problem of detection and identification of leaks in pipelines is handled using the analytical redundancy involved in the input-output parity equations together with a set of dedicated observers. Since the detectability and distinguishability conditions for multi-leak cannot be met, it is established a compromise for each parity equation between the robustness of the residual with respect to the considered leak and the residual sensitivity of the rest. The main contribution in this paper is the formulation of the FDI issue in the continuous time framework for a pipeline discretized in n sections, and the criterium used to select the suboptimal decoupling conditions in the parity equations. It is shown by simulation that the proposed FDI scheme can detect and isolate more than one leak, even if they occur at the same time. The model considered is based on the parameters of a physical installation at the National University of Mexico. The laboratory pilot pipeline is connected to a tank and a valve at its extremes. At the present we are validating our approach in the physical installation.

The layout of the remainder of the paper is as follows. The pipeline model together with the leak characterization is introduced in section 2. The section 3 is dedicated to present the parity equations for the linearized model and to propose a criterium which allows to select the parity equation with the required sensitivity and robustness. In section 4 the performance of the FDI scheme is analyzed when two leaks are introduced in a simulated system and finally the conclusions and future works are introduced in section 5.

2. PIPELINE MODEL

Consider the mathematical description of an unsteady one phase fluid in a one-dimensional pipeline in which it is assumed that the elastic effects of the wall and the term $Q\partial Q/\partial x$ are neglected, the isothermic speed of sound b is constant and the velocity of the flow is small in comparison to the sound speed (Chaudry, 1979)

$$\frac{\partial Q}{\partial t} + gA\frac{\partial H}{\partial x} + \mu Q\left|Q\right| = 0 \qquad (1)$$

$$b^2\frac{\partial Q}{\partial x} + gA\frac{\partial H}{\partial t} = 0 \qquad (2)$$

with H the pressure head $[m]$, Q the flow $[m^3/s]$, x the length coordinate $[m]$, t the time coordinate $[s]$, g the gravity $[m/s^2]$, A the section area$[m^2]$, D is the pipeline diameter $[m]$. Also, $\mu = \frac{f}{2DA}$ where f is the friction coefficient without dimensions.

A leak at point x_f with outflow

$$Q^f \mid_{x_f} = \lambda\sqrt{H} \mid x_f$$

produces a discontinuity in (1-2). As consequence the system must be handled as two pipelines sections with the boundary condition

$$Q^b \mid_{x_f} = Q^a \mid_{x_f} + Q^f \mid_{x_f}$$

in-between, where b and a denote the pipeline sections before and after the leak respectively (Zhidkova, 1973).

Assuming that the pipeline of length l is connected at its extremes to a tank of pressure head H_r and to a valve with open coefficient β, and discretizing the space in n cells of size Δx, the partial differentials with respect to x in (1-2) can be approximated by

$$\frac{\partial H}{\partial x} \simeq \frac{H_{i+1} - H_i}{\Delta x} \quad \forall \quad i = 1, \dots, n$$

$$\frac{\partial Q}{\partial x} \simeq \frac{Q_i - Q_{i-1}}{\Delta x} \quad \forall \quad i = 2, \dots, n+1$$

where the subscript i is associated to the section i of the pipeline and the boundary conditions are given by $H_1 = H_r$ and $Q_{n+1} = \beta\sqrt{H_{n+1}}$. For the purpose of FDI, a leak λ_i for the first $\tilde{n} = n-1$ sections of the pipeline is considered. Under these conditions, the pipeline model can be written as n sets of coupled nonlineal dynamic equations given by

For $i = 1, \dots, n$

$$\dot{Q}_i = a_1(H_i - H_{i+1}) - \mu Q_i\left|Q_i\right| \qquad (3)$$

For $i = 2, \dots, n+1$

$$\dot{H}_i = a_2(Q_{i-1} - Q_i - \lambda_i\sqrt{H_i}) \qquad (4)$$

with $a_1 = gA/\Delta x$, and $a_2 = b^2/gA\Delta x$. These nonlinear equations can be written in the compact form

$$\dot{x} = Ax + B(x)u - \mu nl(x) - a_2 nl_f(x, \lambda) \quad (5)$$

where the state vector is given by

$$x = \begin{bmatrix} Q_1 & H_2 & Q_2 & \cdots & H_{n+1} \end{bmatrix}' \epsilon\, \Re^{2n}$$

the input vector is

$$u = \begin{bmatrix} H_r & \beta \end{bmatrix}' \epsilon\, \Re^2$$

the leaks vector is

$$\lambda = [\lambda_1 \ \lambda_2 \ \cdots \ \lambda_{\tilde{n}}]' \in \Re^{\tilde{n}}$$

the matrix of the linear system part is

$$A = \begin{bmatrix} 0 & -a1 & 0 & 0 & \cdots & 0 & 0 \\ a2 & 0 & -a2 & 0 & \cdots & 0 & 0 \\ 0 & a1 & 0 & -a1 & \cdots & 0 & 0 \\ \vdots & \vdots & \ddots & \ddots & \ddots & \vdots & \vdots \\ 0 & 0 & 0 & \ddots & \ddots & -a2 & 0 \\ 0 & 0 & 0 & 0 & \ddots & 0 & -a1 \\ 0 & 0 & 0 & 0 & \cdots & a2 & 0 \end{bmatrix}$$

and the nonlinear terms are

$$B(x) = \begin{bmatrix} a_1 & 0 \\ \vdots & \vdots \\ 0 & -a_2\sqrt{x_{2n}} \end{bmatrix}$$

$$nl(x) = [x_1 \mid x_1 \mid 0 \ \cdots \ x_{2n-1} \mid x_{2n-1} \mid 0]'$$

$$nl_f(x,\lambda) = [0 \ \lambda_1\sqrt{x_2} \ 0 \ \lambda_2\sqrt{x_4} \ \cdots \ 0 \ 0]'$$

If only the inflow Q_1 and the pressure head H_{n+1} at the valve are measured, the output equation takes the form

$$y = Cx = \begin{bmatrix} 1 \ 0 \cdots 0 \ 0 \\ 0 \ 0 \cdots 0 \ 1 \end{bmatrix} x \qquad (6)$$

Now, (5-6) can be used as the basic relations for the leak detection method. Note from (5) that the leaks can be seen as multiplicative uncertainties in the nonlinear model.

3. MULTI-LEAK DETECTION METHOD

Considering the leak vector λ as the faults to detect and the model (5-6) at the operation point (x_o, u_o), the system deviations can be written by

$$\dot{x} = A_0 x + B_0 u + E\lambda \qquad (7)$$

$$y = Cx \qquad (8)$$

with matrices $B_0 = B(x)\mid_0$

$$A_0 = A + diag(\tilde{\mu}\, x_1, 0, \tilde{\mu}\, x_3, ..., 0, \frac{a^*\beta_0}{\sqrt{x_{2n}}})\mid_0$$

with $\tilde{\mu} = -2\mu$, $a^* = -a_2/2$ and

$$E = a^* \begin{bmatrix} 0 & 0 & \cdots & 0 \\ \frac{1}{\sqrt{x_2}} & 0 & \cdots & 0 \\ 0 & 0 & \cdots & 0 \\ 0 & \frac{1}{\sqrt{x_4}} & \cdots & 0 \\ \vdots & \vdots & \ddots & \vdots \\ 0 & 0 & \cdots & \frac{1}{\sqrt{x_{2n-2}}} \\ 0 & 0 & \cdots & 0 \\ 0 & 0 & \cdots & 0 \end{bmatrix}_0 \in \Re^{2n\times\tilde{n}}$$

Now, the FDI issue is to design a set of dynamic residual generators for the system (7-8) that takes u and y as inputs and produces a residual vector $r \in \Re^{\tilde{n}}$ with the properties:

(1) If the leak vector λ is zero, the residual vector r must decay asymptotically to zero.

(2) If $\lambda_i \neq 0$ and $\lambda_j = 0 \quad \forall j \neq i$, the component $r_i = 0$ and the rest $r_j \neq 0$.

Note that this objective demands sensitivity of the residual generator to $\tilde{n}-1$ leaks and a robust residual again only one leak. This formulation is more realistic to achieve from practical point of view than the usually one in which the robustness again $\tilde{n}-1$ leaks is demanded.

Isolating the fault λ_i, the model of the system for a time horizon $s \leq 2n - 1$ can be transformed to the form

$$\begin{bmatrix} y \\ py \\ \vdots \\ p^s y \end{bmatrix} = H_x x + H_u \begin{bmatrix} u \\ pu \\ \vdots \\ p^s u \end{bmatrix} + \qquad (9)$$

$$+ \bar{H}_i \begin{bmatrix} \Lambda \\ p\Lambda \\ \vdots \\ p^s\Lambda \end{bmatrix} + H_i \begin{bmatrix} \lambda_i \\ p\lambda_i \\ \vdots \\ p^s\lambda_i \end{bmatrix}$$

with the derivative operator $p = d/dt$, the vector $\Lambda = [\lambda_1 \ \cdots \ \lambda_{i-1} \ \lambda_{i+1} \ \cdots \ \lambda_{n-1}]'$ and the matrices

$$H_x = \begin{bmatrix} C \\ CA_0 \\ \vdots \\ CA_0^s \end{bmatrix} \in \Re^{\tau\times 2n}$$

with $\tau = 2(s+1)$

$$H_u = \begin{bmatrix} 0 & 0 & 0 & 0 \\ CB_0 & 0 & 0 & 0 \\ CA_0B_0 & CB_0 & 0 & 0 \\ \vdots & \vdots & \vdots & \vdots \\ CA_0^{s-1}B_0 & \cdots & \cdots & CB_0 \end{bmatrix} \in \Re^{\tau\times 2s}$$

$$\bar{H}_i = \begin{bmatrix} 0 & 0 & \cdots & 0 \\ CE_i & 0 & \cdots & 0 \\ CA_0E_i & CE_i & \cdots & 0 \\ \vdots & \vdots & \ddots & 0 \\ CA_0^{s-1}E_i & \cdots & \cdots & CE_i \end{bmatrix} \in \Re^{r \times s(\bar{n}-1)}$$

with $E_i = E$(deleting i column) and the matrix

$$H_i = \begin{bmatrix} 0 & 0 & \cdots & 0 \\ C\Xi_i & 0 & \cdots & 0 \\ CA_0\Xi_i & C\Xi_i & \cdots & 0 \\ \vdots & \vdots & \ddots & 0 \\ CA_0^{s-1}\Xi_i & \cdots & \cdots & C\Xi_i \end{bmatrix} \in \Re^{r \times s}$$

with Ξ_i = Column(i of E). The structure of matrix H_i takes a down triangular form and is the key to find a robust residual with respect to the leak λ_i (Ibinarriaga-Celis, 1998). The unknown input fault detectability condition for the parity vector (9) requires to select a residual generator vector v_i, such that $v_i\prime H_x = 0$, $v_i\prime H_i = 0$ and $v_i\prime\ \bar{H}_i \neq 0$. The fulfill these conditions allows to use the scalar residual

$$r_i = v_i' \begin{bmatrix} y \\ py \\ \vdots \\ p^s y \end{bmatrix} - v_i' H_u \begin{bmatrix} u \\ pu \\ \vdots \\ p^s u \end{bmatrix}$$

as a leak detector (Chow and Willsky, 1984). This means r_i will be zero in normal conditions and when only the leak λ_i is present; otherwise it will be no zero. However, there is not a vector v_i for the pipeline, that satisfies the detectability conditions assuming one leak as unknown input and the rest as faults.

To overcome this problem, it is suggested to weaken the detectability conditions. Since the multi-leak detection is the main objective here, the state constraint given by $v_i\prime H_x = 0$ will be removed. The effect of this modification is that the residuals will not be exactly zero in normal conditions.

In order to reduce as much as possible the effect of the state in the residual for each leak λ_i, the following optimization problem is minimized

$$\min \frac{\| v_i' H_x \|_2}{\| v_i' \bar{H}_i \|_2} \text{ with constraint } \| v_i' H_i \|_2 = 0$$

Following the procedure reported by Wünnenberg (Wünnenberg, 1990) to generate closed loop residuals, a set of observers is designed on the base of the observer canonical form of (7-8). For the pipeline leaks this form can be reduced to

$$\dot{z} = A_i z + B_i u + E_z \Lambda + O\lambda_i \qquad (10)$$

$$y_2 = C_i z$$

where O denotes zero vector and the system matrix has the form

$$A_i = \begin{bmatrix} 0 & 0 & 0 & 0 & 0 & 0 \\ 1 & 0 & 0 & 0 & 0 & 0 \\ \vdots & \ddots & \ddots & \vdots & \vdots & \vdots \\ 0 & 0 & 1 & a_{ii} & 0 & 0 \\ 0 & 0 & 0 & \ddots & 0 & 0 \\ 0 & 0 & 0 & 0 & 1 & 0 \end{bmatrix}$$

with coefficient $a_{ii} \neq 0$, since $\| v_i' H_x \|_2 \neq 0$; the largest this norm is, the largest will be the residual in absence of leaks.

To obtain the closed loop residual, an estimator of the output error of (10) for each leak is designed by a Kalman Filter. The advantage of this estimator is its robustness again noise and model errors and that it can been extended for the nonlinear case (Gelb, 1979). The selected covariance matrices for the Kalman Filter approach can be found in the reference (Ibinarriaga-Celis, 1998).

4. NUMERICAL EXAMPLE

The above procedure is applied here to detect and isolate leaks in a water pipeline of $132m$ divided in 7 sections with $D = 0.1m$, head pressure of $10m$ and $b = 1200m/s$. The numerical results presented here are obtaining by simulation. The nonlinear continuous model (3-4) was integrated using the Adams-Gear algorithm. The operation

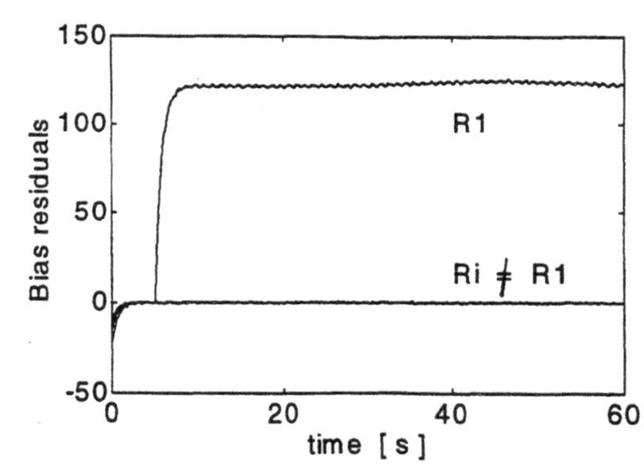

Fig. 1. Time evolution of the bias residuals for a leak in the first section at 5s.

point used in the design is $H_r = 11m$ for the inflow-tank $\beta = 0.5$ for the valve at the tip and $Q = 7.6 \times 10^{-3} m^3/s$ for the flow. To analyze the performance of the proposed approach under real conditions, the tank level and the valve coefficient are simulated adding to the nominal values deviations with sine functions; variations of $\pm 7\%$ and $\pm 80\%$ of the nominal values for H_r and β are used respectively. In addition, white noise is added to the measurement vector.

For the leak coefficient $\lambda_i = 1 \times 10^{-4}$ is used; this generates an outlet flow of $Q^f \simeq 1.4 \times 10^{-7}$ m^3/s, that corresponds to a value of 0.0018% of the steady flow.

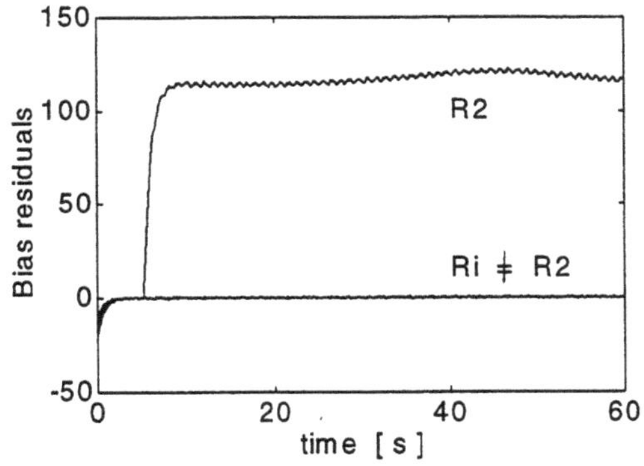

Fig. 2. Time evolution of the bias residuals for a leak in the second section at 5s.

The time response of the bias residuals $R_i = r_i - dc_i$ for the six considered leaks are shown in figures from 1 to 6. The bias dc_i is calculated from the residual value without leak. Figure 7 shows the bias residuals when leaks in section 2 and 4 are introduced at the same time (5s). From these figures it can be seen that the bias residuals are zero in absence of leaks and they respond only to their corresponding leaks. Moreover, the obvious differences between residuals in normal and fault conditions for a loss of 0.0018% of the steady flow indicate that this approach can be applied even for small leakage.

5. CONCLUSIONS

In this paper, the parity vector approach has been used to detect and locate multi-leak in a pipeline with a one phase fluid. The effect of a leak in the mathematical model is introduced by breaking the continuity of the partial differential equations and defining the boundary conditions in term of outflow and pressure at the leak position. Using the model discretized in space, it has been shown that if only measurements of inlet flow and outlet pressure are used the unknown input fault

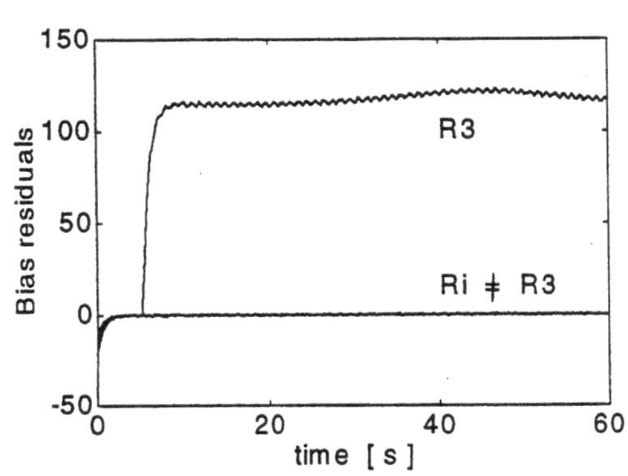

Fig. 3. Time evolution of the bias residuals for a leak in the third section at 5s.

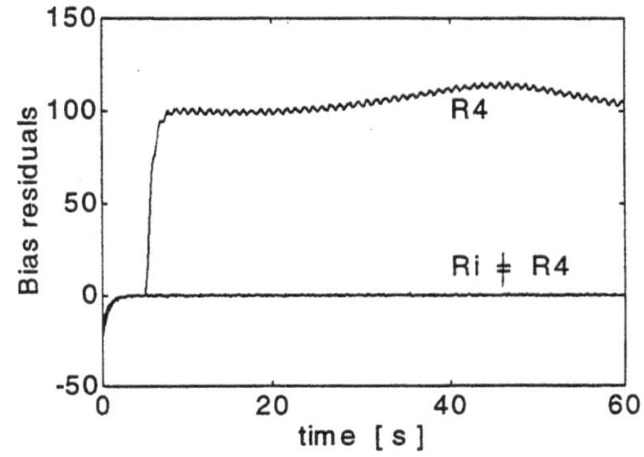

Fig. 4. Time evolution of the bias residuals for a leak in the fourth section at 5s.

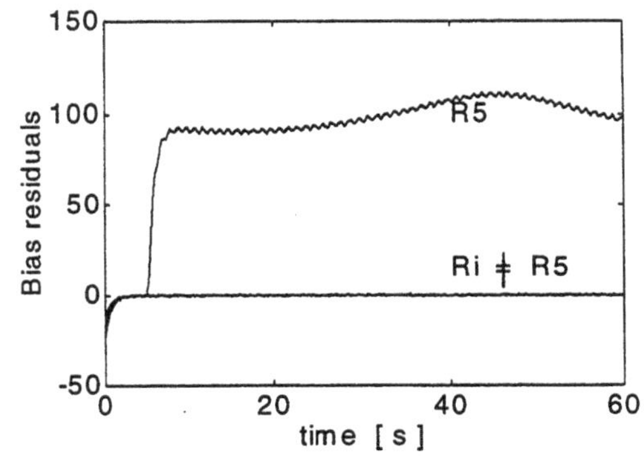

Fig. 5. Time evolution of the bias residuals for a leak in the fiveth section at 5s.

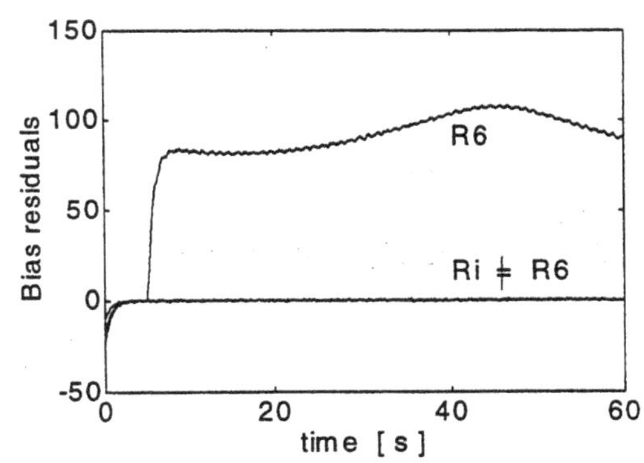

Fig. 6. Time evolution of the bias residuals for a leak in the sixth section at 5s.

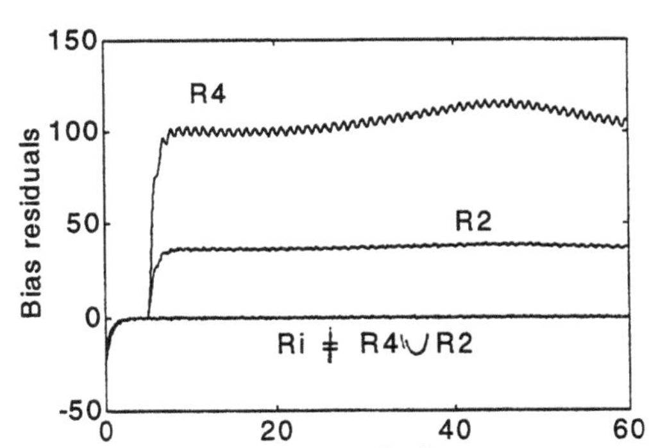

Fig. 7. Time evolution of the bias residuals for leaks in sections two and four at 5s

detectability and distinguishability conditions can not be fulfilled. To overcome this difficulty it is proposed to define a suboptimal residual generator which is robust again a leak and sensitive to the rest. Moreover, to reduce the effect of the model uncertainties and noise in the results, the residuals are calculated via a set of Kalman Filters. The performance of the approach is discussed by simulations. It is illustrated that more than one leak in the pipeline can be detected and isolated at the same time. The results are, even for unsteady condition, satisfactory; despite of the linearized model used in the FDI scheme. Future work is the validation and comparison of our approach with the methods reported in the literature using the pilot pipeline of the National University in Mexico.

6. REFERENCES

Baumoel, J. (1994). Performance of clamp-on ultrasonic flowmeter pipeline leak detection systems. In: *API Pipeline Conference, April.*

Billman, L. and R. Isermann (1984). LEAK DETECTION METHODS FOR PIPELINES. *9th IFAC World Congress, Budapest, Hungary* pp. 1813–1118.

Chaudry, M. H. (1979). *APPLIED HYDRAULIC TRANSIENTS.* Van Nostrand Reinhold Co. New York.

Chow, Y. E. and A.S. Willsky (1984). ANALYTICAL REDUNDANCY AND DESIGN OF ROBUST FAILURE DETECTION SYSTEMS. *TRANSACTIONS ON AUTOMATIC CONTROL* **Vol-AC- 29**(No 7), 603–613.

Frank, P.M. (1996). SUPERVISION, FAULT DETECTION AND DIAGNOSIS OF TECHNICAL SYSTEMS. *IFAC 96 PRE-CONGRESS TUTORIAL.*

Frank, P.M. and X. Ding (1994). FREQUENCY DOMAIN APPROACH TO OPTIMALLY ROBOUST RESIDUAL GENERATION AND EVALUATION FOR MODEL -BASED FAULT DIAGNOSIS. *Automatica* **Vol 30**(5), 789–803.

Gelb, A. (1979). *Applied Optimal Estimation.* MIT Press, USA.

Ibinarriaga-Celis, F. (1998). Deteccion de fugas en tuberias. Master's thesis. Facultad de Ingenieria- UNAM.

Liou, J.C. (1996). PIPELINE INTEGRITY USING SYSTEM IMPULSE RESPONSE. *International Pipeline Conference ASME* **Vol 2**, 1137–1142.

Mactaggart, R. and R. Myers (1996). PC BASED LEAK DETECTION. *International Pipeline Conference ASME* **Vol 2**, 1101–1108.

Massoumnia., M., G.C. Verghese and A.S. Willsky (1989). FAILURE DETECTION AND IDENTIFICATION. *IEEE Trans. Autom. Control* **AC-34**, 316–321.

Patton, R and J. Chen (1991). A REVIEW OF PARITY SPACE APPROACHES TO FAULT DETECTION. *IFAC Safeprocess Conference Baden Baden Germany.*

Wünnenberg, J. (1990). *OBSERVER BASED FAULT DETECTION IN DYNAMICS SYSTEMS.* Reihe 8, Nr. 222. Vereins des Deutscher Ingenieure Verlag.

Zhidkova, M. A. (1973). *Gas Transportation in Pipelines.* Naukova, Dumka, URSS. (in Russian).

COMPARISON OF THE PERFORMANCES OF PARAMETRIC ESTIMATION, TIME-FREQUENCY, WAVELET AND SEGMENTATION WITH DETECTION OF ABRUPT CHANGES IN NON-DESTRUCTIVE EVALUATION

S. Femmam - N.K. M'sirdi

L.R.P-U.P.M.C-U.V.S.Q-C.N.R.S (URA 1778) 10-12 Av. de l'Europe Vélizy 78140 France.
e-mail : {femmam, msirdi}@robot.uvsq.fr

Abstract: Laboratory studies of absorption frequency behavior in rocks often use spectral ratios of digitally recorded ultrasonic signals which have been transmitted through a rock sample and a reference sample of a very low absorption, respectively. It is proposed to treat the digitally recorded signals as autoregressive AR process which, using recursive adaptive filter concepts. The purpose of this paper is to show the advantages of the parametric methods over Fourier analysis method for characterization of materials, and by use of time-frequency, time-scale, and abrupt changes detection (Linear Predictive Error algorithm (LPE)) the choice of the signal length in spectral ratio method is solved by segmentation. Also, as a tool of analysis time-frequency and time-scale are very adapted for faults discrimination. *Copyright © 1998 IFAC*

Key words: Quality factor, parametric spectral estimation, time-frequency, wavelet, LPE, segmentation and fault diagnosis.

1. INTRODUCTION

Recent developments in geosciences have focused attention on the fact that there is more information encoded in seismic and sonic logging data than merely geometrical arrangements of reflection horizon and travel time. Interest in relating wave propagation parameters to porous rock and pore fluid parameters has spurred a considerable amount of research effort, particularly in the area of laboratory measurements and correlation.
One of the most widely used laboratory methods to determine energy loss in rocks is to monitor the amplitude decay of ultrasonic waves which have propagated through rock cores. In most of the past studies, the first techniques involved either comparing amplitudes of first arrivals from successive multiple reflections from the sample faces (Levykin, 1962) or comparing first arrival amplitudes from samples of different lengths (Datta, 1968).
The limiting factor in this approach is that the frequency of measurement is assumed to be the resonant frequency of the transducer. Unfortunately, ultrasonic transducers are not monochromatic and, therefore, the resulting loss determinations refer to a band of frequencies rather than a single frequency. The most promising experimental approach which allows a more complete results is spectral analysis.
The goal of this work is to use parametric spectral analysis rather than non-parametric ones, and to solve the problem of segmentation by use of localization energy techniques (time-frequency and time-scale) and the abrupt changes detection.
Time-frequency and time-scale analysis can be used as a tool for faults discrimination.

1.1. Experimental manipulation.

In the non-destructive evaluation, the ultrasonic signal emission reception technique is widely used for material characterization. This is realized by an apparatus of ultrasonic signal generation, signal recorder, and a set of computing devices. The ultrasonic part is made up with two transducers and the ultrasound signal generator. Tests show the

symmetry of the apparatus, the transmitter transducer is the piston and the receiver transducer is the base. The parametric generator produces electric pulses which excite the emitter sensor. The emitter transducer generates a longitudinal waves with a central frequency of 0.8 MHz for a band pass of 250-750 Khz at 6 dB. The wave travels across the material and received by the receiver transducer.
The characteristics of the signal response and its assumptions are detailed in Guillaume (1992).

2. BACKGROUND TO ACOUSTIC PULSE ATTENUATION

2.1. Amplitude attenuation.

The energy loss due to "internal friction" during a stress-strain cycle is given by:

$$\frac{\Delta E}{E_{max}} = \frac{-2\pi}{Q} \qquad (1)$$

where, ΔE : the energy dissipated per cycle of the periodic oscillation in a certain volume of rock. E_{max} : the peak elastic energy stored in the same volume and Q : the quality factor (Bourbier and Nur, 1989; Kjartansson, 1979).
The propagating wave equation is:

$$\frac{dA}{A} = \frac{-\pi}{Q}\frac{dx}{\lambda} \qquad (2)$$

where, A : wave amplitude, dx : infinitesimal distance along the direction of propagation, and λ : the wavelength of a single propagating frequency.
(2) is the exponential decay relation for peak amplitude with travel distance in a plane wave (Kavestsky et al., 1990).
The present approach is simpler method, based on the concept of an effective frequency.

2.2. Frequency attenuation.

From (2), for a given travel distance, higher frequencies (i.e. shorter wavelength) will be attenuated more rapidly than lower frequencies. This leads to a general broadening of the pulse, and a consequent increase in its first arrival rise-time with increasing travel distance through the rock.

2.3. Derivation of a simple description of acoustic pulse propagation.

As an acoustic pulse propagates through rock, its frequency composition changes by preferential attenuation of the higher frequencies.
From exponential decay (2):
The equation for a reference (index r) sample is:

$$Ln\frac{A_r(\lambda_r)}{A_0} = \frac{-\pi x}{Q_r\lambda_r} + cte \qquad (3)$$

The same equation for the sample to be tested.

$$Ln\frac{A(\lambda)}{A'_0} = \frac{-\pi x}{Q\,\lambda} + cte' \qquad (4)$$

Subtracting (4) from (3):

$$Ln\frac{A(f)}{A_r(f)} = -\pi[\frac{1}{QV} - \frac{1}{Q_rV_r}]xf + cte \qquad (5)$$

where, $\lambda = V/f$, V : the velocity, and f : the frequency.
The Aluminum is considered as a perfect material, its quality factor is very large so (5) becomes :

$$Ln\frac{A(f)}{A_r(f)} = -\pi\frac{1}{QV}xf + cte \qquad (6)$$

The slope of (6) gives the quality factor of the material sample.

$$a = \frac{-\pi x}{QV} \Rightarrow Q = \frac{-\pi x}{aV} \qquad (7)$$

3. SPECTRAL ESTIMATION ANALYSIS

3.1. Classical spectral analysis (Fourier).

The application of the spectral ratio method does not give good results when we use the classical spectral estimation technique (Moreau, 1990; Cadzow, 1982) (figure 1). The main problem is due to resolution of the Fourier transform, and the variations of information in the signal.
In fact, the basic non-parametric analysis method provide a spectral ratio estimation from observed serial time data of the signal. For this estimation the process is supposed to be stationary. However if the characteristics of the process vary with time, the results are not able to follow these variations. Therefore the parametric models to characterize these processes have been chosen to be used.

Figure 1. Original Spectral Ratio (Fourier)

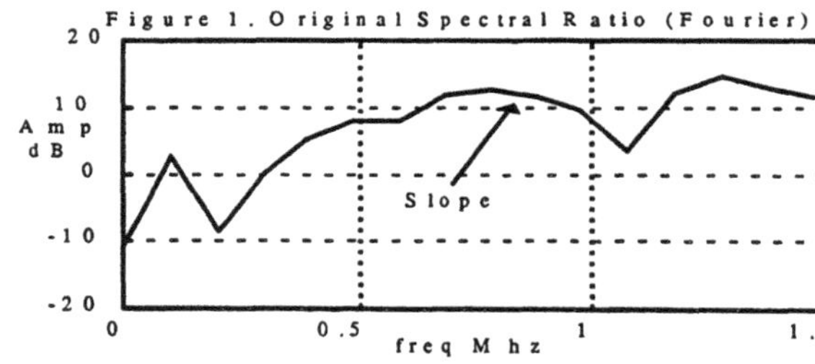

3.2. Parametric spectral estimation methods.

The use of the adaptive parametric spectral estimation methods, for instance the Least Mean Square, the Recursive Least Square, the Burg algorithm, and the Marple algorithm in AR modeling will provide a straight line of the spectral ratio compared to the non-parametric ones (Fourier). This leads an accurate slope estimation and an efficient quality factor (Femmam et al., 1996) (figure 2).

Figure 2. Estimated Spectral Ratio (parametric)

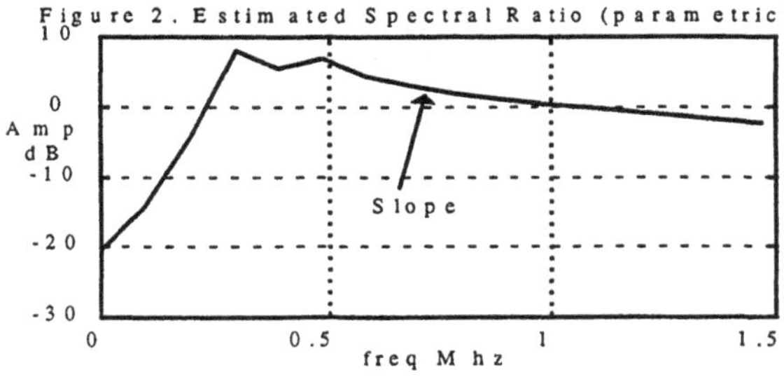

4. ULTRASONIC SIGNAL SEGMENTATION METHODS

4.1. The signal composition and behavior.

Afterward, interest is focused in the segmentation and the choice of the signal length in order to estimate the desired time instant. Preliminary tests realized over different propagated signals across several materials, show the sensitivity of the spectral ratio method towards three parameters :

- the starting point of the signal,
- the choice of the ultrasound signal length,
- the adjustment of the reference signal length with respect to the signal to be characterized.

Refering to the spectral ratio method established by Töksoz et al. (1979), a whole transmitted signal is used. In practice, the samples have a limited size (a sample has a cylindrical shape, its dimension is 80 mm height and 80 mm diameter), the lateral reflected waves make the length of the signal longer. So exploiting the whole transmitted signal is an impossible task.

A first solution used, is to take only the direct arrival signal until the first lateral reflected wave, then calculate its spectrum, and finally the spectral ratio. But this method does not provide a very accurate results because of the loss of transmitted signal information.

After several tests, a threshold value of 2% of the highest peak amplitude of the signal was chosen to be the starting point, since the SNR (signal to noise ratio) is satisfactory (> 100 dB).

The signal length used in spectral ratio is very difficult to determine, because there are no steps to follow. Several research deal with this problem (Kavestsky et al., 1990; Tarif, 1986), and until now to our knowledge there is no way to treat the problem in an acceptable fashion. In this work it is proposed a solution to estimate the signal length without loss of information.

Tests are performed to evaluate the stability of the measured quality factor by spectral ratio method with respect to the signal length. The results show the dependency of the quality factor towards the exploited signal length. To study this problem, the interesting is the calculation of the main signal energy.

Actually, it is clear that, if the signal corresponding to a large energy interval is truncated, the quality factor takes a large and variable values. But, calculating the quality factor with respect to a constant level energy provides a stable results (Tarif, 1986).

This emphasize the need to segment properly the ultrasound signal regarding to energy. If the energy of the whole signal is well localized, the signals lengths which provide the accurate quality factor is well determined.

4.2. Time-frequency and time-scale methods.

To localize the total energy of the ultrasound signal, the time-frequency transforms is applied, which are the spectrogram, the pseudo Wigner-Ville with a smoothing sliding window along the temporally axis, and the smoothing pseudo Wigner-Ville with two smoothing sliding windows one along the temporally axis and the other one along the frequency axis for filtering the interference's (Sessarego et al., 1990; Auger, and Doncarli, 1992; Basseville et al., 1992a; Flandrin, and Martin, 1992; Femmam et al., 1997) (figure 3).

The time-scale transforms are also used for comparison and validation, which are the scalogram (Morlet wavelet), the radiation frequency ω_0 of Morlet wavelet is $5 \leq \omega_0 \leq 6$. The optimal value of ω_0 used is 5,4285) (Lasaygues, 1992; Lasaygues, and Lefebvre, 1995) (figure 4).

Remark that, in all images (figures 3, 4) the main energy is localized around the resonance frequency of the transducer (0.8 MHz), and the main energy corresponds to the main first arrival wave, in Femmam and M'sirdi, (1998a) more details are shown and explained for different ultrasonic signal for various samples.

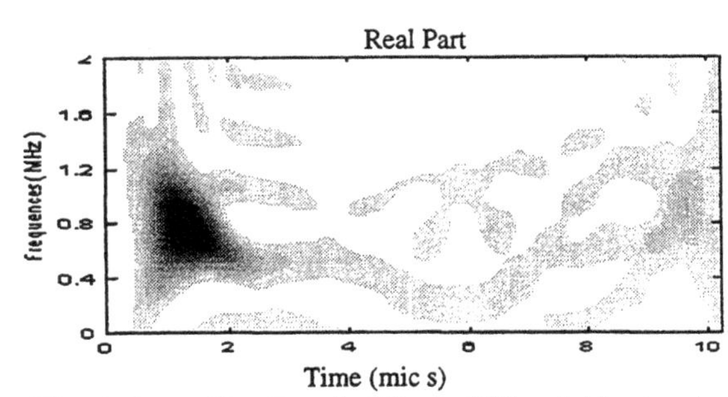

Fig. 3. Smoothing Pseudo Wigner-Ville of Alu signal.

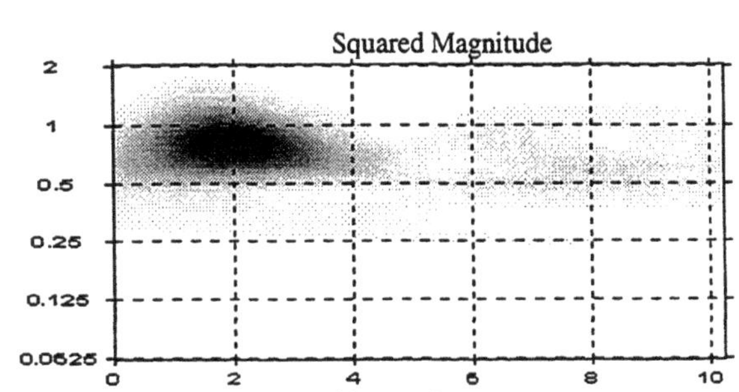

Fig. 4. Morlet Transforms of Granite signal.

4.3. Detection of abrupt changes methods.

4.3.1. Theoretical approach.

General Representation:

Let $X(nT)$ is a discrete signal noted $X(n)$:

$$X(n) = -\sum_{k=1}^{P} a_k X(n-k) + B(n) \qquad (8)$$

Consider the parameter estimation:

$$\underline{\hat{a}} = \left|\hat{a}_1, \ldots.. \hat{a}_p\right|^T \qquad (9)$$

From (8) an LPE is defined (Linear Predictive error).

$$\varepsilon(n) = \sum_{k=0}^{P} \hat{a}_k X(n-k) \qquad a_0 = 1 \qquad (10)$$

$$E[\varepsilon^2(n)] \overset{\Delta}{=} \sigma_\varepsilon^2 \qquad (11)$$

Suppose a good estimation from which $\varepsilon(n)$ is a discrete white noise (gaussian process).

The method principle is as follows: A pointer variable is introduced,

$$P(n) = 1 \qquad for \qquad |\varepsilon(n)| > t \qquad (12)$$

$$P(n) = 0 \qquad for \qquad |\varepsilon(n)| < t$$

where, t: a threshold value which will be chosen as follows:

The "Exceed Word" is defined as:

$$E_l = (P(n-l+1), \ldots., P(n)) \qquad (13)$$

where l is an LPE observation window (Basseville et al., 1992b; Ramdani and M'sirdi, 1994; Doncarli, 1983).

See (Femmam and M'sirdi, 1998b) for segmentation test based on Recursive Likelihood Ratio Deviation (RLRD).

It is demonstrated in Castanie and Soule (1983) That a robust rule is to research l "Successive Exceeds" (with l ~ 3 to 5)

The Neyman-Pearson approach is applied:

* under H_0, a probability of a fixed false alarm $P_{FA} = \alpha$, this determine the "threshold" value.

$$P\left[\left|\varepsilon(n-l+1)\right| > t, \ldots, \left|\varepsilon(n)\right| > t / H_0\right] = \left\{ERF\left(\frac{t}{\sigma_\varepsilon}\right)\right\}^l \qquad (14)$$

under hypothesis that $\varepsilon(n)$ is white and gaussian.

* under H_1, the probability of detection P_D is maximized according to the parameter l.

5. FAULTS DISCRIMINATION

The faults discrimination is one of the famous Problem in non-destructive control.

The reflection signals on the surface of the material are the similar (a few distorted due to the surface irregularity) as the emitted signal, and the diffusion signals by small inclusions (compared to the wave length) are similar to the second derivative of this emitted signal (Morse and Ingard, 1968).

The use of this technique shows that if the signal is a reflection signal or a fault signal due to a hole, a pore, or a fissure.

The detection of a fault inside the material becomes interesting. A hole has been made inside the material to characterize (in our case the Granite) (figure 5(a)). The problem is how the detection of the presence of this hole, which is in fact a fault from an echo of a surface. To solve this problem time-frequency and time-scale analysis are used for this purpose (figure 5(b)).

A three dimensional representation shows the behavior surface of the two main peaks which represent the first arrival reflection (first peak) and the second peak due to the hole made inside the granite sample (figures 6, 7).

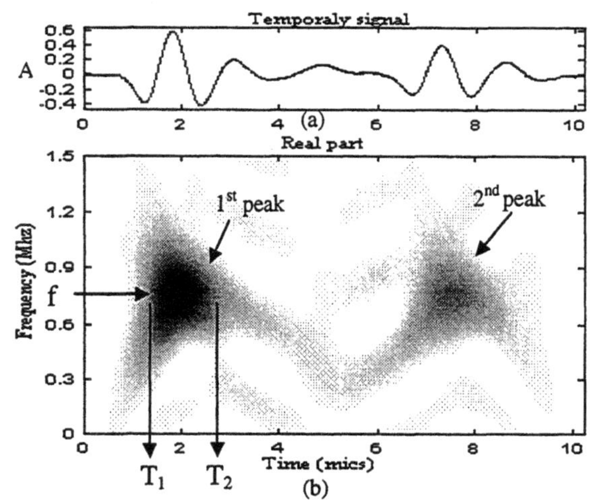

Fig. 5. Granite sample with a hole made inside.

(a). Ultrasonic Granite signal.

(b). Smoothing Pseudo Wigner-Ville Granite signal, f=0.8 MHz, $[T_1, T_2]$ the desired segment.

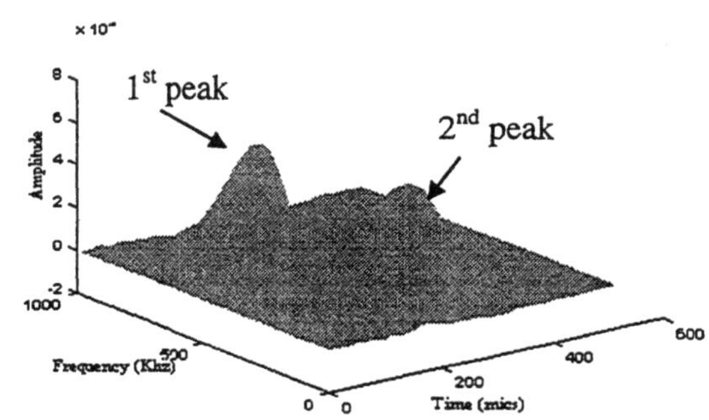

Fig. 6. Three dimensional representation of Granite signal.

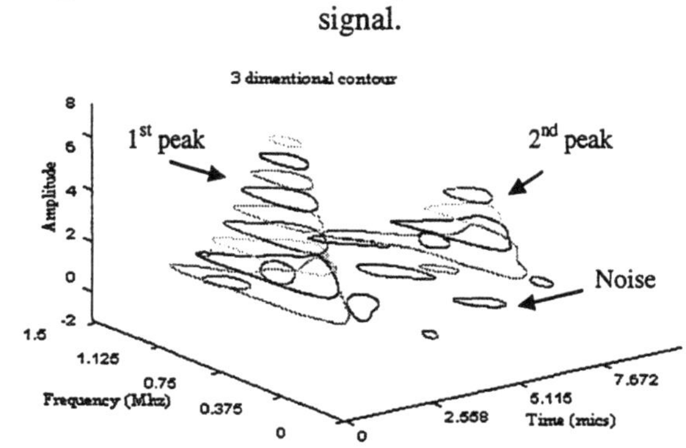

Fig. 7. Three dimensional contour of Granite signal.

6. APPLICATION OF PROPOSED METHOD

6.1. Jumps in variance:

The used model is:

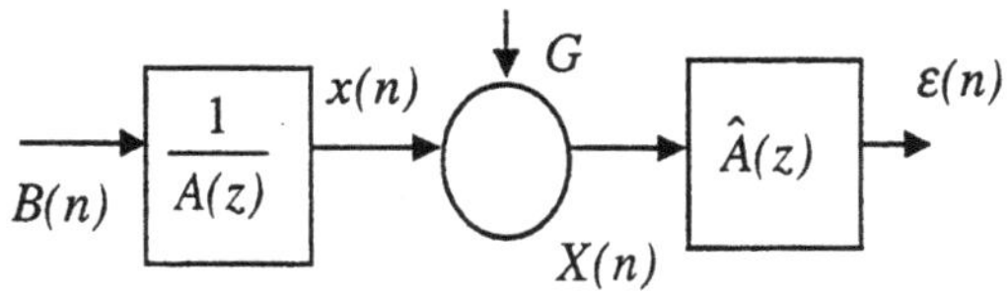

$$X(n) = x(n) \qquad n < n_0 \qquad (15)$$
$$X(n) = G.x(n) \qquad n \geq n_0$$

$X(n)$ models a jump process of a variance.
The variance is given by: (16)

$$\sigma_\varepsilon^2(n) = \sigma_b^2[2G-1] + (G-1)^2 \sum_{i,j=0}^{Inf(p,n-n_0)} a_i a_j C_x(i-j)$$

where $\{a_i\}$ are the AR parameters of $x(n)$ and $C_x(.)$ its covariance. It is clear that:

$$\sigma_\varepsilon^2(n) = G^2 \sigma_b^2 \qquad n \geq n_0 + p \qquad (17)$$

The signature of the non-stationary can not be defined easily. The strategy for that is to keep the same expression for P_{FA}. But P_D is practically difficult to calculate when l >2 or 3.
As a remark, when the detection strategy "exceed word" is changed P_D does not change, but P_{FA} decreases considerably.
The proposed strategy is applied experimentally with l=3 (successive exceeds) $D_3 = 111$ and the variance jumps from σ_x^2 to $G^2\sigma_x^2$. The figures 8 to 13 show the automatic segmentation of the ultrasonic granite signal using the LPE algorithm for obtaining the interesting portion of the signal. Note that there is a delay of segmentation (figure 13), this is due to the abrupt changes detection technique.

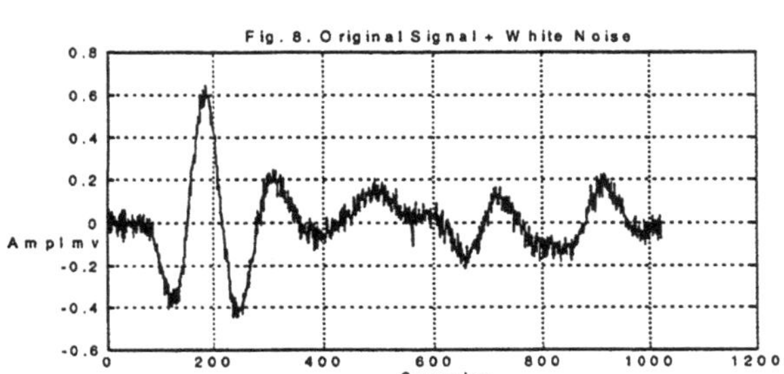

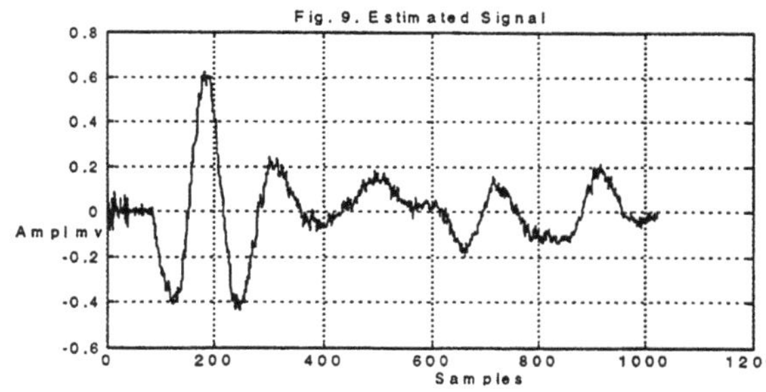

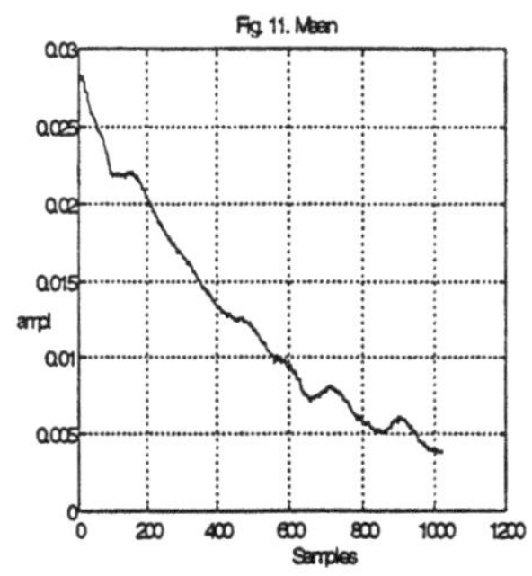

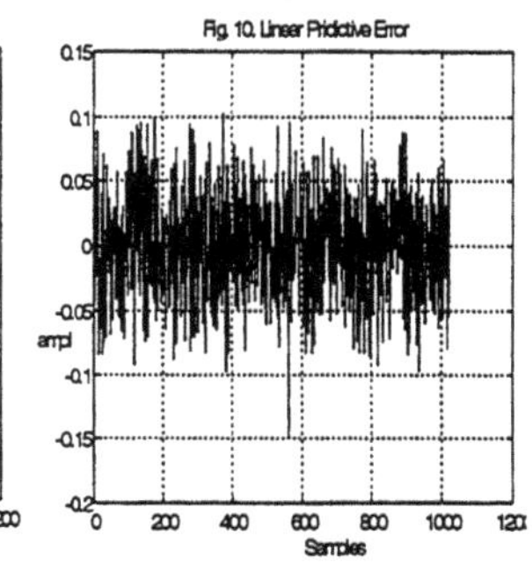

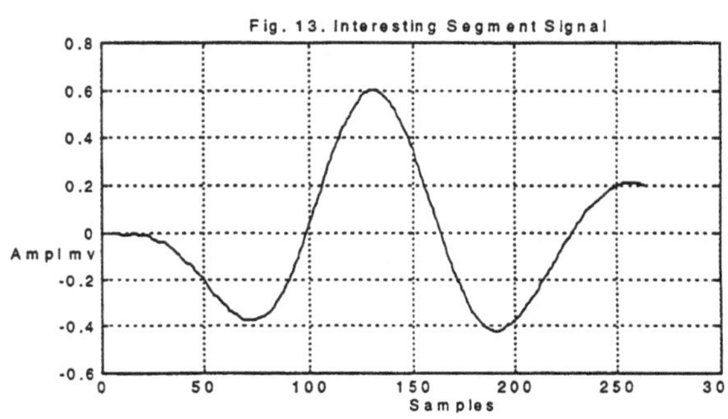

6.2. Experimental evaluation and comparisons.

The results of the quality factor determining by the methods stated above are shown in table I. The quality factor Q_1 is calculated by the non parametric spectral method (classical Fourier), the quality factor Q_2 is calculated by the parametric spectral estimation method (only results of Burg algorithm are given), the quality factor Q_3 is calculated by the use of time-frequency and time-scale transforms (here again only results of smoothing pseudo Wigner-Ville transforms with Burg algorithm are given), and the quality factor Q_4 is calculated by use of automatic segmentation by abrupt changes detection techniques (LPE algorithm with burg algorithm).

Table I, Rocks Description, their characteristics and their quality factors (Q).

Rock nature	Geologic layer	Porosity & Fissure	γ_d (g/cm)	Velocity m/s	Amplitude Volts	Expected values by CGI *	Q_1	Q_2	Q_3	Q_4
Aluminum	/	/	/	3728.908	1.583	reference	/	/	/	/
Granite	Hercynian	Porosity of fissure	2.678	5431.831	0.605	25<Q<40	27.45	36.25	30.92	32.20
Basalt	Pleistocen	Porosity of fissure	2. 836	4170.577	0.814	20<Q<35	16.22	25.17	29.22	31.34
Marble	Aptian superior	1% of porosity less fissure	2.720	5431.831	1.048	10<Q<20	12.14	16.79	16.11	17.62
Marble	Aptian superior	1% of porosity microfissure	2.740	5986.232	1.092	50<Q<100	23.47	87.98	73.17	71.51
Limestone	Oxfordian	Porosity of pore	2.715	2666.667	0.154	35<Q<50	24.35	44.89	42.73	40.82
Limestone	Callovian	Less Porous	2.715	3384.667	0.729	10<Q<15	21.50	11.27	15.42	13.41
Limestone	Batonian	38 % of Porosity	2.715	2993.116	0.636	2<Q<5	26.40	6.71	2.91	4.28

* CGI : Centre de Géologie de l'Ingénieur de l'Ecole de Mines de Paris.

7. CONCLUSION

The superiority of parametric spectral estimation in non destructive control is shown theoretically and experimentally compared to the Fourier analysis, the parametric methods give a very smooth spectrum which is easy to exploit and characterized. The use of time-frequency and/or time-scale are able to detect anomalous behavior inside a material (fissure or pore), in other words discrimination between different signatures.

Segmentation of the ultrasonic signal can be done by use of Linear Predictive Error (LPE) algorithm and the validation is performed by the time-frequency and time-scale energy localization.

The use of detection of abrupt changes with parametric methods are preferable compared to the other methods (see table I). The detection of abrupt changes algorithm provides the desired ultrasonic signals length.

8. REFERENCES

Auger, F. and Doncarli, C. (1992). Quelques commentaires sur des représentations temps-fréquence proposées récemment. *Traitement du signal,* Vol. 9, n° 1, pp. 3-25, 1992.

Basseville, M et al., (1992a). Signaux non-stationnaires, analyse temps-fréquence et segmentation. Fiches descriptives d'algorithmes. *Traitment du signal,* Suppl. au Vol. 9, n° 1, 1992.

Basseville, M. et al. (1992b). Signal segmentation. *Traitement du signal,* 9, Supplément n°1, pp. 115-147.

Bourbier, T. and Nur, A. (1984). Effects of attenuation on reflections: Experimental test. *J. Geo. Re,* Vol. 89, 6197-6202.

Cadzow, J. (1982). Spectral estimation: An overdetermined rational model approach. *Proc. IEEE.* Sept.

Castanie, F. and Soule, P. (1983). An improvment of non stationarity detection by level crossing analysis of linear prediction error. *ICASSP'83 Boston (USA)* 14-16 avril pp. 251-53.

Datta, S. (1968). Absorption of dilatational waves in rocks. *Geoexploration* 6, 127-139.

Doncarli, C. (1983). Sequential detection of abrupt changes in ARMA models. 9ième *Colloque sur le traitement du signal et ses applications,* Nice du 16 au 20 mai.

Femmam, S. et al. (1996). AR processing of ultrasonic signals and its application to determine characteristics for a given material. Submitted 12 july to *Materials Evaluation journal,* 1711 Arlingate Lane, Columbus, USA (reviewed).

Femmam, S. et al (1997). Mise en œuvre et validation d'analyse temps-fréquence. Application à la segmentation d'énergie en contrôle non-destructif par ultrasons. soumis le 23 février au *European Journal of Automation, edition Hèrmes* (reviewed).

Femmam S. and M'sirdi N.K. (1998a). Application of change detection technique to ultrasound signal processing. Use of spectral parametric estimation, time-frequency and wavelets. 30th *IEEE Southeastern symposium on system Theory (SSST),* March 8-10, West Virginia University, Morgantown, west Virginia, USA (Accepted).

Femmam S. and M'sirdi N.K. (1998b). Application of signal segmentation and parametric estimation to determine material characteristics. *IEEE, CESA'98, Computational Engineering in Systems Applic., IMACS Multiconf.* Nabeul-Hammamet, Tunisia, April 1-4 (Accepted).

Flandrin, P. and Martin, N. (1992). Méthodes temps-fréquence. *Traitement du signal,* 9, Suppl. au n°1, pp., 77-113.

Guillaume, S. (1996). Cractéristiques ultrasonores et evolutions texturales du bassin Bressan sous sollicitations thermiques et mécaniques. *Thèse de doctorat de ENSMP, Centre de Géologie de l'Ingénieur de Paris.*

Kavetsky, A. et al (1990). A model of acoustic pulse propagation and its application to determine Q for rock mass. *Int. J. Rock. Mech. Min. Sci. & Geomech. Abstr.* Vol. 27, N°1, pp. 33-41.

Kjartansson, E. (1979). Constant Q-wave propagation and attenuation. *J.Geoph. Res.,* 84, 4737-4748.

Lasaygues, P. (1992). Etude de l'écho d'une fissure. Traitement de signaux par analyse en ondelettes en contrôle non-destructif. *Thèse de l'université d'Aix-Marseille II.*

Lasaygues, P. and Lefebvre, J.P. (1995). Application de l'analyse en ondelettes en tomographie ultrasonore. *Traitement du Signal,* 12, n° 4.

Levykin, A. I. (1962). Damping of ultrasonic waves in rock specimens at various frequencies. *Bulletin Academy of Sciences USSR, Geophysical series* 3, 259-261.

Moreau, N. (1990). Modélisation, estimation spectrale. *ENST, département signal,* 90 sig. 003.

Morse, P. M. and Ingard, K. U. (1968). Theoretical acoustics. *McGraw-Hill Book compagny NY.*

Ramdani, R. and M'sirdi, N.K. Change detection statistical tests in non stationary signal. *European Journal of Diagnosis and safety in Automation,* Volume 4, N° 2, pp. 229-271.

Sessarego, J.P. (1990). Analyse temps-fréquence de signaux en milieux dispersifs. Application à l'étude des ondes de Lamb. *J. Acoust.,* Vol. 3, pp. 273-280, 1990.

Tarif, P. (1986). Mesure de l'atténuation des ondes compressionelles ultrasoniques dans les roches application à la mesure de l'anistropie d'atténuation. *Thèse de doctorat, Université de Paris VII.*

Toksöz, M. N. et al. (1979). Attenuation of seismic waves in dry saturated rocks: I. Laboratory measurements. *Geoph.* Vol. 44, p 681-690, April.

DATA RECONCILIATION AS AN ON-LINE TOOL FOR FAULT DETECTION AND PROCESS SUPERVISION

Bruno Vrielynck, Belsim, Belgium

Abstract: The first part of the article introduces the concepts of data reconciliation. It is a method that uses information redundancy and conservation laws to correct measurements. Its benefits are numerous and range from improved measurements lay-out. The sources of measurement errors are described as well as the fundamental concepts of redundancy and measurement accuracy. Finally the article shows the differences between process simulation and data reconciliation.

The second part of the article deals with practical applications. The present status of data reconciliation is analysed. A few examples are then shown in chemical plants, refineries and power plants. As a conclusion, the future of data reconciliation is investigated. *Copyright © 1998 IFAC*

Keywords: Fault Detection, Validation

1. THE CONCEPT OF DATA RECONCILIATION

1.1 What is data reconciliation ?

Data reconciliation is a method that uses information redundancy and conservation laws to correct measurements and convert them into accurate and reliable information. This method is used in refinery, petrochemical, chemical plants as well as power plants including nuclear power stations. It allows to detect faulty sensors and to pin-point degradation of equipments performance (heat rate, compressor efficiency, etc.)

With tools available today, plant measurements, including lab analyses, can be reconciled in such a way that mass (on a component per component basis) and heat balances are satisfied. When necessary L/V equilibrium and performance constraints can be added. Unmeasured values are calculated and the precision of the reconciled values are quantified. Sensitivity analysis tools can also show the interdependence between the measurements.

1.2 What are the benefits of data reconciliation ?

Benefits from data reconciliation are numerous and include :

- ◊ improvement of measurement lay-out
- ◊ decrease of number of routine analyses
- ◊ reduced frequency of sensor calibration. Only faulty sensors need to be calibrated
- ◊ systematic improvement of process data
- ◊ clear picture of plant operating condition and reduced measurement noise in trends of key variables
- ◊ early detection of sensors deviation and degradation of equipment performance
- ◊ actual plant balances for accounting and performance follow-up
- ◊ quality at process level
- ◊ on-line optimisation tools work with a more accurate information

1.3 The basics of data reconciliation

To understand the basic principles of data reconciliation, one must first recognise that plant measurements (including lab analyses) are not 100% error free. Using these measurements without correction to generate plant balances usually yields incoherence in the balances.

Some sources of errors in the balances directly depend on the sensors themselves:

◊ intrinsic sensor precision,
◊ sensor calibration,
◊ sensor location.

A second source of error when calculating plant balances is the small instabilities of the plant operation and the fact that samples and measurements are not taken exactly at the same time. Using time averages for plant data partly reduces this problem. Lab analyses however cannot be averaged.

Finally, one must also realise that in some cases too many measurements are available whereas in some other cases some measurements are missing and must be back calculated form other measurements. The aspect of data redundancy is here a key factor. It is because too many measurements are available that one can evidence the fact that the measurements are somewhat inaccurate. For example, measuring all inlet and outlet flow rates of a plant (or of part of plant) usually leads to some imbalance between the total inputs and total outputs.

The main idea of data reconciliation is to use the data redundancy of the system as the source of information to correct the measurements. Each measurement is corrected as slightly as possible in such a way that the corrected measurements match all the constraints (or balances) of the process.

Mathematically, this can be expressed as:

$$\text{Min} \sum_i \left(\frac{y_i^* - y_i}{\sigma_i} \right)^2$$

$$\text{subject to} \quad F(x \ , y^*) = 0$$

where y_i^* is the reconciled value of measurement i,

y_i is the measured value of measurement i,

σ_i is the measurement inaccuracy or confidence interval,

x_j denotes unmeasured variables

$F(x \ , y^*) = 0$ corresponds to the process constraints.

1.4 The concept of redundancy

The redundancy is an essential concept of data reconciliation and is the source of information the method uses to correct measurements.

The level of redundancy is defined as the number of measurements which come in addition to the set of measurements needed to be able to calculate the system.

The easiest way to understand this concept of redundancy is to consider an example where only a global mass balance is taken into account.

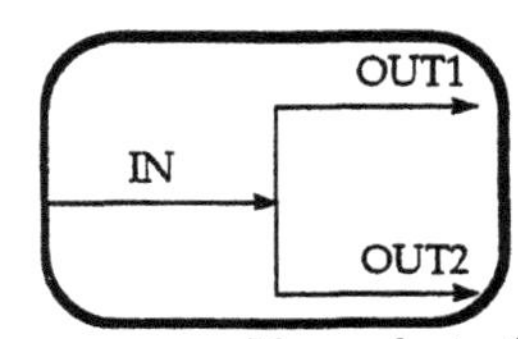

	Measured rates (t/h)		
IN	100	100	100
OUT1	?	50	50
OUT2	?	?	40
Redundancy	-1	0	+1

Fig. 1. The concept of redundancy

If only the rate of stream IN is measured, than there is not enough information to describe the state of the system as nothing can tell how much goes through each outlet lines OUT1 and OUT2. The redundancy is negative.

If the inlet rate and one of the outlet rates are known, that one has just enough information to describe the system and the redundancy level is zero. The rate of OUT2 can be calculated as the difference between the rates of streams IN and OUT1.

Finally, if both outlet rates are measured in addition to the inlet rate, than we have more information than what is necessary to describe the system. The redundancy is of level one.

Of course, in such a trivial example, the correction can be carried out manually. The interest of a software like VALI comes for problems dealing with many variables involved in many constraints where manual corrections cannot be made.

1.5 Process measurements

Measurements are the essential source of information used by data reconciliation (in addition to the process topology) and covers all information coming either from the plant (temperatures, pressures, rates, on-line analyses, etc.) or from the laboratory (GC, densities, etc.)

The concept of measurement can be generalised into pseudo-measurement which represents any numerical information that can be given about the process in addition to actual measurements. For example, turbine efficiencies can be given as pseudo-measurement although this information comes rather from design data than from a real measurement.

The measurement inaccuracy reflects the confidence given to a measurement value. It indicates the expected range of deviation of the measurement.

Assuming a Gaussian distribution of the sensor error, the measurement inaccuracy can be numerically defined on basis of the standard deviation of the distribution. The probability that the real value will be distant from the measured value by less than twice the standard deviation is about 95 %.

Finding the right combination of the inaccuracies is partly a trial and error procedure, because the effect of error sources like for example the improper location of a sensor cannot be predicted. Starting values for the inaccuracies can be the intrinsic (im)precision of the sensor as specified by the manufacturer.

1.6 The difference between process simulation and data reconciliation

Although process simulation and data reconciliation are similar on several aspects, they are as a matter of fact quite different as well, the main difference being the way plant data are handled: in simulation, plant data are used to tune various parameters of simulation models, giving a better model; while in data reconciliation plant data are used to check and correct themselves, giving a better information of the actual state of the plant.

On one hand, both process simulation and data reconciliation are based on a mathematical model of a process. These models are built using predefined objects modelling the process units operations and the products that are handled. Some information must be available about the physical properties of the chemical components present in the system like their molecular weight, their density, etc.

On the other hand, some major differences are to be made on the goal of the modelling, on the way to handle process measurements and on the mathematical methods that must be used.

The primary goal of data reconciliation is to get the most accurate representation of the actual state of an existing process. The primary goal of simulation is to have a model of a process (existing or not) that can be used for various parametric studies and/or process optimization.

In data reconciliation, all measurements are considered simultaneously, each with a given accuracy, and each measurement is corrected. A higher level of redundancy is beneficial as this leads to a higher accuracy of the reconciled measurements. In process simulation, some measurements are considered as specified and some simulation parameters are varied so as to better reproduce the remaining measurements. A higher level of redundancy is generally a source of problems as it is impossible to make the difference between deviations that are due to imperfection of the model and the ones that are due to measurement inaccuracy.

On the mathematical side, process simulation has been traditionally developed on a modular sequential approach. Each process unit is represented by a mathematical model that calculates the state of the output streams knowing the state of the input streams and some unit operation specifications. Each unit is modelled after the other. When some streams are recycled, convergence algorithms are to be used.

Using such an approach is not possible for data reconciliation. Instead the program must generate equations that relate inputs and outputs. An optimization algorithm is then used to find the set of values that are as close as possible to the measurements while satisfying the equations.

It should be noted that while a modular sequential approach is not effective for data reconciliation, on the other side an equation based approach can be used for data reconciliation as well as for process simulation.

2. APPLICATIONS OF DATA RECONCILIATION

2.1 The situation today

As of today, data reconciliation is becoming a mature technology.

Several commercial programs are now available with capabilities ranging from simple mass balances to more sophisticated heat and mass balances, as well as L/V equilibrium equations.

The need for data reconciliation is recognised by an increasing number of companies. With the expansion of computer networks, process plant data become available not only everywhere in the plant but also anywhere in the companies, from the research centres to the headquarters. However, process plant were not initially designed for such use. In particular, when engineers start to calculate mass balances using such data they quickly realise that some unbalances do appear.

The solution to this problem is the use of data reconciliation either off-line or on-line.

In off-line mode, process plant data are reconciled on a case per case basis. The primary benefit is improvement of process data necessary for simulation, de-bottlenecking, retrofit studies and better follow-up of plant operation.

In on-line mode, process data are reconciled on a regular basis (e.g.: each hour, 10 minutes). The primary benefit is process operation improvement thanks to better measurements accuracy, work closer to specifications, early detection of sensor and equipment failures, reduction of number of lab analyses, etc.

In both cases, actual and accurate plant balances are available for process follow-up.

Only a few real on-line applications of data reconciliation are in use today. Most of them are still only overall mass balances.

2.2 Application of data reconciliation in Chemical plants

Vali is used by Wacker Chemie since 1992. Figure 2 shows the evolution of the total sum of errors in one of their plants. This drastic reduction of the sum of errors is due to several factors:

- detection of erroneous measurements
- data reconciliation model improvements: unaccounted by-passes, ...
- process knowledge improvements: unexpected reactions have been identified, better estimations for some unmeasured stream compositions, etc.

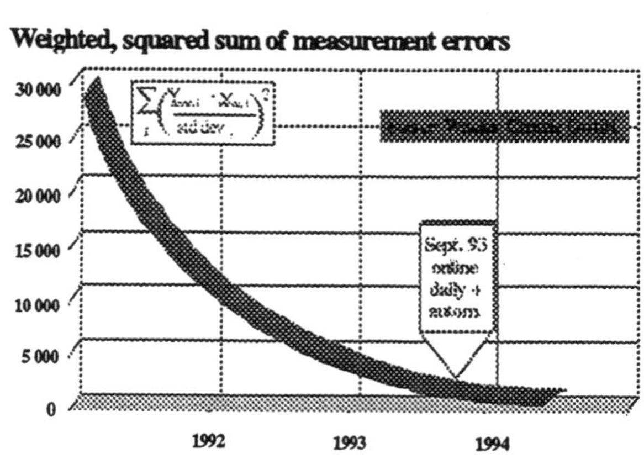

Fig. 2. Reduction in measurements errors

Nowadays the total sum of errors is in the range of 1000, and is the primary factor tracked by the operator. Any significant increase in the sum of errors is an indication of either a process upset or a sensor drift or breakdown.
The main economical benefit in this application has been the reduction by 50% of the number of routine analyses that were necessary to operate the plant.

2.3 Application of data reconciliation in Refining

VALI has been applied to most processes that can be encountered in refineries: distillation units, reformers, FCC, desulfurisation , alkylation, MTBE, etc. Two applications will be further discussed.

In one refinery, the use of VALI for the performance monitoring of the preheat train of a Crude Distillation Unit has pin-pointed a problem with two exchangers of a pump-around system. On all test-run data available on the plant since the last shut-down, VALI systematically indicated that the actual flow rate going through one line of exchangers was virtually null.

On-line cleaning of the exchangers, check of all thermocouples and additional test-runs did not solve the problem. Eventually, the exchangers were opened and the origin of the problem was identified: blinders of the exchangers had erroneously not been removed after the last unit shut-down, 2 years before.

As a second example, the use of VALI on a cyclo-hexane unit showed that the regular cleaning of some of the exchangers was of no use. A few days only after the cleaning, the same fouling level was reached again. The cleaning policy has then been revised.

2.4 Application of data reconciliation in power plants

Data reconciliation is used in a growing number of power plants, including nuclear power plants.

In a nuclear power plant, an accurate measurement of the feed water flow rate is of crucial importance as it is used to calculate the nuclear reactor power, which must never exceed the licensed power. However, due to orifice degradation, one usually observes a slow decay of the actual feed water rate for a constant feed rate measurement. This becomes only obvious at the yearly orifice re-calibration where an expensive Na tracer technology is used. This means that in between two re-calibrations, the actual reactor power is over-estimated, and thus that the plant does not deliver its full load of electricity.

The nuclear power station of Leibstadt (KKL) in Switzerland has suppressed this problem by replacing the Na tracer re-calibration by data reconciliation., bringing a double benefit. First, they don't need the expensive yearly Na tracer re-calibration any more. Secondly, they re-calibrate their orifices on a higher frequency on basis of the reconciled values. This allows them to work permanently at full throughput.

Another typical use of data reconciliation in power plants relates to acceptance test-runs. After any upgrade of a plant, tests are usually performed to prove that the expected benefits are achieved. The expensive measurement devices that are normally used to perform those tests can be avoided thanks data reconciliation. For example, KKL has used this technique after replacement of their LP turbines and identified, at virtually no cost and weeks before the acceptance tests, that the turbine upgrade had increase their production by 46 MW in place of the predicted 23 MW.

Another experience reported by a Chemical company relates to their gas bill. Thanks to the use VALI they have been able to detect a measurement error leading to an over-estimation of a few percent of their gas import. A few percent may seem of low importance, but in this case, this single error was corresponding to an over-cost of 0.5 million DM per year.

3. CONCLUSION: THE FUTURE OF DATA RECONCILIATION

The future of data reconciliation is certainly in on-line applications. Raw process plant data are today available to places where their present quality is no more sufficient. To improve this quality, data reconciliation is the best solution.

Data reconciliation will more and more be coupled with (or will include) the determination of process performance indicators (fouling factors, efficiency losses, etc.). The operators will then be informed simultaneously of the performance of the measurement devices as well as the performance of the plant itself.

We also anticipate that using data reconciliation will become more and more common for acceptance test-runs, suppressing endless discussions on the way to correct the measurements in order to close the balances.

The coupling of data reconciliation with optimisation is another topic. In one hand, using reconciled data to optimise a plant seems to be the proper approach, while on the other hand, time constraints might prevent the use of closed-loop systems.

An area were some improvements have still to be made is the data reconciliation of plants which are not perfectly, or at least not sufficiently, steady, and in particular batch processes. Dynamic data reconciliation is still in the early stages of its development and no real commercial application is currently available.

However the widespread use of data reconciliation will have to over-come several obstacles. One of the major problems is certainly operator acceptance and more generally an acceptance problem at several stages in the companies. Many people will be, one way or the other, affected by the use of data reconciliation in the plant: from process control to maintenance, engineering and even research departments, the systematic use of data reconciliation must be accepted.

Sensor Failures Detection in Steam Condensers

F. Busson[**], A. Aïtouche[**], B. Ould Bouamama[*], M. Staroswiecki[*]

LAIL URA CNRS D1440
[*]Université des Sciences et Technologies de Lille
59655 Villeneuve d'Ascq Cedex, France
Email: Marcel.Staroswiecki@univ-lille1.fr.
Tél: (33) 3 20 43 45 65 Fax : (33) 3 20 33 71 89

[**]ERASM-HEI
13, rue de Toul
59046 Lille Cedex
Email: aitouche@hei.fr
Tél: (33) 3 28 38 48 58 Fax: (33) 3 8 38 48 04

Abstract: This paper is concerned with Fault Detection and Isolation (FDI) in industrial systems such as steam generator processes whose models are non linear. The method is based on Analytical Redundancy Relations. These relations are used to detect and isolate sensors faults. *Copyright © 1998 IFAC.*

Keywords: Fault Detection and Isolation, Bond Graph, Sensors Failures, Residuals, Steam Condensers

1-INTRODUCTION

The increasing importance of condensers in the industrial, current world is not to be demonstrated. It is an essential link of the water cycle currently used in the energetic carriage because of the abundance of this material on our planet and because of its interesting thermodynamic characteristics.
This paper is concerned with model-based sensor failure detection and isolation around the condenser. The interest of such an approach is quickly understandable, since sensors are vital: the system knowledge depends on sights which are given by them. A faded vision - for example a pressure sensor failure - can result in an important increase in the steam inflow, in order to compensate the apparent difference and finally can cause the condenser destruction.
The research has been carried out using a pilot steam generator process available for benchmark experiments (Pollart and Ould Bouamama, 1996). The model we use has been developed using the Bond Graph approach. The Bond Graph model is represented in a vectorial way showing the junction structure and the different fields (Thoma, 1991; Ould Bouamama *et al.*, 1997). The model is decomposed into 33 relations to which 11 measurement relations are added.

The aim of our work is to generate a set of residuals using the structural analysis technique (Staroswiecki and Declerck, 1989; Declerck and Staroswiecki, 1991). This approach consists in finding analytic redundancy relations which contain only known variables. Under normal operating conditions, those relations are satisfied and the residuals are near zero.
If some fault appears in their arguments, the absolute values of some residuals can increase in high proportion. By a binary sequence which codes the result of some decision procedure (0 if the residual is near zero, 1 else), we can theoretically detect which sensor is to be suspected. In the present application, we simulated bias and drift faults of temperature, pressure and level sensors. Our residuals are calculated under the Matlab environment from experimental data. They detect most of the simulated faults.

2-PROCESS DESCRIPTION AND MODELISATION

2.1-Process description

The condenser schematics is shown on figure 1. Steam enters at the left side, condenses on tubes filled with cooling water, falls to the floors and leaves from there to the receiver through control valves. There the water outflow (condensate) is controlled by three valves, in order to keep its level constant.

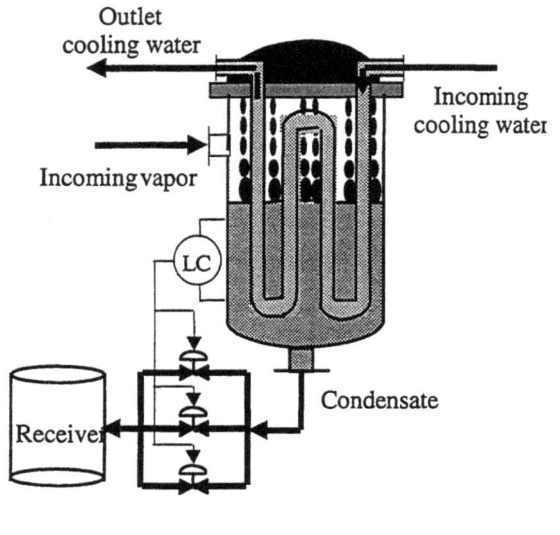

Fig. 1 : Condenser schematics.

2.2-Modelisation

The condenser modelisation has been presented in (Ould Bouamam *et al*, 1997), using a Bond Graph approach. In this paper we only recall the model - not the employed method.

The technological level of the model is represented by the word bond graph given by figure 2.

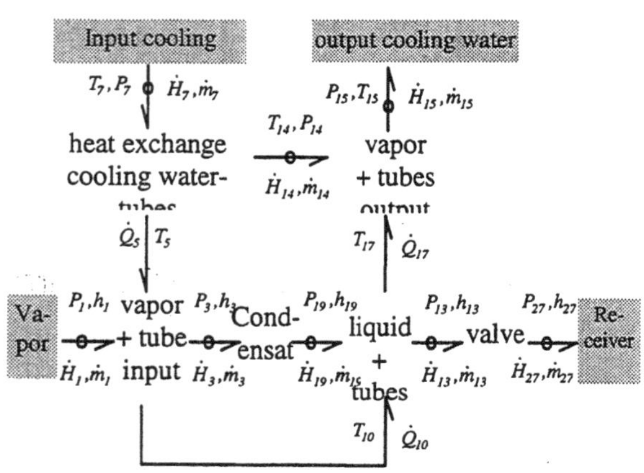

Fig. 2 : Condenser word bond graph.

Figure 3 shows the condenser pseudo Bond Graph model, with pressure and temperature, or specific enthalpy as efforts and mass and enthalpy flows as flows. The reasons for this choice are simpler conservation properties : mass flow is always conserved and enthalpy flow is conserved in a RECO (convection resistor), and obtained by a simple power balance in an HEXA (heat exchanger). This choice allows to reconstruct the other variables, such as the quality of steam (also called dryness fraction) in a two phases mixture and to use thermodynamic functions for simulation.

2.3-Equations

From this model we can deduce the relations between the internal variables of the system. We obtain the following equations:

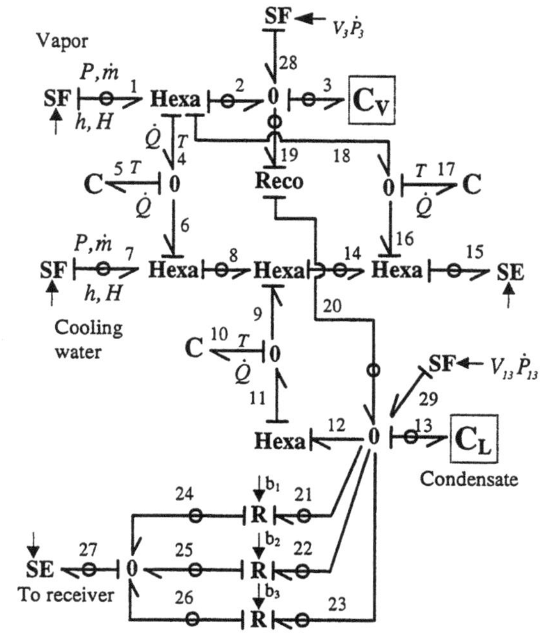

Fig. 3 : Condenser bond graph model.

Partial derivative of saturated steam density is evaluated from saturated steam tables:

$$\frac{\partial \rho_v}{\partial T} = 4.59*10^{-11}*T_{cv}^4 + 1.59159*10^{-9}*T_{cv}^3 + 8.0471*10^{-7}*T_{cv}^2 + 1.1346*10^{-5}*T_{cv} + 1.90366*10^{-4} \quad (1)$$

Saturated steam density and mass of steam are:

$$\rho_v = \frac{\int(\dot{M}_{ec} - \dot{M}_{ct})dt}{\left(\frac{\partial \rho_v}{\partial T}\right)*V_{cv}} + \rho_{v0} \quad (2)$$

$$M_{cv} = \rho_v * V_{cv} \quad (3)$$

Liquid specific enthalpy and steam specific enthalpy are evaluated from thermodynamic tables:

$$h_l = 7.355*10^{-9}*T_{cv}^4 + 1.54722*10^{-6}*T_{cv}^3 - 2.8662*10^{-4}*T_{cv}^2 + 4.20015*T_{cv} - 3.8679 \quad (4)$$

$$h_v = -740 * P_{cv}^2 - 17210 * P_{cv} + 2680000 \quad (5)$$

Liquid specific enthalpy is a function of the pressure:

$$h_l = -24.3*P_{cv}^4 + 8487*P_{cv}^3 - 11900*P_{cv}^2 - 99970*P_{cv} + 347000 \quad (6)$$

Steam saturation specific volume and water specific volume are evaluated from thermodynamic tables:

$$v_v = -5.3*10^{-5}*P_{cv}^5 + 0.0027*P_{cv}^4 - 0.032*P_{cv}^3 + 0.2498*P_{cv}^2 - 1.03*P_{cv} + 2.166 \quad (7)$$

$$v_L = -3.59*10^{-7}*P_{CV}^3 + 1.256*10^{-5}*P_{CV}^2 + 139*10^{-3} \quad (8)$$

The first part and the second part of the mixture equation are:

$$h_{cv} = h_v * X + h_l * (1 - X) \tag{9}$$

$$\frac{1}{\rho_v} = v_v * X + v_l * (1 - X) \tag{10}$$

Latent condensation is evaluated from thermodynamic tables as given by equation (11):

$$L_v = -3.4 * 10^{-11} * T_{cv}^4 + 4.421 * 10^{-11} * T_{cv}^3 - 1.54503 * 10^{-6} * T_{cv}^2 - 2.21406 * 10^{-3} * T_{cv} + 249716 * 10^6$$

The height of the condensing film is extracted from a Nusselt type formula:

$$e_x = \left(\frac{4 * \mu_l * (l - N_{cond}) * (T_{cv} - T_{vt}) * \lambda_l}{g * \rho_l * (\rho_l - \rho_v) * [L_v + 0.68 * c_{tv} * (T_{cv} - T_{vt})]} \right)^{1/4} \tag{12}$$

The relation between saturated pressure and the temperature is obtained from thermodynamic tables (equation 13):

$$P_{cv} = 1.268 * 10^{-22} * T_{cv}^{10} - 2.368 * 10^{-19} * T_{cv}^9 + 1.914 * 10^{-16} * T_{cv}^8 - 8.703 * 10^{-14} * T_{cv}^7 + 2.428 * 10^{-11} * T_{cv}^6 - 4.248 * 10^{-9} * T_{cv}^5 + 4.710 * 10^{-7} * T_{cv}^4 - 2.969 * 10^{-5} * T_{cv}^3 + 9.832 * 10^{-4} * T_{cv}^2 - 1.041 * 10^{-2} * T_{cv} + 2.048 * 10^{-2}$$

The entering steam temperature is:

$$T_{cv} = \frac{\begin{bmatrix} -c_v * M_{cv} * \dot{T}_{cv} + \dot{M}_{ec} * h_{ec} - \dot{M}_{ct} * h_{cv} - M_{ct} * \dot{h}_{cv} \\ + K_{vt} * (T_{vt1} + T_{vt2}) + \dot{M}_{ct} * L_v + M_{ct} * \dot{L}_v + \dot{P}_{cv} * V_{cv} \end{bmatrix}}{c_v * \dot{M}_{cv} + 2 * K_{vt}} \tag{14}$$

The pressure of the liquid is:

$$P_{cv} = \int \dot{P}_{cv} * dt + P_{cv0} \tag{15}$$

The specific enthalpy of the steam in the condenser is obtained from thermodynamic tables:

$$h_{cv} = -3 * 10^{-7} * T_{cv}^5 + 2.0766 * 10^{-4} * T_{cv}^4 - 7.0061 * 10^{-2} * T_{cv}^3 + 8.9607 * T_{cv}^2 + 1.18506 * 10^3 * T_{cv} + 2.5202 * 10^6 \tag{16}$$

The condensate flow is:

$$\dot{M}ct = \frac{\rho_l * (\rho_l - \rho_v) * g * e_x^3 * \pi * d_t * n_t}{3 * \mu_l} \tag{17}$$

The temperature of the incoming cooling tube, the temperature of the tube in the liquid, the temperature of outlet cooling tube, the average temperature of the tubes and the temperature of the outlet cooling water are given respectively by the following equations (18)-(22):

$$T_{vt1} = \frac{K_{vt} * T_{cv} + K_{tv} * T_{ee} - m_{vt1} * c_{tv} * \dot{T}_{vt1}}{K_{vt} + K_{tv}}$$

$$T_{lt} = \frac{K_{lt} * T_{cl} + K_{tl} * \left(1 - \frac{K_{tv}}{\dot{M}_{ee} * c_{ee}}\right) * T_{ee} + \frac{K_{tl} * K_{tv}}{\dot{M}_{ee} * c_{ee}} * T_{vt1} - m_{lt} * c_{tl} * \dot{T}_{lt}}{K_{lt} + K_{tl}}$$

$$T_{vt2} = \frac{1}{K_{vt} + K_{tv}} * \begin{bmatrix} K_{vt} * T_{cv} + K_{tv}^2 * \frac{\dot{M}_{ee} * c_{ee} - K_{tl}}{(\dot{M}_{ee} * c_{ee})^2} * T_{vt1} \\ + \frac{K_{tv} * (\dot{M}_{ee} * c_{ee} - K_{tv}) * (\dot{M}_{ee} * c_{ee} - K_{tl})}{(\dot{M}_{ee} * c_{ee})^2} * T_{ee} \\ + \frac{K_{tl} * K_{tv}}{\dot{M}_{ee} * c_{ee}} * T_{lt} - m_{vt2} * c_{tv} * \dot{T}_{vt2} \end{bmatrix}$$

$$T_{vt} = \frac{T_{vt1} + T_{vt2}}{2}$$

$$T_{se} = \frac{1}{\dot{M}_{ee} * c_{ee}} * \begin{bmatrix} -K_{tv} * T_{vt2} + \frac{(\dot{M}_{ee} * c_{ee} - K_{tl}) * (\dot{M}_{ee} * c_{ee} - K_{tv}) * (\dot{M}_{ee} * c_{ee} + K_{tv})}{(\dot{M}_{ee} * c_{ee})^2} * T_{ee} \\ -K_{tl} * \left(1 - \frac{K_{tv}}{\dot{M}_{ee} * c_{ee}}\right) * T_{lt} - K_{tv} * \frac{(\dot{M}_{ee} * c_{ee} - K_{tl}) * (\dot{M}_{ee} * c_{ee} - K_{tv})}{(\dot{M}_{ee} * c_{ee})^2} * T_{vt1} \end{bmatrix}$$

The mass of liquid in the condenser, the liquid level in the condenser and the liquid pressure in the condenser are:

$$M_l = \int (\dot{M}_{ct} - \dot{M}_{vd}) * dt + M_{l0} \tag{23}$$

$$N_{cond} = \frac{M_l}{\rho_l * S_c} \tag{24}$$

$$P_{cl} = N_{cond} * \frac{g * \rho_l}{10^5} + P_{cv} \tag{25}$$

The outlet liquid flow is:

$$\dot{M}_{vd} = \dot{M}_{ct} - \sum b_i * \sqrt{\rho_c * \frac{P_{cl} - P_{ls}}{K_{dv}}} \tag{26}$$

The volume of steam is:

$$V_{cv} = V - S_c * N_{cond} \tag{27}$$

The outlet liquid temperature and the specific enthalpy of the outlet liquid are given by equations (28) and (29):

$$T_{cl} = \frac{-M_l * c_l * \dot{T}_{cl} + \dot{M}_{ct} * h_l + M_{ct} * c_l * \dot{T}_{cv} + \frac{M_{ct}}{\rho_c} * \dot{P}_{cl} + K_{lt} * T_{lt}}{c_l * \left(\dot{M}_{ct} + \frac{K_{lt}}{c_l}\right)}$$

$$h_{dv} = c_l * T_{cl}$$

The control law of the 3 output valves is defined by :

$$\begin{aligned} &\textstyle\sum b_i = 1 \text{ if } N_{cond} > 1{,}1 * niv \\ &\textstyle\sum b_i = 2 \text{ if } N_{cond} > 1.3 * niv \\ &\textstyle\sum b_i = 3 \text{ if } N_{cond} > 1.5 * niv \\ &\textstyle\sum bi = 0 \text{ if not} \end{aligned} \tag{30}$$

The mass of incoming tubes in the steam, the mass of outlet tubes in the steam and the mass of tubes in the liquid are given respectively by:

$$m_{vt1} = (l - N_{cond}) * \pi * d_t * e * \rho_t * n_t \tag{31}$$

$$m_{vt2} = (l - N_{cond}) * \pi * d_t * e * \rho_t * n_t \tag{32}$$

$$m_{lt} = (2 * N_{cond} + d) * \pi * d_t * e * \rho_t * n_t \tag{33}$$

The following equations describe the links between the sensors and the measured variables.

niv = water level defined by the operator (34)

$$N_{cond}=L_{18} \tag{35}$$

$$N_{cond}=L_{19} \tag{36}$$

$$T_{ee}=T_{21} \tag{37}$$

$$T_{se}=T_{22} \tag{38}$$

$$P_{cv}=P_{16} \tag{39}$$

$$T_{cv}=T_{17} \tag{40}$$

$$T_{cl}=T_{20} \tag{41}$$

$$\dot{M}_{ee}=F_{23} \tag{42}$$

$$\sum b_i=b_1+b_2+b_3 \tag{43}$$

where b_i = 1 if the valve i is open and equal to 0 if not.

$$P_{ls}=P_{27} \tag{44}$$

3-RESIDUALS GENERATION BY STRUCTURAL ANALYSIS

Structural analysis only makes use of information about the constraints which link the variables. This is interesting for large scale systems described by a wide variety of types of relationships: qualitative or quantitative, linear or non linear, static or dynamic.

The system is described by a set of constraints F which apply to a set of variables Z, known or unknown. The subset C of known variables contains the control variables, the variables whose value is measured by a sensor and the supervised parameters. The subset X of unknown variables contains the remaining variables:

$Z=X\cup C$.

Let $F=(f_1,f_2,\ldots,f_n)$ be the set of constraints and $Z=(z_1,z_2,\ldots,z_p)$ be the set of variables.

The structure is described by the following binary relation: $s:F\times Z\rightarrow\{0,1\}$:

$(f_i,z_j)\rightarrow s(f_i,z_j)=1$ if and only if the constraint f_i applies to the variable z_j, $s(f_i,z_j)=0$ otherwise.

The structure leads to a digraph whose incidence matrix represents links between variables and constraints. The problem of finding the redundancy relations which can be obtained in a given system is equivalent to the problem of finding the over-determined sub-systems in the structure of the system. An over-determined sub-system is characterised by a complete matching with respect to X only, such that the unknown variables can be computed in different ways using a subset of the constraints and the known variables C (Staroswiecki and Declerck, 1989 ; .Declerck and Staroswiecki, 1991). In order to take into account the fact that some variables cannot be computed from some relations (due to non linearities or derivations), constraints on the matching are introduced into the structure. In the steam condenser application, the structural analysis of the model showed the existence of 12 residuals.

The first residual expresses the hardware redundancy of the L_{18} and L_{19} sensors:

$$r_1=L_{18}-L_{19} \tag{45}$$

Residual 2 is the derivative of the liquid level measured by L_{19} with time, taken from (23), (24), (25) and (26) equations:

$$r_2=\dot{L}_{19}-\frac{\left(P_{27}-P_{16}-L_{19}*\frac{g*\rho_l}{10^5}\right)^2*\rho_c*\sum b_i}{K_{dv}^2*\rho_l*S_c} \tag{46}$$

Residual 3 is the difference between the P_{16} value and the theoretical expression taken from the thermodynamic table and the T_{17} value:

$$\begin{aligned}r_3=&1.268*10^{-22}*T_{17}^{10}-2.368*10^{-19}*T_{17}^{9}+1.914*10^{-16}*T_{17}^{8}\\&-8.703*10^{-14}*T_{17}^{7}+2.428*10^{-11}*T_{17}^{6}-4.248*10^{-9}*T_{17}^{5}\\&+4.710*10^{-7}*T_{17}^{4}-2.969*10^{-5}*T_{17}^{3}+9.832*10^{-4}*T_{17}^{2}\\&-1.041*10^{-2}*T_{17}+2.048*10^{-2}-P_{16}\end{aligned} \tag{47}$$

Residual 4 is the difference between the T_{22} value and the theoretical expression of the temperature of outlet cooling water. This expression is taken from equations (18), (19), (20), (22), (31), (32) and (33); we obtain equation(48):

$$\begin{aligned}r_4=&T_{22}-\frac{1}{F_{23}*c_{ee}}*(-K_{tv}*T_{vt2}-K_{tv}*\frac{(F_{23}*c_{ee}-K_{tl})*(F_{23}*c_{ee}-K_{tv})}{(F_{23}*c_{ee})^2}\\&+T_{vt1}\frac{(F_{23}*c_{ee}-K_{tl})*(F_{23}*c_{ee}-K_{tv})*(F_{23}*c_{ee}+K_{tv})}{(F_{23}*c_{ee})^2}*\\&*T_{21}-K_{tl}*\left(1-\frac{K_{tv}}{F_{23}*c_{ee}}\right)*T_{lt})\end{aligned}$$

Residual 5 is the difference between the T_{20} value and the theoretical expression of the temperature of outlet liquid. This expression (49) is deduced from (5), (6), (7), (8), (9), (10), (11), (12), (16), (17), (21), (25), (26), (28):

$$r_5=T_{20}-\frac{-M_l*c_l*\dot{T}_{20}+\dot{M}_{ct}*h_l+M_{ct}*c_l*\dot{T}_{17}+\frac{M_{ct}}{\rho_c}*\dot{P}_{cl}+K_{lt}*T_{lt}}{c_l*\left(\dot{M}_{ct}+\frac{K_{lt}}{c_l}\right)}$$

Residual 6 is deduced from the valves control law:

$$r_6=\sum b_i-f(L_{19},niv) \tag{50}$$

Residual 7 is the difference between the P_{11} value and the theoretical expression of the pressure of the outlet liquid. In this case, we supposed that $P_{ls}\approx P_{cl}$:

$$r_7=P_{27}-P_{16}-\rho_l*g*L_{18} \tag{51}$$

Residual 8 is the L_{18} value and the water level defined by the operator:

$$r_8=niv-L_{18} \tag{52}$$

Residual 9 is the difference between the L_{19} value and the water level defined by the operator:

$$r_9=niv-L_{19} \tag{53}$$

Residual 10 considers that h_{dv} is constant, this is the case as long as the system operates around the same operating point.

This expression is deduced from thermodynamic tables.

$$r_{10}=T_{20}-\frac{h_{dv}-0.098343401*P_{11}+0.29025619}{-0.00024118*P_{11}+4.181057148} \tag{54}$$

Residual 11 is nearly the same as residual 4, but its sensitivity to the sensor F_{23} is largely better.

$$\begin{aligned}r_{11}=&F_{23}-\frac{1}{T_{22}*c_{ee}}*(-K_{tv}*T_{vt2}-K_{tl}*\left(1-\frac{K_{tv}}{F_{23}*c_{ee}}\right)*T_{lt}\\&+\frac{(F_{23}*c_{ee}-K_{tl})*(F_{23}*c_{ee}-K_{tv})*(F_{23}*c_{ee}+K_{tv})}{(F_{23}*c_{ee})^2}*T_{21}\\&-K_{tv}*\frac{(F_{23}*c_{ee}-K_{tl})*(F_{23}*c_{ee}-K_{tv})}{(F_{23}*c_{ee})^2}*T_{vt1}\end{aligned} \tag{55}$$

Residual 12 considers that T_{vt1} is constant, this is the case as long as we keep the same operating point.

$$r_{12} = T_{21} - \frac{(K_{vt} + K_{tv}) * T_{vt1} - K_{vt} * T_{17} + m_{vt1} * c_{tv} * \dot{T}_{vt1}}{K_{tv}} \qquad (56)$$

However, in real situations, the operating point may fluctuate, according to the operator's will.
So in order to completely determine the system we have to do experiments using a selection of different operating points.
Analysing the equations of the model, together with the expressions for different residuals, one can deduce the effects of faults on the residuals at steady state. They are summarised by means of the matrix shown in Table 1. A "1" entry in the *i*th row and the *j*th column of the table indicates that the residual r_i is sensitive to the *j*th fault.

Table 1: Theoretical signatures

	L18	L19	P16	P11	T17	T21	T20	F23	T22
r1	1	1							
r2		1	1	1					
r3			1		1				
r4		1			1	1	1	1	1
r5		1	1	1	1	1	1	1	
r6		1							
r7	1		1	1					
r8	1								
r9		1							
r10				1			1		
r11		1			1	1	1	1	1
r12					1	1			

4- RESULTS

To analyse the sensitivity of the residuals to different faults, a number of simulations have been performed. Two situations were considered separately: the case of abrupt faults, which can be simulated by step-like additive signals, and incipient faults, which are simulated by ramp-like additive signals.

We first simulate a bias on each sensor. Experimental data of a condenser without any fault were used. We have added a bias on each sensor average value : 10% from point 30 to point 50, +50% from point 60 to point 80, and +100% from point 90 to point 100.

The residuals are calculated using the Matlab-Simulink software. These residuals were normalised. The decision procedure consists to compare the current value with the fixed threshold. If this value exceeds the threshold, we can consider that a fault has occurred. For example we show the reaction of the residuals 1, 7 and 8 in response to a L_{18} sensor fault.

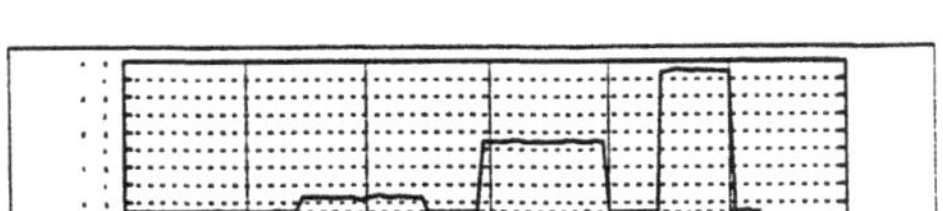

Fig. 4: Evolution of residual 1

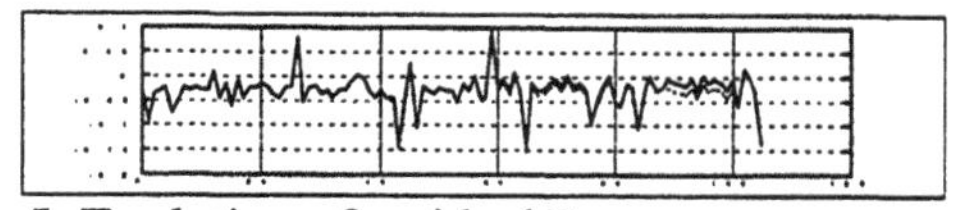

Fig. 5: Evolution of residual 7

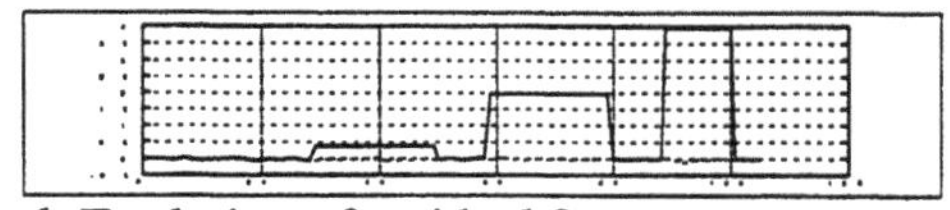

Fig. 6: Evolution of residual 8

The residual 1 and 8 have a good reaction which can be simply detected. The faults were detected by the test as soon as they appeared. As we can see on figure 5, the residual 7 is insensitive to the fault.

We now simulate a drift on the L_{19} sensor from point 50 to point 100. The residuals 1, 2, 6 and 9 are:

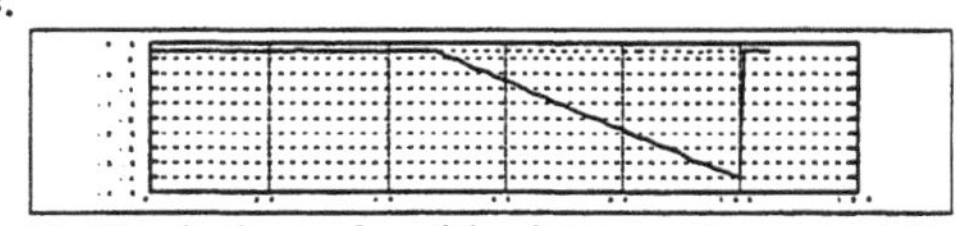

Fig. 7: Evolution of residual 1

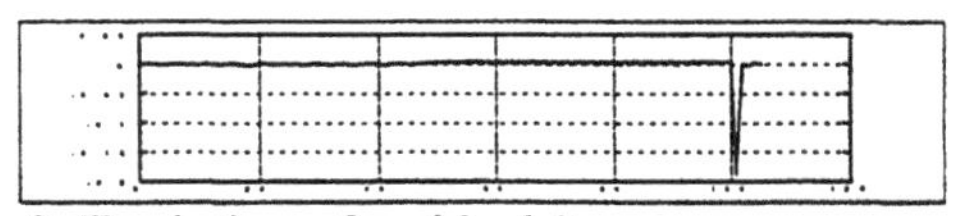

Fig. 8: Evolution of residual 2

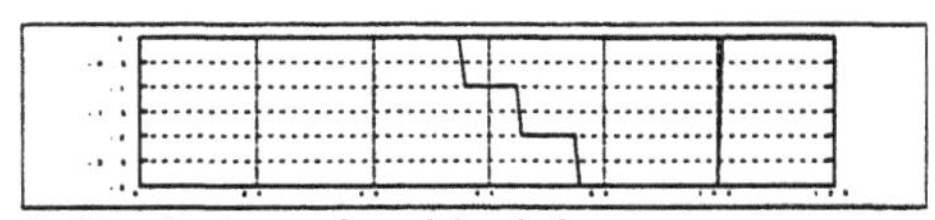

Fig. 9: Evolution of residual 6

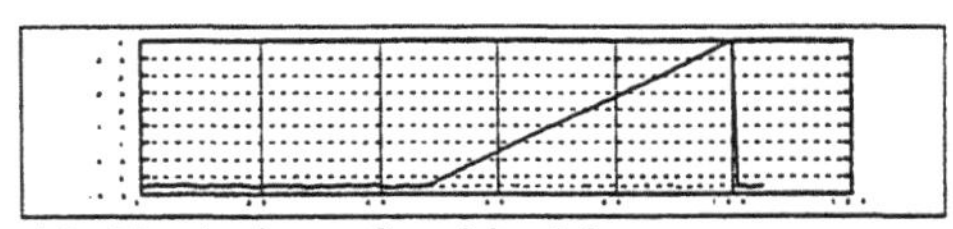

Fig. 10: Evolution of residual 9

Residual 1 follows correctly the sensor drift.
Residual 2 has a reaction only if the sensor variation is high. It cannot be used for an early drift detection.
Residual 6 has a good reaction. As we can notice, it can only take discrete values, because of the valves states (open or close).
Residual 9 like residual 1 follows correctly the sensor drift.

Resulting from these experiments, we obtain actual failure signatures for each sensor. Table 2 gives the experimental signatures for the nine considered sensors.

table 2: Experimental signatures

	L18	L19	P16	P11	T17	T21	T20	F23	T22
r1	1	1							
r3			1		1				
r4						1			1
r6		1							
r7			1	1					
r8	1								
r9		1							
r10							1		
r11						1		1	1
r12					1	1			

This paper illustrates the application of Analytical Redundancy to an industrial non linear system: the steam condenser. Residuals were generated through the structural analysis of the system equations, issued from a Bong Graph modelisation. The detection method was very simple and however very efficient in our case. We normalized the residuals and fixed a detection threshold. If the value of the residual exceeds the threshold, we consider that it contains some faulty element. The normalisation and the threshold depend on the operating point.

The analytical redundancy relations can be complex, and the sensitivity of each of them is difficult to determine. Nevertheless it seems that the faults can be correctly detected in our case.

Different extensions should now be considered, taking into account the variations of the system parameters (in particular h_{dv} and T_{vt1}) with the operating point, and the eventual presence of simultaneous faults.

REFERENCES

Declerck, P. and M. Staroswiecki (1991), Characterisation of the canonical components of a structural graph. application to fault detection in large scale industrial plants. ECC'91, pp. 289-303, Grenoble (France).

Ould Bouamama, B., J. U. Thoma and J.P. Cassar (1997), Bond graph modelisation of steam condensers. In: *IEEE International Conference on Systems, Man, and Cybernetic,* pp. 490-494, Orlando (USA).

Pollart, M. and B. Ould Bouamama (1996), Générateur de vapeur, mise au point et interconnexion des sous ensembles mécanique, informatique, commande, modélisation. *Technical report*, LAIL, Lille (France).

Staroswiecki, M. and P. Declerck (1989), Analytical redundancy in nonlinear interconnected systems by means of structural analysis. In: *IFAC Symposium in Advanced Information Processing in Automatic control.* pp. 23-27, Nancy (France).

Thoma, J.U. (1991). Simulation by bond graphs. *Springer Verlag*, p.180.

NOMENCLATURE

T_{cv} : Temperature entering steam (°C)
ρ_v : Mass density of steam (kg/m^3)
$\dot{M}_{ec}$: Entering steam flow (kg/s)
$\dot{M}_{cl}$: condensate flow(kg/s)
V_{cv} : Volume of steam (m^3)
M_{cv} : Mass of steam (kg)
h_l : Specific enthalpy of the liquid phase (J/kg)
h_v : Specific enthalpy of the steam phase (J/kg)
P_{cv} : Pressure of steam in the condenser (bar)
v_v : Specific volume of steam phase (m^3/kg)
v_l : Specific volume of liquid phase (m^3/kg)
h_{cv} : Specific enthalpy of the steam (J/kg)
X : Title of steam
L_v : Latent condensation heat (J/kg)
e_x : Height of the condensing film (m)
T_{vt} : Average temperature of the tubes (°C)
T_{ee} : Temperature of the incoming cooling water (°C)
N_{cond} : Level of liquid in the condensor (m)
T_{vt1} : Temperature of the incoming cooling tube (°C)
T_{lt} : Temperature of the tube in the liquid (°C)
$\dot{M}_{er}$: Mass flow of cooling water (kg/s)
T_{vt2} : Temperature of the outlet cooling tube (°C)
T_{se} : Temperature of the outlet cooling water (°C)
M_l : Mass of liquid in the condensor (kg)
$\dot{M}_{vd}$: Flow of outlet liquid (kg/s)
P_{cl} : Pressure of liquid in the condensor (bar)
$\sum b_i$: Coefficient of opened valves (1, 2 or 3)
T_{cl} : Temperature of outlet liquid (°C)
h_{dv} : Specific enthalpy of outlet liquid (J/kg)
P_{ls} : Pressure of outlet liquid (bar)
m_{vt1} : Mass of the incoming tubes in the steam (kg)
m_{vt2} : Mass of the outlet tubes in the steam (kg)
m_{lt} : Mass of the tubes in the liquid (kg)

The constants are:
g : Gravity (m/s^2)
K_{tv} : Global heat transfer coefficient from metal tube (in vapor section) to cooling water (W/(m^2.°C))
K_{vt} : Global heat transfer coefficient from the vapor to the metal tube (W/(m^2.°C))
K_{lt} : Global heat transfer coefficient from the condensate to the metal tube (W/(m^2.°C))
K_{tl} : Global heat transfer coefficient from metal tube (in liquid section) to cooling water (W/(m^2.°C))
c_{tv} : Specific heat of the metal tube in vapor section (J/(kg.°C))
c_{tl} : Specific heat of the metal tube in liquid section (J/(kg.°C))
c_l : Specific heat of the liquid (J/(kg.°C))
c_{ee} : Specific heat of the cooling water (J/(kg.°C))
d : Condenser's diameter (m)
S_c : Horizontal section of the condenser (m^2)
a : Coefficient of charge losing for the outlet condensate
V : Total volume of the condenser (m^3)
l : Length of the condensor (m)
ρ_{v0} : First mass density of the steam (kg/m^3)
h_{ec} : Specific enthalpy of the incoming vapor (J/kg)
λ_l : Thermal conductibility of the condensate (W/(m.°C))
μ_l : Dynamic viscosity of the liquid (kg/(m.s))
ρ_l : Mass density of the liquid (kg/m^3)
d_t : Tubes' diameter (m)
n_t : Number of tubes
c_v : Specific heat of the steam (J/(kg.°C))
M_{ct0} : First mass of the condensate (kg)
ρ_t : Mass density of the metal tube(kg/m^3)
e : Height of tubes (m)
ρ_c : Mass density of the condensate (kg/m^3)
K_{dv} : Coefficient of charge loosing of the valves (Pa/(m^3/s))

DETECTION AND ISOLATION OF PROCESS FAULTS IN AN INDUSTRIAL BOILER

B. Ould Bouamama M. Staroswiecki and R. Litwak
Laboratoire d'Automatique et d'Informatique Industrielle de Lille URA CNRS D1440
USTL, 59655 Villeneuve d'Ascq cedex, FRANCE
belkacem.bouamama@.univ-lille1.fr

Abstract: The present paper deals with modeling and monitoring of a steam generator pilot process. Based on the structural analysis of the obtained nonlinear model, the analytical redundancy relations are determined, and the isolation and detection properties are evaluated. In order to test the proposed monitoring system, two faults scenari have been realized on the actual process: a water leak in the boiler, and a blocked output of the steam pipe. The calculated residuals show out to have a good reaction to all the process faults listed in the FDI system specifications. *Copyright © 1998 IFAC*

Keywords: Redundancy relations, Fault Detection and Isolation, Steam generators, structural analysis, residuals.

INTRODUCTION

Steam generators are widely used in chemical industry. The complexity of such plants makes them very sensitive to faults. The improvement of their safety essentially rests on fault detection and isolation procedures, which mainly consist in the comparison of the actual behavior of the system with reference behaviors describing the normal operation (for fault detection) or different kinds of faulty ones (for fault isolation). Different approaches for the design of FDI procedures have been developed, depending on the kind of knowledge which is used to describe the plant operation. One of them rests on the use of quantitative dynamic models, which lead to the determination of Analytic Redundancy Relations, allowing the real time monitoring of this type of process (Staroswiecki, 1989).
The present paper deals with modeling and monitoring a part of a steam generator pilot process (Pollart and Ould Bouamama, 1997).
In the first part of the paper, the specifications of the global performances of the fault detection and isolation system will be defined to determine the cause of the faults taken into account. For each cause, a knowledge relation which involves a known variable or parameter is associated in the system model.
The second part deals with the equations describing the process thermodynamics. The obtained model is under nonlinear form.
In the third part, based on the structural analysis of the obtained model, the analytical redundancy relations are determined, and the isolation and detection properties are evaluated.
Finally in the last part, in order to test the proposed monitoring system, two process faults scenari have been realized on the installation. The residuals are calculated using the Matlab Simulink software. They show out to have a good reaction to all the process faults listed in the FDI system specifications.

1. SPECIFICATION OF THE GLOBAL PERFORMANCES OF THE FAULT DETECTION AND ISOLATION SYSTEM

The FDI system is aimed at the detection and isolation of process failures in the steam generator. The water level in the boiler is important since the water absorbs the energy from the heating device:

insufficient water level could result in the boiler destruction, with catastrophic consequences (in the nuclear process for example the primary and the secondary circuit could mix). Furthermore the steam flow controls the whole electrical subsystem. This is why a water leak in the boiler, and a blocked output of the steam pipe have been chosen as our FDI specifications. Both faults are process faults.
Taking into account the different closed loop controls, the validity of the measurements in the steam generator is also of primary importance, since a sensor failures might have serious consequences. Sensors failures are the second element of the specifications we consider.

2. PROCESS DESCRIPTION AND MODELING

2.1 Description of the pilot process

The test plant represented by the general scheme in figure 1, is designed to be a scale-model of part of a power station.
It is a complex non linear system, which is mainly constituted of four subsystems: a receiver tank with the supply system, a boiler heated by a 60 kW resistor, a steam flow system and a complex condenser coupled with a heat exchanger.
The present work is only concerned with a part of this installation which is constituted of the boiler, the heating resistor, the feed water supply system (which consists of two pumps and the pipe) and the steam flow system. Sensors failures detection and isolation in the condenser - heat exchanger subsystem has also been implemented. The design procedure and some experimental results are reported in (Busson, *et al.*,1998).
In the following, mass flows (kg/s), temperatures (°c), pressures (bar), levels (m) and masses (kg) will be respectively represented by the letters *F, T, P, M* and *L*. The thermal flow (j/s) transferred by convection and by conduction will be indicated respectively by $\dot{H}$ and $\dot{E}$. The letters with a numerical suffix represent the sensor outputs (as indicated in figure 1) and the letters with a literal suffix indicate the variables of the model. For example, P7 is the output pressure sensor and P_{GV} the corresponding model variable.

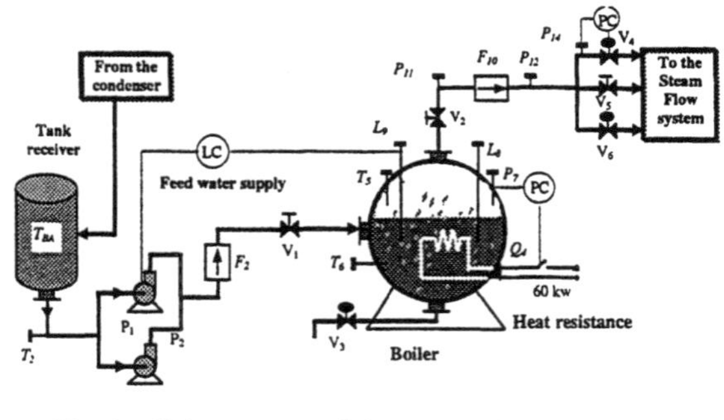

Fig.1 Schematics of the studied process

The feed water flow F_{AL} taken from the receiver is pressurized via the feed pump which is controlled by a relay to maintain a constant water level inside the steam generator. The heat power Q_{TH} is determined based on the available accumulator pressure P_{GV} . The expansion of generated steam is realized by three valves in parallel connection. In figure 1, PC and LC represent the pressure and the level controllers.

2.2. Process model

♦ *The feed water system*

It consists of a pump, a pipe and a tank receiver. The mass flow refueled by the pump F_{AL} is determined as the intersection between the pump and the pipe characteristics (F_{PA} and F_{AL}) which are given by the following equations:

$$\begin{cases} \dfrac{F_{PA}}{\rho_{AL}} = \left(-8{,}4948.0^{-10}\Delta P_{PA} + 9{,}722 * 10^{-4}\right) b_1 \\ \dfrac{F_{AL}}{\rho_{AL}} = \sqrt{\dfrac{(P_{PA} - P_{GV}).10^5}{K_D(z_{V1})}} \end{cases} \quad (1)$$

ρ_{AL} is the water density equal to 1000kg/m^3. ΔP_{PA} is the difference between the output pressure P_{PA} and the input pressure P_{SB} of the pump:

$$\Delta P_{PA} = (P_{PA} - P_{SB}).10^5$$

$K_D(z_{V1})$ is the flow nonlinear characteristic of the pipe identified experimentally and depending on valve V1 position Z_{V1}. b_1 is a Boolean variable based on he reference (L_{GV_ref}) and the actual (L_{GV}) water level inside the boiler. It depends also on the dead zone Δ of the "on-off" LC controller.

$$b_1 = F(sign(L_{GV\_ref} - L_{GV}), \Delta) \quad (2)$$

L_{GV} is based on the mixture quality X (determined by equation (16)), on the fluid mass m_{GV} and on the geometry of the steam generator :

$$L_{GV} = Fg(M_{GV}.(1-X)) \quad (3)$$

We assume that heat losses in the feed water system are negligible because the temperature is ambient. The enthalpy flow convected by the feed water at T_{AL} is given by:

$$\dot{H}_{AL} = F_{AL} h_{AL} = 0{,}420 F_{AL} T_{AL} \quad (4)$$

h_{AL} is the water specific enthalpy (j/kg)

♦ *The steam generator*

We assume that the fluid in the steam generator exits as a saturated, homogeneous mixture of vapor and

liquid at uniform pressure P_{GV}, specific enthalpy h_{GV}, mass M_{GV}, and total energy $\dot{H}_{GV}$. Furthermore, the boiler is bad calorifuged, thus, some heat losses exist.
The mass and energy conservation are given by:

$$\frac{d}{dt}(M_{GV}) = F_{AL} - F_{VD} \quad (5)$$

$$\dot{H}_{GV} = Q_{TH} + \dot{H}_{AL} + \dot{E}_W - \dot{H}_{VD} - \dot{E}_{PG} \quad (6)$$

Where F_{VD} is the mass steam flow through valve V1, determined by:

$$F_{VD}^2 = K_D(z_{V2}).\frac{(P_{GV} - P_{VD}).P_{VD}}{T_{GV}}. \quad (7)$$

in equation (7) P_{VD} is the valve V1 downstream pressure. This variable is imposed by the steam flow system. T_{GV} is the saturated temperature in the steam generator determined by the thermodynamic functions of pressure :

$$T_{GV} = -0{,}4594.P_{GV}^2 + 12{,}7243.P_{GV} + 99{,}003 \quad (8)$$

$\dot{H}_{VD}$ represent the enthalpy flow convected out of the steam generator to the steam flow system.

$$\dot{H}_{VD} = F_{VD} h_{GV} \quad (9)$$

$K_D(z_{V2})$ is the flow nonlinear characteristic of the steam pipe depending on valve V2 position Z_{V2}.
The heat power delivered by the thermal resistance is given by:

$$Q_{TH} = b_2 Q_4 = b_2 6000. \quad (10)$$

b_2 is a Boolean variable based on the reference (P_{GV_ref}), the actual (P_{GV}) boiler pressure and the dead zone Δ of the "on-off" PC controller.

$$b_2 = F(sign(P_{GV\_ref} - P_{GV}), \Delta) \quad (11)$$

The term $\dot{E}_w$ is the thermodynamic work done in pressurizing the volume of generator V_{GV}:

$$\dot{E}_W = V_{GV}.10^5 \frac{dP_{GV}}{dt}. \quad (12)$$

♦ *Heat losses*

The temperature (T_{MG}) stored by the metal can be calculated from the following equation:

$$\rho_{MG} V_{MG} C_{MG} \frac{dT_{MG}}{dt} = \frac{\dot{E}_{PG} - \dot{E}_{EX}}{\rho_{MG} V_{MG} C_{MG}} \quad (13)$$

where

$$\dot{E}_{PG} = K_{GM}(T_{GV} - T_{MG}), \quad (14)$$

and

$$\dot{E}_{EX} = K_{EX}(T_{MG} - T_{EX}) \quad (15)$$

are the energy flows transferred by conduction respectively to the metal and to the atmosphere. K_{GM} and K_{EX} represent the corresponding global heat transfer. C_{MG} , ρ_{MG} and V_{MG} designate respectively the specific heat capacity, the density and the volume of the steam generator body. T_{EX} is the given ambient temperature.

♦ *Mixture equation*

The pressure P_{GV} and the steam quality X (dryness fraction) in the boiler are determined solving the mixture equation:

$$\begin{cases} h_{GV} = \dfrac{H_{GV}}{M_{GV}} = h_v(P_{GV})X + h_L(P_{GV})(1-X) \\ v_{GV} = \dfrac{V_{GV}}{M_{GV}} = v_v(P_{GV})X + v_L(P_{GV})(1-X) \end{cases} \quad (16)$$

where v_L and v_V represent water and steam saturation specific volumes and h_l and h_V specific enthalpy. All are thermodynamic functions of pressure:

$$h_v(P_{GV}) = -0{,}74P_{GV}^2 - 17{,}21P_{GV} + 2680$$

$$h_l(P_{GV}) = -0{,}0243P_{GV}^4 + 0{,}8487P_{GV}^3 - 11{,}9P_{GV}^2 - 99.97P_{GV} + 347$$

$$v_v(P_{GV}) = -5{,}3.10^{-5}P_{GV}^5 + 0{,}00207P_{GV}^4 - 0.032P_{GV}^3 + 0{,}2498P_{GV}^2 - 1{,}03P_{GV} + 2{,}166$$

$$v_l(P_{GV}) = -3{,}59.10^{-7}P_{GV}^3 + 1{,}24.10^{-5}P_{GV}^2 + 1{,}039.10^{-3}$$

3. RESIDUALS GENERATION BY STRUCTURAL ANALYSIS

The structural analysis only keeps information about constraints acting on variables. This method is interesting for large scale systems described by a great variety of types of relationships: qualitative or quantitative, linear or non linear, static or dynamic. The theoretical basis and how it is used for the FDI system design are given in (Staroswiecki,1989), (Declerck,1991).
The analytical redundancy approach consists in finding relations between the known variables (inputs-outputs) of the system. These relations are satisfied if the actual behavior of the system is coherent with the reference behavior, not satisfied otherwise. Many approaches develop the analytical redundancy relations for linear models and additive faults (Patton,1989]. They need some linearization around given operating points and are not well

adapted to complex non linear systems, where the faults are deviations of process parameters. Structural analysis easily allows to take into account system non linearities, and non additive faults. It exhibits the residuals computation scheme, which results from the elimination of the unknown variables from the system equations using elimination theory. The system equations describe on one hand the constraints which link the process variables and on another hand the knowledge we have about these variables (the measurements).

- *Constraint relations*

$$\text{RT1}: \ T_{GV} = -0{,}4594 P_{GV}{}^2 + 12{,}7243 P_{GV} + 99{,}003$$

$$\text{RP1}: \ h_{GV} = h_V X + h_L(1-X)$$

$$\text{RP2}: \ v_{GV} = v_V X + v_L(1-X)$$

$$\text{RP3}: \ h_L = -2{,}43.10^{-2} P_{GV}{}^4 + 0{,}8487 P_{GV}{}^3 - 11{,}9 P_{GV}{}^2 + 99{,}97 P_{GV} + 347$$

$$\text{RP4}: \ h_V = -0{,}74 P_{GV}{}^2 + 17{,}21 P_{GV} + 2680$$

$$RP5: \ v_L = -3{,}59.10^{-7} P_{GV}{}^2 + 1{,}2456.10^{-5} P_{GV} + 1{,}039.10^{-3}$$

$$RP6: v_V = -5{,}3.10^{-5} P_{GV}{}^5 + 2.07.10^{-3} P_{GV}{}^4 - 3{,}2.10^{-2} P_{GV}{}^3 + 0{,}2498 P_{GV}{}^2 - 1{,}03 P_{GV} + 2{,}166$$

$$\text{RP7}: \ v_{GV} = \frac{V_{GV}}{M_{GV}}$$

$$\text{RP8}: \ h_{GV} = \frac{H_{GV}}{M_{GV}}$$

$$\text{RP9}: \ F_{VD}{}^2 = K_D(z_{V2}).\frac{(P_{GV} - P_{VD}).P_{VD}}{T_{GV}}$$

$$\text{RM1}: \ M_{GV} = \int \dot{M}_{GV} dt + M_{GV}(0)$$

$$\text{RM2}: \ F_{GV} = F_{AL} - F_{VD}$$

$$\text{RE1}: \ H_{GV} = \int \dot{H}_{GV} dt + H_{GV}(0)$$

$$\text{RE2}: \ \dot{H}_{GV} = Q_{TH} + \dot{H}_{AL} + \dot{E}_W - \dot{E}_{PG} - \dot{H}_{VD}$$

$$\text{RE3}: \ \dot{H}_{AL} = 0.42 F_{AL} T_{AL}$$

$$\text{RE4}: \ \dot{H}_{VD} = F_{VD} h_{GV}$$

$$\text{RE5}: \ \dot{E}_{PG} = K_{GM}(T_{GV} - T_{MG})$$

$$\text{RE6} \quad \rho_{MG} V_{MG} C_{MG} \dot{T}_{MG} + T_{MG}(K_{EX} + K_{GM}) = K_{GM} T_{GV} + K_{EX} T_{EX}$$

$$\text{RE7}: \ T_{MG} = \int \dot{T}_{MG} dt + T_{MG}(0)$$

- *Knowledge relations*

M1 :	$T_2 = T_{AL}$	(T_2, T_{AL}°C)
M2 :	$\frac{F_3}{3600} = F_{AL}$	(F_3 kg/h, F_{AL} kg/s)
M3 :	$Q_4.1000 = Q_{TH}$	(Q_4 kW, Q_{TH} j/s)
M4 :	$T_5 = T_{GV}$	(T_5, T_{GV}°C
M5 :	$T_6 = T_{GV}$	(T_6, T_{GV}°C)
M6 :	$P_7 = P_{GV}$	(P_7, P_{GV} bar)
M7:	$\frac{F_{10}}{3600} = F_{VD}$	(F_{10} kg/h, F_{VD} kg/s)
M8:	$P_{11} = P_{GV}$	(P_{11}, P_{GV} bar)
M9:	$P_{12} = P_{VD}$	(P_{12}, P_{VD} bar)
M10:	$P_{14} = P_{VD}$	(P_{14}, P_{VD} bar)

4. ANALYTICAL REDUNDANCY RELATIONS

The system is described by a set of *F* variables which can be split into known variables (measurements *C*) and unknowns ones *X:*

$$F = \{RT1, RP1\ RP2, RP3, RP4, RP5, RP6, RP7, RP8\ RP9, RM1, RM2, RE1, RE2, RE3\, RE4, RE5, RE6, RE7, RE8, M1, M2, M3, M4, M5, M6\ M7, M8, M9, M10\}$$

$$X = \{P_{GV}, T_{GV}, h_{GV}, h_V, X, h_L, v_V, v_L, v_{GV}, M_{GV}, P_{VD}, F_{VD}, H_{GV}, F_{VD}, H_{GV}, F_{GV}, F_{AL}, \dot{H}_{GV}, Q_{TH}, \dot{H}_{AL}, \dot{E}_{PG}, \dot{H}_{VD}, T_{AL}, T_{MG}, \dot{T}_{MG}, \dot{E}_W\}$$

$$C = \{T_2, F_3, Q_4, T_5, T_6, P_7, F_{10}, P_{11}, P_{12}, P_{14}\}$$

Let $Z = X \cup C$
The structure is described by the following binary relation:

$$s: F \ x \ (X \cup C) \ \rightarrow \{0, 1\}$$

$(f, z) \ \rightarrow \ s(f, z) = 1$ iff the constraint *f* applies to the variables z,

$s(f, z)=0$ otherwise.

The structure leads to a digraph whose incidence matrix represents the links between variables and constraints (fig.2) and whose analysis shows the existence of six redundancy relations (thus six residuals r_1 to r_6) which are now analyzed.

$$r1 = T_6 + 0{,}4594 P_7{}^2 - 12{,}7243 P_7 - 99{,}003$$

This residual is the difference between the T6 sensor value and the theoretical value calculated from thermodynamic dependence of the pressure. This expression is deduced from equations RT1, M5 and M6.
Residual 2 represents the relation between the boiler pressure P7 and the valve V2 downstream pressure

$$r2 = F_{10}{}^2 - \frac{(P_7 - P_{12}).K_D.P_{12}}{(-0.4594 P_7{}^2 + 12.7243 P_7 + 99.003)}$$

	P_{GV}	T_{GV}	h_{GV}	h_V	x	h_L	v_V	v_L	v_{GV}	M_{GV}	P_{VD}	F_{VD}
RT1	1	1										
RP1			1	1	1	1						
RP2					1		1	1	1			
RP3	1					1						
RP4	1			1								
RP5	1							1				
RP6	1						1					
RP7									1	1		
RP8			1							1		
RP9	1	1									1	1
RM1										1		
RM2												1
RE1												
RE2												
RE3												
RE4			1									1
RE5		1										
RE6		1										
RE7												
RE8												
M1												
M2												
M3												
M4		1										
M5		1										
M6	1											
M7												1
M8	1											
M9											1	
M10											1	

	H_{GV}	F_{GV}	F_{AL}	$\dot{H}_{GV}$	Q_{TH}	$\dot{H}_{AL}$	$\dot{E}_{PG}$	$\dot{H}_{VD}$	T_{AL}	T_{MG}	$\dot{T}_{MG}$	$\dot{E}_W$
RT1												
RP1												
RP2												
RP3												
RP4												
RP5												
RP6												
RP7												
RP8	1											
RP9												
RM1		1										
RM2		1	1									
RE1	1			1								
RE2				1	1	1	1	1				1
RE3			1			1			1			
RE4								1				
RE5							1			1		
RE6										1	1	
RE7										1	1	
RE8										1	1	
M1									1			
M2			1									
M3					1							
M4												
M5												
M6												
M7												
M8												
M9												
M10												

	T_2	F_3	Q_4	T_5	T_6	P_7	F_{10}	P_{11}	P_{12}	P_{14}
RT1										
RP1										
RP2										
RP3										
RP4										
RP5										
RP6										
RP7										
RP8										
RP9										
RM1										
RM2										
RE1										
RE2										
RE3										
RE4										
RE5										
RE6										
RE7										
RE8										
M1	1									
M2		1								
M3			1							
M4				1						
M5					1					
M6						1				
M7							1			
M8								1		
M9									1	
M10										1

Fig.2 : Incidence matrix of the FDI

The third residual represents the energy. This residual is complex because it is obtained from some non measurable variables such as the enthalpy, steam quality, density...,

$$r_3 = ((\frac{\frac{0.175}{[\int (F_3 - F_{10})dt + M_{GV}(0)]} - v_L(P_7)}{(v_V(P_7) - v_L(P_7))} h_V(P_7)$$
$$- h_L(P_7)) + h_L(P_7)) * \int (F_3 - F_{10})dt + M_{GV}(0))$$
$$-(\int (Q_4 + 0.420 T_2 F_3 - K_{GM}(T_5 - T_{MG}) - F_{10} h_V(P_7))dt$$
$$+ H_{GV}(0))$$

Residuals 4, 5 and 6 are the residuals which correspond to hardware redundancy.

$$r_4 = T_5 - T_6$$

The sensors T5 and T6 are both situated in the boiler. T6 indicate the steam pressure and T5 the liquid pressure. But, since the mixture is homogeneous, T5 and T6 are equal.

P7 and P11 are also the same because the hydraulic losses through the V2 valve is low and the steam regime is isenthalpic.

$$r_5 = P_7 - P_{11}$$

P12 and P14 are in a parallel connection, thus the pressure is the same.

$$r_6 = P_{12} - P_{14}$$

5. SIMULATIONS RESULTS

In order to test the monitoring system, simulations were performed. Two situations were considered:
a water leak in the boiler, and a blocked output of the steam pipe. Both faults are faults whose effect cannot be represented using an additive model. The FDI scheme for these two faults can be performed using only two out of the six existing residuals: the first one gives a relation between the steam flow and the output steam pressure and temperature; the second one expresses the boiler power balance. The residuals are calculated using the Matlab Simulink software. The model equations were programmed on a digital computer for simulation. Input data includes polynomial fits to various steam properties, valves characteristics, pump curves, empirical heat and hydraulic coefficients .
The blocked output of the steam pipe was realized by opening valve V2 from 20 sec up to 100 sec.
The second process fault: water leak in the boiler was realized opening valve V3. The scenario is dangerous, since the boiler operates at 150 °C. This is why the disturbance didn't last long: from sec. 125 to sec. 128 only. The simulation results are shown on figure 4.
The loss of water in the boiler caused a water level drop inside the steam generator. Following this, after few seconds, the pump was turned on (b_1=1) with a subsequent water level increase.

6. CONCLUSION

The present paper deals with modeling and monitoring of a steam generator pilot process using a structural analysis.

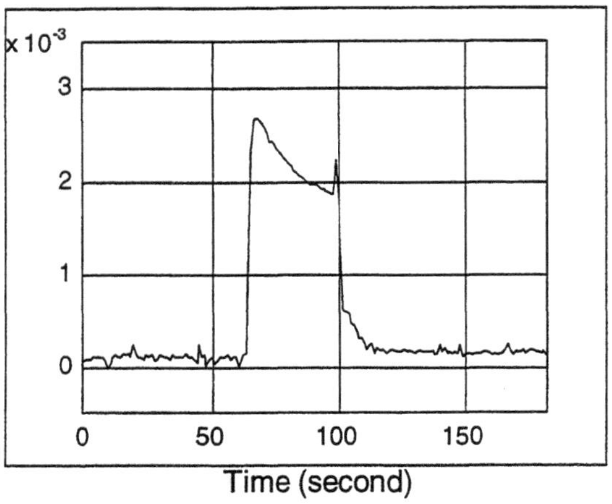

Fig. 3 Evolution of r2 in $(kg/s)^2$
(Blocked output of the steam pipe)

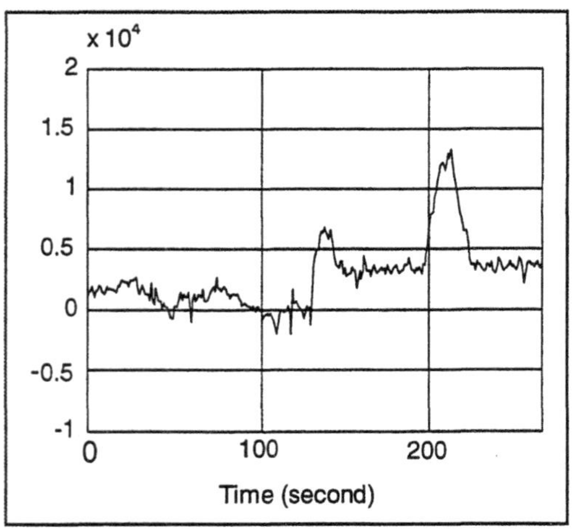

Fig. 4 Evolution of r3 in joule(water leak)

This method, well adapted to complex non linear systems, easily allows to take into account system nonlinearities, and non additive faults. It exhibits the residuals computation scheme, which results from the elimination of the unknown variables from the system equations.
Two faults scenari have been realized on the pilot process: a water leak in the boiler, and a blocked output of the steam pipe. Both parameters are important since in the real power station the insufficient water level could result in the boiler destruction, with catastrophic consequences and the second controls the whole electrical subsystem.
The simulated monitoring system show out to have a good reaction to both process faults listed in the FDI system specifications.

REFERENCES

Busson et al. (1998). Sensor Failures Detection in Steam Condensers. In: *3rd IFAC Workshop* 4- june 1998, IFP, Solaize (Lyon), France.

Chow, E. Y. and A. S. Willsky, (1984) Analytical Redundancy and the Design of Robust Failure Detection System. In: *IEEE Trans. on Automatic Control A.C.*, AC 29, **n°7**.

Declerck, P. and M. Staroswiecki (1991), Characterisation of the canonical components of a structural graph. application to fault detection in large scale industrial plants. In: *ECC'91 European Control Conference* , pp. 289-303, Grenoble (France).

Patton R. J., Frank, P. M. and Clark, R. N. (1989) Fault Diagnosis. In: *Dynamic Systems, Theory and Applications*. Prentice Hall, Englewoods Cliffs, NJ.

Pollart, M. and B. Ould Bouamama (1996), Générateur de vapeur, mise au point et interconnexion des sous ensembles mécanique, informatique, commande, modélisation. *Technical report*, LAIL, Lille (France).

Staroswiecki, M. and P. Declerck (1989), Analytical redundancy in nonlinear interconnected systems by means of structural analysis. In: *IFAC Symposium in Advanced Information Processing in Automatic control.* pp. 23-27, Nancy (France).

PROCESS SUPERVISION BASED ON FUZZY REASONING APPLICATION TO A URBAN WATER SUPPLY NETWORK

D. Mandel, D. Maquin, J. Ragot

Centre de Recherche en Automatique de Nancy
2, avenue de la Forêt de Haye
54516 Vandoeuvre-lès-Nancy
Tel. : 03 83 59 59 59 - Fax : 03 83 59 56 44
e-mail : {dmandel, dmaquin, jragot}@ensem.u-nancy.fr

Abstract : A diagnosis strategy based on fuzzy reasoning is presented in this paper. The first stage of the suggested approach consists in generating residuals revealing of the process functioning state. Two methods of residuals generation for linear-dynamic systems, based on the exploitation of analytical redundancy, are proposed. Then, a procedure of fault isolation using these residuals and based on fuzzy concepts is explained and implemented. To illustrate and show the validity of this approach, it's applied to experimental signals collected on a urban water supply network. *Copyright © 1998 IFAC*

Keywords : diagnosis, fuzzy supervision, fault isolation, sensor failures

1. INTRODUCTION

The unceasing development of computer and instrumentation technology allows the implementation of more and more developed diagnosis methods to be considered. That's what this paper enforces, through the application of diagnosis methods to a urban water supply network. This study has been lead in the framework of a collaboration with the Urban Hydraulic department of the Urban Community of Nancy.

To begin, in a preliminary section, state space representation is briefly recalled. Then a description of the considered water network is carried out, in order to clearly put into relief the supervision's problematic. A mathematical model of the water network is as well established. In the following, two methods of residual generation are presented. After that, these residuals are used for fault isolation. A preliminary fuzzification of the residuals enables to implement an isolation method using fuzzy concepts. The application of the suggested approach to the water network enables to illustrate and test it by using experimental data.

2. PRELIMINARIES : STATE SPACE REPRESENTATION

To describe the model of a process, various representations can be chosen. The methods proposed in this paper are applied to models described in state space form. In this section that kind of representation is succinctly recalled.

For a given system the state space representation can be written :

$$\begin{cases} EX(k+1) = FX(k) \\ Z(k) = HX(k) \end{cases} \quad (1)$$

with $\quad Z(k) = \begin{pmatrix} z_x(k) \\ z_u(k) \end{pmatrix}; \; X(k) = \begin{pmatrix} x(k) \\ u(k) \end{pmatrix}$

$$E = (I_n \;\; 0); \; F = (A \;\; B)$$

$$H = \begin{pmatrix} C & 0 \\ 0 & D \end{pmatrix}; \; W = \begin{pmatrix} V_{z_x} & 0 \\ 0 & V_{z_u} \end{pmatrix}; \; \varepsilon_Z(k) = \begin{pmatrix} \varepsilon_{z_x}(k) \\ \varepsilon_{z_u}(k) \end{pmatrix}$$

where x(k) corresponds to the (n,1) state vector at time k, u(k) is the (m,1) input vector at time k, and X(k) is called generalised state vector. A and B are matrices of appropriate dimension. I_n designates an (n,n) identity matrix. $z_x(k)$ is the (p,1) vector of measured sates, $z_u(k)$ is the (q,1) vector of measured inputs. $\varepsilon_{z_x}(k)$ is the (p,1) vector of state measurement noise and $\varepsilon_{z_u}(k)$ is the (q,1) vector of input measurement noise. $\varepsilon_{z_x}(k)$ and $\varepsilon_{z_u}(k)$ are assumed zero mean Gaussian vectors with covariance V_{z_x} and V_{z_u}. C and D are respectively selection matrices of state and input measures.

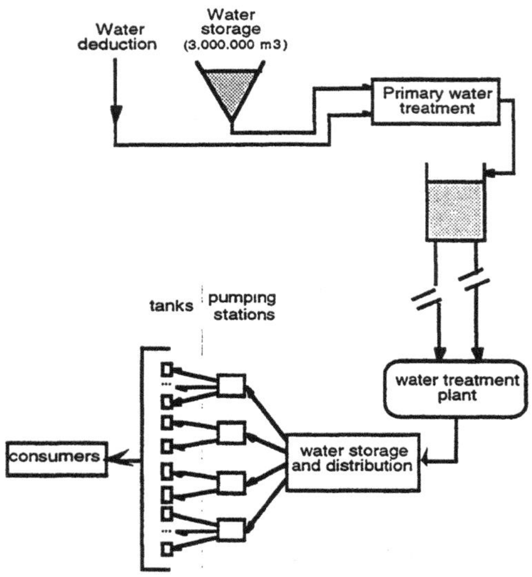

Fig. 1. Scheme of the urban water supply network

3. SUPERVISED PROCESS DESCRIPTION

3.1 - General presentation of the process

The process considered throughout this work is a urban water supply network. This process is depicted schematically in figure 1. The part of the water network considered in this study is the storage network. The function of this network consists, with help of pumping station, in sharing out water reserves among the whole agglomeration, before its distribution to the consumers. More precisely, the storage network is composed of pumping stations that make up an arborescent network regrouping about thirty tanks. On account of its importance for the water distribution reliability, the managers pay a great attention to the supervion of the storage network. Consequently, this part of the network has been supplied with numerous measuring devices. Indeed, to each tank corresponds a water level measurement and almost all the flow-rate variables, roughly a hundred, are measured.

For safe of clarity, the different methods presented in this document are applied to only one of the pumping stations. But the extension to the whole network is practically immediate thanks to the utilisation of state space representation (1). Indeed, whatever the dimension of the considered part of the network may be, the model formulation remains similar. Only the matrix dimensions and entries change. Thus the suggested methods can independently be applied to some parts of the network or to the whole network.

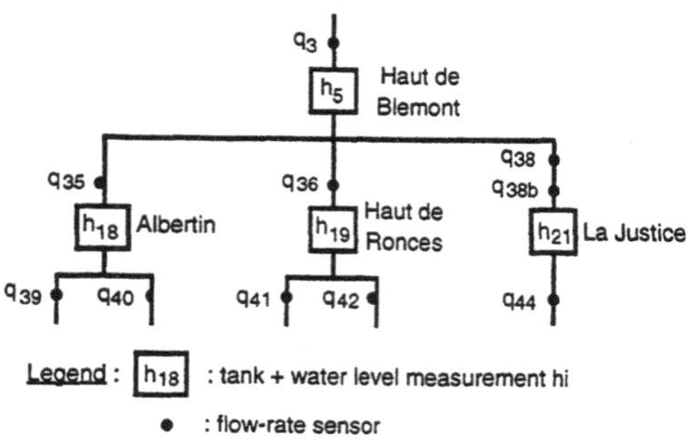

Fig. 2. Scheme of pumping station Haut de Blemont

The mentioned pumping station, called "Haut de Blemont", is composed of four tanks. Its scheme appears on figure 2. For this pumping station the water level measurements of the four tanks, as well as ten flow rate measurements are available.

3.2 Water network modelling

The data available for the supervision of the water network are water level measurements and flow-rate measurements. A model involving these variables can be established on the basis of the material-balance equations expressed for each tank.

With a sampling period ΔT the discrete form of the material-balance equation, for one tank, is :

$$\frac{S}{\Delta T}\left[h(k+1)-h(k)\right]=\Sigma\, q_{in}(k)-\Sigma\, q_{out}(k) \qquad (2)$$

where h is the water level in the tank, S its section and q_{in} and q_{out} are respectively the output and input flow rates. k indicates the sampling time.

More generally, for the whole of the tanks, the model can be described by using state space representation (1). In the case of the pumping station "Haut de Blemont" (Fig. 2), the matrices of system (1) are :

$$X(k) = [\, h_5\ h_{18}\ h_{19}\ h_{21}\ q_3\ q_{35}\ q_{36}\ q_{38}\ q_{39}\ q_{40}\ q_{41}\ q_{42}\ q_{44}\,]^T$$

$$E=(\,I_4\ \ 0\,) \qquad F=(\,A\ \ B\,) \qquad A=I_4 \qquad B=SB_0$$

$H = I_{13}$ since all the variables of X(k) are measured

$$S=\begin{bmatrix}\Delta T/S_5 & 0 & 0 & 0\\ 0 & \Delta T/S_{18} & 0 & 0\\ 0 & 0 & \Delta T/S_{19} & 0\\ 0 & 0 & 0 & \Delta T/S_{21}\end{bmatrix}$$

$$B_0=\begin{bmatrix}1 & -1 & -1 & -1 & 0 & 0 & 0 & 0 & 0\\ 0 & 1 & 0 & 0 & -1 & -1 & 0 & 0 & 0\\ 0 & 0 & 1 & 0 & 0 & 0 & -1 & -1 & 0\\ 0 & 0 & 0 & 1 & 0 & 0 & 0 & 0 & -1\end{bmatrix}$$

with the notation I_n for an (n,n) identity matrix.

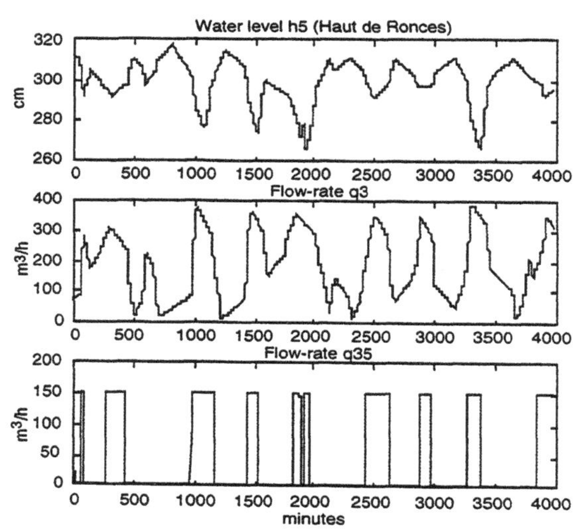

Fig. 3. Experimental signals (filtered)

3.3 Experimental signals

As an example, the filtered signals of measurements h_5, q_3 and q_{35}, collected on the pumping station "Haut de Blemont", appear figure 3. Let's note that the measurement signal of flow rate q_{35} has a particular shape because it corresponds to a pump delivery.

4. SUPERVISION PROBLEMATIC

In a general way, the supervision objective consists in detecting and isolating each anomaly that can occur on a process but as well on its instrumentation. Here, model-based diagnosis methods are considered. The organisation of a model-based supervision procedure has been described by Chow and Willsky (1984). The first stage of the proposed procedure is the generation of residuals (Gertler, 1991), that are revealing of the concordance between the model corresponding to the normal functioning state and the measures. Theoretically residuals are zero mean in normal functioning state and different from zero when they are affected by a fault. Then on the basis of the residuals analysis, faults can be detected and isolated. Finally, a diagnosis can be posed and in function of this diagnosis, actions of reconfiguration or maintenance can be started.

Supervision is a problem that has preoccupied the managers of water supply networks for a long time. They need to detect quickly all the unexpected events that could occur on the networks, such as leakage or damage. That can be achieved through the analysis of the measured values coherence, for instance, by verifying the material balancing. The supervision problem concerns simultaneously the process, here the water supply network, and the sensors. From the manager's point of view, supervision is essential because it leads to provide the state of the network's variables, to locate failures and to provide consistent data. That increases the process safety and the water distribution reliability. These points are crucial since interruptions of the water distribution are less and less tolerated, especially by companies, moreover the damage caused by leaks are often expensive.

5. RESIDUALS GENERATION

In the following two methods of residuals generation for linear dynamic systems are presented. They are both based on the exploitation of a model of the process in normal functioning. The residuals obtained with these methods can be used in a complementary way for fault isolation.

5.1 Direct residuals

The first method of residuals generation is based on the direct exploitation of the analytical redundancy relations. By definition these relations involve only measured variables and thus enable to compute, at each sampling time k, a vector of direct residuals $R_d(k)$. If all the variables are measured, the equations of the model are all redundancy relations. The calculation form of the residuals is then :

$$R_d(k) = -EZ(k+1) + FZ(k) \quad (3)$$

or else $$R_d(k) = MZ(k, k+1) \quad (4)$$

with $$M = (F \quad -E)$$

If all the variables aren't measured, the equations of the model must be analysed in order to eliminate the unmeasured variables (Maquin and Ragot, 1991).

5.2 Estimation residuals

Estimation residuals are generated by comparing the measures and their estimation. So, this method requires a preliminary stage of state estimation.

State estimation consists in adjusting the measures so that they verify exactly the normal functioning model, while minimising the measure corrections. State estimation ensures the obtainment of data that are more accurate than the measures. Beside it enables the value of unmeasured variables to be deduced. If the observability conditions aren't respected, the observable part of the system must be extracted (Ragot, et al., 1990).

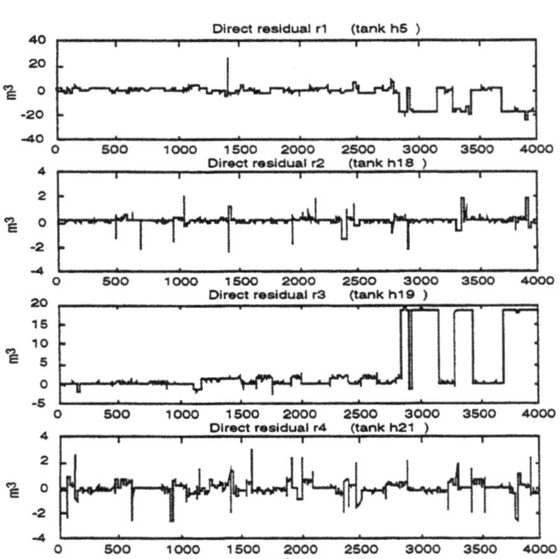

Fig. 4. Experimental direct residuals

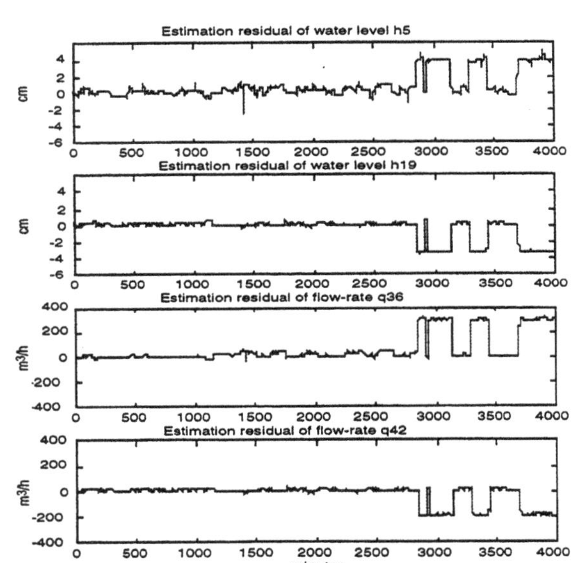

Fig. 5. Experimental estimation residuals

A formulation of generalised state estimation is proposed. The interest of this approach arises from the simultaneous estimation of state and input variables. That means, in the case of this application, the estimation of water levels and flow-rates. Moreover for dynamic systems the quality of the estimation increases clearly if it's worked out by taking into account several sampling times. Therefore the estimation is carried out on a gliding window of width N. The value of N results from a compromise between estimation accuracy and fault sensitivity.

Let's consider a system which model is described in the state space, according to definition (1). On the observation horizon [k, k+N], this system can be written :

$$\begin{cases} \mathcal{M}X(k,\ k+N) = 0 \\ Z(k,\ k+N) = \mathcal{H}X(k,\ k+N) \end{cases} \quad (5)$$

with

$$X(k,k+N) = \begin{pmatrix} X(k) \\ X(k+1) \\ \vdots \\ X(k+N) \end{pmatrix};\ Z(k,k+N) = \begin{pmatrix} Z(k) \\ Z(k+1) \\ \vdots \\ Z(k+N) \end{pmatrix}$$

$$\mathcal{M} = \begin{pmatrix} F & -E & 0 & . & . & . & 0 \\ 0 & F & -E & 0 & \vdots & \vdots & 0 \\ \vdots & \vdots & \vdots & \vdots & \vdots & \vdots & \vdots \\ 0 & . & . & . & 0 & F & -E \end{pmatrix};\ \mathcal{H} = \begin{pmatrix} H & 0 & . & 0 \\ 0 & H & \ddots & . \\ . & \ddots & \ddots & 0 \\ 0 & 0 & 0 & H \end{pmatrix}$$

The variance-covariance matrix of Z(k,k+N) is :

$$\mathcal{W} = \begin{pmatrix} W & 0 & . & 0 \\ 0 & W & \ddots & . \\ . & \ddots & \ddots & 0 \\ 0 & 0 & 0 & W \end{pmatrix}$$

On the horizon [k, k+N], carrying out generalised estimation of system (5) amounts to find the value of $\hat{X}(k,\ k+N)$ solution of the following optimisation problem.

$$\min_{\hat{X}(k,k+N)} J = \frac{1}{2}\left\| Z(k,k+N) - \mathcal{H}\hat{X}(k,k+N) \right\|^2_{\mathcal{W}^{-1}} \quad (6)$$

with respect to : $\mathcal{M}\hat{X}(k,\ k+N) = 0$ (7)

The resolution of this problem leads to the result (Ragot, et al., 1990) :

$$\hat{X}(k,\ k+N) = PR^{-1}\mathcal{H}^T\mathcal{W}^{-1}Z(k,\ k+N) \quad (8)$$

with $R = \mathcal{H}^T\mathcal{W}^{-1}\mathcal{H} + \mathcal{M}^T\mathcal{M}$ (9)

$$P = I - R^{-1}\mathcal{M}^T\left(\mathcal{M}R^{-1}\mathcal{M}^T\right)^{-1}\mathcal{M} \quad (10)$$

With a view to real time implementation, a recursive formulation of this solution is as well given. Let's first define the notation $\hat{X}_N(k)$ that designates the state estimation at time k computed by taking into account an observation horizon of length N. A recurrence between the estimation $\hat{X}_N(N)$ and $\hat{X}_{N-1}(N-1)$ can be established. To find this recurrence two state estimation problems must be considered, on the one hand on the horizon [1, N] and on the other hand on the horizon [1, N+1]. The reader can verify, see (Kratz, 1994), that in this way a recursive formulation defined by expressions (11) and (12) can be obtained.

$$\hat{X}_N(N) = R_N^{-1}(N)\left(QF\hat{X}_{N-1}(N-1) + TZ(N)\right) \quad (11)$$

$$R_N(N) = TH + QE \quad (12)$$

with $Q = E^T\left(FR_{N-1}^{-1}(N-1)F^T\right)^{-1}$

$$T = H^TW^{-1}$$

$R_N^{-1}(N)$ is the variance-covariance matrix of the estimation $\hat{X}_N(N)$.

So, at each sampling time the generalised state of the system can be estimated.

The second method of residuals generation uses the state estimation results. It consists in generating residuals by comparing each measure and its estimation. The calculation form of the estimation residual vector $R_e(k)$ is given below.

$$R_e(k) = Z(k) - H\hat{X}(k) \quad (13)$$

5.3 Example of experimental residuals

A model of the pumping station "Haut de Blemont" has been established (section 3.2) and the measurements of all its variables are available. Hence, on the basis of this knowledge, the direct residuals (Fig. 4) and the estimation residuals (Fig. 5) can be generated.

Let's note that at time 2850 (Fig. 4 and 5) a fault appears. It's now a question of making systematic the detection and isolation of such faults.

6. FAULT SIGNATURE MATRIX

The objective of the residual generation is to provide signals revealing of the process functioning state. However the residuals values aren't sufficient for carrying out fault isolation. Indeed, fault isolation methods require as well the knowledge of the influence of each variables on the residual values. In other words the relation between the symptoms and their causes must be known. This information can be coded with a boolean matrix called fault signature matrix. The aim of this section is to describe the construction of this matrix.

Let's consider the case of the direct residuals which mathematical expression is recalled below.

$$R_d(k) = MZ(k,\ k+1)$$

The fault signature matrix S is then deduced from matrix M. The elements S_{ij} (where i indicates the row number and j the column number) are defined according to the following principle :

$$\begin{cases} S_{ij} = -1 \text{ if } M_{ij} < 0 \\ S_{ij} = 0 \text{ if } M_{ij} = 0 \\ S_{ij} = +1 \text{ if } M_{ij} > 0 \end{cases}$$

The jth column of matrix S represents the theoretical signature of a fault affecting the measure that corresponds to the jth element of Z(k, k+1).

The determination of the signature matrix associated to estimation residuals isn't detailed here. The principle is the same that for the direct residuals but by using, instead of matrix M, the matrix that links the estimation residuals to the measurements.

7. FUZZY-BASED FAULT ISOLATION

7.1 Principle of the method

The suggested method uses fuzzy concepts for fault isolation. This method relies upon the residuals fuzzification. Three fuzzy sets r_j^-, r_j^0 and r_j^+ are defined. The fuzzy set r_j^0 corresponds to the case of consistency of the residual r_j, whereas r_j^- and r_j^+ correspond respectively to the case of negative and positive inconsistency.

The membership functions $\mu_{r_j^-}$, $\mu_{r_j^0}$ and $\mu_{r_j^+}$, respectively associated to the fuzzy sets r_j^-, r_j^0 and r_j^+, are defined by expressions (14), (15) and (16). These membership functions appear figure 6.

$$\mu_{r_j^0}(k) = 1 - \frac{\left(r_j(k)/\tau_j\right)^{\rho}}{1 + \left(r_j(k)/\tau_j\right)^{\rho}} \qquad (14)$$

$$\mu_{r_j^-}(k) = \begin{cases} 1 - \mu_{r_j^0}(k) & \text{if } r_j(k) \le 0 \\ 0 \text{ si } r_j(k) > 0 \end{cases} \qquad (15)$$

$$\mu_{r_j^+}(k) = \begin{cases} 1 - \mu_{r_j^0}(k) & \text{if } r_j(k) \ge 0 \\ 0 \text{ si } r_j(k) < 0 \end{cases} \qquad (16)$$

where $\mu_{r_j^0}(k)$ refers to the value of the membership function associated to r_j^0, at time k. ρ is a form factor that must be even and τ_j the tolerance associated to residual r_j.

So, for each residual and at each time k the value of the membership functions can be computed. The following vector can then be built for each residual :

$$\mu_{r_j}(k) = \begin{pmatrix} \mu_{r_j^-}(k) & \mu_{r_j^0}(k) & \mu_{r_j^+}(k) \end{pmatrix}^T \qquad (17)$$

Fault isolation is based, on the one hand, on the observation of the symptoms affecting the residuals and, on the other hand, on the analysis of the fault signature matrix.

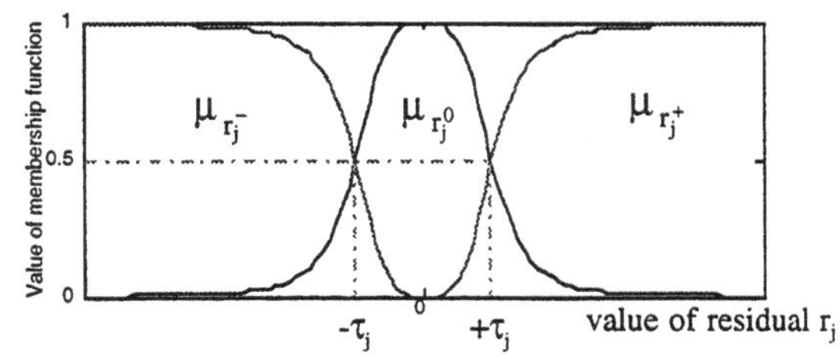

Fig. 6 - Membership functions

Let's consider, for instance, a variable which theoretical fault signature is $(-1\ 1\ 0)^T$. A positive fault affecting this variable involves a value of the membership function $\mu_{r_1^-}$ close to 1 and values of membership functions $\mu_{r_1^0}$ and $\mu_{r_1^+}$ close to 0. With regard to the second and third residuals, the values of the membership functions $\mu_{r_2^+}$ and $\mu_{r_3^0}$ are close to 1 where as the values of the other membership functions are close to 0.

Fault isolation consists in solving the reverse problem. That means deducing, from the value of the residuals, the variable affected by the fault. So, for each fault situation a rule is established.

Let's consider again the previous example, by calling f the envisaged fault. The rule corresponding to f is :

$$\text{if } \mu_{r_1^-} \approx 1 \ \underline{\text{and}}\ \mu_{r_2^+} \approx 1 \ \underline{\text{and}}\ \mu_{r_3^0} \approx 1 \quad \text{then f}$$

The symbol and appearing in this rule refers to a conjunction operator defined subsequently. In the same way two rules can be associated to each variable, one for positive fault and the other for negative fault. A rule-base is then obtained.

After that, the truth value μ_f of each rule, that means of each failure hypothesis, is computed. To compute these values, conjunction operators must be defined. Various conjunction operators can be used (Bouchon-Meunier, 1995). The "minimum" (18) and "mean" (19) operators are in current use. Another operator (20) corresponding to a combination of the "mean" and "minimum" operators, is defined. It increases the robustness of the proposed fault isolation method.

$$\mu_f = \min\left(\mu_{r_1^{A1}}, \mu_{r_2^{A2}}, \ldots, \mu_{r_p^{Ap}}\right) \qquad (18)$$

$$\mu_f = \frac{1}{p} \sum_{j=1}^{p} \mu_{r_j^{Aj}} \qquad (19)$$

$$\mu_f = \min\left(\frac{\sum_{j \in \{j | Aj \in \{0\}\}} \mu_{r_j^{Aj}}}{\left|\{j | Aj \in \{0\}\}\right|}, \frac{\sum_{j \in \{j | Aj \in \{-,+\}\}} \mu_{r_j^{Aj}}}{\{j | Aj \in \{-,+\}\}} \right) \qquad (20)$$

where $A_j \in \{-, 0, +\}$ according to the influence of f on r_j.

The isolation result is based on the analysis of these truth values. Indeed, the rule which truth value is the highest designates the most likelihood fault.

7.2 Application to experimental data

An application of the proposed fault isolation method to the experimental direct residuals of figure 4 is now presented.

Firstly, the fault signature matrix (21) corresponding to the envisaged faults is built. The matrix S is then used for the generation of a rule-base. This rule-base is described by table 1, where q_i^- and q_i^+ designate respectively the case of negative and positive faults affecting the variable q_i.

$$S = \begin{array}{c|ccccccccc} & q_3 & q_{35} & q_{36} & q_{38} & q_{39} & q_{40} & q_{41} & q_{42} & q_{44} \\ \hline r_1 & 1 & -1 & -1 & -1 & 0 & 0 & 0 & 0 & 0 \\ r_2 & 0 & 1 & 0 & 0 & -1 & -1 & 0 & 0 & 0 \\ r_3 & 0 & 0 & 1 & 0 & 0 & 0 & -1 & -1 & 0 \\ r_4 & 0 & 0 & 0 & 1 & 0 & 0 & 0 & 0 & -1 \end{array} \qquad (21)$$

Next step concerns the residual fuzzification. The values of the three membership functions $\mu_{r_1^-}$, $\mu_{r_1^0}$ and $\mu_{r_1^+}$ are computed for each residual. The fuzzification of the experimental residual r_1 is illustrated by figure 7.

Table 1 Rule-base used for fault isolation

	r_1			r_2			r_3			r_4		
Fault	r_1^-	r_1^0	r_1^+	r_2^-	r_2^0	r_2^+	r_3^-	r_3^0	r_3^+	r_4^-	r_4^0	r_4^+
q_3-	X				X			X			X	
q_3+			X		X			X			X	
q_{35}-			X	X				X			X	
q_{35}+	X					X		X			X	
q_{36}-			X		X		X				X	
q_{36}+	X				X				X		X	
q_{38}-			X		X			X		X		
q_{38}+	X				X			X				X
q_{39}-		X				X		X			X	
q_{39}+		X		X				X			X	
q_{40}-		X				X		X			X	
q_{40}+		X		X				X			X	
q_{41}-		X			X				X		X	
q_{41}+		X			X		X				X	
q_{42}-		X			X				X		X	
q_{42}+		X			X		X				X	
q_{44}-		X			X			X				X
q_{44}+		X			X			X		X		
none		X			X			X			X	

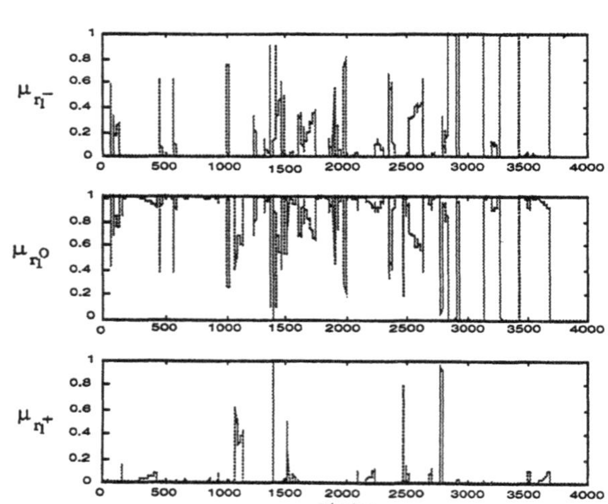

Fig. 7. Fuzzification of direct residual r_1

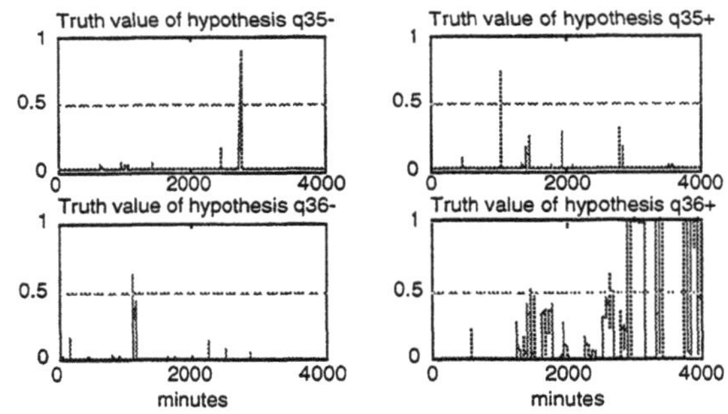

Fig. 8. Truth value of hypothesis q_{35}^-, q_{35}^+, q_{36}^-, q_{36}^+

As an example, the truth value evolution of failure hypothesis q_{35}^-, q_{35}^+, q_{36}^- and q_{36}^+ are shown figure in 8. The analysis of all the truth values leads to conclude that the failure appearing at time 2850 (figure 4) is a positive fault affecting the variable q_{36}. The validity of this result has been confirmed by the supervisors. Moreover, this fault isolation procedure has been tested for other faults and the results obtained attest its efficiency.

8. CONCLUSION

A diagnosis strategy based on fuzzy reasoning has been elaborated and implemented for the supervision of a water distribution network. This study was very fruitful because it put into relief the difficulties linked to the implementation of a supervision procedure for a real process. Let's note that a supervision software has been developed on the basis of the diagnosis strategy presented in this paper.

The suggested approach offers many prospects since it enables, thanks to the use of fuzzy concepts, an easy integration of additional information. This point is very important since our aim is to propose a diagnosis procedure that can take into account all the available data. In this way, the supervision efficiency could still be increased.

REFERENCES

Bouchon-Meunier, B. (1995). *La logique floue et ses applications.* Collection Vie Artificielle, Addison-Wesley, Paris.

Chow, E. Y. and A. S. Willsky (1984). Analytical redundancy and the design of robust failure detection system. *IEEE Transactions on Automatic Control*, **AC-29 n°7**, pp. 603-614.

Gertler, J. (1991). Analytical redundancy methods in fault detection and isolation. *IFAC Symposium Safeprocess'91*, Baden-Baden, vol. **1**, pp. 9-21.

Kratz, S. (1994). *Diagnostic de fonctionnement de procédés continus par réconciliation d'état généralisé - application à la détection de pannes de capteurs et d'actionneurs*, 150 pages. Thèse de Doctorat de l'université de Nancy I.

Maquin, D. and J. Ragot (1991). Comparison of gross errors detection methods in process data. *Proceedings of the 30th IEEE Conference on Decision and Control*, p. 2253-2261, Brighton.

Ragot, J., M. Darouach, D. Maquin and G. Bloch (1990). *Validation de données et diagnostic*, 430 pages. Traité des nouvelles technologies, série diagnostic et maintenance, Hermès.

ON-LINE PERFORMANCE ESTIMATION AND CONDITION MONITORING USING NEURO-FUZZY TECHNIQUES

Geert Uytterhoeven, Jean-Michel Renders, Françoise de Viron, Michel De Vlaminck

TRACTEBEL ENERGY ENGINEERING, BELGIUM

Paolo F. Fantoni

IFE OECD Halden reactor Project, Norway

Abstract : Condition monitoring and performance estimation of a system or component means to assess the current state and estimate the future state of the system or component, using real time measurements and calculations. the difference between current and optimal conditions, caused by degradation, are used to evaluate the need and the time for maintenance, where the cost-benefit ratio for the operation is the lowest possible. Increase of availability, reliability and life time are the reasons why condition monitoring should be the preferred maintenance strategy.
This paper shows the advantages of neuro-fuzzy based systems for plant monitoring and estimation, by presenting three state-of-the-art applications: ANNODIN (an early fault detection system), OCEAN (Optimisation of combustion and gas emission) and PEANO (Process signal validation and estimation).

1. INTRODUCTION

Conventional process monitoring & control systems provide the operator with information based on simple signals issued from the process instrumentation. Such system can react only when an alarm threshold has been exceeded, and only after this it could assist the operator make a diagnosis and initiate the appropriate corrective action.
Since "preventing is better than curing", the necessity was perceived for systems that would be capable of detecting a potential problem well before the conventional alarm system operates.

The concept looks simple : build a highly-realistic yet affordable process simulator which, based on known input parameters, can mimic in real-time the entire process (without its possible shortcomings and deviations), and compare this simulator's output to the output of the actual process. If the discrepancy between the two gets significant, then a warning is given to the operator.
The advantages of neural network technology, which include:

- no need for "perfect" analytical process-modelling
- learning by examples

- quick response time
compared to conventional analytical simulation methods, resulted in the decision to build the affordable highly-realistic model with the help of neural networks.

2. DETECTION OF INCIPIENT FAULTS

ANNODIN (Artificial Neural Networks fOr Detection of Incipient Nuisance) (Renders, *et al.*, 1995 ; de Viron, *et al.*, 1995) is an early detection system aimed at supervision of complex industrial processes or equipment, capable of "feeling" all kinds of nascent symptoms enough time before they materialise and before they can be detected by the operator or give rise to a conventional alarm. The system is aimed primarily at detecting malfunction, allowing predictive maintenance to be done, thanks to timely warning about abnormal wear, fouling, leakage, measurement drift, ...

2.1. The principle and the architecture of the system

The fault detection methodology is based on the following principles:

- The detection of faults is based on the modelling of the physical (thermodynamic, chemical or nuclear) processes occurring in the plant and on the analysis of the discrepancy between the real behaviour and the "normal models" by an appropriate decision logic
- The neural networks are used for modelling physical relationships (for instance mass balances, energy balances, ...) which govern the evolution of some state variables of the plant. In a certain sense, one attempts to exploit the *a priori* knowledge of the system, even if it is rough and incomplete, by isolating physical subsystems and by associating to each of them a set of variables which one hopes to be necessary and sufficient to describe the subsystem behaviour with a satisfactory accuracy.
- The choice of the relationships to be modelled follows our principle of isolating several subsystems of the complex plant and of determining in a suitable way the state variables which allow a correct description (prediction) of the behaviour of each subsystem to be made. The separation into independent subsystems provides also the advantage of making easier the isolation of the fault and, therefore, its diagnosis. Indeed, the analysis of the residuals associated with each network, and consequently with each subsystem, allows to isolate the faulty component. The selection of variables is of course not unique, but it must adhere as much as possible to the following guidelines: : (i) the variables used must be present in the training data base; (ii) ideally, the set of chosen variables should be complete (i.e. sufficient to describe the behaviour of the subsystem) and minimal (i.e. without variable which linearly or nonlinearly depends on already chosen variables); (iii) the subsystem and the associated state variables must react significantly to certain types of faults.

The choice of these relationships will be based on physical and experimental considerations: starting from a particular subsystem, we will isolate some state variable (y_i) which we hope to be sensitive both to certain variables ($\vec{x}^i$) and to certain faults. Qualitative physical reasoning allows us to discern variables $\vec{x}^i$ necessary to predict the behaviour of y_i, but often these variables are not present in our available database. Moreover, the roughness of the qualitative reasoning results in neglecting or overlooking some phenomena. In other words, it is not always possible in practice to ensure the properties of completeness, minimality and sensitivity of the chosen variables and subsystems. Consequently, a complementary approach, more heuristic or more experimental, could be necessary in order to find a satisfactory (as far as completeness, minimality and fault sensitivity are concerned) set of variables $\vec{x}^i$, which allows to correctly predict y_i. In a practical way, we have chosen to add to the state variables already chosen by physical considerations, other state variables which produced a significant decrease of the prediction error on y_i.

Then, according to the principles of the "analytical redundancy" methodology, the discrepancies between the networks' predictions and the actual process measurements result in "residuals" whose values are the contribution of the process noise, the measurement errors, the modelling errors and the possible faults. A suitable local decision logic is developed, which combines statistical tests and filters, in order to discriminate the incipient fault from the other components of the residual. Additional information regarding the confidence level and the validity of each ANN-model for the current state of the process are also integrated in the decision logic, resulting in a better processing of the "ignorance" of neural networks and a more reliable localisation of the fault. Then, based on the entire set of local decisions (one for each subsystem), a pre-diagnosis is established, taking into account the physical relationships between the subsystems (this is a benefit of our system having been split in several small physical subsystems). The following diagram briefly summarises the different steps of the method.

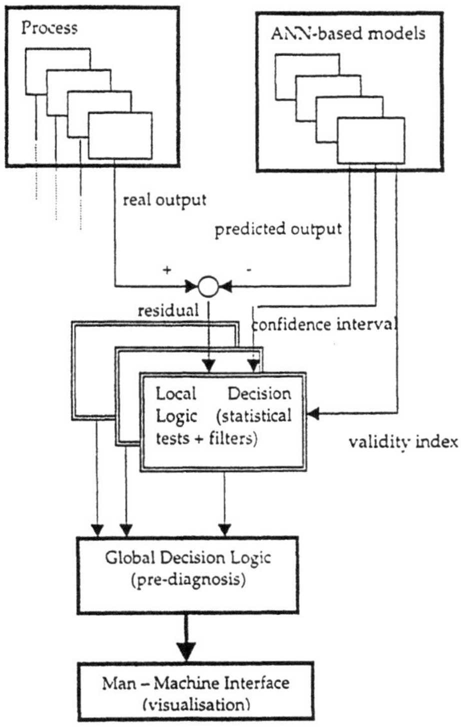

Figure 1. Schematic diagram of the method.

2.2. *ANNODIN tested on the nuclear plant simulator*

First, ANNODIN was tested on the full-scope simulator of DOEL 4 at the Scaldis Training Center near Antwerp in Belgium. ANNODIN had been trained to model the secondary system of Unit 4 of Doel nuclear power plant. With 94 neural networks some 30 components are simulated, including heat exchangers, a turbine, a condenser, ... These models were trained on the data-network at nominal power under normal plant operation conditions, i.e. without failures, incidents or accidents. After this a variety of scenarios were enacted that included simulation of small local leaks and gradual clogging of filters. The results were found good to excellent: ANNODIN detected small leaks and fouling long before the conventional alarms went off, and there were not too many false alarms.

2.3. *ANNODIN on the real power station*

During three months some 160 process variables were recorded, with a sampling frequency of 12 per hour. ANNODIN was re-trained on part of these recorded measurements, and tested on another part.

The main purpose was to demonstrate that the ANNODIN methodology, which showed excellent results when using the simulated data, was still valid when used on the real process, despite the many differences this entailed (measurement noise, complex phenomena not accounted for in the simulator, missing data, ...).

First a set of tools was developed for automating as much as possible the tasks relating to extraction of all the data, pre-processing, building of the models and evaluating them. In particular, the following tools were developed :

- optimised filtering tools that can attenuate as much as feasible the noise in the measurements, affecting as little as possible the useful signal;
- data selection tools enabling to select among all the available data a sufficiently small number of representative samples, in order to provide the learning set;
- statistical processing tools for the «outliers»;
- model construction tools that facilitate the selection of inputs/outputs and the structure of the model.

Thanks to these tools some sixty models were identified and validated on the basis of real process data, taking the same breakdown into subsystems and functional relations (thermal inventory, mass inventory, ...) as was previously adopted when using the simulated data.

The results obtained so far are encouraging : the first conclusions as to the feasibility, drawn from experiments with simulation scenarios, seem to demonstrate that extrapolation to real scenarios is possible despite the difficulties encountered during this change (simulation → reality).

2.4. *Interface*

A special Engineer-Machine Interface has been developed, in order to validate the above methodology (this is obviously not the final MMI presented to the operator). The normalised residuals (=residual/confidence interval) of the 94 models, as well as the validity indices of these models, are displayed at each time (see Figure 2).

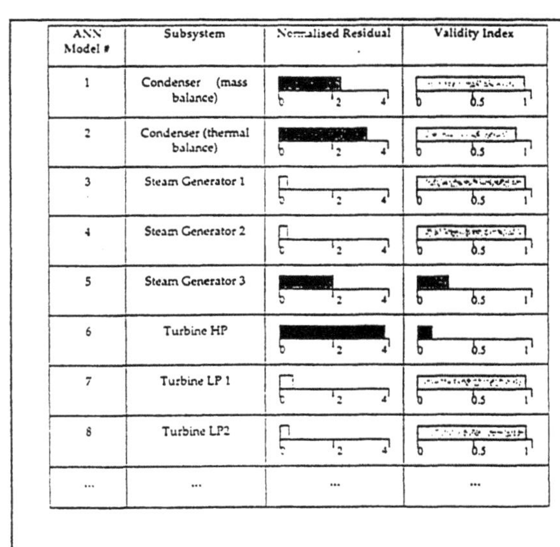

Figure 2. Schematic representation of the model results

Then, one particular model can be selected in order to see which subsystems it is dealing with (see figure 3) and what is the evolution of the model results (residuals, confidence interval, validity index) in order to validate the sensitivity to known incidents and the insensitivity to other "normal" perturbations (see figure 4).

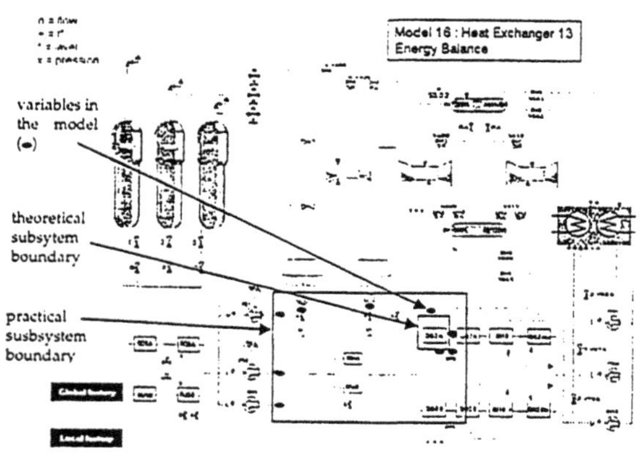

Figure 3. Visualisation of the subsystem and the associated variables.

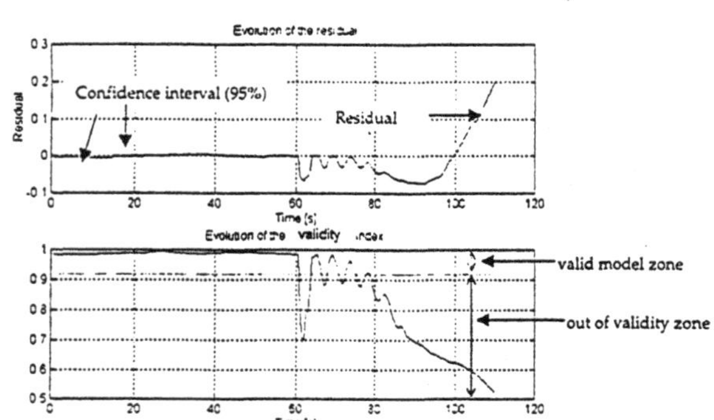

Figure 4. Typical evolution of the residual and model validity at the beginning of the fault

3. PROCESS OPTIMISATION

OCEAN (Optimisation of Combustion and Emissions with Artificial Neural Networks) is a project aimed at optimising combustion in fossil-fired power stations with a view to reducing the NOx emissions, taking into account the limit conditions regarding fuel quality, efficiency and fuel residue in the fly-ash. This project is being developed with the collaboration of two Belgian companies : Electrabel and Laborelec (the Laboratory of Electrabel).

3.1. The principle

In a fossil-fired power station there exist a few hundred parameters with which the combustion can be adjusted. Until now, of all these parameters only some 30 are actually used to control the combustion process in normal operation (mainly for adapting to load change) or to adjust to a variation in fuel quality. Increasingly strict environmental regulations and fiercer competition have forced the operators to react as quickly and as adequately as possible to any external variations, so as to retain the energy-efficiency and meet at any time all the relevant regulations.
OCEAN, which is a neural-network based model, supplies advice to the operator so that he may optimise the process.
Just like with ANNODIN, a neural model will be built that mimics the combustion process. Downstream of this model a number of optimisation algorithms are installed so that the system can always be used in two ways :

- either as a 'process-optimiser': OCEAN supplies the optimal position of all the combustion-air control valves, the speed of the coal mills, the fuel feed rates, ... depending on what the operator intends to optimise more particularly in function of outside constraints or policy with respect to fuel quality, load, ... The first target is the reduction in NOx emission, but this can be extended to optimisation of the energy efficiency and/or the quality of the fly-ash
- either as a simulator, allowing a "what-if" analysis, to investigate the effects of various factors on the combustion efficiency and emissions (influence of the coal quality, register positions, burner configuration, ...).

3.2. The preliminary study

The preliminary study was aimed at checking whether a neural network based optimisation system is feasible both technically and financially, forming an idea of the possible problems that could arise, taking a look at other fossil-fired power stations where development of similar projects is in progress and, finally, performing a market survey concerning existing software packages that could be suitable.

4. SIGNAL VALIDATION

Artificial Neural Networks and Fuzzy Logic models can be combined to exploit learning and generalization capability of the first technique with the approximate reasoning embedded in the second approach. Real-time process signal validation is an application field where the use of this technique can improve the diagnosis of faulty sensors and the identification of outliers in a robust and reliable way.

The PEANO system implements a fuzzy and possibilistic clustering algorithm to classify the operating region where the validation process has to be performed. The possibilistic approach (rather than probabilistic) allows a "don't know" classification that results in a fast detection of unforeseen plant conditions or outliers.
Specialized Artificial Neural Networks are used for

the validation process, one for each fuzzy cluster in which the operating map has been divided. They work concurrently on each signal pattern presented to the system and the overall contribution is weighted according to the membership function of the pattern in each cluster.

There are two main advantages in using this technique: the accuracy and generalization capability is increased compared to the case of a single network working in the entire operating region, and the ability to identify abnormal conditions, where the system is not capable to operate with a satisfactory accuracy, is improved.

Figure 5.1 shows a functional diagram of PEANO, with the bank of ANN and the reliability model driven by the classifier.

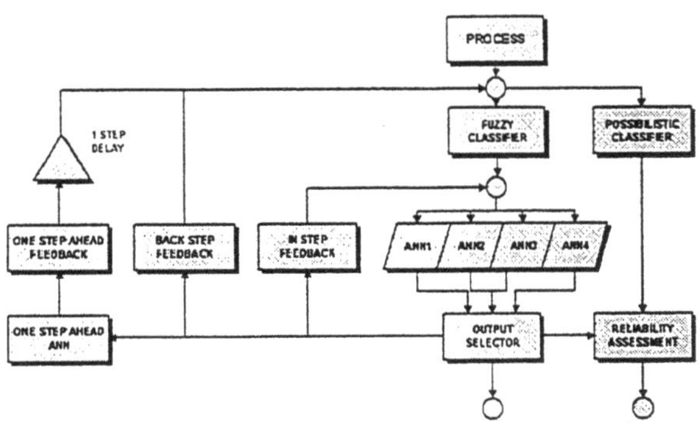

Figure 5.1 PEANO functional diagram

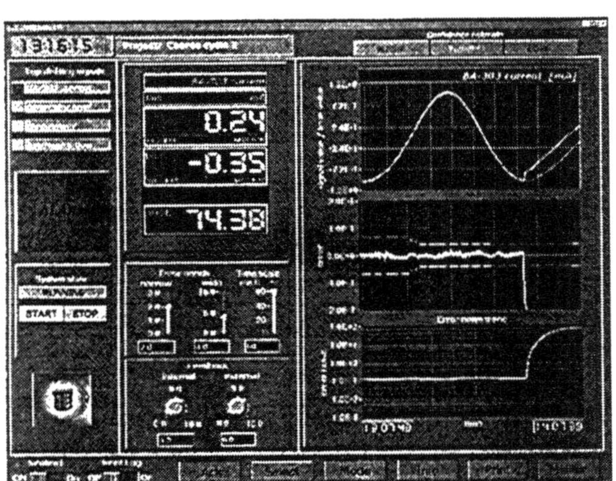

Figure 5.2 The PEANO display in monitor mode

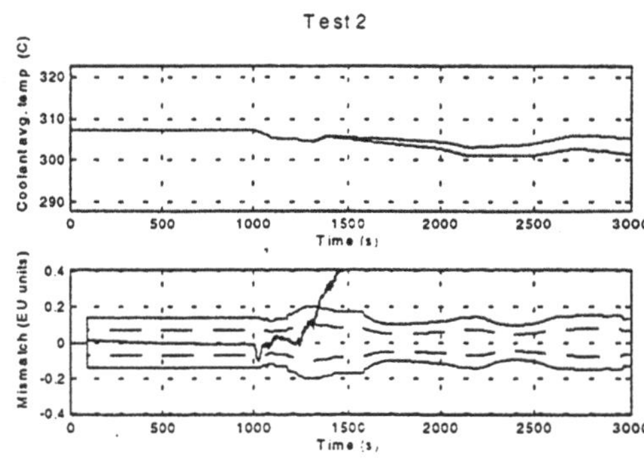

Figure 5.3 Detection of a small signal drift in PEANO

PEANO has been recently tested (P.F. Fantoni, S.Figedy, B.Papin, 1997 ; P.F. Fantoni, S.Figedy,A.Racz, 1998) successfully using data from a Pressurized water nuclear reactor in France. Figure 5.3 shows how a drift in the Primary loop water temperature is detected, as soon as the drift exceeds 0.2 C.

The same figure gives an idea of the accuracy of the system (represented by the the two bands in red), which is a consequence of the use of specialized ANN ineach operating region of the process.

PEANO will be installed in a nuclear full-scope training simulator in the second half of 1998, where tests in real-time will be performed.

5. CONCLUSIONS

Neural Networks and Neuro-Fuzzy systems for process monitoring and diagnosis have been the object of an extensive research and testing, in the last ten years.

The results achieved with emerging system implementations, like those described in this paper, suggest that this technology is mature for industrial implementations.

The following two considerations, in the authors opinion, will encourage and booster industrial applications of this technology in the near future:

5.1. The technical aspect

The experiments and experience with neural network applications in complex industrial environments lead us to the following conclusions :

- the approach based on using *a priori* knowledge (of the physical laws of the process) seems sound and quite effective;

- the development methodology for process supervision, diagnosis or optimising systems, using neuro-fuzzy techniques, which the authors developed and experimented, is feasible and reliable and is suitable in many industrial contexts, such as petro-chemical industries.

5.2. The point of view of the process operator or the plant maintenance planner

In an environment where competition gets ever fiercer and where safety and economic criteria are paramount, tools that can help optimise the process during normal operation or tools that can help timely planning of preventive maintenance, are in great demand and prove more and more valuable.

REFERENCES

J.M. Renders, A. Goosens, F. de Viron en M. De Vlaminck, "A prototype neural network to perform early warning in nuclear power plant" Fuzzy Sets and Systems 74 (1995) 139-151

F. de Viron, M. De Vlaminck, J.M. Renders, "Artificial Neural Networks in process monitoring: a practical experience with a prototype to perform early warning in nuclear plants", paper presented at the International Conference on Artificial Neural Networks, Paris, October 9-13, 1995, France

P.F. Fantoni, S.Figedy, B.Papin, "*A* Neuro-Fuzzy Model Applied to Full *Range* Signal Validation of PWR *Power Plant Data*", Second OECD Specialist Meeting on Operator Aids for Severe Accident Management (SAMOA-2), Lyon, 8-10 September 1997

P.F.Fantoni,S.Figedy,A.Racz, "PEANO: A TooBox for Process *Signal Validation and Estimation*", HWR-515, OECD Halden Reactor Project, February 1998

TOWARDS SELF-VALIDATING CONTROL LOOPS

Jun Chen and John Howell

Department of Mechanical Engineering, University of Glasgow, Glasgow G12 8QQ, UK
Email: j.howell and jchen @mech.gla.ac.uk

Abstract: The possibility of developing self-validating control loops is examined for those loops that inherently eliminate steady state error, for instance by the incorporation of integral action. In these situations it should be possible to reason about abnormal, but steady state, deviations in measured, control loop related variables. The approach discussed here involves a search and test strategy which is based on a variant of the signed directed graph and on knowledge pertaining to control performance. The approach is applied to a simulated CSTR process successfully. *Copyright © 1998 IFAC*

Keywords: fault diagnosis, process control, validation, supervision

1. INTRODUCTION

Conventionally, signed-directed-graph(SDG)-based fault diagnosis has focused on the process plant, with its associated control and sensory systems, as a single entity (e.g. Kramer & Palowitch, 1987; Mo et al,1997). Although functional decomposition has been discussed (Finch & Kramer, 1988), the focus has still remained on process performance as a whole. The advent of the 'smart' sensor, actuator and so on, appears to go against this philosophy and to raise the question as to whether there should be 'smart' fault diagnosis in general. Thus the notion of the self-validating control loop is timely (Clarke, 1995): the focus would be on individual loops and the process would only be of interest in as far as it affects these loops. In-process faults would still be diagnosed however because it is likely that most of the physical processes involved would form parts of the control loops i.e. the process would effectively be under observation. If the focus is on the individual control loop, then one approach to fault diagnosis is to depict pathways of causality by SDGs pertaining to individual control loops connected by an SDG of the process. If the plant then meets a number of fairly general criteria, a straightforward reasoning process can be applied using knowledge of how faults in control loops cause deviations in measured variables. These criteria are as follows:

(1) all control loops are able to guarantee zero steady state error e.g. they have integral action;
(2) the fault remains until a new steady state is reached;
(3) controller outputs, together with measurements of the control variables and valve-openings are available as *observations*.

If these criteria are met throughout the plant, then fault diagnosis can be achieved by a method that combines functional decomposition(FD), SDGs and controller-based knowledge(CBK); self-validating control loops (SVCLs) can then be viewed as the intersection of these three components (figure 1).

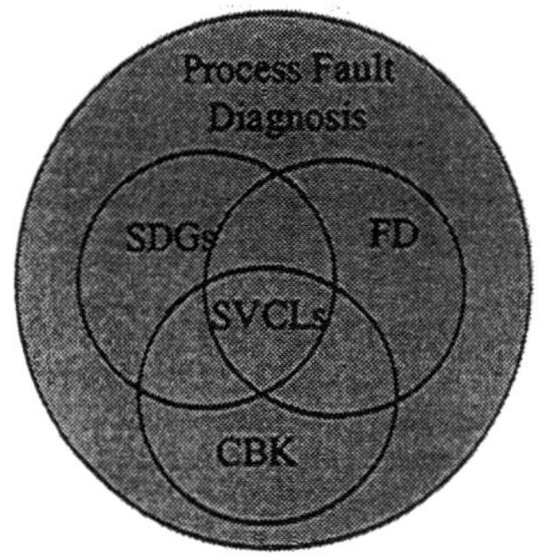

Fig. 1. Process fault diagnosis scheme

2. CONTROL LOOPS: THEIR SDG REPRESENTATION AND CAUSE-EFFECT KNOWLEDGE

The SDG representation for the single loop control system shown in Figure 2, is given in Figure 3 (N.B. +/- signs have been omitted for convenience): components enclosed by the dashed circle can be viewed as one vertex (or node) with external influences applied through E. It is possible to construct a table (Table 1) to describe the various effects that individual faults would have on the observations available. A similar approach can be applied to a cascade control system: its SDG is presented in Figure 4 and cause-effect knowledge pertaining to the resultant vertex is given in Table 2. Faults like a dead sensor, or a sticking valve or a large exogenous fault (or disturbance) are not considered in Tables 1 and 2 because, in these circumstances the steady state is unlikely to be obtained. Some appropriate simple heuristic rules can be applied to isolate these faults.

The directions of the deviations in the various observations can provide additional information with which to infer the 'direction' of the various fault hypotheses e.g. "fails-high" or "fails-low". The following sub-sections discuss the determination of "fails-high" or "fails-low" for a sensor bias, valve bias and exogenous/ancestor fault.

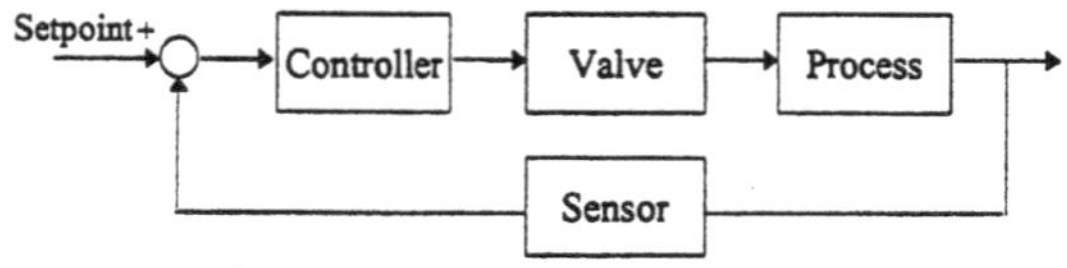

Fig. 2. Single loop control system

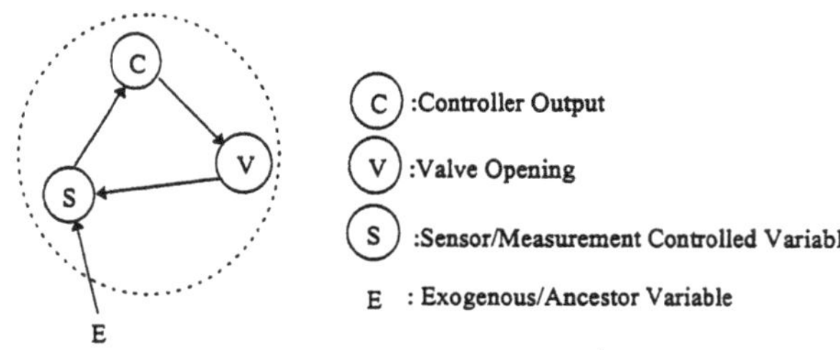

Fig. 3: SDG of the single loop control system in Figure 2

A sensor bias in a single loop control system or in the outer loop of a cascade control system. If the sensor biases, the controller will take action to compensate for this with the net effect that there will be a deviation in the controller output and the sensor measurement will return to its normal value. The direction of the sensor bias can then be determined by looking at the following (Figure 5):

- R-sc: the relation (i.e. forward or reverse acting) between the sensor measurement and the controller output;
- Dc: the deviation of the controller output.

The various possibilities are listed in Table 3.

A sensor bias in the inner loop of a cascade control system. Both the inner and the outer loop controllers will attempt to compensate (Figure 6) with the net effect that the sensor deviation observed (Ds) will have the same direction as the (inner) sensor bias. Table 4 summarises this.

An exogenous/ancestor fault or disturbance. In both single and cascade loop cases, such occurrences will be compensated by the controllers with the net effect that there will be a deviation in the controller outputs. The direction of the exogenous/ancestor fault or disturbance can then be determined by looking at the following:

- R-sc: the relation (i.e. forward or reverse acting) between the sensor measurement and the controller output;
- R-ex: the relation (i.e. forward or reverse acting) between the exogenous/ancestor variable and the sensor measurement;
- Dc: the deviation in the controller output.

Although these 3 factors refer to the various loops differently (see Figures 7-9) the various outcomes can be represented in a single table, Table 5.

A Valve bias. If the valve biases, there will be a deviation in the controller output with the effect that the sensor measurement will return to its normal value. The direction of the valve bias can then be determined by looking at the following factors (Figure 10):

- *R-cv*: relation between the controller output and the valve opening;
- *Dc*: the deviation in the controller output.

Table 6 describes the possible outcomes.

Table 1: Cause-effect knowledge pertaining to a single loop control system in steady state

observations		fault candidate
sensor reading = setpoint	controller output ≠ normal	sensor bias valve bias exogenous/ancestor fault or disturbance
	controller output = normal	no fault

Table 2: Cause-effect knowledge pertaining to a cascade control system in steady state

observations			fault candidate
Inner loop	sensor 2 reading = setpoint 2 = normal	controller 2 output ≠ normal	valve bias exogenous/ancestor fault or disturbance 2
		controller 2 output = normal	no fault
	sensor 2 reading = setpoint 2 ≠ normal	controller 2 output = normal	sensor-2 bias
		controller 2 output ≠ normal	outer loop fault
Outer loop	sensor 1 reading = setpoint 1	controller 1 output ≠ normal	sensor 1 bias exogenous/ancestor fault or disturbance 1
		controller 1 output = normal	no fault

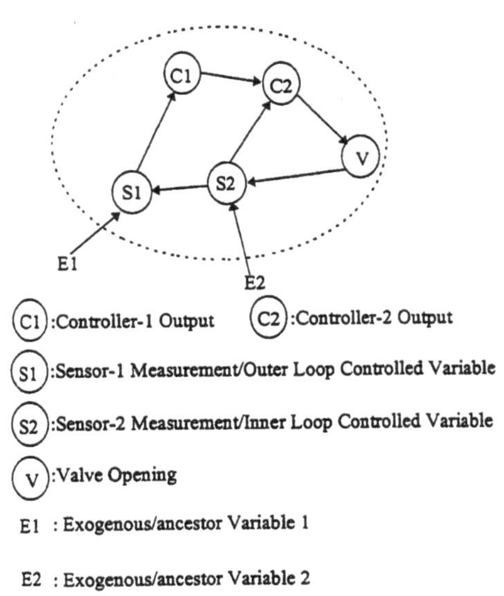

Fig. 4: SDG of a standard cascade control system

Table 3: A Sensor bias in a single loop or the outer loop of a cascade control system

Dc	*R-sc*	Sensor bias
+	+	fails-high
-	-	fails-high
+	-	fails-low
-	+	fails-low

Table 4: A Sensor bias in the inner loop of a cascade control system

Ds	Sensor bias
+	fails-high
-	fails-low

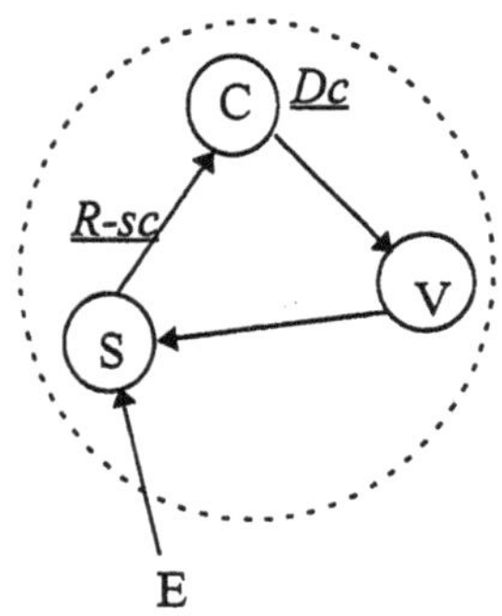

Fig. 5: Sensor bias factors: single loop

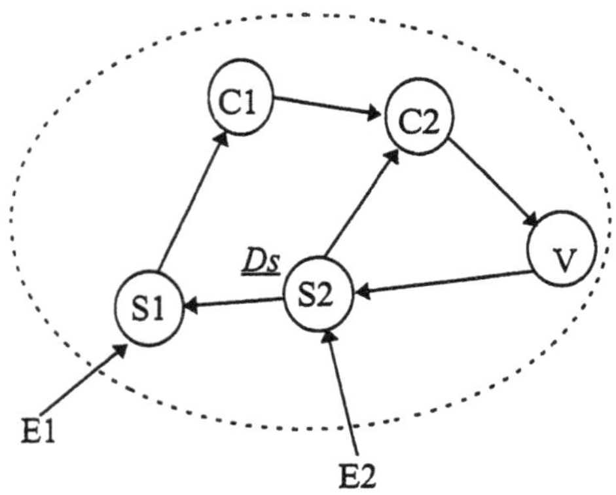

Fig. 6: Sensor bias factors: inner loop

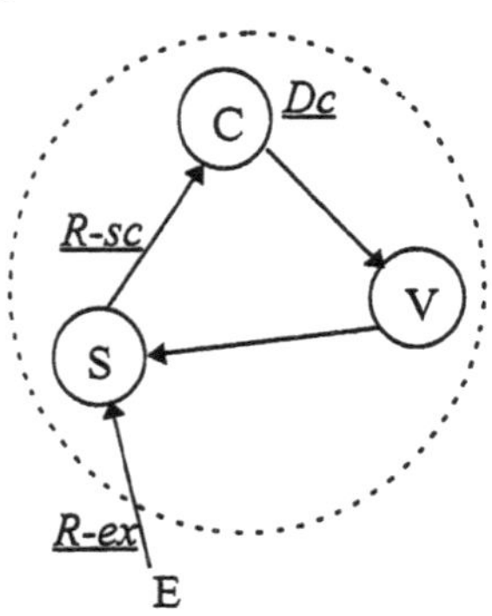

Fig. 7: Effect of an exogenous/ancestor fault or disturbance on a single loop

Table 5: Exogenous/ancestor fault or disturbance

Dc	R-sc	R-ex	Exogenous/Ancestor fault or disturbance
+	+	+	fails-high
+	-	-	fails-high
-	+	-	fails-high
-	-	+	fails-high
+	+	-	fails-low
+	-	+	fails-low
-	+	+	fails-low
-	-	-	fails-low

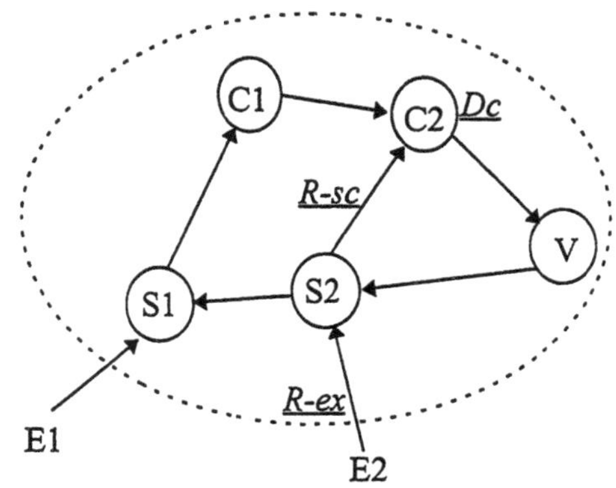

Fig. 8: Effect of an exogenous/ancestor fault or disturbance on the inner loop

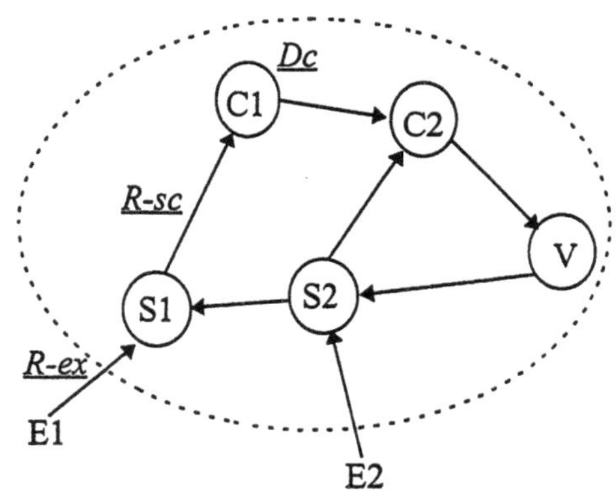

Fig. 9: Effect of an exogenous/ancestor fault or disturbance on the outer loop

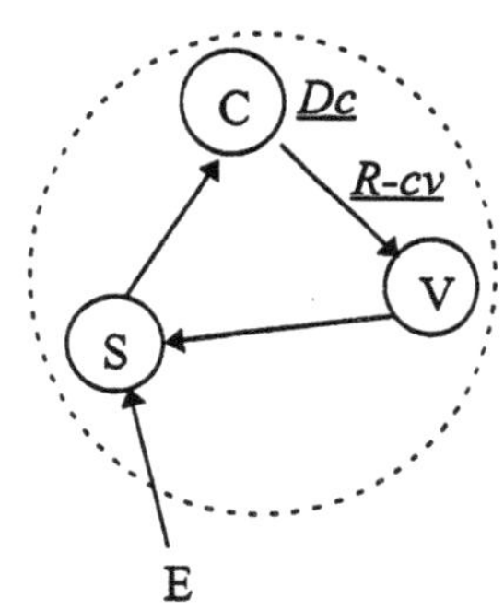

Fig. 10: Effect of a valve bias on a single loop

Table 6: A valve bias

Dc	R-cv	Valve bias
-	+	fails-high
+	-	fails-high
+	+	fails-low
-	-	fails-low

3. FAULT ISOLATION

Any steady state deviation in any control loop measurement must be caused by either an exogenous fault or disturbance or by a fault in a loop element such as in the sensor or in the valve. Although, by the application of appropriate simple heuristic rules, it is very easy to isolate faults like a dead sensor, or a large exogenous fault (or disturbance), or a sticking valve, it is more difficult to isolate faults like a sensor bias, or a valve bias, or a small exogenous fault (or disturbance). In these circumstances, the controlled variable, and its effect on descendants, is arguably the key to fault isolation: with a sensor bias, the controlled variable will deviate from its nominal value and descendants of the controlled variable will be affected; with a valve bias, or with an exogenous fault or disturbance, the controlled variable will remain at its nominal value and its descendants will not be affected. This leads to a search and test strategy composed of a number of sub-strategies, two of which are described below.

Case 1. Consider the case in Figure 11: S1 is a measured controlled variable, S2 is a measured descendant of S1, E1 is an ancestor of both S1 and S2, it affects S1 and S2 simultaneously, and E2 is an ancestor of S1 and only affects S1. Suppose S1 loop controller deviates; then, initially, the fault candidate will be {S1 sensor bias, E1, E2, valve bias in S1 control loop}. There are now two possibilities:

- S2 is affected: the fault candidate set shrinks to {S1 sensor bias, E1} because E1 is the common ancestor of S1 and S2; if the direction of the deviation of S2 contradicts that expected from a S1 sensor bias, then {E1} is the only fault root; if there is no common ancestor between S1 and S2, then {S1 sensor bias} is the only root fault.
- S2 is not affected: the fault candidate set shrinks to {E2, valve bias in the control loop}. If there is no more information about E2, then these two possibilities cannot be separated; otherwise, if E2's descendants deviate, E2 will be the only root fault.

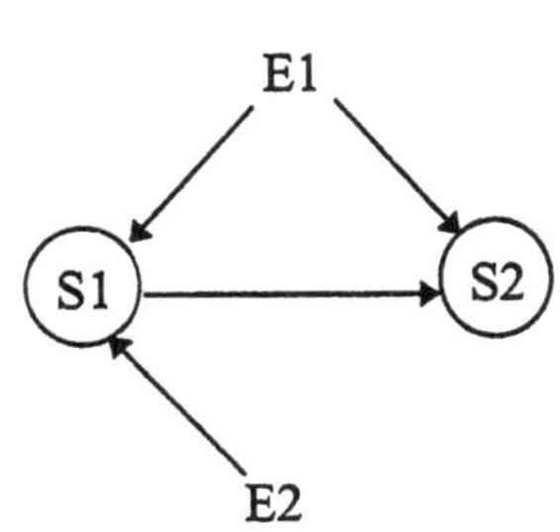

Fig. 11. Case 1

Case 2. Consider the case in Figure 12: S1 and S2 are measured controlled variables, they affect each other. If only one of the control loops deviates, the fault can be separated by the procedure of case 1. If both control loops deviate, there are two possibilities:

- S1 and S2 have at least one common ancestor, the fault candidate set is {S1 sensor bias, S2 sensor bias, common ancestors fault};
- S1 and S2 have no common ancestor, the fault candidate set is {S1 sensor bias, S2 sensor bias}.

Fig. 12. Case 2

4. APPLICATION TO A CSTR SIMULATION

A simulated CSTR process (Figure 13) controlled by a single loop level control system and a cascade temperature control system was investigated. Its SDG is shown in Figure 14. Table 7 shows the results obtained when 9 different faults were applied to the process. Although it is difficult to isolate between a valve bias and an exogenous fault/disturbance, sensor biases or exogenous fault/disturbances pertaining to the outer loop can all be isolated.

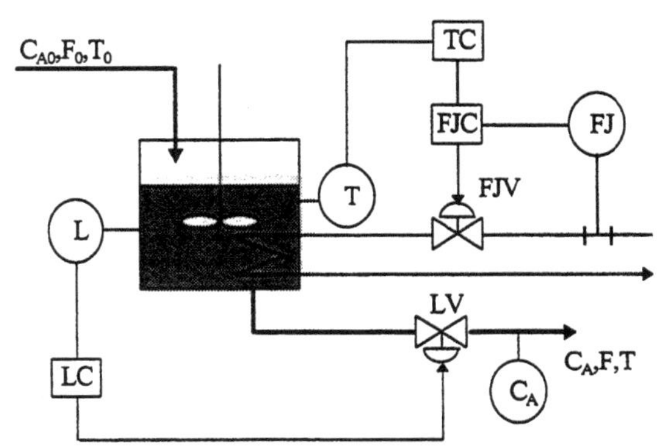

Fig. 13. A simulated CSTR process

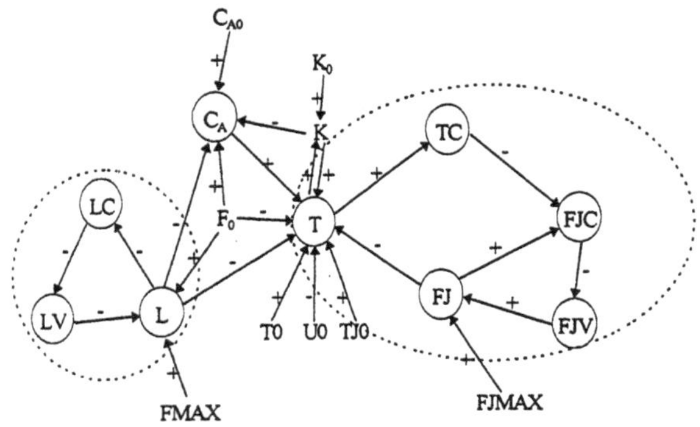

Fig. 14. SDG of the CSTR example (circled variables are measured)

Table 7: CSTR results

true fault	fault diagnosis candidate
L sensor-fails-high	{L sensor-fails-high}
T sensor-fails-high	{T sensor-fails-high}
FJ sensor-fails-low	{FJ sensor-fails-low}
LV fails-high	{LV fails-high, FMAX is low}
FJV fails-low	{FJV fails-low, FJMAX is low}
FMAX decreases	{LV fails-high, FMAX is low}
FJMAX decreases	{FJV fails-low, FJMAX is low}
F_0 decreases	{F_0 is low}
K_0 decreases	{K_0 is low}

5. CONCLUSIONS

It has been shown how faults can be isolated by viewing a plant from a control loop perspective; both single loop and cascade control systems have been examined. The isolation process, which consists of the application of a number of tables and search processes, has been tested on a simple CSTR simulation.

REFERENCES

Clarke, D.W. (1995). Sensor, actuator, and loop validation. *IEEE Control Systems*, Aug, 39-45.

Finch, F.E. and M.A. Kramer (1988). Narrowing Diagnostic Focus Using Functional Decomposition, *AIChE Journal* 34(1), 25-36.

Henry, M.P. and D.W. Clarke (1993). The self-validating sensor: rationale, definitions and examples. *Control Engineering Practice* 1(4), 585-610.

Kramer, M.A. and B.L. Palowitch (1987). A rule-based approach to fault diagnosis using the signed directed graph. *AIChE Journal* 33(7), 1067-1078.

Mo, K.J., Lee, G., Nam, D.S., Yoon, Y.H. and E.S. Yoon (1997). Robust fault diagnosis based on clustered symptom trees, *Control Eng. Practice* 5(2), pp.199-208

Yang, J.C. and D.W. Clarke (1997). The self-validating actuator. *IFAC Symposium Safeprocess'97*, Hull, UK, Vol 2, 579-584.

INTEGRATING ON-LINE AND OFF-LINE FAULT DETECTION FOR PROCESS SAFETY

J. Cao, J.A. Romagnoli, H.M. Tweeddale and P. K. Walsh*

Department of Chemical Engineering
**Graduate School of Business*
The University of Sydney, NSW 2006, Australia

Abstract: A systematic approach to integration of on-line and off-line fault detection systems for process safety assurance has been developed. It comprises three sequential computer-based stages: a) hazard identification; b) design of the on-line and off-line monitoring scheme to match those hazards; and c) design of the protocol for periodic auditing of the effectiveness and quality of the monitoring scheme. At each stage there is provision for review and editing by experienced staff. As the nature of auditing does not suit on-line operation, it is not included in this paper. *Copyright © 1998 IFAC*

Keywords: computer aided system design, fault detection, human factors, monitoring, rule-based system, variables, safety-critical

1. INTRODUCTION

The range of faults found in a process plant includes hardware faults and human errors. "Hardware" includes all physical features of the plant such as pipes, pumps, vessels, valves and instrument systems. The nature of hardware faults depends on the nature of the process, and the equipment used. In general, hardware faults are defined as 'synonyms to designate the departure from an acceptable range of an observed variable or calculated parameter associated with the equipment' (Himmelblau, 1978) e.g. hardware faults may result from failures of equipment and instrumentation systems. Faults resulting from human errors may be due to such causes as a slip or a momentary lapse of attention, poor training or instructions, a lack of physical or mental ability, a lack of motivation, and management error (Kletz, 1991). Effective monitoring for on-line or off-line detection of hardware faults would prevent or minimise the cost of downtime from malfunction of equipment and instrumentation, and reduce the frequency or likelihood of occurrence of a serious incidents such as fire, explosion or release of toxic materials. Provision of monitoring for off-line detection of faults due to human errors would not only improve operational reliability, and thus enhance efficiency and effectiveness of the on- or off- line hardware fault detection, but also improve process safety.

In design of on-line fault detection schemes for process plant, most attention has usually been given to those variables which affect production rate, product quality, process efficiency and cost, and reliability. Less attention has usually been directed toward systematic design of monitoring schemes for early detection of the onset of faults which affect process safety, and the potential for major incident. It is postulated here that the early warning signs of an increasing risk of serious safety-critical incident on process plant are often inadequately represented in these monitoring schemes.

An approach is outlined here for integrating design of plant specific monitoring and auditing protocols for early detection of the onset of faults, matching the plant-specific hazards. The approach comprises three sequential computer-based stages: i) computerised hazard identification, ii) process safety monitoring schemes, and iii) process safety auditing protocols. The third stage is an essential part of the safety assurance programme, but is not directly related to on-line detection of faults, and so is not included in the scope of this paper. The approach integrates on-line and off-line monitoring of both hardware and human factor faults as an essential component of process safety assurance.

2. INTEGRATED APPROACH

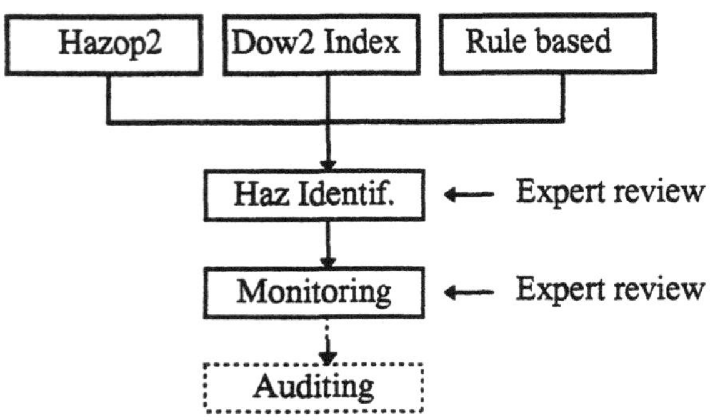

Fig. 1. Integrated Approach to Hazard Identification and On-line and Off-line Fault Monitoring Schemes (note: auditing is not discussed in this paper)

The first stage (hazard identification) aims to identify those variables where early response to a detected fault is necessary to prevent a major mishap. This stage comprises three parallel modules: an extension of computerised Hazop records, a reworking of the Dow Fire and Explosion Index, and a rule-based expert system. These three modules provide redundancy and diversity of approach to maximising the reliability of detection of hazards, and hence of minimising the risk of overlooking a significant variable. For each identified hazard, the likelihood and severity of the possible consequences are assessed, on logarithmic scales from 1 to 5, enabling preparation of a ranked and prioritised list of variables for monitoring.

The second stage, for each identified variable, accesses a data base and selects one or more approaches to monitoring for faults, using the nature of the process materials being handled, the process conditions and the equipment types being used. A recommended frequency of monitoring is calculated, based on the expected time from fault detection to the onset of mishap, the expected reliability of the monitoring method, the assessed consequence and likelihood (or frequency) of the incidents, and the cost of monitoring. It is important that incidents with high assessed consequence and likelihood be monitored on-line. Where the frequency is low, either on-line or off-line monitoring may be used, depending on the technology available and cost of use. This second stage covers monitoring of both hardware and human factors variables. (In practice, human factors are monitored off-line.)

Detailed methods for design of hazard identification and monitoring protocols are described below.

2.1 Hazard Identification

Three methods are used in parallel to identify possible hazardous incidents. The output of all three methods is in an identical format, to enable consolidation, comparison, sorting and analysis. The following principles are used:

- a team is used of with varied experience in process design and process operation;
- adoption of a systematic and detailed approach,
- comprehensively covering on deviations from normal condition;
- risks to the safety of people and the plant is calculated as the product of the consequence and frequency of occurrence.

These three methods are described as follows.

Computerised Hazop. Hazop is a widely used approach for identifying hazardous scenarios in process plants. Commercial computer software is widely available to aid administration of the studies. Hardy (1994) developed a similar system ("Hazcom"). Hazcom was adopted here because it has accessible computer coding, enabling addition of further features to suit the integrated approach. The revised Hazcom computer system, named Hazcom2, has following important additions:

- keyword field 1: for the types of equipment involved in the incidents identified;
- keyword field 2: for the types of potential incident;
- frequency and consequence fields: for assigned numerical magnitude scores;

When using the Hazcom2 system in deriving monitoring (and auditing) protocols, the report is sorted as required, e.g. by type of processed material, type of equipment, type of incident, including assessments of the frequencies and consequences of the incidents, etc.

Reworked Dow Index. The Dow Index (Dow Chemical Co., revision 1994) is an index of fire and explosion risk which is calculated from factors for the hazards of the materials handled and the inherent hazards of the process and equipment used. In this research, the method used in determining the Dow Index has been modified, separating the factors according to whether they are frequency- or consequence-related, to enable determination of separate consequence and frequency indices, from which a combined risk index is then calculated. This enables merging of the output from this method with the output of the other two hazard identification methods, each of which generates both consequence and frequency indices.

Other modifications include extension of the list of special process hazard factors to include new factors covering equipment size, equipment lifetime, operational status, the reliability of preventive systems (e.g. automatic and manual control systems) the reliability of protective systems. The modified Dow Index system used here is termed "Dow2".

The fields in the output from Dow2 are the same as those from Hazcom2. This has necessitated normalising the assessed frequency and consequence scores from Dow2 to the same scale of 1 to 5 for comparability.

Rule-Based Computer System. The rule-based computer system is a machine-driven approach using information about the plant operation (eg. types and features of equipment used, properties and quantities of hazardous materials used, types of potential initiating causes, etc.) to generate a list of potential incidents by systematically exploring the various possible combinations. The rule-based approach is a objective and comprehensive technique that has the potential to generate a wider range of possible incidents than either the Hazcom2 or the Dow2 approaches although, in practice, some of these incidents may not be significant or feasible.

The concept of 'dimensionless consequence' was developed to simplify the quantification of the magnitude of impact from the identified incidents on people, property and the environment. Similar to the Dow2 approach, the frequency and consequence of an identified incident are normalised to a logarithmic scale of 1 to 5.

Human factors. The 'human factor' has been regarded as one of major causes of hazardous incidents and hardware faults in process plants. Ultimately, progressively penetrating layers of the root causes of most process plant incidents reaches a universal root cause of "management failure", a fault in the human factor category (Tweeddale, 1996). Clearly, high standards of management of human factors relating to organisational structure, functional systems, responsibilities, attitudes and "culture" etc. are essential. Rasmussen (1989) suggests that it may be more fruitful to focus on the allocation of responsibilities, improvements in systems, and changing the situation so as to change the attitudes and behaviour of people, rather than on specific departures from normal operating practice. Unfortunately, it was not practicable to incorporate human factor in a quantitative manner in any of these three methods of hazard identification because of the difficulty of assessing the frequencies of the various types of human error and of the resulting incidents. It was therefore concluded that preferable to use the results of three hazard identification methods for design of only the hardware fault monitoring scheme, and to approach automatic generation of human factors monitoring requirement differently, using a generic list of classes and types of human factor weaknesses.

Expert Review. The results of the hazard identification stage are to be reviewed by experts, making use of the same philosophy as was postulated by Romberg et al (1996) that "the insights, gained from experience, are generally more valuable than those relying solely on the unthinking application of advanced signal processing strategies. Combining this insight with comprehensive signal processing strategies can often turn the detection of an 'anomalous event' into meaningful information about the process under examination".

2.2 Process Safety Monitoring Protocols for Fault Detection

Design of protocols for fault monitoring is based on application of 'performance indicators' and the derived concept of 'risk drivers' (Cao et al, 1996). The monitoring stage of the approach generates a list of possible performance indicators based on hazards identified, equipment used, etc. from the first stage. The list is then reviewed and edited by plant experts and managers so as to incorporate insights gained by experience. Monitoring for hardware and human factor faults detection are described separately below.

Hardware fault monitoring. The monitoring topics are generated according to types of equipment involved in the incidents identified using a data base of the most practicable methods for those types of equipment. There are several factors which determine the monitoring frequency with which any particular intermediate safety variable should be monitored. These include: the rate at which the fault condition develops from when it could be first detected until occurrence of the incident; the probability that the fault condition would not be detected per monitoring occasion; the frequency with which the fault condition is likely to occur; the all-inclusive cost of monitoring; the severity of the consequences of the fault condition if it is not adequately responded to; the feasibility and probability of effective response if onset is detected; the safety of undertaking the monitoring; the level of technical or other expertise needed to do the monitoring, etc. An initial estimate of the optimal frequency can be derived from these factors, and refined in the light of on-going plant experience. The broad relationship between these variables is illustrated by Equation 1 below, noting that the detailed relationships would need to be defined with extensive empirical review.

$$f_m = f\left\{\left(\frac{1}{T}, P_f, \frac{1}{C_m}, f_r, c_r\right)\right\} \quad (1)$$

where,

f_m - annual frequency of monitoring
T - mean time from onset of detectable condition to the incident
P_f - probability of failure of monitoring to detect the onset of the condition
C_m - cost of monitoring
c_r - normalised consequence of an incident
f_r - normalised frequency of an incident

Human factor faults monitoring. The design of this module is based on six requirements for safety assurance postulated by Hawksley (1987). This can be used as the framework for categorising which all human factors into six classes. Simplified here from Hawksley's original form, they are:

- good understanding of hazards, consequences, safeguards;
- provision of appropriate equipment and facilities;
- provision of appropriate procedures and systems;
- provision of appropriate organisation, training, supervision and competent staff;
- provision of adequate emergency capability; and
- active promotion of safety and development of strong safety climate and culture.

These six requirements are used as the basis for structuring a process safety measurement system. According to the nature and characteristics of the human factors, the initial off-line monitoring frequencies should be decided by a team of appropriately experienced people, and regularly updated on the basis of observations over time, both in the continuing monitoring programme and the periodic auditing programme.

2.3 Integration of Hazard Identification and Monitoring of Fault Detection Protocols

Integration was approached through:

- structuring the output of the three hazard identification modules in Stage 1 in an identical format, and combining them into one data base ready for sorting, ranking and reviewing;
- using the results from Stage 1 (after review by line management) as the input to Stage 2 for generation of monitoring topics by use of a combination of automated rules such as criteria for selection of the risk drivers and quantitative prioritisation criteria, together with manual input from experienced and expert insight.

3. DETERMINATION OF TOPICS TO BE MONITORED ON-LINE OR OFF-LINE

As mentioned before, on-line and/or off-line could be designed for faults detection monitoring. These are discussed in sections below.

3.1 On-line Fault Monitoring

Monitoring for hardware fault detection is more likely to be an on-line operation than for human fault detection. However, while it is technically practicable to monitor most process variables on-line, it is not currently practicable to monitor all variables of equipment condition on-line. Further, the cost of establishing on-line monitoring facilities, and of continued operation of them, may not be economically justified, depending on the likelihood and consequences of the incidents which could occur in the event of a fault occurring. On-line monitoring has most priority where the equipment is most important to plant production and safe operation, and the likelihood of occurrence is at least credible. There are two main safety monitoring schemes. Firstly, the generated monitoring topics may have already been implemented in the plant control system, possibly for control of production rate, quality or cost, but with a higher or lower frequency. In this case, the frequencies should be compared, and an appropriate frequency selected which meets the safety requirements as well as the normal production requirements. Secondly, the monitoring topics have not been currently included in the control system, in which case additional on-line or off-line monitoring would be considered. Normally, on-line monitoring would be recommended where the monitoring frequency (f_m) calculated by the computer model is higher than is conveniently undertaken manually, or where the consequences could be serious and where instrumented methods are available which are more reliable than manual methods. Otherwise, an off-line monitoring with an appropriate interval may be scheduled.

3.2 Off-line Fault Monitoring

Clearly, some faults related to hardware and all human factors are likely to need to be monitored off-line. Such monitoring is likely to be mainly infrequent (e.g. not more often than once per shift, or perhaps once per hour. This monitoring is still very important: it often relates to the condition of critical equipment (e.g. integrity of pipework, quality of adherence to defined operating and maintenance procedures etc.) and much of the monitoring on a plant is of this type. Hence determination of an appropriate monitoring frequency is important, and will probably rely on people with experience in both line operation and line management.

4. CASE STUDY

This computer-based approach has been implemented in a preliminary form integrating hazard identification and on- or off- line fault monitoring (and off-line auditing) into a single package. It has been trialed in several chemical installations and petrochemical plants, and is still under modification in the light of the findings. A case study carried out in a petrochemical plant (called Plant A) is briefly described below, with details omitted or changed for reasons of commercial confidentiality.

Plant A consists of five sections namely: light hydrocarbon vaporisation, cracking, quenching, caustic scrubbing and distillation, and manufactures a range other hydrocarbons with a range of molecular weights. The scope of this case study was confined to the vaporisation section only.

It should be noted that use of changed data has altered the assessed consequences, frequency and hence risk, and so altered the computed monitoring frequencies. However, the identified types of incident, the types of equipment involved, etc. have not been affected, thus the broad nature of the findings has not been affected.

5. RESULTS AND DISCUSSION

5.1 Hazard Identification

Three modules were in turn applied for hazard identification. Fire and confined space vapour cloud explosion (VCE) were identified as potential incidents by all three hazards identification modules. The results are summarised in Table 1.

Table 1 Results of Hazard Identification

Incident	Number of Incidents Identified			Max Normalised Consequence Score			Max Normalised Risk Score		
	H	***D***	***R***	***H***	***D***	***R***	***H***	***D***	***R***
Fire	2	9	17	3.5	3.5	2.1	6.5	7.3	5.8
VCE	0	9	17	-	3.5	2.7	-	7.3	5.5

Key: ***H***azop; ***D***ow; ***R***ule based.

In the computerised Hazop studies, five lines were considered. Thirteen scenarios were examined by team members, among which two would lead to a major fire. There was no vapour cloud explosion identified. The estimated consequences of each postulated fire were 3.5 (out of 5) on the severity scale, resulting from a) failure of a level controller; and b) opening at low pressure of a safety valve on the light hydrocarbon feed pipeline. The likelihoods of the two incidents were assessed as 4 and 2 (out of 5), respectively.

The scenarios involving failure of the level controller was assessed as having a risk score of 6.5 (out of 10). The predicted potential consequences were loss of light hydrocarbon, physical damage to the vaporising heater and to an adjacent furnace.

In the modified Dow Index module, eighteen hazardous scenarios were identified. Fires and vapour cloud explosions could result from failure of pipelines with a high consequence score of 3.49 (out of 5) and a maximum risk score of 7.29 (out of 10).

In the rule-based module, thirty-four scenarios were identified relating to the types of equipment involved and the potential initiating causes. The assessed maximum consequence (2.71 out of 5) resulted from a postulated fire following a leak or joint failure at the light hydrocarbon feed pump and the maximum risk score was 5.79 (out of 10).

It is evident that the three modules postulated different incidents, and generated different consequence and frequency assessments for them. The explanation is: a) the Hazop module generates a list of results as determined by the study team, and thus represents the knowledge and practical experience of the team members. Hence, the number of hazardous scenarios discussed is limited only by their knowledge and experience; b) the modified Dow Index module, because of its structure, generates a single output for each specified equipment item, and does not indicate the nature of the initiating event; and c) the Rule-based module is entirely machine-driven, and the potential hazardous incidents are derived mechanically from the inherent hazards, with the relationships between these factors being structured in the database, and the actual operational conditions being supplied by the user. In the case study, some high severity incidents were assessed as having low frequencies, and hence as presenting only a moderate risk.

5.2 Hardware Fault Monitoring

Hardware monitoring topics were generated from the identified hazards, and reviewed using expert judgement. An example of the monitoring scheme generated for equipment is shown in Table 2 for the liquid hydrocarbon delivery pipeline from a feed drum.

Table 2. Hardware Fault Monitoring

Eq't	Monit Topic	Freq	Review*
Pipe	Thickness test	1 pa	Yes
Pipe	Leak detection	on-line	Yes
Pipe	Magnetic particle test	1 pa	Yes
Pipe	Radiography	1 pa	No
Pipe	Acoustic emission	1 pa	No
Pump	Corrosion test	1 pa	Yes
Pump	Vibration	4 pa	Yes
Pump	Visual inspection	2 /day	Yes, pst
Pump	Throttle bush check	when pump open	Yes
Pump	Pressure difference	on-line	dbpc
Pump	Visual inspection	2/day	Yes, pst

*dbpc: done by process control;
pst : part of shift tour

Table 2 illustrates that not all equipment requires on-line monitoring; and how the computer-generated monitoring topics and frequencies are subjected to expert review.

5.3 Human Faults Off-line Monitoring

As mentioned earlier, the topics for off-line monitoring of human factors are based on the requirements for process safety assurance postulated by Hawsley. These have been compiled into a data base with six broad categories, each category comprising a checklist, possible fault modes, monitoring frequencies and tick boxes for adoption of the checklisted items. An example of the category of "Provision of sufficient training and supervision" is presented in Table 3.

Table 3. Off-Line Monitoring of Human Faults

Monit Topics	Fault Mode	Monit Freq	To be adopted?
Training	Poor training	2 pa	Yes
Supervision	Lack of supervision	anytime	Yes
Staffing	Improper assignment	4 pa	Yes, but 2 pa
...			

As in the case of hardware monitoring, Table 3 indicates that expert judgement would play an important role in deciding whether a specific topic needs to be monitored and, if so, how often.

6. CONCLUSION

Protocols for monitoring and detection of faults have commonly been designed, installed and implemented for principally meeting routine production requirements. The integrated approach described above broadens the coverage to include the requirements for process safety assurance. The approach is structured to use both the benefits of automatic generation of information derived by rules using information from data bases, together with systematic expert review, so as to gain insights which would not be available from either source in isolation. The output provides a valuable aid in determining which plant variables should be monitored on-line, and assists in justifying the associated expense.

REFERENCES

Cao, J.D., H.M. Tweeddale and K.P. Walsh (1996). An Approach to Integrating Hazard Identification with the Design of Risk-based Process Safety Monitoring and Auditing Systems. *Proceedings of CHEMECA96*, Vol. 6, pp. 13-18, Australia.

Dow Chemical Company (1994). Fire and Explosion Index Hazard Classification Guide. 7th edition. Dow Chemical Co., Miami.

Hard, H. (1994). Development of a Computerised Hazop Database Package. *Undergraduate thesis*, Dept of Chem Eng, Univ of Sydney.

Hawksley, J.L. (1987). Strategy for Safety Assurance for Existing Installations Handling Hazardous Chemicals. *Proceedings of WHO Conference on Chemical Accidents*, Rome.

Himmelblau, D.M. (1978). Introduction In: *Fault Detection and Diagnosis in Chemical and Petrochemical Processes*. pp.1-2. Elsevier Scientific Publishing Company, Amsterdam.

Kletz, T. (1991). An Engineer's View of Human Error. 2nd Edition. pp.1-7. Institution of Chemical Engineers, UK.

Rasmussen, J. (1989). The Rule of Error in Organising Behaviour. *The CEC Workshop on Errors in Operation of Transport Systems*, MRC Applied Psychology Unit, Cambridge.

Romberg, T.M., L.L. Black and T.J. Ledwidge (1996). Overview and Principles In: *Signal Processing for Industrial Diagnostics*. p.11. John Wiley & Sons Ltd., Chichester, England.

Tweeddale, H.M. (1996). The Intangible Dimension in Risk Assessment. Proc Chemeca96, Sydney.

ANALYSIS OF POST-UPSET PLANT DATA FOR FAULT DETECTION

Thomas J. English *
Lars Kaiser *
Frank Doyle **
P. S. Dhurjati *

* University of Delaware, Newark Delaware, USA
** Star Enterprises, Delaware City, Delaware, USA

Abstract: A reboiler attached to a debutanizer column allowed lighter than normal hydrocarbons in a stream supplying equipment not designed to handle the light hydrocarbons. Models of pieces of plant equipment were developed based on first principles and simple mathematical representations of valve behavior. The models were capable of indicating fault periods and of indicating equipment that was not functioning properly. This is the first step in a fault detection system designed for use in an actual chemical plant. *Copyright © 1998 IFAC*

Keywords: data mining, detection algorithms, distillation columns, fault detection, refinery upsets

I. BACKGROUND

1.1 *Introduction*

Faults occur in chemical plants. These incidents can cause anything from poor quality product to a plant shutdown. A plant not operating within its design specifications costs companies large amounts of money.

Currently many types of fault detection have been developed. However, industry uses few of these techniques due to the fact that currently plants are not equipped with all the sensors necessary to utilize such techniques.

Plant data taken during fault periods contains characteristics of the fault. Analysis of these data provides indicators of the fault. These indicators can be looked for in future faults to determine if a similar fault has occurred. This research focused on finding these indicators, mainly in one unit, Polymerization unit (26), in the Star Refinery, located in Delaware City, DE.

The polymerization unit consists of four reactors and several columns – important ones being the depropanizer and the debutanizer. Focus for this work was on the debutanizer. For the debutanizer, an accumulator temporarily stores some of the tops until a fraction of them are returned to the column, and a reboiler which returns a fraction of the bottoms from the column back to the column. The remaining fraction of the bottoms is sent to an olefin unit. Lastly, there are three pumps. Two pump fluid from the accumulator to the top of the column. One of these pumps is steam powered and the other electric. The steam pump is the main one used in the process and is operated off a 175 psi steam line. The third pump pumps liquid from the accumulator to butane storage. The valve controlling the steam flow to the reboiler, which is supplied from a 600 psi steam line, has a manual bypass valve connected with it. This bypass valve is open to some unknown extent during operation. The measurement of the flow is made after the bypass valve and the control valve. As shown in figure 1. The steam flows in the direction of the arrow and the flow measurement taken at the double hash mark. The control valve is pictured above the manual valve.

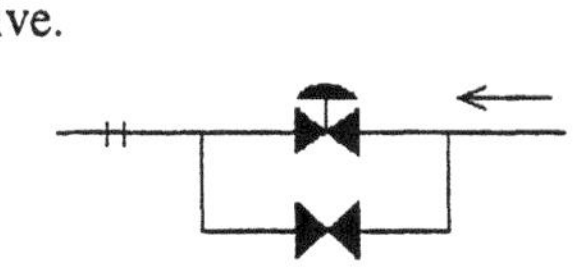

Fig. 1 Schematic of control valve, manual bypass valve, and flow measurement

For a simplified schematic of the portion of the unit focused on in this research refer to figure 2.

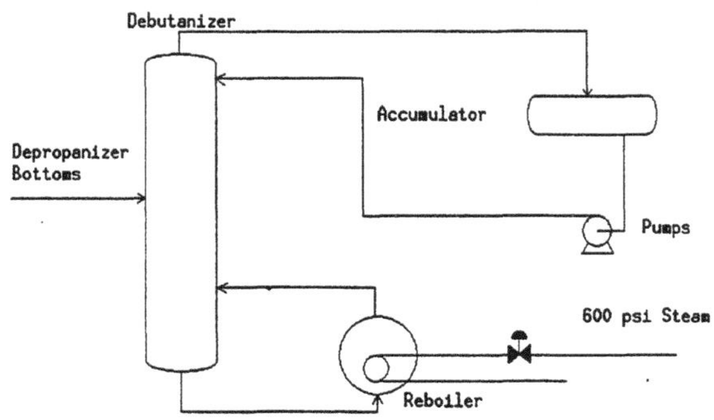

Fig. 2 Schematic of portion of unit

The Star Refinery encountered a problem with its debutanizer column in December 1996, which persisted into January 1997. This problem – lighter hydrocarbons than expected in the stream from the reboiler to the olefin unit – resulted in the unscheduled shutdown of the debutanizer unit and disassembly of the reboiler. Initially, fouling was suspected in the reboiler, causing poor heat transfer and resulting in lighter hydrocarbons not being boiled off allowing them to be passed through in the stream to the olefin unit – which is not equipped to handle lighter hydrocarbons. However, when the reboiler was opened for cleaning very little deposits were found, certainly not enough to cause the apparent poor heat transfer that was occurring. The reboiler was reassembled with nothing else done to correct the problem, yet, the plant returned to normal operation. In this paper, operation referred to as faulty is operation in which lighter hydrocarbons go to the olefin unit. Operation in which there were no reported problems is referred to as normal.

A unit shutdown is the last resort in industry when trying to fix a fault in plant operation. Costs of a one day shut down can easily surpass one million U.S. dollars. Thus, when the unscheduled shutdown on 14 January 1997 yielded no explanation as to the problem or its solution, interest was renewed in fault detection and diagnosis systems.

In the polymerization unit itself 522 measurements are made every two seconds. Twenty-seven of these measurements were chosen for the focus on the debutanizer column. However, storing the mass of data that is generated is impractical. Thus, once a minute the measurements are recorded. Also, due to problems with the historian, at times data are lost. Typically the length of time in which data is lost are only a few days – times when the historian is taken off line for upgrades or maintenance. These periods are infrequent and may only occur three to four days a year.

1.2 *Objective*

The point was to develop a tool that could point either process engineers or operators in the direction of the fault in the process. Using available plant data, a methodology had to be developed to find indicators of faults, and indicate which piece of process equipment was functioning or not functioning properly, thus indicating a starting point for a search in the cause of a fault. These indicators would be just the very beginning of a system for fault detection designed around an existing plant without requiring any new sensors or equipment being installed in the plant itself.

II. METHODOLOGY

Although a considerable amount of data was available and a large number of measurements are taken in the plant, the available data were not sufficient to complete simple mass and energy balances, to apply dynamic simulations, or to check the results of these calculations. This is due mainly to the fact that the data are recorded only once a minute, and only the process variables that the operators use to supervise the process are available. It is not possible to make a distinction between some process faults and sensors faults because the measurements taken are not redundant. In addition to this, items as essential as the compositions of the streams entering and leaving the column are unknown – only once an eight hour shift a sample is taken and sent to the lab for analysis and these values were not available for this project. Also, plant specifics such as the efficiency of the column, the types of valves used, piping sizes, and maintenance logs were rarely available.

Given these constraints, whatever methodology was developed it would indicate a group of possible faults – such as equipment malfunction, sensor malfunction, or controller malfunction – and not narrow the source down to one specific cause of the fault.

Without critical constants, calculating residuals could not be done. However, if these constants are grouped in such a way as that they appear only on one side of the equation and measured values appear on the other, a calculation of the group of constants can be made and these values are used to determine the relative operation of the plant.

2.1 *Varying Valve Coefficient*

Detailed study led to examine the behavior of the valve on the steam line supplying the reboiler. Using the plant data, and equation 1 a valve coefficient is calculated using two hour data segments and a least squares approach.

$$W = c_v \cdot C \cdot \sqrt{\Delta P \cdot P_1 / R \cdot T_1} \quad (1)$$

The flow is recalculated using the obtained valve coefficient. With data from periods of normal operation, the recalculated flow matched closely with the measured flow and the calculated constant remained steady. With data from periods of faulty operation, the recalculated flow did not match the actual flow and the coefficient varied frequently. This approach was taken before it was known that there was a manual bypass valve associated with the control valve (fig. 1). This equation assumes that the controller could cut off the flow of steam completely, which is not true with the bypass valve inline. Though a good starting point, a better model needed to be found.

2.2 *Pseudo Energy Balance*

With a focus on the reboiler, an energy balance was performed, that indicates proper or improper operation of the reboiler. A traditional energy balance appears as:

$$dU/dt = f_b \cdot (\Delta T_b \cdot Cp_b + X \cdot \Delta H_{vap}) - f_s \cdot \Delta T_s \cdot Cp_s \cdot E \quad (2)$$

Assuming steady state and that the heat capacities, fraction vaporized, the efficiency, and the heat of vaporization are constants, the equation can be reduced to:

$$\frac{f_b(\Delta T_b \cdot Cp_b + X \cdot \Delta Hvap)}{f_s \cdot \Delta T_s \cdot Cp_s \cdot E} = 1 \quad (3)$$

Because equation 3 is a linear equation the fraction vaporized and heat of vaporization, both considered constants, the heat of vaporization term may be dropped resulting in only a change of the value of the right hand side. Now, by grouping the remaining constants on the right side and the remaining variables on the left the following result is obtained, known as the pseudo energy balance:

$$\frac{f_b \cdot \Delta T_b}{f_s \cdot \Delta T_s} = k \cdot \frac{E \cdot Cp_s}{Cp_b} \quad (4)$$

The left hand side consists entirely of measured quantities, the right hand side consists of constants, both sides are dimensionless. The condition of the reboiler can be deduced by the calculation of the left hand side of equation 4. A normal value for reboiler operation can be determined by calculating the left hand side using data from a time period in which the reboiler is known to be operating properly. Since the numerator is related to the bottoms flow and the denominator is related to the steam flow, a higher than normal value indicates that too little steam is being supplied to the reboiler to boil off the lighter hydrocarbons for the amount of bottoms that are being sent through the reboiler. A lower than normal value indicates too much steam being fed to the reboiler for the bottoms flow. This case would be an indication of a decrease in the efficiency of the reboiler. By examining the right hand side of equation 4 confirmation is obtained that a decrease in the reboiler's efficiency results in a decrease from the normal value obtained in the calculations. All of these conclusions assume that the sensors are functioning properly.

2.3 *Linearized Valve Ratio*

Modeling the flow through the valve as a linear function of the control output and the bias allowed by the manually operated bypass valve results in:

$$\gamma \cdot C + \beta = f_s \quad (5)$$

Using the same idea as in the pseudo energy balance of dropping the intercept term in a linear relationship the following equation resulted:

$$C/f_s = K/\gamma \quad (6)$$

The left hand side of equation 6 consists of values measured and recorded. Valve behavior can be deduced by the calculation of the left hand side of equation 6. Normal behavior of the values calculated from the left hand side of equation 6 can be determined by calculating the values with data from a period of normal operation in the plant. Unlike the pseudo energy balance which has one set normal range of operation, the linearized valve ratio depends more on the way in which the values change. During periods of normal operation the ratio remains a relatively quiet signal, which is indicated by the standard deviation of data groups of 60 values, equivalent to the standard deviation for one hour of operation. An upward or downward shift in value followed by another period of quiet operation, only in this case a different range indicates a change in either the gain or bias. A change in gain would occur if the controller had been tuned. A change in bias would occur if the manual bypass valve had been opened or closed. To determine which change had occurred, data from before and after the change in ranges of operation would be used with equation 5 to find a least squares fit of β and γ. A fault which would somehow be related to the operation of the valve would be indicated by continuous shifts in the value of the ratio, which may appear as a wider range in the value of the ratio such a change would also be evident in the increase of the standard deviation of the data. Again, when making these conclusions it was assumed that the sensors were functioning properly.

III. PROCESS DATA ANALYSIS

Figures 3, 4, and 5 are examples of values generated by the pseudo-energy balance. Figure 3 illustrates values obtained from the pseudo energy balance while the unit was functioning properly. The values for the ratio remain between 75 and 225. These values were used as a basis for determining normal operation of the reboiler. For ease in comparing figures 3, 4, and 5 dotted lines are drawn in the figures at 75 and 225.

Figure 4 is an example of values obtained using the pseudo energy balance prior to and just after a reboiler cleaning on 8 November 1995 in which the reboiler was found to have "extensive heavy deposits" and reduced efficiency. For the time just before cleaning the reboiler the ratio was found to drop and remain below 50. This is an indication – as explained in the methodology section – of decreased reboiler efficiency. This conclusion would make sense due to the deposits found in the reboiler. After the cleaning took place the ratio returned to its normal range as determined by results in figures such as and including figure 2.

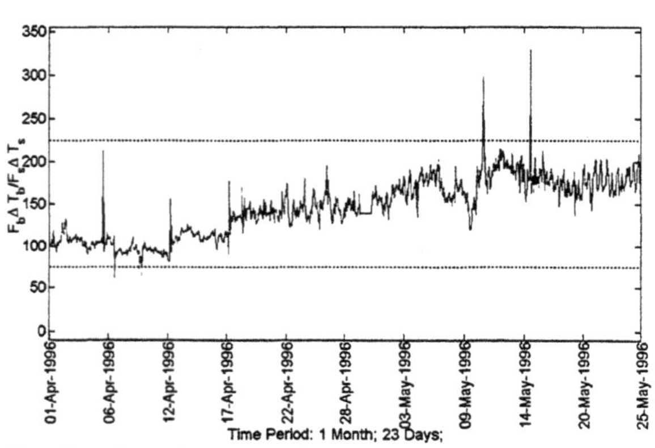

Fig. 3 Pseudo energy balance results during normal operation.

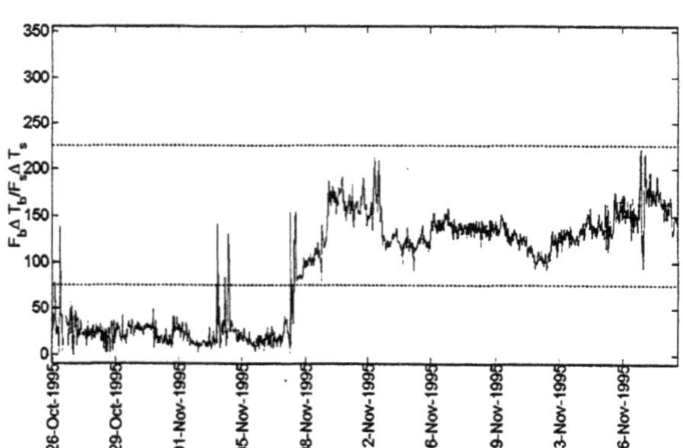

Fig. 4 Pseudo energy balance prior to and following a reboiler cleaning.

Figure 5 is a graph of values obtained using the pseudo energy balance during the fault period from 3 December 1996 to 14 January 1997. This fault led to a shutdown of the unit to examine the reboiler. No deposits were found in the reboiler to indicate that the efficiency of the reboiler had dropped or that the reboiler was operating improperly. The figure indicates that for the period of the fault the ratio had increased to a range of 125 to 350. For ease in seeing this new range, solid lines were drawn at 125 and 350. This is an indication – as explained in the methodology section – of too little steam supplied to the reboiler for

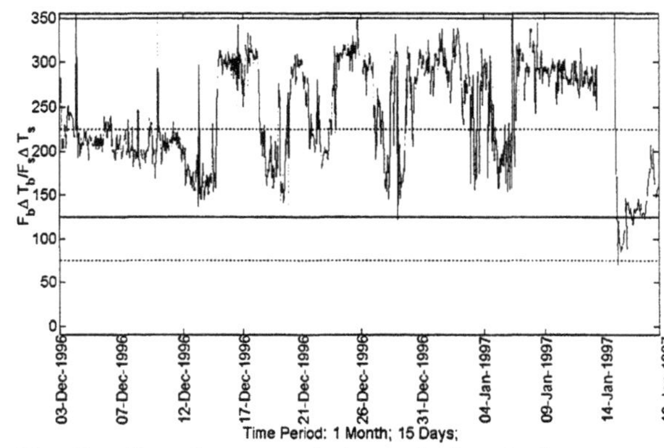

Fig. 5 Pseudo energy balance during fault.

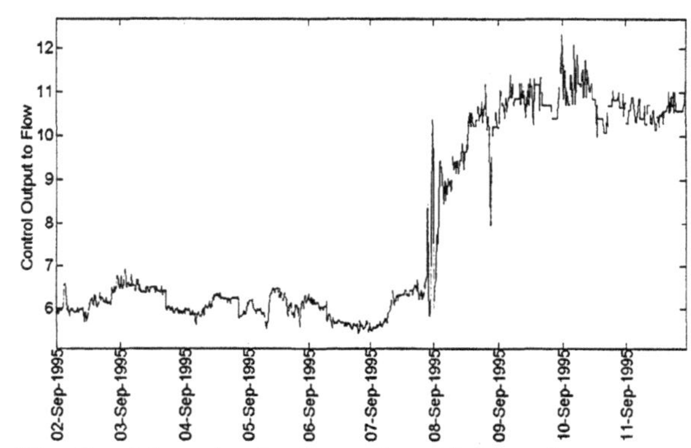

Fig. 6 Linearized valve ratio with bias change. Note bias changes on 8 September.

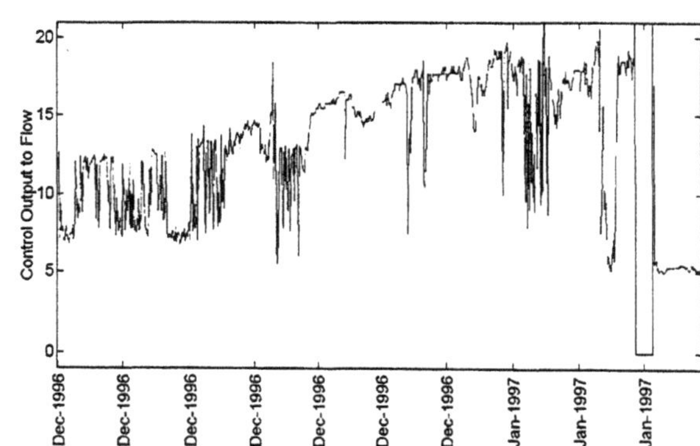

Fig. 7 Linearized valve ratio during fault. Note periods of widened range on 6-10, 12-15, 17-20 December and 4-6 January. Also note the return to a quieter lower value after the restart and return to normal operation on 14 January.

the feed of bottoms to the reboiler, and absolves the reboiler of suspicion in being the cause of the fault.

Figures 6 through 9 deal with the results from applying the linearized valve ratio to the steam valve supplying the reboiler. In figure 6, normal operation of the steam valve is illustrated along with a bias change. The bias change was determined using a least squares analysis of the data, as explained in the methodology section. The ratio remains within a tight band throughout normal operation and is a relatively quiet signal. These qualities helped define normal operation.

Figure 7 shows linearized valve ratios during the fault period which led to the shutdown of the unit in January 1997. In this case the value shows a larger range and an increase in the value itself up to the point of shutdown. The increase in the value itself may be
attributed to the manual opening of the bypass valve in a desperate attempt to supply more steam to the reboiler –

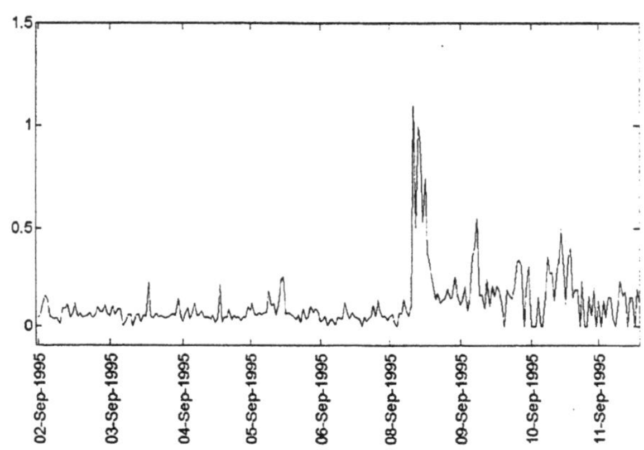

Fig. 8 Standard deviation of the linearized valve ratio taken using 60 minute groupings of data, during normal operation.

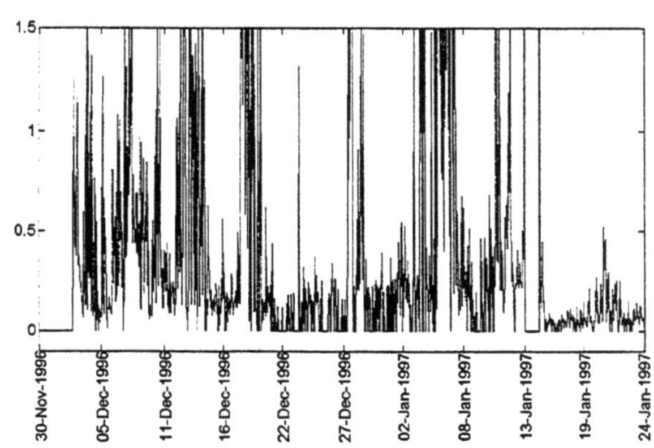

Fig. 9 Standard deviation of the linearized valve ratio taken using 60 minute groupings of data, during a fault period. Note the return to a normal value after the unit was restarted on 14 January.

this would result in a bias gain. The increased range demonstrates the breakdown in the ability of the controller to control the flow.

To indicate the relative quietness of normal operation compared to faulty operation, as determined by the linearized valve ratio, plots of the standard deviation of groups of 60 minute data of the valve ratio values were generated in figures 8 and 9. Sixty minute data groupings were chosen because the Star Refinery control software is already set up to take the standard deviation of 60 minute data groupings. Visual inspection illustrates that the standard deviation of the normal operating period is smaller that that of the fault period. It is also important to note in figure 9 that after the plant was restarted on 14 January, the standard deviation returned to a normal operating range, just as the operation of the unit returned to normal. These figures make clearer the point of increased activity of the linearized valve ratio during a fault when compare with a period of normal operation.

IV. RESULTS

The purpose of this work was to create simple mathematical models that take real plant data and manipulate it so that problems within the process can be detected and that the source of these problems could be either located or other parts of the plant eliminated as the source of the problem. These models are very similar to what the operators already intuit in the everyday operation of a chemical plant. Thus, the numbers generated by these models would have a physical meaning readily understood by the operators and process engineers. The first model developed – one that calculates the valve coefficient for the steam supply valve indicates if the unit is in a fault or not. One model developed – the pseudo energy balance – indicates if the reboiler is or is not a possible cause of operational problems. Another model, valve linearization ratio, coupled with statistical analysis of its results, is capable of indicating that the valve operation had changed in some way, and – during a known fault – that the valve was involved in a fault in some way.

The first model calculated a valve coefficient for the steam supply to the reboiler. After this value was calculated by a least squares procedure using two hours of data at a time, the flow was recalculated using the valve coefficient. At times when the plant was in fault the recalculated flow did not match the measured flow. When the plant was operating properly the calculated flow matched the measured flow. This valve coefficient method was used to indicate fault periods within the unit, but not to indicate the valve being the source of the fault due to the fact that no changes were made to the valve during the unscheduled shut down, and upon restart the valve functioned properly.

The second model developed was the pseudo energy balance – named because of the dropping of some terms to simplify the model. This indicator calculated from the model points at a drop in efficiency of the reboiler in the debutanizer column by a marked decrease in its normal value – determined from plant data during normal operation. Drops in efficiency were indicated before cleanings in which extensive deposits were found by the decrease in the value of the ratio generated by pseudo energy balance.

The indicator also points to a problem in the steam supply to the reboiler via a marked increase in its normal value. This symptom is evident at the time before a fault which caused an unscheduled shut down of the unit to inspect the reboiler. The model indicates that the efficiency, and hence the operation of the reboiler was not a problem. When the reboiler was opened no deposits were found to inhibit efficiency. Of the twenty-one months of data the model did not indicate any non-existent faults. With three known faults, the indicator showed that the reboiler was functioning properly.

The third model was a simple linearization ratio of the control valve that was supplying steam to the reboiler. The model related the controller output to the flow through the valve. This model was capable of indicating either a change in bias or change in gain in the relationship between the controller output and the flow by a steady shift in the value of the linearized model ratio from the prior steady value of the linearized ratio. However, determining if the bias or the gain changed requires a more in-depth analysis of the data utilizing least squares approaches and finding times in which flow rates were purposely changed. Faults related to the steam supply, and not necessarily the valve operation itself, were indicated by an increased range of the linearized ratio during operation. This increasing range over a short amount of time indicates a loss of control over the flow. This shift in range is also observable by the increase of the standard deviation of the linearized ratio. In the twenty-one months of data, this model did not indicate any non-existent faults, but did indicate bias or gain changes and three known faults related to the steam supply.

These models provide simple results using real plant data that allow an operator to better understand the current state of the plant, and allow engineers to locate possible sources of a fault quickly. The models also work around the fact that not all numbers needed for formal textbook analysis are available. In the future, similar simplistic models of the components of a chemical plant need to be developed with the idea in mind of using numbers that are actually recorded in chemical plants so that these models can be used to detect abnormal operation of the plant without requiring additional sensors and equipment for the plant. Placing these models online, a system can be designed that indicates components of the plant that are not operating properly before alarm systems indicate the fault. Such a system would indicate which piece of equipment is cause of the fault, reducing or eliminating downtime to correct the fault, saving large amounts of money for the company.

V. APPENDIX

5.1 Acknowledgments

We acknowledge Robert Brendel and Dennis Michaud for discussions about plant operation and their own industrial experience.

We acknowledge Dam Nguyen of Star Enterprises for locating and downloading data, Don Nolte of Star Enterprises for providing hands on information about the unit, and to Dave Hardin of Star Enterprises for providing a direction for this project.

We acknowledge Star Enterprises for financial support of this research.

5.2 Notation

C	controller output, in percent
c_v	valve coefficient
$Cp_{b,s}$	Heat capacity of bottoms and steam respectively
E	Efficiency of heat transfer from steam to bottoms. Defined as percentage of heat removed from the steam actually transferred to the bottoms in the reboiler.
$f_{b,s}$	flow rate of bottoms and steam respectively
ΔH_{vap}	Heat of vaporization of the bottoms
k	proportionality constant for pseudo energy balance
K	proportionality constant for valve linearization
ΔP	pressure drop across valve
P_1	pressure at valve inlet
R	individual gas constant
T_1	absolute temperature at valve inlet
$\Delta T_{b,s}$	change in temperature of bottoms and steam respectively
U	energy
W	mass flow through the valve
X	fraction of bottoms vaporized in the reboiler

Greek letters

β	bias of controller
γ	Gain of controller

ON-LINE EVENT-BASED SUPERVISION OF A BIOTECHNOLOGICAL PROCESS

Ramon Sarrate-Estruch[☆], **Josep Aguilar-Martin**[★†] **and Julio Waissman-Vilanova**[★]

[☆]*Universistat Politècnica de Catalunya*
Edifici Vapor Sala, Rambla de Sant Nebridi, 10
08222 Terrassa, Catalonia, sarrate@esaii.upc.es
[★]*Laboratoire d'Analyse et d'Architecture des Systemes, Toulouse, France*
[†]*Universitat de Girona, Girona, Catalonia*

Abstract: This paper describes an event-based approach to the design and implementation of a supervisory system applied to a biotechnological process. Process variables are continually analysed in order to identify signal trends which conform significative events. An heuristic model of the process under supervision is characterised as a discrete-event system (DES). Plant events generate transitions on the DES model, thus reflecting the dynamics of the continuous system underneath. This DES model is used as an state observer of the plant. This approach is applied to a yeast production batch process monitoring. *Copyright © 1998 IFAC*

Keywords: Intelligent control, Supervisory control, Discrete-event systems, Automata Theory, Signal-processing algorithms, Knowledge-based systems, Biotechnology.

1. INTRODUCTION

Nowadays chemical process industries are among the most important costumers on demand of an efficient supervisory system. Process supervision involves tasks such as control, monitoring or diagnosis of the plant. Situation assessment and fault prevention are of prime interest in order to increase productivity which, afterwards, equates to benefit.

According to Kotch (1993), two main tasks are to be considered: process monitoring and supervisory control. Process monitoring comprises data collection and processing in order to have an updated plant state knowledge. At this stage the system must be able to decide wether the process is in an abnormal state and wheter a corrective action should be undertaken. Such intervention along with diagnosis is regarded under the supervisory control task.

Chemical processes tend to be quite complex MIMO systems which become an interesting challenge for developing supervisory strategies. Among them, biotechnological processes are particularly interesting due to the inexistance of a satisfactory mathematical model on which to implement classical model-based supervisory techniques. However, daily the operator interacts with these complex systems, identifying normal and abnormal situations and reacting accordingly. This fact shows that knowledge-based approaches seem appropriate to develop and design so-called intelligent supervisory systems (ISS). That is, systems which intend to implement human reasoning mechanisms in an automated way to carry out supervisory tasks.

A great deal of expert reasoning analysis is required in order to be able to abstract these mechanisms into the ISS. Many times, events are the main class of observations that experts are able to understand. For instance, as experts on physics, we know that after throwing a ball upwards it will return to the floor in a few seconds. Failure to do so will mean something abnormal has succeded, which will make us look upwards in case someone has got it. The ball throwing, a timeout, the ball return, etc., all are events, upon which we reason and draw conclusions.

In fact event-based monitoring task involves two main intrinsic stages. On the one hand, the expert is able to detect significative events from the process under supervision. Significative events carry useful information for the expert. On the other hand, he reasons based on these observations, carrying out his supervisory rol. That is, monitoring and supervisory control. Following the previous example, a timeout can be indicative of an abnormal situation whereas an increase in ambient temperature has nothing to do with the ball not returning back.

This paper deals with the design and developement of event-based supervisors based on a two layered hierarchical structure, see Fig. 1. Under this scheme, in the event generation layer, process signals are continually monitored and processed in order to abstract events which are relevant to the reasoning layer. This second layer handles those events to draw conclusions about the state of the plant (i.e., process monitoring), with the possibility of reacting accordingly (i.e., supervisory control). Heuristic knowledge of the plant is used in order to construct both layers, as it will be shown in the application example.

The sliding-window approach will be followed in order to implement the event generation layer. This has been proved a convinient way to abstract qualitative information from continuous signals (Sarrate and Aguilar, 1996).

The reasoning layer must deal with events. According to the literature such a system is known as a discrete-event system (DES). Automata theory is used to express the DES model of the plant upon which to make decisions. Automata theory (Hopcroft and Ullman, 1979) is a convinient formalism to express DES models and to carry out supervisory control (Ramadge and Wonham , 1989). Since this DES is interacting with a continuous process, the interaction of both yields what is known as an hybrid system (HS), see (Antsaklis and Koutsoukos, 1997). So, this paper is a contribution to both paradigms, DES and HS.

The methodology described in this paper can be partially regarded as intelligent control as described by Antsaklis, *et al.*, (1993), in the sense that it could be used in decision-making tasks needed in autonomous systems.

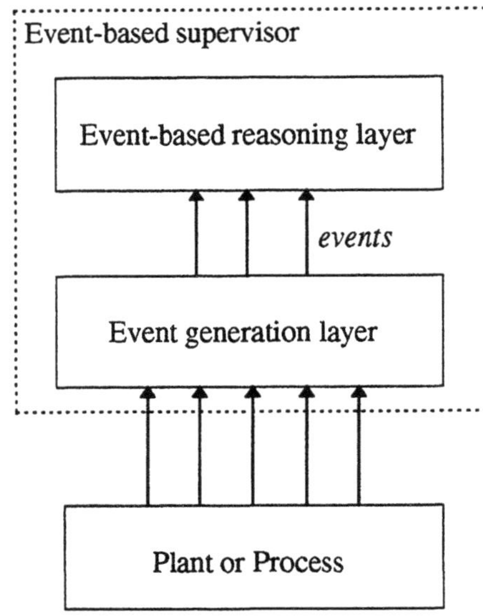

Fig. 1. Monitoring task of an intelligent supervisory system based on a two layer hierarchical scheme.

These ideas has been applied to a fermentation process. Measured signals available from the fermentation tub are analysed through the sliding-window methodology in order to obtain relevant trends. A DES model of the biotechnological plant is used to determine which batch stage is in progress.

A developing environment has been implemented under SIMULINK™. Results shown in this paper has been obtained using this software.

2. ISS DESIGN ISSUES

In the developement of an ISS,, the hierachical structure showed on Fig. 1 is to be followed. The ISS must be design according to the specificacions stated by the expert (i.e., what he expects the ISS to do). Based on this design goals, the engineer will have to decide:

- which is the reasoning strategy most appropiate to achieve such goals, and
- which are the relevant events that better abstract useful information for the reasoning layer.

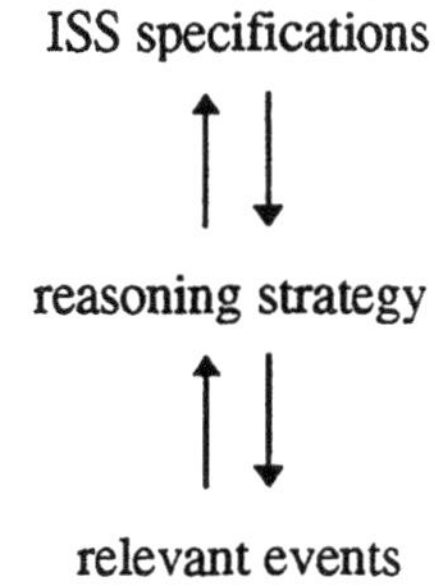

Fig. 2. Main interactions in the ISS developement.

Such selection it is not so straihgtforward and the developement of both layers should be done jointly. In fact, an interative design process will be necessay according to the interactions showed on Fig. 2: reasoning will be carried out based on the events available, and specifications will accomodate feasible goals which the ISS can accomplish.

In a second developing phase, the engineer will design the event generation layer using the sliding-window approach, and the event reasoning layer applying the DES formalism. Those issues are developed next.

2.1 Event Generation Layer.

This layer is in charge of detecting relevant events from measured signals. In fact, building this layer is not an easy task that the engineer must confront. Dialog with the expert will be of prime importance, and a great deal of process knowledge and signal processing.

Upon the whole set of possible events, Σ, the engineer must decide which are the relevant ones for the ISS, $\Sigma_R \subseteq \Sigma$. Relevant events will be transferred to the reasoning layer, so they must abstract useful information from plant signals. Such events can be associated to thresholds, trends, frequency content or temporal shapes (a pulse, a step, a ramp outliers, oscillations).

Commonly, relevant events will be obtained by the combination of other events (not necessarily relevant) which will be generated from measured signals. So, the event generation layer can be subdivided in two sublayers, see Fig. 3: the generation of atomic events and the combination of those events.

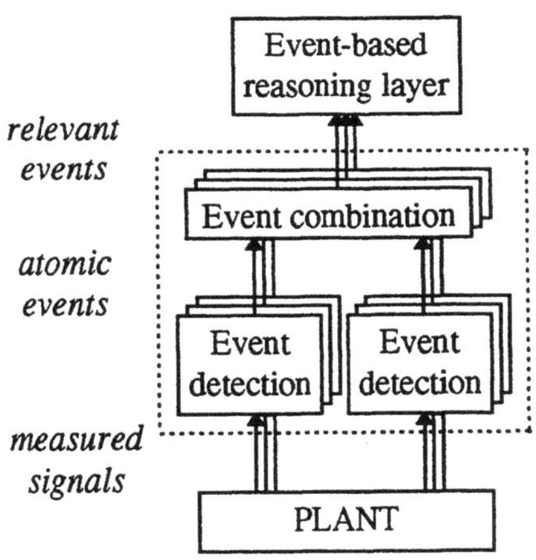

Fig. 3. Internal sub-layered architecture of the event generation layer.

A set of atomic events, $\Sigma_A \subseteq \Sigma$, are directly obtained by event detection algorithms such as those based on the sliding-window paradigm.

Under this approach, data is processed in sets comprising a time window. Each window is analyzed in order to produce an attribute which constitues new data. Appliying this algorithm recursively with different kinds of analysis, the required events can be robustly generated. For instance, signal trends can be generated by conviniently chaining a regression analysis and an statistical analysis with histograms. This method was successfully applied to gas turbine supervision for process monitoring and situation assessment, under the ESPRIT project TIGER, see (Sarrate, *et al.*, 1996).

The event detection block can be seen as a quantitative to qualitative interface which abstracts simbols (i.e., the atomic events) from continuous data. Next, those atomic events are to be combined in order to obtain the desired relevant events, Σ_R, with possibly $\Sigma_A \cap \Sigma_R \neq \emptyset$. This process can be conviniently described as a Moore machine (Hopcroft and Ullman, 1979) by a six-tuple $(Q, \Sigma_A, \Sigma_R, \delta, \lambda, q_0)$, where:

- Q is a set of states,
- Σ_A is the input alphabet (i.e., the atomic events),
- Σ_R is the output alphabet (i.e., the relevant events) with $\varepsilon \in \Sigma_R$ (ε is the silent or null event),
- $\delta : Q \times \Sigma_A \rightarrow Q$, is the transition mapping which is allowed to be partially defined[1],
- $\lambda : Q \rightarrow \Sigma_R$, is the output mapping, and
- q_0 is the initial state.

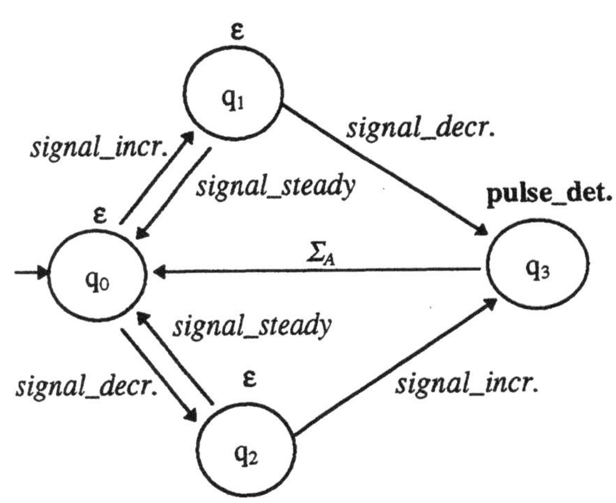

Fig. 4. Transition diagram associated with M to detect a pulse from input data.

[1]Transitions not defined are selfloops, by default.

For instance, to detect a narrow pulse in a measured signal, a Moore machine, M, can be used, with Σ_R = { **pulse_detected** , ε } and Σ_A = { *signal_increased* , *signal_decreased* , *signal_steady* }, see Fig. 4. So, whenever the input signal increases after decreasing, or decreases after increasing, M generates the event **pulse_detected.**

The Moore machine helps in generating complex events from atomic events, with the latter being easier to detect.

Beside the sliding-window approach, other event detection techniques are available in the literature, which could be integrated to this architecture. Cheung and Stephanopoulos (1990) introduced triangular descriptions of trends. Triangular descriptions are based upon the first and second derivative of data. Then a set of triangular simbols is defined which embeds slope and concaveness information about the input signal. Rengaswamy and Venkatasubramanian (1995) provided robust algorithms under noisy environments. Colomer, *et al.* (1997), increased the triangular simbols by defining qualitative interval descriptions for the first and second derivative.

2.2 Event-Based Reasoning Layer.

This layer receives relevant events from the event generation layer and must assess the operator in their decisions, with possibly automatically reacting as needed. This paper just deals with the monitoring task, leaving the supervisory control task as future work.

The process under supervision is described by a logical DES model. It is logical in the sense that it includes no explicit timing information. It just expresses state sequencing based on relevant events observed in the plant. Again a Moore machine is used to formalise the plant model. Now the six-tuple is $(P, \Sigma_R, \Sigma_O, \phi, \mu, p_O)$, where:

- P is a set of states,
- Σ_R is now the input alphabet (i.e., the relevant events),
- Σ_O is the output alphabet (i.e., the output events) with $\varepsilon \in \Sigma_O$,
- $\phi : P \times \Sigma_R \rightarrow P$, is the transition mapping which is allowed to be partially defined,
- $\mu : P \rightarrow \Sigma_O$, is the output mapping, and
- p_O is the inital state.

States are associated to the output alphabet through the mapping μ. The output events, $e_O \in \Sigma_O$, are messages or information on the present process state that the operator (i.e., the expert) needs to know. For instance **temperature_alarm_on_thermocouple_A**. Every output event has a relevant meaning for the expert, not only as a diagnostic message, alarm or faulty situation, but also as a normal physical process state. In that sense, this Moore machine can be regarded as an observer of the process under supervision. The silent simbol must be considered as an output event with no relevant content for the expert.

The most difficult task is to define the transition mapping, ϕ. As in the event generation layer design, the engineer needs a great knowledge of the process. Often, only through exhaustive dialog with the expert, the heuristic knowledge which he possesses on the process can be transferred by the engineer to this Moore machine. Transitions, governed by relevant events, express the dynamics of the plant. In that sense, relevant events change the process state. Such mechanism reflects the reasoning scheme followed by the expert.

So, this Moore machine is used to describe a qualitative model of the process behaviour under normal, or even abnormal situations.

3. YEAST PRODUCTION APPLICATION EXAMPLE

The supervisory architecture of Fig. 1 has been applied to a biotechnological process. *Sccharomyces Cerevisiae* is studied under oxidative regime (i.e., no ethanol production) to produce yeast under a laboratory environment in a fermentating reactor as shown in Fig. 5.

Two different procedures are applied: a batch procedure is followed by a continuous procedure. The batch procedure is composed by a sequence of biological stages. By this procedure the inoculum is introduced into the reactor and, progressively, biomass concentration (i.e., the yeast) increases to a maximum. This can be thought as an start-up procedure.

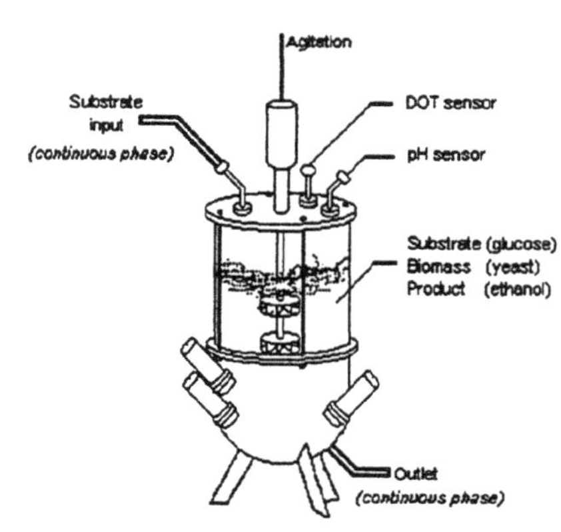

Fig. 5. Fermentation plant.

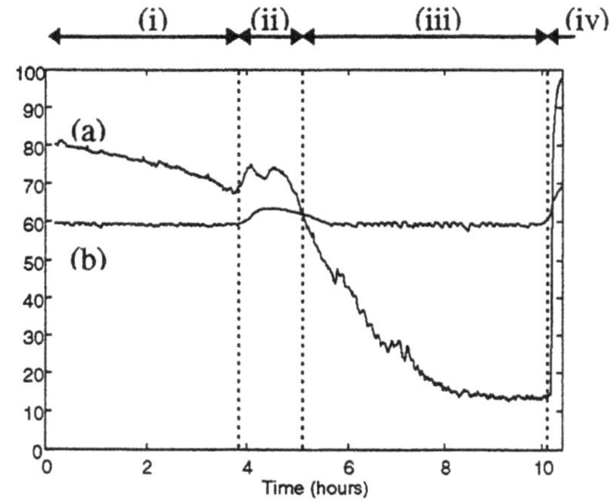

Fig. 6. Batch procedure stages with (a) pO_2 and (b) pH. For (i) to (iv) see text below. Vertical axis scale is irrelevant.

Following, there is the continuous procedure, where the outlet is opened and yeast is obtained by separating it from the substratum while the inlet biomass concentration is kept constant. It can be though as a production procedure.

This paper deals with the batch procedure. Just a subset of two signals, which are available from the plant, are relevant: pO_2 and pH. Those signals will be processed through the sliding-window approach to detect trends as events. Reasoning is carried out on those trends in order to identify batch stages. According to the biotechnologist, four main stages compose the batch procedure, see Fig. 6:

i. The inoculant is introduced into the fermenting tub along with glucose, which constitutes the substratum. During the first stage, biomass concentration increases, while glucose is consumed and ethanol produced. This stage is characterized by pO_2 decreasing.
ii. As soon as it has run out of glucose, there is an enzymatic transition. pH and pO_2 experiment a transitory dynamic.
iii. During the third stage, ethanol is consumed while biomass concentration increases. Again pO_2 decreases.
iv. Finally, as soon as it has run out of ethanol, there is a big transitory change. Normally, at this stage the biotechnologist initiates the continuous procedure. So this is the end of the batch procedure. This stage is characterized by a sudden increase in pO_2 and pH.

Although being a difficult process to model, the biotechnologist has a quite clear heuristic model of the batch procedure. As seen in Fig. 6, trends in the available measured variables can be used as a way to detect stage transitions. So, the following alphabets are defined:

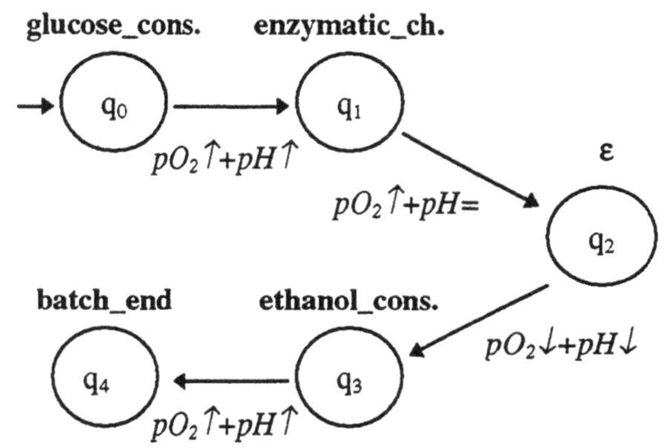

Fig. 7. Moore machine M_R which constitutes the event-based reasoning layer.

- $\Sigma_A = \{ pO_2\uparrow , pO_2\downarrow , pO_2= , pH\uparrow , pH\downarrow , pH= \}$, a set of atomic events which can be detected using the sliding-window approach ('↑' means increased, '↓' means decreased and '=' means steady).
- $\Sigma_R = \{ pO_2\uparrow+pH\uparrow , pO_2\uparrow+pH= , pO_2\downarrow+pH\downarrow , \varepsilon \}$, a set of relevant events according to Fig. 6 ('+' means that both atomic events are simultaneously detected).
- $\Sigma_O = \{$ glucose_consumption , enzymatic_change , ethanol_consumption , batch_end_stage , $\varepsilon \}$

Two Moore machines are defined:

- M_C, which generates Σ_R from Σ_A. This machine composes (i.e., '+' operation defined above) trend events and produces relevant events.
- M_R, which generates Σ_O from Σ_R. This mahine reproduces the process behaviour, identifying the batch stages, see Fig. 7.

The whole ISS architecture has been implemented in SIMULINK™, see Fig. 8. Data collected from the real plant is processed on-line by this ISS SIMULINK™ model. Both Moore machines, M_C and M_R, have been implemented through *if-then* rules. So the DES block constitutes the event combination sublayer along with the event-based reasoning layer of Fig. 3.

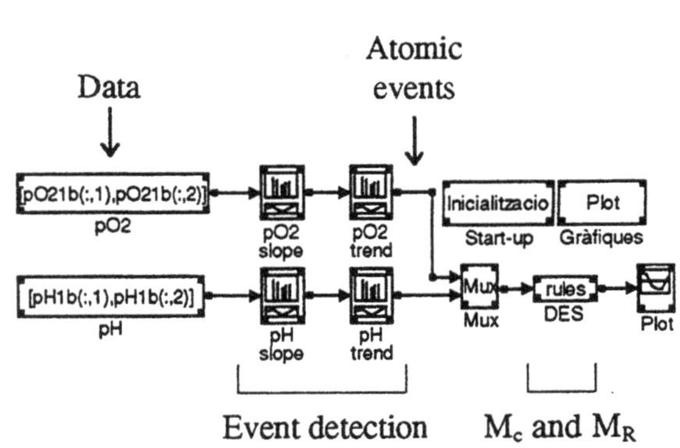

Fig. 8. On-line ISS SIMULINK™ model.

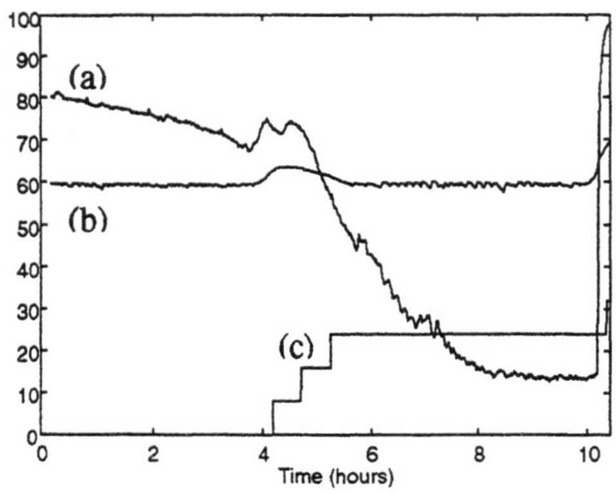

Fig. 9. ISS output: (a) pO_2, (b) pH and (c) the observed batch stages. Vertical axis scale is irrelevant.

The results obtained are showed in Fig. 9. The ISS is able to correctly identify the four batch stages declared before. Every step change in (c) is univocally related to a process stage transition. There is a delay due to the causal scheme used in the event detection algorithms. Sliding-window algorithms have several parameters which have been manually tuned by checking the consistency of the results obtained running the whole system on a set of several scenarios available. So, the same ISS is able to correctly supervise the plant under different conditions, due to its robustness

4. CONCLUSIONS

An efficient ISS architecture has been proposed in this paper. It continually monitors the plant detecting relevant events which are later used to observe the state of the plant under supervision. Automatical knowledge of the plant state is important for the operator job. This prevents human errors and increases the efficiency of the whole plant.

The ISS layered structure implements the sliding-window approach to detect atomic events from measured signals. Those events are processed by hierarchised Moore machines which intend to mimic expert reasoning schemes.

The proposed architecture is implemented under SIMULINK™ and applied to the supervison of a fairly complex biotechnical process. Results are presented which confirm its validity.

Future work will involve new applications in other domains which will allow the ISS architecture to progress towards supervisory control, extending and refining its present capabilities.

ACKNOWLEDGEMENTS

This paper is partially supported by the Spanish Science and Technology Board (CICYT) under contract TAP96-1114-C03-04 and by the University and Research Comission of the Generalitat de Catalunya (ref. 1997SGR 00098).

The authors wish to thank also the support received, as members, by LEA-SICA (European Associated Laboratory on Intelligent Systems & Advanced Control) and CERCA (Collective for the Study and Research in Control and Automation)

REFERENCES

Antsaklis, *et al.* (1993). *Defining Intelligent Control*. Technical Report of the Task Force on Intelligent Control. University of Notre Dame, Notre Dame.

Antsaklis, P.J. and X.D. Koutsoukos (1997). *On Hybrid Control of Complex Systems: A Survey*. Technical Report of the ISIS Group. University of Notre Dame, Notre Dame.

Cheung, J.T. and G. Stephanopoulos (1990). Representation of process trends, part I. A formal representation framework. *Computers in Chemical Engineering*, **14**, 495-510.

Colomer J., J. Meléndez, J. Ll. de la Rosa and J. Aguilar (1997). A qualitative/quantitative representation of signals for supervision of continuous systems. In: *European Control Conference. ECC'97*, cd-rom. Brussels.

Hopcroft, J.E. and J.D. Ullman (1979). *Introduction to Automata Theory, Languages, and Computation*. Addison-Wesley, London.

Kotch, G.G. (1993). *Modular Reasoning. A New Approach Towards Intelligent Control*. PhD Thesis. Swiss Federal Institute of Technology, Zurich.

Ramadge, P.J.G. and W.M. Wonham (1989). The control of discrete event systems. *Proceedings of the IEEE*, **77**, 81-97.

Rengaswamy, R. and V. Venkatasubramanian (1995). A syntactic pattern-recognition approach for process monitoring and fault diagnosis. *Engineering Applications in Artificial Intelligence*, **8**, 35-51.

Sarrate, R., and J. Aguilar (1996). Window-based quantitative-to-qualitative interface for intelligent supervisory systems. In: *Modelling and Simulation. ESM'96*, pp. 724-728. Budapest.

Sarrate, R., J. Aguilar and Ch. Nicol (1996). Eat, a real-time window-based event generation tool: an industrial application example. In: 2nd *Portuguese Conference on Automatic Control. CONTROLO'96*, pp. 37-42. Oporto, Portugal.

SIMULTANEOUS FAULT DETECTION OF HEAT EXCHANGERS

Gábor Szederkényi * Erik Weyer ** Katalin M. Hangos ***

* *Dept. of Computer Science, University of Veszprém, H-8201 Veszprém, Egyetem u. 3., Hungary*
** *Dept. of Electrical and Electronic Engineering, The University of Melbourne, Parkville VIC 3052, Australia*
*** *Systems and Control Research Laboratory, Computer and Automation Institute HAS, H-1518 Budapest, PO Box 63, Hungary*

Abstract: A grey box model-based method for fault diagnosis is proposed in this paper. The method is based on a first principle model of the process unit: a heat exchanger, and on a grey box model of the faults: the deterioration of the heat transfer surface by aging and the leaking of the outer container. The deterioration of the heat transfer surface is due to material settling, and in old heat exchangers pieces of this material can break off and cause damage in the equipment. A recursive least squares estimator with forgetting factor is used to track both the heat transfer coefficients and the cold side volume. The heat transfer coefficients are estimated both from the hot and cold side equations. The settled material breakage fault is detected via detection of abrupt positive jumps in the estimated heat transfer coefficients using a detector based on a cumulative sum (CUSUM) test. The proposed method enables simultaneous detection of the two fault types considered. Fault localization along the equipment length is also possible when temperature measurements are available along the length.

Keywords: Fault detection, Fault isolation, Fault location, Process model, Grey box modelling, Identification

1. INTRODUCTION

Fault detection and diagnosis are playing an increasingly important role in the process industries. Dynamic modelling of the equipment, called operating units in the process industries is a well developed area, and model based diagnostic methods can therefore efficiently and relatively easily be applied. From a modelling point of view heat exchangers are the simplest and most investigated operating units for which we have both reliable dynamic models with known properties and accumulated operational experience. This experience can be used in formulating suitable dynamic models of the typical faults which can be encountered in heat exchangers.

In this paper we extend our previous grey box model-based fault detection method (Weyer and Hangos, 1997), (Weyer *et al.*, 1998) for detecting the deterioration of the heat transfer surface to also include detection of leaking in the outer container. By a grey box model we understand a black box model in which we have incorporated prior physical knowledge.

The paper is organized as follows. In the next section models for the heat exchanger and the faults are presented. The identification and detection algorithms are given in section 3, and they are illustrated on a simulation example in section 4.

2. MODELS

Next we present a dynamic model based on first engineering principles of the heat exchanger, and semi-empirical models of the faults.

2.1 *Heat exchanger model*

A countercurrent heat exchanger is modelled as a sequence of cascaded cells (Varga *et al.*, 1995). Each cell consists of two perfectly stirred tanks with in- and outflows. The two tanks are connected by a heat transfer area between them.

The three cell model A three cell model is used (see Fig. 1) because it can describe the dynamics of the heat exchanger well enough for fault diagnosis purposes for a majority of the industrial equipments.

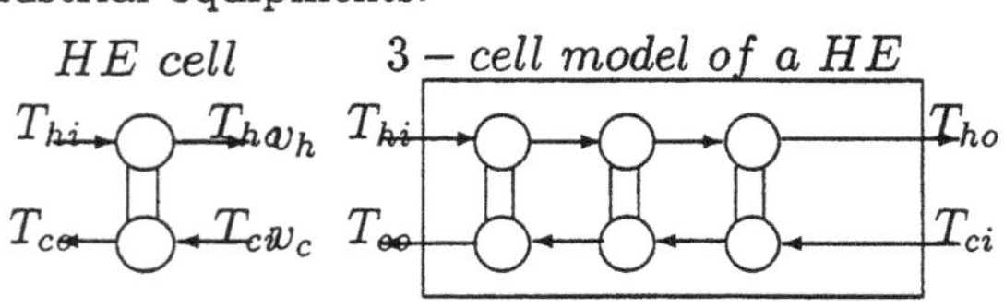

Fig. 1. Cascade model of a countercurrent heat exchanger

The model equations for a single cell consist of the total mass and energy balances for the hot and cold sides. Constant volume (and hence mass) is assumed on the hot side. The total mass balance is therefore only presented for the cold side. The equations for the jth cell are as follows:

$$\begin{aligned}\frac{dT_{jc}}{dt} &= \frac{v_c}{V_{jc}}(T_{(j+1)c} - T_{jc}) + \\ &\frac{U_j A}{c_{pc}\rho_c V_{jc}}(T_{jh} - T_{jc})\end{aligned} \tag{1}$$

$$\begin{aligned}\frac{dT_{jh}}{dt} &= \frac{v_h}{V_{jh}}(T_{(j-1)h} - T_{jh}) + \\ &\frac{U_j A}{c_{ph}\rho_h V_{jh}}(T_{jc} - T_{jh})\end{aligned} \tag{2}$$

$$\frac{dV_{jc}}{dt} = v_{ci} - v_{co} - v_{jl} \tag{3}$$

$$\begin{aligned}j = 1,2,3,\ T_{4c} = T_{ci},\ T_{1c} = T_{co}, \\ T_{0h} = T_{hi},\ T_{3h} = T_{ho}\end{aligned} \tag{4}$$

where the subscripts h and c denote the hot and cold sides respectively. T_{hi}, T_{ci} and T_{ho}, T_{co} are the inlet and outlet temperatures, v_h and v_c are flowrates, v_{jl} is the leaking flowrate, V_{jc} and V_{jh} are the volumes, A the heat transfer surface area, c_{pc} and c_{ph} the specific heats, ρ_c and ρ_h the densities and U_j is the heat transfer coefficient.

The above model can be regarded as a lumped version of a distributed parameter model of the heat exchanger. The cells are therefore associated with a particular spatial location along the equipment length. The state variables of the above model are the temperatures of the cold and hot sides and the cold side volume in the cells, i.e. $\{T_{1h}, T_{1c}, V_{1c}; T_{2h}, T_{2c}, V_{2c}; T_{3h}, T_{3c}, V_{3c}\}$. If an equidistant space lumping is performed, i.e. the parameters A and V_{jh} are the same for all cells, then the heat transfer coefficients U_1, U_2, U_3 carry spatial information.

The simplified model A simplified dynamic model containing only a single cell can be safely used for fault detection *not requiring spatial information* of the equipment. With only a single cell as in Fig. 1.a. describing the heat exchanger the outlet temperatures T_{ho} and T_{co} of the heat exchanger are the same as the outlet temperatures of the cell. In this case we have only three state variables $\{T_{co}, T_{ho}, V_c\}$ and a single triplet of equations (1)-(3).

2.2 *Discrete time model*

We first convert the model (1)-(3) into a discrete time model suitable for tracking of the heat transfer coefficient. The cell index j is omitted from the equations for simplicity. Let t_s denote the sampling interval and introduce the variables

$$\bar{v}_c(k) = \frac{v_c(kt_s)}{V_c},\quad \tau_c(k) = \frac{U(kt_s)A}{c_{pc}\rho_c V_c} \tag{5}$$

and similar for the hot side. Furthermore let (with some abuse of notation)

$$T_{co}(k) = T_{co}(kt_s),\ T_{ci}(k) = T_{ci}(kt_s)$$

and again similar for the hot temperatures. We approximate the derivative using a simple Euler approximation. Using the δ operator (Middleton and Goodwin, 1990) which is defined as

$$\delta T(k) = (T(k+1) - T(k))/t_s$$

we can write the discrete time model as

$$\begin{aligned}\delta T_{co}(k) &= \bar{v}_c(k)(T_{ci}(k) - T_{co}(k)) + \\ &\tau_c(k)(T_{ho}(k) - T_{co}(k)) \\ \delta T_{ho}(k) &= \bar{v}_h(k)(T_{hi}(k) - T_{ho}(k)) + \\ &\tau_h(k)(T_{co}(k) - T_{ho}(k)) \\ \delta V_c(k) &= v_{ci}(k) - v_{co}(k) - v_l(k)\end{aligned} \tag{6}$$

In order to avoid numerical differentiation of the temperature measurements we filter them through the linear filter $e_0/(\delta + e_0)$, i.e.

$$\delta T^f(k) = -e_0 T^f(k) + e_0 T(k)$$

Using the filtered temperature measurements the above equations take the form (assuming $\bar{v}_c, \bar{v}_h, \tau_c$ and τ_h constant)

$$\begin{aligned}\delta T_{co}^f(k) &= \bar{v}_c(k)(T_{ci}^f(k) - T_{co}^f(k)) + \\ &\tau_c(k)(T_{ho}^f(k) - T_{co}^f(k)) \\ \delta T_{ho}^f(k) &= \bar{v}_h(k)(T_{hi}^f(k) - T_{ho}^f(k)) + \\ &\tau_h(k)(T_{co}^f(k) - T_{ho}^f(k))\end{aligned} \tag{7}$$

The above equations are also good approximations when $\bar{v}_c, \bar{v}_h, \tau_c$ and τ_h are slowly varying. For the purpose of estimating the heat transfer coefficient we introduce the signals

$$\begin{aligned}y_c(k+1) &= \delta T_{co}^f(k) - \\ &\bar{v}_c(k)(T_{ci}^f(k) - T_{co}^f(k)) \\ y_h(k+1) &= \delta T_{ho}^f(k) - \\ &\bar{v}_c(k)(T_{hi}^f(k) - T_{ho}^f(k)) \\ u_c(k) &= T_{ho}^f(k) - T_{co}^f(k) \\ u_h(k) &= T_{co}^f(k) - T_{ho}^f(k)\end{aligned} \tag{8}$$

The discrete time model can now be written as

$$\begin{aligned}y_c(k+1) &= \tau_c(k)u_c(k) \\ y_h(k+1) &= \tau_h(k)u_h(k)\end{aligned} \tag{9}$$

where $y_c(k+1), y_h(k+1), u_c(k)$ and $u_h(k)$ can be computed by assumption and $\tau_c(k)$ and $\tau_h(k)$ are proportional to the heat transfer coefficient with known proportionality constants. Hence in order to estimate the heat transfer coefficient we can estimate $\tau_c(k)$ and $\tau_h(k)$.

2.3 *Fault modelling*

Deterioration of the heat exchanger surface Under normal operating conditions the heat transfer coefficient is constant or slowly decreasing due to a layer of settled material building up on the heat transfer surface. In old heat exchangers large pieces of the settled material can break away from the surface causing damage in the equipment. When the settled material breaks off, the heat transfer coefficients will undergo abrupt positive jumps. These jumps can be grouped into two qualitatively different types of jumps. In the beginning when the jumps start to occur they are small and infrequent, while they are large and frequent in the later stages.

Leaking When the outer container is leaking, there is an unmeasurable outflow v_{jl} from the heat exchanger which may result in a slow decrease of the mass in the outer phase, usually the cold side, or - if the levels are controlled, in the decrease of the outflow from the equipment.

Here it is assumed that the volume flow rates v_h and v_c are constant, leaking can be therefore detected via detection of a slow decrease of the cold side volume V_{jc}.

3. ALGORITHMS

3.1 *Tracking of the heat transfer coefficient*

To track the heat transfer coefficients we make use of a simple recursive least squares method with forgetting factor $\lambda < 1$ (see e.g. (Ljung and Soderstrom, 1983)). Using the model (9) the estimates of τ_c and τ_h are given by

$$\begin{aligned}\hat{\tau}_c(k+1) &= \hat{\tau}_c(k) + \\ p_c(k+1)u_c(k)&(y_c(k+1) - \hat{\tau}_c(k)u_c(k)) \\ p_c(k+1) &= \frac{p_c(k)}{\lambda + u_c^2(k)p_c(k)}\end{aligned} \tag{10}$$

$$\begin{aligned}\hat{\tau}_h(k+1) &= \hat{\tau}_h(k) + \\ p_h(k+1)u_h(k)&(y_h(k+1) - \hat{\tau}_h(k)u_h(k)) \\ p_h(k+1) &= \frac{p_h(k)}{\lambda + u_h^2(k)p_h(k)}\end{aligned} \tag{11}$$

The choice of λ is a trade off between good tracking capabilities of the heat transfer coefficients and sensitivity to noise and unmodelled dynamics. A small value of λ means that the estimate will track variations in the heat transfer coefficients quicker, but on the other hand the estimate will also become more sensitive to noise and unmodelled dynamics. The actual estimates of the heat transfer coefficients can be computed either by

$$\hat{U}_c(k) = \frac{\hat{\tau}_c(k)c_{pc}\rho_c V_c}{A} \tag{12}$$

or

$$\hat{U}_h(k) = \frac{\hat{\tau}_h(k)c_{ph}\rho_h V_h}{A} \tag{13}$$

or a convex combination

$$\hat{U}(k) = \alpha\hat{U}_c(k) + (1-\alpha)\hat{U}_h(k), \quad 0 \le \alpha \le 1 \tag{14}$$

where α should reflect the relative confidence we have in the two estimates.

3.2 *Jump detection*

Under normal operations the heat transfer coefficient will either be constant or slowly decreasing. This means that after the estimates have reached a "steady state" the prediction error $\epsilon_c(k) = y_c(k) - \hat{\tau}_c(k-1)u_c(k-1)$ will fluctuate around zero (constant heat transfer coefficients) or be negative (decreasing heat transfer coefficients). The prediction error $\epsilon_h(k) = y_h(k) - \hat{\tau}_h(k-1)u_h(k-1)$ will similarly either fluctuate around zero or be positive. However, after an abrupt positive jump in the heat transfer coefficient $\epsilon_c(k)$ will become positive and $\epsilon_h(k)$ will become negative. This makes the prediction errors or a scaled version of them a well suited signal to be monitored in order to detect a jump since we are back to the standard situation of detecting a change in the mean value of the signal, and in this application we have prior knowledge of the sign of the change. In this paper we have chosen to monitor $X(k) = \hat{U}(k) - \hat{U}(k-1)$ which is a scaled version of the prediction errors. Because of the above properties of $X(k)$ we propose to use a cumulative sum (CUSUM) detector for jump detection (see e.g. (Baseville and Nikiforov, 1993)). Let

$$S(k) = \sum_{t=1}^{k} X(t), \quad S(0) = 0 \tag{15}$$

and

$$m(k) = \min_{0 \le t \le k} S(t) \tag{16}$$

Since $X(t)$ is either fluctuating around zero or negative during normal operation, we expect $m(k)$ and $S(k)$ to be close. However, after a jump $S(k)$ will be considerably larger than $m(k)$. A jump is therefore detected if

$$S(k) - m(k) > \gamma \tag{17}$$

where γ is the threshold level. Again the value of γ is a trade off between quick detection and sensitivity to noise and unmodelled dynamics. A small value of γ will give quick detection, but will also give more false alarms (detects jumps when no jump occurred).

Here we will employ two detectors, one with a low threshold value to detect the small jumps which occur when the jumps start, and one with a large threshold to detect the large jumps during the final stage.

When a jump is detected the forgetting factor is reduced to allow the estimate in question to reach its new value faster. The forgetting factor is gradually increased back to its normal value according to the formula

$$\lambda(t) = \lambda_0 \lambda(t-1) + \lambda(1 - \lambda_0) \tag{18}$$

where λ is the normal value of the forgetting factor and λ_0 is the rate at which $\lambda(t)$ approaches λ. When a jump is detected the detection mechanism is switched off for a while in order to avoid multiple detection of the same jump. Also should several jump occur within a short time interval, it is only necessary to detect the first one since the reduced forgetting factor will ensure quick adaptation when the other jumps occur.

3.3 *Tracking of the cold side volume*

In order to track the cold side volume the discrete time model is rewritten as

$$z(k+1) = \beta(k)W(k) \tag{19}$$

where

$$z(k+1) = \delta T^f_{co}(k) \tag{20}$$

$$\beta(k) = \frac{1}{V_c(k)} \tag{21}$$

$$\begin{aligned} W(k) &= v_c(k)(T^f_{ci}(k) - T^f_{co}(k)) + \\ &+ \frac{UA}{c_{pc}\rho_c}(T^f_{ho}(k) - T^f_{co}(k)) \end{aligned} \tag{22}$$

Again a recursive least squares algorithm with forgetting factor $\lambda_V < 1$ is used, and the estimate of β is given by

$$\begin{aligned} \hat{\beta}(k+1) &= \hat{\beta}(k) + \\ K(k+1)&(z(k+1) - W(k)\hat{\beta}(k)) \end{aligned} \tag{23}$$

$$K(k+1) = \frac{P(k)W(k)}{\lambda_V + W^2(k)P(k)} \tag{24}$$

$$\begin{aligned} P(k+1) &= \frac{1}{\lambda_V}(P(k) - \\ &\frac{P^2(k)W^2(k)}{\lambda_V + W^2(k)P(k)} \end{aligned} \tag{25}$$

Since our aim is to detect slow changes in the cold side volume λ_V should be chosen close to 1 in order to reduce the sensitivity to noise.

Detection of leaking The leaking can be detected via detection of the slow decrease of the cold side volume applying a CUSUM detector. It is important to note that the steady state values for the temperatures are independent of the volumes, therefore we can only estimate the volumes accurately when we have variations in the temperatures, e.g. changing inlet temperatures

3.4 *Location of faults along the equipment length*

The possibility to detect faults in any industrial equipment is heavily dependent on the availability of suitable measurements. For heat exchangers the variables related to the in- and outflows of the equipment (flow rates and temperatures) are usually measured but measurements along the equipment length are rarely available. Therefore the possibilities of fault location in space are rather limited. However, simplified models can be used for fault detection in this case.

All measurements available In this case it is assumed that all the constant parameters of the heat exchanger are known and that the flow rates and both the hot and cold side temperatures in each cell are measured in addition to the inlet temperatures. Changes either in the heat transfer coefficients or in the cold side volume can be detected using (10)-(11) or (23)-(25).

Simultaneous tracking of the heat transfer coefficient and the cold side volume is possible since the estimation of the heat transfer coefficient from the hot side equations is practically unaffected by a slow change in the cold side volume. It is because the effect of V_c appears in the estimation only through the cold side temperature T_{co}. In this case U is substituted by $\hat{U}_h$ in (22), i.e.

$$W(k) = v_c(k)(T^f_{ci}(k) - T^f_{co}(k)) + \\ + \frac{\hat{U}_h(k)A}{c_{pc}\rho_c}(T^f_{ho}(k) - T^f_{co}(k)) \qquad (26)$$

In order to obtain suitable estimates for the cold side volume quick tracking of the heat transfer coefficient is required. After a jump in the heat transfer coefficient is detected λ should be decreased appropriately to let the estimate reach its new value quickly.

Simplified model In this case the constant parameters of the heat exchanger are assumed known and only the hot and cold side inlet and outlet temperatures are assumed to be measured. In this case a simplified dynamic model containing only one cell can be used. By using a simplified model either jumps in the heat transfer coefficient or leaking can be detected, but the simultaneous use of the two methods does not give good results.

The reason for this is that (26) makes use of the estimate of U, and with fewer measurements this estimate becomes much less accurate which again is reflected in the estimate of the cold side volume. The jumps in the heat transfer coefficients would still be detected since the estimates of them are rather insensitive to the cold side volume, and moreover since we detect jumps we can tolerate biased estimates as long as they behave qualitatively in the same manner as the real heat transfer coefficients.

4. SIMULATION RESULTS

In this section we illustrate the proposed method on a simulated example.

During the simulations the heat transfer coefficients were estimated and jumps in them as well as leaking were detected in the two different measurement situation above, i.e. all measurements available and only inlet and outlet temperatures available.

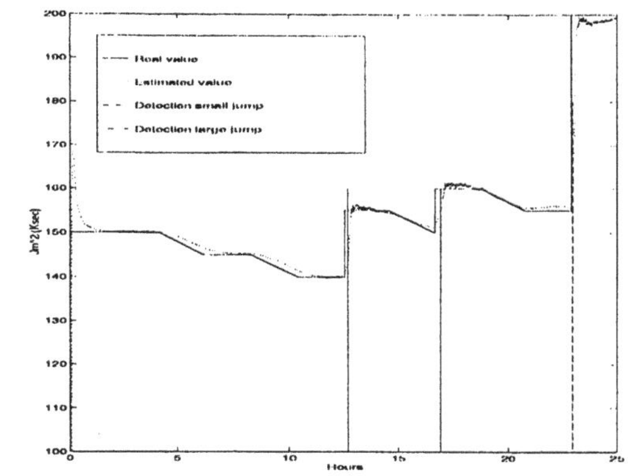

Fig. 2. Real and estimated heat transfer coefficient in the second cell (all measurements available)

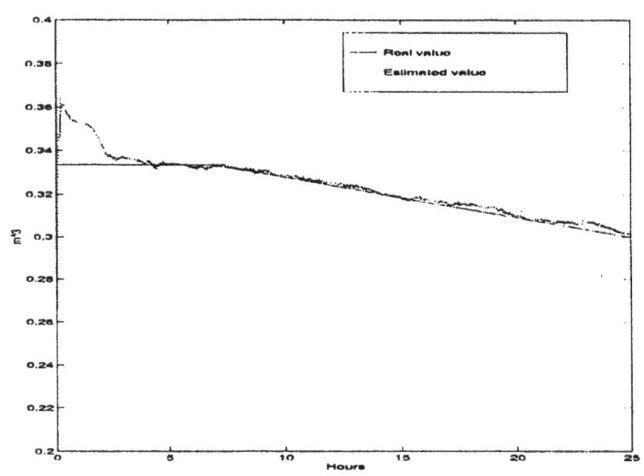

Fig. 3. Real and estimate volume on the cold side in the second cell (all measurements available)

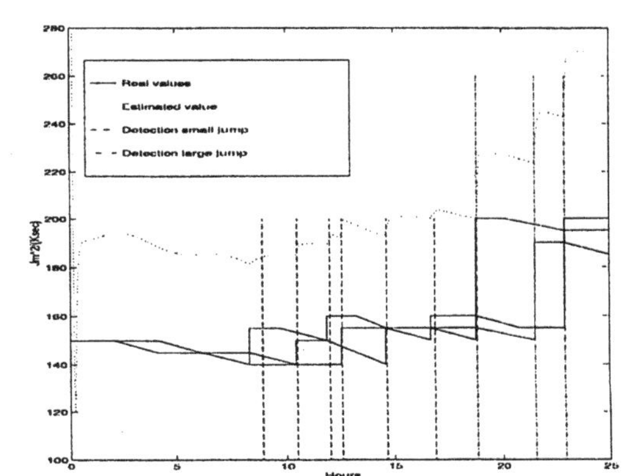

Fig. 4. Real and estimated heat transfer coefficient (simplified model)

4.1 *The simulated heat exchanger*

A model consisting of three cascaded cells was simulated using equations (1)-(3). The same parameter values were used for each cell and can be found in Table 1. The system was sampled every 30 seconds. Gaussian measurement and process noise were added at each sampling instant with variances 0.04 and 0.005 respectively. The profile of the heat transfer coefficient was different in each cell in both cases. The filter coefficient e_0 corresponding to the model was 0.0063.

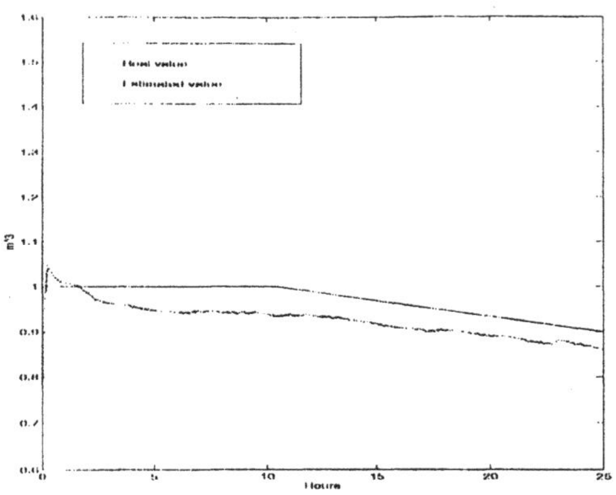

Fig. 5. Real and estimated cold side volume (simplified model)

Table 1. Parameter values of the simulated heat exchanger

v_c	$2.0 \cdot 10^{-3} m^3/sec$
v_h	$0.7 \cdot 10^{-3} m^3/sec$
V_c	$0.333 m^3$
V_h	$0.166 m^3$
c_{pc}	$1910 J/K \cdot kg$
c_{ph}	$1590 J/K \cdot kg$
ρ_c	$1000 kg/m^3$
ρ_h	$1000 kg/m^3$
A	$1.333 m^2$

4.2 *All measurements available*

The two fault types (i.e. jumps in the heat transfer coefficients and leaking) were simulated and detected simultaneously. The threshold levels for the CUSUM detectors were $\gamma = 7Jm^2/(Ksec)$ (small jumps) and $\gamma = 20Jm^2/(Ksec)$ (large jumps). After a jump was detected the forgetting factor was reduced to 0.8 and restored to $\lambda = 0.99$ according to (18). The forgetting factor λ_V used in the tracking of the cold side volume was chosen to be 0.99. The real and estimated values of the heat transfer coefficient and the cold side volume in the second heat exchanger cell are shown in figures 2 and 3 respectively. As can be seen the two fault types can be detected at the same time in this case.

4.3 *Simplified model*

In the algorithm for tracking the heat transfer coefficient the threshold levels for the CUSUM detectors were $\gamma = 5Jm^2/(Ksec)$ (small jumps) and $\gamma = 20Jm^2/(Ksec)$ (large jumps). After a jump was detected the forgetting factor was reduced to 0.95 and restored to $\lambda = 0.99$ according to (18). Leaking was also simulated but it did not affect the estimation of the heat transfer coefficient. The results are shown in figure 4. Note that in this case only an average heat transfer coefficient can be estimated. From figure 4 we see that good fault detection was achieved even with the simplified model at the price of a small detection delay. During the simulation of leakage the heat transfer coefficients were kept constant in each cell. Their values were $198.8Jm^2/(Ksec)$. The forgetting factor in the algorithm was $\lambda_V = 0.99$ The results are shown in figure 5. The estimate is less accurate but leaking of the outer tank can be detected also in this case.

5. CONCLUSION

In this paper we have considered detection of settled material breaking off from the heat transfer surface and leaking of the outer container in a heat exchanger. The methods we have proposed are based on the use of recursive least squares estimators with forgetting factors to track the heat transfer coefficients and the cold side volume. The settled material breakage fault is detected via detection of abrupt positive jump in the estimated heat transfer coefficients using a CUSUM test. Leaking in the outer tank is also detected by recursive least squares method in such a way that the detection of surface aging is not disturbed by the presence of leaking. The proposed method enables simultaneous detection of the two faults when all measurements are available.

Acknowledgement This work is partially supported by the Hungarian National Research Fund (OTKA) through grant T017 108 and by the EC through the COPERNICUS project grant CT94-0237.

6. REFERENCES

Baseville, M. and I.V. Nikiforov (1993). *Detection of abrupt changes. Theory and practice.* Prentice Hall. London.

Ljung, L. and T. Soderstrom (1983). *Theory and practice of recursive identification.* The MIT Press. Massachusetts.

Middleton, R.H. and G.C. Goodwin (1990). *Digital control and estimation. A unified approach.* Prentice Hall. London.

Varga, E.I., K.M. Hangos and F. Szigeti (1995). Controllability and observability of heat exchanger networks in the time varying parameter case. *Control Engineering Practice* **3**, 1409–1419.

Weyer, E. and K.M. Hangos (1997). Grey box fault detection in heat exchanger networks. In: *IFAC Symposium on Fault Detection, Supervision and Safety for Technical Processes - SAFEPROCESS'97* (R.J. Patton and J. Chen, Eds.). Vol. 1. pp. 187–192. Pergamon. Hull.

Weyer, E., G. Szederkenyi and K.M. Hangos (1998). Grey box fault detection of heat exchangers. *Control Engineering Practice* , (submitted).

LINEAR MATRIX INEQUALITY FORMULATION OF FAULT-TOLERANT CONTROL SYSTEMS DESIGN

Jie Chen

Department of Mechanical Engineering, Brunel University, Uxbridge, UB8 3PH, UK
Email: Jie.Chen@brunel.ac.uk

Ron J. Patton, Zheng Chen

Department of Electronic Engineering, University of Hull, Hull, HU6 7RX, UK
Email: R.J.Patton@e-eng.hull.ac.uk, Z.Chen@e-eng.hull.ac.uk

Abstract: This paper presents a new approach for designing fault-tolerant control systems. This approach is based on the linear matrix inequality (LMI) framework. The design objectives of fault-tolerance and robust performance and stability requirements are formulated in a number of linear matrix inequalities. The satisfaction of design requirements leads to a multi-objective optimisation problem which is solved using LMI tool. The paper studies the flight control problem of an unmanned aircraft in the case of suffering battle damage on its wing. The simulation results shows that the robust stability and satisfactory performance can be maintained in the presence of faults. *Copyright © 1998 IFAC*

Keywords: Fault-tolerant Control, Robust Control, LMI, Multi-objective Optimisation

1. INTRODUCTION

There is an increasing demand for man-made dynamic systems to become safer and more reliable. These requirements extend beyond normally accepted safety-critical systems of nuclear reactors, chemical plants and aircraft, to advanced systems employed in cars, rapid transit trains, banking systems, etc. Faults (serious or tolerant) are difficult to avoid due to unexpected factors in operating conditions. Therefore, some fault-tolerance measures should be taken in control systems design to ensure reliable and efficient system operation. A fault-tolerant controller engages suitably fixed or varied structure to guarantee stability and satisfactory performance when all components are in normal condition, as well as the cases when some components are of malfunction. Although advanced fault-tolerant control is important in safety-critical systems, the technique can also be used to improve plant efficiency, maintainability, availability and reliability, thereby saving both cost and mantainance time.

There have been considerable amounts of research in the field of fault-tolerant control (Rauch, 1995; Patton, 1997). However, the effective approaches for designing fault-tolerant control systems are awaited. In this paper, the robust fault-tolerant control problem is formulated using the Linear Matrix Inequality (LMI) framework. With this formulation, the satisfactory performance level which also guarantees robust stability can be achieved under faulty conditions. The fault-tolerant control system design method proposed in this paper is based on the LMI formulation established by the Bounded Real Lemma (Gahinet & Apkarian, 1994) and the parameter-dependent controller derived by Apkarian & Gahinet (1995).

Compared with the robust control system design, the fault-tolerant control system design should consider the imperfect estimation of fault effects which provided by fault diagnosis mechanism. The estimation error is due to exogenous disturbances and modelling uncertainty which are always present in real control systems. To ensure desirable control performance, both robust performance and robust stability with respect to modelling uncertainty should be considered in robust fault-tolerant control systems design. The simultaneous consideration of modelling uncertainty and performances will introduce a rank constraint in the linear matrix inequality formulation. This may lead to a non-convex problem and its numerical tractability is discussed in this paper. In this paper, a multi-objective approach is used to build up a matrix inequality formulation for fault-tolerant control system synthesis which is based on the assumption that both fault effect factors and uncertainties can be of linear-fractional-transformation (LFT) parameter dependence (actually, many other kinds of practical systems, such as process control etc., satisfy this kind of assumption).

An example of flight control system design demonstrates that whilst faults occur in the dynamic system which cause big changes in system parameters, the closed-loop system is still stable and has satisfactory performance.

2. PRELIMINARY AND NOTATION

For appropriately dimensioned matrices K and M with partition

$$M = \begin{bmatrix} M_{11} & M_{12} \\ M_{21} & M_{22} \end{bmatrix}$$

and assuming the inverses exist, the lower LFT is defined as

$$F_l(M,K) = M_{11} + M_{12}K(I - M_{22}K)^{-1}M_{21}$$

and the upper LFT is defined as

$$F_u(M,K) = M_{22} + M_{21}K(I - M_{11}K)^{-1}M_{12}$$

For a stable real-transfer function matrix M, the H_∞ norm is defined as

$$\|M(s)\|_\infty = \sup_{\omega \in \mathbf{R}} \bar{\sigma}(M(j\omega))$$

For real symmetric matrices M, $M > 0$ means that it is a positive definite matrix, i.e. all the eigenvalues of M are positive. Similarly, $M < 0$ means negative definite, i.e. all the eigenvalues of M are negative. For $M > 0$, $M^{\frac{1}{2}}$ denotes the unique positive definite square root.

Here, we also give the structure definition matrices for:

Fault factor matrix

$$\Lambda_\xi := \left\{ diag\left(\xi_1 I_{r_1}, \cdots, \xi_k I_{r_k}\right) \ : \ \xi_i(t) \in \mathbf{R}, r = \sum_{i=1}^{k} r_i \right\}$$

Uncertainty matrix

$$\Lambda_\Delta := \{diag\left[\delta_1 I_{r_1}, \ldots, \delta_S I_{r_S}, \Delta_{S+1}, \ldots, \Delta_{S+F}\right] \ : \ \delta_i \in \mathbf{C}, \Delta_{S+j} \in \mathbf{C}^{m_j \times m_j}, 1 \le i \le S, 1 \le j \le F\}$$

Symmetric scaling matrix for fault factors

$$S_{\Lambda_\xi} := \left\{ S > 0 \ : \ S \in \mathbf{R}^{r \times r}, S\Xi = \Xi S, \forall \Xi \in \Lambda_\xi \right\}$$

Symmetric scaling matrix for uncertainties

$$S_{\Lambda_\Delta} := \{diag\left[S_1, \ldots, S_s, s_{s+1} I_{m1}, \ldots, s_{S+F} I_{mF}\right] \ : \ S_i \in \mathbf{C}^{ri \times ri}, \ S_i = S_i^* > 0, \ s_{s+j} \in \mathbf{R}, \ s_{s+j} > 0 \}$$

Mixed scaling matrix

$$S_{\Lambda_M} := \left\{ \begin{bmatrix} S_{i,1} & S_{i,2} \\ S_{i,2}^T & S_{i,3} \end{bmatrix} > 0 \ : \ S_{i,1}, S_{i,3} \in S_{\Lambda_\xi}; \ and \ S_{i,2}\Xi = \Xi S_{i,2}, \forall \Xi \in \Lambda_\xi \right\}$$

<u>Scaled Bounded Real Lemma</u> (Apkarian *et al*, 1995): Consider a parameter structure $\Delta \in \Lambda_\Delta$, $\|\Delta\|_\infty < 1$, the associated scaling set S_{Λ_Δ}. The following statements for a system with the transfer function $T(s) = D + C(sI - A)^{-1}B$ are equivalent.

1) A is stable and there exists $S \in S_{\Lambda_\Delta}$ such that $\left\| S^{\frac{1}{2}} T(s) S^{-\frac{1}{2}} \right\|_\infty < 1$.

2) There exist positive definite solutions X_{cl} and $S \in S_{\Lambda_\Delta}$ to the matrix inequality

$$\begin{bmatrix} A^T X_{cl} + X_{cl} A & X_{cl} B & C^T \\ B^T X_{cl} & -S & D^T \\ C & D & -S^{-1} \end{bmatrix} < 0 \qquad (2.1)$$

⊠

3. PROBLEM FORMULATION

During the operation, the effects of a fault on the controlled system could be the changes in signals and parameters. Whilst just considering small parameter and signal changes, we can use a robust controller to achieve a passive form of fault tolerance. However, in many practical cases, the changes in system parameters caused by faults are large. It is even impossible to stabilise such a system with very big changes in system parameters just by means of a fixed robust controller. Here a fault detection and isolation (FDI) mechanism is engaged to provide diagnostic information which is used to supervise the controller reconfiguration or restructure via active fault tolerance. The scheme for the fault-tolerant control system (FTCS) with supervision and FDI is given in Fig.3.1, where Δ reflects uncertainties, $\xi(t)$ means the factors of fault effects, $P(s,\xi)$ denotes the generalised plant, $K(s,\hat{\xi})$ represents the feedback controller.

The manifestation of faults varies in different systems, with different components, and with different kinds of faults. However, for those kinds of faults which introduce changes in the system's dynamic parameters, it is convenient to describe the effects of faults in LFT forms as one of following:

$$F_l(P,\Xi) = P_{11} + P_{12}\Xi(I - P_{22}\Xi)^{-1} P_{21} \quad (3.1)$$

$$F_u(P,\Xi) = P_{22} + P_{21}\Xi(I - P_{11}\Xi)^{-1} P_{12} \quad (3.2)$$

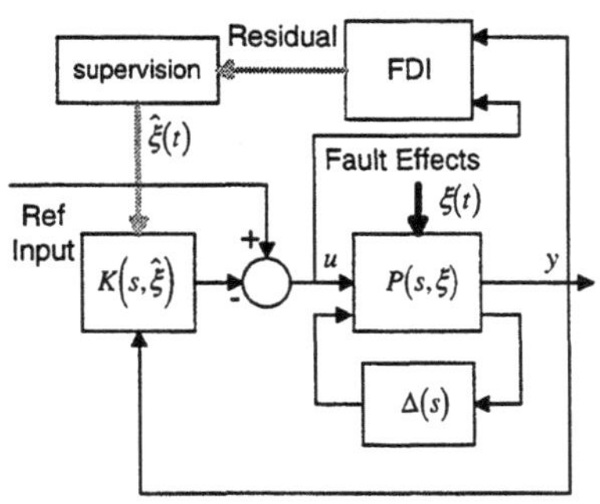

Fig. 3.1 General scheme of fault-tolerant control system

Based on the above assumption, the fault-tolerant control system shown in Fig. 3.1 can be re-structured in Fig. 3.2. In Fig. 3.2, the thick lines represent the FDI signal paths, and the thin lines give the scheme of connection for control system design.

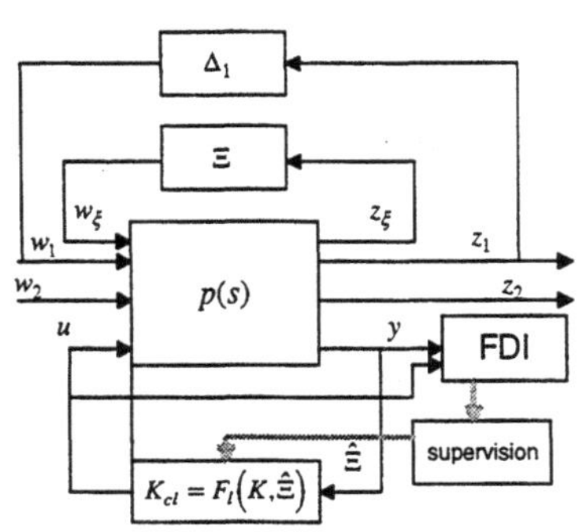

Fig. 3.2 Signal connection scheme in FTCS

where, $\Xi \in \Lambda_\xi$ is the fault factor matrix, $\Delta_1 \in \Lambda_\Delta$ is the disturbance matrix, and $\hat{\Xi} \in \Lambda_\xi$ is the matrix for estimated fault factors. Here, it has already been defined that the relationship between controller and fault factor matrix can also be described by a LFT connection:

$$K_{cl} = F_l\left(K(s), \hat{\Xi}\right) = F_l\left(K(s), \Xi\right) \quad (3.3)$$

In a practical situation, it is impossible to express fault effects precisely. All inaccuracies, modelling errors, and effects of non-linearity should be considered as uncertainties to achieve controller robustness. Meanwhile, we can assume $\hat{\Xi} = \Xi$. As for uncertainties, it is well known that in many cases, it would be convenient to describe them in LFT forms (Packard et al, 1993).

Considering the use of H_∞ optimisation for fault-tolerant controller synthesis, the problem now can be expressed as:

Problem 3.1: Find an internally stabilising controller $K(s)$ such that the closed-loop system $T_{z_2 w_2}$ is internally stable for all fault factors $\|\Xi\|_\infty \le 1$ and uncertainties $\|\Delta_1\|_\infty \le 1$, and so that the H_∞ norm of closed-loop transfer function matrix satisfies

$$\left\|T_{z_2 w_2}\left(P, K, \Xi\right)\right\|_\infty < 1 \quad (3.4)$$

for all $\|\Xi\|_\infty \le 1$ and $\|\Delta_1\|_\infty \le 1$. ⊠

4. LMI SOLUTION FOR ROBUST FTCS

When the scaled bounded real lemma is applied directly to the system shown in Fig 3.2 for finding a controller, a rank constraint has to be involved in the matrix inequality system. This rank constraint can be expressed as:

$$rank\left(\begin{pmatrix} S_3 & 0 \\ 0 & S_4 \end{pmatrix}\begin{pmatrix} S_{\hat{3}} & 0 \\ 0 & S_{\hat{4}} \end{pmatrix} - I\right) \le rank(S_1)$$

where $S_1, S_3, S_{\hat{3}} \in S_{\Lambda_\xi}$ and $S_4, S_{\hat{4}} \in S_{\Lambda_\Delta}$.

Because of the rank constraints, the convexity property of the problem is not guaranteed, and the numerical tractability of the matrix inequality constraints becomes unclear.

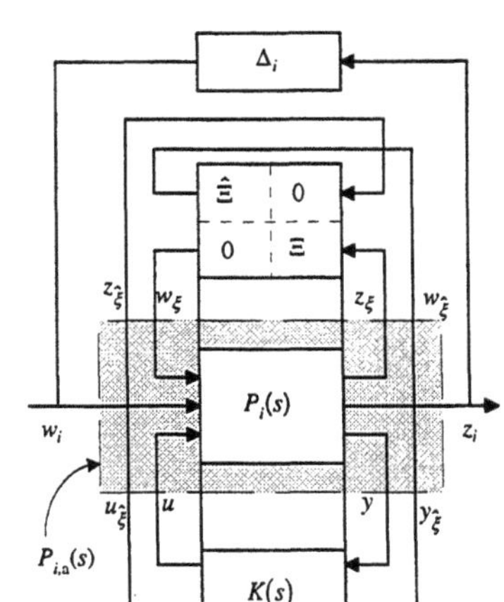

Fig. 4.1 Equivalent structure for robust FTCS, the i_{th} channel

In order to recover the convexity of the optimisation problem, a constrained multi-objective approach is introduced. Here, for the system shown in Fig 3.2, one objective is robust stability in the presence of fault effects. In this case, only the w_i to z_i,

$(i = 1, 2)$ channel is considered independently, and an equivalent structure of Fig. 3.2 is given in Fig. 4.1.

For the augmented plant $P_{i,a}(s)$, $(i = 1, 2)$, the output vector is $\begin{bmatrix} z_{\xi}^T & z_{\xi}^T & z_i^T & y^T & y_{\xi}^T \end{bmatrix}^T$, the input vector is $\begin{bmatrix} w_{\xi}^T & w_{\xi}^T & w_i^T & u^T & u_{\xi}^T \end{bmatrix}^T$, Δ_1 represents uncertainties, Δ_2 gives the measure of robust performance, and the realisation of the generalised plant could be:

$$P_{i1,a}(s) = \begin{bmatrix} 0 & 0 & 0 & 0 & I_r \\ 0 & D_{i,\xi\xi} & D_{i,\xi 1} & D_{i,\xi 2} & 0 \\ 0 & D_{i,1\xi} & D_{i,11} & D_{i,12} & 0 \\ 0 & D_{i,2\xi} & D_{i,21} & D_{i,22} & 0 \\ I_r & 0 & 0 & 0 & 0 \end{bmatrix} + \begin{bmatrix} 0 \\ C_{i,\xi} \\ C_{i,1} \\ C_{i,2} \\ 0 \end{bmatrix} (sI - A)^{-1} \begin{bmatrix} 0 & B_{i,\xi} & B_{i,1} & B_{i,2} & 0 \end{bmatrix} \tag{4.1}$$

Similarly, the realisation of a dynamic controller K can be describe by

$$K(s) = \begin{bmatrix} D_{K11} & D_{K1\xi} \\ D_{K\xi 1} & D_{K\xi\xi} \end{bmatrix} + \begin{bmatrix} C_{K1} \\ C_{\xi 1} \end{bmatrix} (sI - A_K)^{-1} \begin{bmatrix} B_{K1} & B_{K\xi} \end{bmatrix} \tag{4.2}$$

Hence, the closed-loop description from w_i to z_i, $(i = 1, 2)$, can be denoted as

$$T_{zwi} := F_u\left(F_l\left(P_{i,a}(s), K(s)\right), \begin{bmatrix} \hat{\Xi} & 0 \\ 0 & \Xi \end{bmatrix} \right) \tag{4.3}$$

According to the small gain theory, a sufficient condition for robust stability of the closed-loop connection closed by Δ_i with $\|\Delta_i\|_\infty \le 1$ in the face of the fault effects $\begin{bmatrix} \hat{\Xi} & 0 \\ 0 & \Xi \end{bmatrix}$ could be expressed as:

If there exists a mixed scaling matrix $S_i = \begin{bmatrix} S_{i,1} & S_{i,2} \\ S_{i,2}^T & S_{i,3} \end{bmatrix} \in S_{\Lambda_M}$ and a controller (4.2) such that the nominal closed-loop system $F_l\left(P_{i,a}(s), K(s)\right)$ is internally stable and satisfies

$$\left\| \begin{bmatrix} S_i^{\frac{1}{2}} & 0 \\ 0 & I \end{bmatrix} F_l\left(P_{i,a}(s), K(s)\right) \begin{bmatrix} S_i^{-\frac{1}{2}} & 0 \\ 0 & I \end{bmatrix} \right\|_\infty < 1 \tag{4.4}$$

Then, the closed-loop controller (3.3) is a fault-tolerant controller with stability robustness in the sense of H_∞ optimisation.

For the FTCS shown in Fig. 4.1 and expressed in (4.1), (4.2) and (4.3), assume that 1). $(A, B_{i,2}, C_{i,2})$ is stabilisable and detectable; 2). $D_{i,22} = 0$. Then, $F_l\left(P_{i,a}(s), K(s)\right)$ is given by $\begin{bmatrix} A_{i,cl} & B_{i,cl} \\ C_{i,cl} & D_{i,cl} \end{bmatrix}$ with

$$A_{i,cl} = \begin{bmatrix} A & 0 \\ 0 & 0 \end{bmatrix} + \left[\begin{array}{c|cc} 0 & B_{i,2} & 0 \\ \hline I_k & 0 & 0 \end{array}\right] F \left[\begin{array}{c|c} 0 & I_k \\ \hline C_{i,2} & 0 \\ 0 & 0 \end{array}\right] = \underline{A} + \underline{B}_{i,2} F \underline{C}_{i,2} \tag{4.5}$$

$$B_{i,cl} = \left[\begin{array}{ccc} 0 & B_{i,\xi} & B_{i,1} \\ \hline 0 & 0 & 0 \end{array}\right] + \left[\begin{array}{c|cc} 0 & B_{i,2} & 0 \\ \hline I_k & 0 & 0 \end{array}\right] F \left[\begin{array}{ccc} 0 & 0 & 0 \\ \hline 0 & D_{i,2\xi} & D_{i,21} \\ I_r & 0 & 0 \end{array}\right] = \underline{B}_{i,1} + \underline{B}_{i,2} F \underline{D}_{i,21} \tag{4.6}$$

$$C_{i,cl} = \left[\begin{array}{c|c} 0 & 0 \\ C_{i,\xi} & 0 \\ C_{i,1} & 0 \end{array}\right] + \left[\begin{array}{c|cc} 0 & 0 & I_r \\ 0 & D_{i,\xi 2} & 0 \\ 0 & D_{i,12} & 0 \end{array}\right] F \left[\begin{array}{c|c} 0 & I_k \\ \hline C_{i,2} & 0 \\ 0 & 0 \end{array}\right] = \underline{C}_{i,1} + \underline{D}_{i,12} F \underline{C}_{i,2} \tag{4.7}$$

$$D_{i,cl} = \begin{bmatrix} 0 & 0 & 0 \\ 0 & D_{i,\xi\xi} & D_{i,\xi 1} \\ 0 & D_{i,1\xi} & D_{i,11} \end{bmatrix} + \left[\begin{array}{c|cc} 0 & 0 & I_r \\ 0 & D_{i,\xi 2} & 0 \\ 0 & D_{i,12} & 0 \end{array}\right] F \left[\begin{array}{ccc} 0 & 0 & 0 \\ \hline 0 & D_{i,2\xi} & D_{i,21} \\ I_r & 0 & 0 \end{array}\right] = \underline{D}_{i,11} + \underline{D}_{i,12} F \underline{D}_{i,21} \tag{4.8}$$

$$F = \left[\begin{array}{c|cc} A_K & B_{K1} & B_{K\xi} \\ \hline C_{K1} & D_{K11} & D_{K1\xi} \\ C_{K\xi} & D_{K\xi 1} & D_{K\xi\xi} \end{array}\right] \tag{4.9}$$

By applying the *Scaled Bounded Real Lemma* to $F_l\left(P_{i,a}(s), K(s)\right)$, a set of matrix inequalities with the variables $X_{i,cl}$ and L_i is obtained:

$$\Psi_i + P_{i,X}^T F Q_i + Q_i^T F^T P_{i,X} < 0, \quad (i = 1, 2) \tag{4.10}$$

where

$$\Psi_i = \begin{bmatrix} \underline{A}^T X_{i,cl} + X_{i,cl}\underline{A} & X_{i,cl}\underline{B}_{i,1} & \underline{C}_{i,1}^T \\ \underline{B}_{i,1}^T X_{i,cl} & -L_i & \underline{D}_{i,11}^T \\ \underline{C}_{i,1} & \underline{D}_{i,11} & -L_i^{-1} \end{bmatrix}$$

$$P_{i,X} = \begin{bmatrix} \underline{B}_{i,2}^T X_{i,cl} & 0 & \underline{D}_{i,12}^T \end{bmatrix}$$

$$Q_i = \begin{bmatrix} \underline{C}_{i,2} & \underline{D}_{i,21} & 0 \end{bmatrix}$$

$$L_i = diag(S_i,\, I) = diag\left(\begin{bmatrix} S_{i,1} & S_{i,2} \\ S_{i,2}^T & S_{i,3} \end{bmatrix}, I\right)$$

$$L_i^{-1} = diag\left(S_i^{-1},\, I\right) = diag\left(\begin{bmatrix} S_{i,\hat{1}} & S_{i,\hat{2}} \\ S_{i,\hat{2}}^T & S_{i,\hat{3}} \end{bmatrix}, I\right)$$

and the definition for $\underline{A}$, $\underline{B}_{i,1}$, $\underline{C}_{i,1}$, etc. ., is given by equation (4.5), (4.6), (4.7), and (4.8).

Partition $X_{i,cl}$ as

$$X_{i,cl} = \begin{bmatrix} Y_i & N_i \\ N_i^T & R_i \end{bmatrix},\ X_{i,cl}^{-1} = \begin{bmatrix} X_i & M_i \\ M_i^T & T_i \end{bmatrix}$$

Then, the objectives expressed in Eq. (4.4) can be achieved if there exist positive definite matrices $X_i \in R^{n\times n}$, $Y_i \in R^{n\times n}$, $S_{i,3}, S_{i,\hat{3}} \in S_{\Lambda_\xi}$ such that:

$$N_{i,X}^T \left[\begin{array}{c|cc|cc} AX_i + X_i A^T & B_{i,\xi} & B_{i,1} & X_i C_{i,\xi}^T & X_i C_{i,1}^T \\ \hline B_{i,\xi}^T & -S_{i,3} & 0 & D_{i,\xi\xi}^T & D_{i,1\xi}^T \\ B_{i,1}^T & 0 & -I & D_{i,\xi 1}^T & D_{i,11}^T \\ \hline C_{i,\xi} X_i & D_{i,\xi\xi} & D_{i,\xi 1} & -S_{i,\hat{3}} & 0 \\ C_{i,1} X_i & D_{i,1\xi} & D_{i,11} & 0 & -I \end{array}\right] N_{i,X} < 0 \qquad (4.11)$$

$$N_{i,Y}^T \left[\begin{array}{c|cc|cc} A^T Y_i + Y_i A & Y_i B_{i,\xi} & Y_i B_{i,1} & C_{i,\xi}^T & C_{i,1}^T \\ \hline B_{i,\xi}^T Y_i & -S_{i,3} & 0 & D_{i,\xi\xi}^T & D_{i,1\xi}^T \\ B_{i,1}^T Y_i & 0 & -I & D_{i,\xi 1}^T & D_{i,11}^T \\ \hline C_{i,\xi} & D_{i,\xi\xi} & D_{i,\xi 1} & -S_{i,\hat{3}} & 0 \\ C_{i,1} & D_{i,1\xi} & D_{i,11} & 0 & -I \end{array}\right] N_{i,Y} < 0 \qquad (4.12)$$

$$\begin{bmatrix} X_i & I \\ I & Y_i \end{bmatrix} \geq 0, \text{ and } \begin{bmatrix} S_{i,3} & I \\ I & S_{i,\hat{3}} \end{bmatrix} \geq 0 \qquad (4.13)$$

where , $N_{i,X}$ and $N_{i,Y}$ are arbitrary bases of the *Null Spaces* of $\begin{bmatrix} B_{i,2}^T \mid 0 \mid D_{i,\xi 2}^T & D_{i,12}^T \end{bmatrix}$ and $\begin{bmatrix} C_{i,2} \mid D_{i,2\xi} & D_{i,21} \mid 0 \end{bmatrix}$, and $(i = 1, 2)$ for all the equations above.

Now, all specifications and objectives are expressed in terms of linear matrix inequality constraints (4.11), (4.12), and (4.13) with $(i = 1, 2)$. Multi-objective optimisation can be achieved by joining all these matrix inequality constraints together. In order to recover the convexity of the problem, X_1 and X_2 are forced to be identical as $X_1 = X_2 = X$, as are also well as Y_1 and Y_2 with $Y_1 = Y_2 = Y$. This actually means that an identical solution for equation (4.10) with (i=1 , 2) is guaranteed, i.e. $X_{1,cl} = X_{2,cl} = X_{cl}$. Here, the LMI Toolbox in MATLAB (Gahinet *et al*, 1995) can be used for the computation of the controller matrices.

From the linear matrix inequalities (4.11), (4.12) and (4.13), (i=1 and 2), a solution of a set of positive definite matrices $X \in R^{n\times n}$, $Y \in R^{n\times n}$, $S_{i,3}, S_{i,\hat{3}} \in S_{\Lambda_\xi}$, (i=1 and 2), are given. Based on X and Y, X_{cl} can be derived, and based on $S_{1,3}$, $S_{1,\hat{3}}$, $S_{2,3}$ and $S_{2,\hat{3}}$, L_1 and L_2 can be derived. Finally, from linear matrix inequalities (4.10), the controller parameter matrix F is given.

5. AN EXAMPLE

Here we use an example of unmanned aircraft, suffering wing impairment, as a design example. At a particular operating point, the linearised dynamic equation for a second order short period mode could be $\dot{x} = Ax + B\delta_e$ with $x = \begin{bmatrix} \alpha & q \end{bmatrix}^T$, where α is angle of attack, q is the pitch rate, and δ_e is the elevator input. Any minor battle damage on the aircraft wing will cause changes in aerodynamic coefficients C_L, C_{MW} etc., which will bring changes in some elements of the system dynamic matrix A . Using ξ_1 and ξ_2 to denote the normalised deviation caused by the changes of C_L and C_{MW} , a system description could be:

$$A = \begin{bmatrix} (-2.3298 + 0.8032\xi_1)(1 + 0.1\delta_1) & 1 \\ (-14.9358 + 5.4563\xi_2)(1 + 0.05\delta_2) & -0.0499(1 + 0.1\delta_3) \end{bmatrix}$$

$$B = \begin{bmatrix} 0.1611 \\ -13.5789 \end{bmatrix}$$

here, $\delta_i, (i = 1,2,3)$ reflect parametric modelling uncertainties.

The LFT structure of the control system is shown in Fig. 5.1. In this structure, δ_4 and δ_5 are introduced to reflect the estimation errors in factors ξ_1 and ξ_2 . G_m is the ideal model to be followed, which is used to achieve the performance requirements.

$$G_m = \frac{16}{s^2 + 5.656s + 16} .$$

W_1 and W_2 are weighting functions.

$W_1 = \frac{s+15}{4s+8}$ is a low-pass function which enforces the tracking ability at lower frequencies. $W_2 = \frac{s+10}{s+60}$ is a high-pass function used suppress the effects of uncertainties and disturbances at high frequencies. Using the process described in the last section, a 7th order controller is developed.

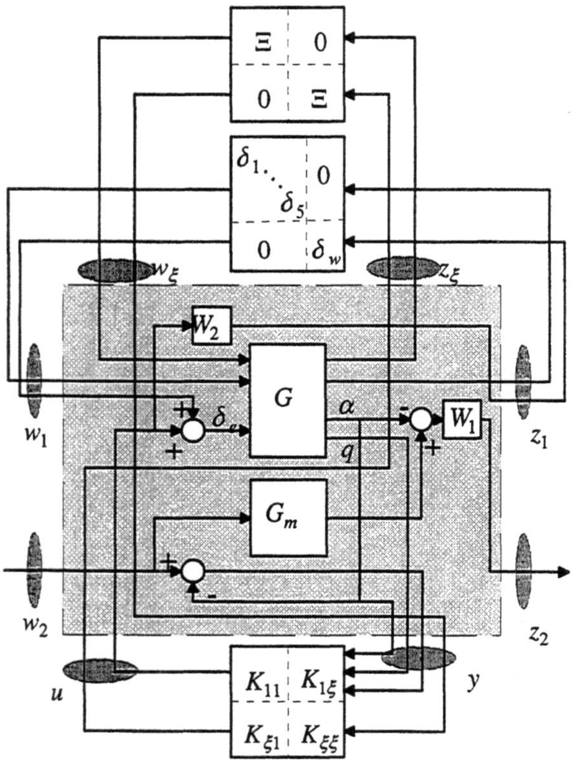

Fig. 5.1. Augmented plant and control system structure

Fig 5.2 shows the closed-loop unit step input responses. At the time $t = 5\sec$, both ξ_1 and ξ_2 are changed from the initial value 0 to the value 1 in a steep ramp. The system keeps stable whilst still has good performance, as shown in the responses to the second unit step input at t=12 sec.

6. CONCLUSION

This paper has presented an LMI-based synthesis scheme for robust fault-tolerant control system design. In order to eliminate the rank constraints whilst considering the robust performance in the presence of faults, uncertainties and estimation errors of the fault effects, a multi-objective approach has been introduced. The simulation results of design example indicate that the fault-tolerant property with the guarantee in performance and stability has been achieved. However, in this scheme, only the unstructured uncertainty block has been considered, and the knowledge of fault effect factors has not been fully utilised. Further development in these aspects is expected.

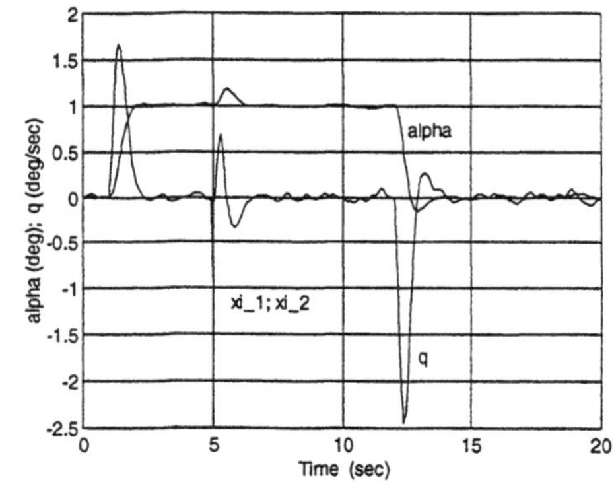

Fig 5.2 Unit step input responses of the closed-loop system

REFERENCES

Apkarian, P. and P. Gahinet (1995), A Convex Characterisation of Gain-Scheduled H_∞ Controllers, *IEEE Transactions on Automatic Control*, **AC-40** (5), May 1995, 853-864.

Gahinet, P., A. Nemirovski, A. J. Laub and M. Chilali (1995), *LMI Control Toolbox*, The Math Works Inc., May, 1995.

Zhou, K., J. C. Doyle, and K. Glover (1995), *Robust and Optimal Control*, Prentice Hall, 1995.

Boyd, S., L. El Ghaoui, E. Feron, and V. Balakrishnan (1994), *Linear Matrix Inequalities in System and Control Theory*, Vol. 15 of SIAM Studies in Applied Mathematics, SIAM, 1994.

Gahinet, P. and P. Apkarian (1994), A Linear Matrix Inequality Approach to H_∞ Control, *Int. Journal of Robust and Non-linear Control*, **4**, 421-448.

Packard, A and J. C. Doyle (1993), The Complex Structured Singular Value, *Automatica*, **29** (1), 1993, 71-109.

Patton, R. J. (1997), Fault-tolerant control: the 1997 situation (survey), *Preprints of "IFAC Sympo. on Fault Detection, Supervision and Safety for Technical Processes: SAFEPROCESS'97"*, Vol.2, 1033-1055, The University of Hull, August 26-28.

HYBRID FDI ON CHEMICAL PLANTS

Ferreiro García, Ramón; Pardo Martínez, Xoán C. & Vidal Paz, José

Grupo de Enxeñeria de Sistemas e Control Automático.
Departamento de Electrónica e Sistemas. Universidade da Coruña.
E.S. da Mariña Civil. Paseo de Ronda, 51. 15071, A Coruña. Spain.
Email: {ferreiro, pardo, vidal}@des.fi.udc.es

Abstract: This paper illustrates an alternative implementation to automatic protection and supervision task using the basic principles of model-based fault-detection and isolation associated to the rule-based FDI with subsequent appropriate corrective actions. Hybrid systems may be precisely described by sequential function charts (SFC). Operation supervision of hybrid systems may be carried out simultaneously on the basis of such SFC description. As discrete event and continuous processes operate simultaneously in close cooperation, a strategy to diagnose such hybrid system by means of a structured methodology under SFC description tool on the IEC-848 standard is developed.

Keywords: Hybrid model based diagnosis, FDI by SFC, Parity equations, Supervisory control.

1. INTRODUCTION

The operation of technical processes requires increasingly advanced supervision and fault diagnosis to improve safety and production costs. With increasing process monitoring associated to an ever-higher level of automation to achieve desired quality, plants have become more vulnerable to faults in its instrumentation and consequently plant availability.

Most production plants operate as a combination of both discrete event and continuous processes (Blanke et all, 1997). Consequently, a structured methodology to solve the supervision task of hybrid processes is strongly demanded. The proposed methodology uses the association of plant operation description under SFC to the classical model-based approaches such as parameter estimation, state estimation (Gertler, 1997; Iserman, 1997) and a novel parity equation procedure (Patton and Chen, 1997). This work describes the basic approach for the engineering design of control systems or plants supervisor. It outlines how fault detection and isolation (FDI) can be associated to decision making procedures achieving a class of fault tolerant control by means of a hybrid process control system description and programming tool: the SFC according standard IEC-848.

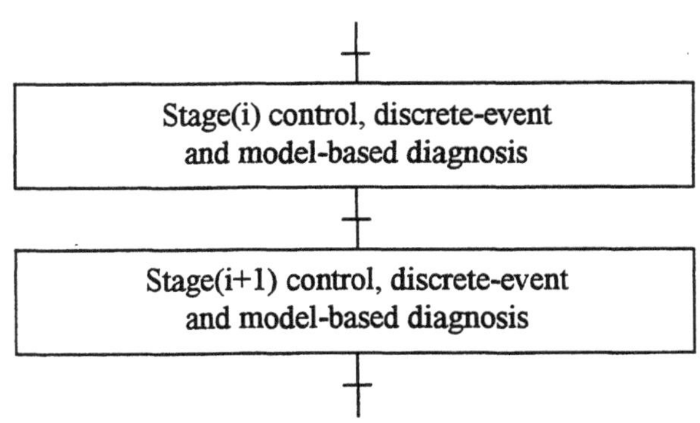

Fig. 1. Task description for control and supervision using SFC tools.

Figure 1 shows the basic approach of a partial sequential procedure illustrating a multitask structure as shown at (Blanke et all, 1997) with a difference that is, this implementation is carried out under the IEC-848 standard, in which both the control and supervision tasks are executed under the same tool.

Decisions on control sequence are taken by event following for plant degradation or fault detection and they depend on the state of plant and type of fault. Classically, the so-called diagnostic reasoning approach is performed by means of a rule base in which the single logical operations are state as IF THEN ELSE rules. SFC description tool contains all necessary knowledge and an inherent manager for its rule base capable to solve a discrete event control problem and consequently to solve any rule based control system. The major disadvantage of this approach has always been that binary logical decisions with Boolean operators do not reflect the gradual nature of many real world diagnostic problems. Fortunately SFC tool is capable to support both the discrete and the continuous event supervision under a unique representation frame

2. THE APPROACH TO DISCRETE EVENT DIAGNOSTICS

Such partial method consist in a control supervisor capable for evaluating, in any stage of the control process, the condition of the actuators (effectors), sensors and/or detectors involved on previous and present stage of the sequential control path, and consequently decides which action should be appropriate in order avoid system damages or system failure (Ferreiro, 1992). Let's consider part of a flow control sequence described by means of a SFC, where CT(i) is the condition for transition from stage S(i-1) to stage S(i) as shown in figure 2.

For every stage (i) of a discrete event control sequence the time to perform activities is a key in diagnostic tasks. There are stages for which such time is a constant, while for others time is a variable that depends on process characteristics.

For stages in which the time necessary to complete the activities is a constant, such as on-off valves, actuator cylinders at constant speed and many other periodic activities, the evidence of a fault in an activity is the time T(i) the action takes to reach the next condition for a normal transition. With such information available, the problem to be solved so as to validate the acquired data from detectors /or sensors and process supervision can be formulated by means of a true table shown at table 1. The application of true tables logic supply the same conclusions as rule bases.

From the knowledge stored in true table it must be remarked that:

Row 000, there is no evidence of fault,

Row 001, there is no evidence of fault

Row 010, there is at least one fault, but it will be required further analysis to conclude the search in the direction (symptom, event, and fault).

Row 011, the activity is performed perhaps with symptoms of degradation or overload, but extensive analysis is also necessary to conclude with a deterministic solution.

Row 100, suggest that detectors/sensors are set from last stage. Further analysis is necessary. Obviously in such discrete event analysis it is clear that do not reflect the gradual nature of real world diagnostic problems. More information can be achieved by adding to the true table another limit time To, that is a low limit time, for which an activity could never be performed under normal conditions.

Table 1 The basic FDI true table

t<To	t>T	CT	Case	The cause of fault
0	0	0		Normal operation
0	0	1		Normal operation
0	1	0	A	Detectors and/or Sensors, variables, process.
0	1	1	B	Slow actuation (degradation or load excess)
1	0	0	C	Detectors/Sensors set from last cycle (MBA required)

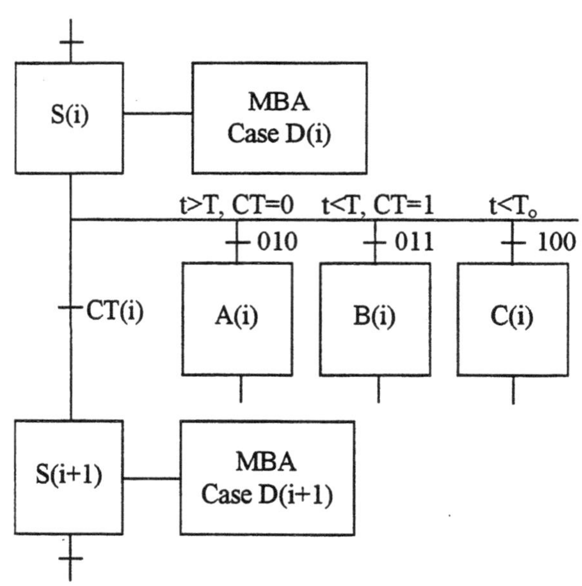

Fig. 2. The principle of hybrid supervision multitask.

It is clear that further effort is necessary to conclude the cause of faults. In this way discrete event search is useful as a management tool in scheduling model-based analysis. Figure 2 illustrates the diagnostic task added to a partial SFC control sequence.

Detection of the variable that do not satisfy the condition for transition is carried out by comparing the logic value of every actual variable involved in the transition function with the true value of same variable that satisfy the condition for transition. The conclusion is straightforward applied to a decision-making procedure. Decision-making through the conclusions given by the proposed true table, is to be implemented by SFC and depends on the desired action to be performed, which will be programmed by a human expert in the particular problem-solving task under the criteria of avoiding critical or ambiguous situations. In the case a condition for transition CT(i) fails, then SFC flow enter in supervision task as shown in figure 2. The result of supervision may be the detection of a single fault (non-critical fault) in one or more detectors involved in such transition function. If plant can operate or complete this operating cycle without forcing the control flow to a shutdown, then, flow should be transferred to next safety condition for transition by passing control flow to transition CT(i+1). In this case, the architecture belongs to fault tolerant control (FTC) for which SFC is able.

Generally, the time to complete an activity is not constant, it depends on process variables, internally coupled variables and process interrelated variables. Prediction of normal activity time may not be possible with required precision. Consequently, discrete analysis can not be carried out with a deterministic conclusion. The unique solution to this problem is to perform a model-based analysis for which if processes are well understood, dynamic models for fault detection and isolation using the well-established continuous-time modelling can be used. Such alternative is shown at figure 2 as case D.

3. HYBRID FDI

The most common case in FDI tasks is that one in which model based analysis must be associated to fault tree analysis to conclude and decide the required safe action.

Case A(i) in figure 2 is a poor or ambiguous conclusion if no further search were performed. The first conclusion is that at least a fault exists in one or both: detectors and/or process parameters and/or process variables.

In order to perform the isolation fault task for the case A(I), the worst case, a new fault tree analysis by means of a rule-based procedure is presented. Let's consider that if a fault is present in the case A(i) then A(i)=1 and if the fault is present in Model-Based Analysis (MBA), then MBA=1. Table 2 shows the sequence to be performed in the analysis approach.

Table 2 True table for analysis sequence

A(i)	MBA	Detectors sensors	plant params.	plant vars.	state event
0	0				
0	1		X	X	plant fault
1	0	X			
1	1	X	X	X	critic state

At above table, the cluster with X means that at least a fault is present. So that from row 2 in the above table it is concluded that plant parameters and/or plant variables state is abnormal. At this state an MBA task on plant parameters is useful to discriminate faults in plant variables. That means, if a after a MBA operation plant parameters are normal, then it is to be concluded that the fault is isolated in plant variables. Last row in table 2 indicates that a critic fault is present and normal sequence operation can not continue. Consequently, heuristic reasoning is strongly demanded for deciding under human inspection.

4. APPLICATION OF THE DIAGNOSTIC TASK ON A TANK LEVEL PROBLEM

Let's consider a subtask or fragment of a process based in the tank level control by means of an on-off control valve. In order to denote plant variables and parameters, it is shown in figure 3. The control sequence consists in open a valve when level is low limit, (CT1=1), and close the valve when level is at high limit, (CT2=1). The SFC for control and supervision is shown in figure 4.

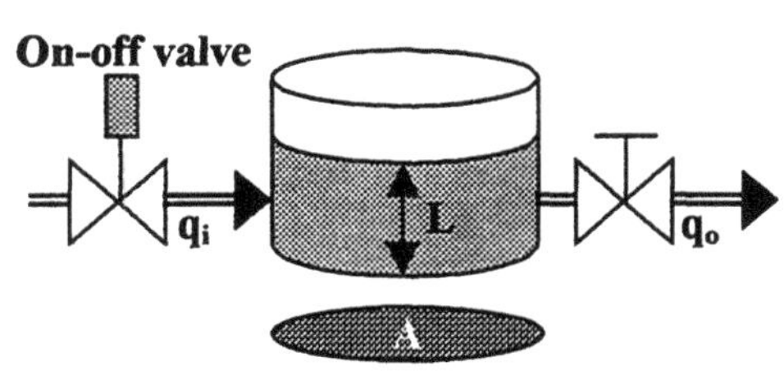

Fig. 3. The process to be supervised.

The supervision task or decision making procedure is carried out by processing MBA and true table logic with the conditions of table 1. If both results indicate plant faults, discrimination by means of another true table logic with the conditions of table 2 (tab2) is executed. Finally, a decision effector is activated to jump towards a safety stage. Figure 5 shows the sequence for decision making task.

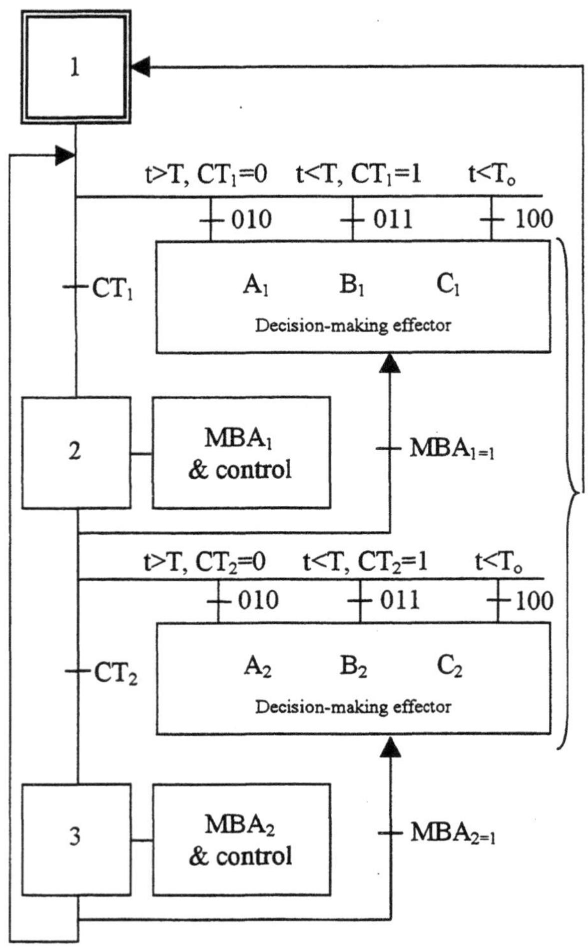

Fig. 4. The hybrid supervision multitask applied to level control sub-plant.

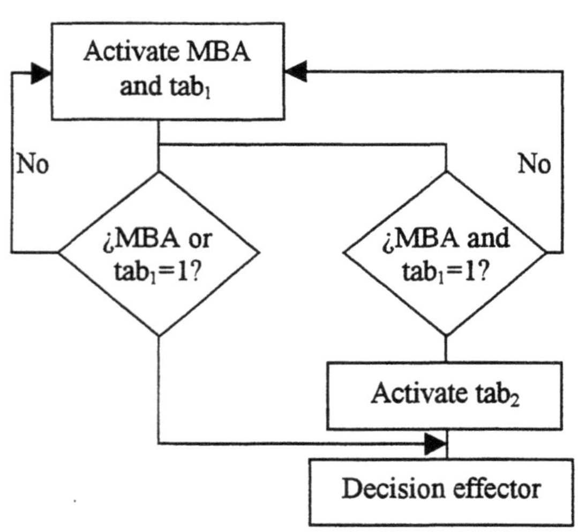

Fig. 5. The decision-making task sequence.

Such control system is implemented by means of a PLC programming language, the Pl7-Micro, a language of last generation, being applicable to several different PLC manufacturers such as Square-D, Modicon, AEG-Schneider and Telemecanique. Model based diagnosis is restricted in this application to compare the actual tank filling velocity with a nominal value. That is, a time series function is compared with the actual measured level, which supply the necessary information to conclude that filling and discharge events are performing according desired models.

4.1. The model-based analysis for level tank dynamics with parity equations

The used models then,

$$q_i - q_o = A\frac{dL}{dt} \tag{1}$$

The transfer functions are

$$G_{ML} = \frac{L}{q_i - q_o} = \frac{1}{AS} \tag{2}$$

According residual generation models for level

$$r_L = G_{ML} f(q_i - q_o) + fL \tag{3}$$

4.2. Fault detection procedure

At least one fault exists if residuals are different from zero. It is possible also the existence of fault when for short periods of time the terms of right hand in equation (3) cancel each other. The detection of faults is straightforward from expression (3).

4.3. Isolation

The isolation of faults in the tank level process requires defining its time domain model, the model for plant parameters is

$$L(t) = \frac{1}{A}\int f(q_i - q_o)dt + fL \tag{4}$$

for application purposes it becomes

$$r_{L(t)} = \frac{1}{A} f(q_i - q_o)t + fL \tag{5}$$

Consequently, an equation system may be achieved

$$r_{L(t_1)} = \frac{1}{A} f(q_i - q_o)t_1 + fL \tag{6}$$

$$r_{L(t_2)} = \frac{1}{A} f(q_i - q_o)t_2 + fL \tag{7}$$

which in matrix form is

$$\begin{bmatrix} r_{L(t_1)} \\ r_{L(t_2)} \end{bmatrix} = \begin{bmatrix} \frac{1}{A}t_1 & 1 \\ \frac{1}{A}t_2 & 1 \end{bmatrix} \begin{bmatrix} f(q_i - q_o) \\ fL \end{bmatrix} \quad (8)$$

whose solution for plant parameters is given as

$$\begin{bmatrix} f(q_i - q_o) \\ fL \end{bmatrix} = \begin{bmatrix} \frac{1}{A}t_1 & 1 \\ \frac{1}{A}t_2 & 1 \end{bmatrix}^{-1} \begin{bmatrix} r_{L(t_1)} \\ r_{L(t_2)} \end{bmatrix} \quad (9)$$

The task of FDI requires plant simulation.

4.4. Rule-based diagnosis by Boolean Logic

Fault detection is based on a discrimination searching. It is considered that if measured velocities are outside the limits given by proposed plant operation safety limits (Low, High), then plant is running under normal conditions. That is,

$$VL_L < VL < VL_H \quad (10)$$

VT_L, VT_H, VL_L, VL_H are defined from process models with safety tolerances

Isolation is concluded by means of heuristic reasoning as

$$\begin{aligned} VL < VL_L &\Rightarrow \text{Pump Efficiency} \\ VL > VL_H &\Rightarrow \text{Level Sensor} \end{aligned} \quad (11)$$

5. CONCLUSION

Some particular characteristics distinguish the SFC based supervision system from others. The PLC exerts its computational effort on the stage in which process is operating. That means, is centred in the active stage for control and supervision. No inactive parts of plant overload the PLC with useless computations. The structured nature of SFC helps in supervision tasks as a straightforward application tool.

Logic and analog process supervision, can be performed simultaneously by means of the same PLC control program. The advantage of such procedure, is plant reliability and safety included its control system, avoiding ambiguity and dangerous actuations after some faults were detected. The indication of the stage in which the system fails, helps in minimise shut-downs, reducing maintenance cost and enlarging plant control system life.

REFERENCES

Alcorta Garcia, E. and Frank, P.M. (1997). *Deterministic Non-linear Observer-Based Approaches to Fault Diagnosis: A Survey.* Control Engineering Practice. May 1997. pp. 663-670.

Blanke, M., Izadi, Zamanabadi, R., Bogh, S. and Lunau (1997). *Fault-tolerant Control Systems. A Holistic View.* Control Engineering Practice. May 1997.

Basseville, M. and Nikiforov, I. (1994). *Statistical Change Detection.* Ed. Prentice Hall.

Ferreiro Garcia, R. (1992). *Real-Time Supervision by Grafcet Algorithm.* Postprint edition of IFAC Symposium on Low Cost Automation (LCA'92) pp. 299-304. Ed. Pergamon. U.K.

Gertler. J (1997). *Fault Detection and Isolation Using Parity Equations.* Control Engineering Practice. May 1997. pp 652-670.

Iserman, R (1997). *Supervision, Fault Detection and Fault-Diagnosis Methods. An Introduction.* Control Engineering Practice. May 1997. pp 639-652.

Patton, R. and Chen, J. (1997). *Observer-Based Fault Detection and Isolation: Robustness and Applications.* Control Engineering Practice. May 1997. pp. 671-682.

NEW INTRINSIC RUNAWAY CRITERIA AND THEIR APPLICATION TO THE MODEL-BASED ON-LINE-MONITORING OF CONTINUOUS AND DISCONTINUOUS REACTORS

T. Obertopp [†], M. A. Alós [‡], J. M. Zaldívar [§], F. Strozzi [§] and E. D. Gilles [†]

[†] *Institut für Systemdynamik und Regelungstechnik, Universität Stuttgart; Germany*
[‡] *Institut Químic de Sarrià; Barcelona, Spain*
[§] *Institute for Systems, Informatics and Safety, Joint Research Centre; Ispra, Italy*

Abstract: In this paper two intrinsic runaway criteria are applied to the early detection of hazardous states in batch, semibatch and continuous chemical reactors. The new criteria are based on parametric sensitivity and the divergence. The evaluation of the criteria is done by means of a model-based measuring technique, namely a Kalman filter. After a comparison with runaway criteria known from literature, the power of the early warning system is shown for two examples, a batch esterification and a continuous oxidation. *Copyright © 1998 IFAC*

Keywords: chemical, safety, model-based recognition, sensitivity, chaos theory

1. INTRODUCTION

Highly exothermic processes can lead to the loss of thermal control if operational faults occur. The on-line-monitoring of chemical reactors is therefore of great importance.

Model-based warning systems have proven useful in the detection of hazardous states and the elucidation of their primary causes at an early stage (King and Gilles, 1990). In these systems, a mathematical model of the process and easily accessible measurements are used in the method of Kalman filtering to provide on-line estimates of all the state variables and their derivatives. The development of the process in the near future can be predicted with the mathematical model and the actual state estimation. The estimations or their predictions are applied to criteria defining the safe operation of the reactor.

However, none of the on-line applicable runaway criteria known from literature are intrinsic, i. e. they are only applicable to a certain type of reactor, as in the case of the inflexion criterion for the temperature reading of a batch process (Adler and Enig, 1964) . In addition, other criteria call for the definition of a critical value depending on the type of reaction, e. g. the maximum accumulation of the dosed reactant in a semibatch reactor. Consequently, these criteria are not easy to handle, as they cannot be used without taking into account the specific characteristics of the process considered.

The situation is even more severe in the case of continuously operated tank reactors. Despite the economical importance of these processes, the research has been almost limited to the study of the steady state. Therefore it is not surprising that the constraint values for the safe operation of continuous reactors are typically based on the steady state. However, according to overall chemical plant accident statistics the incident frequency in continuous plants is much higher during transients, e. g. during start-up and shut-down.

To make the assessment of process safety more generic, a runaway criterion must be based on an intrinsic property of the system. In this contribution we will discuss the application of two new intrinsic runaway criteria to the model-based on-line-monitoring of continuous and discontinuous reactors.

The first criterion uses the well-known high sensitivity that chemical reactors exhibit with respect to small changes of the operating conditions. Morbidelli and Varma (1988) presented an intrinsic off-line criterion based on a generalized behaviour of the sensitivities. This criterion was extended to an on-line application: a set of conditions belongs to the supercritical region if all the sensitivities show the maximum at the same time with the temperature. The criterion is evaluated using sensitivities predicted by the state estimator.

The second criterion originates from chaos theory and it is based on the divergence of the reaction system (Strozzi, 1997), which is a measure of the behaviour of nearby orbits in the phase space. To prevent a runaway, the divergence of the system has to be negative on the whole segment of the reaction path. The divergence is easily calculated on-line using the Jacobian estimated by the Kalman filter.

The fundamentals of the new criteria are explained studying reactions of first and second order. Both criteria are applied to batch, semibatch and continuous reactors and a comparison with respect to restrictivity is done. The early-warning system has been checked for two processes: the continuous oxidation of ethanol by hydrogen peroxide and the batch esterification of 2-butanol with propionic anhydride.

2. CRITERIA FOR THERMAL RUNAWAY

Several criteria are known from literature to recognize if a reaction process is becoming dangerous. Basically, these criteria check values of the state variables or their derivatives. In the case of a semibatch reactor a well-known example is the test of the accumulation of the dosed reactant. However, this criterion is not intrinsic, because a safety limit has to be determined for every type of reaction. An intrinsic criterion must not need the definition of a set value and it should be applicable to any type of reaction. In addition, for a general assessment of process safety, it would be desirable if a criterion could be defined that is applicable to any type of reactor, i. e. not only to the semibatch reactor (SBR), but also to the batch reactor (BR) and the continuous stirred tank reactors (CSTR).

To the knowledge of the authors, none of the existing runaway criteria has proven to fulfill these conditions. This holds especially for the independence of the type of reactor.

2.1 Criterion Based on Parametric Sensitivity

To make the assessment of process safety more generic, a runaway criterion must be based on an intrinsic property of the reaction system. Morbidelli and Varma (1988) presented a runaway criterion in the context of parametric sensitivity. It is based on a generalized behaviour of the sensitivities of the state variables with respect to the operating conditions, i. e. the parameters p_1, p_2, ... of the process. The boundary between sub- and supercritical behaviour is defined as the set of critical parameters $\mathbf{p_{crit}}$ where the maximum value T_{max} of the temperature has a maximum sensitivity $S_{(T_{max},p_i)}$ with respect to a selected operating condition p_i:

$$S_{(T_{max},p_i)}(\mathbf{p_{crit}}) \quad = \quad \max_{p} \left.\frac{\partial T}{\partial p_i}\right|_{t|T_{max}} \qquad (1)$$

In general, the critical condition $\mathbf{p_{crit}}$ depends on the parameter p_i used to study the influence on the maximum temperature. Morbidelli and Varma found, that if a supercritical condition leads to a thermal runaway, the critical condition $\mathbf{p_{crit}}$ is independent of the parameter p_i used to define the sensitivity. This means that the maxima of the sensitivities with respect to all the operating conditions are located in the same place of the parametric space. However, the criterion of Morbidelli and Varma cannot be applied on-line, because it requires to find the maximum of sensitivities in the parametric space. The criterion of an on-line detection system has to be able to assess criticality attending to a unique parametric condition, namely to the one that corresponds to the operating condition in which the process is running.

Nevertheless, the criterion of Morbidelli and Varma is useful for off-line studies, because it is generally accepted to discern between safe and runaway conditions at least for batch reactors, plug flow reactors and continuous stirred tank reactors without taking into account the characteristics of the reaction.

In this work the behaviour of the sensitivities was studied not only as a function of the parameters, but also as a function of the time. Along the line of the criterion of Morbidelli and Varma it is reasonable to postulate the following on-line criterion based on parametric sensitivity: a condition belongs to the runaway region if the sensitivities of the temperature with respect to any parameter show their maxima at the time where also the temperature has a maximum. This criterion turns out to be applicable to several kinds of reactions and reactors despite their different characteristics.

To quantify the closeness of the M times $t|S_{i,max}$ of maximum sensitivity to the time $t|T_{max}$ of maximum temperature the following measure is used:

$$\sigma \quad = \quad \sqrt{\sum_{i=1}^{M} \frac{(t|S_{i,max} - t|T_{max})^2}{M}} \qquad (2)$$

This measure is a kind of standard deviation of the times of maximum sensitivity from the time of maximum temparature. If it decreases drastically towards zero, the process is in a runaway condition.

The following considerations are based on three dimensionless models for the reactor types BR and CSTR with a reaction of first order (A $\rightarrow$ products) and a SBR with a reaction of second order (A+B $\rightarrow$ products) and stoechiometric dosing of B stopped at $\tau = 1$. The material and energy balances for the dimensionless conversion z of A and the dimensionless temperature $\Theta = T/T_{ref}$ read

$$\frac{dz}{d\tau} = \begin{cases} r_1 & \text{(BR)} \\ r_2 + (1-z)\dfrac{d\ln V}{d\tau} & \text{(SBR)} \\ r_1 - z & \text{(CSTR)} \end{cases} \quad (3)$$

$$\frac{d\Theta}{d\tau} = \begin{cases} Bo\, r_1 + Q_c & \text{(BR)} \\ Bo\, r_2 + Q_c + Q_d\dfrac{d\ln V}{d\tau} & \text{(SBR)} \\ Bo\, r_1 + Q_c + Q_d & \text{(CSTR)} \end{cases} \quad (4)$$

$$\frac{d\ln V}{d\tau} = \begin{cases} \dfrac{\epsilon}{1+\epsilon\tau} & \tau \le 1 \\ 0 & \tau > 1 \end{cases} \quad (5)$$

$$Q_c = St(\Theta_c - \Theta) \qquad Q_d = (\Theta_d - \Theta) \quad (6)$$

with the dimensionless reaction rates

$$r_1 = Da_1(1-z)f_\gamma \quad (7)$$

$$r_2 = Da_2(1-z)\left(1 - z - \frac{1-\tau_d}{1+\epsilon\tau_d}\right) f_\gamma \quad (8)$$

$$\tau_d = \min(1,\tau) \quad (9)$$

$$f_\gamma = \exp\left[\frac{1}{\gamma}\left(1-\frac{1}{\Theta}\right)\right] \quad (10)$$

The functions $d\ln V/d\tau$ and f_γ account for the change of the volume with time and the change of the reaction rate with temperature (Arrhenius law), respectively. For a detailed derivation of e. g. the SBR model the reader is referred to the PhD of Steensma (1990) .

For the BR, the SBR, and the CSTR with a kinetic of first and second order figure 1 shows the dimensionless times, when the temperature and the sensitivities of the temperature show the maxima, both as a function of a dimensionless parameter. To demonstrate the intrinsic nature of the criterion, for each type of reactor another parameter was changed. The parameters kept constant are depicted in table 1. In addition, $T_0 = T_d = T_c$ was assumed.

Table 1. Conditions of simulations in figure 1 (var/–): varied/irrelevant parameter

	T_{ref}	T_c	ϵ	γ	Da	Bo	St
BR	298	298	-	0.1	var	2.85	1
SBR	300	var	0.4	0.0238	1.8	0.7	4
CSTR	298	298	-	0.05	0.0818	var	1

Obviously, after passing a critical condition (BR: $Da_1 \approx 0.02$, SBR: $T_c \approx 284$ K, CSTR: $Bo \approx 0.6$), the maxima of the temperature and the sensitivities appear at the same time. In case of the BR and the CSTR a single boundary discerns between a safe process and a runaway. For the CSTR a range of conditions was selected, where periodic behaviour, which is beyond the scope of this paper, cannot occur. Therefore, the possibility of a runaway in the CSTR is restricted to the start-up.

It is well-known, that in a SBR two runaway boundaries have to appear. At sufficient exothermicity of the process the first boundary separates non-ignition from runaway profiles and the second boundary separates runaway from so-called QFS (quick onset, fair conversion, smooth temperature) profiles. The QFS represents the ideal op-

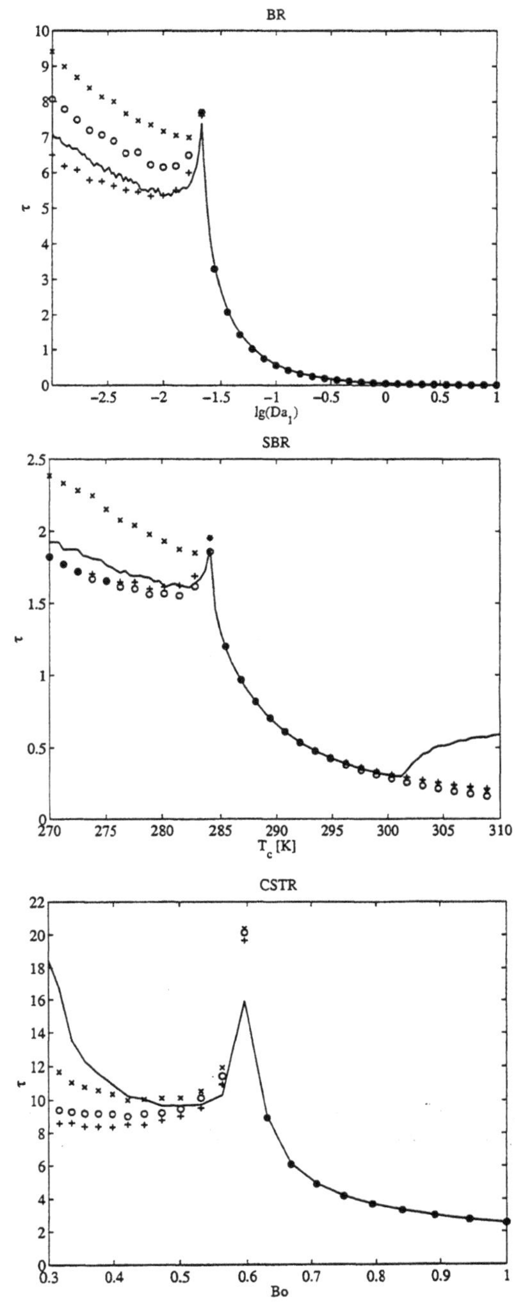

Fig. 1. Times of maximum temperature [–] and times of maximum sensitivity of the temperature with respect to $Da_{1/2}$ [+], St [x] and γ [o] as a function of Da_1 (BR), T_c (SBR) and Bo (CSTR)

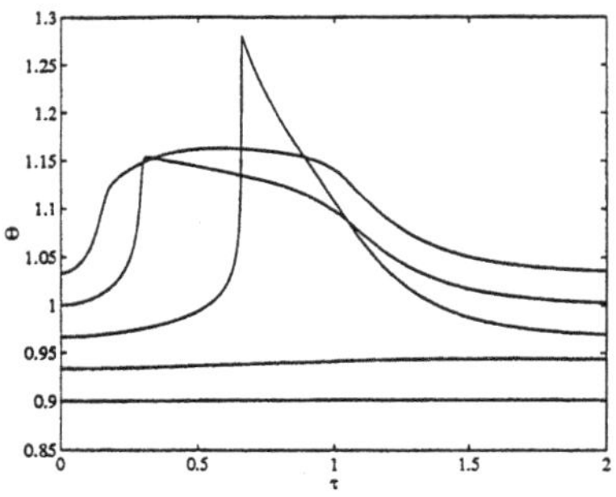

Fig. 2. Temperature readings of SBR processes at different $T_c = T_0 = T_d$, conditions see table 1

eration of a SBR (Steensma, 1990), where the accumulation of the dosed reactand stays small. As a consequence, after surpassing a second critical parameter ($T_c \approx 301$ K in figure 1), the times of maximum temperature and sensitivity start to diverge, defining a second boundary. Figure 2 shows the transition of SBR processes from non-ignition via runaway to QFS changing simultaneously the cooling, dosing and initial temperature from 270 K to 310 K, i. e. the corresponding dimensionless temperatures from 0.9 to 1.033.

2.2 Criterion Based on the Divergence

Strozzi (1997) developed a runaway criterion based on the divergence of the system. The criterion states, that an operating condition belongs to the supercritical region if the divergence becomes positive. For the example of a reaction system with two state variables

$$\frac{dz}{d\tau} = f_z(z,\Theta) \qquad \frac{d\Theta}{d\tau} = f_\Theta(z,\Theta) \tag{11}$$

the divergence is defined by:

$$\mathrm{div}_\tau = \frac{\partial f_z}{\partial z} + \frac{\partial f_\Theta}{\partial \Theta} \tag{12}$$

The subscript τ indicates, that the time is used as independent variable. The divergence criterion is shown in (Strozzi, 1997) to be equivalent to a runaway criterion presented by van Welsenaere and Froment (1970) for tubular reactors, if instead of the time the conversion is used as independent variable:

$$\frac{d\Theta}{dz} = \frac{f_\Theta(z,\Theta)}{f_z(z,\Theta)} \qquad \mathrm{div}_z = \frac{\partial}{\partial\Theta}\left(\frac{f_\Theta}{f_z}\right) \tag{13}$$

The two expressions of the divergence are related by:

$$\mathrm{div}_\tau = \frac{1}{f_z}\left(\mathrm{div}_z - \frac{d\dot{z}}{dz}\right) \tag{14}$$

However, as in critical conditions the temperature becomes the dominant variable and therefore the reaction rate is mainly altered by changes of the temperature, the order of magnitude of $d\dot{z}/dz$ is negligible and hereby $\mathrm{div}_\tau = 0$ and $\mathrm{div}_z = 0$ are giving a similar runaway boundary.

2.3 Comparison with Previous Runaway Criteria

Hub and Jones (1986) developed an early warning criterion for batch reactors using the increase of the heat evolution:

$$\frac{d(T-T_c)}{dt} > 0 \quad \text{and} \quad \frac{d^2T}{dt^2} > 0 \tag{15}$$

The evaluation of this criterion can be done by means of numerical differentiation and filtering to avoid amplification of measurement noise or by means of a model-based estimator. A similar criterion was presented by Adler and Enig (1964) :

$$\frac{dT}{dz} > 0 \quad \text{and} \quad \frac{d^2T}{dz^2} > 0 \tag{16}$$

For a first order batch process a comparison of the new criteria with the criteria of Hub and Jones (1986), Adler and Enig (1964) and Morbidelli and Varma (1988) is given in figure 3 on the top. The boundary between a safe process and a runaway is shown as a function of the Damköhler number Da_1 and the dimensionless adiabatic temperature rise Bo. In case of safety relevant processes with a significant exothermicity Bo all the criteria predict a quite similar boundary. All of them are therefore useful for this kind of processes.

It is easily shown that the divergence criterion can be extended to autocatalytic reactions in BR. The same holds for the sensitivity based criterion. However, the criteria of Hub and Jones (1986) and Adler and Enig (1964) turn out to give always an alarm or to be too restrictive, respectively.

The off-line criterion of Morbidelli and Varma (1988) discerns in a SBR only between non-ignition and runaway conditions. An off-line criterion for a distinction of runaway and QFS processes was presented by Alós (1997). This criterion defines the boundary corresponding to the condition, where the time τ_Θ of maximum temperature has a minimum, see also figure 1. Along the same line, the maximum time of maximum temperature defines a boundary between non-ignition and runaway. The satisfying agreement with the sensitivity based on-line criterion is shown in the middle of figure 3. The divergence criterion turns out to be too restrictive, excluding an obviously safe process, see the QFS process at $T_c = 310$ K in figure 2. It is anyway surprising that both new criteria give the second boundary, although a high sensitivity is not involved in the transition from runaway to QFS.

At low exothermicity, the transition from non-ignition to QFS is done via the so-called marginal ignition region, where the maximum temperature is reached close to the end of the dosing. In fact, a third boundary, namely between marginal ignition and runaway has to be considered in a SBR. Also here, the divergence is more restrictive than the on-line sensitivity criterion. E. g. a process with $Bo = 0.32$ at $T_c = 297.5$ K is safe.

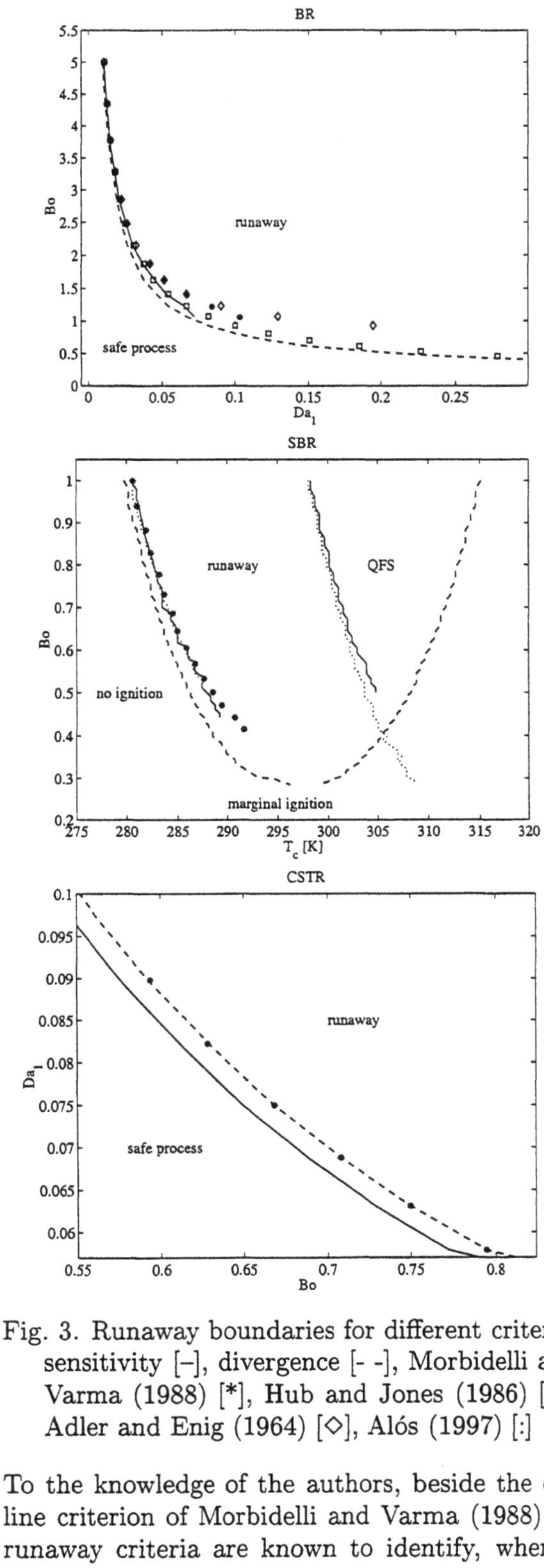

Fig. 3. Runaway boundaries for different criteria: sensitivity [–], divergence [- -], Morbidelli and Varma (1988) [*], Hub and Jones (1986) [□], Adler and Enig (1964) [◇], Alós (1997) [:]

To the knowledge of the authors, beside the off-line criterion of Morbidelli and Varma (1988) no runaway criteria are known to identify, when a CSTR is running into a hazardous condition during a transient. Figure 3, bottom, shows a good agreement with the new criteria.

3. MODEL-BASED ON-LINE-MONITORING

A model-based measuring technique, e. g. a Kalman filter can be used to provide on-line estimations not only of unmeasurable concentrations, but also of the divergence or the sensitivities. For details concerning the filter equations we refer the reader to the textbook of Gelb (1992) .

Given a state estimation a prediction of the parametric sensitivities can be done by solving the underlying variational equations, where $\mathbf{F}$ denotes the Jacobian:

$$\frac{d\mathbf{S}}{dt} = \mathbf{F}\cdot\mathbf{S} + \frac{\partial \mathbf{f}}{\partial \mathbf{p}} \qquad \mathbf{S}(0) = \mathbf{0} \qquad (17)$$

The estimated divergence is simply given by the sum of the diagonal elements of the estimated Jacobian:

$$\hat{\mathrm{div}}_{\tau} = \sum_{n=1}^{N} \mathrm{diag}_{n}(\hat{\mathbf{F}}) \qquad (18)$$

Obviously, using a Kalman filter, a minimum calculation effort is needed to check the divergence criterion.

4. CASE STUDIES

The power of the model-based warning system will now be demonstrated for a continuous and a batch process using the sensitivity and the divergence criterion, respectively.

As a first process the oxidation of ethanol by hydrogen peroxide with iron-III-nitrate as catalyst has been studied in a CSTR. The kinetical scheme of the process reads (Hafke, 1972) :

$$\begin{aligned}
CH_3CH_2OH + H_2O_2 &\overset{Cat}{\rightarrow} CH_3CHO + 2\,H_2O \\
CH_3CHO + H_2O_2 &\overset{Cat}{\rightarrow} CH_3COOH + H_2O \\
H_2O_2 &\overset{Cat}{\rightarrow} 0.5\,O_2 + H_2O \\
Cat + CH_3COOH &\rightleftharpoons Cat - Complex
\end{aligned}$$

Due to two consecutive, one parallel and an equilibrium reaction, this is a suitable example to check the intrinsicity of the new runaway criterion. During the start-up, for an immediate initiation of hydrogen peroxide consumption, a certain initial amount of iron-III-nitrate is needed, leading to an achievement of the steady-state temperature in a smooth way, see solid line in figure 4(a). Here, the initial amount of iron-III-nitrate required for a safe start-up has been forgotten, resulting in an accumulation of hydrogen peroxide. This accumulation is consumed after enough catalyst has been provided by the dosing, that was started at $t = 24$ min. The ignition of the process is accompanied by a sharp increase of the temperature, see dashed line in figure 4(a).

The concentration of iron-III-nitrate estimated by the Kalman filter is given in figure 4(b). In figures 4(c) and (d) the predicted sensitivities based on the estimation at $t = 24$ min and $t = 25$ min are shown. Clearly, after a correction of the concentration of catalyst, the sensitivities indicate with a change of the deviation from $std = 0.16$ to $std = 9.3 \cdot 10^{-4}$ that the reactor is running into a hazardous condition. The runaway is detected 9 minutes before the thermal excursion takes place.

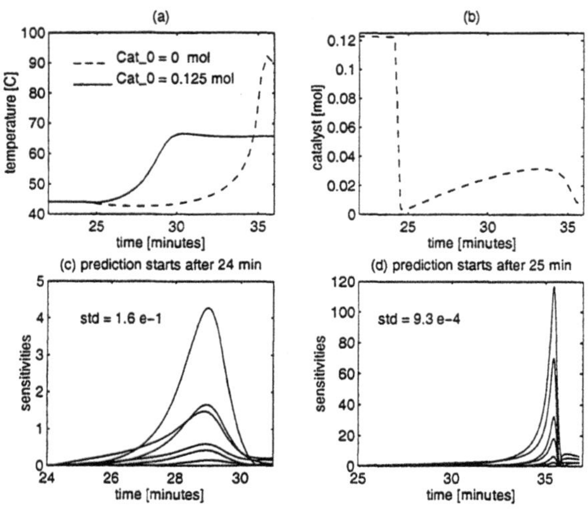

Fig. 4. Simulated [-] and filtered [- -] readings of (a) the temperature and (b) the catalyst concentration for the ethanol oxidation in a CSTR. Plots (c) and (d): predicted sensitivities after 24 and 25 min, respectively

The second case study refers to the batch esterification of 2-butanol with propionic anhydride. If sulphuric acid is used as catalyst, the reaction reveals an autocatalytic mechanism. Detailed kinetical aspects as well as the mathematical model have been described elsewhere (Alós, 1997).

The power of the model-based evaluation of the divergence will be shown for the example of an initial mischarge of sulphuric acid in an isoperibolic experiment with $T_0 = T_c = 25°$C. The reactor was initially filled with 6.86 mols of 2-butanol. Instead of the foreseen initial amount of sulphuric acid of $w_S = 0.5$ weight-% a bigger amount, namely $w_S = 0.7$ weight-% were filled in. The corresponding temperature readings are given in figure 5(a) with the dashed and the solid line, respectively. In this case neither the criterion of Hub and Jones (1986) nor the criterion of Adler and Enig (1964) is applicable. The estimated divergence in figure 5(b) shows, that the runaway is detected at $t = 1.53$ h, about 9 minutes before the excursion takes place. If, in this moment, emergency cooling at $T_c = 15°$C is activated, the process follows the dotted line in Figure 5(a) and the runaway is stopped.

CONCLUSION

The aim of this work was to apply two new intrinsic runaway criteria to the model-based online detection of hazardous states in BR, SBR and CSTR. The criteria are based on the parametric sensitivity and the divergence of the system. The sensitivity criterion reveals a good agreement with previous criteria for different kinds of reactors, whereas the divergence criterion turned out to be too restrictive in case of a SBR. The intrinsic power of the criteria was shown using a continuous oxidation and an autocatalytic batch esterification.

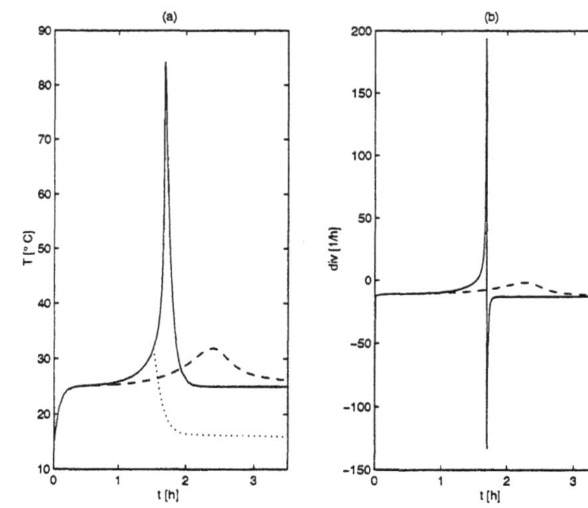

Fig. 5. Readings of (a) temperature and (b) divergence in case of [–] a hazardous and [– –] a safe batch esterification. [:]: Temperature after an emergency cooling

REFERENCES

Adler, J. and Enig, J. W. (1964). The critical conditions in thermal explosion theory with reactant consumption. *Combustion and Flame*, **8**:97–103.

Alós, M. A. (1997). *Generalized criteria for assessing thermal stability of discontinuous reactors.* PhD thesis, Institut Químic de Sarrià, Barcelona, Spain.

Gelb, A., editor (1992). *Applied Optimal Estimation*, Cambridge, M.I.T. Press.

Hafke, C. (1972). *Theoretische und experimentelle Analyse des dynamischen Verhaltens eines kontinuierlich betriebenen Rührkesselreaktors.* PhD thesis, Universität Stuttgart.

Hub, L. and Jones, J. D. (1986). Early on-line detection of exothermic reactions. *Plant Operation Progress*, **5**:221.

King, R. and Gilles, E. D. (1990). Multiple filter methods for detection of hazardous states in an industrial plant. *AIChE J.*, **36**:1697–1706.

Morbidelli, M. and Varma, A. (1988). A generalized criterion for parametric sensitivity: application to thermal explosion theory. *Chem. Eng. Sci.*, **43**:91–102.

Steensma, M. (1990). *Runaway and thermally safe operation of batch and semi-batch reactors.* PhD thesis, Twente University, Enschede.

Strozzi, F. (1997). *Runaway prevention in chemical reactors using chaos theory techniques.* PhD thesis, Twente University, Enschede.

van Welsenaere, R. J. and Froment, G. F. (1970). Parametric sensitivity and runaway in fixed bed catalytic reactors. *Chem. Eng. Sci.*, **25**:1503–1516.

IDENTIFICATION AND CONTROL OF AN INDUSTRIAL DISTILLATION COLUMN : A CASE STUDY

C. Defranoux *, H. Garnier *, P. Sibille * and P. Martinez **

** Centre de Recherche en Automatique de Nancy (CRAN), ESA CNRS 7039*
Université Henri Poincaré, Nancy 1,
BP 239, F-54506 Vandœuvre Cedex, France,
Téléphone : (33) 3 83 91 20 69 - Fax : (33) 3 83 91 20 30
hugues.garnier@cran.u-nancy.fr

*** Elf Atochem,*
Etablissement de Carling - Saint-Avold - Marienau,
BP 1005, F-57501 Saint-Avold Cedex, France,
Téléphone : (33) 3 87 91 74 22 - Fax : (33) 3 87 91 75 31

Abstract: The challenging problem of identification and control of an industrial distillation column is addressed in this paper. It aims at presenting a process analysis, an identification procedure and a control design for a binary distillation column. Process identification, which represents an alternative to modeling is shown to be the appropriate procedure for predictive control design. The predictive controllers based on the identified model indeed ensure stability and a high performance for a wide operating range of the distillation column.

Keywords: Industrial application, identification, predictive control, distillation column

1. INTRODUCTION

Industrial distillation column control (Buckley *et al.*, 1985) (Skogestad and Morari, 1988), (Vester *et al.*, 1993) is a well-known problem but the diversity of columns makes a general approach impracticable. The asymetric nature, nonlinearities, multivariable character and time delays are the source of identification and control design problems. To obtain an accurate mathematical representation, classical approaches use modeling (Loffler, 1990), (Skogestad *et al.*, 1990), (Kooijman and Taylor, 1995), which requires good knowledge about the process. They are based on balanced equations and lead to complex, nonlinear and high order models inappropriate for control design. The identification approach presented in this paper consists in reduced order model determination of the distillation column from input/output measurements. A Predictive Functional Control (PFC) (Richalet, 1993) has then been developed based on the identified model. This type of control has been chosen due to the simplicity of its mathematical formality and implementation. Moreover it makes possible the design of a simple feedforward control.
This paper is organized as follows. The distillation column and its phenomenological study are discribed in section 2. Section 3 is devoted to the identification procedure which is split into experiment design and result description. At last, before the conclusion, the control design and results are presented.

2. PROCESS DESCRIPTION

2.1 *Column description*

The industrial distillation column is equipped with 48 trays, a steam-heated reboiler and a total condenser, as depicted in figure 1. The column is fed in at the eighteenth tray with a binary mixture

Table 1. Standard operating conditions

Variable	Nominal set-points	Normal operation	Operating objective	Unit
F_r	28.5			t/h
T_{12}	−10			$°C$
T_{40}	64.5			$°C$
F_f	43			t/h
X_t	500	$500 \leq X_t \leq 800$	$X_t \leq 1000$	vpm
X_b	2000	$1700 \leq X_b \leq 2100$	$X_b \leq 1500$	vpm
P	25			$bars$

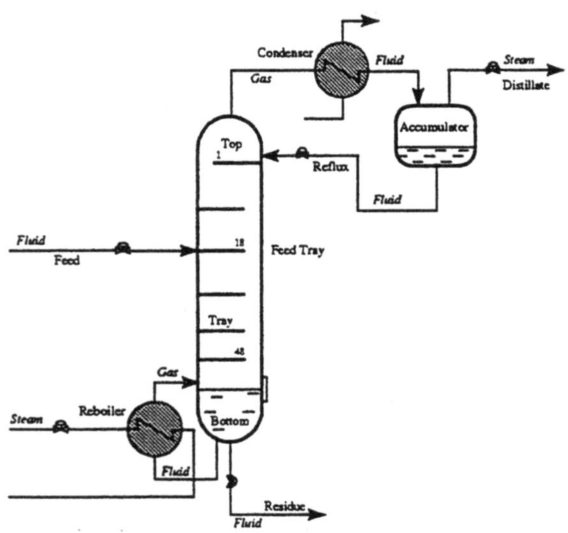

Figure 1. Binary distillation column.

of carbonate components. The separation of component takes place under controlled pressure. The objectives are to control the impurity of the top product or distillate X_t and the impurity of the bottom product or residue X_b with respect to changes on reflux flow F_r and heating power Q while preventing influence of changes on feed flow F_f and feed composition. The compositions X_t and X_b are measured by analysers and expressed in volume per million (vpm).

Table 1 summarises the most important operating conditions of the column, where T_{12} and T_{40} are the temperature of tray number 12 (sensitive tray) and 40 respectively, and P is the column pressure.

2.2 *Phenomenological study*

The main objective is to control two variables : the top and the bottom product quality X_t and X_b respectively. After a careful examination of the first data sets, two elementary systems were defined : the top system and the bottom system. Moreover, as illustrated on figure 2, the temperature measurement of the sensitive tray T_{12} has been found to be a continuous image of the distillate that reacts quickly towards changes :

$$\tilde{X}_t(t) = 233T_{12}(t - 840) + 19 \quad (1)$$

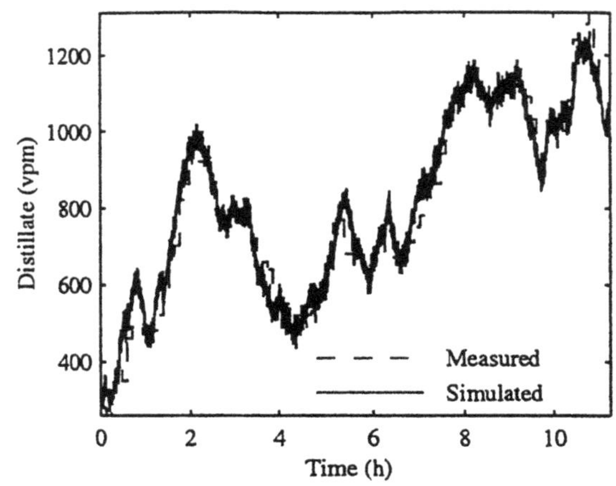

Figure 2. Measured X_t and simulated $\tilde{X}_t$ distillate

A temperature model has then been considered with T_{12} as output variable instead of the distillate for the top system, whereas a product composition model with X_b as output variable has been considered for the bottom system. Classicaly used (Uribe and Gallegos, 1986), (Andersen *et al.*, 1989), (Yin, 1992), (Sriniwas *et al.*, 1995), the reflux flow F_r and heating power representing by the controlled temperature T_{40} are input variables for the top and bottom systems respectively. The most important disturbance entering this distillation column is a change in the feed flow rate. This feed flow rate F_f being measured, it has therefore been used as an input variable for both subsystems. Finally, an acute analysis showed a coupling between the bottom and the top of the column. A transfer function between the input variable T_{40} and the output variable T_{12} was thus considered. On the other hand, no coupling between the top and the bottom of this column could have been demonstrated. Therefore no transfer function between X_b and F_r was considered in the final model structure shown in figure 3. This one may then be represented by the following transfer function matrix form :

$$\begin{pmatrix} T_{12}(s) \\ X_b(s) \end{pmatrix} = \begin{pmatrix} H_{11}(s) & H_{12}(s) & H_{13}(s) \\ 0 & H_{22}(s) & H_{23}(s) \end{pmatrix} \begin{pmatrix} F_r(s) \\ T_{40}(s) \\ F_f(s) \end{pmatrix} \quad (2)$$

3. IDENTIFICATION

The process identification problem may be summarised by the following steps (Ljung, 1987) :

(1) collecting appropriate data from the process for a phenomenological study,
(2) determining the model structure,
(3) estimating the model orders and parameters, which includes the design of appropriate excitation signals,
(4) validating the final estimated models,
(5) evaluating the models in their intended application, i.e. for process control.

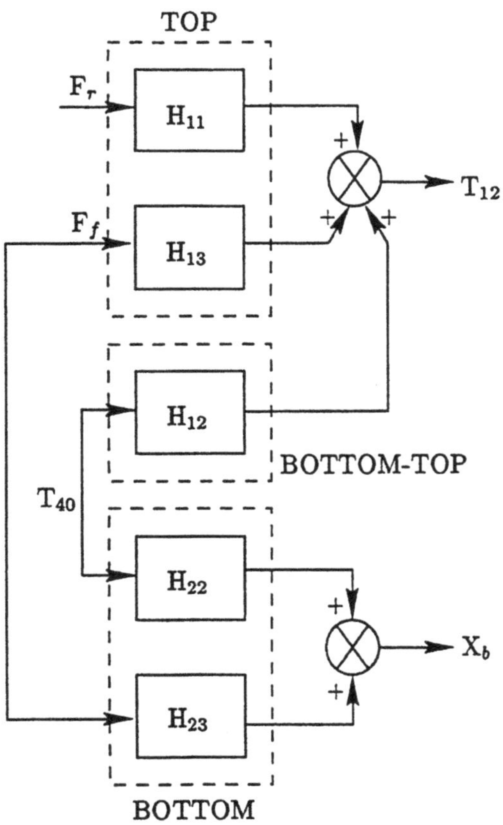

Figure 3. Model structure of the distillation column

The first two steps have already been addressed in the previous section. The last three steps are discussed in the following sections.

3.1 *Experiment design*

Two kinds of experiment were carried out. They consisted in manipulating separately the set-point of the reflux flow and the temperature of tray 40 around their desired operating point ; the other variables being locally controlled. The sampling time was set to 10 seconds. The experiment lasted between 10 and 17 hours. The inputs were zero-mean Random Binary Signals (RBS) with suitable magnitude for staying around the operating point. Two RBS with a magnitude of 0.3 t/h and of 1.5°C were separately applied on the reflux flow F_r and the temperature T_{40} respectively. These RBS generated temperature changes from 1 to 2°C on T_{12} and composition changes from 500 to 700 vpm on X_b. Portions of input-output signals used for identifying all transfer functions H_{ij} defined in equation 2 are presented in figures 4 and 5.

Before executing the estimation procedure, a classical data processing was carried out on the raw data sets, which consisted in filtering and decimating the input-output signals.

Table 2. Model performance in crossvalidation

Subsystem	Top	Bottom	Bottom-Top
$\|R\|$	0.91	0.94	0.87

3.2 *Experiment results*

A Prediction Error Method (Ljung, 1987) was used to estimate all transfer functions H_{ij} defined in equation 2. Different structures of model were tested and reduced order models were chosen. An increase of the model order led indeed to an unsignificant residual variance decrease and a parameter variance deterioration. The identified transfer function matrix is :

$$\begin{pmatrix} T_{12}(s) \\ X_b(s) \end{pmatrix} = H(s) \begin{pmatrix} F_r(s) \\ T_{40}(s) \\ F_f(s) \end{pmatrix} \tag{3}$$

with :

$$H(s) =$$

$$\begin{pmatrix} \frac{-2.5e^{(-200s)}}{1+721s} & \frac{12se^{(-100s)}}{1+1346s} & \frac{1.3e^{(-240s)}}{1+650s} \\ 0 & \frac{-377e^{(-1800s)}}{1+955s} & 0 \end{pmatrix} \tag{4}$$

The feed flow changes did not disturb explicitly the bottom product composition of the column. The transfer function H_{23} was therefore not considered in the estimation procedure and was set to zero. To quantify the model quality, the following performance index was considered :

$$R^2 = 1 - \frac{\sum_{i=1}^{N}(y_m(i) - y(i))^2}{\sum_{i=1}^{N}(y(i) - \bar{y})^2} \tag{5}$$

where y is the output measurement, $\bar{y}$ is the mean of y and y_m is the simulated output. Table 2 shows the performance index on the data sets used for crossvalidation, *i.e.* on the data sets that were not used for identification.

Crossvalidation results are presented in figure 6. These figures show the good agreement between the model and the real plant.

4. CONTROL

Two control loops, using a PFC strategy (Richalet, 1993) as well as a feedforward control on the feed flow rate, have been developed. The bottom product composition X_b and the top product composition X_t, via the temperature T_{12} of tray 12, are controlled by two primary loops. In addition,

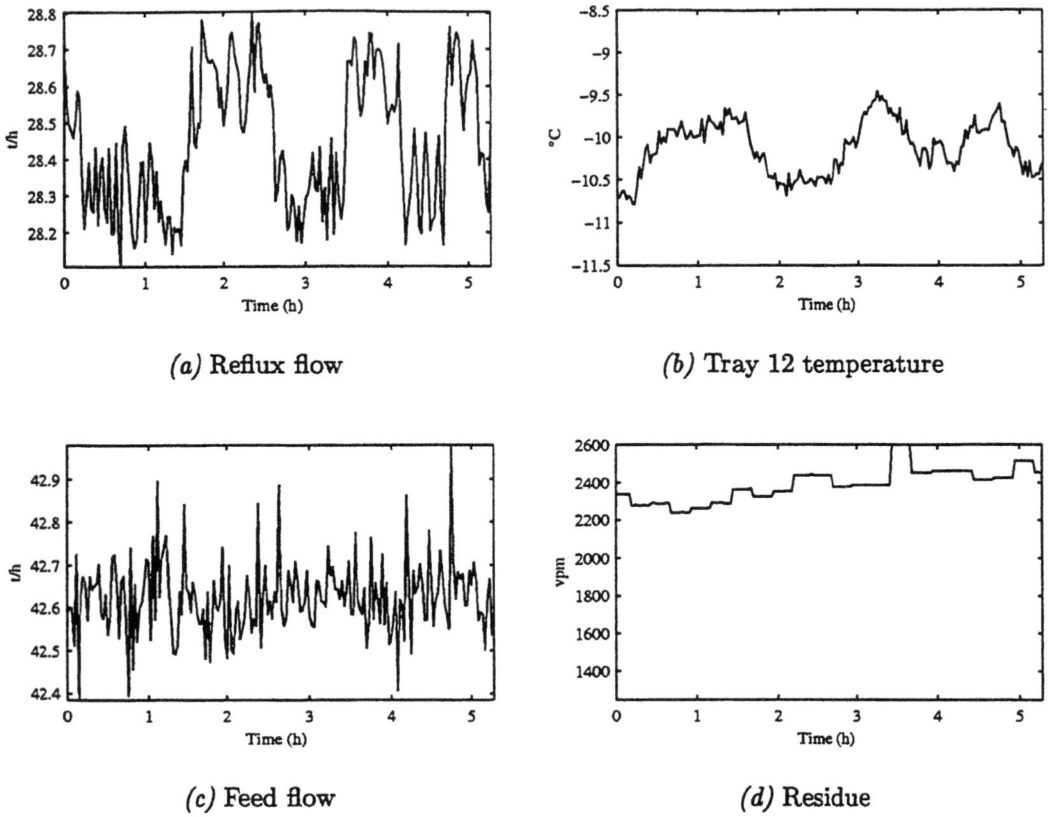

(a) Reflux flow (b) Tray 12 temperature

(c) Feed flow (d) Residue

Figure 4. Portion of the raw input-output data used for top model identification

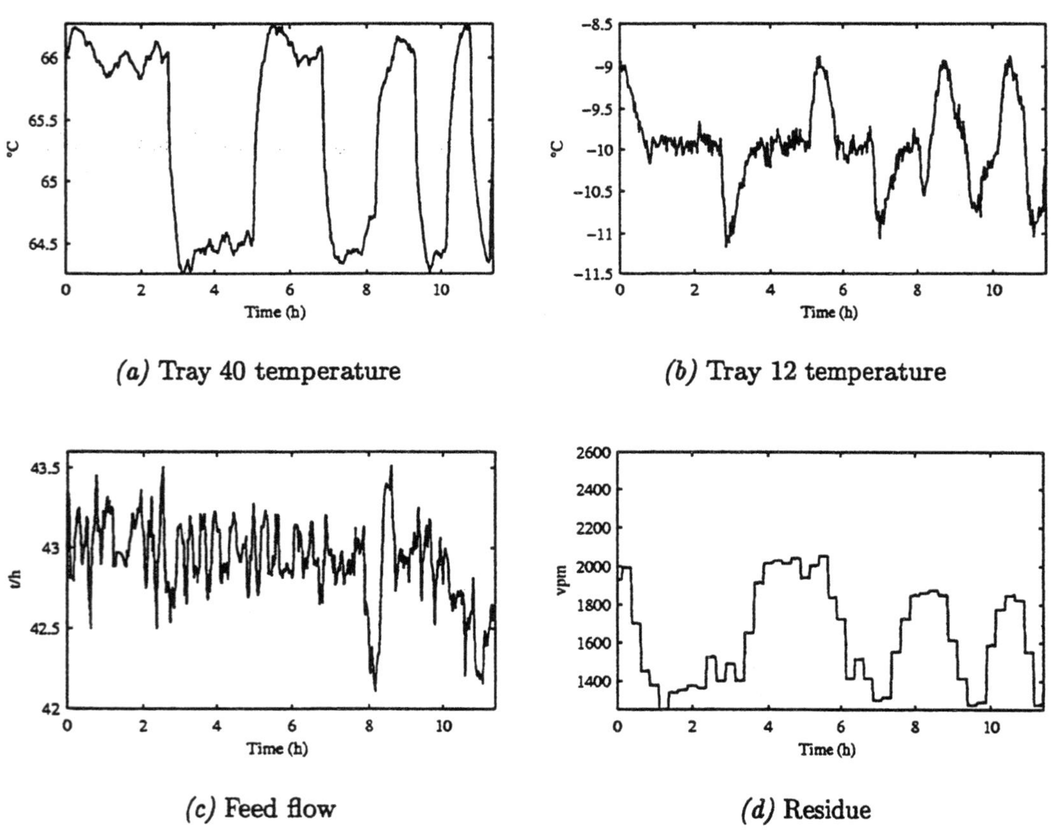

(a) Tray 40 temperature (b) Tray 12 temperature

(c) Feed flow (d) Residue

Figure 5. Portion of the raw input-output data used for bottom and bottom-top model identification

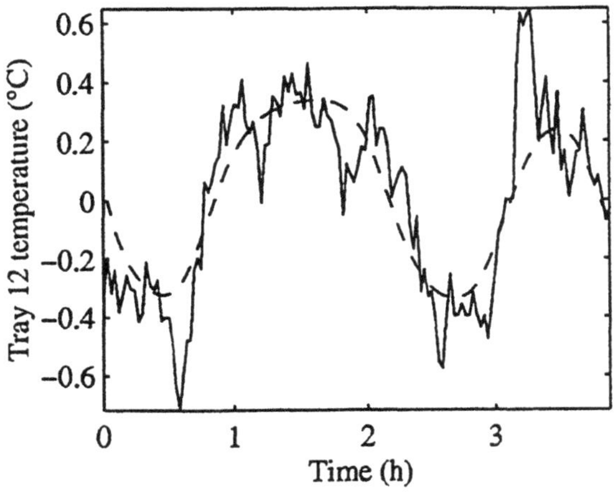

(a) Top system

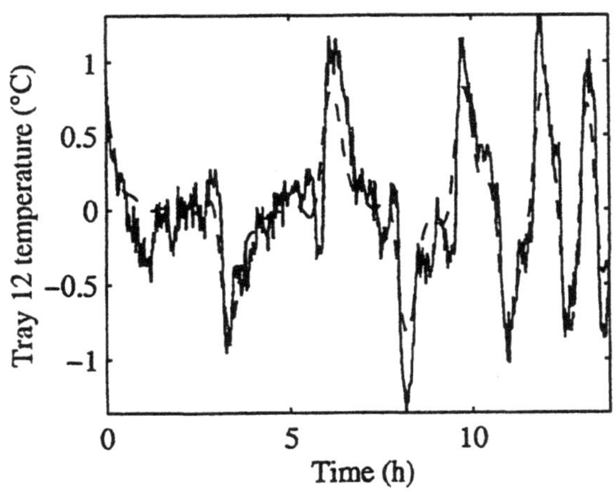

(b) Bottom-Top system

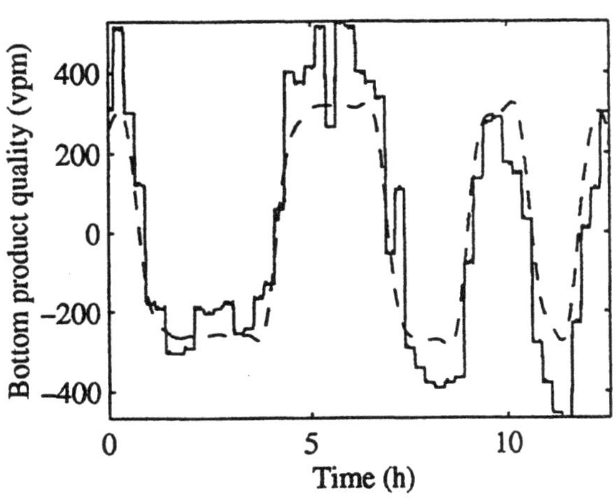

(c) Bottom system

Figure 6. Crossvalidation results. Measured (—) and simulated (- -) outputs.

four secondary variables, namely, the column pressure, the bottom accumulator level, the bottom column level and the temperature of tray 40 have to be maintained at their level for steady operation of the column. Since the secondary loops are noninteracting, simple PID type controllers were already used to control the secondary variables. PFC control was used for the primary loops. Reflux flow F_r has been used as the manipulated variable for the top product composition, via temperature T_{12}, while temperature T_{40} has been used for the bottom product composition. The balance sheet of this new control strategy on the top and bottom product quality is presented in figure 7 under histogram form. Before the predictive control implementation, the column ran with local simple PID control loops around $X_t = 500$ vpm (figure 7(a)) while the desired value was near $X_t = 800$ vpm. Since then, the set point for the distillate has been changed to $X_t = 1400$ vpm and the predictive control, based on models identified around the previous operating point, has been keeping very high performances (figure 7(b)).

5. CONCLUSION

Only a few applications of the modern control theories have been reported so far. The main objective of this paper is to contribute to fill in the gap between theory and practice. Therefore, the application results of a predictive control to an industrial distillation column has been presented. This kind of modern control requires a mathematical description of the process. Identification of linear transfer function matrix model was consequently first carried out from input/output data measurements. The implementation on the real plant of the predictive control has significantly improved the performances for a wide operating range of the distillation column and fully justifies the extensive efforts required in the identification stage.

6. ACKNOWLEDGEMENT

The authors wish to thank J. Richalet, Director of ADERSA for his constructive comments during the study.

7. REFERENCES

Andersen, H.W., M. Kummel and S.B. Jorgensen (1989). Dynamics and identification of a binary distillation column. *Chemical Engineering Science* 44(11), 2571–2581.

Buckley, P.S., W. L. Luyben and J.P. Shunta (1985). *Design of distillation column control systems.* Edward Arnold.

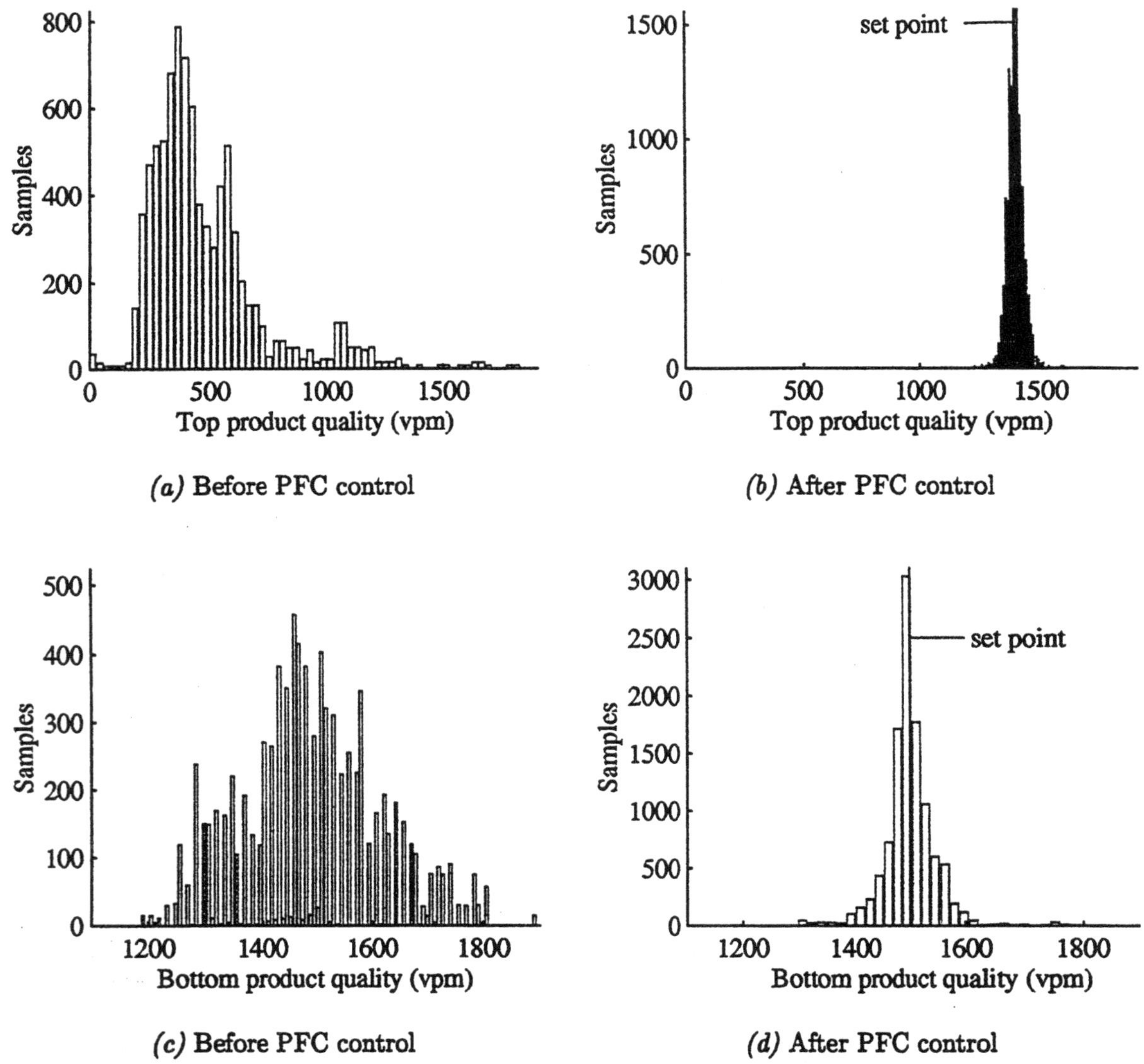

Figure 7. Top and bottom product quality histogram before and after predictive control implementation

Kooijman, H.A. and R. Taylor (1995). A nonequilibrium model for dynamic simulation of tray distillation columns. *AIChE Journal* **41**(8), 1852–1862.

Ljung, L. (1987). *System Identification. Theory for the user.* Prentice Hall.

Loffler, C. (1990). Modélisation, simulation et conduite d'un atelier de distillation extractive. PhD thesis. Université de Nancy 1.

Richalet, J. (1993). Industrial applications of model based predictive control. *Automatica* **29**(5), 1251–1274.

Skogestad, S., E.W. Jacobsen and P. Lundström (1990). Modelling requirements for robust control of distillation columns. In: *IFAC 11th Triennal World Congress, Tallinn - USSR.* pp. 191–197.

Skogestad, S. and M. Morari (1988). Lv-control of a high-purity distillation column. *Chem. Eng. Sci.* **43**, 33–48.

Sriniwas, G.R., Y. Arkun, I.L. Chien and B.A. Ogunnaike (1995). Nonlinear identification and control of a high-purity distillation column : a case study. *J. of Process Control* **5**(3), 149–162.

Uribe, L.B. and J.A. Gallegos (1986). Identification of a binary distillation column. *Int. J. Systems Sci.* **17**(2), 295–304.

Vester, M.T., R.J.P. Van Der Linden and J.L.A. Pangels (1993). Control design for an industrial distillation column. *Computers chemical Engineering* **17**(5/6), 609–615.

Yin, K. (1992). Order determination and online estimation - application to distillation process. In: *American Control Conference, Chicago - USA.* pp. 112–116.

FAULT DETECTION FOR CONTINUOUS-TIME SYSTEMS, THE P.M.F IDENTIFICATION APPROACH

B. Bensaker and F. Ouchene

Institute of electronics, University of Annaba
B.P.12 23000, Annaba, Algeria

Abstract: In this paper, we present an application of Poisson Moment Functionals (P.M.F) identification method in fault detection for continuous-time systems. The application concern an electro-mechanical system which can be used in chemical industry as a mixer. A continuous-time model is derived in the form of differential equation.

The fault detection procedure uses the parameter estimation approach via the implementation of the P.M.F identification method. The PMF method is implemented for determining the nominal model and the corresponding bias of the ordinary least-squares method by the use of a set of training data. Based upon the bias analysis of the observed model we have detected the faulty system behaviour.

The advantages of the PMF identification approach is that it does not need any direct transformation of the continuous-time model and permits to handle the derivative terms occurring in the continuous-time model easily and directly.

Keywords: Fault detection, continuous-time systems, parameter estimation, PMF approach, least-squares, bias analysis.

1. INTRODUCTION

Systems have generally a continuous-time behaviour and their control and supervision requires often more accurate and rapid algorithms. On this viewpoint, several approaches based on parameter estimation of continuous-time models have been developped. These approaches include transformation methods from discrete-time equivalent models or use the sampled input-output data or the filtered input-output signals (Strmcnik and Bremsak, 1979), (Young, 1981), (Sinha and Lastman, 1982), and (Isermann et al., 1993).

In this paper, we consider the continuous-time systems modeled by the following general linear time-invariant differential equation.

$$\sum_{i=0}^{n} a_i \frac{d^i y(t)}{dt^i} = \sum_{j=0}^{m} b_j \frac{d^j u(t)}{dt_j} \quad m<n \qquad (1)$$

Where u(t) and y(t) are the input-output signals and a_i, b_j are the parameters to be estimated. The initial conditions are considered null.

Based on the parameter estimation approach the fault detection procedure uses the P.M.F identification method which converts the continuous-time input-output system signals into a set of real numbers and uses these real numbers for estimating the parameters via the ordinary least-squares method (Saha and Rao, 1980), (Garnier et al., 1994).

The paper is organised as follows: The P.M.F method is presented in the second section, the procedure of fault detection is presented in the third section. In the forth section we present the model of the considered system. The simulation results are presented in the fifth and final section.

2. TH P.M.F APPROACH

Let s(t) be a continuous-time signal, the k-th order P.M.F of this signal is defined by:

$$S(k)=P_k[s(t)]=\int_0^{\tau} s(t)p_k(\tau-t)dt$$

$$=TL^{-1}[P_k(s)S(s)]$$

$$k=0,1,2,\ldots \qquad (2)$$

where $p_k(t)=(1/k!)t^k \exp(-\alpha t)$ is the Poisson pulse function.

A simple calculus shows that $P_k[.]$ acts as a linear operator which converts the continuous-time signal s(t) over the interval $[0,\infty]$ into a set of real numbers.

This property of the operator $P_k[.]$ leads us to write the relation (1) into the following linear in the parameter form.

$$\sum_{i=0}^{n} a_i[(-1)^i \sum_{p=0}^{i}(-1)^p C_i^p \alpha^{i-p} Y(k-p) =$$

$$= \sum_{j=0}^{m} b_j[(-1)^j \sum_{p=0}^{j}(-1)^p C_j^p \alpha^{j-p} U(k-p) \quad (3)$$

Where U(k) and Y(k) are the P.M.F's of the input-output system signals, and $C(i,p)=i!/(i-p)!p!$.

If we assume that the P.M.F of the input-output system signals are measured with a certain error v(k) which is a combination of noise terms affecting the system such as unmeasurable disturbances, measurements noise, and modelling errors the relation (3) can be rewritten in the following form.

$$Y(k) = X(k)\theta + v(k) \quad (4)$$

The ordinary least-squares method can be then applied for estimating the parameters a_i and b_j (Garnier et al., 1994).

3. BIAS ANALYSIS

The bias of the least-squares algorithm can be determined by the use of a set of training data at the same time that the nominal model or estimated as follows (Garnier et al., 1994):

$$\hat{\theta}(N) = (X^T X)^{-1} X^T Y \quad (5)$$

where N is the number of training data.

Using the stationary and ergodicity properties we can write:

$$\lim_{N \to \infty} \hat{\theta}(N) = \theta + (R_{xx})^{-1}(R_{xv}) \quad (6)$$

Where $\hat{\theta}(N)$ is the estimated parameter vector, θ is the true parameter vector, and R_{xx}, R_{xv} are respectively the autocorrelation matrix and the intercorrelation vector.

The term $d\theta = (R_{xx})^{-1}(R_{xv})$ represents the bias, it can be reduced to $d\theta = R_{yv}$ if we assume the measurements noise independent of the input signal (Garnier et al., 1994).

Let $\delta\theta = d\theta_o - d\theta_N$ be the difference between the bias of the observed and nominal model. Based upon the bias analysis of the observed model and the nominal model one can detect the faulty behaviour of the system. The observed model parameters are expressed as a function of physical system parameters which leads to isolate the faulty part of the system (Isermann, 1994), (Basseville, 1988), (Frank, 1996).

4. THE SYSTEM MODEL

The electromechanical system considered consists essentially of a d.c generator-motor commonly called the Ward-Leonard system. In this system, generator G serves to supply d.c directly to motor M to drive the mixture contained in the vessel V.

The linearised dynamic model of this system in the differential equation representation is (White and Daniels, 1985):

$$\frac{d^2w(t)}{dt^2} + \left[\frac{1}{J}\left(f + \frac{K_m^2}{r}\right) + \frac{R}{L}\right]\frac{dw(t)}{dt} + \frac{R}{JL}\left(f + \frac{K_m^2}{r}\right)w(t) = \frac{K_m K_G}{JLr}u(t) \quad (7)$$

Where u(t) is the applied voltage [V], R the armature resistance [Ω], L the armature inductance [H], r the G-M circuit resistance [Ω], w(t) angular velocity [rd/s], J the moment if inertia [Kg.m^2], f the friction coefficient [Kg.m^2], and K_m, K_G are the motor and generator coefficient [wb].

Based on the relationships between model parameters and physical parameters one can detect a significant changes in the command circuit (R, L) via the parameters a_0 and a_1 or in the G-M circuit (r) via the parameter b_0 if we assume the remain physical parameters are not affected by any type of failure.

5. SIMULATION RESULTS

The nominal model is obtained via a set of training data at the same time that the correponding bias for two the different values of the parameter table 1.

Simulation 1: we assume a changes of 5% has occured in the physical command parameter R and we implement the Least-squares algorithm for estimating the observed model parameter and the corresponding bias, the results are gathered in table 2.

Simulation 2: we assume now that the same changes of 5% has occured in the physical G-M circuit via the parameter r and we estimate the observed model parameter and the bias as simulation 1. The results are gathered in table 3.

Based on the bias analysis table 4 one can detect the faulty behaviour system. In order to isolate the faulty part of the system we have introduced a threshold $T=2d\theta_N$ and $T=3d\theta_N$, and we have considered the difference between the bias the observed model and nominal model parameters. If this difference is strictly superior to the threshold T then the corresponding parameter gives the faulty part of the system table 5.

CONCLUSION

In this paper we have presented an application of the PMF approach in fault detection for continuous-time system.

The first advantage of the PMF identification approach in the procedure of fault detection is that does not need any direct transformation to discrete-time equivalent model.

The second and important advantage is that we can handle the derivative terms of the continuous-time model easily..

On this viewpoint the PMF approach is considered to be one of the most efficient technique for continuous-time model identification.

REFERENCES

Basseville M. (1998): Detecting changes in signals in signals and systems. A survey. Automatica, vol.24, n°3, pp.309-326.

Frank P.M. (1996): Analytical and qualitative model-based fault diagnosis. A survey and some new results. European Jour. of control, vol.2, pp.6-28.

Garnier H. Sibille P. and Nguyen H.L. (1994): A new bias-compensating least-squares method for continuous-time MIMO system identification applied to laboratory-scale process. Proc. 3-th IEEE Conf. Cont. Appl., pp.1711-1716.

Iserman R. (1984): Process fault detection based on modeling and estimation methods. A survey. Automatica, vol.20, n°4, pp.387-404.

Isermann R., Voigt K.U. and Pfeiffer K. (1993): Parameter estimation and digital control with continuous-time models and application to a combustion-engine dynamometer. 12-th IFAC World Congress Sidney, Australia.

Saha D.C. and G.P. Rao (1980): Identification of lumped linear time-varying parameter systems via Poisson moment functionals. Int. Journal of Control Vol.32, pp.707-721.

Sinha N.K. and Lastman G.J. (1982): Identification of continuous-time multivariable systems from sampled data. Int. Jour. of Cont. Vol.35, pp.117-126.

Strmcnik S. and Bremsak F. (1979): Some new transformation algorithms in the identification of continuous-time multivariable systems using discrete identification methods. Proc. 5th symp. on "identif. and syst. param. estim." Darmstadt. pp.397-405.

White B.A. and Daniels A.R. (1985): Digital control of d.c motors. North-Holland systems and control series Vol.7, pp.253-272.

Young P. (1981): Parameter estimation for continuous-time models. A survey Automatica, vol.17, pp.23-39.

Nominal Model Parameters			
	b_0	a_0	a_1
0.50	3.92	50.53	236.77
0.75	3.79	50.06	227.66
Aver. Val.	3.85	50.29	232.21

Table 1: Nominal Model

Observed Model Parameters			
	b_0	a_0	a_1
0.50	1.42	18.01	84.49
0.75	1.91	24.32	115.03
Aver. Val.	1.66	21.16	99.76

Table 2: Observed Model 1

Observed Model Parameters			
	b_0	a_0	a_1
0.50	2.34	27.66	123.89
0.75	2.98	38.31	158.20
Aver. Val.	2.64	32.98	141.04

Table 3: Observed Model 2

Bias Analysis			
	b_0	a_0	a_1
0.B1	0.24	3.15	15.27
0.B2	0.30	5.32	17.15
N.B. Max	0.06	0.23	4.56

Table 4: Bias Analysis

	Obs. Model 1	Obs. Model 2	$T=2.d\phi_N$	$T=3.d\phi_N$	Decision	
db_0	0.18	0.24	0.12	0.18		F/r
da_0	2.92	5.09	0.46	0.69	F/R	F/R
da_1	10.71	12.59	9.12	13.68	F/R	F/R

Table 5: Isolation of the faulty part of the system

PROCESS INTEGRITY : A FRAMEWORK FOR EXTENDED AND SAFE PROCESS OPERATION

Tomas B. Co

Michigan Technological University
Houghton, MI 49930, USA
email: tbco@mtu.edu

Abstract: A mathematical theory of process integrity is developed as a tool for safe and extended operation of critical processes. Process integrity becomes a characteristic measure of the ability of the system to function within prescribed operating region. The proposed framework bridges the tools of process reliability with process dynamics and control, thereby allowing improved decision to either perform online process integrity control, and in some cases help determine when a shutdown is needed to avert costly failure situations. *Copyright © 1998 IFAC*

Keywords: Failure detection, fault-tolerant systems, process safety, process integrity

1. INTRODUCTION

Failure and loss prevention are very crucial tasks in the hydrocarbon processing and other chemical processing industries. Productivity is significantly reduced due to unscheduled downtimes. In some cases, damages to the environment can be irreversible and involve costly cleanups.

Current methods of fault supervision begins with fault detection. Once identified, the fault is diagnosed by isolating the cause and location of the fault. The fault is then evaluated to determine the potential hazards. Finally, a decision is made to perform necessary response (Himmelblau, 1978). Several methods for fault supervision include state and parameter estimation (Willsky, 1976), (Iserman, 1984), and other model-based techniques (Kim and Bien, 1991), (Ray, 1990).

It is generally accepted that early fault detection is needed in order to allow appropriate responses to commence in a timely fashion. However, the issue of timeliness should not be a function of fault supervision alone. It should encompass the needs for possible on-line recovery as well. In some cases, failure occurs almost at the instant of the fault. Thus to avoid the failure condition in these cases, monitoring the process for undesirable trends or situations are just as critical as assessing the presence of faults.

To address the need for timely prevention of failure conditions, fault supervison methods can be enhanced with a measure referred to in this paper as process integrity. This measure is developed to determine the capacity of a system to prevent failure conditions within a set time horizon. This measure can further be used to provide optimal tradeoffs between primary operational goals of a process and the goal of avoiding a failure condition.

To help clarify, consider the process in a distillation column. The primary goal of a distillation column is to produce distillate of prescribed purity. For optimal productivity, most columns are likely to operate close to the flooding constraints. The issue of process integrity then becomes how close to the constraints should the process be allowed. The various flow rates, differential pressure as well as holdup information are needed. In addition, the availability of control actions plus available response times are needed to evaluate whether a timely movement away from the flooding conditions is possible. These factors are all significant and do not necessarily describe faults. A measure which combines these pieces of information into one convenient characteristic function can be very useful for monitoring the safe operation of the process.

In this paper, the definition and construction of process integrity is developed. Then a discussion is included to explain how the measure can be used to improve the sustainability of a process through a new objective which is referred to as process integrity control. These concepts are then concretized through a simulation example of integrity control of a surge tank level.

2. PROCESS INTEGRITY

Process integrity is a metric for assessing the capacity of a process to maintain its designed function in the presence of changes in either the external influence (e.g. manipulated and disturbance variables) or internal characteristics (e.g. state variables and system parameters). To this end, the process integrity measure will be defined to be sum of two terms

$$I = \mu + \Delta\mu \tag{1}$$

The first term is the instantaneous measure of the distance from failure or safety index, denoted by μ. The second term is the predicted change in the safety index $\Delta\mu$ within a prescribed time Δt.

2.1 *The Safety Index, μ*

The safety index $\mu(\alpha, x, u, \eta)$ is a fuzzy (or belief) measure indicating membership in the acceptable regions of operation. Its range is from 0 to 1, with 0 indicating a failure condition and 1 indicating a safe condition. The safety index represent a static map (i.e. not explicity dependent on time) of the process in the space spanned the process parameters α, state variables x, manipulated variables u and disturbance η. Methods and tools from order statistics and reliability theory can be used on historical data to determine the contour of this function.

2.2 *Change in Safety Index, $\Delta\mu$*

The second term in the definition of process integrity, I, contains the predictive characteristic of the measure,

$$\Delta\mu = \int_t^{t+\Delta t} \left(\frac{d\mu}{dt}\right) dt \tag{2}$$

The safety index, μ, becomes a function of time only implicitly through the dynamic nature of α, x, u and η. The application of the chain rules exposes the underlying structure that will incorporate the various dynamic models

$$\begin{aligned}\frac{d\mu}{dt} &= \nabla_\alpha \mu \frac{d\alpha}{dt} + \nabla_x \mu \frac{dx}{dt} \\ &+ \nabla_u \mu \frac{du}{dt} + \nabla_\eta \mu \frac{d\eta}{dt}\end{aligned} \tag{3}$$

This expansion exhibits the main feature of the process integrity measure. The predictive models enter through the time derivatives dx/dt, $d\alpha/dt$, du/dt and $d\eta/dt$.

The first time-derivative, $d\alpha/dt$, represents the model for time-varying parameters. It needs to incorporate the nominal "*wear-and-tear*" (off-line) models plus monitored (on-line) parameter drifts. Various on-line parameter estimation techniques, (Molle and Himmelblau, 1987),(Co and Ungarala, 1997), could be used to improve the estimates as well as the prediction of parameter values.

The second time-derivative is the familiar state space model

$$\frac{dx}{dt} = f(\alpha, x, u, \eta) \tag{4}$$

Various methods exist for obtaining theses models ranging from using first principles to using empirical methods.

The third time-derivative, du/dt, can be assessed from the feedback control laws

$$\frac{du}{dt} = g(\alpha, x, u, \eta, y_{\text{set}}) \tag{5}$$

where y_{set} is the target output trajectory. This term, together with its gradient, represents the availability of on-line recovery. Since some of the manipulated variables will saturate within the time horizon Δt, the time derivative has to be replaced by an approximate: $\Delta u/\Delta t$, where

$$\Delta u = \begin{cases} u_{\min} - u(t) & \text{if } u_{t+\Delta t} < u_{\min} \\ u_{\max} - u(t) & \text{if } u_{t+\Delta t} > u_{\max} \\ u_{t+\Delta t} - u(t) & \text{otherwise} \end{cases} \tag{6}$$

and

$$u_{t+\Delta t} = \int_t^{t+\Delta t} \left(\frac{du}{dt}\right) dt \tag{7}$$

The fourth time-derivative represents the disturbance model. It incorporates both measured and unmeasured disturbances. For the case of measured disturbances, online correction is possible.

For the unmeasured disturbances, existing observers are needed. Note however that it is the rate of change in disturbance and not the actual disturbance measure that is needed for this last derivative term.

These derivatives are then weighted by their corresponding gradients shown in equation (3). The weights hold important implications for process integrity. First, it shows that not all changes move toward decreasing the integrity of the system. Sometimes, a parameter change can improve integrity. Second, the gradients determine the sensitivity weights among the different classes of variables, e.g. whether the effects of changes in the state variables are more significant than parameter changes or vice verse. Third, it determines whether the available change for the manipulated variables Δu, after weighted by the gradient $\nabla_u \mu$, is sufficient to move the process away from failure. Fourth, the gradients depend on the location in the α-x-u-d space and reflects the need for different modes of operation.

2.3 *Time Horizon, Δt*

The time horizon, Δt, is another important term in the definition of process integrity. It should accomodate for physical constraints such as available response times and time delays. In some cases, operational constraints determine the allowed rates of changes, e.g. quenching rates, and should be represented in Δt. In short, Δt should reflect process demands for recovery.

If the value of Δt is underestimated, the utility of the process integrity measure would be significantly reduced. This could result in making correct decisions but not having enough time for implementation. However, a gross overestimate of Δt could likewise reduce the reliability of the prediction. In some cases, it may be preferrable to obtain a map, $\Delta t = \Delta t(\alpha, \mu, u, \eta)$ which corresponds to the mode of operation. This map would then show recovery time constraints in the α-x-u-d space.

3. PROCESS INTEGRITY CONTROL

Once process integrity measure has been obtained, it could be used to extend the lifetime of a process. This is true when there exists manipulated variables which could move the process away from the failure boundaries. At a higher decision level, the various terms in Δu (cf. equation (3)) can be used to determine whether the process needs to be shutdown for necessary repair. This shutdown may be partial or complete. Nontheless, this type of process downtime is still more preferrable, as opposed to unexpected failure events.

At a lower level, process integrity measures, $I(t)$, can be used to override the primary control objectives in a feedback scheme as shown in Figure 1.

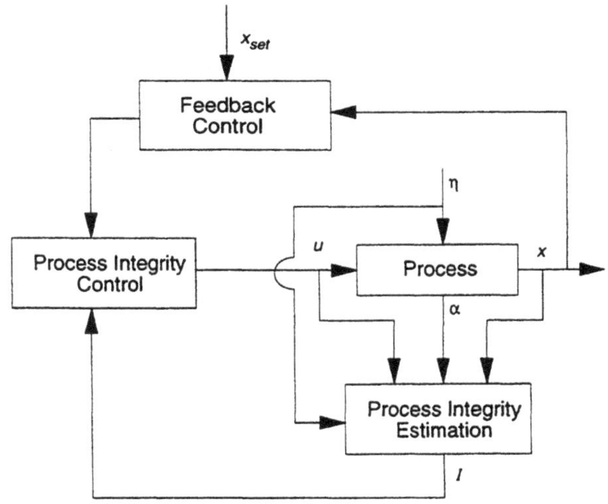

Fig. 1. Structure for incorporation of Process Integrity Control.

This override control will be referred to as process integrity control. As mentioned in the introduction, most processes are actually driven to the boundaries of allowed operation to maximize productivity. In these cases, the objectives of optimal operation are at odds with process safety. With the quantification of process integrity, mathematical methods can be used to obtain a tradeoff between these two objectives. For instance, in optimal decision control, weights (including economic indices) on both integrity and productivity can be used to decide changes in the manipulated variables.

In some cases, the state of the process could be moved to a better position in the α-x-u-d space without necessarily sacrificing the primary control objectives. This is usually true when there are some degrees of freedom in the process. For example, in controlling liquid flow through a pipe, valves and pumps are used to manipulate differential pressures. To prevent cavition, head needs to be kept below vapor pressure. Thus in this case, the process integrity measure needs to be a function of head, and process integrity control can be used to prevent cavitation, without necessarily having to sacrifice flow control.

4. SURGE TANK EXAMPLE

To illustrate the implementation of process integrity control, consider the surge tank system as shown in Figure 2. The example is chosen primarily to show how the different concepts discussed above can be implemented.

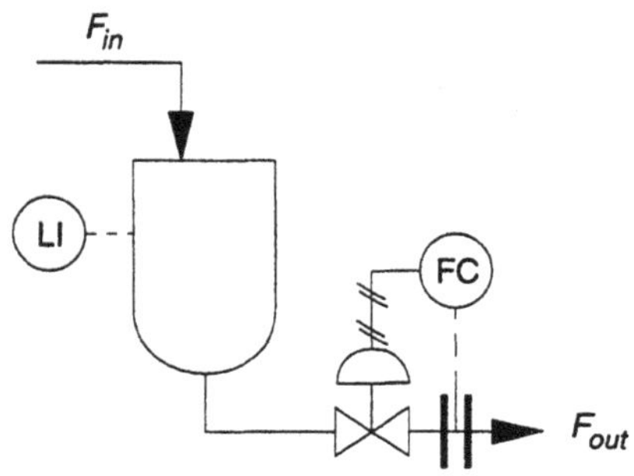

Fig. 2. Flowsheet for flow system with a surge tank.

The surge tank is used to allow for flow rate control downstream. The level inside the tank is not crucial until it is about to overfill or be empty. At those boundaries, process integrity control needs to override flow control to prevent spills or vapor entry. The model to be used for simulation is given by

$$\frac{dh}{dt} = \frac{1}{A}\left(F_{\text{in}} - ku\sqrt{h}\right) \tag{8}$$

where A is the tank cross-section area, F_{in} is the volumetric flow rate in, k is the valve coefficient, u is the valve opening and h is the liquid level. In this process, k is the parameter, u is the manipulated variable, h is the state, and F_{in} is the process disturbance.

4.1 *Construction of Safety Index, μ.*

To construct the safety index, we start by identifying the failure boundary. For simplicity, define this boundary as the surface of an ellipsoid in the k-h-u space in which one of the principal axis has a negative slope in the h-u plane. The rationale for this relationship between level h and valve opening u is that when the level is low, a small outlet valve opening is not a desireable operating condition. Likewise, when the level is high, a wide open outlet valve can not increase outflow much further, and can not be considered a safe operating condition.

The ellipsoid can be described as a unit sphere, v, deformed by matrix operator Q^{-1},

$$\begin{pmatrix} k - k_o \\ h - h_o \\ u - u_o \end{pmatrix} = Q^{-1}v \tag{9}$$

where $(k_o,\ h_o,\ u_o)$ is the center of the ellipsoid.

Inside the ellipsoid, the gradation from the center of the ellipsoid to the surface is given by a sigmoid function, resulting in the following potential function for the safety index,

$$\mu = \left[1 + \exp\left(\frac{5}{1-\epsilon}(2\|v\| - 1 - \epsilon)\right)\right]^{-1} \tag{10}$$

where $0 < \epsilon < 1$ determines the gradation of the index.

A cross-section view of the safety index in the h-u plane is shown in Figure 3.

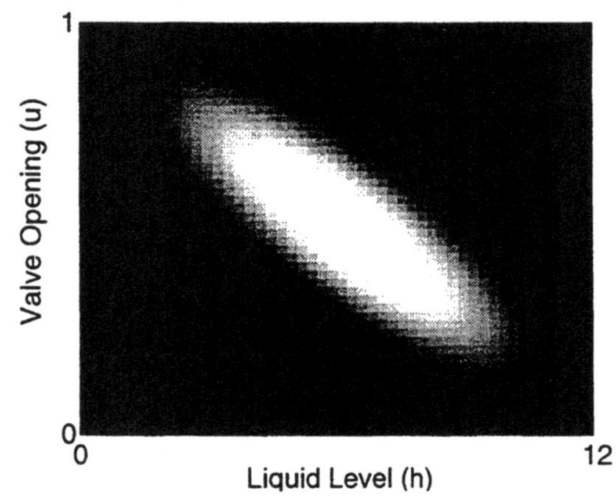

Fig. 3. A crossection projection of safety index in the h-u plane.

4.2 *Construction of Process Integrity, I.*

To further simplify implementation, the definition of $\Delta\mu$ given in equation (3) will be approximated to the first-order using the following:

$$\Delta\mu = \beta\zeta^T\xi\Delta t \tag{11}$$

where

$$\zeta^T = [\nabla_k\mu\ ,\ \nabla_h\mu\ ,\ \nabla_u\mu] = \left[\frac{10\mu(\mu-1)}{1-\epsilon}\right]\left(\frac{v^T}{\|v\|}\right)Q \tag{12}$$

$$\xi = \begin{pmatrix} dk/dt \\ dh/dt \\ du/dt \end{pmatrix} \tag{13}$$

and $1 \leq \beta \leq 2$ is a correction term to improve the approximation. Thus, the integrity is given by

$$I = \mu + \beta\zeta^T\xi\Delta t \tag{14}$$

with μ having the function described in equation (10). Note that if $\Delta\mu$ has been accurately determined, integrity $I(t)$ will be $\mu(t + \Delta t)$. In Figure 4, this approximation shows process integrity appropriately leading safety index by a time horizon of $\Delta t = 10$ sec.

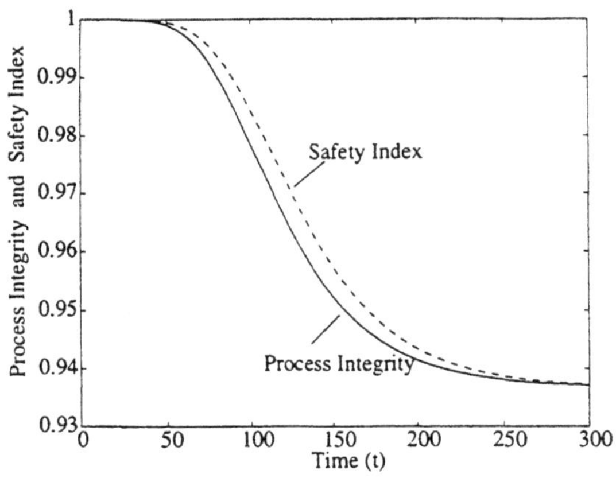

Fig. 4. Plot of Process Integrity (I) and Safety Index (μ) .

4.3 *Process Integrity Control*

Consider the scenario in which the flow rate into tank increases from 2.45 to 2.8 in^3/sec for a duration of 200 secs. The height of the tank is only a foot high. If only flow control is applied, a spill occurs as shown in Figure 5.

One way of applying the process integrity measure is to use the measure to blend (and ultimately switch) between flow control and level control, i.e.

$$u = (I)\ u_{\text{FLOW}} \quad + \quad (1 - I)\ u_{\text{LEVEL}} \qquad (15)$$

where u_{FLOW} and u_{LEVEL} are the PI-flow control and PI-Level control, respectively. Even though level control normally implements a proportional control strategy, the PI control improves process integrity faster. And when I approaches the value of 1, flow control will dominate again.

Using this override strategy avoids a spill, as shown in Figure 5. However, once the flow rate into the tank (F_{in}) reaches its initial value of 2.4 in^3/sec, the flow rate out (F_{out}) is allowed to reach its set point, as shown in Figure 6. The process integrity for the inclusion of the override scheme is shown in Figure 7.

By plotting the changes in manipulated variables (cf. Figure 8) note that the value of u does not move back to its original value of 0.5 even though F_{in} has returned to 2.4 in^3/sec. This is due to the extra degree of freedom existing in this process due to the term $u\sqrt{h}$ in equation (8). This represents a cumulative deterioration of process integrity, i.e. every increase in the flow rate into the tank decreases the steady state outlet valve opening. The proposed integrity control scheme improves the situation by moving u closer to 0.5. But it still suffers from cumulative decrease of process integrity as shown in Figure 8.

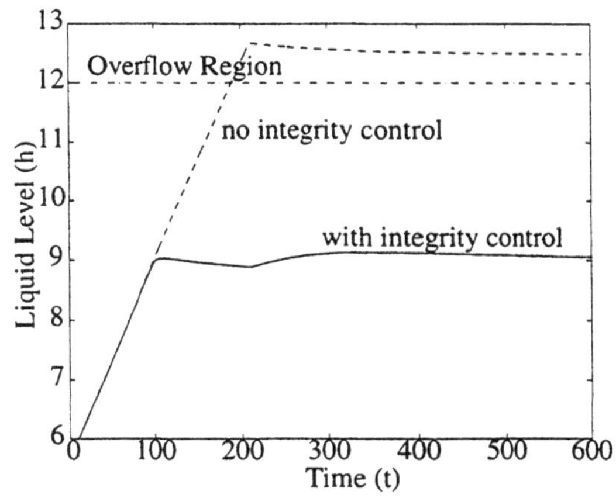

Fig. 5. A comparison of liquid level response to a strategy with and without process integrity control.

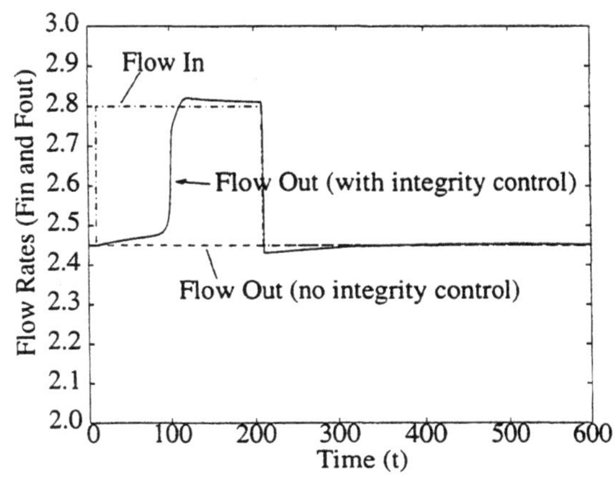

Fig. 6. A comparison of flow rates out in response to a strategy with and without process integrity control.

This implies that two modes of integrity control are needed. One mode is the aggresive override mode which immediately moves the system towards acceptable operating conditions. The other mode is to put integrity control in the background, but reduced activity, mode. In this mode, the system would slowly reduce the accumulated damage of previous events. [1]

5. CONCLUSION

In this paper, the concepts and definition of process integrity are discussed. The use of process integrity can help in faster monitoring of the conditions of a process. It evaluates how far a process is to a failure condition and it combines existing methods of reliability theories with methods

[1] One analogy to the background mode is the "*savings and investment*" routine of a conservative business approach.

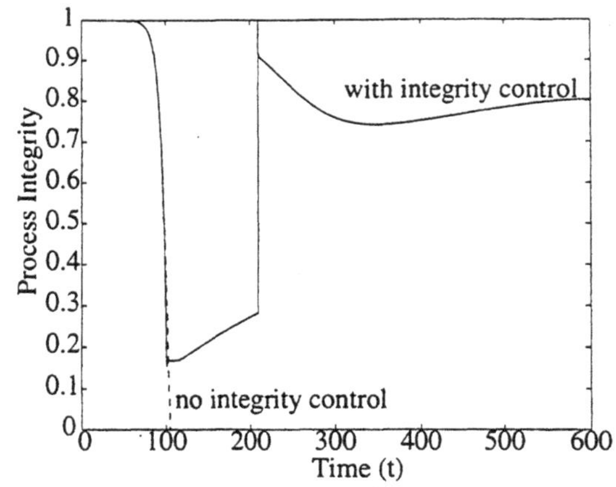

Fig. 7. A comparison of changes in process integrity in response to a strategy with and without process integrity control.

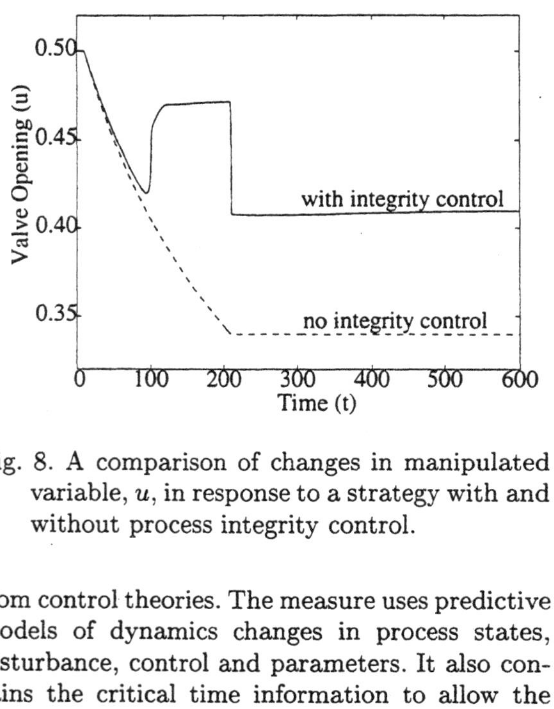

Fig. 8. A comparison of changes in manipulated variable, u, in response to a strategy with and without process integrity control.

from control theories. The measure uses predictive models of dynamics changes in process states, disturbance, control and parameters. It also contains the critical time information to allow the implementation of appropriate recovery actions to occur in a timely manner.

Aside from the development of the measure, we have explored how process integrity can be used to improve the operation of processes by avoiding situations that can lead to failure conditions. As shown in a simulation example, process integrity measures can be used to construct online override schemes. The scheme used in this example was a basic blending of different control objectives. A more advanced scheme should be able to further improve the response by incorporating the measure in the decision algorithm, e.g. in the objective function of model predictive controllers.

Several issues need further investigation. First, the construction of the safety index is crucial since it is the basis for the process integrity measure. Reliability theory, Monte Carlo simulations and other methods can help in identifying the boundaries. Inside the boundaries, the gradients will need other tools such as order statistics, cell-to-cell mapping and fuzzy set modeling. Second, the issue of process integrity control, including stability and robustness analysis has to be established. Third, advanced methods for integrity control are needed. For instance, an optimal control strategy may be able to use the various terms in equation (3) for a better response. Fourth, an extension from process integrity to a system-wide implementation, called system integrity, is needed for large scale implementation.

6. REFERENCES

Co, T.B. and S. Ungarala (1997). Batch scheme recursive parameter estimation of continuous-time systems using the modulating functions method. *Automatica* **33**, 1185–1191.

Himmelblau, D.M. (1978). *Fault Detection and Diagnosis in Chemical and Petroleum Processes.* Elsevier Scientific Pub. Co.. Amsterdam.

Iserman, R. (1984). Process fault detection based on modeling and estimation methods. *Automatica* **20**, 387–404.

Kim, J.H. and Z. Bien (1991). Geometric approach for fault diagnosis. *IEE Proc. Pt D* **138**, 293–302.

Molle, D.T. Dalle and D.M. Himmelblau (1987). Fault detection in a single-stage evaporator via parameter estimation using kalman filter. *Ind. Engr. Chem. Res.}* **138**, *2482–2489.*

Ray, A. (1990). Multi-level hypothesis testing for fault detection in continuous process. Journal of Dynamical Systems, Measurement and Control **112**, *787–790.*

Willsky, A.S. (1976). A survey of design methods for failure detection in dynamic systems. Automatica **12**, *601–611.*

7. ACKNOWLEDGEMENTS

The author wishes to acknowledge and thank the support of this research through a grant provided by the Petroleum Research Fund of the American Chemical Society. Part of this research was done during the author's sabbatical leave at Honeywell Technology Center which is partially supported by Honeywell Inc. and Michigan Technological University and is greatly appreciated.

AUTHOR INDEX

www.ingramcontent.com/pod-product-compliance
Ingram Content Group UK Ltd.
Pitfield, Milton Keynes, MK11 3LW, UK
UKHW052301280926
15471UKWH00006B/1049

9 780080 432335